2006 IBC® Handbook

Structural Provisions

S.K. Ghosh, Ph.D.
R. Chittenden, S.E.
John R. Henry, P.E.

2006 IBC Handbook

Structural Provisions

ISBN-978-1-58001-522-6

Cover Design:	Mary Bridges
Publication Manager:	Mary Lou Luif
Project Editor:	Roger Mensink
Illustrator/Interior Design:	Mike Tamai
Typesetting:	Mike Tamai
Manager of Development:	John Henry

First Printing: July 2008

Printed in United States of America

Preface

Internationally, code officials recognize the need for a modern, up-to-date building code addressing the design and installation of building systems through requirements emphasizing performance. The 2006 *International Building Code* (IBC) is designed to meet those needs through model code regulations that safeguard the public health and safety in all communities, large and small.

This comprehensive code establishes minimum regulations for building systems using prescriptive and performance-based provisions. It is founded on broad-based principles that make possible the use of alternate materials as well as new, improved design methodologies.

The changes in the structural provisions of the 2006 IBC when compared to previous editions of the IBC and legacy codes (*Uniform Building Code, National Building Code, Standard Building Code*) creates the need for a comprehensive work that discusses the changes and expands on many of the new or improved provisions. This book, the *2006 IBC Handbook—Structural Provisions,* is intended to do just that.

By helping code users understand and properly apply the structural provisions in Chapters 16 through 23 of the 2006 IBC, this handbook is a valuable resource for those who design, plan review, inspect or construct buildings or other structures regulated by the 2006 IBC. Although it will prove useful to a broad range of individuals, it was written primarily so that architects, engineers and code officials can understand the IBC's provisions and gain insight into their underlying basis and intent. To that end, the handbook's numerous figures, tables and examples help clarify and illustrate the proper application of many code provisions.

One of the significant differences between the structural provisions in the UBC and the IBC is that the IBC adopts national (structural) standards by reference rather than transcribing the structural provisions of the standards into the code itself. This is true for structural loads as well as structural materials. This trend has continued with each subsequent edition of the IBC to the extent that the 2006 IBC relies on the referenced standards even more than the 2000 and 2003 editions. Therefore, in many cases the discussion in this handbook pertains to the provisions in the referenced standard rather than the IBC itself.

The *2006 IBC Handbook—Structural Provisions* covers four major structural categories:

- Structural load effects and design provisions of Chapter 16
- Special inspection, structural testing and structural observation provisions of Chapter 17
- Foundation and soil provisions of Chapter 18
- Specific structural material provisions for concrete, masonry, steel and wood in Chapters 19 through 23

As an added benefit to readers, ICC has included a CD-ROM containing the complete structural handbook as well as a variety of helpful resource documents such as 2003

NEHRP *Recommended Provisions for Seismic Regulations for New Buildings and Other Structures with Accompanying Commentary* (FEMA 450 parts one and two), NEHRP *Recommended Provisions: Design Examples* (FEMA 451), NEHRP *Recommended Provisions for New Buildings and Other Structures: Training and Instructional Materials* (FEMA 451B), *Seismic Considerations for Steel Storage Racks Located in Areas Accessible to the Public* (FEMA 460), *Homebuilders' Guide to Earthquake-Resistant Design and Construction* (FEMA 232), *Communicating with Owners and Managers of New Buildings on Earthquake Risk: A Primer for Design Professionals* (FEMA 389), *Designing for Earthquakes: A Manual for Architects* (FEMA 454) and *CodeMaster - 2006 IBC Seismic Design - 2006 IBC*, 2003 NEHRP, ASCE 7-05.

Foreword

The era of regional model codes in the United States is over. The 2006 *International Building Code* (IBC) is the official model code of the land. Understanding its provisions and the intent behind them is of critical importance for practicing engineers. This is particularly true for those who have practiced for a long time using one of the three earlier regional model codes (*Uniform Building Code* (UBC), *National Building* Code (NBC) or *Standard Building Code* (SBC).

The *2006 IBC Handbook—Structural Provisions* contains the most recent information on the structural provisions as they pertain to the 2006 IBC and by implication to local codes that are based on the 2006 IBC. The scope and treatment of the various provisions of the code presented in this handbook will benefit designers, students and code officials alike. In addition to explaining code provisions, the handbook chronicles the history behind many of these provisions. The *2006 IBC Handbook—Structural Provisions* represents the most thorough, comprehensive coverage of the IBC structural provisions to date.

A unique feature of this handbook is the inclusion of a CD-ROM containing essential references that support the IBC. Although no single reference can hope to replace the code or substitute for on-the-job experience, the *2006 IBC Handbook—Structural Provisions* comes as close as possible, providing in-depth coverage in a consistent and coherent manner.

As a member of the design community and one who is intimately familiar with structural codes and standards, I am pleased to be a part of the historic movement toward code consolidation across the United States and applaud the efforts of the authors in producing a comprehensive handbook to help practicing structural engineers in the understanding and correct application of various structural provisions of the 2006 IBC.

Farzad Naeim, Ph.D., S.E., Esq.
Editor of the *Seismic Design Handbook*

Acknowledgements

Appreciation goes to the original authors of the 2000 edition of this handbook, S.K. Ghosh, Ph.D., and Robert Chittenden, S.E., both of whom have extensive knowledge, expertise and experience in the development of many of the structural provisions of the IBC. Dr. Ghosh authored the introductory chapter, Chapters 16 and 19, and the appendices of this handbook; Mr. Chittenden authored Chapters 17, 18, 20, 21, 22 and 23 of the 2000 edition. John Henry updated Chapters 17, 18, 20, 21, 22 and 23 to the 2006 IBC and expanded the discussion where appropriate.

Dr. Ghosh would like to acknowledge the significant contributions of Prabuddha Dasgupta, Ph.D., his colleague at S. K. Ghosh Associates, Inc. Dr. Dasgupta was involved in the update of Chapters 16 and 19 of the *2000 IBC Handbook—Structural Provisions* from the inception of the project until the final manuscripts were sent to the ICC. He contributed to every aspect of the update effort. The project indeed could not have been completed without his help. Dr. Ghosh is also indebted to John Henry, P.E., for his thorough and constructive review of Chapters 16 and 19. His review comments led to significant improvements in those chapters.

This publication was produced by a team of talented and highly qualified people who put in many hours of effort. Thanks to Roger Mensink for editing; and to Mike Tamai for designing both the publication and CD-ROM, and for typesetting and creating the illustrations and figures.

Contents

INTRODUCTION

This introduction provides a brief background on the current (2008) situation concerning building codes in this country, including the development of the *International Building Code*® (IBC®).

Legacy Model Codes

The building codes of most legal jurisdictions (cities, counties or states) within the United States used to be based on one of three model codes that are now being called the legacy model codes: the *BOCA National Building Code* (BOCA/NBC)[1] published by the Building Officials and Code Administrators International, Country Club Hills, Illinois; the *Uniform Building Code*™ (UBC)[2] published by the International Conference of Building Officials (ICBO), Whittier, California; and the *Standard Building Code* (SBC)[3] published by the Southern Building Code Congress International (SBCCI), Birmingham, Alabama. Figure 1 shows the approximate areas of influence of each legacy model building code—areas in which each was typically adopted.

The 1993 edition of the BOCA/NBC, the 1994 edition of the UBC and the 1994 edition of the SBC were organized into a common code format established by the Council of American Building Officials (CABO). The new format established common chapter designations for the three model building codes published in the United States. For example, structural design requirements were placed in Chapter 16 of all three model codes so they would all be organized by subject in a similar manner, which would then facilitate the comparison process in creating a single code.

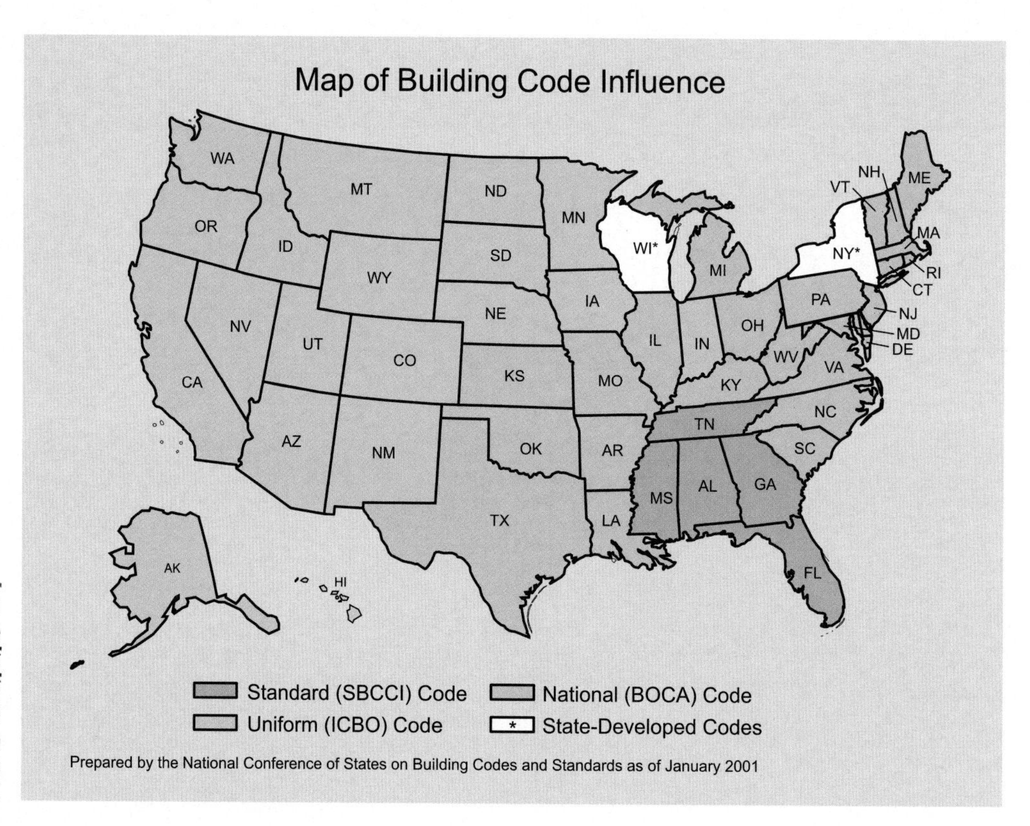

Figure I-1 **Approximate areas of influence of each model code prior to publication of the IBC**

Traditionally there existed a difference in the way standards were adopted into the UBC and the other two model codes. The BOCA/NBC and SBC adopted standards by reference, reproducing only such portions as were likely to be frequently used by building officials, making few, if any, modifications in the process. The UBC, on the other hand, adopted standards by both transcription and by reference. Frequently used standards such as material design standards were reproduced in the code itself, while the other standards were transcribed as UBC Standards. Modifications, sometimes significant ones, were often made during this process of adoption.

International Building Code

At the end of 1994, the three existing model code groups together formed the International Code Council (ICC) with the express purpose of developing a single set of construction codes for the entire country. Included in this family of *International Codes®* is the *International Building Code* (IBC), which represents a major step in a cooperative effort to bring national uniformity to building codes. The development of the IBC is schematically illustrated in Figure 2.

Five technical subcommittees (Fire Safety, Structural, Occupancies, Means of Egress and General) were formed in 1996 by the International Code Council®. The intent of the technical subcommittees was to develop a working draft of a comprehensive building code

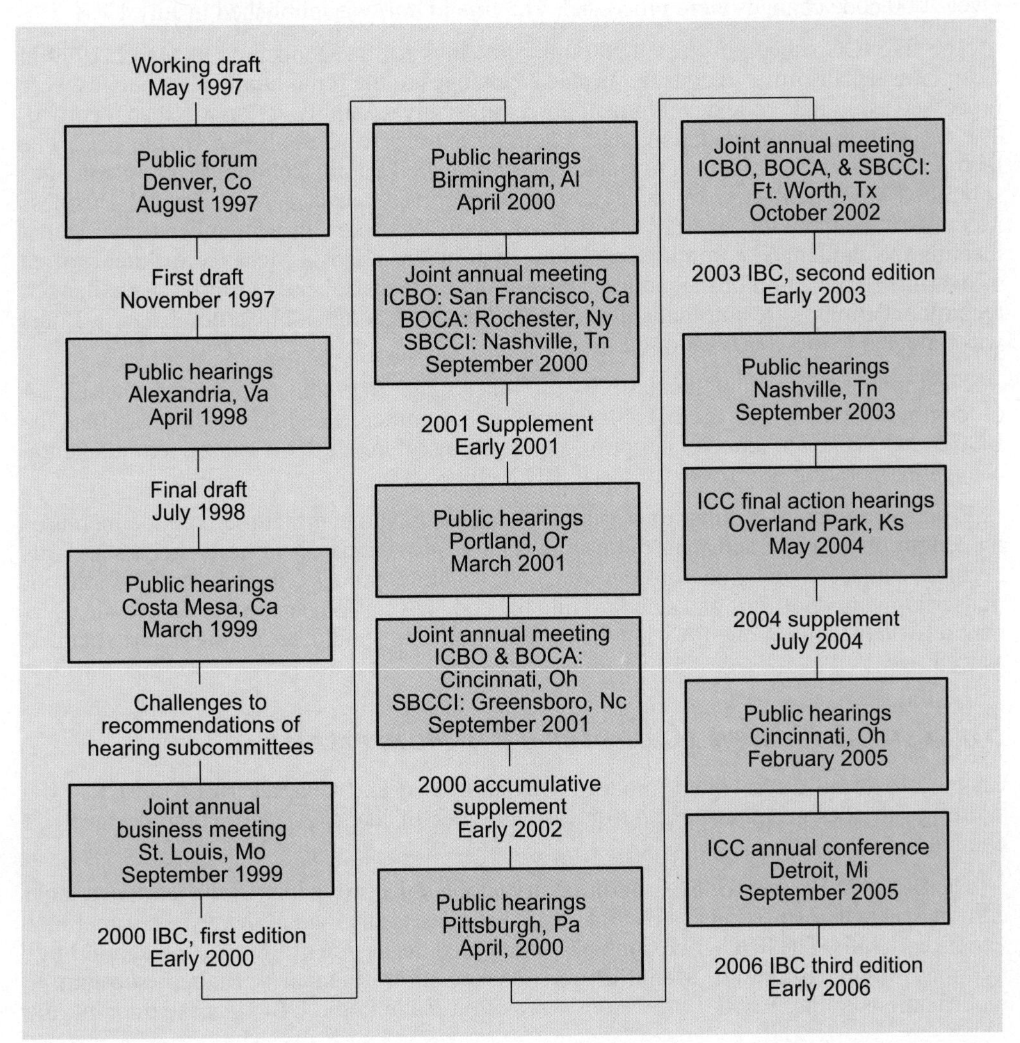

Figure I-2 Development of IBC

that would be consistent with and inclusive of the scope and content of the legacy model building codes.

The technical content of the latest building codes promulgated by BOCA, ICBO and SBCCI was the nucleus for the development of the working draft. Other sources of relevant technical information were also considered. Although there were a great many similarities among the three codes, careful consideration was given to identified differences. The technical subcommittees used certain principles as guidance in the resolution of technical differences. The principles were based on the intent to establish provisions consistent with the scope of a building code that adequately protects public health, safety and welfare; provisions that do not increase construction costs unnecessarily; provisions that do not restrict the use of new materials, products or methods of construction; and provisions that do not give preferential treatment to particular types or classes of materials, products or methods of construction. The Working Draft was published in May 1997.

To assess the views of building code users, the building industry and all other interested parties, a public forum was held in August 1997 in Denver, Colorado, where the technical subcommittees considered close to 3000 submitted public comments to determine the content of the First Draft, which was published in November 1997. The ICC subsequently appointed five hearing subcommittees responsible for the same IBC chapters as the five technical subcommittees. These five subcommittees held public hearings in April 1998 in Alexandria, Virginia, using ICC code development procedures, except that successful assembly votes replaced committee recommendations for text revisions in the Final Draft. Over 2000 code changes were processed. The Final Draft was published in July 1998.

The five ICC subcommittees held public hearings for a second time in March 1999 in Costa Mesa, California, to consider proposed changes to the Final Draft. The changes were processed using ICC code development procedures; any assembly action was also recorded. The committee recommendations and assembly actions were published by the ICC. The recommendations were subject to challenge (or so-called Public Comment). The challenges submitted were published by the ICC and then considered at a Joint Annual Business Meeting of the three model code groups in St. Louis, Missouri, in September 1999. At that meeting the challenged committee recommendations from Costa Mesa were either ratified or overturned by action of the memberships of the three model code groups. Unchallenged recommendations were automatically ratified. The final membership action from St. Louis was reflected in the 2000 edition of the IBC.

The topic of adoption of standards by reference rather than transcription was the subject of lengthy debates within the IBC Structural Subcommittee. The IBC chose to emulate the BOCA and SBCCI practice of adopting standards by reference. This caused a fundamental change in operating procedure for the ICBO membership.

Primary arguments against transcribing standards have been that the practice increases the length of the code and that, more importantly, provisions of the referenced standards become subject to change through the code change process. Legal ramifications of transcribing only portions of a standard into the code have also been pointed out: code users may rely on the code alone for what they need when they should be relying on the standard as well.

Structural Design (Chapter 16): Nonseismic

Chapter 16 of all three legacy model codes addressed and Chapter 16 of the IBC now addresses the design requirements for various types of loads, as well as the design load combinations.

The BOCA/NBC and the SBC traditionally adopted the provisions of the latest available edition of ASCE 7 (previously ANSI A58.1)[4] for nonseismic loads. There were sometimes variations to this. The SBC, for example, also allowed the design of low-rise metal buildings by the Metal Building Manufacturers Association manual.[5] Rather substantive modifications to the ASCE 7 provisions were often made in the UBC. A case in point: the

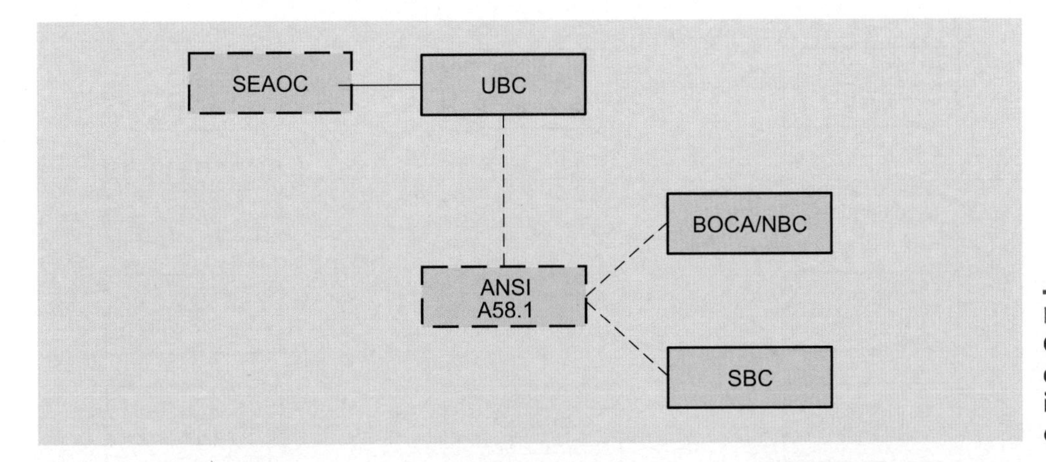

Figure I-3
Origin of seismic design provisions in U.S. building codes

wind-load provisions of the 1997 UBC, which were simplified versions of the corresponding provisions of ASCE 7-88. The 2000 IBC chose to make ASCE 7-98 the basis of all provisions related to nonseismic loads, making only relatively rare exceptions. The 2003 IBC updated to ASCE 7-02. The 2006 IBC updated to ASCE 7-05. Much of the text was also dropped from the various sections of Chapter 16. The provisions covered by the deleted text are now to be found only in ASCE 7-05. Simplified wind design provisions, for instance, are now found only in Chapter 6 of ASCE 7-05.

Structural Design (Chapter 16): Seismic

Until the very beginning of the 1990s, seismic design provisions in U.S. building codes followed a certain pattern (Figure 3). Provisions were first proposed by the Structural Engineers Association of California (SEAOC) in its *Recommended Lateral Force Requirements*[6] (commonly referred to as the *Blue Book*), then adopted by ICBO in its *Uniform Building Code*. The UBC provisions were then adopted (often with modifications) by the American National Standards Institute (ANSI) Standard A58.1 (later to become ASCE 7), which, in turn, was adopted by the other two model codes: the BOCA/NBC and the SBC. Thus, the seismic design provisions of all three model codes were based on the SEAOC *Blue Book* provisions, although a time lag was sometimes involved. The latest available editions of the BOCA/NBC and the SBC at a particular time might be based on an older edition of the *Blue Book* than the then-latest edition of the UBC.

A departure from the above pattern was initiated in 1972 when the National Science Foundation and the National Bureau of Standards (now the National Institute of Standards and Technology) decided to jointly sponsor a Cooperative Program in Building Practices for Disaster Mitigation. Under that program, the Applied Technology Council (ATC) developed a document entitled *Tentative Provisions for the Development of Seismic Regulations for Buildings*.[7] This document, published in 1978 and commonly referred to as ATC 3-06, underwent a thorough review by the building community in ensuing years. Trial designs were conducted to establish the technical validity of the new provisions and to assess their impact. A new entity, the Building Seismic Safety Council (BSSC), was created under the auspices of the National Institute of Building Sciences (NIBS) to administer and oversee the trial design effort. The trial designs indicated the need for certain modifications to the original ATC 3-06 document. The modifications were made. The resulting document was the first edition, dated 1985, of the NEHRP (National Earthquake Hazards Reduction Program) *Recommended Provisions for the Development of Seismic Regulations for New Buildings*[8] (*and Other Structures* was added to the title starting with the 1997 edition). Under continued federal funding, this document has been updated every three years; the 1988, 1991, 1994, 1997, 2000 and 2003 editions of the NEHRP *Provisions* have been issued by the Building Seismic Safety Council (Figure 4). The 2009 edition of the NEHRP

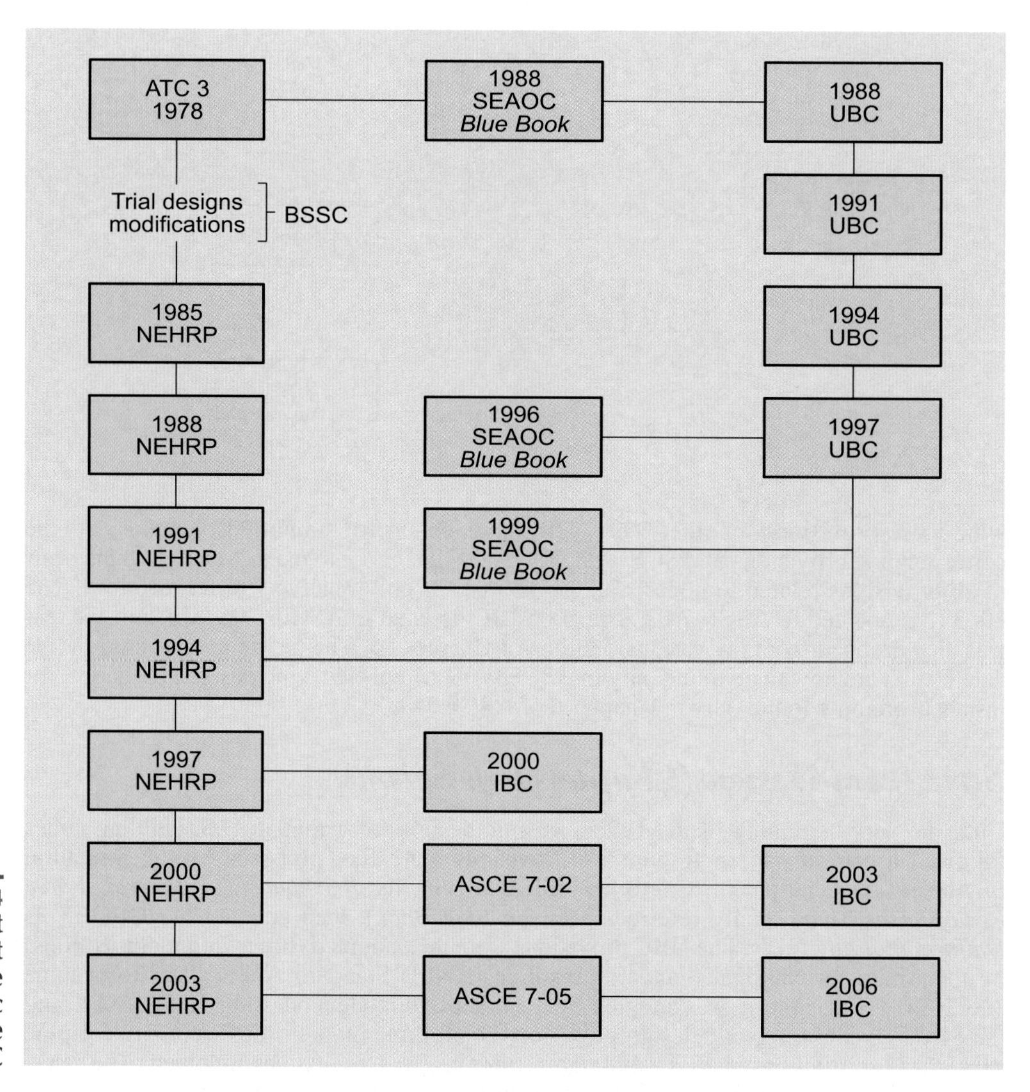

Figure I-4 **Development and subsequent assimilation of the NEHRP Seismic Provisions in the UBC and IBC**

Provisions is currently under development. This edition obviously represents a change in the three-year cycle on which this document has been published in the past.

In 1980, the SEAOC Seismology Committee undertook the task of developing an ATC-based revision of their *Blue Book*. This extensive effort resulted in the 1988 edition of the SEAOC *Blue Book*, which was then adopted into the 1988 edition of the UBC (Figure 4). Changes from the 1988 to the 1991 and from the 1991 to the 1994 editions of the UBC were minor. There were major revisions to the seismic code provisions between the 1994 and the 1997 editions of the UBC, which were due to a code change under development since 1993 by the SEAOC Seismology Committee and its ad hoc subcommittees on strength design and strong ground motions. Numerous significant revisions were incorporated into their code change proposal to reflect the lessons learned from the Northridge and Kobe earthquakes. Also, recognizing that the IBC would be based on existing model codes, SEAOC evaluated the basis of the seismic provisions in the BOCA/NBC and the SBC to determine if convergence of the seismic provisions was possible. As a result, significant concepts consistent with the seismic provisions in these other two model codes were incorporated into SEAOC's code change proposal. To put it more directly, the 1997 UBC included a number of significant features out of the 1994 NEHRP *Provisions*—most notably, the six-tier soil classification that had been introduced in the 1994 *Provisions*. These changes, sometimes in modified forms, were included in Appendix C to the 1996 SEAOC *Blue Book* and in the 1999 SEAOC *Blue Book*.

The BOCA/NBC and the SBC, which traditionally adopted the seismic provisions of the ANSI A58.1 (now ASCE 7) Standard, adopted seismic design provisions based on the 1991 NEHRP *Provisions* in 1993 and 1994, respectively. This was because the seismic provisions of the national loading standard had fallen behind the times. The 1988 edition of the ASCE/ANSI Standard, published in 1990, still had seismic design provisions based on the 1979 UBC. Requirements based on the 1991 NEHRP *Provisions* were retained in the 1996 edition of the BOCA/NBC and the 1997 edition of the SBC. The 1994 NEHRP *Provisions* did not become the basis of the seismic design provisions of the 1996 BOCA/NBC or the 1997 SBC. However, seismic design provisions based on the 1994 NEHRP *Provisions* were adopted into the 1995 edition of the ASCE 7 Standard. Both the 1996 edition of the BOCA/NBC and the 1997 edition of the SBC permitted seismic design by ASCE 7-95. The seismic design provisions of the 1999 edition of the BOCA/NBC remained the same as those of the 1996 edition. Similarly, the seismic design provisions of the 1999 edition of the SBC remained the same as those of the 1997 edition. Table 1 summarizes seismic design provisions of the model codes.

The seismic design provisions of the IBC were treated separately from the rest of the structural provisions in the code development process. In 1996, the IBC Code Development Committee agreed in concept for the IBC to be based on the 1997 edition of the NEHRP *Provisions*, which was being developed at the time the last edition of the UBC (1997) was published. A Code Resource Development Committee (CRDC), funded by the Federal Emergency Management Agency (FEMA), was formed under the direction of the Building Seismic Safety Council (BSSC) to generate seismic code provisions based on the 1997 edition of the NEHRP *Provisions*, for incorporation into the 2000 IBC. This effort was successful and the CRDC submittal was accepted by the IBC Code Development Committee for inclusion in the IBC. The seismic design provisions of the Working Draft were thus based on the 1997 NEHRP *Provisions*, but also included a number of features of the 1997 UBC that were not included in the 1997 NEHRP *Provisions*. Many changes were made to the provisions of the Working Draft through the public forum and the two sets of public hearings (see Figure 2). BSSC's Code Resource Support Committee (CRSC), a successor group to the CRDC, played an active role in this development by sponsoring changes of their own and by taking positions on other submitted changes—positions that were carefully considered by the IBC Structural Subcommittee at the public forum and at the public hearings. The seismic design provisions of the first edition of the IBC were based on the 1997 NEHRP *Provisions*, with some of the features of the 1997 UBC also included.

The seismic design provisions of ASCE 7-02 were based on the 2000 NEHRP *Provisions*. The 2003 IBC gave the designer two specific options: Seismic design could be done by the provisions of ASCE 7-02, ignoring Sections 1613 through 1623 of the 2003 IBC, or seismic design could be done by 2003 IBC Sections 1613 through 1623, which referenced specific sections within ASCE 7-02 and often made modifications to them. Thus, the two choices were: seismic design entirely by ASCE 7-02, or seismic design by ASCE 7-02, as modified by the 2003 IBC.

The seismic design provisions of ASCE 7-05 are based on the 2003 NEHRP *Provisions*. These have also undergone a thorough reorganization. The seismic design provisions are now found in Chapters 11 through 23, and Appendices 11A and 11B of ASCE 7-05. The seismic design provisions of the 2006 IBC are now found in just one section, 1613. That section adopts, by reference, Chapters 11, 12, 13 and 15 through 23 of ASCE 7-05 as well as Appendix 11B. Chapter 14, Material Specific Requirements, is not adopted, because the design of concrete, aluminum, masonry, wood and steel structures must be done by Chapters 19 through 23 of the 2006 IBC, respectively, and not by Chapter 14 of ASCE 7-05. Also not adopted is Appendix 11A on inspection and testing, because inspection and testing of a structure designed by the 2006 IBC must be by Chapter 17 of that code and not by Appendix 11A of ASCE 7-05.

Table I-1. Key to seismic design provisions of model codes

Resource Document	Edition	Model Code
SEAOC *Blue Book*[a]	1998	1994 UBC
	1996 (App. C), 1999	1997 UBC
NEHRP	1991	1993, 1996,1999 BOCA/NBC
		1994, 1997, 1999 SBC
	1994	See ASCE 7-95 below
	1997	2000 IBC
	2000	See ASCE 7-02 below
	2003	See ASCE 7-05 below
ASCE 7	1995 (1994 NEHRP adopted)	1996[b], 1999[b] BOCA/NBC
		1997[b], 1999[b] SBC
	2002 (2000 NEHRP adopted)	2003 IBC
	2005 (2003 NEHRP adopted)	2006 IBC

a. Recommended Lateral Force Requirements.

b. The 1991 NEHRP *Provisions* formed the basis of the seismic design requirements of the 1996 and 1999 editions of the BOCA/NBC and the 1997 and 1999 editions of the SBC. However, both codes allowed the use of ASCE 7-95, which adopted seismic design requirements based on the 1994 NEHRP *Provisions*. ASCE 7-98, ASCE 7-02 and ASCE 7-05 have adopted seismic design provisions based on the 1997, the 2000 and the 2003 NEHRP *Provisions*, respectively.

BOCA Building Officials and Code Administrators International
NHRP National Earthquake Hazards Reduction Program
SEAOC Structural Engineers Association of California
ASCE American Society of Civil Engineers
IBC *International Building Code*
SBC *Standard Building Code*
NBC *National Building Code*
UBC *Uniform Building Code*

The changes from the 1994 to the 1997 edition of the *Uniform Building Code* took it in the direction of the 1997 NEHRP *Provisions* and IBC. However, there are still substantial differences between the seismic design provisions of the 1997 UBC and those of all three editions of the IBC. The treatment of ground motion parameters in seismic design is entirely different in the two documents. Another major difference is discussed below.

The IBC uses Seismic Design Categories (SDC) to determine permissible structural systems, limitations on height and irregularity, the type of lateral force analysis that must be performed, the level of detailing for structural members and joints that are part of the lateral-force-resisting system and for the components that are not. The 1997 UBC, as in prior editions of the code, used the seismic zone in which a structure was located for all these purposes. The SDC is a function of occupancy (called Seismic Use Group in the 2000 and 2003 IBC and the 1997, 2000 and 2003 NEHRP *Provisions*) and of soil-modified seismic risk at the site of the structure.

In 1978, ATC 3-06, the predecessor document to the NEHRP *Provisions*, made the level of detailing (and other restrictions concerning permissible structural system, height, irregularity and analysis procedure) a function of occupancy. That was a major departure from UBC practice, which was continued in all the NEHRP *Provisions* through the 1994 edition. Now, in the IBC and the 1997 and subsequent NEHRP *Provisions*, the level of detailing and the other restrictions have been made a function of the soil characteristics at the site of the structure. This is a further major departure from UBC practice and indeed from current or recent seismic design practice across the country.

Structural Design (Chapter 16): Load Combinations

Design load combinations in all three of the legacy model codes progressively became inordinately complex, primarily because of an emphasis on modernizing the seismic regulations in each of those codes over their last decade of existence. Advancements in the understanding of seismic structural response brought about changes in design philosophy which, in turn, resulted in changes in load combinations that must be used when earthquake forces are considered. The design load combinations of ASCE 7, if and when adopted, were subject to so many modifications that confusion was commonplace. A reader seeking clarity may benefit from consulting References 9-11 (also see Appendix 1 to this publication).

The complexity continued into the 2000 IBC, but it has since dissipated some. As the more recent editions of the IBC are increasingly adopted across the country, the situation is bound to improve.

REFERENCES

[1]Building Officials & Code Administrators International, The BOCA National Building Code, Country Club Hills, IL, 1993, 1996, 1999, copyright held by International Code Council.

[2]Pacific Coast Building Officials Conference (later the International Conference of Building Officials) Uniform Building Code, Long Beach, CA (later Whittier, CA), 1935,1952, 1967, 1970, 1973, 1976, 1979, 1982, 1985, 1988, 1991, 1994, 1997, copyright held by International Code Council.

[3]Southern Building Code Congress International, Standard Building Code, Birmingham, AL, 1994, 1997, 1999, copyright held by International Code Council.

[4]American Society of Civil Engineers, ASCE *Standard Minimum Design Loads for Buildings and Other Structures*, ASCE 7-88, ASCE 7-93, ASCE 7-95 (also ANSI A58-55, ANSI A58.1-72, ANSI A58.1-82), New York, NY , 1990, 1993, 1995; ASCE 7-98, ASCE 7-02, ASCE 7-05, Reston, VA, 2000, 2002, 2005.

[5]Metal Building Manufacturers Association, *Low-Rise Building Systems Manual*, Cleveland, OH, 1986.

[6]Seismology Committee, Structural Engineers Association of California, *Recommended Lateral Force Requirements and Commentary*, San Francisco (later Sacramento), CA, 1974, 1988, 1996, 1999.

[7]Applied Technology Council, *Tentative Provisions for the Development of Seismic Regulations for Buildings*, ATC Publication ATC 3-06, NBS Special Publication 510, NSF Publication 78-8, U. S. Government Printing Office, Washington, DC, 1978.

[8]Building Seismic Safety Council, NEHRP (National Earthquake Hazards Reduction Program) *Recommended Provisions for the Development of Seismic Regulations for New Buildings (and Other Structures)*, Washington, DC, 1994 (1997, 2000, 2003).

[9]Ghosh, S.K., "Design of Reinforced Concrete Buildings under the 1997 UBC," Building Standards, International Conference of Building Officials, May-June 1998, pp. 20 – 24.

[10]Ghosh, S.K., "Needed Adjustments in 1997 UBC," Proceedings, 1998 Convention, Structural Engineers Association of California, October 7 – 10, 1998, Reno-Sparks, NV, pp. T9.1 – T9.15.

[11]*Strength Design Load Combinations for Concrete Elements*, IS 521, Portland Cement Association, Skokie, IL,1998.

CHAPTER

16

Part 1 STRUCTURAL DESIGN

This chapter explains and provides background on the development of the structural design requirements of Chapter 16 of the 2006 *International Building Code* (IBC). In a significant change from previous editions of the IBC, large sections of Chapter 16 that were related to the determination of snow, wind and seismic loads have been removed in order to eliminate the possibility of error and confusion arising from the IBC's practice of adopting and transcribing the provisions related to snow, wind and seismic loads from the ASCE 7 Standard *Minimum Design Loads for Buildings and Other Structures*.[1] This practice often required designers and building officials to refer to both the building code and the ASCE 7 standard simultaneously. Further, there was a risk that when transcription into the building code was made, inadvertent errors and omissions might have taken place. Thus, to ensure that the design requirements of the building code are succinctly and unambiguously specified, a coalition of the Structural Engineering Institute (SEI) of ASCE and the National Council of Structural Engineers Associations (NCSEA) developed a proposal whereby all technical specifications relating to snow, wind and seismic loading are incorporated through reference to the 2005 edition of the ASCE 7 standard. This results in substantial reduction in material actually contained in the 2006 IBC. Sections that are still left in the IBC relate to the local geologic, terrain or other environmental conditions that many building officials will wish to specify when adopting the model code by local ordinance. Also, seismic provisions left in the 2006 IBC are now in a single section, Section 1613, as opposed to multiple sections (1613 through 1623) in previous editions of the IBC. As indicated below, many of the provisions are derived from two major sources: ASCE 7-05 and the 2003 NEHRP *Provisions*.[2] The seismic design provisions of ASCE 7-05 are in fact adopted from those of the 2003 NEHRP *Provisions*. Fortunately, both these documents come with detailed commentaries. Sections of these commentaries have at times been paraphrased in an attempt to make this handbook reasonably self-contained. In many instances, fairly detailed background to certain provisions has been provided in appendices, so as not to interfere with the flow of the handbook.

Numerical examples have been included where they serve to illustrate the design requirements.

Section 1601 *General*

Chapter 16, Structural Design, governs the structural design of IBC-regulated buildings, nonbuilding structures and portions thereof. A building is defined in Section 202 as any structure used or intended for supporting or sheltering any use or occupancy.

Chapter 16 provides requirements for minimum structural loads as well as criteria or methods of load application to be used in the design of buildings and other structures. The various types of structural loads specified by Chapter 16 are either gravity loads or lateral loads. Gravity loads specifically addressed are dead loads, live loads, snow loads and rain loads. Lateral loads specifically dealt with are those that are due to wind, earthquakes, soil pressure or flood. Loading conditions, such as uniformly distributed and concentrated live loads, impact loads, and most important, the design load combinations are also regulated by the provisions of Chapter 16. Section 1601 presents the scope of Chapter 16. Section 1602 defines terms that are commonly used in the structural design requirements. Section 1603 specifies the minimum information that must be provided on the construction documents. Section 1604 gives general design requirements and specifically addresses strength criteria and serviceability criteria, including deflection limitations, structural analysis, occupancy categories and load tests. Section 1605 addresses the vital topic of design load combinations. Strength design load combinations and allowable stress design load combinations, as well as special seismic design load combinations, are given. Section 1606 specifies design dead loads. Section 1607 specifies the minimum uniformly distributed live loads and minimum concentrated live loads for various types of occupancies. This section also permits a reduction of design live loads under certain conditions. Section 1608 specifies the design snow loads, largely by reference to Chapter 7 of ASCE 7-05. Section 1609 contains structural design requirements for wind loads, largely by reference to Chapter 6 of ASCE 7-05. Sections 1610, 1611 and 1612 treat soil lateral loads, rain loads and flood loads, respectively. The rain load provisions of Section 1611 are by reference to Chapter 8 of ASCE 7-05. Section 1613 is devoted to the seismic design of buildings and other structures and references Chapters 11 through 23 (excluding Chapter 14) and Appendix 11B of ASCE 7-05.

Section 1602 *Definitions*

The IBC defines general structural terms that are used in Chapter 16 in Section 1602.

Although some notations are given in Section 1602, symbols and notations are typically defined throughout the IBC following the equation(s) in which they are used. The ICC Structural Subcommittee thought this to be more user-friendly than requiring the reader to refer back to the beginning of a chapter each time an equation appears.

Note that definitions related to earthquake loads and seismic force-resisting systems have been deleted from the 2006 IBC because those terms are not specifically used in the 2006 IBC text, because of the total reliance on referencing ASCE 7-05.

Section 1603 *Construction Documents*

This section details the items to be shown on construction documents.

Construction documents are defined in Section 202. Note that the loads are not required to be on the construction drawings but must be included within the construction documents in such a way that the design loads are clear for all parts of the structure. Of course, the indicated loads are required to be equal to or greater than the loads required by the code. The information required to be included in the construction documents is useful to the building official in performing plan review and field inspection. It is also typically found to be useful if additions or alterations are made to a structure at a later date. Each of the items indicated in Section 1603.1.4 is an important parameter in the determination of the wind resistance that is required in the structural system of the building.

The exception to Section 1603.1 simplifies the structural design information required for buildings constructed to the conventional light-frame construction provisions of Section 2308. A registered design professional is not required for such buildings. However, the requirements of Sections 1603.1.1 through 1603.1.6 clearly would require the services of such a professional in order to provide the specified design data. The requirements in the exception provide adequate information for the building official to verify the structural design basis of buildings built to these conventional construction provisions. Note that flood design data has been added in the 2006 IBC to the list of items that must be included in the construction documents for buildings to be constructed in accordance with the requirements of Section 2308.

The 2006 IBC uses "Occupancy Category" as part of the earthquake design data to be included on the construction documents instead of "Seismic Use Group" used in prior editions of the 2003 IBC. The term "Seismic Use Group" (SUG) is omitted from the 2006 IBC altogether, to make the code consistent with ASCE 7-05, which no longer uses SUG as a determining factor for the importance factor, the seismic design category and drift limits. SUG was originally adopted from the NEHRP *Provisions* (preceding the 2003 edition), which is a document intended for seismic design purposes only. On the other hand, ASCE 7-05 links the importance factors for snow, wind and seismic designs directly to the occupancy category of the building, so that uniformity is maintained across the design processes for different loading types. The SUG was simply a consolidation of the occupancy categories; Occupancy Category I and II together constituted SUG I, Occupancy Category III was SUG II and Occupancy Category IV was SUG III. This was felt to be unnecessary.

Section 1603.1.6 provides a pointer to Section 1612.5, which is entitled "Flood Hazard Documentation." Section 1612.5 requires statements to be included on the construction documents if certain situations exist. By including this pointer in Section 1603.1.6 to Section 1612.5, the likelihood is enhanced that these statements will be included on the plans.

Sections 1603.1.8 and 1603.2 through 1603.4 go beyond items to be shown on construction documents. To ensure that seismic-resistant systems and components are properly constructed and/or installed, special inspection for seismic resistance is required in Section 1707.1. To do this inspection, approved construction documents for these systems and components are required. The requirement in Section 1603.1.8 for construction documents is found in Sections 2.6 and 6.2.9 of the 2003 NEHRP *Recommended Provisions*.

Section 1604 *General Design Requirements*

1604.1 General. This section requires buildings and other structures and all portions thereof to be designed and constructed in accordance with the general requirements contained in Section 1604.

1604.2 Strength. Most of this section is a reproduction of ASCE 7-05[1] Section 1.3.1. The basic requirement is that buildings and structures be able to support the factored loads without exceeding their strength. For structural elements of a building or other structure designed using nominal rather than factored loads, the actual design stresses are not to exceed the applicable allowable stresses.

1604.3 Serviceability. The basic requirement for serviceability of buildings, nonbuilding structures and all parts thereof is that they be designed and constructed with adequate stiffness so as to limit deflection and lateral drift to an appropriate degree based on the intended use. The general statement is adopted with modification from Section 1.3.2 of ASCE 7-05. Specific requirements are given in Sections 1604.3.1 through 1604.3.6. Table 1604.3 contains deflection limits of structural members as a function of span and load type.

ASCE 7-05 Section 1.3.2 reads as follows: "Structural systems, and members thereof, shall be designed to have adequate stiffness to limit deflections, lateral drift, vibration or any other deformations that adversely affect the intended use and performance of buildings and other structures." The IBC has omitted any reference to vibration or any other deformations that have an adverse impact on intended use and performance of structure because:

1. The code has no objectively defined standard for structural vibration. Acceptable limits are frequently subjective and highly dependent on the specific requirements of occupants of a building. This information is not necessarily available to the building official.
2. It is impossible for the building official to anticipate everything that can "adversely affect the intended use and performance" of a building. Sections 1604.3.1 through 1604.3.6 provide objectively defined deflection limits for structural members. Limits more restrictive than these should properly be a matter of negotiation

between the design professional and the client and should not be part of a life-safety building code. For example, there are situations in which sensitive computer, optical and mechanical equipment require extraordinary measures to provide adequate support for their intended use. These measures are often very complex and well beyond the life-safety requirements of most structures.

1604.4

Analysis. The first two paragraphs are reproduced with minor modifications from Section 1.3.4 of ASCE 7-05. The third paragraph can be traced back to 1997 *Uniform Building Code*[3] (UBC) Section 1605.2, Rationality. Similarly, the fourth paragraph is mostly from 1997 UBC Section 1605.2.1, Distribution of horizontal shear, and the last paragraph is from UBC Section 1605.2.2, Stability against overturning.

Structural analysis is to be based on fundamental principles of structural mechanics. These principles are equilibrium, stability, geometric compatibility and material properties. Although the code in general does not intend to specify the design method used by the engineer, it does intend that the design method be rational and in accordance with well-established principles of mechanics. Departures from this latter requirement can still be made based on the provisions of Section 104.11 when approved by the building official. For example, the structural adequacy of a building may not admit to a rational analysis; a program of full-scale testing may be the only reasonable way to determine its structural behavior. If the testing program shows that a certain building can safely resist the loads required by the code, the building official may approve the construction of the building.

The basic requirement for self-straining forces is that those that are due to moisture, movement, creep, shrinkage, temperature changes and foundation settlement be adequately accounted for. Thus, a structure is to be designed to resist the forces resulting from self-straining, or alternatively, expansion joints must be provided.

The requirement for stability against overturning makes reference to the following specific sections: 1609 for wind, 1610 for lateral soil loads, and 1613 for earthquake.

The requirement that provisions be made for the increased forces induced in resisting elements of the structural system, resulting from torsion that is due to eccentricity between the center of application of the lateral forces and the center of rigidity of the lateral-force-resisting system has been changed slightly from the 2003 IBC to the 2006 IBC. The 2006 IBC requires the same, "Except where diaphragms are flexible, or are permitted to be analyzed as flexible," because flexible diaphragms cannot transmit torsion.

1604.5

Occupancy category. This section and Table 1604.5 make a combined presentation of the occupancy categories of buildings and other structures. The occupancy category (O.C.) reflects the relative anticipated seriousness of consequence of failure from lowest hazard to human life (O.C. I) to highest (O.C. IV), and is used to relate the criteria for maximum environmental loads and distortions specified in the code to the consequence of the loads being exceeded for the structures and its occupants.

Occupancy Category I contains buildings and other structures that represent a low hazard to human life in the event of failure either because they have a small number of occupants or because they have a limited period of exposure to extreme environmental loading.

Occupancy Category II contains all occupancies other than those in Occupancy Category I, III and IV, and is sometimes referred to as *ordinary* for the purpose of risk exposure.

Occupancy Category III contains those buildings and other structures that have large numbers of occupants, are designed for public assembly, or in which physical restraint or other incapacity of occupants hinders their movement or evacuation. Therefore, these structures represent a substantial hazard to human life in the event of a failure. Occupancy Category III also includes important infrastructures such as power generating stations, water treatment facilities, etc., where a failure may not create an unusual life-safety risk, but can cause large-scale economic impact and/or mass disruption of day-to-day civilian life.

In an effort to improve clarity, the wording of the first criterion for Occupancy Category III is revised in the 2006 IBC. As per 2003 IBC Table 1604.5, buildings where more than 300 people congregate in one area are to be assigned Occupancy Category III. However, it was not clear if the term *one area* meant a single room, a number of connected rooms or a complete floor, etc., which led to inconsistent code enforcement. The statement also seemed to include a large number of commercial buildings where an occupant load greater than 300 is not unusual, which was not the intent. The similar requirements in *The BOCA National Building Code* (NBC)[4] and the *Uniform Building Code* (UBC) were clearer and were limited to public assembly occupancies (i.e., Group A) having an aggregate occupant load greater than 300. Thus, 2006 IBC Table 1604.5 is revised to apply Category III designation to "covered structures whose primary occupancy is public assembly with an occupant load greater than 300."

Occupancy Category IV contains buildings and other structures that are designated as essential facilities and are intended to remain operational in the event of extreme loading such as hospitals, fire stations, etc. Also included are the structures that are supplementary to Occupancy Category IV structures and are required for the operation of Occupancy Category IV facilities during an emergency, e.g., facilities to maintain water pressure for fire suppression. Furthermore, structures holding extremely hazardous materials are also included in Occupancy Category IV because of the potentially devastating effect of a release of those materials in the environment.

In the case where there are multiple occupancies in a structure, the highest (or most restrictive) Occupancy Category is to be assigned to the structure unless the portions are structurally separated. In other words, when a lower group impacts a higher group, the higher group must either be seismically independent of the other, or the two must be in one structure designed seismically to the requirements of the higher group. In the case in which the two uses are seismically independent but are functionally dependent, both portions are required to be assigned to the higher Occupancy Category.

The 2006 IBC references ASCE 7-05 for assigning Importance Factors to structures based on their occupancy categories for the purpose of determining design snow, wind and seismic loads.

1604.6 **In-situ load tests.** This is adapted from Section 1.7 of ASCE 7-05. Whenever there is reasonable doubt as to the stability or load-bearing capacity of a completed building, structure or portion thereof for the expected loads, an engineering assessment may be required by the building official. The engineering analysis may involve either a structural analysis or an in-situ load test, or both. See IBC Section 1713 for more details.

1604.7 **Preconstruction load tests.** In evaluating the physical properties of materials and methods of construction that are not capable of being designed by approved engineering analysis, or which do not comply with applicable material design standards listed in Chapter 35, the structural adequacy must be predetermined based on the load test criteria given in Section 1714.

1604.8 **Anchorage.** This section addresses the anchorage of the various components of a building to resist the uplift and sliding forces that result from the application of the code-prescribed lateral forces. It intends that all members be tied together or anchored to resist the uplift and sliding forces. The section differentiates between the uplift and sliding forces to be resisted in general (Section 1604.8.1) and the lateral support required for concrete and masonry walls (Section 1604.8.2).

Many observed failures of concrete or masonry walls in the 1971 San Fernando and 1994 Northridge earthquakes were attributable to inadequate anchorage between the walls and the roof system. Although requirements for anchorage to prevent the separation of heavy masonry or concrete walls from floors or roofs have been common (although often insufficient) in highly seismic areas, they have been minimal or nonexistent in most other parts of the country. This section requires that anchorage be provided in any location to the tune of 280 pounds per linear foot of wall (Figure 16-1). This requirement alone may not provide complete earthquake-resistant design, but observations of earthquake damage indicate that it can greatly increase the earthquake resistance of buildings and reduce hazards in those localities where earthquakes may occur but are rarely damaging.[2]

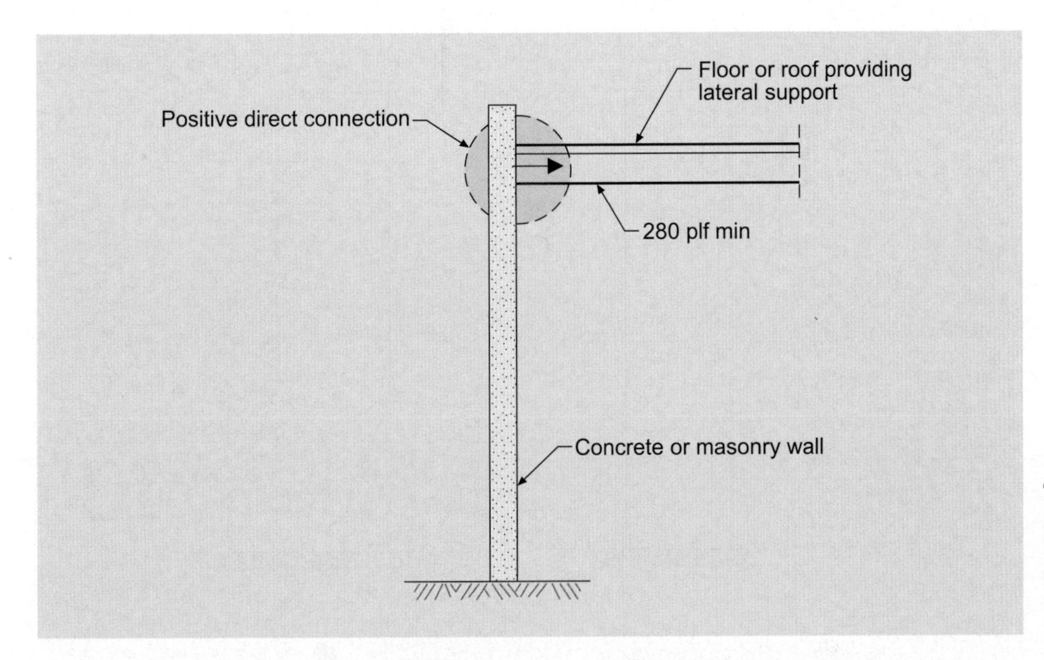

Figure 16-1
Minimum anchorage of concrete and masonry walls

Where the spacing between anchors exceeds 4 feet, the bending of a wall between such anchors must be considered in the design of the wall. If the wall being laterally supported is of hollow-unit masonry or a cavity wall, the required anchors must be embedded in a reinforced grouted structural element of the wall.

1604.9

Counteracting structural actions. This contains a very basic and general requirement that structural members and systems, and components and cladding in a building or other structure be anchored to resist wind- or earthquake-induced overturning, uplift and sliding, and to provide continuous load paths for those forces to the foundation. This section expands the provisions regarding design against overturning, uplift and sliding found in 2003 IBC Section 1609.1.3, so that they apply not just to wind design, but to seismic design as well. It was felt that those requirements are, in fact, applicable to all buildings in general and hence are more appropriately written and located as a general design requirement.

1604.10

Wind and seismic detailing. The forces that a structure subjected to earthquake motions must resist result directly from the distortions induced by the motion of the ground on which it rests. The response (i.e., the magnitude and distribution of forces and displacements) of a structure resulting from such a base motion is influenced by the properties of both the structure and the foundation, as well as the character of the exciting motion.

A simplified picture of the behavior of a building during an earthquake may be obtained by considering Figure 16-2. As the ground on which the building rests is displaced, the base of the building moves with it. However, the inertia of the building mass resists this motion and causes the building to suffer a distortion (greatly exaggerated in the figure). This distortion wave travels along the height of the structure in much the same manner as a stress wave in a bar with a free end. The continued shaking of the base causes the building to undergo a complex series of oscillations.

It is important to draw a distinction between forces that are due to wind and those produced by earthquakes. Occasionally, even engineers tend to think of these forces as belonging to the same category just because codes specify design wind as well as earthquake forces in terms of equivalent static forces. Although both wind and earthquake forces are dynamic in character, a basic difference exists in the manner in which they are induced in a structure. Whereas wind loads are external loads applied and, therefore, proportional to the exposed surface of a structure, earthquake forces are essentially inertial forces. The latter result from the distortion produced by both the earthquake motion and inertial resistance of the structure. Their magnitude is a function of the mass of the structure

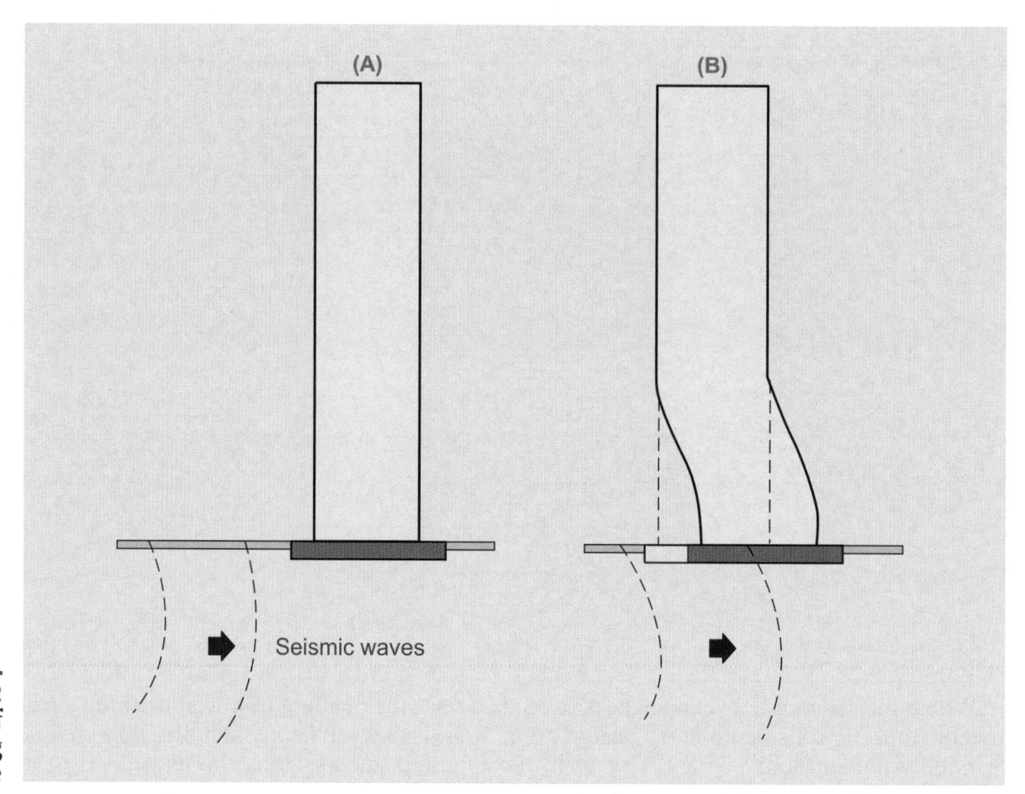

Figure 16-2
Behavior of building during an earthquake

rather than its exposed surface. Also, in contrast to the structural response to essentially static gravity loading or even to wind loads, which can often be validly treated as static loads, the dynamic character of the response to earthquake excitation can seldom be ignored. Thus, where in designing for static loads one would feel greater assurance about the safety of a structure made up of members of heavy section, in the case of earthquake loading, the stiffer and heavier structure does not necessarily represent the safer design.

When a structure responds elastically to ground motions during a severe earthquake, the maximum response accelerations may be several times the maximum ground acceleration and may depend on the mass and stiffness of the structure and the magnitude of the damping. It is generally uneconomical and also unnecessary to design a structure to respond in the elastic range to the maximum earthquake-induced inertia forces. Thus the design seismic horizontal force recommended by codes, including the 2006 IBC, are generally less than the elastic response inertia forces induced by a major earthquake (in the case of the 2006 IBC, the design earthquake ground motion is defined in Section 1613).

Experience has shown that structures designed to the level of seismic horizontal forces recommended by codes can survive major earthquake shaking. This is because of the ability of well-designed structures to dissipate seismic energy by inelastic deformations in certain localized regions of certain members. Decrease in structural stiffness caused by accumulating damage and soil-structure interaction also helps at times. It should be evident that the use of the level of seismic design forces recommended by codes implies that the critical regions of inelastically deforming members should have sufficient inelastic deformability to enable the structure to survive without collapse when subjected to several cycles of loading well into the inelastic range. This means avoiding all forms of brittle failure and achieving adequate inelastic deformability by the yielding of certain localized regions of certain members (or of connections between members) in flexure, shear or axial action. This is precisely why the materials chapters of codes (Chapters 19, 21, 22 and 23 of the 2006 IBC) contain detailing requirements and other limitations that go hand-in-hand with the code-prescribed seismic forces. The design earthquake forces of the code (Chapter

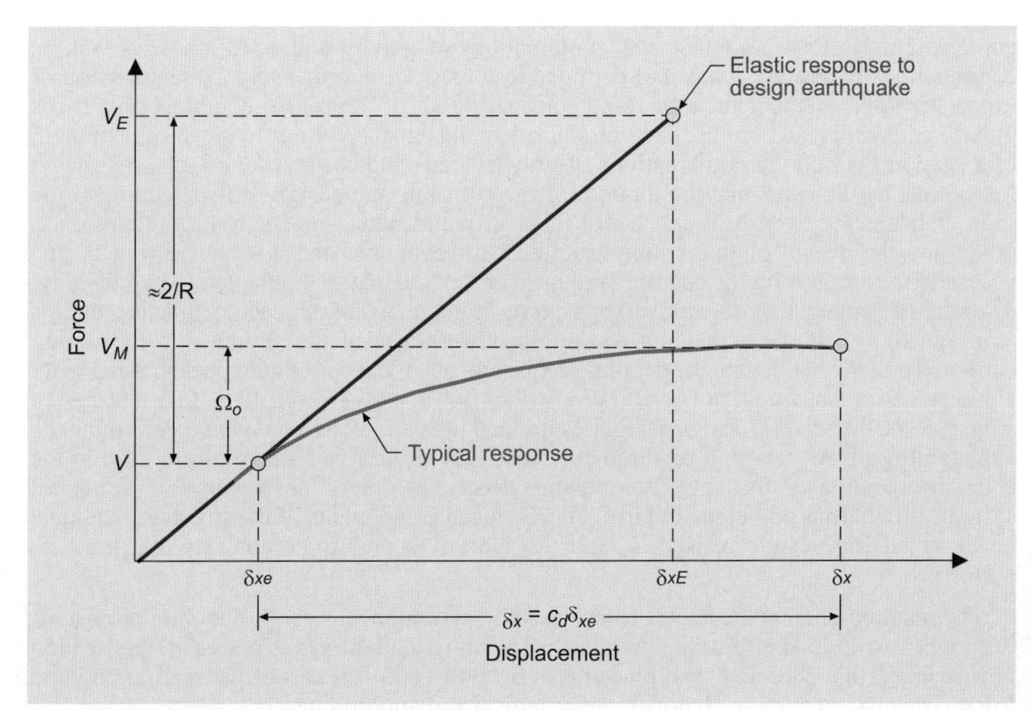

Figure 16-3
Idealized force-displacement relationship of a building subjected to the design earthquake of the IBC

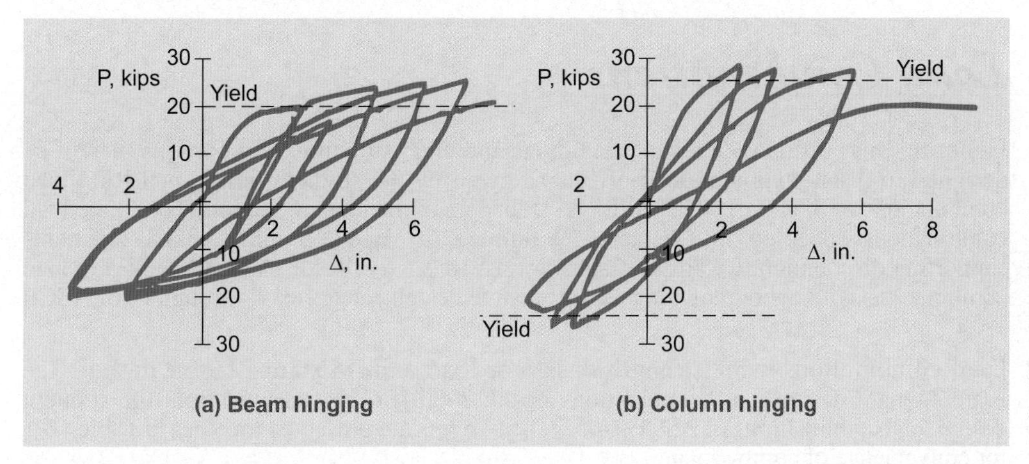

Figure 16-4
Load-deflection curves of structural subassemblies subjected to reverse cyclic displacements

16 of the IBC) and the detailing requirements and other restrictions of the materials chapters are an integral package.

Figure 16-3 shows the idealized force-displacement relationship of a particular structure subject to the design earthquake of the 2006 IBC, as defined in Section 1613. On the y-axis are the earthquake-induced forces; along the x-axis are the earthquake-induced displacements. The curve may be thought of as the envelope or the backbone curve of hysteretic force-displacement loops that describe the response of a structure subjected to reversed cyclic displacement histories of the type imposed by earthquake ground motion (see Figure 16-4).[5]

It should be obvious from the above that as long as seismic design is done using code-prescribed forces, which are reduced below the level that would have produced elastic structural response to the design earthquake of the IBC, the detailing requirements of the materials chapters must be conformed with, irrespective of how high the wind effects might be in comparison with the earthquake effects. Wind and earthquake effects are not considered to occur simultaneously on the structure in U.S. design practice. Section 1605 contains gravity load combinations, combinations of gravity and wind loads, as well as combinations of gravity loads and earthquake forces. Design of every critical section of every structural member must be done considering all of these load combinations. If the gravity and wind load combinations produce demands that are closer to the design strength of a section than do the combinations of gravity loads and earthquake effects, then wind rather than earthquakes may be thought of as governing the design of that section. If the same happens for every critical section of a structure, then wind may be thought of as governing the design of that entire structure. However, this fact has no bearing on the necessity to comply with the detailing requirements of the materials chapters. Theoretically, it could be argued that if wind effects were larger than unreduced earthquake effects (earthquake effects corresponding to the elastic response of the structure to the design earthquake of the IBC), then the detailing requirements of the code could be dispensed with. However, even that would not be allowed by the IBC. Totally irrespective of the severity of wind effects, the seismic design categories defined in Section 1613.5.6 would determine the applicability of the detailing requirements. The seismic design categories are used in the code, irrespective of the severity of wind effects, to determine permissible structural systems, limitations on height and irregularity, those components of the structure that must be designed for seismic resistance, and the type of lateral force analysis that must be performed.

The requirement that the lateral-force-resisting system meet seismic detailing provisions even when wind load effects are greater than seismic load effects is placed in this section now, whereas the 2003 IBC had placed it in Section 1609.1.5, part of the section on wind loads. The new location is obviously more logical and appropriate.

Section 1605 *Load Combinations*

1605.1 General. This section requires that buildings and other structures and portions thereof be designed to resist combined load effects as given by the strength design or LRFD load combinations of Section 1605.2 or the ASD load combinations of Section 1605.3, the load combinations specified in Chapters 18 through 23, and Section 12.3.3.3 (elements supporting discontinuous walls or frames) or 12.10.2.1 (collector elements requiring load combinations with overstrength factor for seismic design categories C through F) of ASCE 7-05.

1605.2 Load combinations using strength design or load and resistance factor design. The basic strength design load combinations of the 2006 IBC are adapted from the strength design load combinations of ASCE 7-05.[1] The two sets are presented together in Table 16-1 for convenience of comparison.

Revising the load combination equations in the 2003 IBC, the effects of fluid load (F), lateral earth pressure (H) and self-straining force (T) are directly incorporated in the load combination equations, as is done in ASCE 7-05. Thus, designers are no longer required to refer to a set of equations in ASCE 7-05 when these load effects are present, as was the case in the 2003 IBC. Also, the ponding load effect (P) is no longer considered separately. When ponding load acted on a building, the 2003 IBC referred to ASCE 7 for the correct load combinations. However, from the 1995 edition onward, ASCE 7 has addressed the issue of ponding instability within the provisions for snow loads and rain loads and stopped using it as a separate load effect in the load combination equations.

It must be noted that differences still exist between the load combination equations in ASCE 7-05 and those in the 2006 IBC. Factors f_1 and f_2 are still used with live load effect (L) and snow load effect (S) in 2006 IBC Eqs. 16-3, 16-4 and 16-5. Although ASCE 7-05 accomplishes the same effect as factor f_1 through Exception 1 to the load combinations, a substantial difference can be seen in the case of snow load. ASCE 7-05 uses a fixed snow load factor of 0.2 in Eq. 5. The 2006 IBC uses a variable snow load factor, f_2, which equals

0.7 for roof configurations (such as saw tooth) that do not shed snow off the structure and 0.2 for other roof configurations.

Where F_a (flood load) is to be considered in design, the load combinations of Section 2.3.3 of ASCE 7-05 are to be used.

Table 16-1. Strength design load combinations of the 2006 IBC and ASCE 7-05

2006 IBC, Section 1605.2.1	
$1.4(D+F)$	(Equation 16-1)
$1.2(D+F+T) + 1.6(L+H) + 0.5(L_r \text{ or } S \text{ or } R)$	(Equation 16-2)
$1.2D + 1.6(L_r \text{ or } S \text{ or } R) + (f_1L \text{ or } 0.8W)$	(Equation 16-3)
$1.2D + 1.6W + f_1L + 0.5(L_r \text{ or } S \text{ or } R)$	(Equation 16-4)
$1.2D + 1.0E + f_1L + f_2S$	(Equation 16-5)
$0.9D + 1.6W + 1.6H$	(Equation 16-6)
$0.9D + 1.0E + 1.6H$	(Equation 16-7)
ASCE 7-05, Section 2.3.2	
$1.4(D+F)$	(1)
$1.2(D+F+T) + 1.6(L+H) + 0.5(L_r \text{ or } S \text{ or } R)$	(2)
$1.2D + 1.6(L_r \text{ or } S \text{ or } R) + (L \text{ or } 0.8W)$*	(3)
$1.2D + 1.6W + L + 0.5(L_r \text{ or } S \text{ or } R)$*	(4)
$1.2D + 1.0E + L + 0.2S$*	(5)
$0.9D + 1.6W + 1.6H$	(6)
$0.9D + 1.0E + 1.6H$	(7)

*The load factor on L is permitted to equal 0.5 for all occupancies in which L_0 in Table 4-1 is less than or equal to 100 psf, with the exception of garages or areas occupied as places of public assembly.

Load combinations using allowable stress design. The IBC contains two alternative sets of ASD load combinations. The basic ASD load combinations are the same as those in ASCE 7-05, whereas the alternative basic ASD load combinations are adapted from the 1997 UBC. These two sets of ASD load combinations are presented in Table 16-2 along with the basic combinations in ASCE 7-05. **1605.3**

The 2006 IBC includes F (load due to fluids), H (load due to lateral pressure of soil and water in soil) and T (self-straining force arising from contraction or expansion resulting from temperature changes, shrinkage, moisture change, creep in component materials, movement that is due to differential settlement, or combinations thereof) in the basic ASD load combinations. However, where F_a (flood load) is to be considered in design, the load combinations of Section 2.4.2 of ASCE 7-05 are to be used.

In the alternative basic ASD load combinations, load effects F, H and T are dealt with under Section 1605.3.2.1, whereby 1.0 times each applicable load (F, H or T) must be added to the combinations specified in Section 1605.3.2 of the 2006 IBC.

The two sets of ASD load combinations of the 2006 IBC are based on different philosophies and are not specifically intended to be equivalent to each other. The basic set of ASD load combinations adopted from ASCE 7 is based on the premise that the design strength resulting from the allowable stress method should, in general, not be less than that resulting from the basic strength design method. The alternate basic set of ASD load combinations is based on the premise that the designs should be about the same as those resulting from the 1994 UBC.

Table 16-2. **Allowable stress design load combinations of the 2006 IBC and ASCE 7-05**

2006 IBC, Section 1605.3.1		2006 IBC, Section 1605.3.2	
$D + F$	(Eq. 16-8)	$D + L + (L_r \text{ or } S \text{ or } R)$	(Eq. 16-16)
$D + H + F + L + T$	(Eq. 16-9)	$D + L + (\omega W)$	(Eq. 16-17)
$D + H + F + (L_r \text{ or } S \text{ or } R)$	(Eq. 16-10)	$D + L + \omega W + S/2$	(Eq. 16-18)
$D + H + F + 0.75(L+T) + 0.75(L_r \text{ or } S \text{ or } R)$	(Eq. 16-11)	$D + L + S + \omega W/2$	(Eq. 16-19)
$D + H + F + (W \text{ or } 0.7E)$	(Eq. 16-12)	$D + L + S + E/1.4$	(Eq. 16-20)
$D + H + F + 0.75(W \text{ or } 0.7E) + 0.75L + 0.75(L_r \text{ or } S \text{ or } R)$	(Eq. 16-13)		
$0.6D + W + H$	(Eq. 16-14)		
$0.6D + 0.7E + H$	(Eq. 16-15)		
ASCE 7-05, Section 2.4.1			
$D + F$	(1)		
$D + H + F + L + T$	(2)		
$D + H + F + (L_r \text{ or } S \text{ or } R)$	(3)		
$D + H + F + 0.75(L + T) + 0.75(L_r \text{ or } S \text{ or } R)$	(4)		
$D + H + F + (W \text{ or } 0.7E)$	(5)		
$D + H + F + 0.75(W \text{ or } 0.7E) + 0.75L + 0.75(L_r \text{ or } S \text{ or } R)$	(6)		
$0.6D + W + H$	(7)		
$0.6D + 0.7E + H$	(7)		

In the 1997 UBC, the strength-level earthquake effect, E, was brought down to service level through division by a factor of 1.4 before it was combined with the effects of unfactored gravity and other loads in the basic as well as the alternative basic load combinations. The same approach is adopted in the 2006 IBC as well as in ASCE 7-05.

The IBC specifically states that increases in allowable stresses "specified in the appropriate materials section of this code or referenced standard" shall not be used with the basic ASD load combinations, except that a duration of load increase shall be permitted in accordance with Chapter 23. Chapter 23 of the 2006 IBC simply adopts 2005 NDS by reference. On the other hand, when using the alternate basic load combinations that include wind or seismic loads, allowable stresses are permitted to be increased or load combinations reduced, "where permitted by the material section of this code or referenced standard." Thus, to summarize:

1605.3.1 Basic load combinations. One-third allowable stress increase is *not permitted*. See Example 16-2, Part 2 of this chapter. For wood, load duration increases are *permitted*.

1605.3.2 Alternative basic load combinations.

- One-third allowable stress increase is *permitted* (where permitted by the material section of the code or the referenced standard) for all materials other than wood. For wood, load duration increases are *permitted*.
- In load combinations that include the counteracting effects of dead loads and wind loads, only two-thirds of the minimum dead load that is likely to be in place during a design wind event shall be used.
- For evaluating sliding, overturning and soil bearing at the soil-structure interface, the reduction of foundation overturning from ASCE 7-05 Section 12.13.4 is *not*

permitted. A detailed discussion can be found at http://www.gostructural.com/article.asp?id=1275.

- For the purpose of proportioning foundations for seismic loadings, the vertical seismic load effects, E_v, is *permitted* to be taken as zero.

The basic ASD load combinations and the alternate basic ASD load combinations are subject to an exception. Flat roof snow loads of 30 psf (1.44 kN/m^2) or less need not be combined with seismic loads. Where flat roof snow loads exceed 30 psf (1.44 kN/m^2), only 20 percent of the flat roof snow load is required to be combined with seismic loads. This is probably an attempt to be consistent with the definition of *W*, the effective seismic weight of the structure, in Section 12.7.2 of ASCE 7-05, Effective Seismic Weight. See Example 16-3, Part 2 of this chapter.

1605.4

Special seismic load combinations. ASCE 7-05 Section 12.3.3.3 (Elements supporting discontinuous walls or frames) and Section 12.10.2.1 (Collector elements requiring load combinations with overstrength factor for seismic design categories C through F) require such elements (Figure 16-5) to be designed by the special seismic load combinations of ASCE 7-05 Section 12.4.3.2. When the definition of maximum earthquake effect, E_m, is incorporated from ASCE 7-05 Section 12.4.3, IBC Equations 16-22 and 16-23 become:

$$(1.2 + 0.2S_{DS})D + f_1L + \Omega_0 Q_E \qquad (16\text{-}22)$$

$$(0.9 - 0.2S_{DS})D + \Omega_0 Q_E \qquad (16\text{-}23)$$

where Ω_0 is the system overstrength factor given in ASCE 7-05 Table 12.2-1, and mostly varies between 2 and 3.

Collectors and elements supporting discontinuous shear walls are designed for magnified forces (the estimated maximum axial forces that can realistically develop in these elements in an earthquake situation) so that they will not fail before the vertical resisting elements. This would ensure that they can deliver earthquake forces to or support the vertical resisting elements so that the vertical elements can dissipate energy through inelastic deformation. See similar discussion under ASCE 7-05 Section 12.10. An analogy is provided by an electrical circuit in which the wire (collector) is sized to safely carry more current than the capacity of the fuse (shear wall) to ensure that the fuse blows before the wire melts.

1605.5

Heliports and helistops. In this section, the IBC intends that the design of touchdown areas be regulated only when the touchdown area is on a building. The minimum loading criteria consist of the following:

1. The weight of the helicopter plus the snow load, *S*.
2. The landing impact effect of the helicopter.
3. A uniform live load of 100 psf.

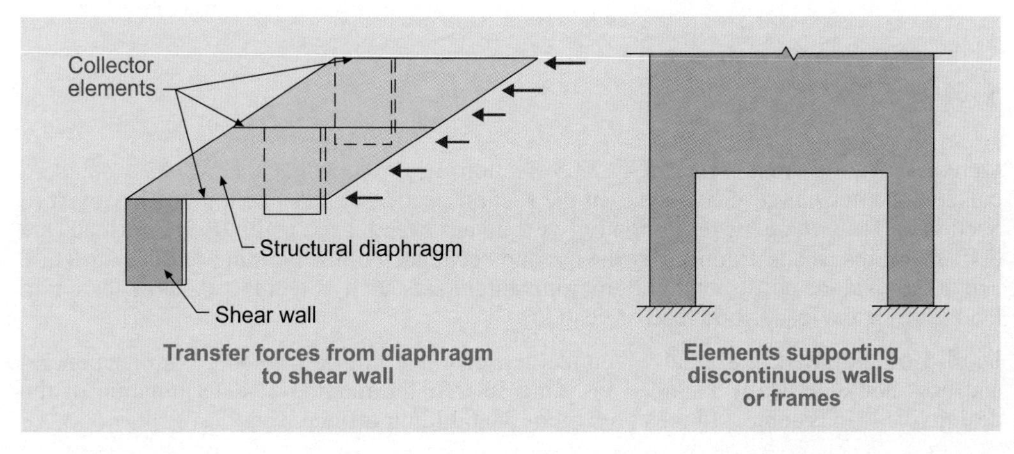

Figure 16-5 Collector elements for diaphragms and discontinuous shear walls

An exception to Item 3 reduces the uniform live load to 40 psf when the landing area is used by helicopters with gross weight not exceeding 3,000 pounds. A study pointed out that about 56 percent of all registered helicopters in the U.S. weigh less than 3,000 pounds, which is comparable to the weight of a small automobile. This, and the large area required by a helicopter to land make the equivalent uniform load in the range of only 2.1 – 4.3 psf. In those cases, a 100 psf design live load would result in excessively conservative design.

Section 1606 *Dead Loads*

This section is similar to ASCE 7-05 Section 3.1. According to the ASCE 7/IBC definition, dead loads consist of the weight of materials of construction incorporated into the building, including but not limited to walls, floors, roofs, ceilings, stairways, built-in partitions, finishes, cladding and other similarly incorporated architectural and structural items, and fixed service equipment, including the weight of cranes.

According to the commentary on ASCE 7-05, to establish uniform practice among designers, it is desirable to present a list of materials generally used in building construction, together with their proper weights. The solution chosen by Committee ASCE 7 has been to present, in the ASCE 7-05 Commentary, an extended list that will be useful to the designer and building official alike. This table, Table C3-2, is truly a valuable resource. It is conceded that special cases will unavoidably arise. Authority is therefore granted in both ASCE 7-05 and the IBC for the building official to deal with such cases.

Engineers, architects and building owners are advised in the ASCE 7-05 Commentary to consider factors that result in differences between actual and calculated loads. Conditions have been encountered in the past which, if not considered in design, may reduce the future utility of a building or reduce its margin of safety. The ASCE 7-05 Commentary points out two such conditions:

1. There have been numerous instances in which the actual weights of members and construction materials have exceeded the values used in design. Care is advised in the use of tabular values. Also, allowances should be made for such factors as the influence of formwork and support deflections on the actual thickness of a concrete slab of prescribed nominal thickness.
2. Allowance should be made for the weight of future wearing or protective surfaces where there is a good possibility that such may be applied. Special consideration should be given to the likely types and position of partitions, as insufficient provision for partitioning may reduce the future utility of a building.

The ASCE 7-05 Commentary also directs attention to the possibility of temporary changes in the use of a building, as in the case of clearing a dormitory for a dance or other recreational purposes.

Section 1607 *Live Loads*

1607.1 General. This is similar to ASCE 7-05 Section 4.1, except that the ASCE 7 standard contains the definition of live loads in the section itself, whereas Section 1607.1 refers to Section 1602.1, where the definition of live load has been placed in the IBC. Live loads are defined as those loads produced by the use and occupancy of the building or other structure and do not include construction or environmental loads such as wind load, snow load, rain load, earthquake load, flood load or dead load.

1607.2 Loads not specified. Essentially identical to Section 4.5 of ASCE 7-05. For occupancies and uses not specifically included in Table 1607.1, the method of determination of the design live load is subject to the approval of the building official.

Extremely valuable information is provided in the Commentary to ASCE 7-05, Tables C4-1 and C4-2, concerning the determination of design live loads for occupancies not listed in Table 1607.1.

1607.3

Uniform live loads. This charges the designer to use the unit live loads set forth in Table 1607.1 and specifies that these loads must be considered minimum live loads. In other words, floors must be designed for the maximum live loads to which they are likely to be subjected during the life of the building based on its intended use, but in no case should the design loads be less than those given in Table 1607.1.

Table 1607.1 is the same as Table 4-1 of ASCE 7-05, but for a few exceptions. The exceptions are detailed in Table 16-3.

The Commentary to ASCE 7-05 advises that in selecting the occupancy and use for the design of a building, the owner should consider the possibility of later changes of occupancy

Table 16-3. **Comparison between IBC Table 1607.1 and ASCE 7-05 Table 4-1**

IBC			ASCE 7		
Occupancy or Use	Min. Uniform Load (psf)	Min. Conc. Load (lb)	Occupancy or Use	Min. Uniform Load (psf)	Min. Conc. Load (lb)
4. Assembly areas and theaters			Assembly areas and theaters		
Follow spot, projection and control rooms	50		—		
—			Lobbies	100	
Stages and platforms	125		Platforms (assembly)	100	
			Stage floors	150	
12. Cornices	60		—		
13. Corridors, except as otherwise indicated	100		Corridors		
			First floor	100	
			Other floors, same as occupancy served except as indicated		
17. Garages: Trucks and buses	Section 1607.6		Garages: Trucks and buses	(a)	
18. Grandstands	(a)		Grandstands	100 (b)	
29. Reviewing stands, grandstands and bleachers	(a)		Reviewing stands, grandstands and bleachers	100 (b)	
33. Skating rinks	100		—		
34. Stadiums and arenas			Stadiums and arenas		
Bleachers	100 (a)		Bleachers	100 (b)	
Fixed seats (fastened to floor)	60 (a)		Fixed seats (fastened to floor)	60 (b)	

IBC:
(a) Design in accordance with the ICC *Standard for Bleachers, Folding and Telescopic Seating and Grandstands.*[6]

ASCE 7:
(a) Garages accommodating trucks and buses must be designed in accordance with an approved method, which contains provisions for truck and bus loadings.
(b) In addition to the vertical live loads, the design shall include horizontal swaying forces applied to each row of the seats as follows: 24 pounds per linear foot of seat applied in a direction parallel to each row of seats and 10 pounds per linear foot of seat applied in a direction perpendicular to each row of seats. The parallel and perpendicular horizontal swaying forces need not be applied simultaneously.

involving loads heavier than originally contemplated. The lighter loading appropriate to the first occupancy should not necessarily be selected.

Footnotes i, j and k of 2006 IBC Table 1607.1 provide the criteria as to when an attic can be considered to have limited or no storage, or when it can be considered habitable. The storage condition is ascertained from the open space available within the roof trusses based on their web configuration. These criteria are not found in ASCE 7-05. The three footnotes are reproduced below. Figure 16-6 illustrates the requirements.

Footnote i: Attics without storage are those where the maximum clear height between the joist and the rafter is less than 42 inches, or where there are not two or more adjacent trusses with the same web configuration capable of containing a rectangle 42 inches high by 2 feet wide, or greater, located within the plane of the truss. For attics without storage, this live load need not be assumed to act concurrently with any other live load requirement.

Footnote j: For attics with limited storage and constructed with trusses, this live load need only be applied to those portions of the bottom chord where there are two or more adjacent trusses with the same web configuration capable of containing a rectangle 42 inches high by 2 feet wide, or greater, located within the plane of the truss. The rectangle shall fit between the top of the bottom chord and the bottom of any other truss member, provided that each of the following criteria is met:

i. The attic area is accessible by a pull-down stairway or framed opening in accordance with Section 1209.2,

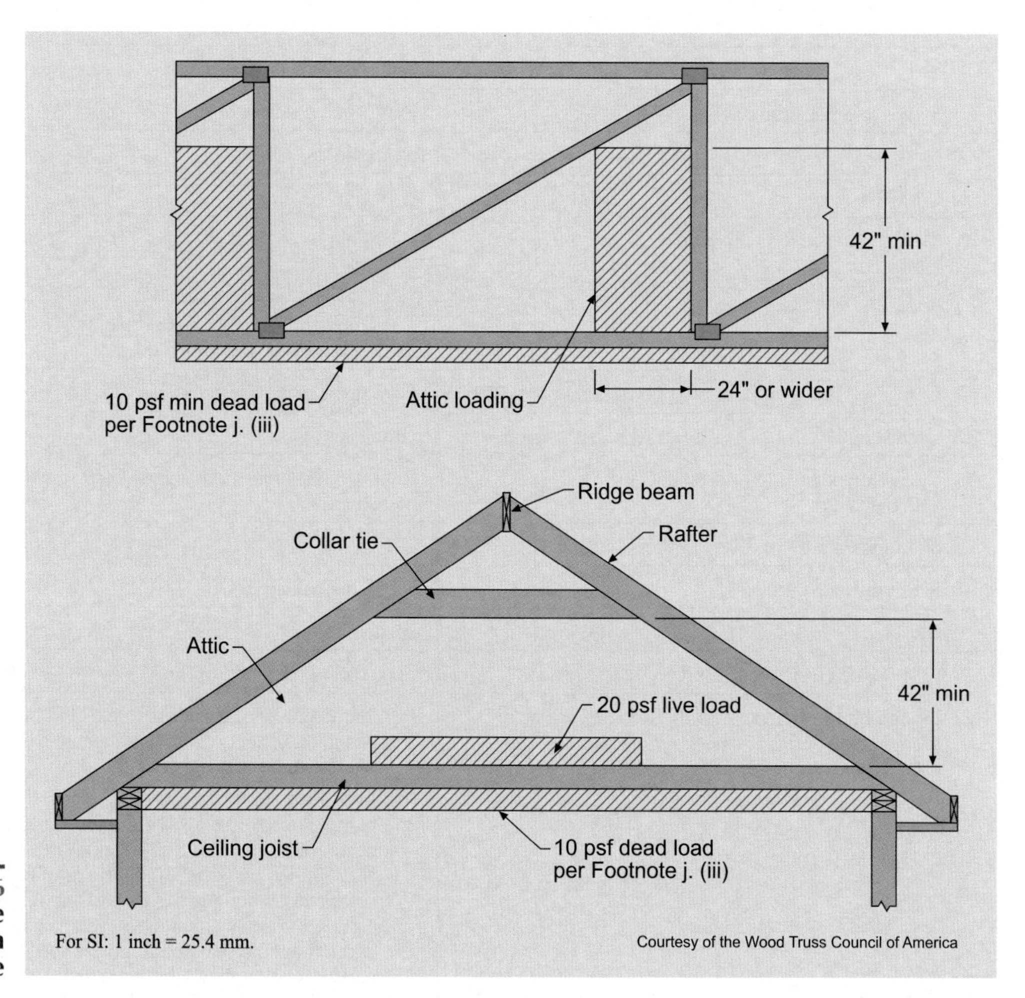

Figure 16-6 Unhabitable attics with limited storage

ii. The truss shall have a bottom chord pitch less than 2:12, and

iii. Bottom chords of trusses shall be designed for the greater of actual imposed dead load or 10 psf, uniformly distributed over the entire span.

Footnote k: Attic spaces served by a fixed stair shall be designed to support the minimum live load specified for habitable attic and sleeping rooms."

The Commentary to ASCE 7-05 also advises that "the building owner should ensure that a live load greater than that for which a floor or roof is approved by the authority having jurisdiction is not placed, or caused, or permitted to be placed, on any floor or roof of a building or other structure." The intent, of course, is to prevent the overloading of floors and roofs.

The Commentary to ASCE 7-05 provides background to the development of the design values in Table 4-1 of that standard. To solicit specific informed opinion concerning the design loads in the corresponding table of ANSI A58.1-1972, a panel of 25 distinguished structural engineers was selected. Design values and supporting reasons were requested of each panel member for each occupancy type. The information was summarized and circulated back to the panel members for a second round of responses; those occupancies for which previous design loads were reaffirmed, as well as those for which there was consensus for change, were included.

Many surveys of live loads in buildings, particularly office buildings, have been conducted over the years.[7,8,9,10] Buildings must be designed to resist the maximum live loads they are likely to be subjected to during some reference period, frequently taken as 50 years. Table C4-2 of the ASCE 7-05 Commentary briefly summarizes how load survey data are combined with a theoretical analysis of the load process for some common occupancy types and illustrates how a design load might be selected for an occupancy not specified in ASCE 7-05 Table 4-1.

Concentrated loads. **1607.4** Many uses are susceptible to the movement of equipment, files, machinery, etc. Therefore, the code requires that floors for these uses, which are listed in Table 1607.1, be designed for the indicated concentrated load placed on a space $2^1/_2$ feet (750 mm) square whenever this load, on an otherwise unloaded floor, produces stresses greater than those caused by the uniform loads required by the code. As this concentrated load can take many forms in the real world, and as the design structural engineer usually does not know in advance what form the load will take and how it will be applied, the best compromise to cover most situations is to consider the concentrated load to be applied through a rigid base $2^1/_2$ feet (750 mm) square.[11]

Footnote a to Table 1607.1 requires garages and other areas where motor vehicles are stored to be designed for concentrated loads or, alternatively, the uniform loads as specified in Table 1607.1. In the case of garages for storage of private pleasure-type vehicles, the note prescribes a single 3000-pound load over a $4^1/_2$-inch by $4^1/_2$-inch area. In 2000 IBC Table 1607.1, this concentrated load used to be 2000 pounds, based on the provisions of ASCE 7-98. In the 2003 IBC, it was increased to 3000 pounds because the uniform live load was reduced from 50 psf to 40 psf. The rationale behind the reduction of uniform live load was that unlike live load in a building, such as an office, where the loads are located randomly, a garage is loaded with vehicles in a regular pattern. For mechanical parking structures without slab or deck for the storage of passenger cars only, 2250 pounds of load per wheel is specified. For sidewalks, vehicular driveways and yards, subject to trucking, the concentrated wheel load is required to be applied over an area of 20 square inches. On stair treads, a minimum concentrated load of 300 pounds is required to be applied over an area of 4 square inches.

For all the concentrated loads specified in this section, the intent of the code is that each concentrated load shall be placed on the floor in such a position as to create maximum stresses in the structural members. Therefore, the loading condition, either uniform or concentrated, which provides the maximum stresses in the structural members, would be used for the design of the floor system.

1607.5 Partition loads. In those uses where the partitions are subject to change in locations, such as office buildings and flexible-plan school buildings, the code requires that the floor system be designed to support a uniformly distributed live load of 15 psf (0.74 kN/m^2). This requirement is irrespective of whether partitions are shown on the construction documents. The exception is where the specified live load exceeds 80 psf (3.83 kN/m^2). The 15 psf represents a reduction from the 20 psf specified in the 2003 IBC, based on ASCE 7-98. As explained in the Commentary to ASCE 7-05, the 15 psf value is arrived at by assuming 10-foot-high partition walls of wood or steel stud wall construction with $^1/_2$-inch gypsum board on each side, and arranged in a square grid of 10-foot sides. This assumption was thought to provide a fairly conservative estimate. However, the ASCE 7-05 Commentary also advises designers to consider a larger partition load if a higher density of partition is anticipated. It should be noted that the uniformly distributed load of 15 psf (0.74 kN/m^2) is considered by the code to be a live load. Thus, it should be included with other live loads in factored load combinations.

1607.6 Truck and bus garages. The live load values used in the design of structural floors for garages storing trucks or buses are to conform to the lane load requirements provided in Table 1607.6 or to a uniformly distributed live load of 50 psf (2.40 kN/m^2) of floor area, whichever produces the greater effect.[1] Table 1607.6 reproduces the lane loading requirements in AASHTO's Bridge Specifications.[12] The lane loads specified are based on truck trains or the concentrated wheel loads from standard load trucks with gross weight limitations. Truck train loads are uniform loads resulting from a series of closely spaced trucks, one after another. Section 1607.6.1 specifies the application of live loads so as to produce the maximum stresses.

1607.7 Loads on handrails, guards, grab bars and vehicle barriers. The entire section is mostly adopted from ASCE 7-05 Section 4.4, except that Section 1607.7.1.3, Stress increase, is not part of ASCE 7-05, while ASCE 7-05 Section 4.4.4, Loads on Fixed Ladders, is not found in the 2006 IBC. The requirements of this section are intended to provide an adequate degree of structural strength and stability to handrails, guards, grab bars and vehicle barriers.

1607.7.1 Handrails and guards. In the second exception to the requirement for the application of the 50 plf load to handrails and guards in areas not accessible to the general public with occupancy of less than 50, the 2006 IBC reduces the minimum load from 50 plf to 20 plf. In contrast, ASCE 7-05 completely exempts these buildings from the design distributed load and requires only the application of a single concentrated load of 200 pounds. These are the same occupancies listed in IBC Section 1013.3, which are allowed to have larger openings in the guard.

Also, in the first exception, the 2006 IBC clearly states that whereas one- or two-family dwellings are exempted from the requirement for a minimum distributed load of 50 plf, the requirement for a single concentrated load of 200 pounds, specified in Section 1607.7.1.1, still applies. The same is also implied in ASCE 7-05 Section 4.4.1, but not stated clearly.

The Commentary to ASCE 7-05 points out that loads that can be expected to occur on handrail and guardrail systems are highly dependent on the use and occupancy of the protected area. It further points out that for cases in which extreme loads can be anticipated, such as long straight runs of guardrail systems against which crowds can surge, appropriate increases in loading need to be considered.

1607.7.2 Grab bars, shower seats and dressing room bench seats. It is noted in the Commentary to ASCE 7-05 that when grab bars are provided for use by persons with physical disabilities, the design is governed by CABO A117 *Accessible and Usable Buildings and Facilities* Standard.[13]

1607.8 Impact loads. These provisions are also adopted from ASCE 7-05. Where unusual vibration or impact forces are likely to occur, their effect may be to produce additional stresses and deflections in the structural system. This section requires that the structural design takes these effects into account. Typically, the dynamic effects are approximated through the application of a static load equal in effect to the dynamic loads. In most cases, this is sufficient. However, in certain situations, a dynamic analysis may be necessary to properly consider the natural frequencies of vibration of a structure.

1607.9

Reduction in live loads. The live load reduction provisions of Section 1607.9.1 are based on Section 4.8.1 of ASCE 7-05, whereas the alternate floor live load reduction provisions of Section 1607.9.2 are those of Section 1607.5 of the 1997 UBC. The alternate floor live load provisions of 1997 UBC Section 1607.6, on the other hand, are those of Section 4.8.1 of ASCE 7-95.

The alternate floor live load reduction provisions of Section 1607.9.2 or the reduction of live load provisions of Section 1607.5 of the 1997 UBC, based on tributary floor area (Figure 16-7), represent the *original* live load reduction provisions that used to be in older editions of the ANSI A58.1 standard (predecessor to ASCE 7) and in all three model codes. Note that in the 2003 IBC, the alternative floor live load reduction equation is used for reducing floor live loads only, whereas in the 1997 UBC, it was used to reduce floor live loads as well as roof live loads.

The concept of, and method for, determining member live load reductions as a function of a loaded member's influence area, A_I, was first introduced into ANSI A58.1 in 1982 and was the first such change since the concept of live load reduction was introduced over 40 years ago. The revised method was the result of more extensive survey data and theoretical analyses. Figure 16-8, reproduced from the Commentary to ASCE 7-95, illustrates the

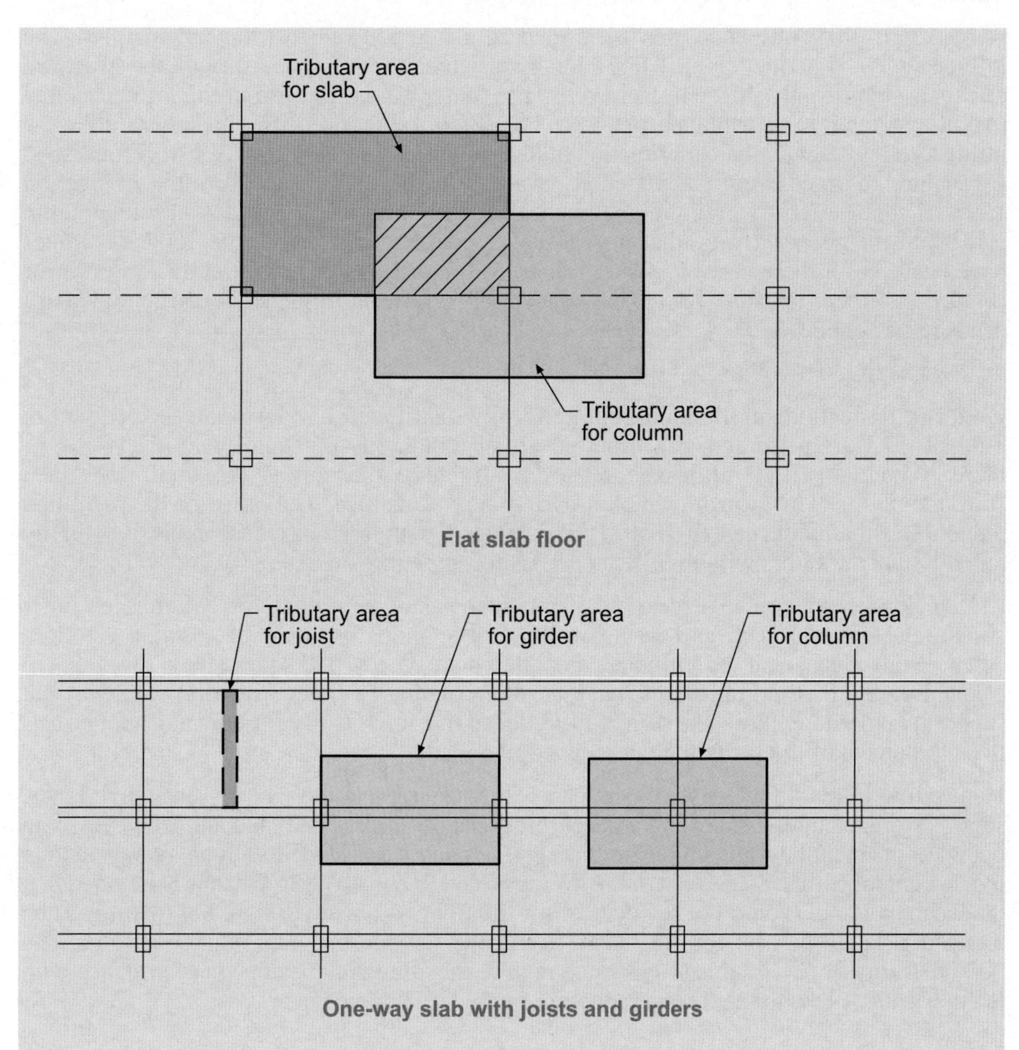

Figure 16-7
Tributary areas

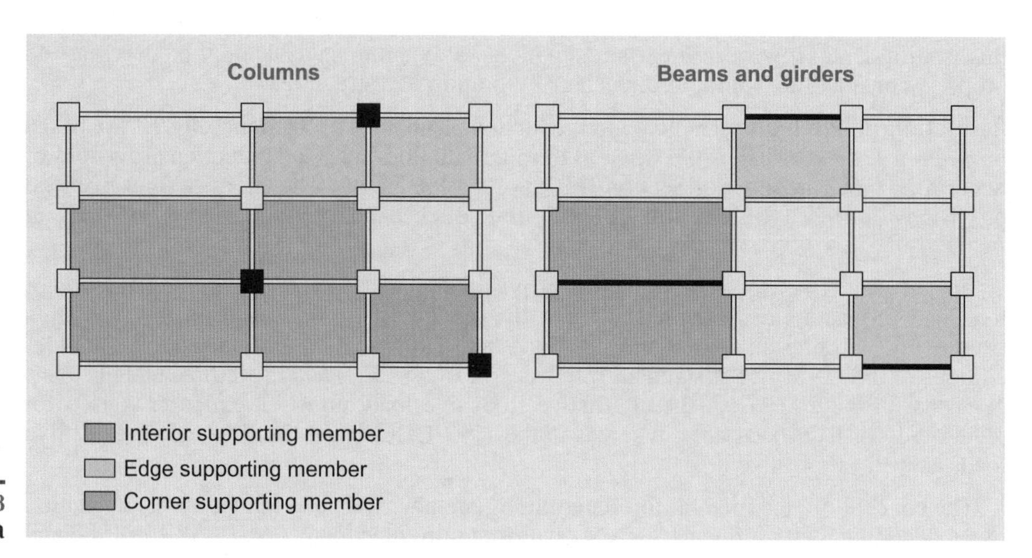

Figure 16-8
Influence area

influence area concept. The influence area-based live load reduction formula of ANSI A58.1-82, ASCE 7-88, ASCE 7-93 and ASCE 7-95 is used in the alternate floor live load reduction provisions of Section 1607.6 of the 1997 UBC.

In ASCE 7-05, influence area is defined as a function of the tributary area, A_T. The influence area is defined as that floor area over which the influence surface for structural effects is significantly different from zero. The factor K_{LL} is the ratio of the influence area (A_I) of a member to its tributary area (A_T), i.e., $K_{LL} = A_I/A_T$ and is used to better define the influence area of a member as a function of its tributary area. Table 1607.9.1 has established K_{LL} values (derived from calculated K_{LL} values) to be used in the formula of Section 1607.9.1 for a variety of structural members and configurations. K_{LL} values vary for column and beam members having adjacent cantilever construction, and Table 1607.9.1 values have been set for these cases to result in live load reductions that are slightly conservative (Figure 16-9). For unusual shapes, the concept of significant influence for structural effect needs to be applied.

See Example 16-4, Part 2 of this chapter.

1607.9.1 **General.** Reductions in the minimum uniformly distributed live load values specified in Table 1607.1 are permitted, based on an influence area of 400 square feet (37.16 m^2) or more. Essentially, the influence area of a structural element is the total floor area surrounding the element from which it derives any of its load. The basis for the permitted reduction is that in the design of structural elements with large influence areas, it is highly unlikely that the floors will be fully loaded over their entire area.

Live load reduction in excess of 50 percent is not permitted for columns or other structural elements (such as bearing walls) that support the loads of a single floor. In essence, this means that the influence area may not exceed 3600 square feet (334.4 m^2) in calculating the reduced unit floor live load. For columns or other structural elements that support two or more floors, the sum of the reduced live loads from all floors must not be less than 40 percent of the sum of the unreduced live loads.

1607.9.1.1 **Heavy live loads.** In the case of occupancies involving relatively heavy basic live loads, such as storage buildings, several adjacent floor panels may be fully loaded. However, data obtained in actual buildings indicate that rarely is any story loaded with an average actual live load of more than 80 percent of the average rated live load.[1] ASCE 7 thus concluded that the basic live load should not be reduced for the floor and beam design, but that it may be reduced a flat 20 percent for the design of members supporting more than one floor. The 2006 IBC further revises this provision to require that reduction be done in accordance with Section 1607.9.1, with maximum reduction limited to 20 percent.

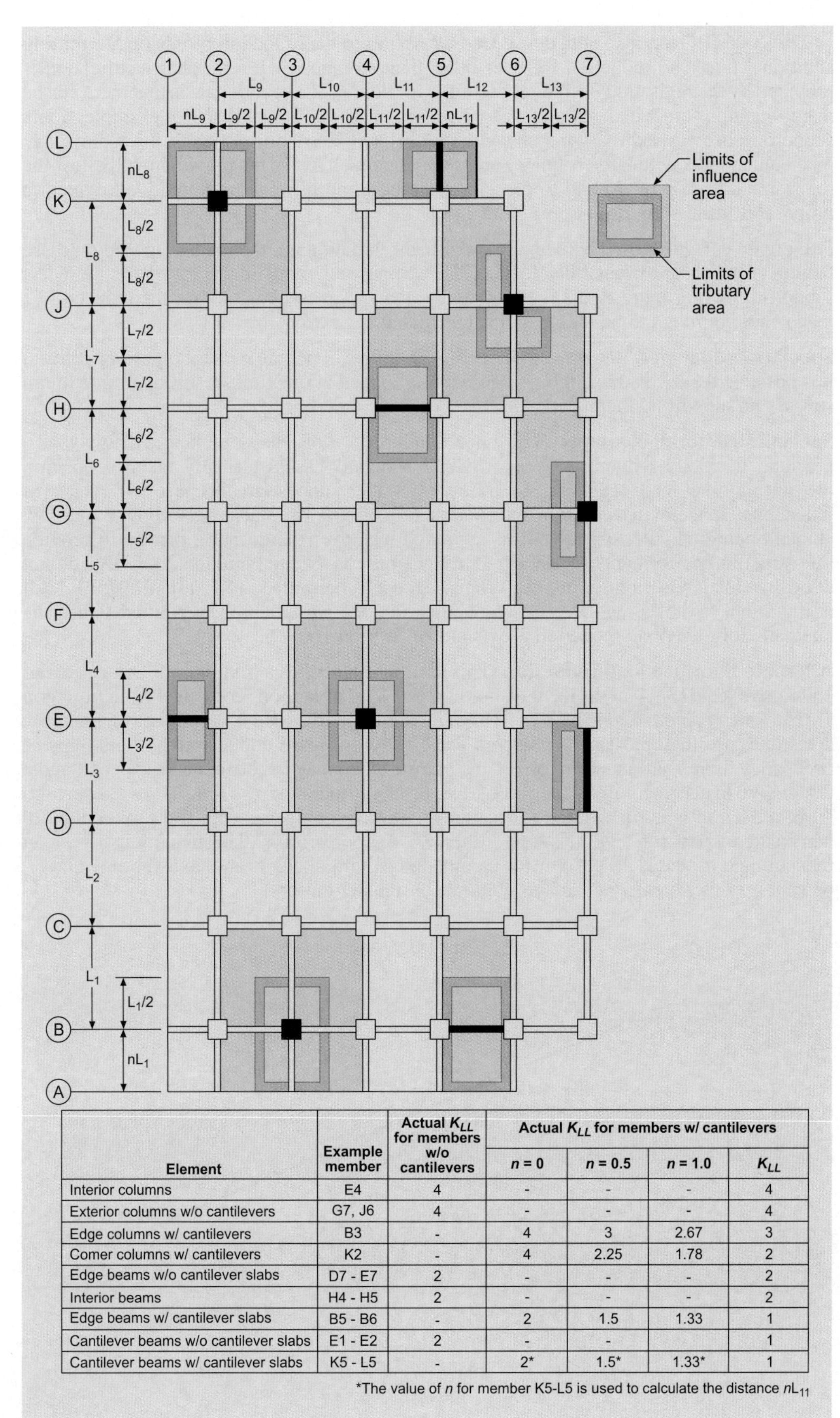

Element	Example member	Actual K_{LL} for members w/o cantilevers	Actual K_{LL} for members w/ cantilevers			
			$n = 0$	$n = 0.5$	$n = 1.0$	K_{LL}
Interior columns	E4	4	-	-	-	4
Exterior columns w/o cantilevers	G7, J6	4	-	-	-	4
Edge columns w/ cantilevers	B3	-	4	3	2.67	3
Comer columns w/ cantilevers	K2	-	4	2.25	1.78	2
Edge beams w/o cantilever slabs	D7 - E7	2	-	-	-	2
Interior beams	H4 - H5	2	-	-	-	2
Edge beams w/ cantilever slabs	B5 - B6	-	2	1.5	1.33	1
Cantilever beams w/o cantilever slabs	E1 - E2	2	-	-		1
Cantilever beams w/ cantilever slabs	K5 - L5	-	2*	1.5*	1.33*	1

*The value of n for member K5-L5 is used to calculate the distance nL_{11}

Figure 16-9
Typical tributary and influence area

The 2006 IBC has also added a second exception to the reduction provision, permitting additional live load reduction for uses other than storage when a rational justification is provided by the registered design professional for doing so. The rationale is that there can be many situations other than storage where live loads may exceed 100 psf. For example, floors supporting heavy machinery may have very localized high uniform loads, but the average load on members with large tributary areas may be much less. This provision will allow the registered design professional to present to the building official a rational load reduction proposal if those scenarios apply.

1607.9.1.2 **Passenger car garages.** There are no significant variations in the loads imposed on these facilities, which are often fully loaded. The 20-percent reduction allowed for members supporting two or more floors is based on the reason given above. Also, the reduced live load is not permitted to be less than that calculated in Section 1607.9.1.

1607.9.1.3 **Special occupancies.** Because of large concentrations of people in these types of facilities, it is possible that two adjacent bays can be fully loaded at one time, resulting in maximum stresses on supporting members.

1607.9.1.4 **Special structural elements.** The uncertainties of roof loads, including snow loads, preclude live load reductions. The ASCE 7 standard has historically taken a position prohibiting live load reduction for one-way slabs, supposedly because of a lack of redundancy inherent in the design and construction of one-way slabs. According to ASCE 7, "though admittedly conservative, this approach has, over the decades, proved to provide safe structural performance." ASCE 7-05 provisions, as adopted into the 2006 IBC, do not allow live load reduction for one-way slabs "except as permitted in Sections 1607.9.1.1." It is important to note that live load reduction for one-way slabs is permitted under the alternate floor live load reduction provisions of Section 1607.9.2.

1607.9.2 **Alternate floor live load reduction.** This section establishes a minimum tributary area of 150 square feet (13.94 m^2) as the threshold for live load reductions computed from Equation 16-25, which is plotted in Figure 16-10. No reduction in live load is permitted for assembly areas, because they must be considered to be fully occupied under normal conditions of occupancy. Live loads in excess of 100 psf (4.79 kN/m^2) may not be reduced, except that the design live loads on columns supporting two or more floors may be reduced by 20 percent. Also, reduction is not permitted for passenger vehicle garages except for a maximum 20 percent reduction for columns supporting two or more floors. The maximum live load reduction permitted is 40 percent for members receiving loads from one level only and 60 percent for other members (such as columns or transfer girders).

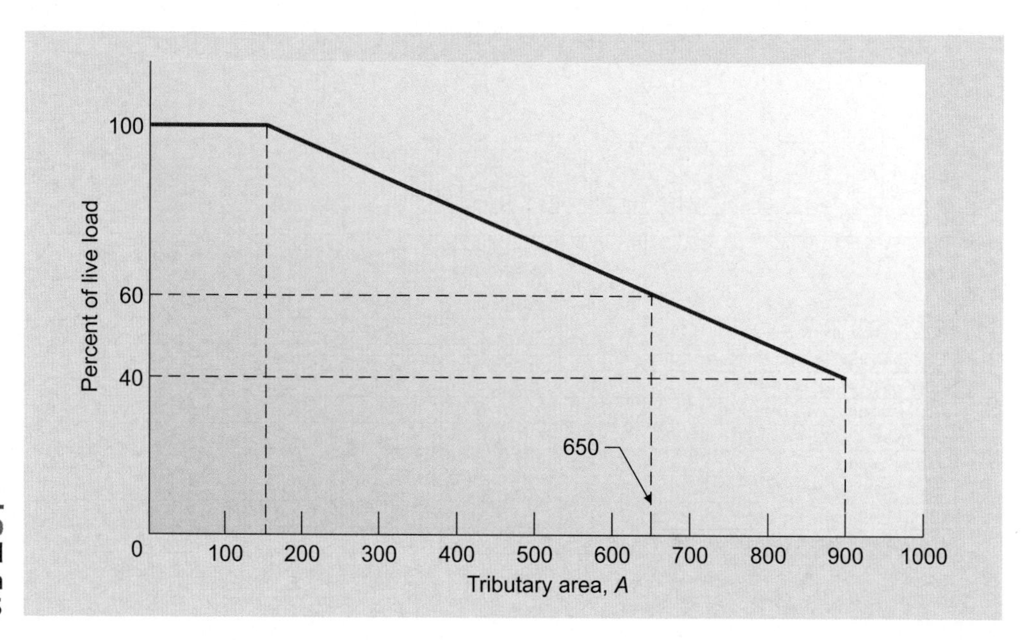

Figure 16-10
Live load reduction Equation 16-25

Note that the prohibition on live load reduction for parking structure deck members is new in the 2006 IBC. This eliminates a provision that existed in the 2003 IBC, which allowed a designer to take advantage of the alternate live load reduction provisions to design the floors and beams of a passenger vehicle parking garage for a uniform live load of only 24 psf, reduced from the 40 psf set forth in Table 1607.1. The 24 psf was inadvertent; a minimum reduced live load of 30 psf was prescribed in the 2004 Supplement to the 2003 IBC. This reduction was not consistent with the provisions of general live load reduction which never permitted reduction for passenger car garages (except for columns supporting two or more floors), but was consistent with the 30 psf lower limit set by two of the three legacy codes (the *Uniform Building Code* and the *Standard Building Code*[14]).

Equation 16-26 was derived so that if a structural member supporting a tributary area of sufficient size to qualify for the maximum reduction allowed by the equation were subjected to the full design live load over the entire area, the overstress would not exceed 30 percent.[15]

It may be noted from Equation 16-26 that the maximum live load reduction is proportional to the ratio of dead load to live load. Therefore, for heavy framing systems, the reduction is permitted to be greater than it would be for lighter framing systems. This is in view of the fact that for a given magnitude of overload on a structural system, the system with the heavier dead load is overstressed proportionately less than one with a lighter dead load. For example, if a floor system weighing 30 pounds per square foot (1.44 kN/m^2) and designed for a live load of 40 pounds per square foot (1.92 kN/m^2) were subjected to a 20-pounds-per-square-foot (0.96 kN/m^2) overload, the amount by which the structural system would be overstressed is about 30 percent, assuming the system was designed to support just the design live load of 40 pounds per square foot (1.92 kN/m^2). If this floor had a dead load of 60 pounds per square foot (2.88 kN/m^2), the overstress would be only 20 percent, again assuming that the system were designed to support just the 40-pounds-per-square-foot (1.92 kN/m^2) live load.

1607.10

Distribution of floor loads. Where loads are uniformly distributed on continuous structural members, they shall be arranged so as to create maximum bending moment in any given critical section. This may require a design to consider so-called skip loading or alternate span loading, as shown in Figure 16-11.

1607.11

Roof loads. In addition to dead and live loads (typically during construction), the structural system of a roof is to be designed and constructed to resist environmental loads caused by wind, snow and earthquakes.

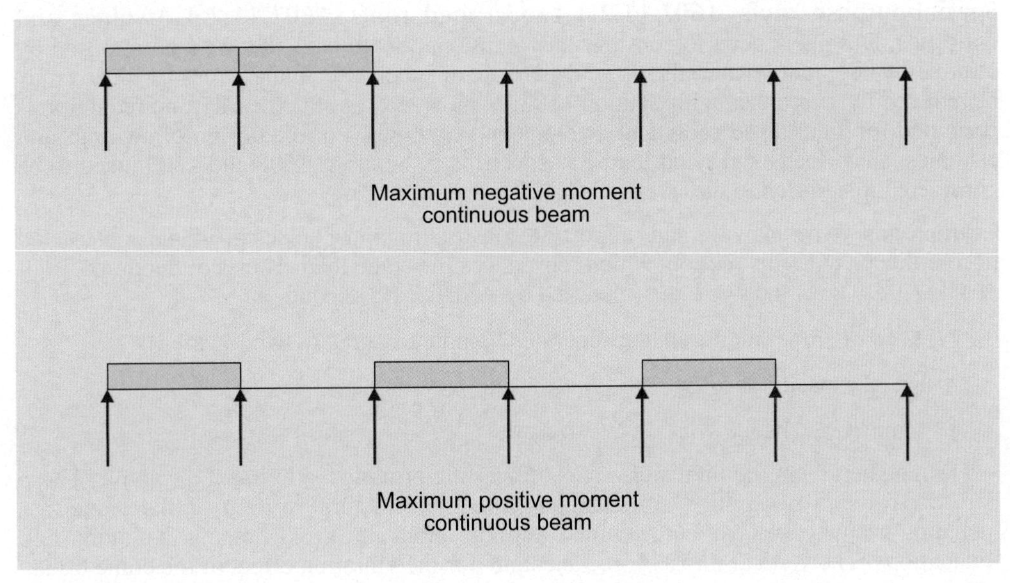

Figure 16-11
Alternate span loading of continuous beams

1607.11.1 Distribution of roof loads. This is nearly identical to Section 1607.10 (see discussion above). However, the provision applies only when the uniform roof live loads are reduced to less than 20 psf. This is in line with the provision in Footnote h to ASCE 7-05 Table 4-1. ASCE 7-05 Section 7.5 is referred to for partial snow load.

1607.11.2 Reduction in roof live loads. Section 1607.11.2.1 is essentially the same as Section 4.9.1 of ASCE 7-05. The reduced load values specified are meant to act vertically upon the projected area and have been selected as minimum roof live loads, even in localities where little or no snowfall occurs. This is because it is considered necessary to provide for occasional loading that is due to the presence of workers and materials during repair operations.[1] The live load reduction for roofs is a function not only of tributary area but also of the slope of the roof. This is because it becomes less probable that the loads on a roof member will be at maximum levels as the roof slope increases.

For *flat, pitched and curved roofs*, the reduction provision of the 2006 IBC does not actually change the final value of the reduced roof live load from that given by the 2003 IBC. In the 2003 IBC, the minimum live load value for these roof types was required to be determined by multiplying the reduction factors R_1 and R_2 by a base value of 20 psf. In the 2006 IBC, the same base value of 20 psf is moved to Table 1607.1 as the minimum uniformly distributed roof live load, and the same R_1 and R_2 factors are used. On the other hand, live load reduction is now possible for special-purpose roof live load values specified in 2006 IBC Table 1607.1. The loads are the same as those specified in the text of the 2003 IBC, but with the allowance of reduction in accordance with the floor live load reduction provisions of Section 1607.9. See also ASCE 7 Section 4.9.2. It is also worth noting that no live load reduction is allowed in the 2006 IBC for landscaped roofs or awnings and canopies.

Where the design roof snow load exceeds the minimum roof live load, the snow load, calculated in accordance with the requirements of Section 1608, is to be used for design of the roof, in accordance with the applicable load combination. Only the greater of the roof load value established by the minimum roof live load or the design snow load determined as indicated above is required to be applied to the roof.

Because the roof loads are handled differently in the 2006 IBC, in that the unreduced values are listed in Table 1607.1, a necessary revision is made to the roof live load reduction equation. In the 2003 IBC, the roof live load reduction equation was $L_r = 20R_1R_2$. In the 2006 IBC, the equation is $L_r = L_0R_1R_2$ where L_0 is the unreduced roof live load specified in Table 1607.1.

1607.11.2.2 Special-purpose roofs, 1607.11.2.3 Landscaped roofs, 1607.11.2.4 Awnings and canopies. Designers need to consider any additional dead loads that may be imposed by saturated landscaping materials. Special-purpose or special-occupancy roof live loads may be reduced in accordance with Section 1607.9.[8] However, as stated earlier, no reduction is permitted for landscaped roofs and awnings and canopies. For the design of awnings and canopies, snow loads and wind loads, as specified in Sections 1608 and 1609, need to be considered in addition to the live loads specified in Table 1607.1.

1607.12 Crane loads. All craneways and supporting construction must be designed and constructed in compliance with this section, which reproduces the crane load criteria of Section 4.10 of ASCE 7-05. The crane live loads specified by ASCE 7-05 depend on:

- Type of crane (monorail, cab-operated, pendant-operated, hand-geared)
- Rated capacity of crane
- Maximum wheel loads

Design lateral, longitudinal and vertical forces are provided in terms of the above. These live loads are to be applied simultaneously to the structural system of the craneway, including runway beams, connections, support brackets, cross-bracing, columns and foundations. The vertical impact force accounts for the vibration effect of the crane bridge movement and the movement of the lifted load. The lateral force (perpendicular to the runway girder) results from acceleration or deceleration of the trolley and the lifted load. The longitudinal force (parallel to the runway girder) results from the acceleration or deceleration of the bridge or the lifted load.

1607.13

Interior walls and partitions. The intent of this section is to provide sufficient strength and durability of wall framing and wall finish, so that a minimum level of resistance would be available to nominal impact loads that commonly occur in the use of a facility and to HVAC pressurization.

Loading provisions for "Fabric partitions" are added in the 2006 IBC as a subsection within the "Interior walls and partitions" provisions. Fabric partitions are defined in Section 1602 as follows:

> FABRIC PARTITIONS. A partition consisting of a finished surface made of fabric, without a continuous rigid backing, that is directly attached to a framing system in which the vertical framing members are spaced greater than 4 feet (1219 mm) on center.

The definition clearly differentiates them from other more traditional type partitions, which include partial-height office partitions that contain rigid panels finished with fabric and attached to a rigid frame. In the case of a fabric partition there is no rigid panel to which the fabric is attached. The fabric simply spans the open space between the rigid frame over which it is stretched and attached. In the definition, it states that the vertical framing members are spaced greater than 4 feet on center, again to differentiate this type of partition from a more traditional partition where the vertical framing members are spaced at 4 feet on center or less. Typically, these partitions are not intended to be full- (ceiling-) height, so they would normally not be attached directly to the ceiling. They are usually supported by the floor, except under conditions where, because of the layout and the height of the ceiling, it may be more appropriate to hang the partition from the ceiling grid or use a combination of floor supports and ties to the ceiling grid or a special structural ceiling grid designed to support such partitions. Figure 16-12 shows some examples of fabric partitions.

The intent behind the concentrated load of 40 pounds was to ensure that the partition not tip over if someone were to inadvertently lean up against the frame or the fabric and that a person inadvertently leaning against the fabric would not cause the fabric to tear and the

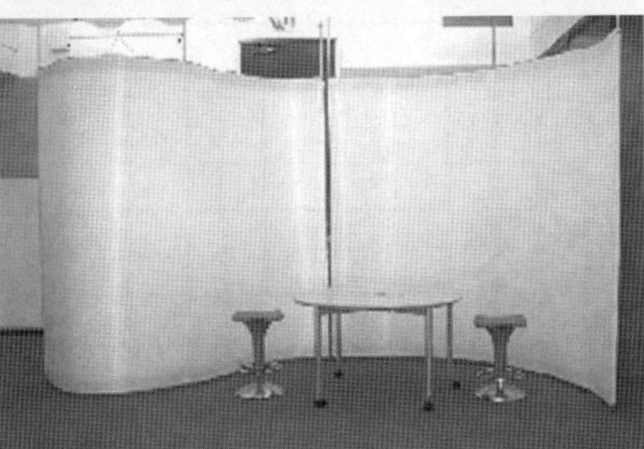

Courtesy of Herman Miller, Inc.

Figure 16-12 Applications of fabric partitions

person to fall abruptly. In the case of the 5 psf horizontal load, it was decided that the 5 psf is to be distributed by calculating the total load based upon the area of the fabric and then having that load distributed proportionally over the horizontal and vertical structural framing members of the partition. Thus, the framing system, in effect, will be resisting the total horizontal distributed load of 5 psf even though the 5 psf is not applied over the field of the fabric between the supports.

Section 1608 *Snow Loads*

The snow load provisions are adopted by reference to Chapter 7 of ASCE 7-05. Only the provisions regarding determination of ground snow loads in the contiguous United States, Alaska and Hawaii are left in the code, which many building officials will wish to specify when adopting the model code by local ordinance. Based on the provisions in the 2006 IBC and ASCE 7-05, Figure 16-13 has been developed to show the organization of the design snow load determination. The variables that must be determined for the calculation of the design snow loads include ground snow load, exposure factor, thermal factor and importance factor. Other considerations include a rain-on-snow surcharge, partial loading and ponding instability from melting snow or rain on snow. The discussions about these provisions are presented in the same order as shown in Figure 16-13.

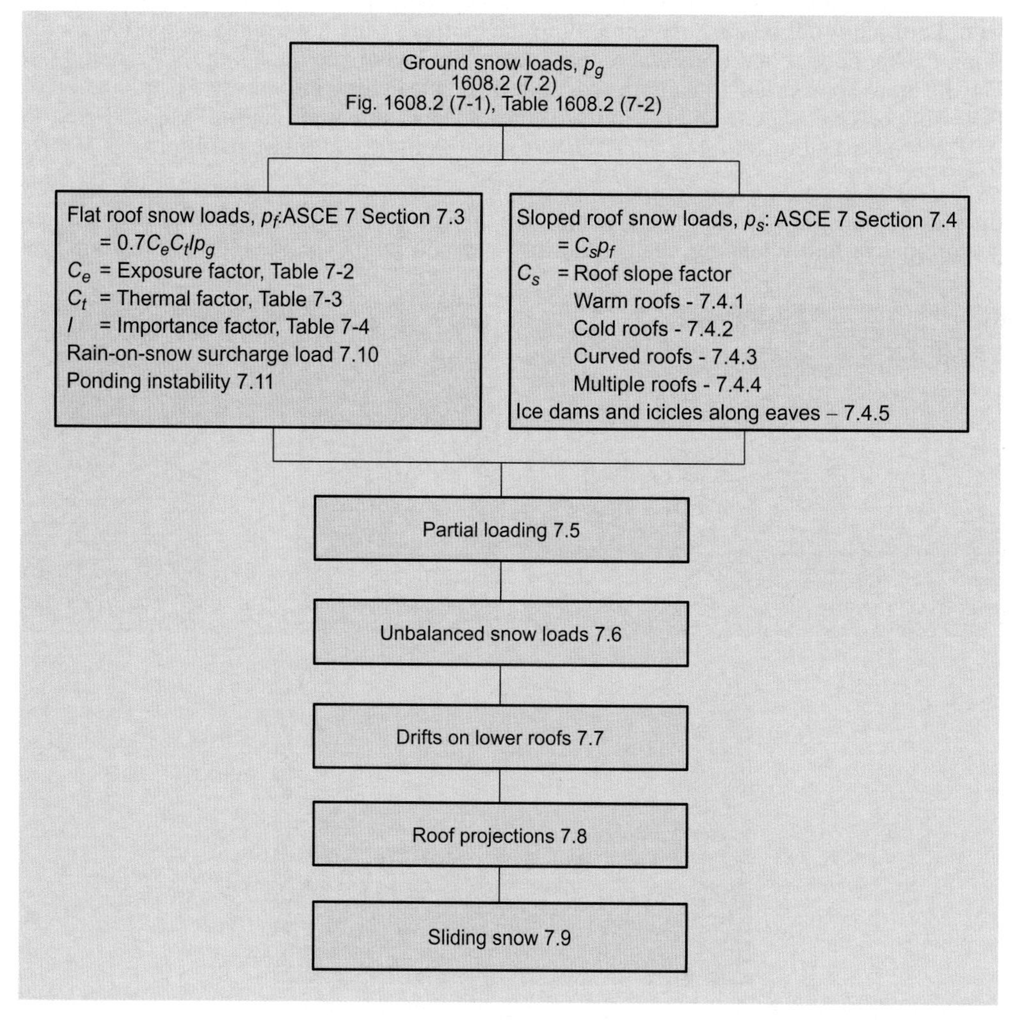

Figure 16-13 Snow loads provisions

General. The IBC stipulates in this section that design snow loads be determined in accordance with Chapter 7 of ASCE 7, but that the design roof load shall not be less than that determined by Section 1607, Live Loads. **1608.1**

Ground snow loads. This is nearly identical to Section 7.2 of ASCE 7-05. Figure 1608.2 and Table 1608.2 are reproductions of Figure 7-1 and Table 7-1 of ASCE 7-05, respectively. Figure 1608.2 for the contiguous United States and Table 1608.2 for Alaska give the ground snow loads to be used in determining the design snow loads for roofs. The 2006 IBC states: "Snow loads are zero in Hawaii, except in mountainous regions as approved by the building official." ASCE 7-05 also contains very similar language. The ground snow loads on the map and the table are generally based on snow depths recorded over a period approaching 50 years. The snow loads on the maps have a 2-percent probability of being exceeded (a 50-year mean occurrence interval). The mapped values indicate the ground snow loads in pounds per square foot (psf). In mountainous areas, the map indicates the highest elevation up to which it is appropriate to use a given snow load. For elevations higher than indicated on the map, a site-specific case study is required to determine the appropriate snow load. Some of the mountainous areas of the western United States require site-specific case studies irrespective of elevation. Assistance in the determination of an appropriate ground snow load for these areas may be obtained from the U.S. Department of Army Cold Regions Research and Engineering Laboratory in Hanover, New Hampshire. **1608.2**

ASCE 7-05 Section 7.3 Flat Roof Snow Loads. This section converts the ground snow load, p_g, to flat roof snow load, p_f. A flat roof is defined as one with a slope equal to or less than 5°. However, even when the roof slope is more than 5°, the flat roof provisions are necessary for the calculation of sloped roof snow load, p_s.

ASCE 7-05 Section 7.3.1 Exposure Factor, C_e. The roof exposure factor depends on the wind exposure category (or terrain category) defined in Chapter 6 of ASCE 7-05 and the exposure of the roof, as described in the footnotes to Table 7-2. The roof snow load is higher for a site in a wooded area than it is for an adjacent site that is flat and open. Attention should be paid to the footnotes to Table 7-2, which requires that the terrain category and roof exposure condition chosen shall be representative of the anticipated conditions during the life of the structure. An exposure is required to be determined for each roof of a structure.

ASCE 7-05 Section 7.3.2 Thermal Factor, C_t. The thermal factor takes into account the heat that is transmitted from the interior of the structure and reduces the snow depth on the roof. Thermal conditions considered in the determination of the thermal factor, C_t, from Table 7-3 are required to be representative of the anticipated conditions during winters for the life of the structure.

ASCE 7-05 Section 7.3.3 Importance Factor, *I*. The snow load importance factor attempts to address the need to relate design loads to the consequences of failure. Roofs of most structures having normal occupancies and functions are designed with an importance factor of 1.0, which corresponds to unmodified use of the statistically determined ground snow load for a 2-percent annual probability of being exceeded (50-year mean recurrence interval). Lower and higher risk situations are established using the importance factors for snow loads, which range from 0.8 to 1.2. The factor of 0.8 corresponds to an annual probability of being exceeded of about 4 percent (about a 25-year mean recurrence interval). The factor of 1.2 is nearly that for a 1-percent annual probability of being exceeded (about a 100-year mean recurrence interval).[1] This factor has been explained further under Commentary Section C7.3.3.

ASCE 7-05 Section 7.3.4 Minimum Value of p_f for Low-Slope Roofs. As explained in the Commentary to ASCE 7-05, this section accounts for a number of situations where the basic exposure factor of 0.7 as well as the factors C_e and C_t do not apply, such as snow load from a single storm in an area where the ground snow load p_g is less than 20 psf.

In ASCE 7-02, for hip and gable roofs the minimum flat roof snow load was applicable when the slope of the roof was less than or equal to (70/*W*)+0.5 degrees, where *W* is measured in feet. This requirement alone made the critical slope very small for very wide roofs. ASCE 7-05 puts a lower limit of 2.38° (1/2 on 12) on this critical slope, a value considered to be reasonable for wide gable roofs from a study of case histories. If the roof slope now is less than the larger of 2.38° and (70/*W*)+0.5 degrees, the minimum flat roof snow load requirement applies. This is illustrated by the shaded region in Figure 16-14. It may be noted that the new 2.38° value governs only when *W* is greater than approximately 37 feet.

ASCE 7-05 Section 7.4 Sloped Roof Snow Loads. The design snow load for a sloped roof is given as the design snow load for a flat roof multiplied by a roof slope factor that is given by Section 7.4.1 for warm roofs, Section 7.4.2 for cold roofs, Section 7.4.3 for curved roofs and Section 7.4.4 for multiple folded plate, sawtooth and barrel vault roofs.

ASCE 7-05 Section 7.5 Partial Loading. In many situations, a reduction in snow load on a portion of a roof by wind scour, melting or snow removal operations may simply reduce the stresses in the supporting members. However, in other cases, removal of snow from an area may induce heavier stresses in the roof structure than can occur when the entire roof is loaded. Cantilevered roof joints are a good example; removing half the snow load from the cantilevered portion increases the bending stress and deflection of the adjacent continuous span. In some situations adverse stress reversals may result.[1] This section requires consideration to be given in design to those adverse situations.

For the same reason as described under Section 7.3.4, the same lower limit of 2.38° on the critical slope expression of (70/*W*)+0.5 degrees is added in ASCE 7-05 for determining if partial loading needs to be considered for a gable roof. The shaded region in Figure 16-14 represents where partial loading needs to be applied.

ASCE 7-05 Section 7.6 Unbalanced Snow Loads. Snow on the roof of a building rarely accumulates evenly. The code intends that the designer investigate conditions of imbalance by requiring that roof designs consider unbalanced snow loading. One case would be the loading of one slope of a gable roof with snow while the other slope is unloaded (Figure 16-15). If this creates a less favorable condition in the design of the roof structural members, then it is the loading to be considered. This type of loading covers the case where the snow may have been removed from one side of the roof due to sliding or melting.[11]

The same lower limit of 2.38°, as discussed under ASCE 7-05 Sections 7.3.4 and 7.5.1, is imposed in the 2006 IBC for evaluating if unbalanced snow loads are required to be considered on hip or gable roofs. Thus, roofs with slopes less than the larger of 2.38° and

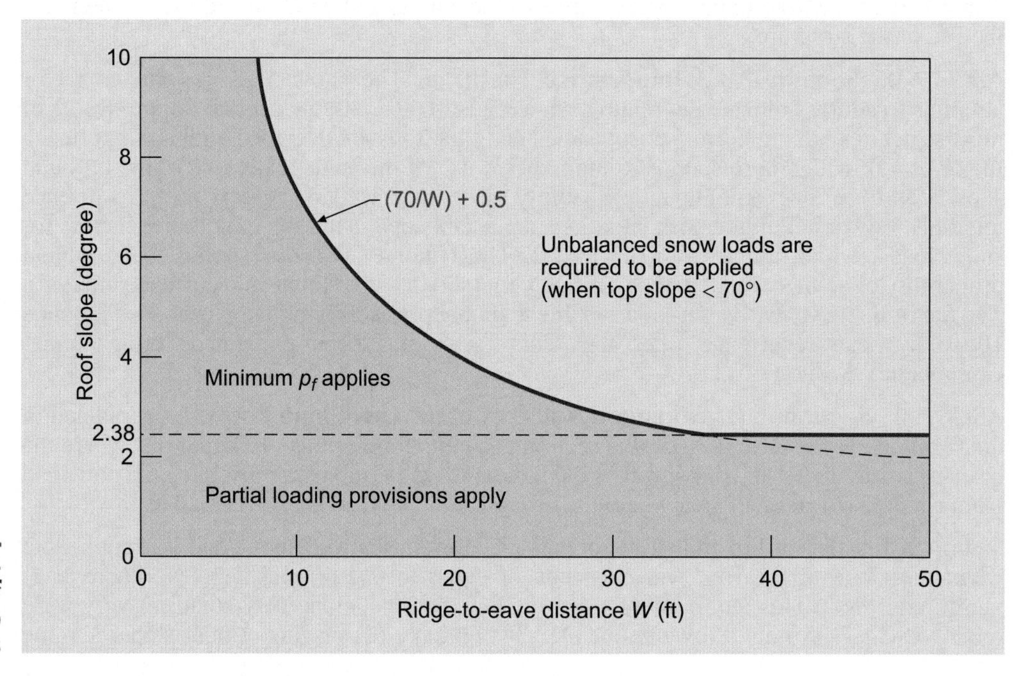

Figure 16-14 Limiting roof slope for hip or gable roofs

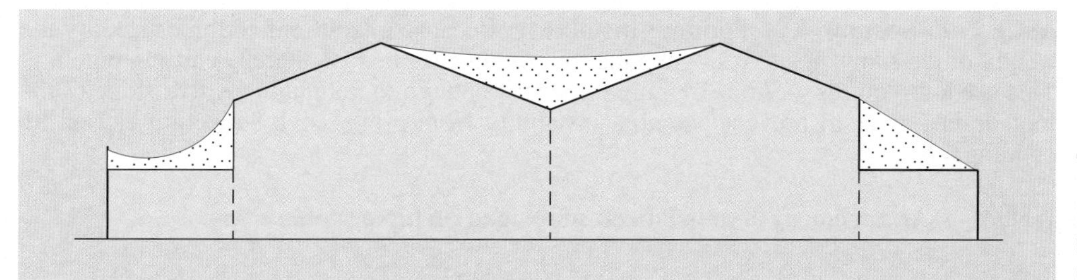

Figure 16-15
Potential snow slope for hip and gable roofs

(70/*W*)+0.5 degrees are not required to be designed for unbalanced snow loads. The other condition for not having to apply this unbalanced snow load, i.e., when slopes are larger than 70°, remains unchanged. The unshaded region in Figure 16-14 indicates roof slopes for which unbalanced snow load needs to be applied.

An additional major revision is made in the values as well as the distribution of the unbalanced snow load on qualifying hip and gable roofs. The change is illustrated in Table 16-4. It has been seen that the load from the triangular unbalanced snow deposited on top of the uniform balanced snow on the leeward side of the roof can be simplified by an equivalent uniform load, which gives conservative estimates of the design moment and shear forces on the rafters over those resulting from the actual triangular loading. However, ASCE 7-02 stipulated that this equivalent uniform load be applied on the entire ridge-to-eave span (W) for roofs with $W < 20$ feet as well as those with $W > 20$ feet, changing only the magnitude of the load for those two categories. ASCE 7-05 applies this equivalent load over only a part of the ridge-to-eave span for roofs with $W > 20$ feet, recognizing that the unbalanced triangular deposit of snow might not cover the entire leeward side. The distance up to which this equivalent load is to be considered can be seen in Table 16-4 and is arrived at by having the centroid of the equivalent rectangle coincide with that of the actual triangular load distribution. The intensity of the equivalent load (height of the rectangular block) is calculated by equating the cross sectional area of the triangular load distribution to that of a rectangular distribution with the same length as the triangular distribution. The load values for both cases are also revised from those given in ASCE 7-02 and can be seen in Table 16-4.

ASCE 7-05 Section 7.7 Drifts on Lower Roofs (Aerodynamic Shade). It is extremely important to consider localized drift loads in designing roofs because drifts onto lower roofs are a common cause of roof failures after a heavy snow (Figure 16-15). This section separately addresses windward drift that forms on the windward side of a high-bay wall area, and leeward drift that forms on the leeward side of a high-bay wall area. Equations and design details for drift loading are given.

ASCE 7-05 Section 7.8 Roof Projections. Solar panels, mechanical equipment, parapet walls and penthouses are examples of roof projections that may cause windward drifts on the roof around them. The drift-load provisions of this section cover most of these situations adequately.

ASCE 7-05 Section 7.9 Sliding Snow. The snow that slides from a higher sloped roof imposes loads that are in addition to the snow load already on the lower roof. This section requires consideration of such additional loads in roof design.

ASCE 7-05 Section 7.10 Rain-on-Snow Surcharge Load. This load accounts for the increased weight of snow after it rains on the snow. It has been shown by O'Rourke and Downey[16] that the major portion of the rain-on-snow surcharge load comes from the rainwater flowing along the slope of the roof by percolating through the snow layer immediately above the roof. Thus, the surcharge load increases with larger ridge-to-eave span W and smaller roof slope. Based on this, the expression for limiting slope has been changed from a constant value of 2.38° ($^1/_2$ on 12) in ASCE 7-02 to W/50 degrees, in ASCE 7-05, where W is expressed in feet.

ASCE 7-05 Section 7.11 Ponding Instability. Because of roof deflection caused by the weight of the snow, positive slope to the roof drains may be lost, thereby causing ponding. This section requires the roof framing to be stiff enough to maintain positive slope to the roof drains, so as to prevent potential instability from progressive deflection caused by ponding.

Table 16-4. **Arrangement of unbalanced snow load on hip or gable roofs**

W	ASCE 7-02	ASCE 7-05
$W \leq 20$ ft	$1.5p_s/C_e$ C_e = Exposure factor from Table 7-2	$I \times p_g$
$W > 20$ ft	$1.2(1 + \beta/2) \times p_s/C_e$ $0.3\ p_s$ β = 1.0 for $p_g \leq 20$ psf; β = 1.5 - 0.025 p_g for $20 \leq p_g \leq 40$ psf; β = 0.5 for $p_g \geq 40$ psf	$8h_d\sqrt{S}/3$ $h_d\gamma/\sqrt{S}$ p_s $0.3\ p_s$ $2h_d/\sqrt{S}$ 1 S γ = snow density; h_d = height of snow drift from Figure 7-9

Section 1609 *Wind Loads*

1609.1 Applications. The following basic requirement is specified: wind loads must not be decreased by considering the effect of shielding by other structures.

1609.1.1 Determination of wind loads. In general, wind pressures must be assumed to come from any horizontal direction and to act normal to the surfaces considered. Section 6 of ASCE 7-05 is to be used to determine wind forces, with two exceptions allowed for building structures:

1. Subject to the limitations spelled out in Section 1609.1.1.1, the SBCCI SSTD 10-99 *Standard for Hurricane Resistant Residential Construction*[17] is permitted to be used for applicable Group R2 and R3 buildings. SSTD 10 is a standard that was originally developed in 1990 with the intent of providing prescriptive provisions for wind-resistant design and construction of one- and two-story residential buildings of wood-framed construction, and one-, two- and three-story residential buildings of concrete and masonry construction sited in high-wind regions.
2. Subject to the limitations of Section 1609.1.1.1, the AFPA *Wood Frame Construction Manual for One- and Two-Family Dwellings*[18] is permitted to be used. This manual gives engineering and prescriptive requirements for the design and construction of one- and two-family dwellings of wood-framed construction.

Figure 16-16 has been prepared to serve as a road map through the wind design provisions of Section 1609.

A background to the wind load provisions of U.S. model codes and standards has been provided in Appendix 2 of this publication. Included is background on the wind design provisions of ASCE 7-05. The ASCE 7 Commentary on the wind design provisions of the 2005 standard is fairly extensive. Also, Example 16-5, Part 2, of this chapter has been prepared to illustrate the application of the analytical design procedure of that standard. The example can serve as a road map through much of Chapter 6 of the ASCE 7-05 standard. The simplified provisions for low-rise buildings of Section 6.4 are also illustrated through a complete design example. The basic steps of the simplified design procedure are discussed below.

Simplified provisions for low-rise buildings. The simplified provisions are applicable to the components and claddings of buildings that satisfy all the following conditions:

1. Enclosed building.
2. Flat roof, gabled roof with $\theta \leq 45°$, or hipped roof with $\theta \leq 27°$.
3. Mean roof height is less than or equal to 60 feet (18.3 m).
4. The building does not have response characteristics making it subject to across wind loading, vortex shedding and instability due to galloping or flutter. Also, the

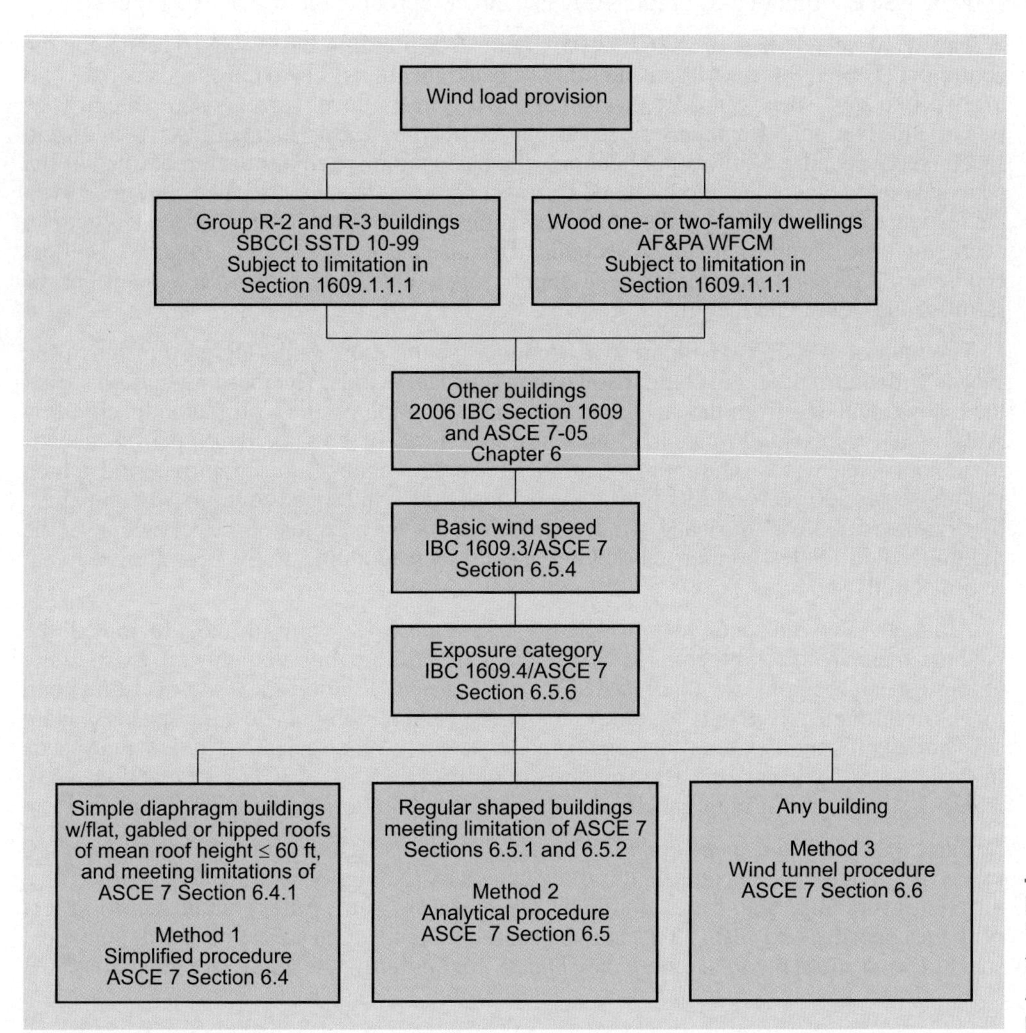

Figure 16-16 "Road map" through the IBC/ASCE 7 wind load provision

building is not to be located on a site for which channeling effects or buffeting in the wake of upwind obstructions warrant special consideration.

5. The building is regular-shaped.

In addition, the building must meet further requirements for its wind-force-resisting system (MWFRSs) to qualify for the simplified design procedure, including simple diaphragm buildings with no expansion joint or structural separation, exempted from torsional load cases, and not classified as a flexible building. Note that if only the components and claddings of a building qualify to be designed by the simplified procedure, then the MWFRSs need to be designed by the analytical procedure or wind-tunnel procedure.

Some examples of simple diaphragm buildings are houses with plywood shear walls, typical CMU (concrete masonry unit) wall buildings, concrete frame buildings and steel frame buildings with vertically spanning walls and floor and roof diaphragms. Metal buildings with horizontal girts that span between frames are not simple diaphragm buildings.

Much of the simplicity of the procedure derives from the fact that internal pressures are not involved in the design pressures for the MWFRS. Because the wind forces are delivered to the MWFRS via floor and roof diaphragms, and the building, by definition, is enclosed, the internal pressures simply do not come into play (or cancel out).

Step 1: Basic Wind Speed. The basic wind speed is to be determined from IBC Figure 1609 or ASCE 7 Figure 6-1. Figure 1609 is a direct reproduction of ASCE 7 Figure 6-1.

The wind speed map of ASCE 7-05 Figure 6-1 presents basic wind speeds for the contiguous United States, Alaska and other selected locations. The wind speeds correspond to 3-second gust speeds at 33 feet (10 m) above ground level for exposure category C. Because the National Weather Service has phased out the measurement of fastest-mile wind speed (see Appendix 2 of this publication), the basic wind speed has been redefined as the peak gust that is recorded and archived for most National Weather Service stations. Given the response characteristics of the instrumentation used, the peak gust is associated with an averaging time of approximately 3 seconds.[1] Because the wind speeds of Figure 6-1 reflect conditions at airports and similar open-country exposures, they account for the effects of significant topographic features.

The map of ASCE 7-05 Figure 6-1 accounts for the more rapid increase of hurricane speeds with return period in comparison to nonhurricane winds. The map is specified so that the loads calculated from the standard, after multiplication by the load factor specified by ASCE 7-05, represent ultimate loads having approximately the same return period as loads for nonhurricane winds. The approach required selection of an ultimate return period, which was chosen as 500 years. A set of design level hurricane wind speed contours were obtained by dividing 500-year hurricane wind speed contours by a factor of 1.225, which is the multiplier that relates nonhurricane wind speeds corresponding to 50-year and 500-year return periods.[1]

Although the wind speed map of ASCE 7-05 Figure 6-1 is valid for most regions of the country, there are special regions in which wind-speed anomalies are known to exist. Some of these special regions are noted in ASCE 7-05 Figure 6-1. Winds blowing over mountain ranges or through gorges or river valleys in these special regions can develop speeds that are substantially higher than the values indicated on the map. When selecting basic wind speeds in these special regions, the IBC requires conformance with ASCE 7-05 Section 6.5.4. ASCE 7 advises use of regional climatic data and consultation with a wind engineer.

Since the basic wind speed of ASCE 7 is now based on 3-second gust speed at 33 feet (10 m) above ground in open country, it is often necessary that regional climatic data based on a different averaging time, for example, hourly mean or fastest mile, be adjusted to reflect peak gust speeds at 33 feet (10 m) above ground in open country. The results of statistical study of wind speed records, reported by Durst[19] for nonhurricane winds, are given in ASCE

7-05 Figure C6-2, which defines the relation between wind speed averaged over t seconds, V_t and over 1 hour, V_{3600}.

Using the Durst curve for the conversion of fastest mile wind speeds (see Table 16-5), which indicates that for the range of fastest-mile wind speeds examined (70 mph to 140 mph), a constant difference of 20 mph can be conservatively assumed for fastest-mile wind speeds of 80 mph or more, and a constant difference of 15 mph for fastest-mile wind speeds of 70 and 75 mph.

When referenced documents are based on fastest-mile wind velocities, the 3-second gust wind velocities of Figure 1609 can be converted to fastest-mile wind velocities using Table 1609.3.1.

Table 16-5. **3-second gust velocity versus fastest mile wind velocity**

V_{fm}	sec./hr.	sec./mile	Ratio of V_{fm}/V_{mh}	V_{mh}	Ratio of V_{3s}/V_{mh}	V_{3s}
70	3,600	51.4	1.27	55.1	1.53	84.5
80	3,600	45	1.28	62.5	1.53	95.6
90	3,600	40	1.29	69.8	1.53	106.7
100	3,600	36	1.34	76.3	1.53	116.8
120	3,600	32.7	1.32	83.3	1.53	127.5
130	3,600	30	1.33	90.2	1.53	138.0
140	3,600	27.7	1.34	97	1.53	148.4

V_{fm} = Fastest mile velocity
V_{mh} = Mean hourly velocity
V_{3s} = 3-second gust velocity

Step 2: Importance factor. The wind load importance factors of Section 6.5.5 of ASCE 7-05 are given in Table 6-1 of that standard based on the occupancy category of the building. It may be recalled that Table 1604.5 of the 2006 IBC defines the four occupancy categories of buildings.

The wind importance factor is used to adjust the level of structural reliability of a building or other structure, depending on its occupancy. The importance factors given in Table 6-1 adjust the wind velocity pressure to different annual probabilities of being exceeded. Importance factors of 0.87, 1.00 and 1.15 are, for nonhurricane winds, associated with annual probabilities of being exceeded of 0.04, 0.02 and 0.01, respectively (mean recurrence intervals of 25, 50 and 100 years, respectively).

In hurricane-prone regions with basic wind speed V exceeding 100 mph, I is changed from 0.87 to 0.77 in Table 6-1. For wind speeds greater than 100 mph, the annual probabilities of exceedance implied by the importance factors of 0.77 and 1.15 will vary along the coast; however, the resulting risk levels associated with the use of these importance factors will be approximately consistent with those applied to wind speeds less than 90 mph.[1]

Further discussion has been provided in ASCE 7-05 Commentary Section C6.5.5.

Step 3: Exposure category. IBC Section 1609.4 requires that an exposure category that adequately reflects the characteristics of ground surface irregularities be determined at the site for each wind direction considered.

In the 2006 IBC, exposure category definitions are restructured to be consistent with the ASCE 7-05 provisions. Thus, instead of determining the exposure category of a building directly from the ground surface irregularities, as was done in the 2003 IBC, a Surface Roughness Category is first determined for the building site, and the Exposure Category of the building is then determined based on the Surface Roughness Category prevailing over specified distances in the upwind direction. The definitions for the three types of surface roughness categories (B, C and D) and the three types of exposure categories (B, C and D) are identical to those given in ASCE 7-05.

Once a surface roughness category has been established for a given wind direction in accordance, an exposure category is determined from the provisions of Section 1609.4.3. Exposure B applies where Surface Roughness B prevails in the upwind direction for a distance of at least 2600 feet (792 m) or 20 times the height of the building, whichever is greater. An exception to this permits the upwind distance to be reduced to 1500 feet (457 m) for buildings with mean roof heights less than or equal to 30 feet (9.1 m). Exposure D is applicable when Surface Roughness D prevails in the upwind direction for a distance of at least 5000 feet (1524 m) or 20 times the building height, whichever is greater. This exposure is to extend into downwind areas of Surface Roughness B or C for a distance of 600 feet (200 m) or 20 times the building height, whichever is greater. Exposure C applies in all cases where Exposures B and D do not apply.

Appendix Section C6.5.6 of ASCE 7-05 contains a detailed discussion on the revisions that were made to the definitions of the exposure categories. It also contains a method to determine surface roughness categories and exposure categories in cases where a more detailed assessment is required or desired.

In 2006 IBC Section 1609.4.1, clear guidelines are provided for determining the governing exposure category whereby two upwind sectors, each enclosing an angle of 45° from the selected wind direction (Figure 16-17), are to be considered separately, and the exposure resulting in the highest wind load governs the design for the selected wind direction. This is particularly helpful in establishing a more objective determination of exposure category for buildings located in a transition zone of two types of exposures.

Step 4: Topographic factor. Topographic factor, K_{zt}, accounts for the increase in wind speeds that are due to an abrupt change in the upwind terrain as a result of the presence of an isolated hill, ridge or escarpment. However, the isolated feature needs to make a significant difference in the general terrain features to have an effect on the design wind speed. ASCE 7-05 Section 6.5.7.1 provides the detailed requirements for the topographic factor to apply. The value of K_{zt} can be calculated by using ASCE 7 Figure 6-4 and Equation 6-3. More discussion can be found in ASCE 7 Commentary Section C6.5.7.

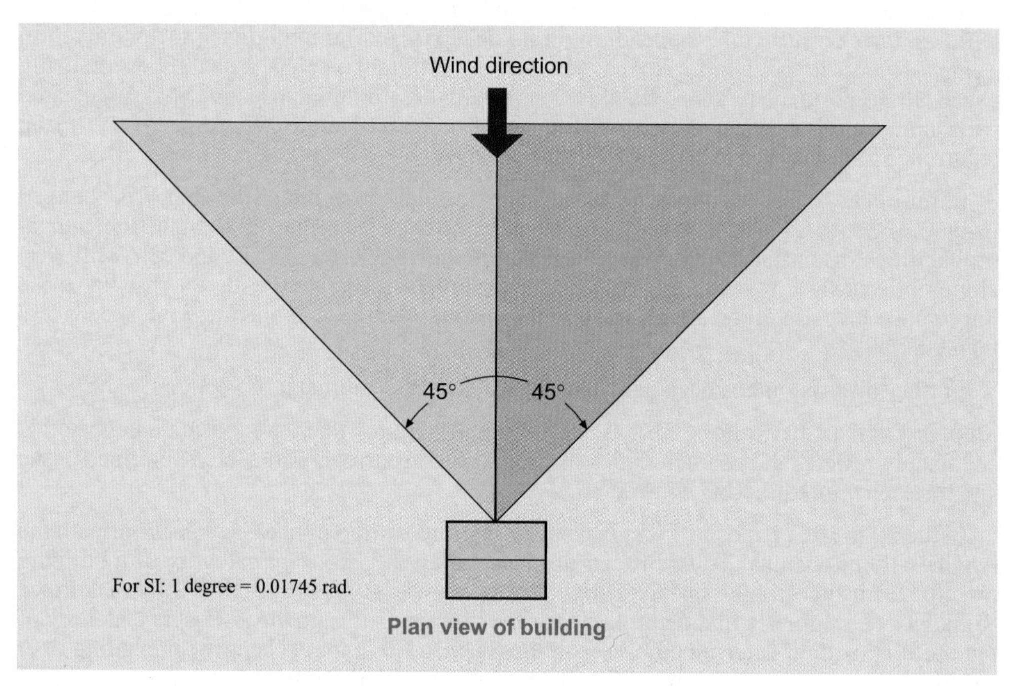

Figure 16-17
Upwind sectors for determination of governing exposure category

Step 5: Design wind pressure: Simplified design wind pressure is calculated by using ASCE 7 Eq. 6-1 for Main Wind Force Resisting Systems (MWFRSs) and Equation 6-2 for Components and Claddings. Figures 6-2 and 6-3 provide the simplified design wind pressures on different pressure szones, designated as A through H, for Exposure Category B and mean roof height of 30 feet. An adjustment factor, λ, is also provided in the same tables to convert the tabulated wind pressure values to those for actual exposure category and mean roof height.

A minimum design wind pressure of 10 psf is also specified in Sections 6.4.2.1.1 and 6.4.2.2.1. For MWFRSs, the minimum pressure is to be applied uniformly over pressure zones A through D, while having zero pressure on zones E through H. For components and cladding of buildings, the pressure is required to be considered in either direction acting normal to the surface.

Additionally, it is also specified that the full design wind pressure is to be applied even for air-permeable cladding, unless a lower value can be justified by approved test data or recognized literature.

Protection of openings. This requires that in wind-borne debris regions, glazing in buildings be made impact-resistant or protected with an impact-resistant covering in compliance with the following: (1) glazed openings located within 30 feet (9.15 m) of grade meet requirements of the Large Missile Test of ASTM E 1996,[20] and (2) glazed openings located more than 30 feet (9.15 m) above grade meet the provisions of the Small Missile Test of ASTM E 1996. The mandatory protection requirements are meant to prevent damages to the building interior and consequent financial as well as functional losses. **1609.1.2**

An exception to the above requirement provides the builder and/or owner of one- or two-story buildings with a low-cost alternative to the installation of permanent shutters or laminated glass meeting the requirements of ASTM E 1996. This exception requires the builder to provide panels precut to fit each glazed opening and further requires that temporary connection hardware be provided that is adequate to withstand the required components and cladding loads of hurricane winds.

The second and third exceptions permit any openings in Occupancy Category I buildings and openings located at least 60 feet above ground in Occupancy Category II, III or IV buildings to remain unprotected. Additionally, the second part of the third exception recognizes that loose roof aggregate that is not protected by a high parapet can act as a potential source of wind-borne debris. When such a source is present within 1500 feet of an Occupancy Category II, III or IV building, an opening needs to be at least 30 feet above the source roof to be exempted from the protection requirements. More discussion can be found in ASCE 7-05 Commentary Section C6.5.9.

Definitions. This section contains the definitions of only two wind-related terms (Hurricane-prone Regions and Wind-borne Debris Region). These are the only terms requiring definition that are left in the few wind design provisions remaining in the 2006 IBC. The definitions are identical to those in ASCE 7-05 Section 6.2. **1609.2**

Roof systems. The roof system consists of the roof deck and the roof covering. Provisions for each are discussed below. **1609.5**

Roof deck. The roof deck is a structural component of the building. Thus, it is required to be designed to resist the wind pressures determined by the provisions of ASCE 7-05, by SBCCI SSTD 10, or by the AF&PA *Wood Frame Construction Manual for One- and Two-Family Dwellings*. The last two procedures have limits on their applicability, as indicated in Figure 16-16. **1609.5.1**

Roof coverings. If the roof deck is relatively impermeable, the wind pressures will act through the roof deck to the building framing system. If the roof covering is also relatively impermeable and fastened to the roof deck, the two components will be subject to the same wind pressures. If the roof covering is air-permeable, wind pressures are able to develop on both the top of the roof covering and underneath the roof covering. This venting action of the roof negates some of the wind pressures on the roof covering. **1609.5.2**

1609.5.3 Rigid tile. As explained above, in certain types of installations, the roof covering is not subject to the same wind pressures as the roof deck. Such installations include concrete and clay tile roof coverings. The gaps occurring at the joints allow some equalization of pressure between the inner and outer faces of the tiles, leading to reduced pressures. The equation given in this section has been developed by research for the determination of wind loads (uplift moments) on loose laid and mechanically fastened roof tiles when laid over sheathing with an underlayment. The procedure is based on practical measurements on real tiles to determine the effect of air being able to penetrate the roof covering.

Section 1610 *Soil Lateral Loads*

1610.1 General. This section reproduces part of ASCE 7-05 Section 3.2, Soil and Hydrostatic Pressure. Table 1610.1, Soil Lateral Load, is a modified version of Table 3-1 of ASCE 7-05. In particular, footnotes C and D to Table 3-1 have been made into a new column in Table 1610.1 entitled "At-rest pressure."

Table 16-6 compares the soil lateral loads given in ASCE 7-95, -98, -02, -05, the last two editions of the BOCA/NBC,[4] the last two editions of the SBC,[14] and the 2000, 2003 and 2006 IBC.

ASCE 7 Table 3-1 includes high earth pressures, 85 pounds per cubic foot or more, to point out that certain soils are poor back-fill material.

In ASCE 7, contrary to what is stated in Footnote a, the soil lateral loads for silts and silt-clay mixtures—unified soil classifications SM-SC, SC, ML and ML-CL—are based on saturated conditions that take hydrostatic pressure into account. Table 16-7, in addition to showing soil lateral loads from the IBC and the ASCE 7 standard, also shows soil lateral load values using the calculation procedure that formed the basis of the values included in BOCA/NBC. The calculated soil lateral loads are presented for both moist and saturated conditions. Table 16-7 shows that for gravels and sands, IBC, ASCE and the calculated soil lateral loads for moist conditions agree. Conversely, for silts and silt-clay mixtures, the IBC values agree with the calculated soil lateral loads for moist conditions, whereas ASCE 7 values are closer to the calculated soil lateral loads for saturated conditions. Footnote a to IBC Table 1610.1 as well as ASCE 7 Table 3-1, states that the design lateral soil loads are given for moist soil conditions. This appears to be true only in the case of the IBC.

ASCE 7 points out in Footnotes c and d to Table 3-1 that when walls are unyielding, the earth pressure is increased from active pressure toward earth pressure at rest, resulting in 60 pounds per cubic foot for granular soils and 100 pounds per cubic foot for silt and clay-type soils. As noted above, these footnotes are incorporated into the last column in Table 1610.1 of the 2006 IBC. According to the exception to Section 1610.1, basement walls extending not more than 8 feet below grade and supporting flexible floor systems are permitted to be designed for active pressure. Examples of light floor systems supported on shallow basement walls, given in the ASCE 7 Commentary, are floor systems with wood joists and flooring, and cold-formed steel joists without cast-in-place concrete floors attached.

Expansive soils exist in many parts of the United States and may cause serious damage to basement walls unless special design considerations are provided.[1] Expansive soils should not be used as backfill, because they can exert very high pressures against walls. Special soil testing is required to determine the magnitude of these pressures. It is preferable to excavate expansive soils and backfill with nonexpansive freely draining sands and gravels. This topic is covered by Footnote b to Table 1610.1.

Table 16-6. Soil lateral loads in model codes and standards

Description of Backfill Material[e]	Unified Soil Classification	Design Lateral Soil Load[a], psf per foot of depth					
		ASCE 7-95, -98, -02, -05	BOCA/NBC 1996, 1999	SBC 1997, 1999	IBC 2000	IBC 2003, 2006	
						Active Pressure	At-rest Pressure
Well-graded clean gravels, gravel-sand mixes	GW	35[c]	30	30	30[c]	30	60
Poorly graded clean gravels, gravel-sand mixes	GP	35[c]	30	30	30[c]	30	60
Silty gravels, poorly graded gravel-sand mixes	GM	35[c]	41	45	40[c]	40	60
Clayey gravels, poorly graded gravel and clay mixes	GC	45[c]	46	45	45[c]	45	60
Well-graded clean sands, gravelly sand mixes	SW	35[c]	30	30	30[c]	30	60
Poorly graded clean sands, sand-gravel mixes	SP	35[c]	30	30	30[c]	30	60
Silty sands, poorly graded sand-silt mixes	SM	45[c]	41	45	45[d]	45	60
Sand-silt clay mix with plastic fines	SM-SC	85[d]	44	45	45[d]	45	100
Clayey sands, poorly graded sand-clay mixes	SC	85[d]	48	60	60[d]	60	100
Inorganic silts and clayey silts	ML	85[d]	45	45	45[d]	45	100
Mixture of inorganic silt and clay	ML-CL	85[d]	44	60	60[d]	60	100
Inorganic clays of low-to-medium plasticity	CL	100	45	60	60[d]	60	100
Organic silts and silt-clays, low plasticity	OL	b	b	b	b	b	b
Inorganic clayey silts, elastic silts	MH	b	55	60	b	b	b
Inorganic clays of high plasticity	CH	b	b	b	b	b	b
Organic clays and silty clays	OH	b	b	b	b	b	b

a. ASCE 7, IBC – Design lateral loads are given for moist conditions for the specified soils at their optimum densities. Actual field conditions shall govern. Submerged or saturated soil pressures shall include the weight of the buoyant soil plus the hydrostatic loads.

b. ASCE 7, IBC – Unsuitable as backfill material. BOCA – Soil shall be replaced with controlled backfill. SBC – Compliance with Section 1804.3 (Expansive Soils) is required.

c. ASCE 7, IBC – For relatively rigid walls, as when braced by floors, the design lateral soil load shall be increased for sand- and gravel-type soils to 60 pounds per square foot per foot of depth. Basement walls extending not more than 8 feet below grade and supporting light floor systems are not considered relatively rigid walls.

d. ASCE 7, IBC – For relatively rigid walls, as when braced by floors, the design lateral load shall be increased for silts and clay-type soils to 100 pounds per square foot per foot of depth. Basement walls extending more than 8 feet below grade and supporting light floor systems are not considered relatively rigid walls.

e. IBC – The definition and classification of soil materials shall be in accordance with ASTM D 2487.

Table 16-7. **Soil lateral loads—calculated versus code and standard values**

Soil Description	Unified Soil Classification	Design Lateral Soil Load (pounds per square foot per foot of depth)		Calculated Lateral Soil Load — Active Lateral Pressure, (pounds per square foot per foot of depth)	
		IBC	ASCE 7	Moist	Saturated
Well-graded clean gravels, gravel-sand mixes	GW	30	35	34	81
Poorly graded clean gravel, gravel-sand mixes	GP	30	35	34	80
Silty gravels, poorly graded gravel-sand mixes	GM	40	35	40	84
Clayey gravels, poorly graded gravel-sand-clay mixes	GC	45	45	44	86
Well-graded clean sand, gravelly-sand mixes	SW	30	35	32	80
Poorly graded clean sands, sand-gravel mixes	SP	30	35	32	79
Silty sands, poorly graded sand-silt mixes	SM	45	45	38	82
Sand-silt clay mix with plastic fines	SM-SC	45	85	40	84
Clayey sands, poorly graded sand-clay mixes	SC	60	85	42	85
Inorganic silts and clayey silts	ML	45	85	39	82
Mixture of inorganic silts and clay	ML-CL	60	85	40	83
Inorganic silts and silt-clay, medium plasticity	CL	60	100	46	86
Organic silts and silt-clays, low plasticity	CL	Unsuitable	Unsuitable	–	–
Inorganic clayey silts, elastic silts	MH	Unsuitable	Unsuitable	–	–
Inorganic clays or high plasticity	CH	Unsuitable	Unsuitable	–	–
Organic clays and silty clays	OH	Unsuitable	Unsuitable	–	–

Section 1611 *Rain Loads*

This reproduces Chapter 8, Rain Loads, of ASCE 7-05, except that Section 8.2, Roof Drainage, is not included, and IBC Section 1611.2, Ponding Instability, simply references the provisions in ASCE 7 Section 8.4. Each portion of a roof is required to be designed to sustain the load of rainwater that will accumulate on it if the primary drainage systems for that portion is blocked, plus the uniform load caused by water that rises above the inlet of the secondary drainage system at its design flow. It should be obvious from the design load combinations of Sections 1605.2 and 1605.3 that the design rain loads would affect design only if they are larger than the design snow loads and the roof live loads.

Section 1612 *Flood Loads*

1612.1 General. Section 1612.1 simply states that buildings and structures in flood hazard areas as established in Section 1612.3 must be designed and constructed to resist the effects of flood hazards and flood loads, as prescribed in this section. It also requires that a building that is located in more than one flood hazard area be designed in accordance with the provisions of the most restrictive flood hazard area. This requirement addresses the situation where a building site is not subjected to high velocity wave action, but is affected by an area designated as floodways.

It is also important to note that Section 5.3 of ASCE 7 presents information on the design of buildings in flood-prone areas. Much of the impetus for flood-resistant design has come from initiatives of flood insurance and flood-damage mitigation sponsored by the federal government.

As indicated in the Commentary to ASCE 7-05, the National Flood Insurance Program (NFIP) is based on an agreement between the federal government and participating communities that have been identified as being flood-prone. The Federal Emergency Management Agency (FEMA), through the Federal Insurance Administration, makes flood insurance available to the residents of communities, provided that the community adopts and enforces adequate flood-plain management regulations that meet the minimum requirements. Included in the NFIP requirements, found under Title 44 of the U. S. Code of Federal Regulations, are minimum design and construction standards for buildings located in special flood hazard areas.

Special flood hazard areas are those identified by FEMA's Mitigation Directorate as being subject to inundation during the 100-year flood (flood having a 1-percent chance of being equaled or exceeded in any given year). Special flood hazard areas are shown on flood insurance rate maps that are produced for flood-prone communities. Special flood hazard areas are identified on such maps as A Zones (A, AE, A1-30, A99, AR, AO or AK) or V Zones (V, VE, VO or V1-30). The special flood hazard areas are those in which communities must enforce NFIP-compliant, flood-damage-resistant design and construction practices.

For the 2000 edition of the IBC, the American Society of Civil Engineers (ASCE) and FEMA proposed modifications to the IBC Final Draft at that time for the purpose of ensuring that the provisions of the IBC and the manner in which they are administered is consistent with those of the NFIP. This proposal was accepted. Much of the above is accomplished through adoption by reference of the ASCE 24-05 *Flood Resistant Design and Construction Standard.*[21] The provisions of ASCE 24-05 meet or exceed the building science provisions of the NFIP and present the consensus state-of-the-art approach to flood-resistant construction.

The first edition of ASCE Standard for flood-resistant construction (ASCE 24-98) represented the culmination of over four years of work by dozens of experts in flood-resistant design and construction from all facets of the land development and regulations community. FEMA was in full support of this effort and provided both financial and technical assistance in this endeavor.

ASCE 24-05 requires that the design of structures within flood hazard areas be governed by the loading provisions of ASCE 7-05. That standard requires that the structural systems of buildings be designed, constructed, connected and anchored to resist flotation, collapse and permanent lateral movement that is due to the action of wind loads and loads from flooding associated with the design flood. Wind loads and flood loads may act simultaneously at coastlines, particularly during hurricanes and coastal storms. This may also be true in some other situations cited in ASCE 7-05 Commentary Section C5.3.1. Flood loads are the loads or pressures on the surfaces of buildings and structures caused by the presence of floodwaters. These loads are of two basic types—hydrostatic and hydrodynamic. Impact loads result from objects transported by floodwaters striking against structures or parts of structures. Wave loads are considered a special type of hydrodynamic load. Hydrostatic loads, hydrodynamic loads, wave loads and impact loads are specified in ASCE 24-05. Wave loads may be determined by using an analytical procedure outlined by more advanced numerical modeling procedures, or by laboratory test procedures or physical modeling.

ASCE 24-05 requires that the design and construction of structures located in flood-hazard areas consider all flood-related hazards; including hydrostatic loads, hydrodynamic loads and wave action, debris impacts, alluvial fan flooding, flood-related erosion, ice flows or ice jams, or mudslides in accordance with the requirements of that standard if specified or, if not specified in that standard, then in accordance with the requirements approved by the authority having jurisdiction. Design documents must identify, and take into account, flood-related and other concurrent loads that will act on the structure. Design documents must include, but not be limited to, the applicable conditions listed below:

1. Wave action
2. High-velocity flood waters
3. Impacts due to debris in the flood waters
4. Rapid inundation by flood waters
5. Rapid drawdown of flood waters
6. Prolonged inundation by flood waters
7. Wave- and flood-induced erosion and scour
8. Deposition and sedimentation by flood waters

1612.2 Definitions. The particular definitions to be noted are Design Flood, Flood-Hazard Area, Flood-Hazard Area Subject to High-Velocity Wave Action, and Special Flood Hazard Area.

1612.3 Establishment of flood-hazard areas. Flood-hazard areas are established by the local jurisdiction through adoption of a flood-hazard map and supporting data. Minimum requirements imposed on the jurisdiction are spelled out.

1612.4 Design and construction. Requires the design and construction of buildings and structures located in flood-hazard areas, including flood-hazard areas subject to high-velocity wave action, to be in accordance with the requirements of ASCE 24.

1612.5 Flood-hazard documentation. This lists a number of documents that must be furnished to the building official. The requirements are different for construction in flood-hazard areas not subject to high-velocity wave action and for construction in flood-hazard areas subject to high-velocity wave action.

Section 1613 *Earthquake Loads*

Overview. In what used to be UBC[3] territory before the *legacy* model codes began to be replaced by the IBC, the State of California and jurisdictions within it, by the time this publication comes out, will have switched to the 2006 IBC directly from the 1997 UBC (on January 1, 2008). The City and County of Honolulu switched to the 2003 IBC from the 1997 UBC on September 17, 2007. The Commonwealth of Puerto Rico, as of the issuance of this publication, is still in the 1997 UBC. The rest of the country, with the exception of some major cities such as Chicago, has adopted the 2000 or the 2003 IBC. Some jurisdictions such as the States of South Carolina and Georgia have already adopted the 2006 IBC. Thus, in the text that follows, references are made to the 1997 UBC and earlier editions of the IBC, but almost none to the BOCA/NBC[4] or the SBC,[14] because hardly any jurisdiction is likely to switch directly from those codes to the 2006 IBC.

Complete seismic design provisions were contained within the 1997 UBC. These were based on what became 1999 SEAOC *Blue Book*,[22] but were also significantly influenced by the 1994 NEHRP *Provisions*. In a deliberate attempt to bring the UBC as close as was considered acceptable at that time to the IBC, the site classification of the 1994 NEHRP *Provisions*, which eventually became the site classification of the IBC, and a number of important features associated with it, were adopted into the 1997 UBC. The 2000 IBC started out the same way as the *legacy* model codes, with complete seismic design provisions in the code itself. These were, by and large, the seismic design provisions of the 1997 NEHRP *Provisions*,[2] but there were modifications.

The 2003 IBC was a transition edition in many ways. It retained eleven sections (1613 through 1623) on the seismic design provisions; however, the sections were largely empty. They simply referred the reader to specifications within ASCE 7-02. Some of the time, in the course of doing that, the 2003 IBC made modifications to the requirements of ASCE 7-02. Thus, when someone designed using the seismic design provisions of the 2003 IBC, he or she was designing by the seismic design provisions of ASCE 7-02, as modified by the 2003 IBC. The 2003 IBC also gave designers the option of designing entirely by ASCE 7-02 and ignoring the seismic modifications made by the 2003 IBC. However, the modifications had to be considered in their entirety or not at all. One could not pick and choose. The seismic design provisions of ASCE 7-02 were those of the 2000 NEHRP *Provisions*,[2] with a number of modifications.

Almost all of the seismic design provisions of the 2006 IBC are adopted through reference to ASCE 7-05. The provisions that remain in the code are related to ground motions and soil parameters as well as definitions of terms actually used within those provisions and the four exceptions under the scoping provisions. See Table 16-8 for a comparison between the organizations of the seismic provisions in the 1997 UBC and the 2000 IBC, the 2003 IBC and the 2006 IBC.

The earthquake regulations of ASCE 7-05 Chapters 11 – 13 and 15 – 23, based on the 2003 NEHRP *Provisions*, are substantially different from the corresponding provisions of the 1997 UBC. This is in large part because the seismic design requirements of the NEHRP *Provisions* underwent major fundamental changes between the 1994 and 1997 editions of that document. For further information regarding the application of the IBC seismic provisions, one should refer to the 2003 NEHRP *Provisions* Part 2 – Commentary. The Provisions (FEMA 450.1), the Commentary (FEMA 450.2) or both and an accompanying set of maps on a CD (FEMA 450-CD) are available free of charge from the FEMA Publication Distribution Facility at 1-800-480-2520. The information on the CD can also be downloaded from the Building Seismic Safety Council website: www.bssconline.org.

The most significant change from the 1997 UBC to the IBC (irrespective of edition) is in the design ground motion parameters, which are now S_{DS} and S_{D1}, rather than Z. S_{DS} and S_{D1} are five-percent-damped design spectral response accelerations at short periods and 1-second period respectively. S_{DS} determines the upper-bound design base shear (the *flat-top* of the design spectrum) used in seismic design (see ASCE 7-05 Figure 11.4-1 and Section 11.4). S_{D1} defines the descending branch or the period-dependent part of the design spectrum (see ASCE 7-05 Figure 11.4-1 and Section 11.4). The seismic zone map of the UBC has been replaced by contour maps giving two quantities from which S_{DS} and S_{D1} are

Table 16-8. Comparison between organizations of 1997 IBC and the 2000, 2003 and 2006 IBC seismic provisions

	1997 UBC	2000 IBC	2003 IBC	2006 IBC
Sections	1626 – 1636	1613 – 1623	1613 – 1623	1613
Number of Pages	30 pages	75 pages (18 pages of maps)	43 pages (18 pages of maps)	24 pages (18 pages of maps)
Location of Simplified Method	Simplified Method included	Simplified Method included	Simplified Method included	Go to ASCE 7-05 for Simplified Method
Reference to ASCE 7	Complete seismic design provisions are contained in Sections 1626 – 1636	Complete seismic design provisions are contained in Sections 1613 – 1623	Other than Simplified Method, ASCE 7-02 is referred to, with modifications, for seismic design provisions	Section 1613.1 references ASCE 7-05, excluding Chapter 14 and Appendix 11A
Modifications to ASCE 7	N/A	N/A	Numerous modifications to ASCE 7-05	Two alternatives to ASCE 7-05

derived. The mapped quantities are the maximum considered earthquake spectral response accelerations S_S (at short periods) and S_1 (at 1-second period).

The maximum considered earthquake (MCE) is the 2500-year return period earthquake (two-percent probability of exceedance in 50 years) in most of the country; however, in coastal California, where the seismic sources and their capabilities are known, it is the largest (deterministic) earthquake that can be generated by the known seismic sources. The design earthquake of the 1997 UBC has an approximate return period of 475 years (ten-percent probability of exceedance in 50 years). The design earthquake of the IBC is two-thirds of the MCE. The two-thirds is the reciprocal of 1.5, which is agreed to be the *seismic margin* built into structures designed by the UBC (1973 edition or later) or older editions of the NEHRP *Provisions* (preceding the 1997 edition). In other words a structure designed by the UBC (1973 edition or later) or older editions of the NEHRP *Provisions* (preceding the 1997 edition) is believed to have a low likelihood of collapse from an earthquake that is up to one and one-half times as large as the design earthquake of those documents. The redefinition of the design earthquake in the IBC, introduced first in the NEHRP *Provisions*, is intended to provide a uniform level of safety across the country against collapse in the maximum considered earthquake. This was not the case before, because the MCE is only 50 percent larger than the design earthquake of the UBC in coastal California, whereas it can be four or five times as large as the design earthquake based on the UBC in the Eastern United States. The mapped MCE spectral response accelerations S_S and S_1 of the 2006 IBC are mapped on Site Class B, soft rock of the western United States, which is equivalent to Soil Profile Type S_B of the 1997 UBC.

S_{DS} and S_{D1} are two-thirds of S_{MS} and S_{M1}, which are the soil-modified MCE spectral response accelerations at short periods and 1-second period, respectively. S_{MS} is obtained by multiplying the mapped MCE spectral response acceleration S_S (at short periods) by F_a, the acceleration-related site coefficient (see ASCE 7-05 Table 11.4-1). S_{M1} is similarly obtained by multiplying the mapped MCE spectral response acceleration S_1 (at 1-second period) by F_v, the velocity-related site coefficient (see ASCE 7-05 Table 11.4-2). F_a and F_v are analogous to C_a/Z and C_v/Z of the 1997 UBC, respectively.

Because the 2006 IBC and the 1997 UBC have both adopted the soil classification and the associated site coefficients first introduced in the 1994 NEHRP *Provisions*, a correlation of ground motion parameters between the two codes is possible. For example, if S_{DS} of the 2006 IBC is equal to $2.5C_a$ and S_{D1} of the 2006 IBC is equal to C_v for a particular location, then the soil-modified seismicity for that site has not changed from the 1997 UBC to the 2006 IBC. Table 16-9 shows a comparison of soil-modified seismicity (for Soil Profile Type S_D or Site Class D) by the 1997 UBC and the 2006 IBC for a number of locations.

The 1997 edition of the UBC for the first time introduced two near-source factors: acceleration-related N_a and velocity-related N_v, the purpose of which was to increase the soil-modified ground motion parameters C_a and C_v when there were active faults capable of generating large-magnitude earthquakes within 15 km or 9 miles of a Seismic Zone 4 site. These factors became necessary in view of the artificial truncation of Z-values to 0.4 in UBC Seismic Zone 4. These near-source factors are not found in the 2006 IBC, because the artificial truncation of ground motion is not a feature of that code. Both S_S and S_1 attain high values in the vicinity of seismic sources that are judged capable of generating large earthquakes. In other words, the near-source factors of the UBC are built into the mapped MCE spectral response acceleration contours of the IBC.

The 2006 IBC maps for S_S and S_1 are based on USGS probabilistic maps and are available on the USGS web site at http://earthquake.usgs.gov/research/hazmaps. Along with the mapped values, the user can also view a design spectrum for a site specified by latitude-longitude or ZIP code. Site coefficients other than Site Class B may be included in the calculations using NEHRP site coefficients or as specified by the designer. It should be noted that the 2006 IBC maps are different from those of the 2000 and the 2003 IBC (maps did not change from the 2000 to the 2003 IBC). Thus, the USGS CD-ROM that was distributed with the 2000 and the 2003 IBC must not be used with the 2006 IBC. USGS is expected to issue a DVD that will enable a code user to establish the S_s- and S_1-values at a particular location by the ZIP code or by the latitude and longitude. As of this writing, the DVD has not been released. Pending the issuance of the DVD, values of S_s and S_1 at a site can also be found in the 2003 NEHRP CD (specifically the maps in that CD).

Table 16-9. Comparison of soil-modified seismicity by the 1997 UBC and the 2006 IBC for site Class D

Site	1997 UBC			2006 IBC	
	Zone	$2.5C_a$	C_v	S_{DS}	S_{D1}
West Los Angeles[1]	4	1.3	0.64	1.11	0.63
Downtown San Francisco (4th & Market)	4	1	0.45	1	0.67
U.C. Berkley Memorial Stadium[2]	4	1.5	0.8	1.32	0.77
Denver	1	0.2	0.08	0.23	0.09
Sacramento	3	0.75	0.3	0.52	0.31
St. Paul	0	0	0	0.06	0.04
Seattle	3	0.75	0.3	0.96	0.49
Portland	3	0.75	0.3	0.73	0.4
Houston	0	0	0	0.09	0.06

1. On Type B Newport-Inglewood Fault
2. On Type A Hayward Fault

One difference really stands out when the seismic design provisions of the 2006 IBC are compared with those of the 1997 UBC. In the UBC, restrictions on building height and structural irregularity, choice of analysis procedures that form the basis of seismic design, as well as the level of detailing required for a particular structure are all governed by the seismic zone in which a structure is located. In the 2006 IBC, all are governed by the seismic design category which combines the occupancy with the soil-modified seismic hazard at the site of the structure.

In seismic design by the provisions of the 2006 IBC, the detailing as well as the other restrictions depend on the soil characteristics at the site of a structure. This is a major departure from UBC design practice—a departure that has significant design implications. Table 16-10 shows a comparison of 1997 UBC seismic zones and seismic design categories of the 2006 IBC for a number of locations across the country.

Finally, ASCE 7-05 Chapter 15 on seismic design requirements for nonbuilding structures, based on the 2003 NEHRP *Provisions*, is much more extensive and detailed than the corresponding 1997 UBC Section 1634 on nonbuilding structures. It is also much more extensive and detailed than 2000 IBC Section 1622 and 2003 IBC Section 1622. ASCE 7-05 Chapter 15 represents a major expansion of the guidance available in the building codes concerning the seismic design of nonbuilding structures.

Scope. Section 1613 adopts seismic design provisions by reference from Chapters 11, 12, 13, and 15 through 23 and Appendix 11B of ASCE 7-05. Chapter 14 of ASCE 7-05 is not adopted because the materials Chapters 19 through 23 of the 2006 IBC must be used instead. Appendix 11A of ASCE 7-05 is also not adopted, because the inspection and testing requirements in Chapter 17 of the 2006 IBC must be used instead. **1613.1**

Table 16-10. Comparison of 1997 UBC Seismic Zones and Seismic Design Categories of the 2006 IBC

Location	UBC	IBC				
	Seismic Zone	Site Class				
		A	B	C	D	E
West Los Angeles	4	D	D	D	D	D
Downtown SFO (4th and Market)	4	D	D	D	D	D
UC Berkeley Memorial Stadium	4	E	E	E	E	E
Denver	1	A	A	B	B	C
Sacramento	3	B	C	D	D	D
St. Paul	0	A	A	A	A	A
Seattle	3	D	D	D	D	D
Portland	3	D	D	D	D	D
Houston	0	A	A	A	A	B
Washington D.C.	1	A	A	A	B	B
Chicago	0	A	A	A	B	C
Baltimore	1	A	A	A	B	B
Boston	2A	B	B	B	B	C
New York	2A	A	B	B	C	D
Cincinnati	1	A	A	B	B	C
Philadelphia	2A	A	B	B	B	C
Richmond	1	A	A	B	B	C
Birmingham	1	A	B	B	C	D
Atlanta	2A	A	A	B	C	C
Orlando	0	A	A	A	A	B
Little Rock	1	B	C	C	D	D
New Orleans	0	A	A	A	B	B
Nashville	1	B	B	C	D	D
Charlotte	2A	B	B	B	C	D
Charleston	2A	D	D	D	D	D

After having stated that every structure and portions thereof shall as a minimum be designed and constructed to resist the effects of earthquake motions and assigned a seismic design category in accordance with Section 1613 or ASCE 7-05, the IBC proceeds to permit a number of important exceptions:

1. Detached one- and two-family dwellings in Seismic Design Category A, B or C or located where S_S is less than 0.4g are exempt from seismic design requirements. The last part of this exception removes the need to determine the Seismic Design Category, which in turn requires the determination of Site Class, in low to moderate hazard areas. Figure 16-18 shows areas of the contiguous United States that have S_S less than 0.4g. 1997 UBC Section 1629.1 exempted one- and two-family dwellings in Zone 1 from seismic design requirements.

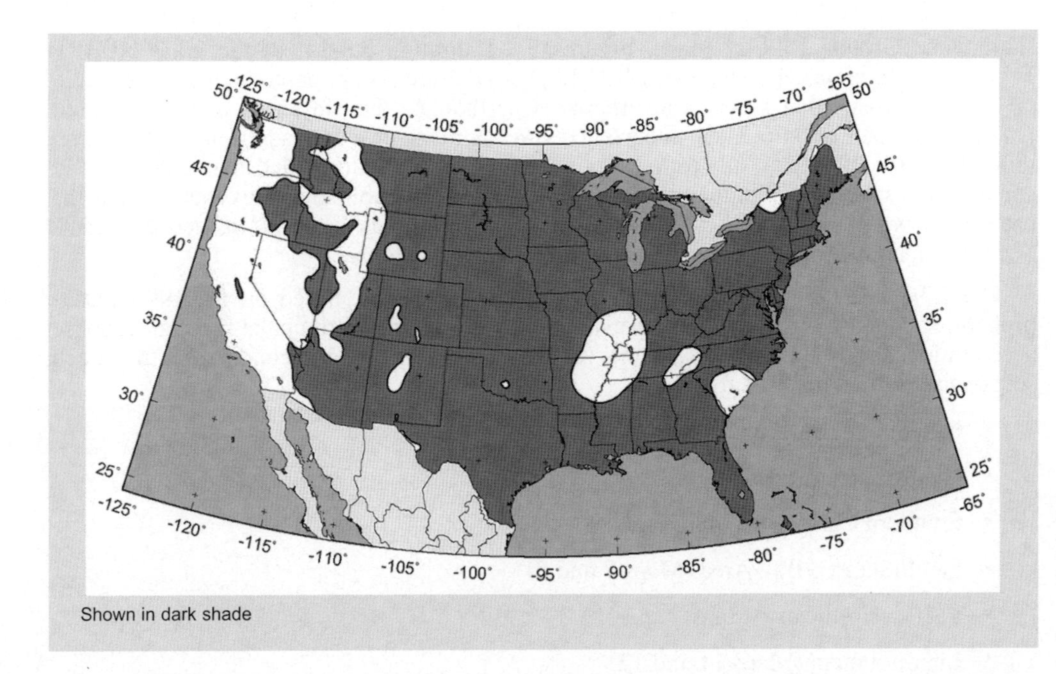

Figure 16-18 Areas of the contiguous United States with $S_S < 0.4g$

2. The seismic force-resisting systems of wood-frame buildings that conform to the provisions of Section 2308 (Conventional Light-Frame Construction) need not be analyzed as required by Section 1613.1. Based on the limitations in 2006 IBC Sections 2308.2 and 2308.11.1, this exception, in effect, increases the allowable height of a wood frame building from two to three stories when the building is assigned to SDC A or B. 1997 UBC Section 1605.1 similarly states that buildings constructed in accordance with the conventional wood frame construction provisions of Section 2320 are deemed to comply with the structural design requirements of Section 1605.
3. Agricultural storage structures intended only for incidental human occupancy are exempt from all seismic design requirements. This is because of the very low risk to life that is involved.
4. Certain special structures such as "vehicular bridges, electrical transmission towers, hydraulic structures, buried utility lines and their appurtenances, and nuclear reactors" are placed outside the scope of Section 1613. These structures require special consideration of their response characteristics and environment, and need to be designed in accordance with other regulations.

The same exceptions are to be found in Section 11.1.2 of ASCE 7-05, except that Exception 2 is somewhat different in ASCE 7-05.

1613.3

Existing buildings. Provisions concerning additions, alterations and change of occupancy of an existing building are located in 2006 IBC Chapter 34. Similar, but not identical, provisions are to be found in Appendix 11B to ASCE 7-05 (Existing Building Provisions).

1613.4

Special inspections. The special inspection requirements of Chapter 17 are cross-referenced.

1613.5

Seismic ground motion values.

1613.5.1

Mapped spectral acceleration parameters. The mapped maximum considered earthquake spectral response accelerations at short periods (S_S) and at 1-second period (S_1) for a particular site are to be determined from Figures 1613.5(1) through 1613.5(14). Where a site is between contours, straight line interpolation or the value of the higher contour may be used. 2006 IBC Figures 1613.5(1) through 1613.5(14) are the same as Figures 22-1 through 22-14 of ASCE 7-05. Both sets are adapted from the Maximum Considered

Earthquake Ground Motion Maps, Figures 3.3-1 through 3.3-14, of the 2003 NEHRP *Provisions* (Reference 4). The 2003 NEHRP *Provisions* maps are based on the 2002 USGS probabilistic maps that incorporate improved earthquake data in terms of updated fault parameters (such as slip rates, recurrence time and magnitude), additional attenuation relations (also referred to as ground motion prediction equations) and data calculated at smaller grid spacing as compared to the 1996 USGS probabilistic maps that formed the basis of the provisions in ASCE 7-98 and ASCE 7-02 as well as the 2000 IBC and 2003 IBC. See Commentary Appendix A to the 2003 NEHRP *Provisions*.

The 1997 NEHRP *Provisions* included two sets of maps, based on the 1996 USGS probabilistic maps—one for the mapped maximum considered earthquake spectral response acceleration at short periods (S_S) and one for the same quantity at 1-second period (S_1). Each set consisted of 12 maps as follows:

- National (Maps 1 and 2)
- California, Nevada (Maps 3 and 4)
- Southern California (Maps 5 and 6)
- San Francisco Bay Area (Maps 7 and 8)
- Pacific Northwest (Maps 9 and 10)
- Intermountain (Maps 11 and 12)
- New Madrid Region (Maps 13 and 14)
- Southeast (Maps 15 and 16)
- Alaska (Maps 17 and 18)
- Hawaii (Maps 19 and 20)
- Puerto Rico, Virgin Islands (Maps 21 and 22)
- Guam, Tutuila (Maps 23 and 24)

It is important to note that these maps are retained in the 2000 and 2003 NEHRP *Provisions*.

Parts 1 through 10 of the 2000 and 2003 IBC Figure 1613.5 correspond to NEHRP Maps 1, 2, 3, 4, 11, 12, 17, 18, 19, 20, 21, 22, 23 and 24, respectively.

The seismic zone map of the 1997 UBC dates back to the 1988 edition of the *Uniform Building Code*. In drawing the zone boundaries, both acceleration and velocity-related maps of the 1994 NEHRP *Provisions* were consulted, and, if they disagreed, the one indicating the higher zone prevailed. The design base shear equation was modified so that the values of Z would correspond to the estimated values of effective peak acceleration.

Given the wide range in return periods for maximum-magnitude earthquakes in different parts of the United States (100 years in parts of California to 100,000 years or more in several other locations), the 1997 NEHRP *Provisions* focused on defining maximum considered earthquake ground motions for use in design. The definition developed for the 1997 NEHRP *Provisions* remains unchanged through the 2003 NEHRP *Provisions*. These ground motions may be determined in different manners depending on the seismicity of an individual region; however, they are uniformly defined "the maximum level of earthquake ground shaking that is considered reasonable to design buildings to resist." This definition facilitates the development of a design approach that provides approximately uniform protection against collapse in the maximum considered earthquake throughout the United States.

The 1997, 2000 and 2003 NEHRP seismic design provisions are based on the assessment that if a building experiences a level of ground motion 1.5 times the design level of the 1994 and prior NEHRP *Provisions*, the building should have a low likelihood of collapse.

Although quantification of this margin is dependent on the type of structure, detailing requirements, etc., the 1.5 factor was felt to be a conservative judgment.

Given that the maximum earthquake for many seismic faults in coastal California is fairly well known, a decision was made to develop a procedure that would use the best estimate of ground motion from maximum magnitude earthquakes on seismic faults with higher probabilities of occurrence. For the purpose of the 1997 and subsequent NEHRP *Provisions*, these earthquakes are defined as *deterministic earthquakes*. Following this approach and recognizing the inherent margin of 1.5 contained in the NEHRP *Provisions*, it was determined that the level of seismic safety achieved in coastal California would be approximately equivalent to that associated with a 2 to 5 percent probability of exceedance in 50 years in areas outside of coastal California. Accordingly, the ground motion corresponding to a 2 percent probability of exceedance in 50 years was selected as the maximum considered earthquake ground motion for use in design where the deterministic earthquake approach discussed above is not used.

2006 IBC Section 1613.5.1 and ASCE 7-05 Section 11.4.1 assign Seismic Design Category A to any structure, regardless of its occupancy category when mapped short-period and 1-second period spectral response accelerations, S_S and S_1, at its site are less than or equal to $0.15g$ and $0.04g$, respectively. In ASCE 7-02 as well as in the 2003 IBC, this provision was included as an exception to the applicability of the respective seismic provisions chapters. It is now moved to this section in the ASCE 7-05, and in Section 1613.5.1 in the 2006 IBC. The areas where $S_S = 0.15g$ and $S_1 = 0.04g$ are shown on the map of the mainland United States in Figure 16-19.

1613.5.2

Site class definitions. The site class definitions of Table 1613.5.2 date back to the 1994 NEHRP *Provisions* in which extensive modifications were made to the consideration of site effects. The 1994 NEHRP classification was also adopted and incorporated into Table 16-J of the 1997 UBC.

1613.5.3

Site coefficients and adjusted maximum considered earthquake spectral response acceleration parameters. Refer to Appendix 4 of this publication. The short- and long-period amplification factors implied by the Loma Prieta strong-motion data and related calculations for the same earthquake by Joyner et al.[23], as well as modeling results at the $0.1g$ ground acceleration level, provided the basis for consensus values of site coefficients F_a and F_v. These were listed, for the first time, in 1994 NEHRP *Provisions* Tables 1.4.2.3a and 1.4.2.3b, reproduced here as Tables 16-11 and 16-12.

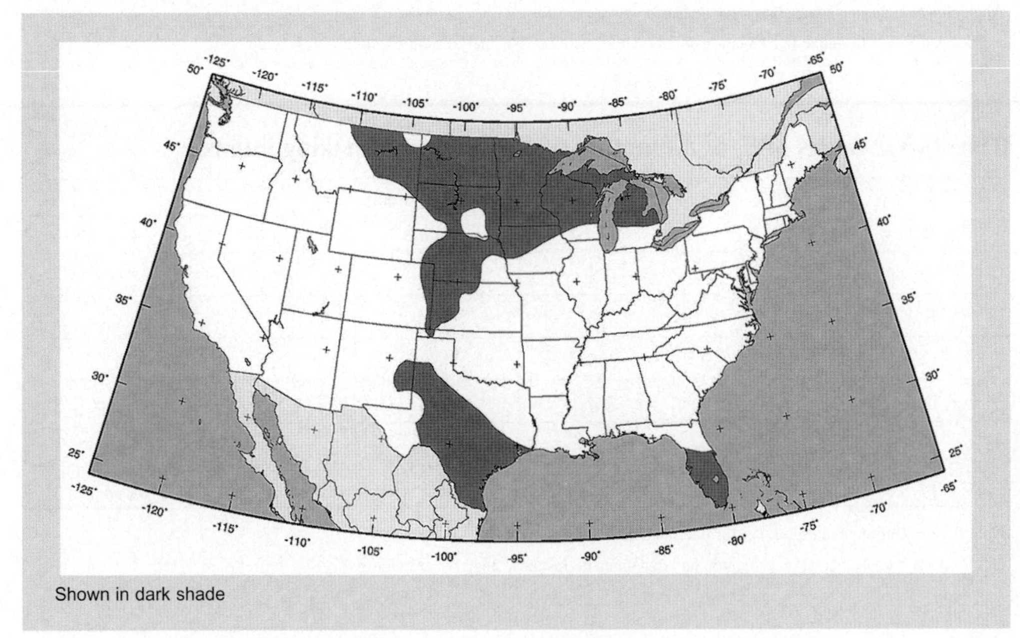

Figure 16-19
Areas of the contiguous United States with $S_S \leq 0.15g$ and $S_1 \leq 0.04g$

Site coefficients F_a and F_v of the 1994 NEHRP *Provisions* are analogous to C_a/Z and C_v/Z, respectively, of the 1997 UBC. The values of C_a/Z and C_v/Z, calculated from 1997 UBC Tables 16-Q and 16-R, respectively, are given in Tables 16-13 and 16-14, respectively. The Seismic Zone 4 values include the near-source factors N_a and N_v that are not part of the 1994 NEHRP *Provisions*, the 1997/2000/2003 NEHRP *Provisions*, or the 2000/2003/2006 IBC. The factors are determined by the closest distance from a site to known seismic sources that are capable of generating large-magnitude earthquakes and the types of those seismic sources. Three types of seismic sources are defined. Seismic Source Type A consists of faults that are capable of producing large-magnitude earthquakes (moment magnitudes ≥ 7.0) and that have a high rate of seismic activity (slip rate ≥ 5 mm/year). Seismic Source Type C consists of faults that are not capable of generating large-magnitude earthquakes (moment magnitude < 6.5) and that have a relatively low seismic activity (slip rate ≤ 2 mm/year). Seismic Source Type B consists of all faults other than Type A and C. The acceleration-dependent Near-Source Factor, N_a, has values of 1.5, 1.2 and 1.0, respectively, at distances of ≤ 2 km, 5 km and ≥ 10 km away from a Seismic Source Type A. It has values of 1.3 and 1.0 at distances of ≤ 2 km and 5 km away from a Seismic Source Type B. The velocity-dependent Near-Source Factor, N_v, has values of 2.0, 1.6, 1.2 and 1.0 at distances of ≤ 2 km, 5 km, 10 km and ≥ 15 km away from Seismic Source Type A. It has values of 1.6, 1.2 and 1.0 at distances of ≤ 2 km, 5 km and 10 km away from a Seismic Source Type B. At distances other than those specifically mentioned, N_a and N_v may be established by linear interpolation of the values given.

Table 16-11. Values of F_a as a function of site class and shaking intensity

Site Class	Shaking Intensity				
	$A_a \leq 0.1$	$A_a = 0.2$	$A_a = 0.3$	$A_a = 0.4$	$A_a \geq 0.5$
A	0.8	0.8	0.8	0.8	0.8
B	1.0	1.0	1.0	1.0	1.0
C	1.2	12	1.1	1.0	1.0
D	1.6	1.4	1.2	1.1	1.0
E	2.5	1.7	1.2	0.9	a
F	a	a	a	a	a

Note: Use straight line interpolation for intermediate values of A_a.

a. Site-specific geotechnical investigation and dynamic site-response analyses are required.

Table 16-12. Values of F_v as a function of site class and shaking intensity

Site Class	Shaking Intensity				
	$A_v \leq 0.1$	$A_v = 0.2$	$A_v = 0.3$	$A_v = 0.4$	$A_v \geq 0.5$
A	0.8	0.8	0.8	0.8	0.8
B	1.0	1.0	1.0	1.0	1.0
C	1.7	1.6	1.5	1.4	1.3
D	2.4	2.0	1.8	1.6	1.5
E	3.5	3.2	2.8	2.4	a
F	a	a	a	a	a

Note: Use straight line interpolation for intermediate values of A_v.

a. Site-specific geotechnical investigation and dynamic site-response analyses are required.

Table 16-13. Ground motion amplification that is due to soil: 1997 UBC Seismic coefficient, C_a/Z

Soil Profile Type	Seismic Zone Factor-Z				
	Z = 0.075	Z = 0.15	Z = 0.2	Z = 0.3	Z = 0.4
S_A	0.8	0.8	0.8	0.8	0.8 N_a
S_B	1.0	1.0	1.0	1.0	1.0 N_a
S_C	1.2	1.2	1.2	1.1	1.0 N_a
S_D	1.6	1.5	1.4	1.2	1.1 N_a
S_E	2.5	2.0	1.7	1.2	0.9 N_a
S_F					

Table 16-14. Ground motion amplification that is due to soil: 1997 UBC Seismic coefficient, C_v/Z

Soil Profile Type	Seismic Zone Factor-Z				
	Z = 0.075	Z = 0.15	Z = 0.2	Z = 0.3	Z = 0.4
S_A	0.8	0.8	0.8	0.8	0.8 N_V
S_B	1.0	1.0	1.0	1.0	1.0 N_V
S_C	1.7	1.7	1.6	1.5	1.4 N_V
S_D	2.4	2.1	2.0	1.8	1.6 N_V
S_E	3.5	3.3	3.2	2.8	2.4 N_V
S_F					

2000 IBC Table 1615.1.2(1) was derived directly from 1994 NEHRP *Provisions* Table 1.4.2.3a by converting the ground motion parameter from A_a to S_S, the mapped MCE spectral acceleration at short periods. The conversion was based on $S_S = 2.5A_a$. 2000 IBC Table 1615.1.2(2)] was likewise derived from 1994 NEHRP *Provisions* Table 1.4.2.3b by converting the ground motion parameter from A_v to S_1, the mapped MCE spectral acceleration at 1-second period. The conversion was based on $S_1 = A_v$. No near-source factors were incorporated, because the mapped values of A_a, A_v in the 1994 NEHRP *Provisions* and those of S_S and S_1 in the 2000 IBC were not artificially truncated. S_s and S_1 are not artificially truncated in the subsequent editions of the NEHRP *Provisions*, the 2003 and the 2006 IBC, or ASCE 7-05. These parameters attain high values close to known seismic sources that are capable of generating large-magnitude earthquakes.

The 2000 IBC required site-specific geotechnical investigation and dynamic site response analysis not only for Site Class F, but also for Site Class E in areas having $S_S \geq 1.25g$ or $S_1 \geq 0.5g$ (this was similar to the situation in Tables 16-11 and 16-12). However, in the 2000 NEHRP *Provisions*, a number of important modifications were made to the equivalent of 2000 IBC Tables 1615.1.2(1) and 1615.1.2(2). First, values of site factors for the highest accelerations for Site Class E soils ($F_a = 0.9$, $F_v = 2.4$) were placed in the tables so that site-specific studies were no longer required for Type E soils regardless of the level of shaking. The values were considered to be reasonably conservative. This was incorporated in the 2003 IBC and continued in the 2006 IBC and ASCE 7-05.

It should be noted that the footnote in ASCE 7-05 Tables 11.4-1 and 11.4-2 requires, through reference to Sections 11.4.7, that the values of F_a and F_v for Site Class F be determined using a site response analysis outlined in Section 21.1. However, an exception is made in Section 20.3.1 for structures with fundamental periods of vibration less than or equal to 0.5 second built on liquefiable soils of Site Class F, whereby F_a and F_v can be directly taken equal to the values that would be determined for these soils in the absence of liquefaction. (Typically, these soils would be in Site Class D or E.)

Commentary guidance was provided for the first time in the 2000 NEHRP *Provisions* on how to determine the response of Type F soils for which values of site factors F_a and F_v were not given in 2000/2003 IBC Tables 1615.1.2(1) and 1615.1.2(2) [and are not given in 2006 IBC Tables 1613.5.3(1) and 1613.5.3(2)] and for which the note to the tables requires that "site-specific geotechnical investigation and dynamic site response analyses shall be performed." This commentary is reproduced, with permission from the Building Seismic Safety Council (BSSC), in Appendix 4 of this publication, for the convenience of the reader. The material was developed by Technical Subcommittee 3 of the Provisions Update Committee of the BSSC for the 2000 NEHRP *Provisions*. The text has been retained in the Commentary to the 2003 NEHRP *Provisions*.

1613.5.4 Design spectral response acceleration parameters. Five-percent damped design spectral response accelerations at short periods, S_{DS}, and at 1-second period, S_{D1}, are equal to two-thirds S_{MS} ($=F_aS_s$) and two-thirds S_{M1} ($=F_vS_1$), respectively. In other words, the design ground motion is 1/1.5 or two-thirds times the soil-modified maximum considered earthquake ground motion. This is in recognition of the inherent margin contained in the NEHRP Provisions that would make collapse unlikely under one and one-half times the design level ground motion (see Appendix 3 of this publication). Table 16-15 summarizes the derivation of the design quantities S_{DS} and S_{D1}.

Table 16-15. Design ground motion of ASCE 7-05 and the 2006 IBC

S_S = MCE spectral acceleration in the short-period range for Site Class B.
S_1 = MCE spectral acceleration at 1-second period for Site Class B.
$S_{MS} = F_a S_S$, MCE spectral acceleration in the short-period range adjusted for site class effects.
$S_{M1} = F_v S_1$, MCE spectral acceleration at 1-second period adjusted for site class effects.
$S_{DS} = \frac{2}{3} S_{MS}$, spectral acceleration in the short-period range for the design ground motion.
$S_{D1} = \frac{2}{3} S_{M1}$, spectral acceleration at 1-second period for the design ground motion.

1613.5.5 Site classifications for seismic design. The definitions of average soil properties of IBC Section 1613.5.5 and the "Steps for Classifying a Site" of IBC Section 1613.5.5.1 first appeared in the 1994 NEHRP *Provisions*. The material was adopted into UBC Section 1636 Site Categorization Procedure. The material has been revised for the 2006 IBC to address a situation where rock layers are present within the upper 10 feet of the soil profile. For determining the average standard penetration resistance, prior versions of the soil classification procedure only specified an upper limit on N of 100 blows/foot for individual soil layers. However, there was no provision for a scenario where penetration refusal is met because of the presence of a rock layer. In the 2006 IBC (as well as ASCE 7-05 and the 2003 NEHRP *Provisions*), it is now stipulated that the N-value for rock layers is to be taken as 100 blows/foot as well. Further, text from Section 20.3.2 of ASCE 7-05 is added in 2006 IBC Section 1613.5.5, which requires that if a site with a total thickness of more than 10 feet of soft clay layers does not qualify for Site Class F, it is to be classified as Site Class E.

The site classification provisions of Section 1613.5.5 are to be found in Chapter 20 of ASCE 7-05. Refer to the discussion on ASCE 7-05 Chapter 20 in this publication for further commentary on site class. Sections 1613.5.1 through 1613.5.4 are the same as Sections 11.4.1 through 11.4.4 of ASCE 7-05.

1613.5.6 Determination of seismic design category. This section creates six design categories that are key to establishing the design requirements for any building based on its occupancy and on the level of expected soil-modified seismic ground motion. The IBC uses Seismic Design Categories (SDC) to determine permissible structural systems, limitations on height and irregularity, the type of lateral force analysis that must be performed, the level of detailing for structural members and joints that are part of the lateral force-resisting system and for the components that are not. The 1997 UBC, as in the prior editions of the UBC, utilized the Seismic Zone (Section 1629.4.1, Figure 16-2, Table 16-I) in which a structure was located for all these purposes. The Seismic Design Category is a function of occupancy and of soil-modified seismic risk at the site of the structure in the form of the design spectral response acceleration at short periods, S_{DS}, and the design spectral response acceleration at 1-second period, S_{D1}.

The 1997 NEHRP *Provisions* introduced a term, Seismic Use Group (SUG), for the purpose of defining SDC. SUG was subsequently adopted into the IBC and ASCE 7. In earlier editions of the NEHRP *Provisions*, the term *Seismic Hazard Exposure Group* was used to describe the categorization of structures by occupancy or use. The name *Seismic Use Group* was adopted into the 1997 NEHRP *Provisions* as being more descriptive of what the categorization is intended to accomplish. Seismic hazard relates to the severity and frequency of ground motion expected to affect a building. Because the buildings contained in the various groups are spread across regions of varying seismicity, from high to low hazard, the groups do not really relate to hazard. Rather, the groups are used to establish design criteria intended to produce the desired levels of performance in design earthquake events, depending on the occupancy or use of a structure. The following excerpt from the Commentary to the 1997 NEHRP *Provisions* is worth reproducing.

> Historically, building code occupancy classifications were based primarily on fire-safety considerations. It was concluded, however, [in the development of the ATC 3-06 document, the predecessor to the NEHRP *Provisions*] that these traditional classifications would at least in part reflect some considerations contrary to good seismic design. Thus, it was decided that a new approach was needed for defining occupancy exposure to seismic hazards . . .

ASCE 7 and the IBC, instead of directly adopting the seismic use group definitions of the NEHRP *Provisions*, chose to refer to occupancy categories of buildings and other structures for defining SUG. Categories I and II constituted SUG I, while Categories III and IV constituted SUGs II and III, respectively. However, in their latest editions, ASCE 7-05 and the 2006 IBC have decided to eliminate the SUG and relate the SDC of a building directly to its occupancy category. The SUG was simply a consolidation of the occupancy categories, and was felt to be just unnecessary. Table 16-16 correlates the occupancy categories of the 1997 UBC, those of the 2006 IBC, and the Seismic Use Groups of the 2000 and the 2003 IBC.

Table 16-16. Occupancy categories and seismic use groups

IBC Occupancy Category	2000, 2003 IBC Seismic Use Group	1997 UBC Occupancy Category	General Description
I	I	5	Low-hazard facility
II	I	4	Standard-occupancy building
III	II	3	High-occupancy building
IV	III	2,1	Essential or hazardous facility

When ATC 3-06 in 1978 made the level of detailing (and other restrictions concerning permissible structural systems, height, irregularity and analysis procedure) a function of occupancy, that was a major departure from prior practice. The departure was continued in all the NEHRP *Provisions* through the 1994 edition. In the 2000 IBC and the 1997 NEHRP *Provisions* (which have been carried through to the 2006 IBC and the 2003 NEHRP *Provisions*), the level of detailing and other restrictions were made a function of the soil characteristics at the site of the structure. This was a further major departure from practice at that time across the country—a move that has had significant impact on the economic and other aspects of earthquake-resistant design.

A structure located where $S_1 \geq 0.6g$ is assigned to SDC E if it is in Occupancy Category I, II or III and to SDC F if it is in Occupancy Category IV. For structures not assigned to Seismic Design Category E or F, the Seismic Design Category needs to be determined twice—first as a function of S_{DS} by Table 1613.5.6(1) (same as ASCE 7-05 Table 11.6-1) and a second time as a function of S_{D1} by Table 1613.5.6(2) (same as ASCE 7-05 Table 11.6-2). The more severe category governs.

Seismic Design Category A applies to structures, irrespective of their occupancy, in regions where anticipated ground motions are minor, even for very long return periods.

Seismic Design Category B includes Occupancy Category I, II or III structures in regions where moderately destructive ground shaking is anticipated.

Seismic Design Category C includes Occupancy Category IV structures where moderately destructive ground shaking may occur as well as Occupancy Category I through III structures in regions with somewhat more severe ground shaking potential.

Seismic Design Category D includes structures of Occupancy Category IV structures located in regions of severe seismicity and above, and Occupancy Category I through III structures located in regions expected to experience destructive ground shaking.

However, as noted above, structures located close to major active faults are directly assigned to SDC E or F outside the provisions of Tables 11.6-1 and 11.6-2. SDC E includes Occupancy Category I, II or III structures in regions located close to major active faults and SDC F includes Occupancy Category IV structures in those locations.

For the purposes of detailing as well as the other restrictions, the UBC Seismic Zones and the IBC Seismic Design Categories that may be considered to be corresponding to them are shown in Table 16-17.

Table 16-17. **Correspondence between UBC Seismic Zones and IBC Seismic Design Categories**

1997 UBC Seismic Zone	0, 1	2A, 2B	3, 4
2006 IBC Seismic Design Category	A, B	C	D, E, F

1613.5.6.1 Alternative seismic design category determination. In view of the fact that it is unnecessary and wasteful to require that the seismic design category of a short-period structure be determined by long-period ground motion, the 2003 IBC permitted the SDC determination to be based on S_{DS} alone [Table 1613.5.6(1)], provided:

1. $T_a = 0.8T_s$ where T_a is the approximate fundamental period of the structure and $T_s = S_{D1}/S_{DS}$,
2. Upper-bound design base shear is used in design (e.g., $V = S_{DS}W/(R/I)$), and
3. Diaphragm is not flexible.

The period $T_s = S_{D1}/S_{DS}$ is the period at which the short-period or constant-acceleration part of the design spectrum transitions into the long-period or velocity-governed part of the spectrum. It is the dividing line between short-period and long-period response. By requiring in Item 1 above that T_a be less than $0.8T_s$ rather than T_s itself, the intent of the code is to minimize the possibility that T might equal or exceed T_s, even though T_a is less than T_s.

Item 2 above requires that the upper-bound design base shear, as given by the constant-acceleration or *flat-top* part of the design spectrum, be used in the design of a structure utilizing the above exception. This requirement is intended to impose a design force penalty on a structure for which T may equal or exceed T_s, while T_a is less than $0.8T_s$.

Item 3 above makes the relaxation in question inapplicable to structures with flexible diaphragms because the flexible diaphragm may dramatically influence the elastic fundamental period of such a structure to the extent that it may exceed the approximate fundamental period, T_a.

The 2003 IBC exception discussed above is now included in 2006 IBC Section 1613.5.6.1. However, the following three additional criteria are included for this relaxation to be applicable, thus making it consistent with the same provision in ASCE 7-05 Section 11.6:

1. 1-second mapped spectral response acceleration, S_1, is less than $0.75g$,
2. In each of two orthogonal directions, the fundamental period of the structure used to calculate story drift is less than the period T_s, and
3. Buildings with flexible diaphragms can now qualify, but only when the distance between vertical elements of the seismic force-resisting system does not exceed 40 feet.

The first additional criterion means that SDC determination cannot be based on S_{DS} alone for structures in the vicinity of known faults that can generate large earthquakes (so-called near-fault structures). It is already required that for the relaxation in question to be applicable, the approximate period T_a of the structure in each of two orthogonal directions be less than $0.8T_s$ and that the upper-bound design base shear (as given by ASCE 7 Eq. 12.8-2) be used in the design of the structure. The second additional requirement would appear to make these requirements superfluous. ASCE 7 does not impose an upper limit on the period used to calculate story drift, and as a result, it probably is a more realistic representation of the actual period of the structure than T_a. This is probably why the second additional requirement was added. The third additional criterion appears to be justified because when the diaphragm span is no more than 40 feet, the dynamics of the diaphragms are unlikely to drastically influence the period of the structure, which is typically the concern with flexible-diaphragm structures.

Whether there is any advantage to be gained from the relaxation in question depends strictly on the relationship between S_{DS} and S_{D1} at the site of a structure. For many locations across the U.S., there is no advantage to be gained, because the soil-modified short- and long-period ground motion parameters yield the same seismic design category. However, in other situations, the relaxation in question may yield a one- or even two-category reduction in the seismic design category. In other words, while the SDC may be D based on S_{D1}, it may be only C on the basis of S_{DS}. In that case, only intermediate, rather than special, detailing would be required for a structure.

1613.6

Alternatives to ASCE 7. The 2006 IBC includes the following two modifications to ASCE 7-05. These modifications are optional, meaning that they may be used if desired; otherwise, the unmodified corresponding ASCE 7-05 provisions are to be used.

1. The 2003 IBC definition for flexible diaphragms stated that a diaphragm is flexible when the computed maximum in-plane deflection of the diaphragm is more than 2 times the average drift of adjoining vertical elements. Section 1617.5.3 additionally stated that diaphragms constructed of untopped steel decking or wood structural panels or similar light-framed construction are permitted to be considered flexible. However, this section only applied to the simplified design method. ASCE 7-05, while retaining the general definition based on the in-plane diaphragm deflection, includes prescriptive provisions that allow certain diaphragms to be idealized as flexible or rigid. A part of the prescriptive provisions is as follows (see later discussion of ASCE 7-05 Section 12.3.1):

   12.3.1.1 Flexible diaphragm condition. Diaphragms constructed of untopped steel decking or wood structural panels are permitted to be idealized as flexible in structures in which the vertical elements are steel or composite steel and concrete braced frames, or concrete, masonry, steel or composite shear walls. Diaphragms of wood structural panels or untopped steel decks in one- and two-family residential buildings of light-frame construction shall also be permitted to be idealized as flexible.

   The 2006 IBC modifies this ASCE 7-05 definition by adding a new set of criteria at the end of ASCE 7-05 Section 12.3.1.1 for idealizing a diaphragm as flexible in light-frame construction. In ASCE 7-05, diaphragms constructed of wood structural panels or untopped steel decking are only permitted to be idealized as flexible when the light-frame construction is a one- or two-family residential building. However, it was realized that an appropriate idealization can only be done based on the nature of the construction itself rather than its occupancy. The modification sets forth four conditions that must be met for other occupancies. These conditions are based on research from the CUREE Caltech Woodframe project, which showed that for regular, light-framed, wood diaphragm buildings, treating the diaphragms as flexible gives a better match with the results of full-size experimental tests.

An exception is added to ASCE 7-05 Section 17.5.4.2, whereby restrictions placed on the use of Ordinary Steel Concentrically Braced Frames (Steel OCBFs) and Ordinary Steel Moment Frames (OMFs of steel) in a seismically isolated SDC D, E or F structure are relaxed significantly. Because of their poor performance during a large earthquake, ASCE 7-05 Table 12.2-1 severely limits the use of these systems in SDC D, E or F structures. However, while these limitations are valid for typical buildings with fixed bases where large ductility demands are expected, seismically isolated structures deserve a different approach, as the ductility demand on them is much smaller, owing to the much smaller value of the response modification factor *R*. Thus, a significant cost reduction is possible by not requiring the implementation of special detailing schemes associated with special moment frames and braced frames. This added exception permits OCBFs and OMFs of steel in a building of SDC D, E or F up to a height of 160 ft, provided the value of R_I is taken as 1 and the structural systems are designed in accordance with the 2005 edition of AISC 341, *Seismic Provisions for Structural Steel Buildings*.[24]

Seismic Design Provisions of ASCE 7-05

Because the seismic design provisions of the 2006 IBC, other than those discussed above, are only in ASCE 7-05, it is now necessary to discuss the seismic design provisions of that document.

The seismic design provisions of ASCE 7-05 have been completely reformatted and reorganized. The goal of the reorganization effort was to present the ASCE 7-05 seismic provisions in such a manner that they are user-friendly and can be understood and interpreted correctly and easily by the intended user.

Chapter 11 Seismic Design Criteria

11.1.2 Scope. After stating that every structure and portion thereof, including nonstructural components, must be designed and constructed to resist the effects of earthquake motions, ASCE 7 proceeds to permit a number of important exemptions:

1. Detached one- and two-family dwellings located where the mapped short-period spectral response acceleration S_S is less than 0.4*g*, or when assigned to Seismic Design Category (Section 11.6) A, B and C, are totally exempt from all seismic design requirements.
2. Detached one- or two-family wood frame dwellings not included in Exception 1 with not more than two stories, and satisfying the limitations and constructed in accordance with the *International Residential Code* (IRC)[25]. The 2006 IBC modifies this exception to require that the seismic-force-resisting systems in wood frame buildings conform to the provisions of Section 2308 of the IBC to qualify for exemption from the seismic design requirements.

3. Agricultural storage structures intended only for incidental human occupancy are exempt from all seismic design requirements. This is because of the very low risk to life involved.

4. Structures such as vehicular bridges, electrical transmission towers, hydraulic structures, buried utility lines and their appurtenances, and nuclear reactors that require special considerations of their response characteristics and environment, are not covered by ASCE 7 or the IBC, and need to be designed in accordance with other regulations.

The 2006 IBC excludes the provisions of ACSE 7-05 Chapter 14 and Appendix 11A from its scope. Chapter 14 of ASCE 7 is "Material Specific Seismic Design and Detailing Requirements." The requirements contained in this ASCE 7 chapter are found in the materials chapters of the 2006 IBC. Appendix 11A is "Quality Assurance Provisions", and those provisions are found in 2006 IBC Chapter 17.

Seismic ground motion values. This section provides a step-by-step procedure for determining the design ground motion acceleration value. Two procedures are provided for the determination of ground motion accelerations, represented by response spectra and coefficients derived from these spectra—the general procedure described in Sections 11.4.1 through 11.4.6 or the site-specific procedure mentioned in Section 11.4.7. The actual provisions of the site specific procedure are contained in Chapter 21 of ASCE 7-05. The provisions of Sections 11.4.1 through 11.4.4 are reproduced in 2006 IBC Section 1613.5.1 through 1613.5.4, respectively, and have been discussed above. **11.4**

Mapped acceleration parameters. The mapped maximum considered earthquake spectral response accelerations at short periods, S_S and at 1-second period, S_1, for a particular site are to be determined from Figures 22-1 through 22-14 of ASCE 7-05, which are also reproduced as Figures 1613.5(1) through 1613.5(14) in the 2006 IBC. Where a site is between contours, straight line interpolation or the value of the higher contour may be used. **11.4.1**

Because the ASCE 7-05 and 2006 IBC maps for S_s and S_1 are new, a DVD has also been prepared (although, as of this writing, not yet released) by the USGS in cooperation with BSSC and FEMA. It can be used to obtain ground motion accelerations for design according to the NEHRP/ASCE 7, the *International Residential Code* (IRC), and the NEHRP Rehabilitation Guidelines.[26] Either the ZIP code or the latitude/longitude of the site is input to the program, and the corresponding S_S and S_1 values are output. Soil-modified accelerations can also be obtained when the site class is input. The USGS maximum considered earthquake, the IBC ground motion maps and the IRC ground motion maps can also be viewed and printed from the DVD.

The same functionality of the DVD can also be obtained from the USGS website at http://earthquake.usgs.gov/research/hazmaps.

This section also includes a provision that assigns Seismic Design Category (SDC) A to any structure regardless of its Occupancy Category when mapped short period and 1-second period response spectral accelerations, S_s and S_1, are less than or equal to 0.15g and 0.04g, respectively. Figure 16-19 shows the geographic region where this provision is applicable.

Site class. Appendix 4 of this publication should be referred to for background on consideration of site soil characteristics in seismic design. This section refers to Chapter 20 of ASCE 7-05 for the definitions of the six soil classes. The 1994 NEHRP *Provisions* defined six soil profile types as follows: **11.4.2**

A. Hard rock with measured shear wave velocity, $\bar{v}_u$ > 5000 ft/sec (1500 m/sec)

B. Rock with 2500 ft/sec < $\bar{v}_s \le$ 5000 ft/sec (760 m/sec < $\bar{v}_s \le$ 1500 m/sec)

C. Very dense soil and soft rock with 1200 ft/sec < $\bar{v}_s \le$ 2500 ft/sec (360 m/sec < $\bar{v}_s \le$ 760 m/sec) or with either $\bar{N}$ > 50 or $\bar{S}_u \ge$ 2000 psf (1000 kPa)

D. Stiff soil with 600 ft/sec < $\bar{v}_s \leq$ 1200 ft/sec (180 m/sec < $\bar{v}_s \leq$ 360 m/sec) or with either 15 < $\bar{N} \leq$ 50 or 1000 psf < $\bar{S}_u \leq$ 2,000 psf (50 kPa < $\bar{S}_u \leq$ 100 kPa)

E. A soil profile with $\bar{v}_s$ < 600 ft/sec (180 m/sec) or a profile with more than 10 feet (3 meters) of soft clay defined as soil with PI > 20, $w \geq$ 40 percent, and $\bar{S}_u$< 500 psf (25 kPa).

F. Soils requiring site-specific evaluations:

1. Soils vulnerable to potential failure or collapse under seismic loading such as liquefiable soils, quick and highly sensitive clays, and collapsible weakly cemented soils.
2. Peats and/or highly organic clays [*H* > 10 feet (3 m) of peat and/or highly organic clay]
3. Very high plasticity clays [*H* > 25 ft (8 m) with *PI* > 75]
4. Very thick soft/medium stiff clays [*H* > 120 feet (36 m)]

In the above,

H = thickness of soil.

$\bar{N}$ = average field standard penetration resistance for the top 100 ft (30 m) of soil.

PI = plasticity index of soil (ASTM D 4318[27]).

$\bar{S}_u$ = average undrained shear strength in top 100 ft(30 m) of soil (ASTM D 2166[28] or ASTM D 2850[29]).

$\bar{V}_u$ = average shear wave velocity in top 100 ft (30 m) of soil.

w = moisture content of soil in percent (ASTM D 2216[30]).

The above classification was adopted and incorporated into Table 16-J of the 1997 UBC. In the process of adoption, one change was made. The above definition of Soil Profile Type E became a footnote to Table 16-J. In the table itself, Soil Profile Type E was defined as a profile with $\bar{v}_s$ < 600 ft/sec or $\bar{N}_{CH}$ < 15 or $\bar{S}_u$ < 1,000 psf, where N_{CH} = average standard penetration resistance for cohesionless soil layers for the top 100 ft (30 m) of soil. Table 20.3-1 of ASCE 7-05 and Table 1613.5.2 of the 2006 IBC are identical to Table 16-J of the 1997 UBC, except that the footnote mentioned above is incorporated into the table itself.

The 1994 NEHRP *Provisions* defined a default site class for use in design by requiring that "when the soil properties are not known in sufficient detail to determine the Soil Profile Type, Type D shall be used. Soil Profile Type E or F need not be assumed unless the regulatory agency determines that Type E or F may be present at the site or in the event that Type E or F is established by geotechnical data." This requirement has been retained in the 1997, 2000 and 2003 NEHRP *Provisions*. ASCE 7-05 Section 11.4.2, as well as Section 20.1, includes a variation of this requirement in that "where the soil properties are not known in sufficient detail to determine the site class, Site Class D shall be used unless the authority having jurisdiction or geotechnical data determines that Site Class E or F soils are present at the site." This is to be compared with the default soil classification in prior design practice, as represented by the 1994 UBC where "in locations where the soil properties are not known in sufficient detail to determine the soil profile type, soil profile S_3 shall be used. Soil profile S_4 need not be used unless the building official determines that soil profile S_4 may be present at the site or in the event that soil profile S_4 is established by geotechnical data, in which case soil profile S_4 will be used."

The default site class of ASCE 7-05 is extremely important because Table 20.3-1 requires 100-foot (30 m) borings as the basis of soil classification.

ASCE 7-05 Section 20.1 also addresses a situation when only a partial soil information is available, stating "where site-specific data are not available to a depth of 100 ft, appropriate soil properties are permitted to be estimated by the registered design professional preparing the soil investigation report based on known geologic conditions."

There is no direct correlation between the site classification given by Table 20.3-1 and the prior S_1 through S_4 soil classifications. An approximate correlation is attempted in Table 16-18. The table is intended solely to give the reader a rough idea of the correlation between the old and the current classifications. It must not be used for any other purpose.

Table 16-18. Soil classifications of ASCE 7-05 versus soil classifications of earlier codes

A	Hard rock	S_1 Rock
B	Rock	S_1 Rock
C	Very dense soil or soft rock	S_1, S_2
D	Stiff soil	S_2, S_3
E	Soft soil	S_3, S_4 Soft soil
F	Soils requiring site-specific evaluations	S_4 Soft soil

Site coefficients and adjusted maximum considered earthquake (MCE) spectral response acceleration parameters. Read the discussion on 2006 IBC Section 1613.5.3. **11.4.3**

Design spectral response acceleration parameters. Read the discussion on 2006 IBC Section 1613.5.4. **11.4.4**

Design response spectrum. This provides a general method for obtaining a 5-percent damped response spectrum from the site design acceleration response parameters S_{DS} and S_{D1}. The design spectrum in ASCE 7-05 is based on that proposed by Newmark and Hall[31] and consists of a series of four curves representing a very short-period region, a short-period region of constant spectral acceleration, a long-period region of constant spectral velocity and a very long-period region of constant spectral displacement (Figure 16-19). **11.4.5**

The spectral response acceleration at any period in the short-period range is equal to the design spectral-response acceleration at short periods, S_{DS}:

$$S_a = S_{DS}$$

The spectral response acceleration at any point in the constant velocity range can be obtained from the relationship:

$$S_a = S_{D1}/T$$

The spectral response acceleration at any point in the constant displacement range can be obtained from the relationship:

$$S_a = S_{D1}T_L/T^2$$

where the long-period transition period T_L is given on new maps in Figures 22-15 through 22-20, which are similar to zone maps, for all fifty states. One must locate one's site on this map to determine T_L, which ranges between 4 and 16 seconds, depending upon the location. The values of T_L are much larger than the corresponding values traditionally used by many engineers in the past.

It should be noted that the constant displacement branch of the design response spectrum is new to ASCE 7-05 and the 2006 IBC and did not exist in ASCE 7-02 or the 2003 IBC. This addition is significant in determining the slosh height in tanks and the design seismic forces of long-period buildings. Because the designs of many tall buildings have assumed displacement cutoff periods of 4 seconds or less, the impact of this change could also be significant for high-rise buildings. As a companion change to the addition of this new branch to the design spectrum, the minimum design base shear of $0.044S_{DS}IW$ of ASCE 7-02 and the 2003 IBC, which was applicable in all seismic design categories, is deleted and is

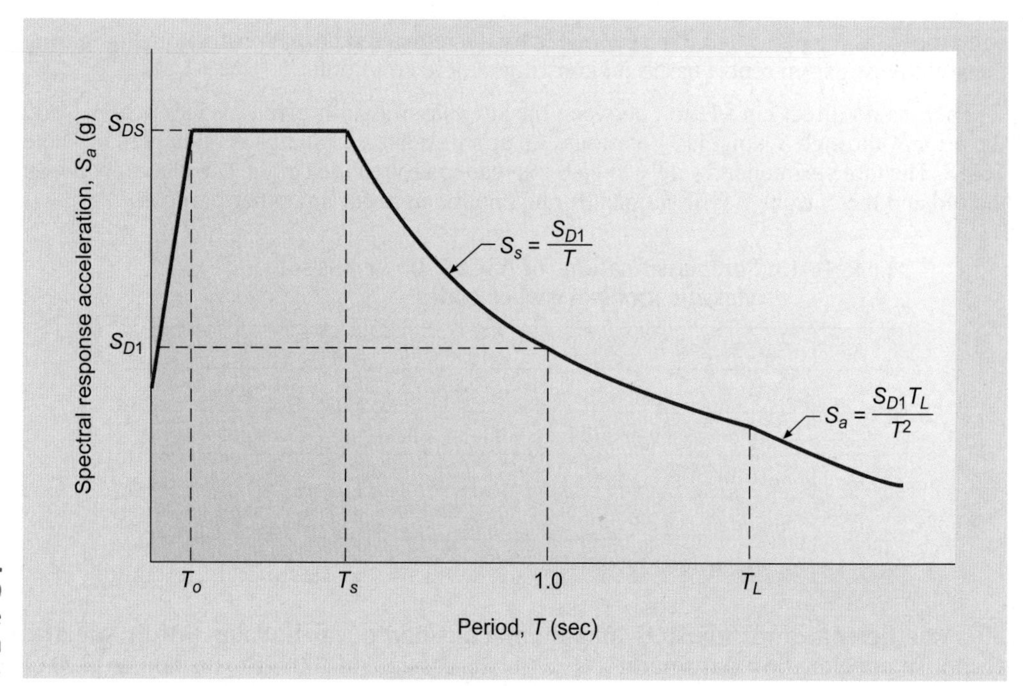

Figure 16-20 Design response spectrum in ASCE 7-05

not part of ASCE 7-05. The deletion of the minimum design base shear of $0.044S_{DS}IW$ may have a significant impact on the seismic design of high-rise buildings.

The ramp building up to the flat top of the design spectrum is defined by specifying that the spectral-response acceleration at zero period is equal to 40 percent of the spectral-response acceleration corresponding to the flat top, S_{DS}, and that the period T_0 at which the ramp ends is 20 percent of the period, T_S, at which the constant acceleration and the constant velocity portions of the spectra meet. That period,

$$T_S = S_{D1}/S_{DS}$$

is solely a function of the seismicity and the soil characteristics at the site of the structure. It also serves as the dividing line between short- and long-period structures.

11.4.6 MCE response spectrum. The maximum considered earthquake spectral response acceleration at any period, S_{aM}, is to be determined by simply multiplying the corresponding ordinate of the design response spectrum by 1.5.

11.4.7 Site-specific ground motion procedures. Whereas the site-specific ground motion procedures, now set forth in ASCE 7-05 Chapter 21, can be used to determine ground motions for any structures, ASCE 7-05 requires "site-specific" analysis to be carried out in two very different situations: (1) for structures located on Site Class F soils and (2) for seismically isolated structures on sites that are *near-source* ($S_1 = 0.6g$). In order to clarify what is required for these two different situations, the site-specific procedure section in ASCE 7-05 is divided into two parts: (1) Section 21.1 Site Response Analysis (applicable to Site Class F soils) and (2) Section 21.2 Ground Motion Hazard Analysis (applicable to seismically isolated structures and structures with damping systems on sites with $S_1 = 0.6g$). The site response analysis is completely new, whereas the ground motion hazard analysis is based on the site-specific procedure in ASCE 7-02 Section 9.4.1.3.

11.5 Importance factor and occupancy category. The seismic importance factor is adopted from the 1997, 2000 and 2003 NEHRP *Provisions*. The importance factor of a structure is determined based on its occupancy category. A detailed discussion on occupancy category is provided under IBC Section 1604.5. ATC 3-06,[32] the predecessor document to NEHRP, deliberately decided to drop the importance factor that had long been used in seismic design by the UBC. Instead, ATC 3-06 chose to institute two other requirements to ensure enhanced performance of structures in higher occupancy categories. First, it made drift limits tighter for structures in higher occupancy categories. Second, it made the level of detailing and other restrictions a function not only of the seismic risk at the site of a structure, but also of the occupancy of the structure. With these provisions still in place, the 1997/2000/2003 NEHRP *Provisions* and the IBC have chosen to include the importance factor. The highest value of the importance factor in the IBC is 1.5, whereas the UBC has used a maximum value of 1.25 since 1988.

According to the Commentary to the 1997/2000/2003 NEHRP *Provisions*, the importance factor represents only one of several ways in which control is attempted of the seismic performance capabilities of buildings in different occupancy categories. Specifically, the occupancy importance factor modifies the R coefficients used to determine minimum base shear forces. Structures assigned occupancy importance factors exceeding 1.0 are, through this mechanism, designed for larger base shear forces. As a result, they are expected to experience lower ductility demands than structures designed with lower occupancy importance factors and, hence, would be expected to sustain less damage.

Importance factor. Table 11.5-1 provides seismic importance factors based on the occupancy category of a building as determined from Table 1604.5. Whereas Occupancy Category I and II buildings are assigned an importance factor of 1.0, Occupancy Category III and IV buildings are assigned 1.25 and 1.5, respectively. The values of 1.25 and 1.5 are to be contrasted with the corresponding values of 1.00 and 1.25, respectively, in the 1997 UBC. **11.5.1**

Seismic design category. The point has been made that it is unnecessary and wasteful to require that the SDC of a short-period structure be determined by long-period ground motion. A relaxation was permitted in the 2003 IBC, which was not part of ASCE 7-02. The 2003 IBC provision, in a modified form, has now been adopted in ASCE 7-05. ASCE 7-05 Section 11.6 permits the SDC determination to be based on S_{DS} (Table 11.6-1) alone, provided all of the following apply: **11.6**

1. 1-second mapped spectral response acceleration S_1 is less than $0.75g$.
2. The approximate fundamental period of the building, T_a, in each of the two orthogonal directions determined in accordance with Section 12.8.2.1 of ASCE 7-05 is less than $0.8T_S = 0.8S_{D1}/S_{DS}$.
3. In each of two orthogonal directions, the fundamental period of the structure used to calculate story drift is less than the period T_s, which marks the end of the constant-acceleration segment of the design response spectrum.
4. Equation 12.8-2 of ASCE 7-05 is used to determine the seismic response coefficient C_S; and,
5. The diaphragms are rigid per the definition given in Section 12.3.1 or for diaphragms that are flexible, distance between vertical elements of the seismic-force-resisting system is not more than 40 feet.

Item 4 above requires that the upper-bound design base shear, as given by the constant acceleration or *flat-top* part of the design spectrum, be used in the design of a structure utilizing the above exception. This requirement is intended to impose a design force penalty on a structure for which T may equal or exceed T_S, while T_a is less than $0.8T_S$.

Item 5 above makes the relaxation in question inapplicable to structures where flexible diaphragms may play a decisive role in determining the elastic fundamental period of the structure, and this period may well be in excess of the approximate fundamental period, T_a. As a result, flexible diaphragms are only permitted when the vertical supports are close enough not to allow the diaphragms to have too much effect on the dynamics of the structure.

It has been demonstrated that this exception can be advantageous in many areas of the United States.[33]

Additionally, this section also stipulates that when a structure is designed in accordance with the simplified design procedure of ASCE 7-05 Section 12.14, SDC can be assigned from Table 11.6-1 alone using the S_{DS} values. The requirements that a structure needs to satisfy to qualify for the simplified design automatically satisfies the five items mentioned above.

11.7 Design requirements for Seismic Design Category A. For structures assigned to SDC A, the IBC requires only that a complete lateral force-resisting system be provided and that all elements of the structure be tied together.

11.7.1 Applicability of seismic requirements for Seismic Design Category A structures. This section contains all the seismic requirements that a SDC A structure needs to meet. The requirements are very basic and are meant to provide a minimum level of structural integrity, avoiding the more complex design and detailing considerations given to SDC B through F.

11.7.2 Lateral forces. The structure needs to be analyzed for static lateral forces equal to 1 percent of the floor weight applied at each floor level, and the resulting load effect E is to be used in the design load combinations. The background to the requirements of Sections 11.7.2 through 11.7.4 is provided in the Commentary to the 2003 NEHRP *Provisions* (Section 1.5). The minimum lateral force requirement is intended to provide a nominal level of structural integrity that will improve the performance of buildings in the event of a possible but rare earthquake and many other types of unanticipated loadings. The 1-percent value has been used in other countries as a minimum value for structural integrity. For many structures, design for the specified wind loads will normally control the lateral-force design when compared to the minimum force specified for structural integrity. However, many low-rise, heavy structures or structures with significant dead loads resulting from heavy equipment may be controlled by the nominal 1-percent acceleration.

ASCE 7 allows orthogonal effects to be neglected in the design of SDC A structures. The design seismic forces may be applied separately in each of the two orthogonal directions.

11.7.3 Load path connections. According to the 2003 NEHRP Commentary (Section 4.6.1.1), probably the most important single attribute of an earthquake-resisting building is that it is tied together to act as a unit. This is important not only for earthquake resistance but also for resisting high winds, floods, explosion, progressive collapse and even such ordinary hazards as foundation settlement. The 2003 NEHRP *Provisions*, as well as ASCE 7-05, require that any smaller portion of a structure (or unit if there are separation joints) be tied to the remainder of the structure with elements having a strength of 5 percent of the smaller portion's weight. In the 2000 NEHRP *Provisions*, ASCE 7-02 and the 2003 IBC, the minimum connection strength was specified as $S_{DS}/7.5$ times the weight of the smaller portion or 5 percent of the portion's weight, whichever is greater. This was revised as the value of $S_{DS}/7.5$ for an SDC A structure is always less than 5 percent because the largest S_{DS} for SDC A is 0.167. Note that this requirement does not apply to the connection between the structure itself and its foundation.

11.7.4 Connection to supports. In addition to the above, a positive connection is to be provided to support each beam, girder or truss to resist a horizontal force acting parallel to the member. The connection must have a minimum strength of 5 percent of the vertical dead and live load reaction applied horizontally.

11.7.5 Anchorage of concrete or masonry walls. As mentioned earlier in the discussion of 2006 IBC Section 1604.8, many failures of concrete or masonry walls in the 1971 San Fernando and 1994 Northridge earthquakes were attributable to inadequate anchorage between the walls and the roof system. This section requires that anchorage be capable of resisting a force that is the larger of the horizontal force specified in Section 11.7.3 and 280 pounds per linear foot of wall (Figure 16-1).

11.8 Geologic hazards and geotechnical investigation.

11.8.1 Site limitations for Seismic Design Categories E and F. It is not possible to reliably design a structure to resist the very large forces it may be subjected to if an active fault were to cause rupture of the ground surface at its location. Consequently, this section prohibits the construction of buildings assigned to SDC E and F at sites where such potential exists.

As pointed out in the 2003 NEHRP Commentary, the effects of landsliding, liquefaction and lateral spreading can also be highly damaging to a building. However, the effects of these site phenomena can more readily be mitigated through the adoption of appropriate measures than can the effects of direct ground-fault rupture. Consequently, construction on sites with these hazards is permitted, if appropriate mitigation measures are undertaken.

Geotechnical investigation report for seismic design categories C through F. This section is the same as 2006 IBC Section 1802.2.6. Please see the discussion of that section in this publication. **11.8.2**

Additional geotechnical investigation report requirements for seismic design categories D through F. This section is the same as 2006 IBC Section 1802.2.7. Please see the discussion of that section in this publication. **11.8.3**

Chapter 12 *Seismic Design Requirements for Building Structures*

Structural design basis. This section provides a general outline of the basic requirements that structures assigned to SDC B through F must meet before going into more detailed seismic design provisions. A building structure shall include complete vertical and lateral force-resisting systems that have adequate strength, stiffness and energy dissipation capacity to withstand design ground motions in any direction, while maintaining much of its gravity load carrying capacity. This adequacy needs to be demonstrated through a mathematical model of the structure with design earthquake loads applied in accordance with the provisions of this standard. Individual members must be capable of carrying the combined effects of axial forces, shear forces and bending moments that are induced in them from the load combinations of this standard, and connections must have adequate strength to transmit the member forces. A continuous load path must be provided for effective transfer of the lateral forces down to the foundation. Foundations themselves need to be designed for the forces imposed by the superstructure with consideration of the dynamic nature of the forces as well as the dynamic properties of the soil. **12.1**

This section also requires that members conform to the material design and detailing requirements of Chapter 14 of this standard. However, the 2006 IBC specifically excludes this requirement, as the material-specific detailing provisions can be found in the materials chapters of the IBC itself.

Structural system selection. **12.2**

Selection and limitations. This section provides a list of seismic force-resisting systems that are permitted by this standard for use individually or in combination. The basic system types and their subdivisions are listed in Table 12.2-1, along with their design parameters as well as limitations. This section also permits the use of seismic force-resisting systems that are not included in Table 12.2-1, but only when it is demonstrated through test data or analytical results that their seismic performance is equivalent to that of the systems included in the table. **12.2.1**

The following general types of structural systems are included in ASCE 7-05 Table 12.2-1.

Bearing wall system. This is a structural system without an essentially complete space frame providing support for gravity loads. Bearing walls provide support for all or most gravity loads. Resistance to lateral load is provided by the same bearing walls acting as shear walls [Figure 16-21(a)].

In SDC D, E and F, the lateral force-resisting shear walls must be special reinforced concrete shear walls, intermediate precast shear walls or special reinforced masonry shear walls; or they may be light-framed walls with shear panels made of wood structural panels, light-framed walls with shear panels made of other materials (SDC D only), or light-framed wall systems using flat strap bracing. In SDC C, lateral force-resisting shear walls may additionally consist of ordinary reinforced concrete shear walls, intermediate reinforced masonry shear walls or ordinary reinforced masonry shear walls. In SDC B, also allowed as lateral force-resisting shear walls are detailed plain concrete shear walls, ordinary plain concrete shear walls, detailed plain masonry shear walls and ordinary plain masonry shear walls. For each type of shear wall listed in ASCE 7-05 Table 12.2-1, an appropriate response modification coefficient, *R*, is specified, with increasing values of *R* reflective of the level of potential inelastic deformability. Each shear wall type must comply with the requirements of the respective section number(s) listed in Column 2 of the table, depending on the assigned seismic design category. The special design and detailing requirements for each seismic design category are stated in the material chapters of the IBC as well as in Chapter 14 of ASCE 7-05.

Building frame system. This is a structural system with an essentially complete space frame providing support for gravity loads. Resistance to lateral loads is provided by shear walls or braced frames [Figure 16-21(b)].

The seismic safety of the building frame system is totally dependent on good-faith satisfaction of the deformation compatibility requirements (ASCE 7-05 Section 12.12.4 for SDC D through F). These recognize that when the designated lateral force-resisting system of a structure deforms laterally under an earthquake of intensity anticipated by the code, the subsystems that have been arbitrarily designated to be outside of the lateral force-resisting system will have no choice but to deform together, because they are connected at every floor level through the floor systems. If in the course of that earthquake-induced lateral displacement, the subsystems designed for gravity loads only are unable to sustain their gravity load-carrying capacity, then life-safety is compromised. It is thus a specific requirement of all seismic codes, including ASCE 7-05, that structural elements or subsystems designated not to be part of the lateral force-resisting system be able to sustain

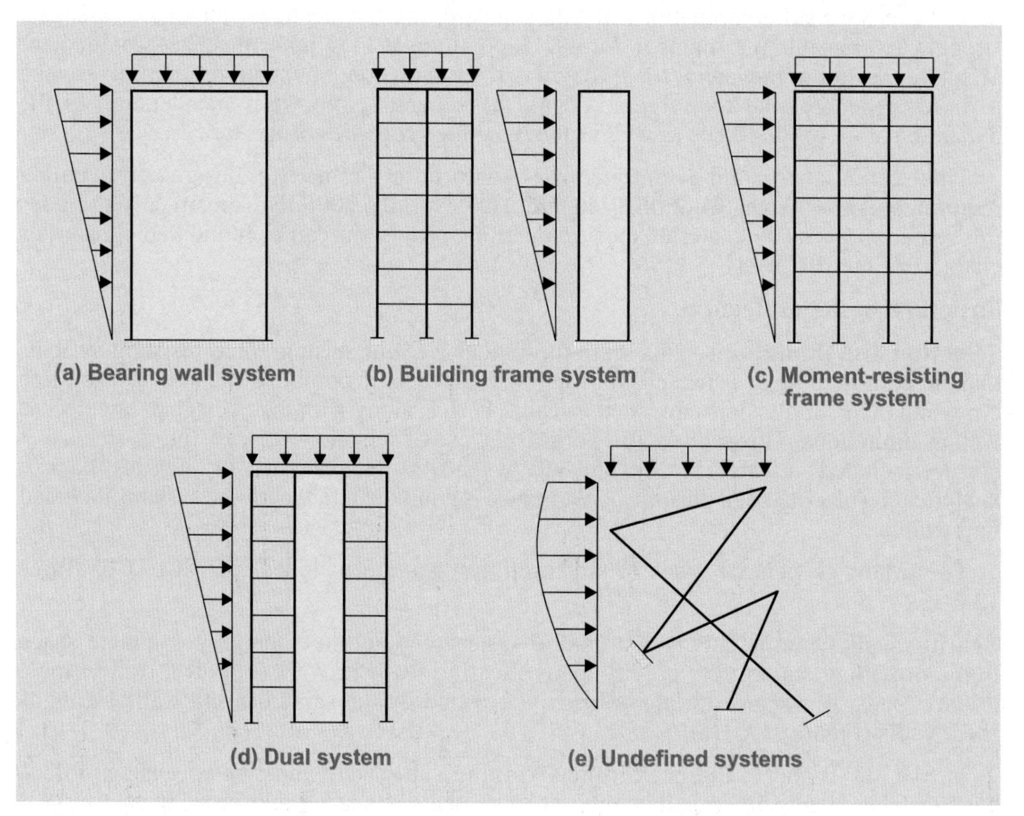

Figure 16-21 Seismic-force-resisting structural systems

their gravity load carrying capacity at a lateral displacement equal to a multiple times the computed elastic displacement of the lateral force-resisting system under code-prescribed seismic design forces. The amplified elastic displacement of the lateral force-resisting system is intended to provide an estimate of the actual displacement of the entire structure caused by an earthquake of intensity anticipated by the code. If, under the estimated earthquake-induced displacements, the gravity loads would cause inelasticity (because of the induced bending moments and shear forces exceeding the design moment strength or the design shear strength, respectively) in any structural element initially designed for gravity only, that structural element should also be detailed for inelastic deformability. If satisfaction of deformation compatibility would require that ductility details be provided in structural members originally designed for gravity only, then the engineer ought to review his or her original decision and consider making such structural elements or subsystems part of the lateral force-resisting system. In other words, an alternative structural system should be considered for the building.

The lateral force-resisting system of a building assigned to SDC D, E or F, and utilizing the building frame system may consist of:

1. Steel eccentrically braced frames with moment-resisting connections at columns away from links.
2. Steel eccentrically braced frames with nonmoment-resisting connections at columns away from links.
3. Special steel concentrically braced frames.
4. Ordinary steel concentrically braced frames.
5. Special reinforced concrete shear walls.
6. Intermediate precast shear walls.
7. Composite eccentrically braced frames.
8. Composite concentrically braced frames.
9. Composite steel plate shear walls.
10. Special composite reinforced concrete shear walls with steel elements.
11. Special reinforced masonry shear walls.
12. Intermediate reinforced masonry shear walls.
13. Light-frame walls with wooden shear panels or steel sheets.
14. Light-frame walls with shear panels of other materials (SDC D only).
15. Buckling restrained braced frames with nonmoment-resisting beam-column connections.
16. Buckling restrained braced frames with moment-resisting beam-column connections.
17. Special steel plate shear walls.

In a building utilizing the building frame system and assigned to SDC C, the lateral force-resisting system may additionally consist of:

18. Ordinary reinforced concrete shear walls.
19. Ordinary composite braced frames.
20. Ordinary composite reinforced concrete shear walls with steel elements.
21. Intermediate reinforced masonry shear walls.
22. Ordinary reinforced masonry shear walls.
23. Detailed plain masonry shear walls.

Each of these shear wall or braced frame types must satisfy the requirements of the corresponding section number(s) listed in column 2 of Table 12.2-1.

SDC B is not mentioned above, because the concept of the building frame system loses its validity for buildings assigned to SDC B, as discussed later.

Two new systems are included in ASCE 7-05: buckling restrained braced frames and special steel plate shear walls. Buckling Restrained Braces (BRBs) are typically made by encasing a steel core brace member in a concrete filled steel tube (Figure 16-22). The steel core is kept separated from the concrete filled tube by a layer of unbonding material applied

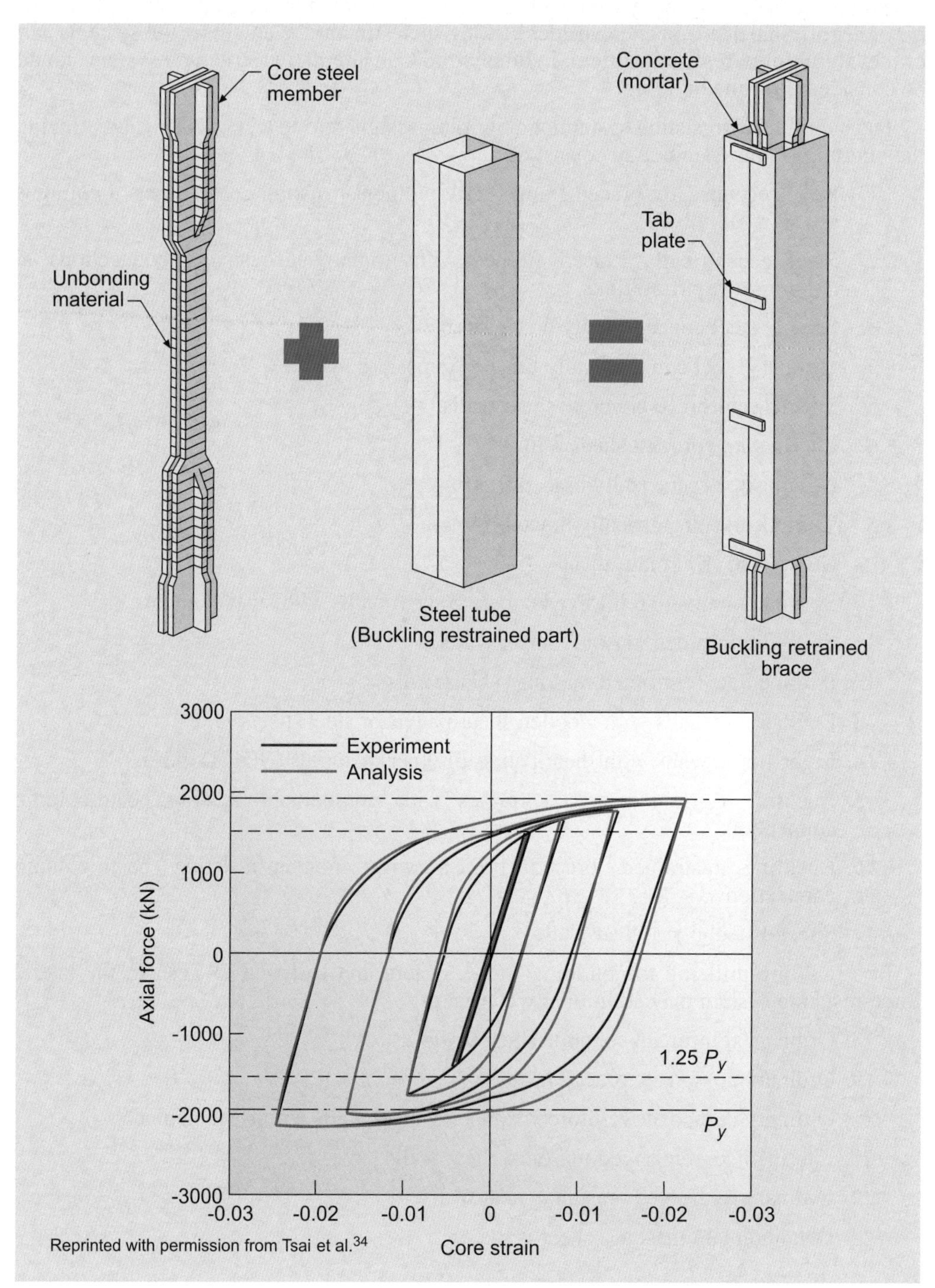

Figure 16-22 Typical configuration of and load-displacement behavior or bucking restrained brace (BRB)

on the surface of the steel core. The role of the concrete encasing and the steel tube is to prevent buckling of the steel core, so that a well formed load-displacement response of the brace is achieved under large displacement reversals. The unbonding material ensures that the force coming into the BRB is carried by the core only, without engaging the encasing material. A load-displacement response obtained from the BRB configuration shown is presented in Figure 16-22. As can be seen, full hysteretic loops and excellent energy dissipation can be achieved.

The provisions for special steel plate shear walls (Figure 16-23) are based on research conducted in the U.S. as well as in Canada. In this system, the post-buckling strength of the web of the steel plate shear wall is utilized to achieve a favorable seismic performance.

Moment-resisting frame system. This is a structural system with an essentially complete space frame providing support for gravity loads [Figure 16-21(c)]. Moment-resisting frames provide resistance to lateral loads primarily by flexural action of the members.

Several aspects of the space frame and the moment-resisting frame require discussion. For a system to qualify as a moment-resisting frame system, it must have a substantially complete vertical load-carrying space frame. The engineer should have the latitude to designate selected portions of the space frame as moment-resisting frames, as long as these portions satisfy the design and detailing requirements and provide the intended behavior of the selected type of lateral force-resisting system. This allows the engineer to select the most effective configuration for the lateral force-resisting system. The need for redundancy should be a primary consideration in this selection process.

For those portions of the space frame that are not part of the designated lateral force-resisting system, the deformation compatibility requirement of ASCE 7-05 Section 12.12.4 must be complied with for SDC D, E or F structures.

Table 12.2-1 recognizes moment-resisting frames of concrete, masonry and steel, as well as of composite steel-concrete construction. As per ASCE 7 Section 14.1 and 2006 IBC Section 2205.2.1, in SDC B and C, steel moment-resisting frames can be ordinary moment-resisting frames (OMRF) proportioned to satisfy the requirements of AISC 360[36] for steel systems not detailed for seismic resistance, using a response modification factor R = 3, or AISC 341[24] using the response factor, *R*, per ASCE 7-05 Table 12.2-1. For concrete moment frames in SDC B, design can be per ACI 318,[37] exclusive of Chapter 21, with no special seismic detailing required. For concrete moment frames in SDC C, an intermediate moment-resisting frame (IMRF) is required, proportioned to satisfy the requirements of ACI 318 and the special design and detailing provisions of Sections 21.2.1.3 and 21.12. In SDC D, E and F, moment frames of steel can be OMRF, IMRF or SMRF, but must be designed and detailed to satisfy the requirements of AISC 341, using the response factors, *R*, per ASCE 7-05 Table 12.2-1, subject to the limitations given in that table. For concrete, special moment-resisting frames (SMRF) must be used in SDC D, E and F, proportioned to comply with ACI 318 and the special design and detailing requirements of Sections 21.2 through 21.5.

Reprinted with permission from American Institute of Steel Construction (AISC)[35]

Figure 16-23
Special steel plate shear wall

There are two lateral force-resisting systems listed under moment-resisting frame systems that are somewhat outside of the regular pattern. These are the special steel truss moment frames (AISC 341 Part I, Section 12), permitted in all seismic design categories other than F, and composite partially restrained moment frames (AISC 341 Part II, Section 8), permitted in seismic design categories other than E and F.

Research work at the University of Michigan has led to the development of special truss girders that limit inelastic deformations to a special segment of the truss.[36,38] As illustrated in Figure 16-24, the chords and web members (arranged in an X pattern) of the special segment are designed to withstand large inelastic deformations while the rest of the structure remains elastic. Special truss moment frames (STMF) have been validated by extensive testing of full-scale subassemblages with story-high columns and full-span special truss girders.

The STMF special segment should be located near mid-span of the truss girder because shear that is due to gravity loads is generally lower in that region. The lower limit on special segment length of 10 percent of the truss span length provides a reasonable limit on the ductility demand, whereas the upper limit of 50 percent of the truss span length represents a more practical limit.

Because the special segment is intended to yield over its full length, no major structural loads should be applied within the length of the special segment. Accordingly, a restrictive upper limit is placed on the axial force in diagonal web members due to gravity loads applied directly within the special segment.

The ductility demand on diagonal web members in the special segment can be rather large. Flat bars have been suggested in the AISC 341 Commentary because of their high ductility. Tests[38] have shown that single angles with width-thickness ratios less than $0.18/\sqrt{E_s/F_y}$ also possess adequate ductility for use as web members in an X configuration. Chord members in the special segment are required to have compact cross-sections to facilitate the formation of plastic hinges.

Composite partially restrained (PR) moment frames (C-PRMF) consist of structural steel columns and composite beams that are connected with partially restrained (PR) moment connections that meet the requirements in AISC 360 Specifications. C-PRMF are to be

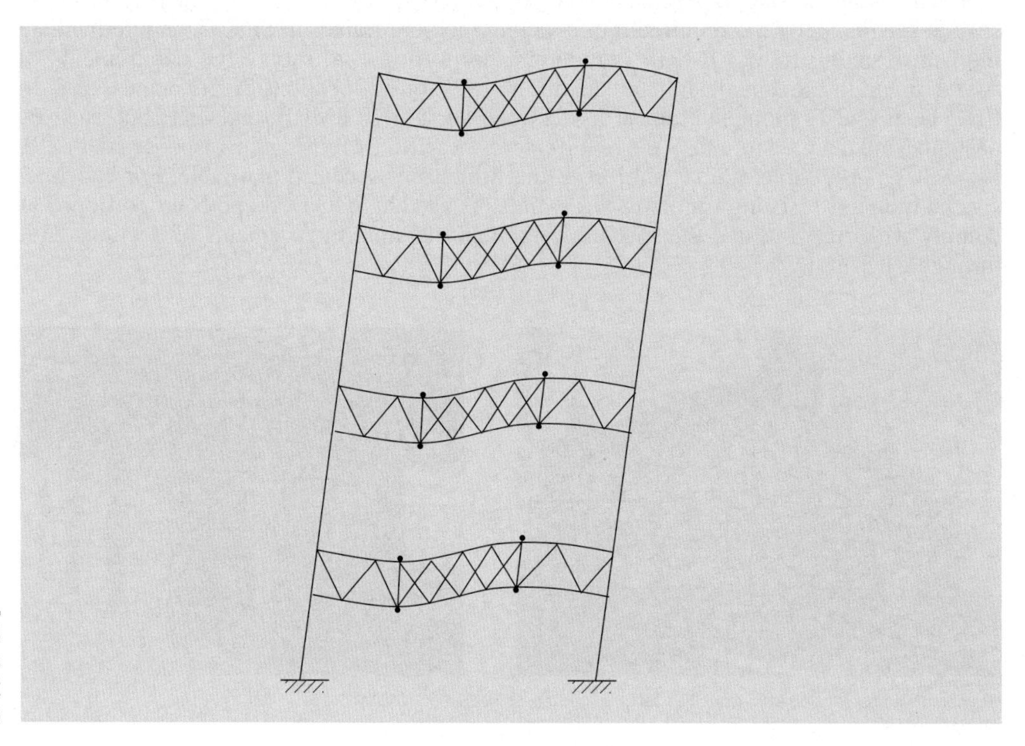

Figure 16-24 Special truss moment frames

designed so that, under earthquake loading, yielding occurs in the ductile components of the composite PR beam-to-column moment connections. Limited yielding is permitted at other locations, such as the column base connection. Connection flexibility and composite beam action are to be accounted for in determining the dynamic characteristics, strength and drift of C-PRMF.

Dual system. A dual system [Figure 16-21(d)] is a structural system with three essential features: (1) An essentially complete space frame provides support for gravity loads, (2) resistance to lateral loads is provided by moment-resisting frames capable of resisting at least 25 percent of the design base shear, and by shear walls or braced frames and (3) the two systems (moment frames and shear walls or braced frames) are designed to resist the design base shear in proportion to their relative rigidities.

In earlier codes, including the 1985 edition of the UBC, a fourth criterion existed, in addition to the three enumerated above, that had to be satisfied for a shear wall-frame system or a moment frame-braced frame system to qualify as a dual system. The shear walls or braced frames acting independently of the ductile moment-resisting portions of the space frame had to be able to resist the total required lateral forces. This requirement was at best counterproductive. It took incentive away from the design engineer who preferred to design the shear wall-frame or moment frame-braced frame system as a dual system. If the shear walls or the braced frames had to resist 100 percent of the design lateral forces, what was the benefit of making the moment-resisting frames part of the lateral-force-resisting system? It is true that the design force level went down some when this was done (because of a decrease in the old *K*-value, which is equivalent to an increase in the *R*-value of the IBC). However, the moment-resisting frames then had to be specially detailed. In most cases, the cost of the increased detailing requirements more than outweighed any savings resulting from the reduced design force level. Also, there were many instances where there simply were not sufficient shear walls in a frame-wall building for the shear walls to carry 100 percent of the design lateral forces. The fourth criterion, in addition to being counterproductive, was also not conservative. The unnecessarily overdesigned shear walls attracted more shear forces to themselves in the event of an actual earthquake, which obviously is not a desirable condition. Fortunately, the undesirable fourth criterion was dropped from the 1988 edition of the UBC and has not reappeared in recent codes.

ASCE 7-05 separately recognizes dual systems in which the moment-resisting frame consists of special moment frames, and dual systems in which the moment-resisting frame consists of intermediate moment frames.

In buildings assigned to SDC D, E and F, which employ dual systems with special moment frames, the shear wall or braced frame part may consist of:

1. Steel eccentrically braced frames.
2. Special steel concentrically braced frames.
3. Special reinforced concrete shear walls.
4. Composite steel and concrete eccentrically braced frames.
5. Composite steel and concrete concentrically braced frames.
6. Composite steel plate shear walls.
7. Special composite reinforced concrete shear walls with steel elements.
8. Special reinforced masonry shear walls.
9. Buckling-restrained braced frame.
10. Special steel plate shear walls.

In SDC C, additionally allowed are:

11. Ordinary reinforced concrete shear walls.
12. Ordinary composite reinforced concrete shear walls with steel elements.

13. Intermediate reinforced masonry shear walls.

ASCE 7-05 requires that each type of shear wall or braced frame must satisfy the requirements of the corresponding section numbers in Chapter 14 listed in Column 2 of Table 12.2-1. However, the 2006 IBC requires that the corresponding provisions in its own materials chapters be used.

In buildings assigned to SDC D, E and F, which employ dual systems with intermediate moment frames, the shear wall or braced frame part may consist of:

1. Special steel concentrically braced frames (not permitted in SDC F).
2. Special reinforced concrete shear walls.
3. Composite steel and concrete concentrically braced frames (not permitted in SDC F).

Additionally permitted in SDC C are:

4. Ordinary reinforced masonry shear walls.
5. Intermediate reinforced masonry shear walls.
6. Ordinary composite braced frames.
7. Ordinary composite reinforced concrete shear walls with steel elements.
8. Ordinary reinforced concrete shear walls.

Each of these shear wall or braced frame types must satisfy the requirements of the corresponding section numbers listed in the second column of Table 12.2-1. IBC users, however, need to refer to the corresponding provisions in the materials chapters in the 2006 IBC.

SDC B is not mentioned above, because the concept of the dual system loses its validity for buildings assigned to SDC B, as discussed later.

Shear wall-frame interactive system with Ordinary Reinforced Concrete Moment Frames and Ordinary Reinforced Concrete Shear Walls. This system was first introduced in the 1997 edition of the UBC, constituting its own category. In the 2000 IBC, it was misplaced under dual systems with intermediate moment frames. This system appears under its own heading in Table 1617.6.2 of the 2003 IBC, but does not appear in Table 9.5.2.2 of ASCE 7-02 (which was an oversight by the ASCE 7 Committee). It is now included in Table 12.2-1 of ASCE 7-05.

Shear walls or braced frames used in conjunction with moment frames to carry lateral loads in buildings assigned to SDC C, D, E or F must be designed as either building frame systems or as dual systems. Central to the concept of the dual system is the backup frame capable of independently resisting at least 25 percent of the design lateral forces. A buildings frame system, on the other hand, utilizes shear walls or braced frames to resist 100 percent of the design lateral forces. The attraction of this system is that the moment frames, because they are not part of the lateral force-resisting system, require only ordinary detailing. The concept of the dual system really loses its validity in buildings assigned to SDC B, because it is questionable whether the moment frames, which are required to have only ordinary detailing, can act as a backup to the ordinarily detailed shear walls or braced frames, the inelastic deformabilities of both subsystems being comparable. The concept of the building frame system also loses its appeal because there is little to be gained from assigning the entire lateral resistance to the shear wall or braced frame subsystem, in the absence of any special detailing requirements for the beam-column or slab-column frames. In fact, over-designing the shear walls or braced frames to resist 100 percent of the design lateral forces makes them uneconomical. Shear walls with excessive flexural strength are also more likely to fail in a brittle shear-governed mode than in a ductile flexural mode. In areas of low seismicity, it has been usual practice to design the shear walls (or braced frames) and frames in a shear wall-frame (or braced frame-moment frame) structure to resist lateral loads in proportion to their relative rigidities, considering interaction between the two subsystems at all levels. Such a system is recognized in the 2000, 2003 and 2006 IBC as a shear wall-frame interactive system, and its use is restricted to an SDC no higher than B.

Cantilevered column systems. In ASCE 7-02 and the 2003 IBC, the title of the category also included inverted pendulum type structures. Inverted pendulum systems are now defined in ASCE 7-05 Section 11.2 as simply a subclass of cantilever column systems where more than 50 percent of the structure's mass is concentrated at the top of the structure and where the stability of the mass at the top relies on the rotational restraint to the top of the cantilever column. As a result, ASCE 7-05 Table 12.2-1 Item G only refers to cantilever columns and does not mention inverted pendulum systems separately. Note that 2003 IBC Table 1617.6.2 actually had cantilevered column systems listed under inverted pendulum systems.

These types of structures are singled out for special consideration because of their unique characteristics. These structures have little redundancy and overstrength and concentrate inelastic behavior at their bases. As a result, they have substantially less energy dissipation capacity than other systems.

A cantilevered column system utilizes cantilevered column elements for resistance to lateral forces. A cantilevered column element is defined in the UBC as a column element in a lateral force-resisting system that cantilevers from a fixed base and has minimal moment capacity at the top, with lateral forces applied essentially at the top. This type of structural system is common for multifamily residential occupancies over carports, strip shopping center storefronts, and single-family dwellings on ocean-side or hillside lots. According to the 1999 SEAOC *Blue Book* Commentary, inclusion of this class of structures in the table of structural systems of the *Blue Book* and the UBC does not imply that this is a preferred type of system, but rather recognizes the existence of the cantilevered column system as a separate category with unique characteristics. Typically, this system does not have redundant elements and multiple load paths, nor does the system have added damping, strength or rotational stiffness developed by the nonstructural elements.

In cantilevered column systems, the column elements acting in cantilever action often provide support for the gravity (vertical) loads in addition to resisting all lateral forces. Hence, there is no independent vertical load-carrying system, and the failure of the primary lateral system compromises the ability of the structure to carry gravity loads.

Overstrength in the cantilevered column system is minimal because the ability to form a progression of plastic hinges is limited. Hence, design for higher strength and stiffness (by use of a low R) is necessary to reduce the high ductility demands imposed on the columns. The 1999 SEAOC *Blue Book* Commentary cautions that this type of system should be used only when the use of more desirable systems is not feasible.

Undefined structural system. Undefined structural systems are those not listed in ASCE 7-05 Table 12.2-1. For such systems [Figure 16-21(e)], the coefficients R, Ω_0 and C_d are to be substantiated by approved cyclic test data and analyses. The following items need to be addressed when establishing R, Ω_0 and C_d (see 1999 SEAOC *Blue Book* Section 104.9.2):

1. Dynamic response characteristics
2. Lateral-force resistance
3. Overstrength and strain-hardening or -softening
4. Strength and stiffness degradation
5. Energy dissipation characteristics
6. System ductility
7. Redundancy

Some of the above terms become clearer in the discussion that follows.

Response modification factor, *R*. The first part of this discussion closely follows the text of the 2000 NEHRP Commentary, Section 5.2.

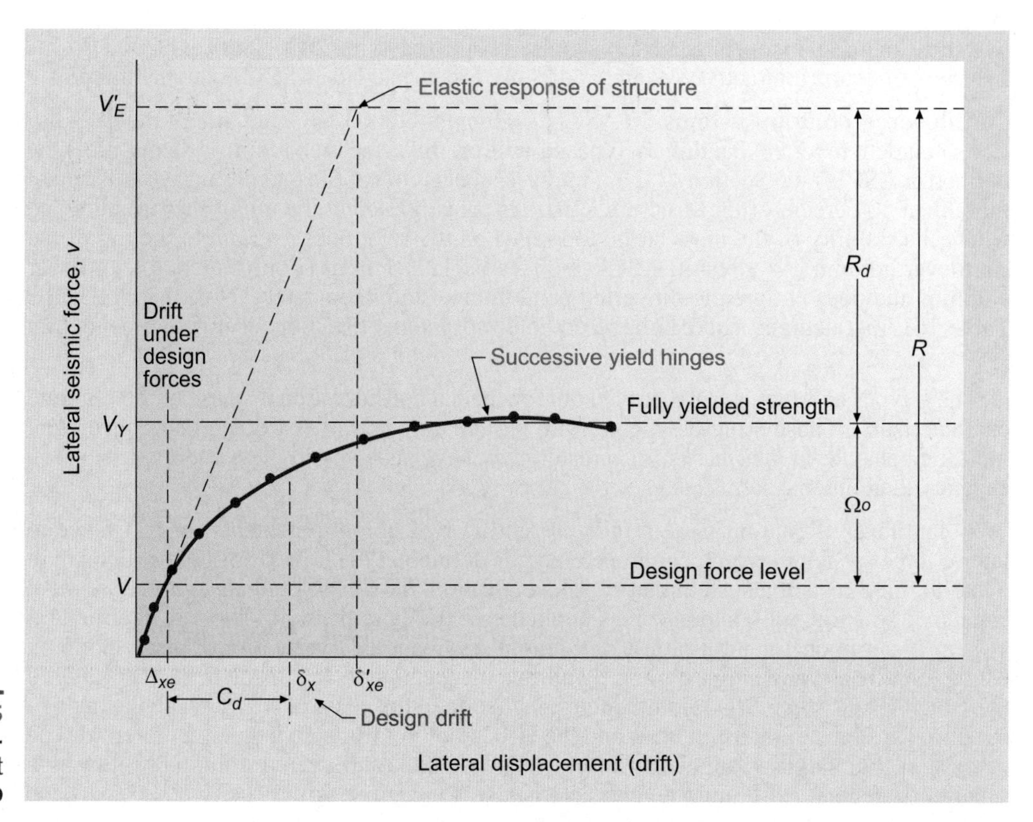

Figure 16-25
Inelastic force-displacement relationship

The response modification factor, *R,* essentially represents the ratio of the forces that would develop under the specified ground motion if the structure had an entirely linear elastic response to the prescribed design forces (Figure 16-25). The reader would benefit from correlating Figure 16-25 with Figures 16-3 and 16-4. The structure is to be designed so that the level of significant yield exceeds the prescribed design force. The ratio, *R,* expressed by the equation:

$$R \quad = V'_E \, / \, V$$

is always larger than 1.0; thus all structures are designed for forces smaller than those the design ground motion would produce in a completely linear-elastic structure. This reduction is possible for a number of reasons. As the structure begins to yield and deform inelastically, the effective period of response of the structure tends to lengthen, which for many structures results in a reduction in strength demand. Furthermore, the inelastic action results in a significant amount of energy dissipation, also known as hysteretic damping, in addition to the viscous damping. The combined effect, which is also known as ductility reduction, explains why a properly designed structure with a fully yielded strength (V_Y in Figure 16-25) that is significantly lower than the elastic seismic force demand (V'_E in Figure 16-25) can be capable of providing satisfactory performance under the design ground motion excitations. A system ductility reduction factor, R_d may be defined as the ratio between V'_E and V_Y.[39]

$$R \quad = V'_E \, / \, V_Y$$

It is then clear from Figure 16-25 that the response modification coefficient, *R,* is the product of the ductility reduction factor and structural overstrength factor:[40]

$$R \quad = R_d \, \Omega_o$$

The energy dissipation resulting from hysteretic behavior can be measured as the area enclosed by the force-displacement curve of the structure as it experiences several cycles of excitation.

The term *significant yield* used above should not be taken as referring to the stage at which first yield occurs in any member. It is defined instead as corresponding to complete plastification of at least the most critical region of the structure (e.g., formation of a first plastic hinge in the structure). A structural steel frame consisting of compact members is assumed to reach this point when a *plastic hinge* develops in the most highly stressed member of the structure. A concrete frame reaches this significant yield when at least one of the sections of its most highly stressed component reaches its design strength as defined in IBC Chapter 19. For other structural materials that do not have their sectional yielding capacities as easily defined, modifiers to working stress values are provided. The code requirements contemplate that the design is based on a seismic force-resisting system with redundant characteristics wherein significant structural overstrength above the level of significant yield can be obtained by plastification at other points in the structure prior to the formation of a complete mechanism. Figure 16-25 shows the lateral load-deflection curve for a typical structure. Significant yield is the level where plastification occurs at the most heavily loaded element in the structure, shown as the lowest yield hinge on the load-deflection diagram. With increased loading, causing the formation of additional plastic hinges, the capacity increases (following the solid curve) until a maximum is reached. The overstrength capacity obtained by this continued inelastic action provides the reserve strength necessary for the structure to resist the actual seismic forces that may be generated by the design ground motion.

Note that the structural overstrength described above results from the development of sequential plastic hinging in a properly designed, redundant structure. Several other sources will further increase structural overstrength. First, material overstrength (i.e., actual material strengths higher than the nominal material strengths specified in the design) may increase the structural overstrength significantly. For example, surveys have shown that the mean yield strength of A36 steel is about 30 to 40 percent higher than the minimum specified strength, nominally used in design calculations. The actual yield strength of Grade 60 reinforcing bars is allowed by ACI 318-05 to be 30 percent higher than their specified yield strength. Second, member design strengths usually incorporate a strength reduction (or resistance) factor, ϕ, to ensure a low probability of failure under design loading. Third, designers themselves introduce additional overstrength by selecting sections or specifying reinforcing patterns that exceed those required by the computations. Similar situations occur when minimum code requirements, for example, minimum reinforcement ratios, control the design. Finally, the design of many flexible structural systems, such as moment resisting frames, are often controlled by the drift rather than strength limitations of the governing code, with sections selected to control lateral deformations rather than provide the specified strength. The result is that structures typically have a much higher lateral resistance than specified as a minimum by design codes. According to the 2003 NEHRP Commentary, first actual significant yielding of structures may occur at lateral load levels that are 30 to 100 percent higher than the prescribed design seismic forces. If provided with adequate ductile detailing, redundancy and regularity, full yielding of structures may occur at load levels that are two to four times the prescribed design force levels.

A detailed history of the evolution of R factors is presented in the ATC-19 report.[41] The R values contained in ASCE 7-05 Table 12.2-1 are largely based on engineering judgment of the performance of various materials and systems in past earthquakes. The factors and characteristics to be considered in the evaluation of R have been determined by SEAOC to include:

1. Observed system performance
2. Level of inelastic response capability
3. Possibilities of failure of vertical load-resisting system
4. Redundancy of lateral force-resisting system
5. Multiplicity of lines of defense, such as backup frames

For ATC 3-06 where the factor R was first introduced, certain agreed-upon reference structures were selected. Two systems having high and low expected levels of performance were chosen to be "a steel ductile frame and a box type masonry or concrete building, respectively." In today's terminology, these would be the special moment frame of steel and a bearing wall system consisting of masonry or concrete shear walls, respectively. The R-values for these two systems were chosen considering the K-values assigned to them by older editions of the UBC. No compelling arguments were offered to change the design basis loads for these systems or to change their interrelationship. The expected performances of other systems were then evaluated relative to these reference systems in order to determine the intermediate R values. The same characteristics discussed above were used, and considerations focused on the following issues:[22]

1. The degree to which the system can be allowed to go beyond the elastic range, its degree of energy dissipation in so doing, and the stability of the vertical load-carrying system during inelastic response that is due to maximum expected ground motion.
2. The consequence of failure or partial failure of vertical elements of the seismic force-resisting system on the vertical load-carrying capacity and stability of the total building system.
3. The inherent redundancy of the system that would allow some progressive inelastic excursions without overall failure. One localized failure of a part must not lead to failure of the system.
4. When dual systems are employed, important performance characteristics include the ability of the secondary (backup) system to maintain vertical support when the primary system suffers significant damage at the maximum deformation response. The backup system can serve to redistribute lateral loads when the primary system undergoes degradation and should stabilize the building in the event that the primary system is badly damaged.

System overstrength factor, Ω_0, and deflection amplification factor, C_d. These factors, also listed in ASCE 7-05 Table 12.2-1, are discussed below.

Figure 16-3 is reproduced here as Figure 16-26 to show the idealized force-displacement relationship of a particular structure subject to the design earthquake of the 2006 IBC, as defined in Section 1613. On the y-axis are the earthquake-induced forces; along the x-axis are the earthquake-induced displacements.

The quantity, V, along the y-axis is the code notation for design base shear, a global force quantity. The quantity δ_{xe} represents the lateral displacements at various floor levels, and Q_E represents the member forces (bending moments, shear forces, axial forces, etc.). As the structure responds inelastically to the design earthquake, the lateral displacements at the various floor levels increase from δ_{xe} to $C_d\delta_{xe}$, and the member forces increase from Q_E to $\Omega_0 Q_E$. Both the deflection amplification factor, C_d, and the system overstrength coefficient, Ω_0, depend on the structural system used for earthquake resistance. Coefficients V_E and δ_{xe} are the base shear and the lateral displacements at various floor levels corresponding to the hypothetical elastic response of the structure to the design earthquake of the 2006 IBC, defined in Section 1613. Figure 16-26 suggests that an R-value of 2 used in design would result in elastic response of a structure to the design earthquake. The basis for this is explained below. The explanation is adapted from the 1999 SEAOC *Blue Book*, Appendix C.

When a structural system is designed for the seismic force level and details established by its R value, there are two behavior properties that allow the structure to perform adequately under the design earthquake ground motion. These are total system resistance and dynamic response modifications. Refer to Figure 16-27 for the following discussion.

Total system resistance. Base shear, V_A, is the required design lateral force level for a given system with structural period, T_A, and R value for $V_A = V'_E/R$. The path from A to B to E represents the key points on the structure pushover curve.

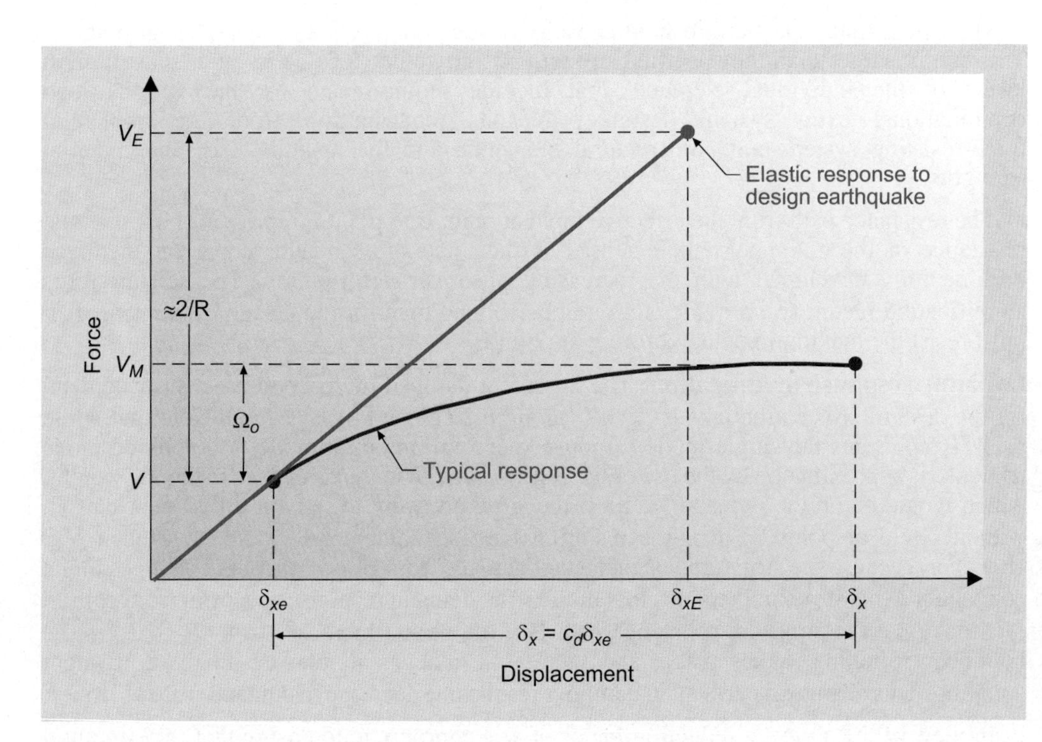

Figure 16-26
Idealized force-displacement relationship of a building subjected to the design earthquake of the IBC

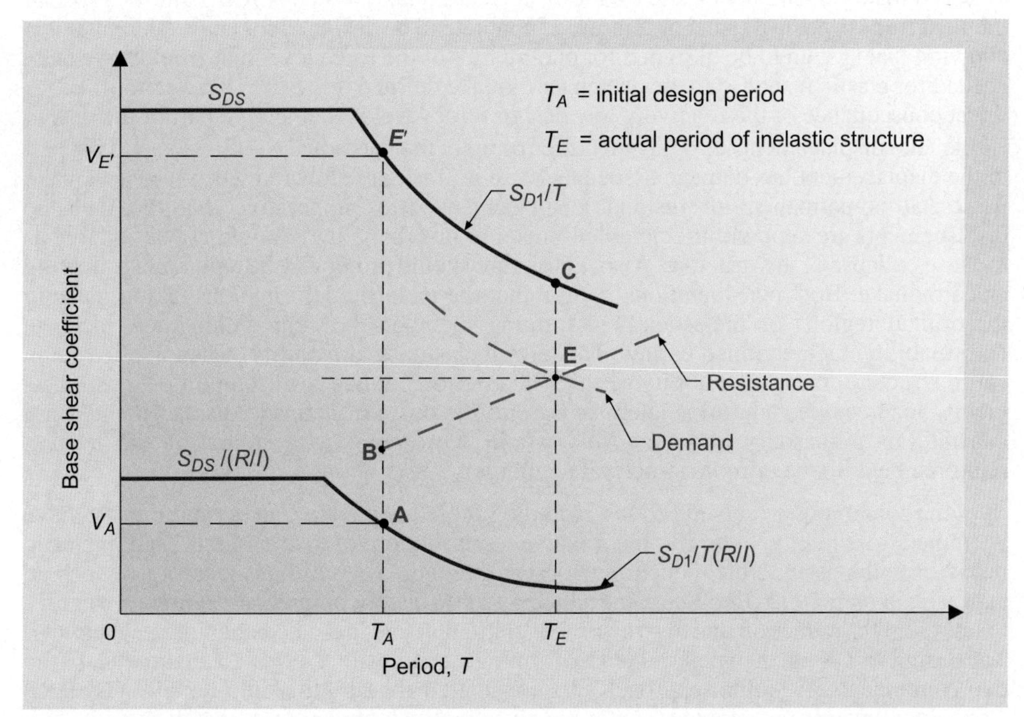

Figure 16-27
Seismic resistance versus seismic demand

The actual linear elastic threshold capacity of the structure is the shear, V_B, at point B. This can be larger than the specified strength design value, V_A, due to the actual strength levels of the individual elements that include requirements for the various load combinations; extra system design provisions; contribution from the nonlateral force-resisting system and nonstructural elements; and the as-built sizes and material strengths.

The resistance path from the effective yield at point B to point E represents how the total resistance of the entire system increases as the understressed and redundant members become fully developed with the increasing pushover deformation. The seismic-force amplification factor, Ω_0, provides for an upper-bound load for the design of elements that must resist the maximum demand forces at point E.

Dynamic response modification. The 5-percent damped multimode base shear demand for the design basis earthquake is $S_{D1}W/T$ along the demand path E' to C. Note that while $S_{D1}W/T$ represents the single-mode response spectrum, the multimode or combined mode response is approximated by the use of the full W rather than the effective first mode weight, which would be on the order of $0.7W$. If the structure were to remain fully linear elastic, without yield at point B, then the resulting base shear demand would have been V_E'. However, because the structure has inelastic behavior from B to E, there are changes in the equivalent dynamic characteristics that modify the demand response spectrum from path E' – C to – E' – E. Along the resistance path BE, the period increases from T_A to T_E as the nonlinear softening takes place, and there is a concurrent increase in the equivalent damping. This change in period and damping results in a decrease in demand from V_E' to V_E

In view of the above, a reduction factor of 2 is appropriate to ensure that the structural system remains essentially elastic for the design earthquake. When ASCE 7 allows a special moment frame system of steel or reinforced concrete to be designed for an R-value of 8, it is allowing that system to be designed for one-quarter of the force level that would have been needed for elastic response to the design earthquake defined in ASCE 7-05 Section 11.2. A direct consequence of the relatively low design force level is that a large part of the design earthquake displacement, $C_d\delta_{xe}$, is inelastic [roughly the part equal to $(C_d - 1)\delta_{xe}$]. This part of the displacement has damage associated with it. It is not recovered when the earthquake passes. It is permanent or residual displacement. It is imperative that the inelastic displacements are kept within tolerable limits. For one thing, they should not be so large as to cause collapse of the structure. Also, a structure should preferably be repairable following an earthquake. Both considerations require that the inelastic deformations taking place in the critical regions of inelastically deforming members be kept within their inelastic deformability, as mentioned earlier. The term inelastic deformability, when applied to an entire structure, means the ability of the structure to continue to sustain all or most of its gravity loads while it deforms laterally beyond the onset of inelastic displacements (in a concrete or a masonry structure, this would correspond to yielding of the tension reinforcement in one or more structural members).

Some contemporary seismic codes, including those adopted in Canada and Europe, have attempted to directly quantify the relative contribution of overstrength and inelastic behavior to the permissible reduction in design strength. SEAOC has incorporated such an approach in their 1999 *Blue Book*, introducing two R-factor components, termed R_o and R_d, to represent the reduction due to structural overstrength and inelastic behavior, respectively. The design forces are then determined by forming a composite R, equal to the product of the two components. A similar approach was considered for adoption into the 1997 NEHRP *Provisions*, but was not adopted for reasons given in the Commentary to that document. As a first step in the direction of split R values, however, the factor, Ω_0 was added to the R values table, to replace the previous $2R/5$ factor used for evaluation of brittle structural behavior modes in prior editions of the NEHRP *Provisions*.

ATC-63 Project. In September 2004 the Applied Technology Council (ATC) was awarded a "Seismic and Multi-Hazard Technical Guidance Development and Support" contract by the Federal Emergency Management Agency (FEMA) to conduct a variety of tasks, including a task entitled "Quantification of Building System Performance and Response Parameters" (ATC-63 Project). The purpose of the ATC-63 project was to establish and document a new, recommended methodology for reliably quantifying building system performance and response parameters for use in seismic design. A key parameter to be addressed on the project was the Structural Response Modification Factor (R-factor), but related design parameters that affect building system seismic response and performance were also addressed. Collectively, these factors were referred to as Seismic Performance Factors.

R-factors, as indicated above, are used to estimate strength demands on systems that are designed using linear methods but are responding in the nonlinear range, and their values are fundamentally critical in the specification of seismic loading. R-factors were initially introduced in the ATC-3-06 report, *Tentative Provisions for the Development of Seismic Regulations for Buildings*, published in 1978. The original R-factors were based largely on judgment and qualitative comparisons of the known response capabilities of relatively few types of lateral force resisting systems that were in widespread use at the time. Since then, the number of systems addressed in seismic codes and in today's NEHRP *Recommended Provisions for Seismic Regulations for New Buildings and Other Structures* has increased dramatically. Many of these recently defined systems have somewhat arbitrarily assigned R-factor and have never been subjected to any significant level of earthquake ground shaking. Their potential response characteristics and ability to meet seismic design performance objectives is both untested and unknown.

The primary objective of the ATC-63 project is the development of a recommended methodology for determining building system performance and response parameters for use in seismic design that, when properly implemented in the design process, will result in:

- *Equivalent safety against collapse in an earthquake for buildings across different seismic force-resisting systems.*

The methodology is intended for use with model building codes and resource documents to set minimum acceptable design criteria for standard code-approved seismic force-resisting systems and to provide guidance in the selection of appropriate design criteria for other systems when linear design methods are applied.

The methodology also provides a basis for re-evaluation of existing tabulations of and limitations on code-approved seismic force-resisting systems for adequacy to achieve the inherent design performance objectives. It is possible that results of future work based on this methodology could be used to modify or eliminate those systems or requirements that cannot reliably meet these objectives.

Height limits. In keeping with the practice incorporated into the *Uniform Building Code* for zones of high seismic risk, ASCE 7-05 continues limitations on the use of certain structural systems over 160 feet in height for buildings assigned to SDC D and E (over 100 feet in height for buildings assigned to SDC F). Although, in view of a lack of reliable data on the behavior of SDC D, E and F buildings whose structural systems include shear walls and/or braced frames, it is considered prudent to retain some height limits, the values of 160 and 100 feet chosen for the 2003 NEHRP *Provisions* and ASCE 7-05 are arbitrary and must be understood to be such. The height limits of ASCE 7-05 Table 12.2-1 are modified by Section 12.2.5.4.

Structural systems of specific materials. The structural systems of steel, concrete, masonry and wood that are recognized in ASCE 7-05 Table 12.2-1 along with the R, Ω_0, C_d values and height limits for various seismic design categories assigned to them, are listed in Tables 16-19, 16-20, 16-21 and 16-22, respectively.

Structural systems for specific seismic design categories. The choice of concrete and masonry structural systems available for buildings assigned to SDC D, E and F is presented in Tables 16-23 and 16-24, respectively. There are very few restrictions on steel and wood structural systems that are related to the seismic design category. With rare exceptions, all steel and wood structural systems listed in Tables 16-19 and 16-22 may be used in all seismic design categories.

Concrete and masonry structural systems restricted in applicability to seismic design categories no higher than C are listed in Tables 16-25 and 16-26, respectively.

Finally, concrete and masonry structural systems restricted in applicability to seismic design categories no higher than B are listed in Tables 16-27 and 16-28, respectively.

Table 16-19. Earthquake force-resisting structural systems of steel – ASCE 7-05

Basic Seismic-Force Resisting System	Detailing Ref. Section	R	Ω_0	C_d	System Limitations and Building Height Limitations (ft) by Seismic Design Category				
					B	C	D	E	F
Bearing Wall Systems									
Light-framed walls sheathed with wood structural panels rated for shear resistance or steel sheets	14.1 and 14.5	$6^1/_2$	3	4	NL	NL	65	65	65
Light-framed walls with shear panels of all other materials	14.1 and 14.5	2	$2^1/_2$	2	NL	NL	35	NL	NL
Light-framed wall systems using flat strap bracing	14.1 and 14.5	4	2	$3^1/_2$	NL	NL	65	65	65
Building Frame Systems									
Steel eccentrically braced frames, moment-resisting connections at columns away from links	14.1	8	2	4	NL	NL	160	160	100
Steel eccentrically braced frames, nonmoment-resisting connections at columns away from links	14.1	7	2	4	NL	NL	160	160	100
Special steel concentrically braced frames	14.1	6	2	5	NL	NL	160	160	100
Ordinary steel concentrically braced frames	14.1	$3^1/_4$	2	5	NL	NL	35[1]	35[1]	NP[1]
Light-framed walls sheathed with wood structural panels rated for shear resistance or steel sheets	14.1 and 14.5	7	$2^1/_2$	$4^1/_2$	NL	NL	65	65	65
Light frame walls with shear panels of all other materials	14.1 and 14.5	$2^1/_2$	$2^1/_2$	$2^1/_2$	NL	NL	35	NP	NP
Buckling-restrained braced frames, nonmoment-resisting beam-column connections	14.1	7	2	$5^1/_2$	NL	NL	160	160	100
Buckling-restrained braced frames, moment-resisting beam-column connections	14.1	8	$2^1/_2$	5	NL	NL	160	160	100
Special steel plate shear walls	14.1	7	2	6	NL	NL	160	160	100
Moment-Resisting Frame Systems									
Special steel moment frames	14.1 and 12.2.5.5	8	3	$5^1/_2$	NL	NL	NL	NL	NL
Special steel truss moment frames	14.1		3	$5^1/_2$	NL	NL	160	100	NP
Intermediate steel moment frames	12.2.5.6, 12.2.5.7, 12.2.5.8, 12.2.5.9, and 14.1	$4^1/_2$	3	4	NL	NL	35[2]	NP[2]	NP[3]
Ordinary steel moment frames	12.2.5.6, 12.2.5.7, 12.2.5.8, and 14.1	$3^1/_2$	3	3	NL	NL	NP[2]	NP[2]	NP[3]

(Continued)

Table 16-19. Earthquake force-resisting structural systems of steel – ASCE 7-05 (Cont'd)

Basic Seismic-Force Resisting System	Detailing Ref. Section	R	Ω_0	C_d	System Limitations and Building Height Limitations (ft) by Seismic Design Category				
					B	C	D	E	F
Dual Systems with Special Moment Frames									
Steel eccentrically braced frames	14.1	8	2 1/2	4	NL	NL	NL	NL	NL
Special steel concentrically braced frames	14.1	7	2 1/2	5 1/2	NL	NL	NL	NL	NL
Buckling-restrained braced frames	14.1	8	2 1/2	5	NL	NL	NL	NL	NL
Special steel plate shear walls	14.1	8	2 1/2	6 1/2	NL	NL	NL	NL	NL
Dual Systems with Intermediate Moment Frames									
Special steel concentrically braced frames[4]	14.1	6	2 1/2	5	NL	NL	35	NP	NP[5]
Cantilevered Column Systems Detailed to Conform to the Requirements for:									
Special steel moment frames	12.2.5.5 and 14.1	2 1/2	1 1/2	2 1/2	35	35	35	35	35
Intermediate steel moment frames	14.1	1 1/2	1 1/4	1 1/2	35	35	35[5]	NP[2]	NP[3]
Ordinary steel moment frames	14.1	1 1/4	1 1/4	2 1/2	35	35	NP[2]	NP[2]	NP[3]
Structural Steel Systems Not Specifically Detailed for Seismic Resistance	14.1	3	3	3	NL	NL	NP	NP	NP

[1]Steel ordinary concentrically braced frames are permitted in single-story buildings up to a height of 60 ft, where the dead load of the roof does not exceed 20 psf and in penthouse structures.

[2]See Sections 12.2.5.6 and 12.2.5.7 for limitations on steel OMFs and IMFs in structures assigned to SDC D or E.

[3]See Sections 12.2.5.8 and 12.2.5.9 for limitations on steel OMFs and IMFs in structures assigned to SDC F.

[4]Ordinary moment frame is permitted to be used in lieu of intermediate moment frame in Seismic Design Categories B and C.

[5]Increase in height to 45 ft is permitted for single-story storage warehouse facilities.

Table 16-20. **Earthquake force-resisting structural systems of concrete – ASCE 7-05**

Basic Seismic-Force Resisting System	Detailing Ref. Section	R	Ω_0	C_d	System Limitations and Building Height Limitations (ft) by Seismic Design Category: B	C	D	E	F
Bearing Wall Systems									
Special reinforced concrete shear walls	14.2 and 14.2.3.6	5	$2^1/_2$	5	NL	NL	160	160	160
Ordinary reinforced concrete shear walls	14.2 and 14.2.3.4	4	$2^1/_2$	4	NL	NL	NP	NP	NP
Detailed plain concrete shear walls	14.2 and 14.2.3.2	2	$2^1/_2$	2	NL	NP	NP	NP	NP
Ordinary plain concrete shear walls	14.2 and 14.2.3.1	$1^1/_2$	$2^1/_2$	$1^1/_2$	NL	NP	NP	NP	NP
Intermediate precast shear walls	14.2 and 14.2.3.5	4	$2^1/_2$	4	NL	NL	40[1]	40[1]	40[1]
Ordinary precast shear walls	14.2 and 14.2.3.3	3	$2^1/_2$	3	NL	NP	NP	NP	NP
Building Frame Systems									
Special reinforced concrete shear walls	14.2 and 14.2.3.6	6	$2^1/_2$	5	NL	NL	160	160	160
Ordinary reinforced concrete shear walls	14.2 and 14.2.3.4	5	$2^1/_2$	$4^1/_2$	NL	NL	NP	NP	NP
Detailed plain concrete shear walls	14.2 and 14.2.3.2	2	$2^1/_2$	2	NL	NP	NP	NP	NP
Ordinary plain concrete shear walls	14.2 and 14.2.3.1	$1^1/_2$	$2^1/_2$	$1^1/_2$	NL	NP	NP	NP	NP
Intermediate precast shear walls	14.2 and 14.2.3.5	5	$2^1/_2$	$4^1/_2$	NL	NL	40[1]	40[1]	40[1]
Ordinary precast shear walls	14.2 and 14.2.3.3	4	$2^1/_2$	4	NL	NP	NP	NP	NP
Moment-Resisting Frame Systems									
Special reinforced concrete moment frames	12.2.5.5 and 14.2	8	3	$5^1/_2$	NL	NL	NL	NL	NL
Intermediate reinforced concrete moment frames	14.2	5	3	$4^1/_2$	NL	NL	NP	NP	NP
Ordinary reinforced concrete moment frames	14.2	3	3	$2^1/_2$	NL	NP	NP	NP	NP
Dual Systems with Special Moment Frames									
Special reinforced concrete shear walls	14.2	7	$2^1/_2$	$5^1/_2$	NL	NL	NL	NL	NL
Ordinary reinforced concrete shear walls	14.2	6	$2^1/_2$	5	NL	NP	NP	NP	NP
Dual Systems with Intermediate Moment Frames									
Special reinforced concrete shear walls	14.2	$6^1/_2$	$2^1/_2$	5	NL	NL	160	60	60
Ordinary reinforced concrete shear walls	14.2	$5^1/_2$	$2^1/_2$	$4^1/_2$	NL	NL	NP1	NP	NP
Shear Wall-Frame Interactive System with Ordinary Reinforced Concrete Moment Frames and Ordinary Reinforced Concrete Shear Walls	12.2.5.10 and 14.2	$4^1/_2$	$2^1/_2$	4	NL	NP	NP1	NP	NP
Cantilevered Column Systems Detailed to Conform to the Requirements for:									
Special reinforced concrete moment frames	12.2.5.5 and 14.2	$2^1/_2$	$1^1/_4$	$2^1/_2$	35	35	35	35	35
Intermediate concrete moment frames	14.2	$1^1/_2$	$1^1/_4$	$1^1/_2$	35	35	NP	NP	NP
Ordinary concrete moment frames	14.2	1	$1^1/_4$	1	35	NP	NP	NP	NP

[1]Increase in height to 45 ft is permitted for single-story storage warehouse facilities.

Table 16-21. **Earthquake force-resisting structural systems of masonry – ASCE 7-05**

Basic Seismic-Force Resisting System	Detailing Ref. Section	R	Ω_0	C_d	System Limitations and Building Height Limitations (ft) by Seismic Design Category				
					B	C	D	E	F
Bearing Wall Systems									
Special reinforced masonry shear walls	14.4 and 14.4.3	5	2 1/2	3 1/2	NL	NL	160	160	100
Intermediate reinforced masonry shear walls	14.4 and 14.4.3	3 1/2	2 1/2	2 1/4	NL	NL	NP	NP	NP
Ordinary reinforced masonry shear walls	14.4	2	2 1/2	1 3/4	NL	160	NP	NP	NP
Detailed plain masonry shear walls	14.4	2	2 1/2	1 3/4	NL	NP	NP	NP	NP
Ordinary plain masonry shear walls	14.4	1 1/2	2 1/2	1 1/4	NL	NP	NP	NP	NP
Prestressed masonry shear walls	14.4	1 1/2	2 1/2	1 3/4	NL	NP	NP	NP	NP
Building Frame Systems									
Special reinforced masonry shear walls	14.4	5 1/2	2 1/2	4	NL	NL	160	160	100
Intermediate reinforced masonry shear walls	14.4	4	2 1/2	4	NL	NL	NP	NP	NP
Ordinary reinforced masonry shear walls	14.4	2	2 1/2	2	NL	160	NP	NP	NP
Detailed plain masonry shear walls	14.4	2	2 1/2	2	NL	NP	NP	NP	NP
Ordinary plain masonry shear walls	14.4	1 1/2	2 1/2	1 1/4	NL	NP	NP	NP	NP
Prestressed masonry shear walls	14.4	1 1/2	2 1/2	1 3/4	NL	NP	NP	NP	NP
Moment-Resisting Frame Systems									
No masonry system included	–	–	–	–	–	–	–	–	–
Dual Systems with Special Moment Frames									
Special reinforced masonry shear walls	14.4	5 1/2	3	5	NL	NL	NL	NL	NL
Intermediate reinforced masonry shear walls	14.4	4	3	3 1/2	NL	NL	NP	NP	NP

Table 16-22. **Earthquake force-resisting structural systems of wood – ASCE 7-05**

Basic Seismic-Force Resisting System	Detailing Ref. Section	R	Ω_0	C_d	System Limitations and Building Height Limitations (ft) by Seismic Design Category				
					B	C	D	E	F
Bearing Wall Systems									
Light-framed walls sheathed with wood structural panels rated for shear resistance or steel sheets	14.1 and 14.5	6 1/2	3	4	NL	NL	65	65	65
Light-framed walls with shear panels of all other materials	14.1 and 14.5	2	2 1/2	2	NL	NL	35	NP	NP
Light-framed wall systems using flat strap bracing	14.1 and 14.5	4	2	3 1/2	NL	NL	65	65	65
Building Frame Systems									
Light-framed walls sheathed with wood structural panels rated for shear resistance or steel sheets	14.1 and 14.5	7	2 1/2	4 1/2	NL	NL	65	65	65
Light frame walls with shear panels of all other materials	14.1 and 14.5	2 1/2	2 1/2	2 1/2	NL	NL	35	NP	NP
Cantilevered Column Systems Detailed to Conform to the Requirements for:									
Timber frames	14.1	1 1/2	1 1/2	1 1/2	35	35	35	NP	NP

Table 16-23. **Concrete structural systems for Seismic Design Categories D, E and F**

Structural System	R	Ω_0	C_d	Height Limit (ft)		
				D	E	F
Bearing Wall Systems						
Special reinforced concrete shear walls	5	2½	5	160	160	100
Intermediate precast shear walls	4	2½	4	40[1]	40[1]	40[1]
Building Frame Systems						
Special reinforced concrete shear walls	6	2½	5	160	160	160
Intermediate precast shear walls	5	2½	4½	40[1]	40[1]	40[1]
Moment-Resisting Frame Systems						
Special reinforced concrete moment frames	8	3	5½	NL	NL	NL
Dual Systems with Special Moment Frames						
Special reinforced concrete shear walls	7	2½	5½	NL	NL	NL
Dual Systems with Intermediate Moment Frames						
Special reinforced concrete shear walls	6½	2½	5	100	100	100
Cantilever Column Systems						
Special reinforced concrete moment frames	2½	1¼	2½	35	35	35

[1]Increase in height to 45 ft is permitted for single-story storage warehouse facilities.

Table 16-24. **Masonry structural systems for Seismic Design categories D, E and F**

Structural System	R	Ω_0	C_d	Height Limit (ft)		
				D	E	F
Bearing Wall Systems						
Special reinforced masonry shear walls	5	2½	3½	160	160	100
Building Frame Systems						
Special reinforced masonry shear walls	5½	2½	4	160	160	100
Dual Systems with Special Moment Frames						
Special reinforced masonry shear walls	5½	3	5	NL	NL	NL

Table 16-25. Concrete structural systems for Seismic Design Category no higher than C

Structural System	R	Ω_0	C_d	Height Limit (ft)
Bearing Wall Systems				
Ordinary reinforced concrete shear walls	4	$2^1/_2$	4	NL
Building Frame Systems				
Ordinary reinforced concrete shear walls	5	$2^1/_2$	$4^1/_2$	NL
Moment-Resisting Frame Systems				
Intermediate reinforced concrete moment frames	5	3	$4^1/_2$	NL
Dual Systems with Special Moment Frames				
Ordinary reinforced concrete shear walls	6	$2^1/_2$	5	NL
Dual Systems with Intermediate Moment Frames				
Ordinary reinforced concrete shear walls	$5^1/_2$	$2^1/_2$	$4^1/_2$	NL
Cantilevered Column Systems				
Intermediate concrete moment frames	$1^1/_2$	$1^1/_2$	$1^1/_2$	35

Table 16-26. Masonry structural system for Seismic Design Category no higher than C

Structural System	R	Ω_0	C_d	Height Limit (ft)
Bearing Wall Systems				
Intermediate reinforced masonry shear walls	$3^1/_2$	$2^1/_2$	$2^1/_4$	NL
Ordinary reinforced masonry shear walls	2	$2^1/_2$	$1^3/_4$	160
Building Frame Systems				
Intermediate reinforced masonry shear walls	4	$2^1/_2$	4	NL
Ordinary reinforced masonry shear walls	2	$2^1/_2$	2	160
Dual Systems with Special Moment Frames				
Intermediate reinforced masonry shear walls	4	3	$3^1/_2$	NL
Dual Systems with Intermediate Moment Frames				
Ordinary reinforced masonry shear walls	3	3	$2^1/_2$	160
Intermediate reinforced masonry shear walls	$3^1/_2$	3	3	NL

Table 16-27. Concrete structural systems for Seismic Design Category no higher than B

Structural System	R	Ω_0	C_d
Bearing Wall Systems			
Detailed plain concrete shear walls	2	$2^1/_2$	2
Ordinary plain concrete shear walls	$1^1/_2$	$1^1/_2$	$1^1/_2$
Building Frame Systems			
Detailed plain concrete shear walls	2	$2^1/_2$	2
Ordinary plain concrete shear walls	$1^1/_2$	$2^1/_2$	$1^1/_2$
Moment-Resisting Frame Systems			
Ordinary reinforced concrete moment frames	3	3	$2^1/_2$
Shear Wall-Frame Interactive System with Ordinary Reinforced Concrete Moment Frames and Ordinary Reinforced Concrete Shear Walls	$4^1/_2$	$2^1/_2$	4

Table 16-28. Masonry structural system for Seismic Design Category no higher than B

Structural System	R	Ω_0	C_d
Bearing Wall Systems			
Detailed plain masonry shear walls	2	$2^1/_2$	$1^3/_4$
Ordinary plain masonry shear walls	$1^1/_2$	$2^1/_2$	$1^1/_4$
Prestressed masonry shear walls			
Building Frame Systems			
Detailed plain masonry shear walls	2	$2^1/_2$	2
Ordinary plain masonry shear walls	$1^1/_2$	$2^1/_2$	$1^1/_4$
Prestressed masonry shear walls	$1^1/_2$	$2^1/_2$	$1^3/_4$

12.2.2 Combinations of framing systems in different directions. This section states that different seismic systems are permitted to be used independently in two orthogonal directions of a structure with their respective R-, C_d-, and Ω_0-values. ASCE 7-02 required that if a building had a seismic force-resisting system with R less than 5, then the seismic force-resisting system in the orthogonal direction needs to be designed with the same R-value as well. This came from an old requirement that was first added to the 1988 UBC for buildings with bearing walls in one direction. However, the current drift limitations and deformation compatibility requirements make this requirement unnecessary, and as a result, it is no longer included in ASCE 7-05.

12.2.2 Combinations of framing systems in the same direction. When different structural systems are used to resist seismic forces in the same direction, the most stringent system limitations would apply. As a result, when different structural systems are combined along the height of the building (Vertical Combination), the rule is that the value of R used in the design of any story must be less than or equal to the value of R used in the given direction for the story above (Figure 16-28). Also, the values of C_d, Ω_0 used in the design of any story must be greater than or equal to the value of C_d, Ω_0 used in the given direction for the story

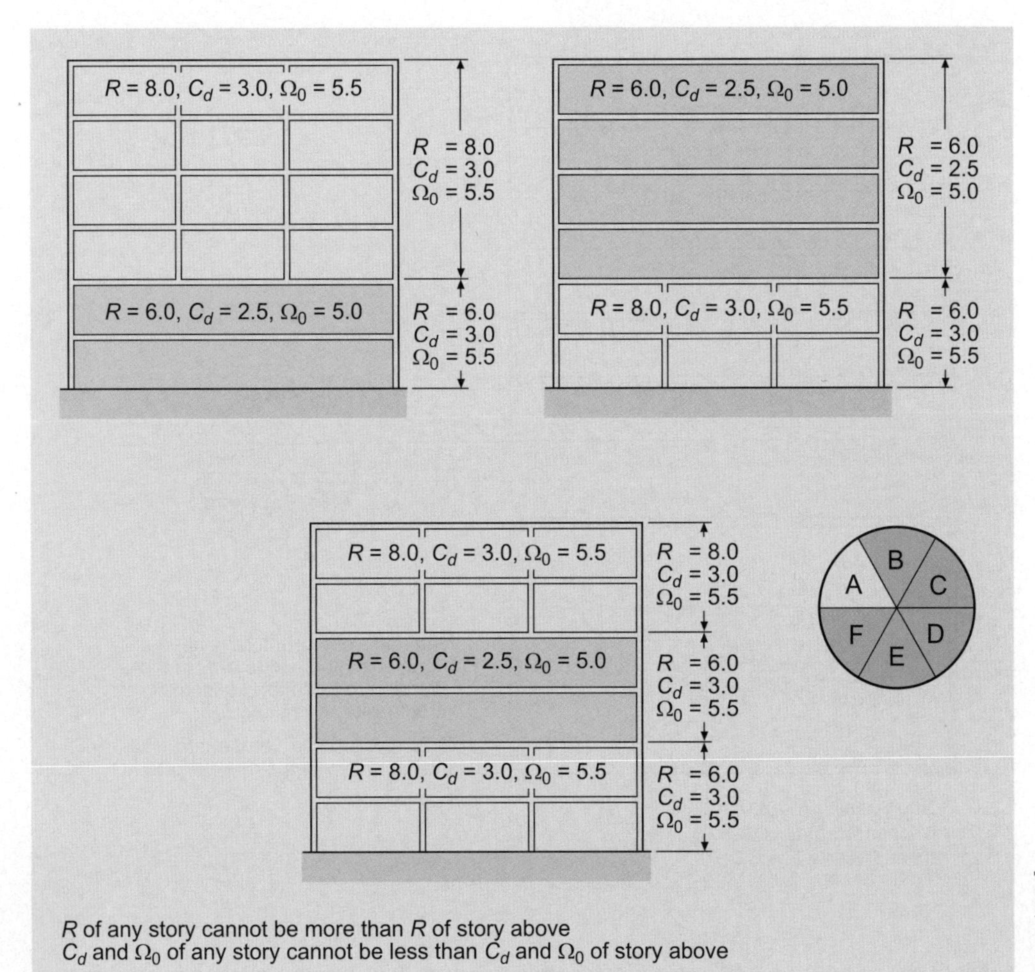

Figure 16-28
Vertical combinations of structural systems

above. However, ASCE makes three exceptions to this requirement for small rooftop structures, other supported small structural systems and detached one- or two-family dwellings of light-frame construction.

ASCE 7-05 also permits a two stage equivalent lateral force procedure for structures having a flexible upper portion above a comparatively stiff lower portion, whereby the two portions can be designed as separate structures with fixed bases with their respective R and ρ, provided:

(a) The stiffness of the lower portion is at least 10 times the stiffness of the upper portion, and

(b) The period of the entire structure is not greater than 1.1 times the period of the flexible upper portion considered as a separate structure with fixed base.

The first criterion is meant to ensure that the lower portion is adequately stiff to act as a fixed support to the upper portion for all practical purposes. The second criterion ensures that the overall structure is not much more flexible than the flexible upper portion.

Additionally, the support reactions from the analysis of the upper portion need to be included in the analysis of the lower portion after scaling them by a factor of $(R_{upper}/R_{lower})(\rho_{lower}/\rho_{upper}) = 1.0$. The factor is intended to prevent the combinations of the response modification factors and the redundancy coefficients for the upper and lower portions to result in a lower force in the lower portion than occurs in the upper portion. This factor adjusts the seismic forces from the upper portion when the lower portion has a lower

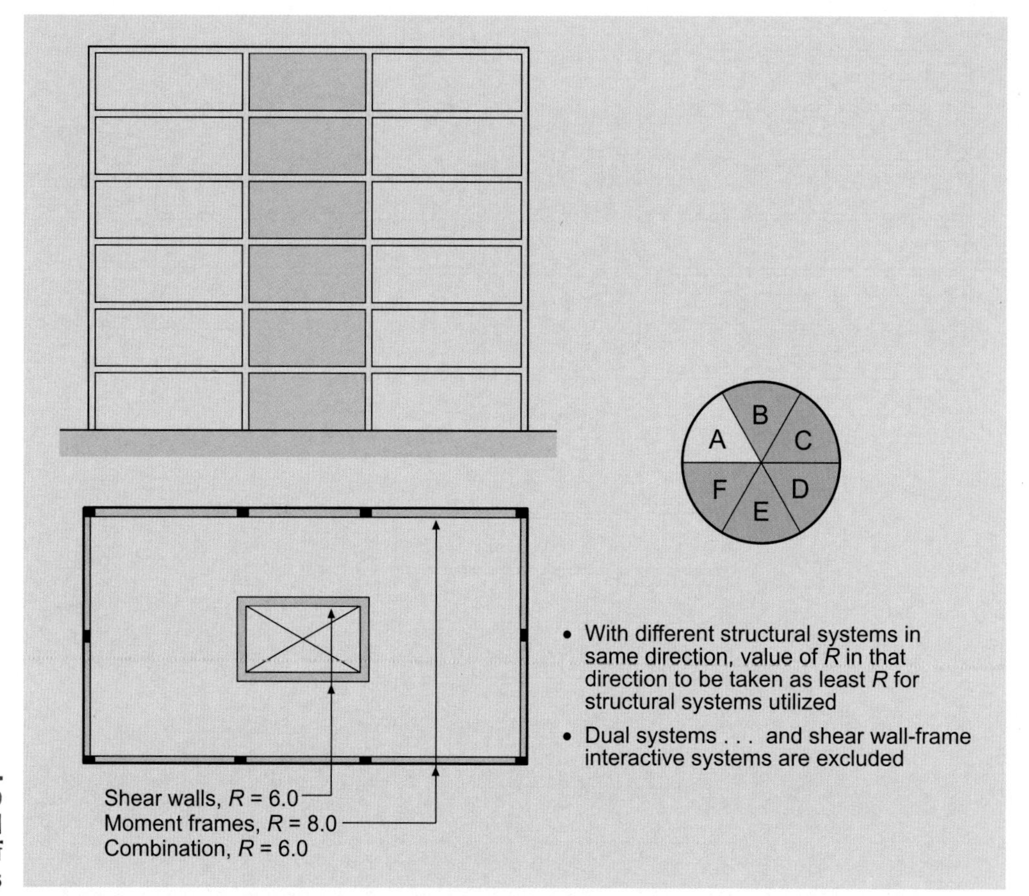

Figure 16-29 Vertical combinations of structural systems

value of the response modification factor (due to lower ductility, overstrength, etc.) and/or is less redundant (thus, has a higher ρ). This makes the reaction forces compatible with the seismic forces obtained from the design response spectrum of the lower portion.

For other than dual systems, where a combination of different structural systems is used to resist lateral forces in the same direction, the value of R used for design in that direction must not be greater than the least value for any of the systems utilized in that same direction (Figure 16-29). However, this requirement is relaxed a little for structures meeting the following three conditions: (1) Occupancy Category I or II, (2) up to two stories in height, and (3) light-frame construction or flexible diaphragms. In these buildings, resisting systems along individual lines of resistance in the same direction can be assigned separate R-values. A given line of resistance is assigned the smallest R-value from all the different structural systems that are part of it. However, R-value used for the design of diaphragms in a particular direction in these buildings must still be lesser or equal to the least R-value of all seismic force-resisting systems used in that direction.

The second paragraph of ASCE 7-05 Section 12.2.3.2 states:

> The deflection amplification factor, C_d, and the system over strength factor, Ω_0, in the direction under consideration at any story shall not be less than the largest value of this factor for the R factor used in the same direction being considered.

The meaning of this provision is far from clear. One possible interpretation is that the design C_d- and Ω_0-values should be those corresponding to the system with the least value of R for any of the systems used in the same direction. In the case where two or more systems have the same least value of R, the largest of the corresponding values of C_d and Ω_0 shall be used.

Combination framing detailing requirements. ASCE 7-05 explicitly states that the detailing required by the higher R-value must be used for structural components common to systems having different R-values. **12.2.4**

System-specific requirements. **12.2.5**

Dual system. This section specifies that the moment frames in a dual system need to be designed for a minimum of 25 percent of the total base shear, while the rest of the seismic force-resisting system needs to be designed for the part of the total base shear that is in proportion to its stiffness as compared to the stiffness of the moment frame. More about the dual system can be found under the discussion of ASCE 7-05, Table 12.2-1. **12.2.5.1**

Cantilever column system. This section requires that the factored axial force on individual cantilever column element be within 15 percent of its axial capacity, owing to the low reliability of these type of systems. As the NEHRP *Provisions* Commentary points out, these structures have little redundancy and overstrength and concentrate inelastic behavior at their bases. As a result, they have substantially less energy dissipation capacity than other systems. A number of buildings incorporating this system experienced very severe damage, and in some cases, collapse, in the 1994 Northridge earthquake. Foundations of such systems need to be designed for load combinations with over strength factor as provided in ASCE 7-05, Section 12.4.3.2. **12.2.5.2**

Inverted pendulum-type structures. As mentioned in the discussion of Table 12.2-1, inverted pendulum-type structure are considered a subset of the cantilever column system where more than 50 percent of the total mass is concentrated at the top of the structure, and the stability of that mass is dependent on the rotational restraint to the top of the cantilevered element. As a result, this section requires that an inverted pendulum-like structure be designed by considering a top moment equal to half the design moment at the base. The design base moment is to be determined from the equivalent lateral force analysis specified in Section 12.8, and a uniform variation of bending moment is to be assumed from the base to the top. **12.2.5.3**

Increased building height limit for steel braced frames and special reinforced concrete shear walls. It may be noted from ASCE 7-05 Table 12.2-1 that buildings in SDC D and E over 160 feet in height and buildings in SDC F over 100 feet in height must have one of the following seismic force-resisting systems: **12.2.5.4**

1. A moment-resisting frame system with special moment frames capable of resisting the total prescribed seismic forces.
2. A dual system with special moment frames wherein the prescribed forces are resisted by the entire system and the special moment frames are designed to resist at least 25 percent of the prescribed seismic forces.

According to this section, for buildings having cast-in-place concrete shear walls or steel braced frames, the height limit prescribed in ASCE 7-05 Table 12.2-1 is allowed to be relaxed from 160 feet to 240 feet for SDC D and E, and from 100 feet to 160 feet for SDC F, provided that: (1) the configuration of the lateral force-resisting system is such that torsional effects do not result in an extreme torsional irregularity (horizontal structural irregularity Type 1b) as defined in table 12.3-1, and (2) the braced frames or shear walls in any plane do not resist more than 60 percent of the seismic design force neglecting torsional effects. The intent is that each of the shear walls or braced frames shall be in a different plane and that the four or more planes shall be spaced throughout the plan or on the perimeter of the building in such a way that the premature failure of one of the single walls or braced frames will not lead to excessive inelastic torsion.

Although a structural system with lateral force resistance concentrated in the interior core is allowed by ASCE 7-05, it is highly recommended that use of such a system be avoided, particularly for taller buildings. Preference should be given to systems with lateral force resistance distributed across the entire building.

See Example 16-9, Part 2 of this chapter.

12.2.5.5 Special moment frames in structures assigned to Seismic Design Categories D through F. This section specifies that where a special moment frame is required by Table 12.2-1 (misprinted as 12.1-1 in the first edition of ASCE 7-05), that is, in buildings exceeding the height limits specified for the structural systems other than moment-resisting frame systems and dual systems with special moment frames, the frame needs to be continued down to the foundation and not be vertically combined with a stiffer system. However, when a special moment frame is used but not required by Table 12.2-1, this section allows it to be discontinued and supported on a more rigid system with a lower *R*-value, provided that the requirements for extreme weak stories (Section 12.3.3.2) and force increase for collector elements (Section 12.3.3.4) are satisfied. These two requirements ensure that the lower rigid portion has the right configuration to provide a continuous load path for the special moment frame above.

12.2.5.6 Single-story steel ordinary and intermediate moment frames in structures assigned to Seismic Design Category D or E. This section relaxes the restrictions imposed by Table 12.2-1 and allows ordinary and intermediate steel moment frames to be used in SDC D and E buildings—however, only when they are single-story buildings with height up to 65 feet and with dead load supported by and tributary to the roof not exceeding 20 psf. In addition, for an exterior wall higher than 35 feet, the dead load of the wall that is carried by the moment frame should not exceed 20 psf.

12.2.5.7 Other steel ordinary and intermediate moment frames in structures assigned to Seismic Design Category D or E. This section specifies the criteria, similar to those above, when multistoried steel ordinary and intermediate moment frames need to satisfy for use in SDC D and E buildings.

12.2.5.8 Single-story steel ordinary and intermediate moment frames in structures assigned to Seismic Design Category F. The conditions under which single-story steel ordinary and intermediate moment frames can be used in an SDC F building are mostly the same as those for SDC D and E buildings, specified in Section 12.2.5.6. However, as an added restriction, the exterior wall dead load limitation now applies to all walls and not just walls more than 35 feet in height.

12.2.5.9 Other steel intermediate moment frames in structures assigned to Seismic Design Category F. Unlike the structures assigned to SDC D or E, multistory steel ordinary moment frames are not allowed in SDC F buildings. For intermediate moment frames, the building needs to be of light-frame construction in addition to satisfying all the conditions for SDC E, as specified in Section 12.2.5.8.

12.2.5.10 Shear wall-frame interactive system. In a shear wall-moment frame interactive system, at least 75 percent of the design story shear in each story shall be capable of being resisted by the shear walls and at least 25 percent of the design story shear in each story shall be capable of being resisted by the moment frames.

12.3 Diaphragm flexibility, configuration irregularities and redundancy. This section defines these three important attributes that have a critical role in determining the behavior of a structure subjected to earthquake ground motions. The first two attributes determine the distribution of seismic forces in different structural elements, whereas the third attribute determines the ability of a structure to go into inelastic deformations without losing significant amounts of vertical or horizontal strength.

12.3.1 Diaphragm flexibility. This section describes a computational method as well as a prescriptive method of determining if a diaphragm can be idealized as flexible or rigid. If the diaphragm cannot be categorized by either of the two methods, then this section requires that structural analysis explicitly include consideration of diaphragm stiffness. By the computational method, which was traditionally used to define diaphragm flexibility, a diaphragm can be considered flexible when the computed maximum in-plane deflection of the diaphragm is more than two times the average drift of adjoining vertical elements (Figure 16-30). This definition has been in the UBC since its 1988 edition. It implies that a diaphragm segment between two vertical lateral force-resisting elements may be modeled as a simple beam spanning between those elements.

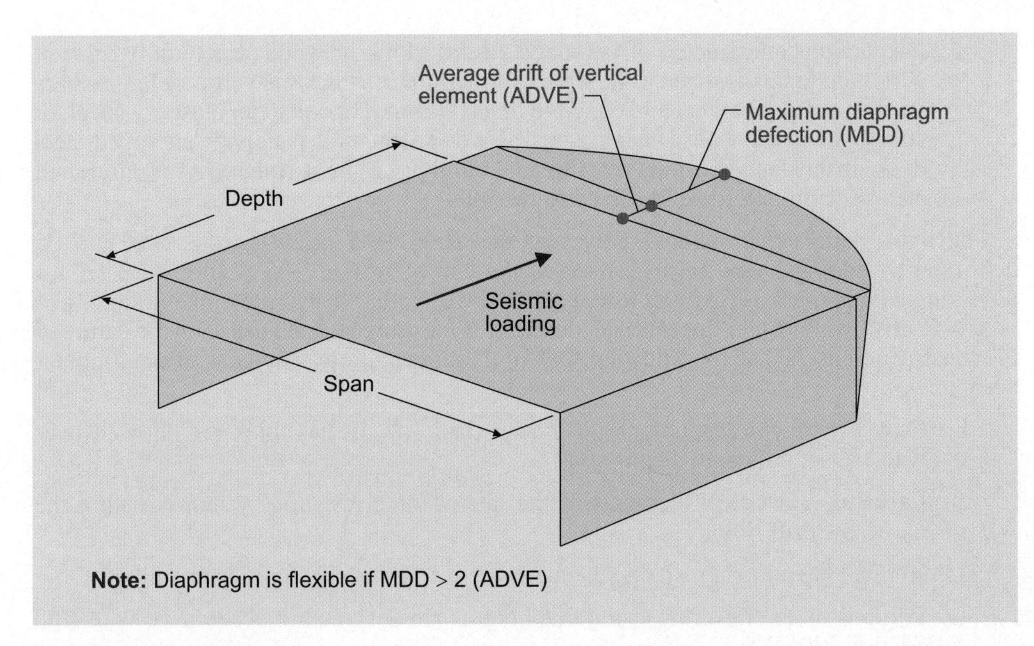

Figure 16-30
Classification of diaphragm based on deflection calculation

The following excerpt from the 1999 SEAOC *Blue Book* Commentary (Section C805.3) is included here as being of interest to a large number of potential readers. Only slight modifications of language have been made in a few places.

> Past and present design practice in California and other areas has almost exclusively used flexible diaphragm assumptions for distribution of seismic forces in structures with wood diaphragms. The assumption of flexible behavior of wood diaphragms conforms quite well with the UBC definition of flexible diaphragms for larger buildings with concrete or masonry walls. Lateral force-resisting systems using wood diaphragms in combination with wood shear walls seldom meet the UBC definition of a flexible diaphragm, suggesting that for many wood shear wall structures, the UBC would require that a rigid diaphragm analysis be performed. The adoption of the 1988 UBC did not seem to prompt any significant change in design practice related to the assumption of diaphragm flexibility for wood framed structures. Most designers continued their past practice and most building officials continued to accept this practice. This is one area where a divergence developed between common design practice and a strict interpretation of the code.

In the wood design examples of the *Seismic Design Manual*,[42] which is in conformance with the 1997 UBC, SEAOC found it necessary to point out that common design practice is not consistent with strict interpretation of the UBC. This caused considerable consternation among designers of residential structures.

The relative flexibility of cantilevered columns compared to parallel shear walls is thought to have been one factor contributing to the poor performance of buildings with tuck-under parking in the Northridge earthquake. A rigid diaphragm analysis performed on this type of a building should have identified the relative flexibility of the columns, and the resulting redistribution of seismic forces might have been accommodated in the design. Although damage resulting from the relative flexibility of vertical elements should theoretically occur to varying extents in many other types of wood buildings, it has not yet been reported as a major cause of damage. Earthquake damage observed to date suggests that life safety has generally been achieved in structures having a fairly regular configuration and redundancy, with wood structural panel diaphragms designed using flexible diaphragm assumptions.

The prescriptive method is new to ASCE 7-05, which states that

> Diaphragms constructed of untopped steel decking or wood structural panels are permitted to be idealized as flexible in structures in which the vertical elements are steel or composite steel and concrete braced frames, or concrete, masonry, steel, or composite shear walls. Diaphragms of wood structural panels or untopped steel decks in one- and two-family residential buildings of light-frame construction shall also be permitted to be idealized as flexible.

For light-frame constructions, however, the 2006 IBC modifies this ASCE 7-05 definition by adding a new set of criteria at the end of ASCE 7-05 Section 12.3.1.1 for idealizing a diaphragm as flexible. It was realized that, rather than simply mentioning one- or two-family dwellings, an appropriate idealization can only be done based on the nature of the construction itself. The modification sets forth four conditions that must be met for other occupancies.

1. Only nonstructural toppings no greater than 1-$^1/_2$ inches thick are allowed over wood structural panel diaphragms.
2. Each line of vertical elements of the lateral-force resisting system is within the allowable drift limits.
3. Vertical lateral load resisting elements are wood or steel sheet shear walls.
4. Portions of cantilevering wood diaphragms are designed in accordance with IBC Section 2305.2.5.

These conditions are in part based on research that was part of the CUREE Caltech Woodframe project, which showed that for regular, light-framed, wood diaphragm buildings, better building performance (lower drift and lower repair cost) resulted when flexible diaphragm force distribution was used. It is expected that these new provisions will allow many more wood framed structures to have the diaphragms idealized as flexible for design purposes. More information is contained in CUREE W30 *Recommendations for Earthquake Resistance in the Design of Construction of Woodframe Building*[43] available through www.curee.org.

ASCE 7-05 Section 12.3.1.2 defines rigid diaphragms as: "Diaphragms of concrete slabs or concrete filled metal deck with span-to-depth ratios of 3 or less in structures that have no horizontal irregularities."

The flow chart in Figure 16-31 is provided to illustrate the use of these diaphragm provisions.

Consideration of rigid diaphragm behavior is recommended where the diaphragms can be judged by observation to be stiff compared to the vertical lateral force-resisting elements, and particularly where one or more lines of resistance are substantially less stiff than the rest. Because of many uncertainties that exist in rigid diaphragm analysis for wood structures, use of an envelope considering some combination of rigid and flexible analysis may be appropriate. For small, substantially regular buildings that have a good distribution of lateral resistance and are well tied together, earthquake performance to date suggests that design using flexible diaphragm assumptions is adequate for life-safety performance.

It may be of interest to note here that 2006 IBC Section 1602 defines a rigid diaphragm as follows: "A diaphragm is rigid for the purpose of distribution of story shear and torsional moment when the lateral deformation of the diaphragm is less than or equal to two times the average story drift." In other words, a diaphragm that is not flexible is rigid. ASCE 7-05 Section 12.3.1 requires that unless a diaphragm can be idealized as either flexible or rigid in accordance with Sections 12.3.1.1 (*prescriptively flexible*), 12.3.1.2 (*prescriptively rigid*), or 12.3.1.3 (*flexible by calculation*), the structural analysis must explicitly include consideration of the stiffness of the diaphragm (i.e., semirigid modeling assumption). However, 2006 IBC Section 1602 definitely implies that unless a diaphragm is flexible by calculation, lateral load distribution can be on the basis of analysis that assumes rigid diaphragm behavior. This is a clear case where the 2006 IBC and ASCE 7-05 are in conflict. 2006 IBC Section 102.4 unequivocally states: "Where differences occur between

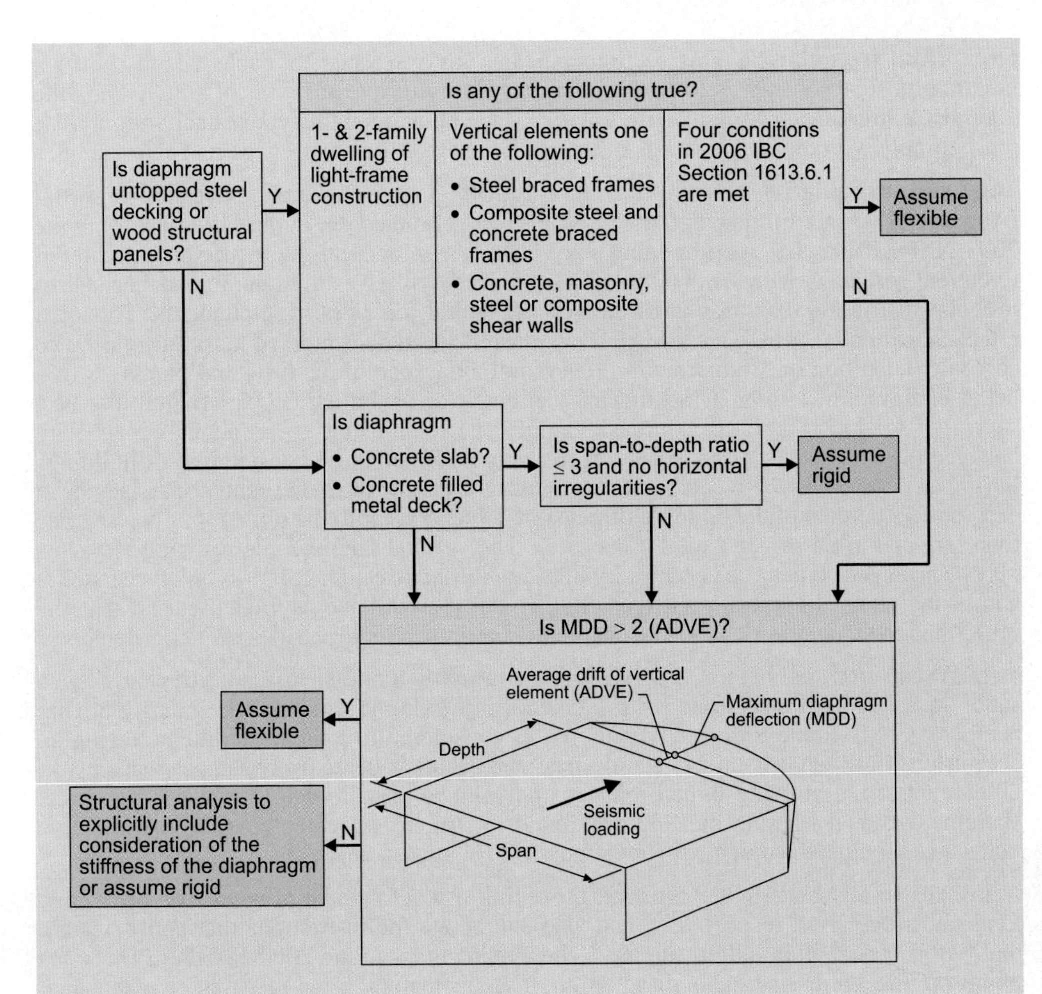

Figure 16-31 **Classification of diaphragm based on ASCE 7-05 Section 12.3.1**

provisions of this code and referenced codes and standards, the provisions of this code shall apply." Section 102.4 can definitely be invoked in this particular case.

12.3.2

Irregular and regular classification. This section requires that buildings be classified as regular or irregular based on the horizontal and the vertical configuration of the building.

Past earthquakes have repeatedly shown that buildings having irregular configurations (in plan or along the height of a structure) suffer greater damage than buildings having regular configurations. This situation prevails even with good design and construction.

The 2003 NEHRP Commentary (Section 4.3.2) has pointed out several reasons for this poor behavior of irregular structures. In a regular structure, inelastic demands produced by strong ground shaking tend to be well distributed throughout the structure, resulting in a dispersion of energy dissipation and damage. However, in irregular structures, inelastic behavior can concentrate in certain localized regions, resulting in rapid deterioration of structural elements in these areas. In addition, some irregularities introduce unanticipated stresses into a structure, which designers frequently overlook when detailing the structural system. Finally, elastic analysis methods typically employed in the design of structures are incapable of accurately predicting the distribution of earthquake demands in an irregular structure, leading to inadequate design in the vicinity of irregular features. For these reasons, the requirements of this section were formulated to encourage that buildings be designed to have regular configurations and to prohibit gross irregularity in buildings located on sites close to major active faults, where strong ground motion and extreme inelastic demands in a structure may be experienced.

12.3.2.1 Horizontal irregularity. This section indicates, by reference to Table 12.3-1, when a building must be considered to have a plan irregularity for the purposes of ASCE 7 and the 2006 IBC. Structures having plan irregularity, depending upon the type of such irregularity, must comply with the requirements of the applicable sections referenced in Table 12.3-1.

Torsional irregularity (Horizontal Irregularity Type 1a) and extreme torsional irregularity (Horizontal Irregularity Type 1b). As pointed out in the Commentary to the 2003 NEHRP *Provisions*, a building may have a symmetrical geometric shape without re-entrant corners or wings, but may still have to be classified as irregular in plan because of irregular distribution of mass or of the vertical lateral force-resisting elements. Torsional effects in earthquakes may and do occur even when the static centers of mass and resistance coincide. Direction of ground motion that is not along one of the principal plan axes of a building may cause torsion. Cracking and yielding in an asymmetrical pattern may also be a source of torsion. These effects may magnify the torsion due to an actual eccentricity between the static centers of mass and resistance. Based on these considerations, buildings having an eccentricity between the static center of mass and the static center of resistance in excess of 10 percent of the building dimension perpendicular to the direction of the seismic force are classified as horizontally irregular. The vertical lateral force-resisting elements may be arranged so that the eccentricity between the static centers of mass and resistance is within the above limitation, but the arrangement may still be asymmetrical so that the prescribed torsional forces would be unequally distributed to the various resisting elements.

In ASCE 7-05, as in the 2003 NEHRP *Provisions*, torsional irregularities have been subdivided into two categories, with a category of extreme torsional irregularity having been created. Extreme torsional irregularity is considered to exist in buildings having an eccentricity between the static center of mass and the static center of resistance in excess of 20 percent of the building dimension perpendicular to the direction of the seismic force. Extreme torsional irregularities are prohibited in structures located close to major active faults and should be avoided, whenever possible, in all structures.

Figures 16-32A and 16-32B illustrate the definitions of torsional irregularity and extreme torsional irregularity, respectively, and also enumerate the restrictions that apply when a building is classified in either category. The Commentary to the 2000 NEHRP *Provisions* points out that there is a type of distribution of vertical lateral force-resisting elements that, while not being classified as irregular, does not produce good structural performance in strong earthquakes. This arrangement produces a core-type building, with the vertical components of the lateral force-resisting system collected near the center of the building. Better performance is observed when the vertical components are distributed near the perimeter of the building.

Re-entrant corners (Horizontal Irregularity Type 2). A building having a regular configuration can be square, rectangular or circular. As the Commentary to the 2003 NEHRP *Provisions* points out, a square or rectangular building with minor re-entrant corners would still be considered regular, but large re-entrant corners creating a crucifix form would cause a building to be classified as irregular in plan. The response of the wings of this type of a building is generally different from the response of the building as a whole, and this produces higher local forces than would be determined from application of the IBC without modification. Other plan configurations, such as H-shapes, that are geometrically symmetrical would also be classified as irregular because of the response of the wings.

The definition of Plan Irregularity Type 2 is illustrated in Figure 16-33. The restrictions that become applicable when a building is classified as having Plan Irregularity Type 2 are also clearly indicated in Figure 16-33.

Diaphragm discontinuity (Horizontal Irregularity Type 3). Significant differences in stiffness between portions of a diaphragm at a level may cause a change in the distribution of lateral forces to the vertical lateral force-resisting elements and create torsional forces not accounted for in the distribution considered for a regular building.

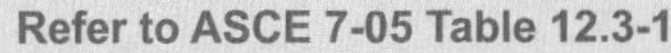

Refer to ASCE 7-05 Table 12.3-1

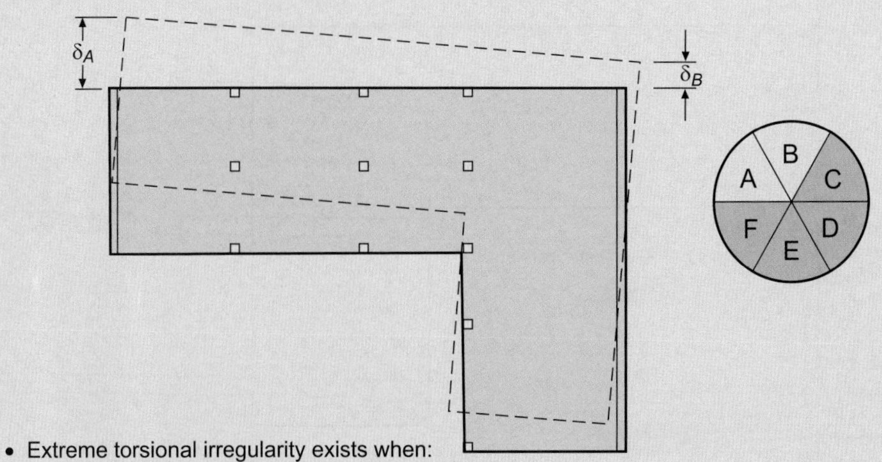

- Extreme torsional irregularity exists when:

$$\delta_A > 1.4\,\frac{\delta_A + \delta_B}{2}$$

- Torsional irregularity to be considered only when diaphragms are not flexible (ASCE 7-05 Table 12.3-1)
- In SDC D, E and F design forces determined from Section 12.8.1 shall be increased 25% for connections of diaphragms to vertical elements and to collectors, and for connections of collectors to vertical elements (ASCE 7-05 Section 12.3.3.4)
- In SDC, D, E and F dynamic amplification of torsion as required by Section 12.8.4.3 shall be applied

Figure 16-32a
Plan irregularity Type 1A, torsional irregularity

Refer to ASCE 7-05 Table 12.3-1

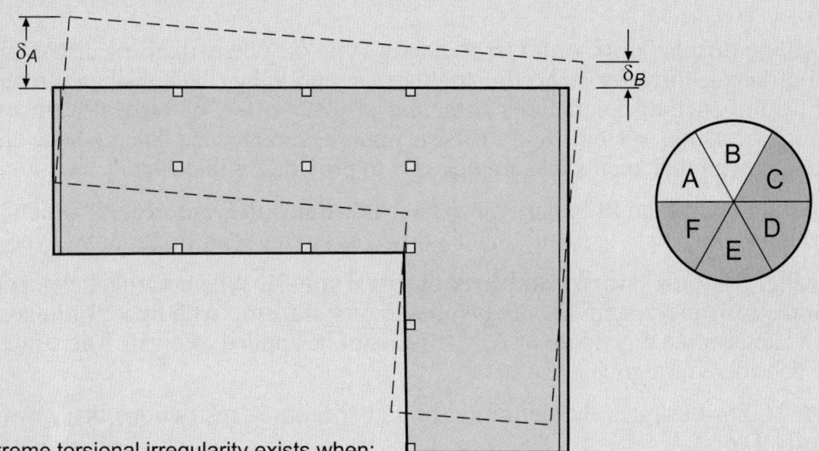

- Extreme torsional irregularity exists when:

$$\delta_A > 1.4\,\frac{\delta_A + \delta_B}{2}$$

- Torsional irregularity to be considered only when diaphragms are not flexible (ASCE 7-05 Table 12.3-1)
- Shall not be permitted in SDC E and F (ASCE 7-05 Section 12.3.3.1)
- In SDC D, E and F design forces determined from Section 12.8.1 shall be increased 25% for connections of diaphragms to vertical elements and to collectors, and for connections of collectors to vertical elements (ASCE 7-05 Section 12.3.3.4)
- In SDC, D, E and F dynamic amplification of torsion as required by Section 12.8.4.3 shall be applied

Figure 16-32b
Plan irregularity Type 1B, extreme torsional irregularity

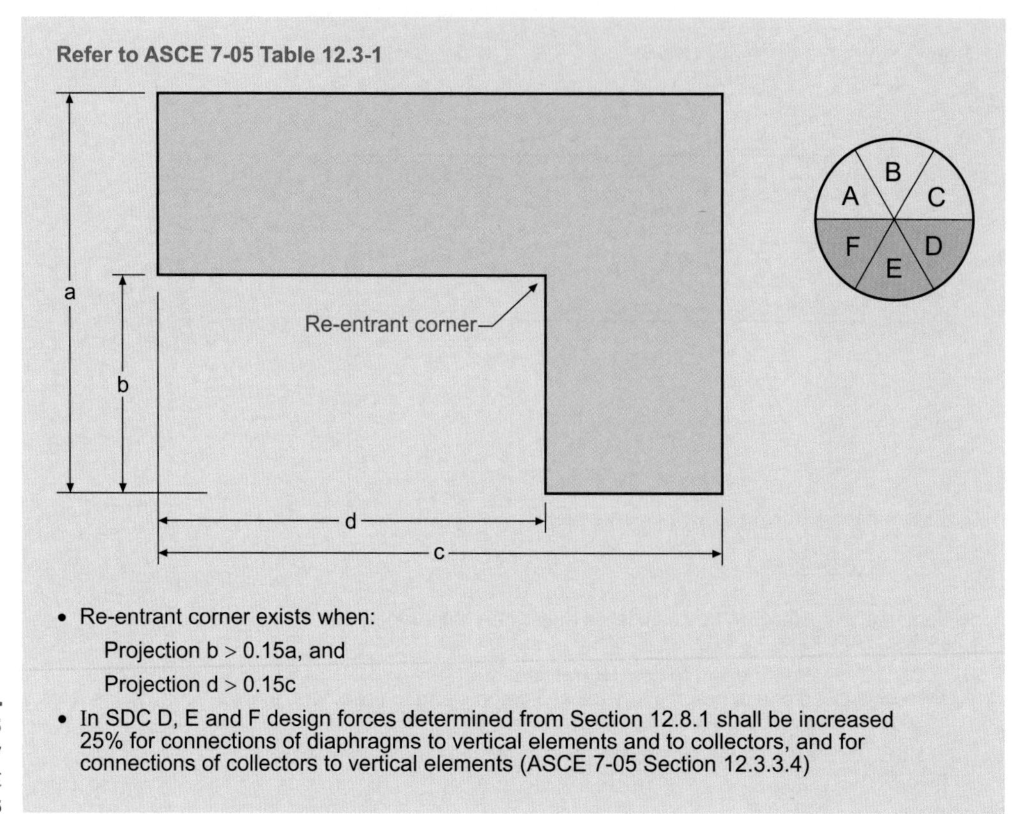

Figure 16-33 **Plan irregularity Type 2, re-entrant corners**

Figure 16-34 illustrates the definition of Plan Irregularity Type 3 and also points out the restrictions that become applicable when a building is classified as having Plan Irregularity Type 3.

Out-of-plane offsets (Horizontal Irregularity Type 4). Where there are discontinuities in the path of lateral-force resistance, a structure may no longer be considered regular. The most critical of such discontinuities is the out-of-plane offset of vertical elements of the lateral force-resisting system. Such offsets impose vertical and lateral load effects on horizontal structural elements that are not easy to provide for in design.

The definition of Plan Irregularity Type 4 is illustrated in Figure 16-35, which also lists the restrictions that apply to a building classified as having Plan Irregularity Type 4.

Nonparallel Systems (Horizontal Irregularity Type 5). Where vertical elements of the lateral force-resisting system are not parallel to or symmetric with major orthogonal axes, the static lateral-force procedure of ASCE 7 cannot be applied as given. The structure then must be considered irregular.

Figure 16-36 illustrates the definition of and shows the restrictions that go with Plan Irregularity Type 5.

12.3.2.2 Vertical irregularity. This section indicates, by reference to Table 12.3-2, when a structure must be considered to have a vertical irregularity. Vertical configuration irregularities affect seismic response at various floor levels and induce loads at these levels that depart significantly from the distribution assumed in the equivalent lateral-force procedure. Structures having vertical irregularity, depending upon the type of such irregularity, must comply with the requirements of the applicable sections referenced in Table 12.3-2.

Stiffness irregularity—soft story (Vertical Irregularity Type 1a), Stiffness irregularity—extreme soft story (Vertical Irregularity Type 1b). A building would be classified in one of these categories if the lateral stiffnesses in adjoining stories differ significantly. This might occur in a moment-resisting frame building if one story were much

Refer to ASCE 7-05 Table 12.3-1

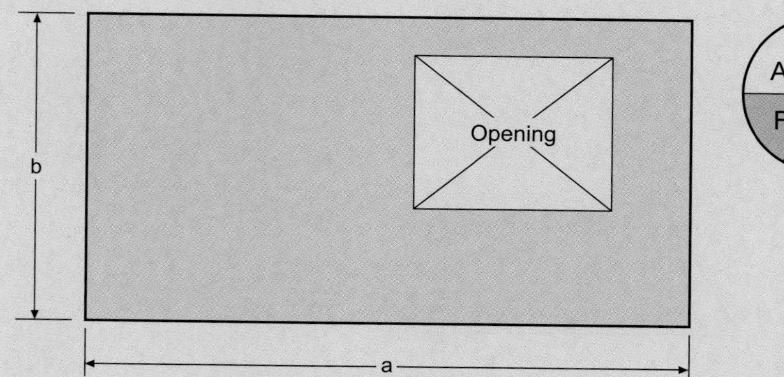

- Diaphragm discontinuity exists when:
  Area of opening > 0.5ab or
  Effective diaphragm stiffness changes more than 50% from one story to the next
- In SDC D, E and F, design forces determined from Section 12.8.1 shall be increased 25% for connections of diaphragms to vertical elements and to collectors, and for connections of collectors to vertical elements (ASCE 7-05 Section 12.3.3.4)

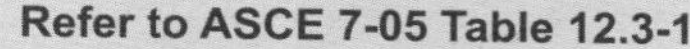

Figure 16-34
Plan irregularity Type 3, diaphragm discontinuity

Refer to ASCE 7-05 Table 12.3-1

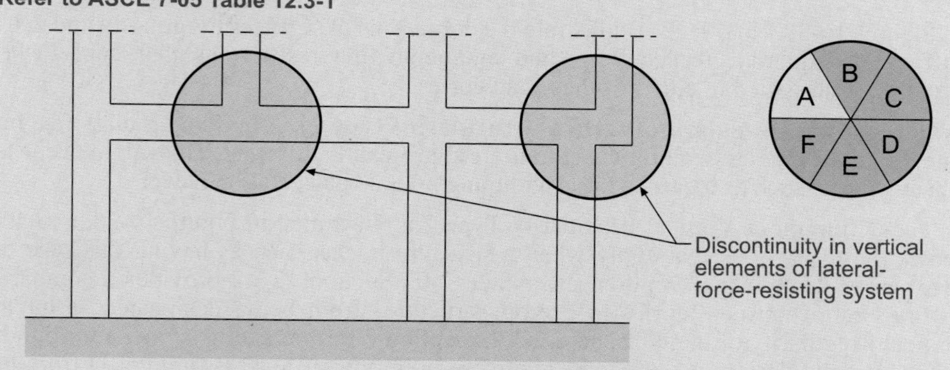

- Required strength of columns supporting discontinuous vertical elements (ASCE 7-05 Section 12.3.3.3):
  Load combinations with overstrength factor (ASCE 7-05 Section 12.4.3.2)
  $(1.2 + 0.2S_{DS})D + \Omega_0 Q_E + L + 0.2S$
  $(0.9 - 0.2S_{DS})D + \Omega_0 Q_E + L + 1.6H$
- Note special detailing requirements in materials chapters.
- In SDC D, E and F, design forces determined from Section 12.8.1 shall be increased 25% for connections of diaphragms to vertical elements and to collectors, and for connections of collectors to vertical elements (ASCE 7-05 Section 12.3.3.4).

Figure 16-35
Plan irregularity Type 4, out-of-plane offsets

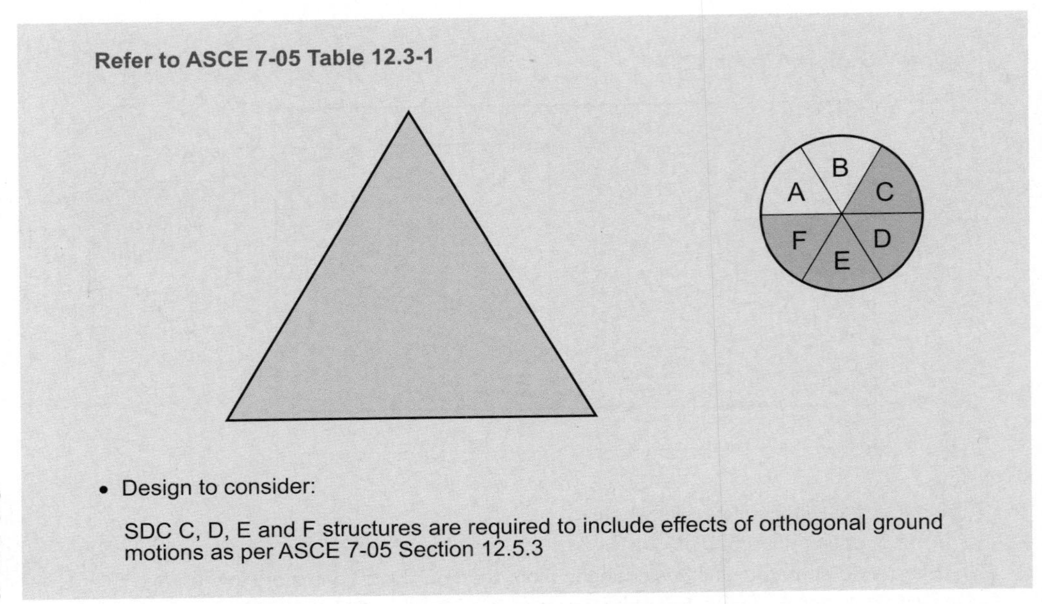

Figure 16-36 Plan irregularity Type 5, nonparallel systems

taller than the adjoining stories and the resulting decrease in stiffness was not, or could not be, compensated for.

In ASCE 7, as in the 2003 NEHRP *Provisions*, stiffness irregularities have been subdivided into two categories, with a category of extreme stiffness irregularity having been created.

Figures 16-37A and 16-37B illustrate the definitions of stiffness irregularity and extreme stiffness irregularity, respectively, and enumerate the restrictions that apply when a structure is classified in either of these categories.

Weight (Mass) irregularity (Vertical Irregularity Type 2). A building is classified in this category if the masses of adjoining stories are significantly different. This might occur when a heavy mass, such as a garden or a swimming pool, is placed at one level.

The definition of Vertical Irregularity Type 2 is illustrated in Figure 16-38, which also shows the restrictions that apply when a structure is classified as having this type of an irregularity. Note that Exception 1 to ASCE 7-05 Section 12.3.2 provides a comparative stiffness ratio between stories to exempt structures from being designated as having a vertical irregularity of the type specified. Exception 2 further exempts one-story buildings in any seismic design category and two-story buildings in SDC B through D from being designated as having a vertical irregularity of the type specified.

Vertical geometric irregularity (Vertical Irregularity Type 3). As illustrated in the Commentary to the 2003 NEHRP *Provisions* (Section 4.3.2.3), one type of vertical irregularity is created by asymmetrical geometry with respect to the vertical axis of the building. A building may have a geometry that is symmetrical about the vertical axis and still be classified as irregular because of significant horizontal offsets in the vertical elements of the lateral-force-resisting system at one or more levels. An offset is considered to be significant if the ratio of the larger dimension to the smaller dimension is more than 130 percent. As pointed out in the 2003 NEHRP Commentary, a building would also be considered irregular if the vertical lateral force-resisting elements having the smaller dimensions supported such elements having the larger dimensions, thus creating an inverted pyramid effect.

Vertical Irregularity Type 3 and the restrictions associated with it are illustrated in Figure 16-39.

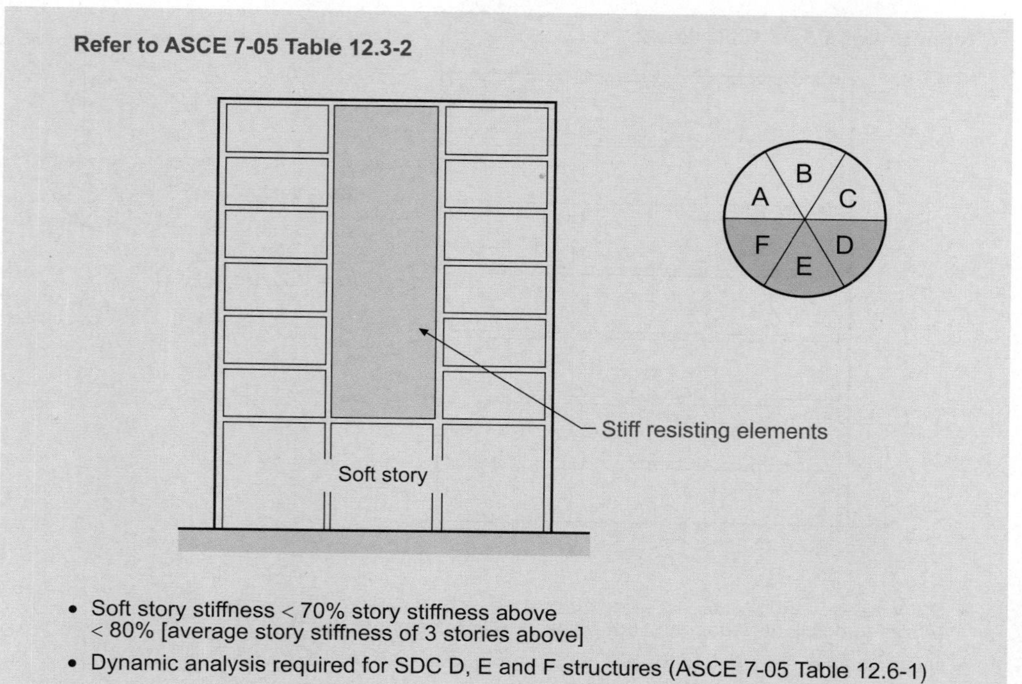

- Soft story stiffness < 70% story stiffness above
  < 80% [average story stiffness of 3 stories above]
- Dynamic analysis required for SDC D, E and F structures (ASCE 7-05 Table 12.6-1)

Figure 16-37a Vertical irregularity Type 1A, soft story

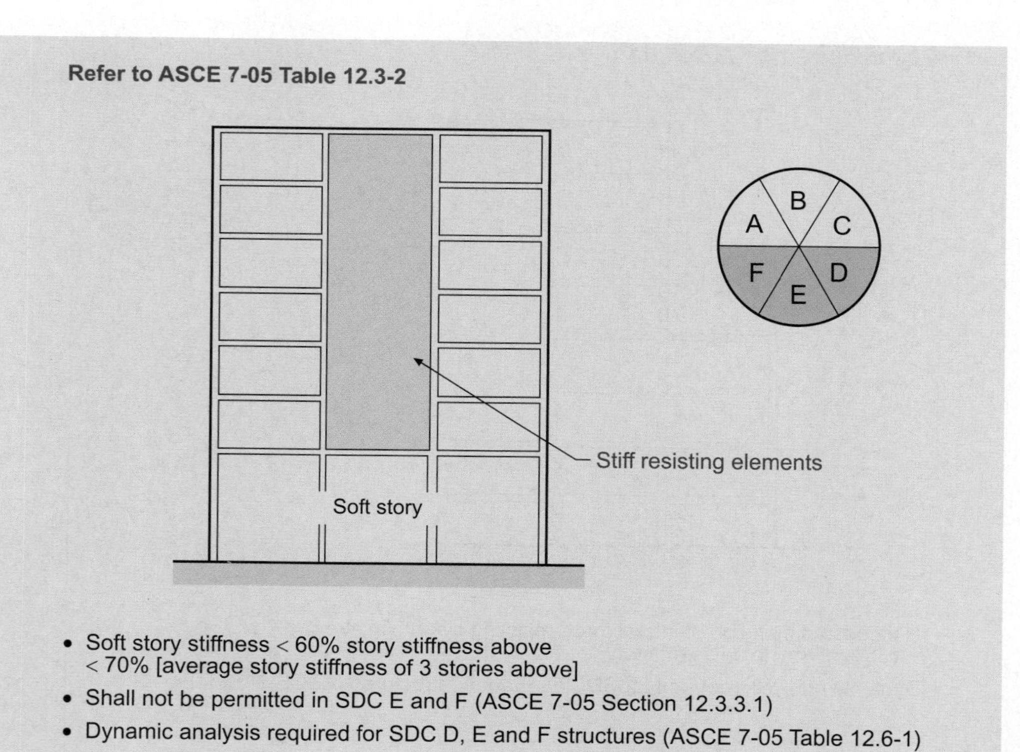

- Soft story stiffness < 60% story stiffness above
  < 70% [average story stiffness of 3 stories above]
- Shall not be permitted in SDC E and F (ASCE 7-05 Section 12.3.3.1)
- Dynamic analysis required for SDC D, E and F structures (ASCE 7-05 Table 12.6-1)

Figure 16-37b Vertical irregularity Type 1B, soft story

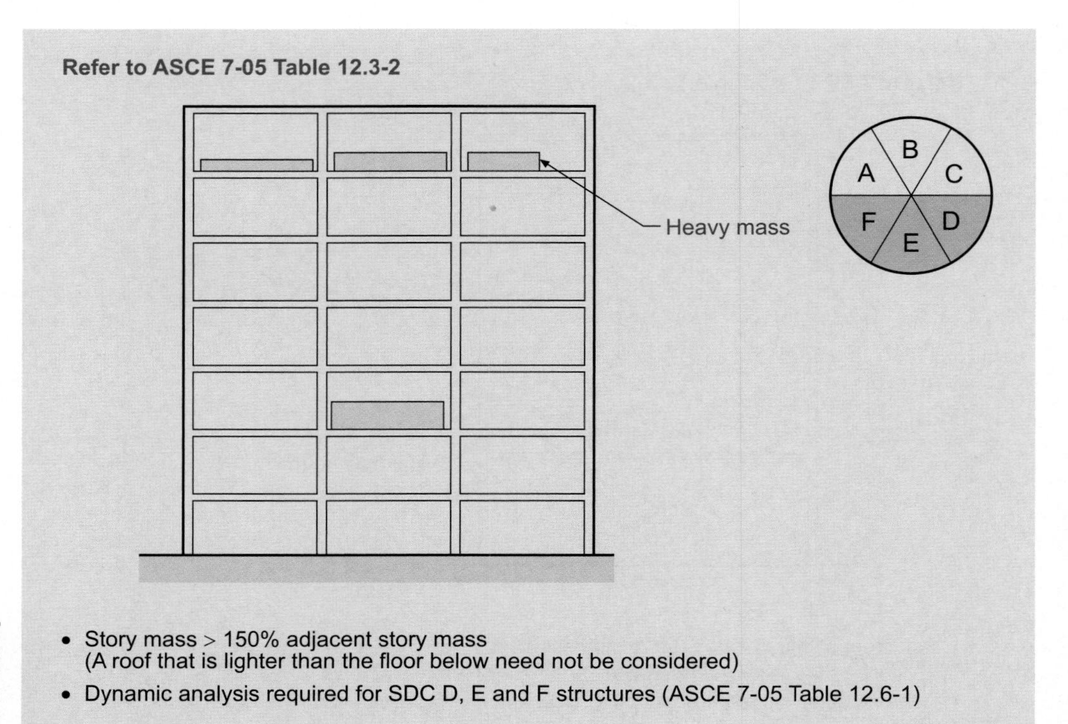

Figure 16-38 **Vertical irregularity Type 2, weight (mass) irregularity**

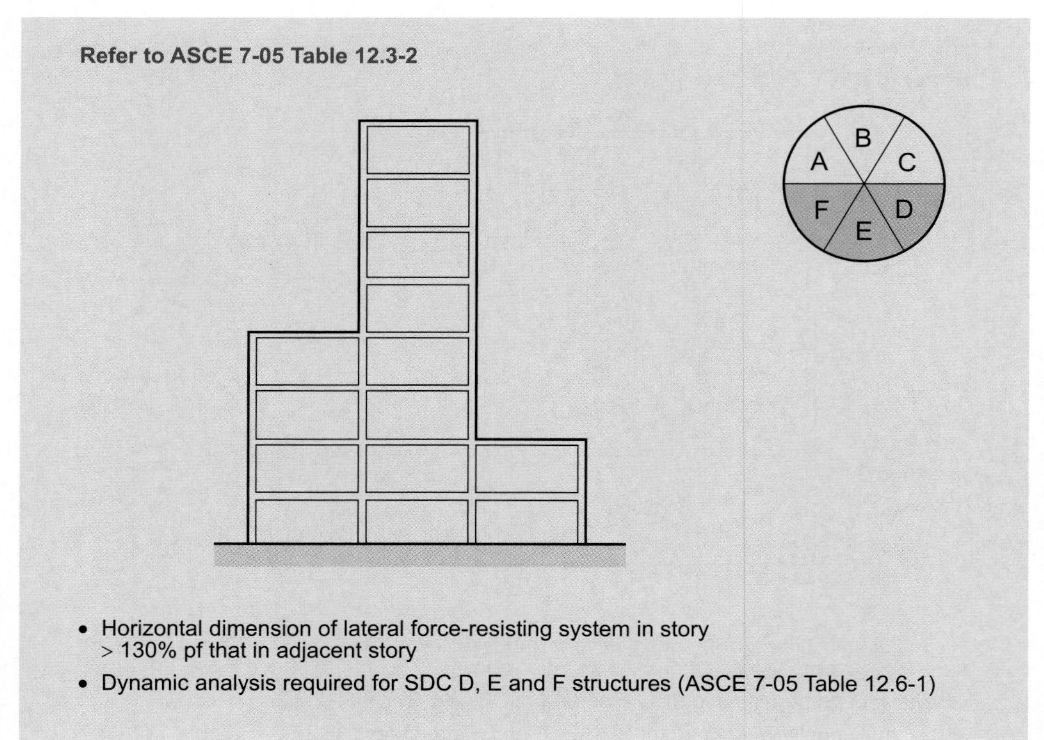

Figure 16-39 **Vertical irregularity Type 3, vertical geometric irregularity**

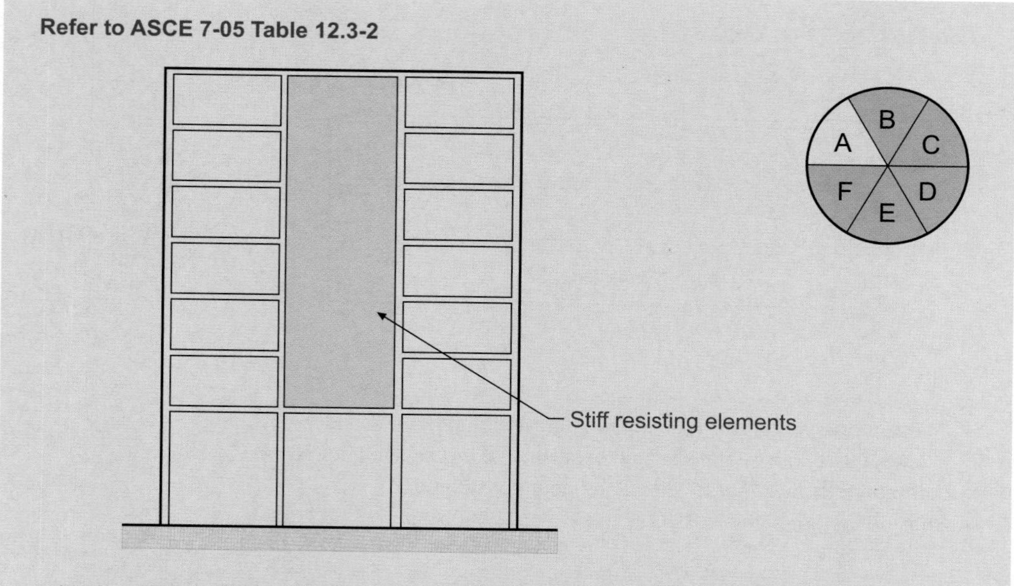

- In-plane offset of lateral-force-resisting elements > lengths of those elements or reduction in stiffness of resisting elements in story below
- Required strength of columns supporting discontinuous vertical elements (ASCE 7-05 Section 12.3.3.3):

  Load combinations with overstrength factor (ASCE 7-05 Section 12.4.3.2) must be used

  $(1.2 + 0.2S_{DS})D + \Omega_0 Q_E + L + 0.2S$

  $(0.9 - 0.2S_{DS})D + \Omega_0 Q_E + L + 1.6H$
- Note special detailing requirements in materials chapters.
- In SDC D, E and F design forces determined from Section 12.8.1 shall be increased 25% for connections of diaphragms to vertical elements and to collectors, and for connectors of collector to vertical elements (ASCE 7-50 Section 12.3.3.4)

Figure 16-40
Vertical irregularity Type 4, in-plane discontinuity in vertical lateral force-resisting element irregularity

In-plane discontinuity in vertical lateral-force-resisting elements (Vertical Irregularity Type 4). This irregularity is not unlike Vertical Irregularity Type 3. The definition and the applicable restrictions are given in Figure 16-40.

Discontinuity in capacity—weak story (Vertical Irregularity Type 5a), extreme weak story (Vertical Irregularity Type 5b). Weak-story irregularities or extreme weak-story irregularities occur whenever the strength of a story to resist lateral demands is significantly less than that of the story above. Buildings with this configuration tend to develop all of their inelastic behavior in the weak story. A significant change in the deformation pattern of the building results, with most earthquake-induced displacement occurring within the weak story. This typically results in extensive damage within the weak story and may even cause instability and collapse.

The definition of Vertical Irregularity Type 5a and 5b are illustrated in Figures 16-41A and 16-41B, respectively. The restrictions applying to a building classified as having Vertical Irregularity Type 5a and 5b are given in the figures.

Limitations and additional requirements for systems with structural irregularities. **12.3.3**

Prohibited horizontal and vertical irregularities for seismic design categories D through F. **12.3.3.1** This section requires that SDC E or F buildings not have extreme torsional irregularity, extreme soft story irregularity or weak story irregularity. For SDC D structures, only extreme weak story is prohibited.

Refer to ASCE 7-05 Table 12.3-2

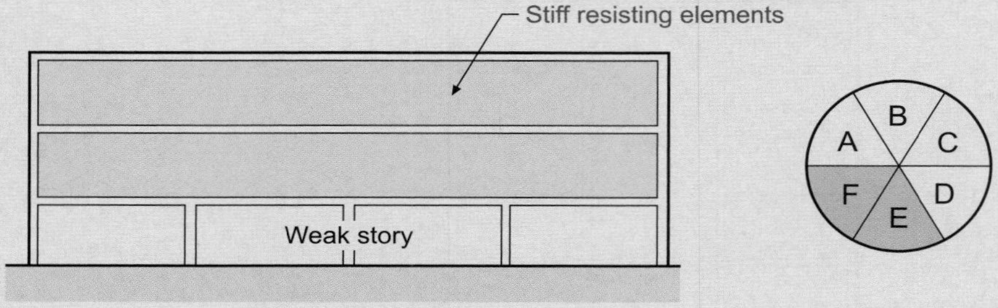

Stiff resisting elements

Weak story

A B C D E F

- "Weak story" lateral strength < 80% laterals strength above story
  Lateral strength = total strength of seismic force-resisting elements
  Shearing story shear for direction under consideration
- Shall not be permitted in SDC E and F (ASCE 7-05 Section 12.3.3.1)

Figure 16-41a Vertical irregularity Type 5A, weak story

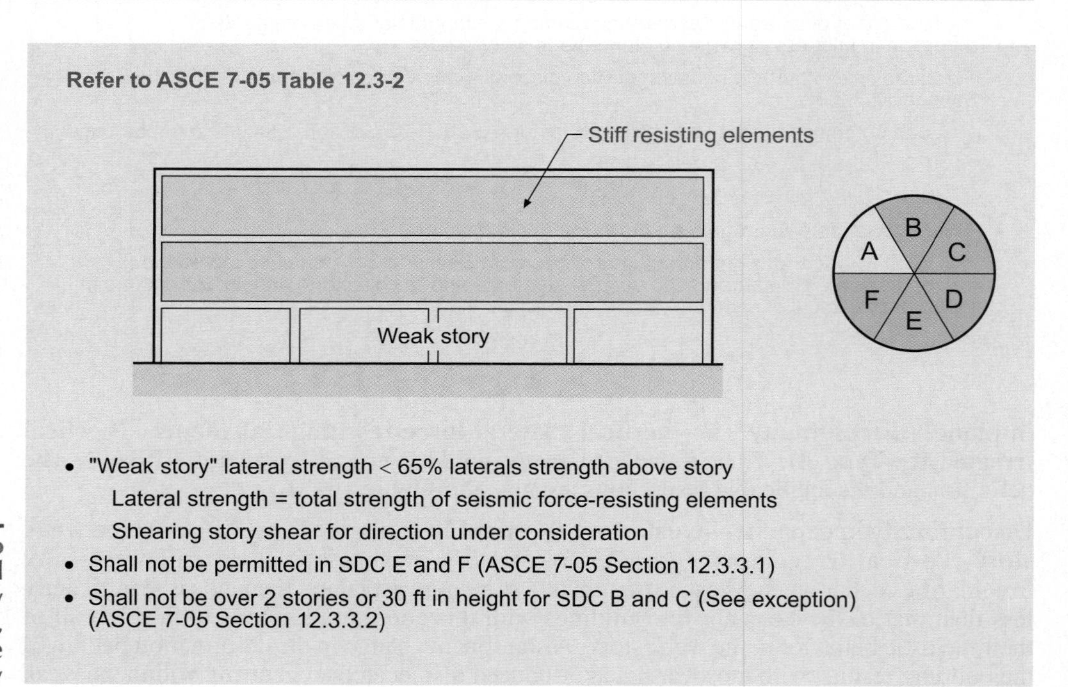

Figure 16-41b Vertical irregularity Type 5B, extreme weak story

12.3.3.2 Extreme weak stories. An extreme weak story is not allowed to be built more than two stories or 30 feet in height in SDC B through F. However, the prohibition does not apply when the story above the weak story is overdesigned, such that even though the strength of the weak story is less than 65 percent of the story above, it can still carry a story shear equal to that obtained from the equivalent lateral force procedure, multiplied by the overstrength factor Ω_o of the system.

12.3.3.3 Elements supporting discontinuous walls or frames. This section requires that, in SDC B through F structures, columns or other elements supporting discontinuous walls or frames in structures having out-of-plane offsets of the vertical elements or in-plane discontinuity of vertical lateral force-resisting elements shall have the design strength to resist the forces that develop in accordance with the seismic load combinations with overstrength factor, as given in ASCE 7-05 Section 12.4.3.2. However, the connections of those discontinuous elements to the supporting members need only be designed for the forces for which the discontinuous elements were designed.

Increase in forces due to irregularities for seismic design categories D through F. **12.3.3.4** Also, for buildings having torsional irregularity, extreme torsional irregularity, re-entrant corners, diaphragm discontinuity, out-of-plane offsets of the vertical lateral force-resisting elements (irregularity Type 1a, 1b, 2, 3 or 4, ASCE 7-05 Table 12.3-1) or in-plane discontinuity in vertical lateral force-resisting elements (irregularity Type 4, ASCE 7-05 Table 12.3-2), the design forces determined from the equivalent lateral force procedure of ASCE 7-05 Section 12.8.1 must be increased 25 percent for connections of diaphragms to vertical elements and to collectors, and for connections of collectors to the vertical elements. It should be noted that collectors and their connectors should also be designed for these increased forces, unless they are designed for the seismic load combinations with overstrength factor of ASCE 7-05 Section 12.4.3.2.

Redundancy. **12.3.4** Redundancy as a concept is simple. It is the concept of the multiple load path—of providing more than one alternative path for every load to travel from its point of application to the ultimate point of resistance. Despite this conceptual simplicity, when the SEAOC Seismology Committee attempted to quantify redundancy for the 1997 UBC, it proved to be a challenging task indeed. The SEAOC formulation, adopted into the 1997 UBC, was adopted into the 2000 IBC with exactly three modifications. The formulation was further refined by introducing half a dozen or so additional modifications in the 2003 IBC and ASCE 7-02.

In ASCE 7-02 and the 2003 IBC, for structures assigned to SDC B or C, the value of the redundancy coefficient, ρ, was 1.0, just as in the 1997 UBC, the value of ρ was 1.0 for structures in Seismic Zones 0, 1 and 2. This reflected the belief that providing multiple load paths for the seismic forces is more important in the higher seismic zones and design categories than in lower ones.

For SDC D through F structures, the redundancy coefficient, ρ, as defined in ASCE 7-02 and the 2003 IBC, depended on three items:

1. The number of vertical lateral force-resisting elements. The term *element* was particularized by the code for different lateral force-resisting systems.
2. Floor areas of the building at different story levels, based on the thinking that a large building would require a larger number of lateral force-resisting elements to be as redundant as a smaller building with fewer such elements.
3. Distribution of lateral forces to the lateral force-resisting elements. This is important because if one of the lateral force-resisting elements at a particular story level dominates and resists most of the story shear, that does not represent a redundant situation, which is to be avoided.

Based on these three factors, the coefficient ρ_x at any story x was computed from the expression

$$\rho_x = 2 - \frac{20}{r_{\max_x}\sqrt{A_x}}$$

Where $r_{\max_x}$ is the ratio of the shear force resisted by the single element carrying the most shear force in the story to the total story shear, for a given direction of loading. A_x is the floor area in ft^2 of the diaphragm level immediately above story x. A minimum value of 1.0 and a maximum value of 1.5 were also specified for ρ_x. The redundancy coefficient, ρ, for the entire building in the direction concerned was the highest value of ρ_x over the height of the building.

In a major departure from this earlier approach, the redundancy provisions have been simplified to a significant extent in ASCE 7-05. The basic premise of the new redundancy provisions that have been adopted into ASCE 7-05 is that the most logical way to determine lack of redundancy is to check whether a component's failure results in an unacceptable amount of story strength loss or in the introduction of extreme torsional irregularity. In ASCE 7-05, the redundancy factor, ρ, is equal to either 1.0 or 1.3, depending upon whether or not an individual element can be removed (deemed to have failed or lost its force-resisting capabilities) from the lateral force-resisting system without causing the remaining structure to suffer a reduction in story strength of more than 33 percent or creating an extreme torsional irregularity (Horizontal Structural Irregularity Type 1b in ASCE 7-05 Table 12.3-1).

Braced frame, moment frame, shear wall and cantilever column systems have to conform to redundancy requirements. Dual systems are included also but are in most cases are inherently redundant. Shear walls with a height-to-length ratio greater than 1.0 are included in redundancy considerations to help ensure that an adequate number of wall elements is included or that the proper redundancy factor is applied. Stories resisting more than 35 percent of the base shear only are considered.

The above way of determining ρ obviously involves a number—possibly a significant number—of structural analyses. There is an alternative way of determining ρ, which is essentially by inspection. If a structure is regular in plan and there are at least two bays of seismic force-resisting perimeter framing on each side of the structure in each orthogonal direction (Figure 16-42) at each story resisting more than 35 percent of the base shear, then ρ = 1. Significantly, the number of bays for a shear wall is to be calculated as the length of the shear wall divided by the story height, or two times the length of shear wall divided by the story height for light-framed construction. Thus, the shear wall length plays a major role, and light-framed construction is given an advantage.

The 1.3 value of ρ is based on the results of reliability and redundancy studies as well as engineering judgment and is considered to provide adequate safeguard for a nonredundant building without causing serious overdesign. The threshold story shear of 35 percent of the base shear was arrived at through parametric studies and roughly includes all stories in buildings up to five or six stories tall and about 87 percent of the stories in taller buildings. The redundancy provision is mostly critical for the design of moment frames and braced frames, although shear walls with height-to-length ratios greater than 1.0, and dual systems,

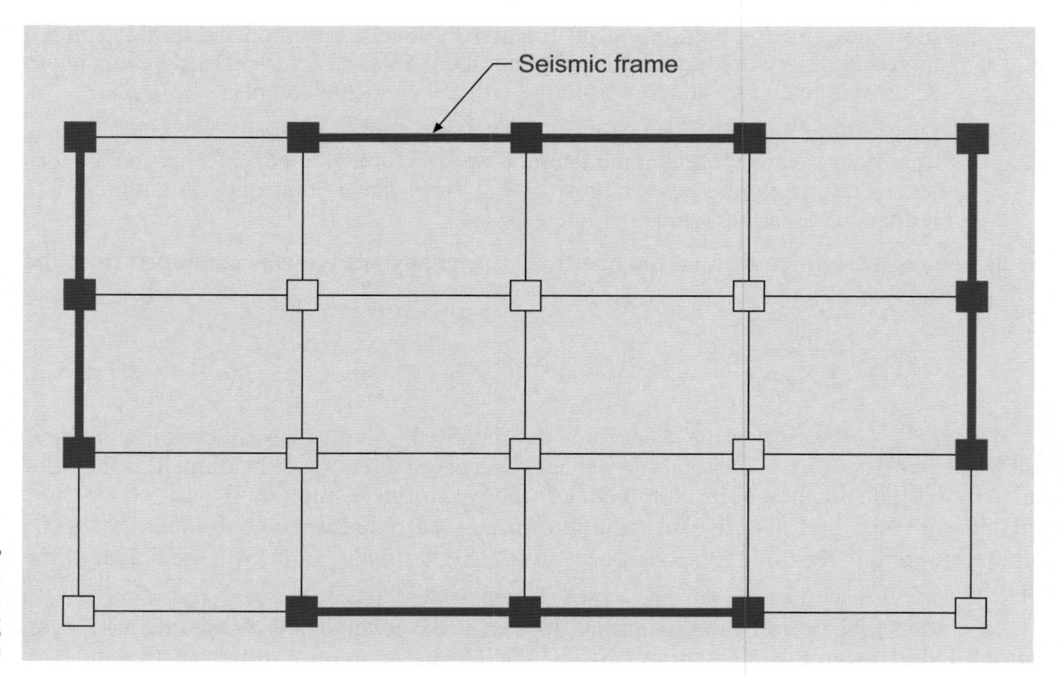

Figure 16-42
An example plan view of building with ρ = 1.0

are also included in the provision. Note, however, that the requirement that the collector elements and their connections be designed for a force that incorporates overstrength generally results in the use of multiple shear walls to reduce the force in the collector elements, making a building with shear walls inherently redundant. Also, dual systems are generally redundant because of the requirements governing their design that are already in place.

ASCE 7-05 includes a new user-friendly feature of conveniently listing situations where the redundancy factor, ρ, may be taken as 1.0. The value of ρ is permitted to equal 1.0 for the following:

1. SDC B or C buildings.
2. Buildings of SDC D through F that are regular in plan and have at least two bays of seismic force-resisting framing on each side of the building perimeter at each story that resists at least 35 percent of the base shear.
3. Drift calculation and *P*-delta effects.
4. Design of nonstructural components.
5. Design of nonbuilding structures that are not similar to buildings.
6. Design of collector elements, splices and their connections, for which the load combinations with overstrength factor, as provided in ASCE 7-05 Section 12.4.3.2 are required to be used.
7. Design of members or connections for which the load combinations with overstrength factor, as provided in ASCE 7-05 Section 12.4.3.2 are required to be used.
8. Diaphragm loads determined from ASCE 7-05 Eq. 12.10-1.
9. Structures with damping systems designed in accordance with ASCE 7-05 Chapter 18.

It can be clarified at this point that the redundancy coefficient, ρ, for structures assigned to SDC D, E and F, is not an element-by-element or story-by-story redundancy factor. It is the redundancy factor for the entire lateral force-resisting system of a building along one principal plan axis. If a structure has lateral force-resisting systems that are not identical along two orthogonal plan axes, then it may have two different ρ-values in the two orthogonal directions. In such cases, if two-dimensional structural analysis is carried out, the effects of the design base shear in the *x*-direction are to be amplified by the ρ-value for that direction, and similarly for the orthogonal direction. However, if a three-dimensional analysis is carried out, it may make sense to amplify the results of the analysis by the larger of the two ρ-values or by an arithmetic or a statistical average of the two ρ-values. No specific guidance is given as to which of these might be the most appropriate.

Example 16-7, Part 2 of this chapter, further illustrate the application of the redundancy coefficient.

Seismic load effects and combinations — 12.4

Applicability. This section requires that *E* for use in the basic load combinations of Sections 2.3 and 2.4 (IBC Sections 1605.2 and 1605.3) be determined from Section 12.4 of ASCE 7-05 and that *E* be modified to include system overstrength for use in the seismic load combinations with overstrength of Section 12.4.3 of ASCE 7-05 when such combinations are specifically required by ASCE 7-05. **12.4.1**

Seismic load effect. In seismic design practice represented by the UBC through its 1994 edition, *E* was simply the effect of the design base shear, *V* (internal forces obtained by elastic analysis of the structure under lateral forces resulting from code-required distribution of *V* along the height of the structure). ATC 3-06 and the NEHRP *Provisions* through its 1991 edition included the effect of the vertical earthquake ground motion in the design load combinations. The 1994 NEHRP *Provisions* modified the definition of *E* to **12.4.2**

include the effect of the vertical earthquake ground motion in addition to the effect of V. The 1997 UBC went a step further and defined E as being made of two parts: the effect of the horizontal earthquake ground motion (same as the effect of the design base shear, V) amplified by a redundancy coefficient, ρ, and the effect of the vertical earthquake ground motion. The 1997 NEHRP *Provisions* adopted the 1997 UBC approach, which in turn was adopted into the 2000 IBC. The 2000 NEHRP *Provisions*, which was adopted into the 2003 IBC and ASCE 7-02, and the 2003 NEHRP *Provisions*, which was adopted into the 2006 IBC and ASCE 7-05, are the same as the 1997 NEHRP *Provisions* in regard to the seismic load effect.

Table 16-29 shows that the consideration of vertical earthquake ground motion, as required by ASCE 7-05 Equations 12.4-1 and 12.4-2, causes an increase in the dead load factor when gravity and earthquake effects are additive (for instance, gravity causes negative bending moments at a particular support section of a particular beam; earthquake forces also cause a negative bending moment at the same support section of the same beam). A decrease in the dead load factor is caused when gravity counteracts earthquake effects (for instance, gravity causes negative bending moments at a particular support section of a particular beam; earthquake forces cause a positive bending moment at the same support section of the same beam). In other words, when gravity effects add to earthquake effects, the direction of vertical earthquake ground motion is taken in the direction of gravity or vertically downward. When gravity counteracts earthquake effects, the direction of vertical earthquake ground motion is taken in the direction opposite to that of gravity or vertically upward.

Table 16-29. Seismic strength design load combinations of ASCE 7-05

Gravity and earthquake effects additive:	
	ASCE 7-05
Required strength	
$U = 1.2D + 1.0E + L + 0.2S$	Eq. 5, Section 2.3.2
$= 1.2D + (\rho Q_E + 0.2S_{DS}D) + L + 0.2S$	Eqs. 12.4-1, 12.4-3 and 12.4-4
$= (1.2 + 0.2S_{DS})D + \rho Q_E + L + 0.2S$	
Gravity and earthquake effects counteractive:	
	ASCE 7-05
Required strength	
$U = 0.9D + 1.0E + 1.6H$	Eq. 7, Section 2.3.2
$= 0.9D + (\rho Q_E - 0.2S_{DS}D) + 1.6H$	Eqs. 12.4-2, 12.4-3 and 12.4-4
$= (0.9 - 0.2S_{DS})D + \rho Q_E + 1.6H$	

The proportioning of shear walls and columns of reinforced concrete as well as masonry is governed by the load combination producing the minimum axial load in compression or the maximum axial load in tension. This is because designing for lower compressive (or higher tensile) axial loads typically causes an increase in the required flexural reinforcement. The governing load combinations of ATC 3-06, the various editions of the NEHRP *Provisions*, the 1997 UBC, the IBC and ASCE 7 are shown in Table 16-30. The dead load factor in these load combinations (denoted as DLF in the following discussion) where the effects of factored gravity counteract the effects of strength-level earthquake forces, is of particular importance—the lower the DLF, the more severe the load combination, and the higher the flexural reinforcement requirement.

It is evident from Table 16-30 that the DLF of ATC 3-06 is a fixed 0.8. In the 1985, 1988 and the 1991 NEHRP *Provisions*, the DLF is equal to $0.9 - 0.5A_v$, the most severe value of which is 0.7 for an A_v of 0.4. In the 1994 NEHRP *Provisions*, the 1997 UBC, and the 1997, 2000 and 2003 NEHRP *Provisions*, as well as the 2000, 2003 and 2006 IBC and the 1998, 2002 and 2005 ASCE 7, the DLFs are $0.9 - 0.5C_a$, $0.9 - 0.5C_aI$ and $0.9 - 0.4F_aS_s/3$, respectively. These values are tabulated for different site classes in Tables 16-31(a), (b) and (c), respectively. The UBC values are for Seismic Zone 4 only. The NEHRP/IBC/ASCE 7 values are for regions of varying seismicity. The UBC shows N_a as the acceleration-related near-source factor. It should be apparent that the consideration of vertical earthquake ground motion is more conservative in the 1997 UBC when compared to the other documents. The DLF values of the 2000, 2003, 2006 IBC/ASCE 7-98, -02, -05 are somewhat less and are more consistent with the values used in the NEHRP *Provisions* through its 1994 edition and in ATC 3-06.

Table 16-30. Consideration of vertical seismic ground acceleration in U.S. codes and resource documents[1,2]

Design Base Shear	
$V = \frac{(1.2A_vS)}{RT^{2/3}}W \le \frac{(2.5A_a)}{R}W$	ATC 3-06, 1985 NEHRP, 1988 NEHRP, 1991 NEHRP, ASCE 7-93
$V = \frac{(1.2C_v)}{RT^{2/3}}W \le \frac{(2.5C_a)}{R}W$	1994 NEHRP, ASCE 7-95
$V = \frac{(C_vI)}{RT}W \le \frac{(2.5C_aI)}{R}W$	1997 UBC
$V = \frac{(S_{D1})}{T(R/I)}W \le \frac{(S_{DS})}{R/I}W$ $\ge 0.044S_{DS}/W$ $\ge \frac{0.5S_1}{R/I}W$ (SDCE, F only) $S_{DS} = 2S_{MS}/3 = 2F_aS_S/3$ $S_{D1} = 2S_{M1}/3 = 2F_vS_1/3$	1997 NEHRP, ASCE 7-98, 2000 IBC, 2000 NEHRP, ASCE 7-02, 2003 IBC
$V = \frac{(S_{D1})}{T(R/I)}W \le \frac{S_{DS}}{R/I}W \ (T \le T_L)$ $V = \frac{S_{D1}}{T^2(R/I)}W \le \frac{S_{DS}}{R/I}W \ (T \le T_L)$ $\ge 0.01W$ $\ge \frac{0.5S_1}{R/I}W \ (S_1 \ 0.6g)$ $S_{DS} = 2S_{MS}/3 = 2F_aS_S/3$ $S_{D1} = 2S_{M1}/3 = 2F_vS_1/3$	2003 NEHRP, ASCE 7-05, 2006 IBC

(Continued)

Table 16-30. Consideration of vertical seismic ground acceleration in U.S. codes and resource documents[1,2] (Cont'd)

Required Strength	
$U = 0.8Q_D \pm 1.0Q_E$	ATC 3-06
$U = (0.9 - 0.5A_V)Q_D \pm Q_E$	1985, 1988, 1991 NEHRP
$E = \pm 1.0Q_E - 0.5A_VD$ $U = 0.9D + 1.0E$ $= (0.9 - 0.5A_V)D \pm 1.0Q_E$	ASCE 7-93
$E = Q_E \pm 0.5C_aD$ $U = 0.9D + 1.0E$ $= (0.9 - 0.5Ca)D + Q_E$	1994 NEHRP, ASCE 7-95
$E = \rho E_h + 0.5C_aID$ $U = 0.9D - 1.0E$ $= (0.9 - 0.5C_a)ID - \rho E_h$	1997 UBC
$E = \rho Q_E - 0.2S_{DS}D$ $U = 0.9D + 1.0E$ $= (0.9 - 0.2S_{DS})D + \rho Q_E$ $= (0.9 - 0.4F_aS_S/3)D + \rho Q_E$	1997 NEHRP, ASCE 7-98, 2000 IBC, 2000 NEHRP, ASCE 7-02, 2003 IBC, 2003 NEHRP, ASCE 7-05, 2006 IBC

Notes:

1. In the 1997 UBC, E_h is the effect of V. In all other documents, the effect of V is Q_E.
2. Load combinations with overstrength are not considered in the table.

Table 16-31(a). 1994 NEHRP Dead Load Factors, including effects of vertical ground acceleration—DLF = 0.9 – 0.5 C_a

Site Class	$A_a = \leq 0.05$		$A_a = 0.1$		$A_a = 0.2$		$A_a = 0.3$		$A_a = > 0.4$	
	C_a	DLF	C_a	DLF	C_a	DLF	C_a	DLF	C_a	DLF
A	0.04	0.88	0.08	0.86	0.16	0.82	0.24	0.78	0.32	0.74
B	0.05	0.88	0.10	0.85	0.20	0.80	0.30	0.75	0.40	0.70
C	0.06	0.87	0.12	0.84	0.24	0.78	0.33	0.74	0.40	0.70
D	0.08	0.86	0.16	0.82	0.28	0.76	0.36	0.72	0.44	0.68
E	0.13	0.84	0.25	0.78	0.34	0.73	0.36	0.72	0.36	0.72

Table 16-31(b). 1997 UBC Dead Load Factors including effects of vertical ground acceleration, for Seismic Zone 4—DLF = 0.9 – 0.5 C_aI

Site Class	C_a	$N_a = 1$		$N_a = 1.2$		$N_a = 1.3$		$N_a = 1.5$	
		$I = 1.0$	$I = 1.25$	$I = 1.0$	$I = 1.25$	$I = 1.0$	$I = 1.25$	$I = 1.0$	$I = 1.25$
S_A	$0.32N_a$	0.74	0.70	0.71	0.66	0.69	0.64	0.66	0.60
S_B	$0.40N_a$	0.70	0.65	0.66	0.60	0.64	0.58	0.60	0.53
S_C	$0.40N_a$	0.70	0.65	0.66	0.60	0.64	0.58	0.60	0.53
S_D	$0.44N_a$	0.68	0.63	0.64	0.57	0.61	0.54	0.57	0.49
S_E	$0.36N_a$	0.72	0.68	0.68	0.63	0.67	0.61	0.63	0.56

Table 16-31(c). **1997, 2000, 2003 NEHRP / 2000, 2003, 2006 IBC / ASCE 7-98, ASCE 7-02, ASCE 7-05 Dead Load Factors, including effects of vertical ground acceleration—DLF = 0.9 - 0.4 $F_aS_S/3$**

Site Class	S_s = 0.25		S_s = 0.5		S_s = 0.75		S_s = 1.00		S_s = 1.25	
	F_a	DLF	F_a	DLF	F_a	DLF	F_a	DLF	F_a	DLF
A	0.8	0.873	0.8	0.847	0.8	0.82	0.8	0.793	0.8	0.767
B	1.0	0.867	1.0	0.833	1.0	0.80	1.0	0.767	1.0	0.733
C	1.2	0.860	1.2	0.820	1.1	0.79	1.0	0.767	1.0	0.733
D	1.6	0.847	1.4	0.807	1.2	0.78	1.1	0.753	1.0	0.733
E	2.5	0.817	1.7	0.787	1.2	0.78	0.9	0.780	0.9*	0.750

*1997 NEHRP/2000 IBC/ASCE 7-98 required site specific geotechnical investigation and dynamic site response analysis; no factor was specified.

Seismic load effect including overstrength factor. As mentioned before, ASCE 7-05 Section 12.3.3.3 requires that elements supporting discontinuous walls or frames in structures assigned to SDC B and higher and with certain irregularity types must have the design strength to resist the seismic load combinations with overstrength factors given in this section. ASCE 7-05 Section 12.3.3.4 likewise requires that diaphragm collector elements, splices and their connections to resisting elements have the design strength to resist the seismic loads with overstrength. The term E_m, maximum seismic load effect, involved in the load combinations with overstrength factor is defined in this section. **12.4.3**

In the maximum seismic load effect, E_m, given by ASCE 7-05 Equations 12.4-5 and 12.4-6, the effect of the horizontal earthquake ground motion, Q_E, which is the same as the effect of the design base shear, V, is amplified by the overstrength factor, Ω_0, of Table 12.2-1, which varies between 2 and 3. In the strength-level seismic load effect, E, given by Equations 12.4-1 and 12.4-2, Q_E, the effect of the design base shear V, is amplified by the redundancy coefficient, ρ, which ranges between 1 and 1.5. This is the sole difference between the two. The consideration of vertical earthquake ground motion is identical in both cases. It may be noted that the consideration of vertical earthquake ground motion was absent in the maximum earthquake effect term, and hence, in the special load combinations of the 1997 UBC. There, E_m was defined simply as equal to $\Omega_0 E_h$, E_h of the UBC and Q_E of ASCE 7 being equivalent terms.

Note that the load combinations with overstrength factor of ASCE 7-05 Section 12.4.3.2 are similar to the special seismic load combinations of 2006 IBC Section 1605.4 and have the same applicability. The two sections, however, are in conflict. First, the special seismic load combinations of 2006 IBC Section 1605.4 do not include the effect of the design snow load, S, while the load combinations with overstrength factor in ASCE 7-05 Section 12.4.3.2 do include the effect of the design snow load, S. Second, 2006 IBC Section 1605.4 contains strength design load combinations only, while ASCE 7-05 Section 12.4.3.2 also contains ASD load combinations with overstrength. There is an allowable stress increase provided for in ASCE 7-05 Section 12.4.3.3, which goes with the ASD load combinations with overstrength of ASCE 7-05 Section 12.4.3.2. In such situations of conflict between the 2006 IBC and ASCE 7-05, 2006 IBC Section 102.4 would normally dictate that the 2006 IBC provisions would apply. However, a change already included in the 2007 Supplement to the 2006 IBC has deleted the special seismic load combinations of 2006 IBC Section 1605.4 in favor of a reference to ASCE 7-05 Section 12.4.3. This change will also be part of the 2009 IBC. Thus, 2006 IBC Section 102.4 will remain the governing provision, but only until a jurisdiction adopts the 2007 Supplement or the 2009 IBC.

Direction of loading. According to this section, the seismic forces on a structure must be applied in a direction so that the most critical load effects are produced in each component. However, how the forces need to be applied varies for structures assigned to different SDCs. For SDC B structures, it is permitted to apply the seismic forces separately and **12.5**

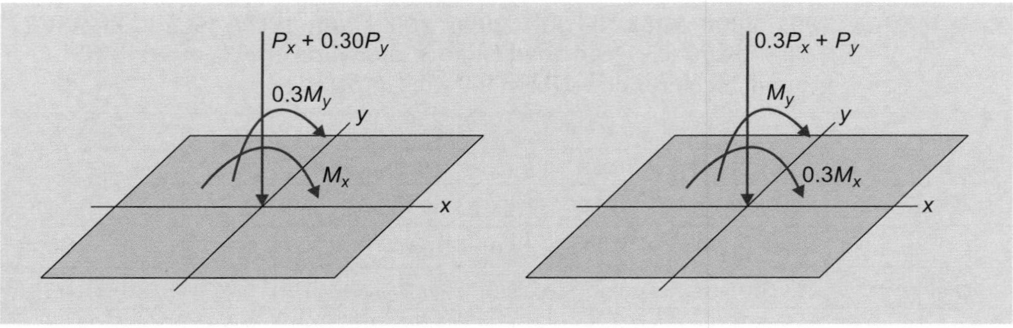

Figure 16-43 Combination of effects of independently applied orthogonal ground motion

independently in each of two orthogonal directions. The same is true for structures assigned to SDC C as well, but with restrictions. When the lateral force-resisting systems are not parallel to or symmetric about the major orthogonal axes of the structure (that is, when Type 5 Horizontal Irregularity exists), an analysis considering orthogonal load effects is required. Two options are provided for this purpose and any one of them can be used. ASCE 7-05 Section 12.5.3, Item a, requires that the seismic forces be applied in two orthogonal directions of a structure independently and elements of the structure be designed for the most severe effect of 100 percent of the seismic forces applied on the structure in one direction plus 30 percent of the seismic forces applied on the structure in the perpendicular direction (Figure 16-43). The foregoing is based on the observation that while earthquake forces act in both principal directions of a building simultaneously, the earthquake effects in the two principal directions are unlikely to reach their maximum values simultaneously. This method provides a reasonable and adequate method of combining the effects. In the second method, ASCE 7-05 Section 12.5.3, Item b, requires that three-dimensional linear or nonlinear dynamic analysis be performed with the simultaneous application of orthogonal pairs of ground motion acceleration time histories.

Orthogonal effects are slight for beams, girders, slabs and other horizontal elements that are essentially one-dimensional in their behavior. They may be significant in columns or other vertical members that participate in resisting earthquake forces in both principal directions of a building. The 2003 NEHRP Commentary advises that for two-way slabs, orthogonal effects at slab-to-column connections can be neglected, provided the moment transferred in the minor direction does not exceed 30 percent of that transferred in the orthogonal direction and there is adequate reinforcement within lines one and one-half times the slab thickness on either side of the column to transfer all the minor direction moment.

12.5.4 Seismic design categories D through F. The requirements are the same as those for SDC C structures, except for an additional requirement for any column or wall that forms part of two or more intersecting seismic force-resisting systems and that is subjected to an axial load due to lateral seismic forces equal to at least 20 percent of its design axial strength. These members are to be designed for the most critical loads obtained after considering the orthogonal load effects by following either Item a or Item b described in the previous section. For structures with flexible diaphragms, 2-dimensional analysis is permitted, unless the structure has any torsional irregularity (horizontal irregularity type 1a or 1b), out-of-plane offsets (horizontal irregularity type 4) or nonparallel lateral force-resisting systems (horizontal irregularity type 5).

12.6 Analysis procedure selection. This section describes the structural analyses permitted as the basis of determining E (the combined effect of horizontal and vertical earthquake-induced forces) and E_m (the maximum seismic load effect). The design load combinations of IBC Section 1605 and ASCE 7-05 Chapter 2 involve the terms E and E_m. Both terms are defined in part by Ω_E (the effect of horizontal seismic forces), the determination of which requires structural analysis in accordance with the requirements of this section. Such analysis also leads to the determination of the design story drift, Δ, which must be kept within the limits prescribed in ASCE 7-05 Section 12.12.

The Commentary to the 2003 NEHRP *Provisions* (Section 4.4.1) lists the standard procedures for the analysis of forces and deformations in structures subjected to earthquake ground motion, in the order of expected rigor and accuracy, as follows:

1. Equivalent lateral force procedure (ASCE 7-05 Section 12.8);
2. Modal analysis procedure (response spectrum analysis) (ASCE 7-05 Section 12.9);
3. Inelastic static procedure, involving incremental application of a pattern of lateral forces and adjustment of the structural model to account for progressive yielding under load application (push-over analysis); and,
4. Inelastic response history analysis involving step-by-step integration of the coupled equations of motion (ASCE 7-05 Section 16.2).

ASCE 7 chose to include a simplified analysis procedure (Section 12.14) that, in the order of expected rigor and accuracy, would precede Item 1 above. Push-over analysis is not formally recognized in the IBC or in ASCE 7-05, although it is high among the recognized analysis procedures in the FEMA 356 *Seismic Rehabilitation Guidelines.*[26]

ASCE 7-05 also recognizes an elastic or linear time history analysis (Section 16.1), which in the order of expected rigor and accuracy, would probably rank the same as Item 2 above. It should be recognized that Items 1 and 3 above and the simplified analysis procedure of Section 12.14 are static procedures, while the other analysis procedures mentioned are dynamic procedures. The philosophy underlying this section is that dynamic analysis is always acceptable for design. Static procedures are allowed only under certain conditions of regularity, occupancy and height.

The equivalent lateral force procedure and the modal analysis procedure are both based on the approximation that inelastic seismic structural response can be adequately represented by linear analysis of the lateral force-resisting system using the design spectrum, which is the elastic acceleration response spectrum amplified by the importance factor, I, and reduced by the response modification factor, R. The effects of the horizontal component of ground motion perpendicular to the direction of analysis, the vertical component of ground motion, and torsional motions of the structure are all considered in the same approximate manner in both cases, if only two-dimensional analysis is used. The main difference between the two procedures lies in the distribution of the seismic lateral forces over the building. In the modal analysis procedure, the distribution is based on the deformed shapes of the natural modes of vibration, which are determined from the distribution of the masses and the stiffnesses of the structure. In the equivalent lateral force procedure, the distribution is based on simplified formulas that are appropriate for regular structures (ASCE 7-05 Section 12.8.3). Otherwise, the two procedures are subject to the same limitations. The total design forces used in the two procedures are also very similar (see ASCE 7-05 Section 12.9.4).

According to the 2003 NEHRP Commentary (Section 5.1), the equivalent lateral force procedure and the modal analysis procedure "are all likely to err systematically on the unsafe side if story strengths are distributed irregularly over height. This feature is likely to lead to concentration of ductility demand in a few stories of the building. The inelastic static (or the so-called push-over) procedure is a method to more accurately account for irregular strength distribution. However, it also has limitations and is not particularly applicable to tall structures or structures with relatively long fundamental periods of vibration."

The least rigorous analytical procedure that may be used in determining the design seismic forces and deformations in a structure depends on the seismic design category and the structural characteristics (in particular, regularity and height). Structural irregularity was discussed under Section 12.3.2.

Table 12.6-1 in ASCE 7-05 clearly summarizes the permitted analytical procedures as a function of the SDC. For Seismic Design Categories B and C, the equivalent lateral force procedure is the minimum level of analysis, except when the simplified analysis procedure of Section 12.14 is applicable. A more rigorous analysis is always permitted. For Seismic Design Categories D, E and F, the least rigorous analysis procedures that may be used are identified in Table 12.6-1. Note that the equivalent lateral force procedure is not permitted to be used for regular structures and for some types of irregular structures where the fundamental computed period, T, is greater than or equal to $3.5T_S = 3.5S_{D1}/S_{DS}$. Historically, the equivalent lateral force procedure has been limited in application in these seismic design categories to regular structures with heights of 240 ft (70 m) or less and irregular structures with heights of 100 feet (30 m) or less. In the 2000 NEHRP *Provisions*, a change in the height limit for regular structures to 100 feet (30 m) was contemplated, acknowledging that the base shear equation with a $1/T$ relationship underestimated the response of structures with significant higher mode participation. It was decided to limit the application of the equivalent lateral force procedure to structures in SDC D, E and F having fundamental periods of response less than 3.5 times the period at which the response spectrum makes a transition from constant response acceleration to constant response velocity. In this way, the importance of higher mode participation in structural response, which is a function of the structure's dynamic properties (height, mass and stiffness of lateral force-resisting elements) and the frequency content of the ground shaking (as represented by the response spectrum) are accounted for when choosing the required analysis method. Assuming that $S_{D1}/S_{DS} = 0.5$, then the limiting period is equal to $3.5 \times 0.5 = 1.75$ seconds, which is approximately the period of a 17 or 18 story frame building. Dynamic analysis is required for such structures to determine the possible effects of higher mode response on force distribution and deformations. In the other cases cited in Table 12.6-1, dynamic analysis is required because the noted irregularities invalidate the assumptions applicable for equivalent static force analysis.

Special consideration of dynamic characteristics in equivalent static analysis is required for a building assigned to SDC D, E or F, if the building has one or more of the plan irregularities of Type 2, 3, 4 or 5, and the building has one or more of the vertical irregularities of Type 4, 5a or 5b. The Commentary to the 2003 NEHRP *Provisions* suggests a procedure for determining whether the modal analysis procedure is needed for a particular structure.

12.7 Modeling criteria. Current professional practice and computational capabilities may lead to the choice of a three-dimensional model for both static and dynamic analyses. Although three-dimensional models are not specifically required for regular structures with independent orthogonal seismic force-resisting systems, they often have important advantages over two-dimensional models.

A three-dimensional model is appropriate for the analysis of torsional effects (actual plus accidental), diaphragm deformability and systems having nonrectangular plan configurations. According to the commentary to the 1999 SEAOC *Blue Book*, when a three-dimensional model is needed for any purpose, "it can also serve for all required loading conditions, including seismic loading in each principal direction, other selected directions and for orthogonal effects."

The actual strength and other properties of the various components of a structure can be explicitly considered only by a nonlinear analysis of dynamic response by direct integration of the coupled equations of motion. If the two translational motions and the torsional motion are expected to be essentially uncoupled, it is sufficient to include only one degree of freedom per floor, in the direction of analysis; otherwise at least three degrees of freedom per floor, two translational and one torsional, need to be included. The 2003 NEHRP Commentary points out—and it cannot be overemphasized—that the results of nonlinear response history analysis of mathematical structural models are only as good as the models chosen to represent the structure vibrating at amplitudes of motion large enough to cause significant yielding at several locations. Proper modeling and proper interpretation of results require background and experience. Also, reliable results can be achieved only by calculating the response to several ground motions—recorded accelerograms and/or simulated motions—and examining the statistics of response.

The effective seismic weight that needs to be included in the model is the total weight of the building and that part of the other gravity loads that might reasonably be expected to be acting on the building at the time of an earthquake. It includes permanent and movable partitions and permanent equipment such as mechanical and electrical equipment, piping and ceilings. "The normal human live load is taken to be negligibly small in its contribution to the seismic lateral forces."[2] Buildings intended for storage or warehouse occupancy must have at least 25 percent of the design floor live load included in *W*. The inclusion of a part of the design snow load in *W* brings up one of the few differences between the requirements of the 2003 NEHRP *Provisions* and those of the IBC.

According to the 2003 NEHRP *Provisions*, in areas where the design flat roof snow load does not exceed 30 psf (1.4 kPa), the effective snow load is permitted to be taken as zero. In areas where the design snow load is greater than 30 psf (1.4 kPa) and where siting and load duration considerations warrant and when approved by the authority having jurisdiction, the effective snow load is permitted to be reduced to no less than 20 percent of the design snow load. The 2003 NEHRP Commentary explains that freshly fallen snow would have little effect on the lateral force in an earthquake, but that ice loading would be more or less firmly attached to the roof of the building and would contribute significantly to the inertia force. For this reason, the effective snow load is taken as the full design snow load for those regions where the snow load exceeds 30 psf, with the provision that the local authority having jurisdiction may allow the snow load to be reduced up to 80 percent. The question of how much snow load should be included in *W* is really a question of how much ice buildup or snow entrapment can be expected for the roof configuration or site topography, and this is a question the 2003 NEHRP *Provisions* leaves to the discretion of the local authority having jurisdiction. ASCE 7-05, however, requires the inclusion of a fixed 20 percent of the flat roof snow load in *W* where the flat roof snow load exceeds 30 psf. When the flat roof snow load is lower, no portion of it need to be included in *W*.

Because moment frames have relatively lower stiffness compared to other seismic force-resisting-systems, rigid elements that are not part of the moment frame may have a significant effect on the total stiffness of the structure and its dynamic behavior. As a result, ASCE 7-05 requires that the effect of these elements be accounted for in the analysis and design.

Equivalent lateral force procedure. The following discussion is on the provisions of Section 12.8. **12.8**

Seismic base shear. The design base shear, as given by ASCE 7-05 Equation 12.8-1, is the starting point in seismic design by the equivalent lateral force procedure. The design base shear is given as a seismic response coefficient, C_s, times the effective seismic weight of the structure, *W*.

Calculation of seismic response coefficient. The seismic response coefficient, C_s, as defined by ASCE 7-05 Equations 12.8-2, 12.8-3 and 12.8-4 describes the design response spectrum of ASCE 7-05 Figure 11.4-1, with two modifications. First, the *ramp* building up to the *flat-top* of the design spectrum is excluded from consideration under Section 12.8.1.1. This is because the equivalent lateral force procedure deals with fundamental mode response. The period of the first mode of a practical building structure is unlikely to be in the very short period range, $T < 0.2T_S$, where the ramp would have been required. In other words, the ramp is excluded from consideration because it is unlikely to be required. Second, the seismic response coefficient defined by Equations 12.8-2 through 12.8-4 is the spectral response acceleration of Figure 11.4-1, amplified by the occupancy importance factor, *I*, and reduced by the response modification factor, *R*.

Equation 12.8-2 represents the constant acceleration portion of the spectrum of Figure 11.4-1, Equation 12.8-3 represents the constant velocity portion and Equation 12.8-4 represents the constant displacement portion. The design force level defined by Equations 12.8-1 through 12.8-4 is based on the assumption that a structure will undergo several cycles of inelastic deformation during major earthquake ground motion and, therefore, the level is related to the type of structural system and its estimated ability to sustain these deformations and dissipate energy without collapse. The force level defined by Equations 12.8-1 through 12.8-4 is used not only for the static lateral force procedure, but also as the lower bound for the modal analysis procedure of ASCE 7-05 Section 12.9. Thus, it is worthwhile to discuss the physical relationship between the design base shear and the spectral representation of major earthquake ground motion, as well as the idealization of the inelastic behavior of the structure that is due to this ground motion.

For a given fundamental period, T, the design base shear Equations 12.8-2 through 12.8-4 provide values that are equal to the acceleration response spectrum ordinate as given in Figure 11.4-1, times the total structure weight, W, multiplied by the importance factor, I, and divided by R. Although only the fundamental mode period is employed, the additional response that is due to higher modes of vibration is represented and approximated by use of the total weight, W. Appendix D to the 1999 SEAOC *Blue Book* points out that the first or fundamental mode base shear is the effective mass of the first mode times the first mode spectral ordinate, S_a. This effective mass is on the order of 0.7 times the total mass, W/g. Therefore the use of W in the base shear equations intends to provide an upper-bound value of the combined mode (SRSS or CQC) base shear response.

Descending branch of the design spectrum. In the 2006 IBC and ASCE 7-05, the elastic acceleration response spectrum for earthquake ground motion has a descending branch for longer values of T that varies roughly as $1/T$. In the NEHRP *Provisions* prior to the 1997 edition and in the UBC prior to its 1997 edition, the actual response spectrum that varies in a $1/T$ relationship was replaced with a design spectrum that varied in a $1/T^{2/3}$ relationship. This was intentionally done to provide added conservatism in the design of tall structures. For the 1997, 2000 and 2003 NEHRP *Provisions*, the true shape of the response spectrum, represented by a $1/T$ relationship, was maintained in the base shear equations. The 1997 NEHRP Commentary explained, "In order to maintain the added conservatism for tall and high occupancy structures, formerly provided by the design spectra which utilized a $1/T^{2/3}$ relationship, the 1997 *Provisions* adopted an occupancy importance factor, I, into the base shear equation. This I-factor, which has a value of 1.25 for SUG II (Occupancy Category III) structures and 1.5 for SUG III (Occupancy Category IV) structures, has the effect of raising the design spectrum for taller, high-occupancy structures to levels comparable to those for which they were designed in previous editions of the *Provisions*." This explanation appears to be less than tenable. The 1999 SEAOC *Blue Book* Commentary offers the right explanation with regard to the 1997 UBC, which also made the descending branch of the design spectrum a function of $1/T$ rather than $1/T^{2/3}$. To paraphrase the explanation for the IBC, in view of the use of the total weight W (in the design base shear equations), along with appropriate S_{DS} and S_{D1} values, and the long-period part of the spectrum, Equation 12.8-4 of ASCE 7-05, the use of the $1/T$ version of the base shear equation provides reasonable parity with prior requirements that used the $1/T^{2/3}$ relation to represent multimode response.

Response modification factor, *R*. The factor, R, in the denominator of Equations 12.8-2 through 12.8-4 is an empirical response reduction factor intended to account for both the damping and the ductility (or inelastic deformability) inherent in a structural system at displacements large enough to surpass initial yield and approach maximum inelastic response displacements.[2] The factor, R, is also intended to account for overstrength, which is partly material-dependent and partly system-dependent. Because design force levels are based on first yield of the highest stressed elements of a system, the maximum force level that the system can resist after the formation of successive hinges, bracing yield, or shear wall yield or cracking is significantly higher than the initial yield value. Designs are also based on minimum expected yield or strength values, whereas the average strength of a material could be significantly higher. The 1999 SEAOC *Blue Book* divided R into two portions: R_o and R_d. While R_o represented the overstrength-dependent part of the R-factor, R_d represented the part dependent upon damping and ductility.

As pointed out in the Commentary to the 2003 NEHRP *Provisions*, for a lightly damped building structure of brittle material that would be unable to tolerate any appreciable deformation beyond the elastic range, the factor, R would be close to 1.0 (that is, no reduction from the force level corresponding to linear elastic response would be allowed). At the other extreme, a heavily damped building structure with a very ductile structural system would be able to withstand deformation considerably in excess of initial yield and would, therefore, justify the assignment of a relatively large response modification factor,

R. ASCE 7-05 Table 12.2-1 stipulates *R*-factors for different types of structural systems using several different construction materials. The coefficient, *R,* ranges in value from a minimum of 1.5 for a bearing wall system consisting of ordinary plain masonry or concrete shear walls to a maximum of 8.0 for a special moment frame system or a dual system consisting of special moment frames of steel. The basis for the *R*-factor values specified in Table 12.2-1 is explained in 2003 NEHRP Commentary Section 4.2.1. See earlier discussion of Section 12.2 in this publication.

Modification by importance factor, *I*. The effective value of *R* used in the base shear equations is adjusted by the occupancy importance factor, *I*. The *I*-value, which ranges between 1.0 and 1.5, has the effect of reducing the amount of ductility a structure will be called upon to provide at a given level of ground shaking, thus also reducing the damage inflicted by the same level of ground motion. However, as pointed out in the 2003 NEHRP Commentary, added strength by itself is not sufficient to produce superior seismic performance in buildings with critical occupancies. "Good connections and construction details, quality assurance procedures, and limitations on building deformation are also important to significantly improve the capability for maintenance of function and safety in critical facilities and those with high-density occupancy."[2] Consequently, the reduction in the damage potential of critical facilities is also handled by using more conservative drift controls (ASCE 7-05 Section 12.12) and by providing special design and detailing requirements (ASCE 7-05 Chapter 12) and materials limitations (2006 IBC Chapters 19, 21, 22 and 23).

Minimum base shear. Following the Long Beach earthquake of 1933, the Riley Act in California required every building to be designed for a minimum service-level seismic force equal to 3 percent of its weight. This minimum requirement eventually became seismic zone-dependent in the UBC. In UBC editions prior to 1997, the minimum was $0.075ZIW$ (service-level) where Z was the Seismic Zone factor (0.4, 0.3, 0.2, 0.15 and 0.075 for Zones 4, 3, 2B, 2A and 1, respectively), I was the importance factor and W was the effective seismic weight. In the 1997 UBC, this minimum design base shear was brought up to strength level and was made soil-dependent; it became: $1.4(0.075)C_aIW = 0.11C_aIW$ where C_a was a function of Z and of the soil characteristics at the site of the structure. The NEHRP *Provisions* prior to the 1997 edition did not include a lower bound on the design base shear. However, in the 1997 edition, in view of the fact that the descending branch of the design spectrum now varied with $1/T$ rather than $1/T^{2/3}$, a minimum value in terms of the design spectral acceleration at 1-second period, S_{D1}, was added. The minimum design base shear of the 1997 NEHRP *Provisions* was $0.1S_{D1}IW$, while in the 2000 NEHRP *Provisions* it became $0.044S_{DS}IW$, which is actually the same as the 1997 UBC minimum if the soil-modified spectral accelerations at short periods and 1-second period in the NEHRP *Provisions* and the 1997 UBC are related as $S_{DS} = 2.5C_a$ and $S_{D1} = C_v$. The minimum design base shear of ASCE 7-02 is also equal to $0.044S_{DS}IW$ (Eq. 9.5.5.2.1-3). This minimum value was included in view of the uncertainty and the lack of knowledge of actual structural response of long-period buildings subject to earthquake ground motion. As mentioned earlier, ASCE 7-05 has introduced a long-period segment in its design spectrum to account for the dynamic response of long-period buildings, and consequently, this minimum base shear has been dropped. The only minimum base shear that ASCE 7-05 still has for all buildings is 1 percent of the seismic weight, which is simply an SDC A minimum based on structural integrity considerations.

In a recent development, however, the general minimum of $0.044S_{DS}IW$ is again being brought back in Supplement II of ASCE 7-05. Results from the Draft ATC-63, *Quantification of Building System Performance and Response Parameters* (read more about ATC-63 project within the discussion on ASCE 7-05 Section 12.2 Structural System Selection) indicate that design base shear in tall buildings may come out to be unacceptably low unless the minimum base shear is increased to the value used in ASCE 7-02.

Following the Northridge earthquake of 1994, a second lower-bound on the design base shear, applicable in Seismic Zone 4 only, was added to the 1997 UBC. This second minimum was in terms of Z, N_v, I and R where N_v was the velocity-dependent near-source factor, and was specifically intended to account for the large displacement and velocity pulses that were observed in near-fault ground motion in the Northridge earthquake. A corresponding minimum was added to the 1997 NEHRP *Provisions*. This minimum was originally applicable to structures assigned to SDC E and F only. In the 2000 IBC, this equation was also applicable to all structures located where the mapped MCE spectral response acceleration at 1-second period, S_1, equaled or exceeded $0.6g$, which is equivalent to Zone 4 of the 1997 UBC. ASCE 7-05 also requires the same, although the 2003 IBC had gone back to SDC E and F only.

Table 16-32 shows a comparison of the design base shear equations of ASCE 7-05 and the 1997 UBC. It can be seen that the two sets of expressions are equivalent, if the following conversions are made: $S_{D1} = C_v$, and $S_{DS} = 2.5C_a$, with the above mentioned exceptions that the 1997 UBC did not have the constant-displacement segment in its design spectrum as ASCE 7-05, and that the UBC minimum base shear that was applicable to all structures is very different from the corresponding ASCE 7-05 minimum (until Supplement II of ASCE 7-05 is published).

Table 16-32. Comparison of design base shear equations of ASCE 7-05 and the 1997 UBC

ASCE 7-05	1997 UBC
$V = \frac{S_{D1}T_L}{T^2(R/I)}W \le \frac{S_{D1}}{T(R/I)}W \le \frac{S_{DS}}{R/I}W$	$V = \frac{C_V}{(R/I)T}W \le \frac{2.5C_a}{(R/I)}W$
$\ge 0.01W$	$\ge 0.11C_aW$
$\ge \frac{0.5S_1}{R/I}W$, where $S_1 \ge 0.6g$	$\ge \frac{32N_V}{(R/I)}W$, in Seismic Zone 4
$S_{DS} = 2/3\ S_{MS} = 2/3\ F_aS_s$ $S_{D1} = 2/3\ S_{M1} = 2/3\ F_vS_1$	

Period determination. In the denominator of Equations 12.8-3 and 12.8-4, T is the fundamental period of vibration of the building. It is preferable that this be determined using the structural properties and deformational characteristics of the resisting elements in a properly substantiated analysis. If a mathematical model has been formulated for the modal analysis procedure of Section 12.9, then the period of the first mode of vibration in a given principal direction can be used for this T-value.

Unfortunately, the methods of structural mechanics cannot be employed to calculate the period of vibration before a building has been designed. Consequently, this section provides an approximate method that can be used to estimate building period, with minimal information available on the building design. It is based on the use of simple formulas that involve only a general description of the building type (such as steel moment frame, concrete moment frame, etc.) and the overall height or number of stories.

Rationally computed period is a function of modeling assumptions and is dependent in particular on stiffness assumptions. The smaller the assumed stiffness, the longer the rationally computed period, which translates directly into a lower design base shear. Rationally computed period is thus open to possible abuse. Although it would have been preferable to prevent such possible abuse by specifying fairly rigid modeling rules, it has so far not been possible to forge consensus behind a complete set of rules to meet all possible variations and conditions encountered in design. The 2000 NEHRP *Provisions* and the IBC, just as their predecessor documents and recent editions of the UBC, chose to impose direct control on rationally computed period. For design purposes, it may not be taken any larger than a coefficient, C_u, times the approximate period given in Section 12.8.2.1. The 1999 SEAOC *Blue Book* Commentary advises that reasonable mathematical rules should be followed such that the increase in period allowed by the C_u coefficient is not taken advantage of when the structure does not merit it.

Table 16-33 expresses the restriction on rationally computed period in terms of design base shear as long as it is proportional to $1/T$ (i.e., $T_s = S_{D1}/S_{DS} = T = T_L$). For example, when T is restricted to be no more than 1.4 times T_a, the design base shear based on T is restricted to be no less than 71.4 percent of the design base shear based on the approximate period of Section 12.8.2.1. Note, importantly, that Section 12.8.6.2 on story drift determination specifies that "for determining compliance with the story drift limits of Section 12.12.1, it is permitted to determine the elastic drifts, (δ_{xe}), using seismic design forces based on the computed fundamental period of the structure without the upper limit (C_uT_a) specified in Section 12.8.2."

It may be noted that larger values of C_u are permitted as the soil-dependent seismic risk of a location decreases. This is because buildings in areas with lower lateral-force requirements are thought likely to be more flexible. Higher values of C_u for lower values of S_{D1} also result in less dramatic changes from prior practice in lower risk areas. As pointed out in the 2003 NEHRP *Provisions* Commentary, it is generally accepted that the equations for T_a are tailored to fit the types of construction common in areas with high lateral-force requirements. It is unlikely that buildings in lower seismic risk areas would be designed to produce as high a drift level as allowed by the 2003 NEHRP *Provisions* or the IBC or ASCE 7, because of stability (*P*-delta) considerations and wind requirements. For buildings with design *controlled* by wind, the use of a large T will not really result in a lower design force.

Table 16-33. Restriction on rationally computed period expressed in terms of design base shear for $T_s = T = T_L$

Design Spectral Response Acceleration at 1-Second Period, S_{D1}	Coefficient C_u = Max. T/T_a	Min. V/V_a
≤ 0.4	1.4	0.714
0.3	1.4	0.714
0.2	1.5	0.667
0.15	1.6	0.625
0.1	1.7	0.588
≤ 0.05	1.7	0.588

Note: V = Design base shear based on rationally computed period
V_a = Design base shear based on approximate period

In the 1997 NEHRP *Provisions*, C_u ranged from 1.2 in zones of high seismicity to 1.7 in zones of low seismicity. In the 2000 NEHRP *Provisions*, C_u in zones of high and medium seismicities were increased to their present values based on an investigation involving extensive instrumentation of real buildings. The results were used to revise the approximate period equations as well the upper limits to the rationally computed periods. This has been described in the next section.

Approximate fundamental period. The approximate fundamental period needs to be conservative in the sense that if this period estimate is later not refined through rational period determination, the resulting design should still be safe. Thus, the approximate period needs to be smaller than the true period of a building.

Taking the seismic base shear to vary as $1/T$ and assuming that the lateral forces are distributed linearly over the height and that deflections are controlled by drift limitations, a simple determination of the period of vibration by Rayleigh's method[28] leads to the conclusion that the period of vibration of moment-resisting frame structures varies roughly with $h_n^{3/4}$ where h_n equals the total height of the building (height above base to the highest level of building). This form of the period was therefore adopted into ATC 3-06[32], where it was originally applicable only to moment-resisting frame systems. Later modifications made it applicable to systems with shear walls and bracing elements. ATC 3-06 originally gave values of C_t equal to 0.035 and 0.025 for steel and concrete frames, respectively. The data upon which the ATC 3-06 values were based were re-examined for concrete frames, and the 0.030 value was judged to be more appropriate. The C_t values for steel and concrete moment frames remained 0.035 and 0.030, respectively, in the 1997 NEHRP *Provisions*, the 1997 UBC and the 2000 IBC.

The values for C_t given are intended to be reasonable lower bound (not mean) values for structures designed according to the 2000 IBC. Surveys and studies of the particular buildings that provided the period data for the ATC 3-06 equations (Bertero et al., 1988) have shown that these original equations, even with the modified $C_t = 0.030$ for concrete frames, provide predictions that are about 80 to 90 percent of the lower bound values of measured periods at deformation values near first yield of the structural elements.

Although this might indicate a large, perhaps excessive degree of conservatism, it is important to recognize that the buildings involved were designed for lateral force requirements prior to those of the 1976 UBC, which were significantly lower than those in the 1976 UBC, as well as those given in more recent seismic codes. Furthermore, the controls on interstory drift for all elements, on irregularity and member detailing provisions, as given in recent seismic codes, are generally more restrictive than those used for the buildings in the period evaluation study. Therefore, given reasonably similar nonstructural elements, the population of structures that conform to the drift provisions of the IBC will have increased stiffness and corresponding lower period values than the structures designed according to previous codes.

According to the 2000 NEHRP *Provisions* Commentary, in recent years, a large number of strong motion instruments have been placed in buildings located within zones of high seismic activity. This program, operated by the California Division of Mines and Geology and the United States Geological Survey (USGS), has allowed a substantial inventory of additional recordings of building response to strong ground shaking to be obtained and the fundamental period of vibration of the buildings to be calculated. Evaluation of the data from this expanded base indicated that adjustments to the approximate period formulas were warranted. The basis for the adjusted formulas was developed in a study by Goel and Chopra.[44] Figures C5.4.2.1-1, C5.4.2.1-2 and C5.4.2.1-1 in the 2000 NEHRP Commentary show plots of these data for moment-resisting concrete frame buildings, moment-resisting steel frame buildings and concrete shear wall buildings, respectively. The approximate period formula was revised in the 2000 NEHRP Provisions to better fit these data. The revised equation appears as Equation 12.8-7 in Section 12.8.2.1 of ASCE 7-05. The values of the coefficients C_t and x in ASCE 7-05 Table 12.8-2 for the moment-resisting frame structures represent the lower-bound fits to the data shown in Figures C5.4.2.1-1 and C5.4.2.1-2. For all other types of structures, the approximate period obtained from Equation 12.8-7 will be the same as those in previous editions of the provisions. Although data were available for concrete shear wall structures, these data do not fit well with an equation of the form of Equation 12.8-7. Thus, Equations 12.8-9 and 12.8-10 were introduced as an alternate means of determining the approximate fundamental period for structures with concrete or masonry shear walls.

Figure 16-44 compares the approximate period obtained from the new formulas (used by ASCE 7-02 and -05) to that obtained from the older formulations (used by the 1997 UBC and the 2000 IBC) for steel and concrete special moment resisting frame structures of different heights. As can be seen, differences are negligible for both steel and concrete midrise moment resisting frame structures. However, as building height exceeds 100 feet, there is a significant increase in the predicted period for moment-resisting frame structures of concrete, as compared to previous predictive equations, with the difference in period between taller steel and concrete structures reduced significantly.

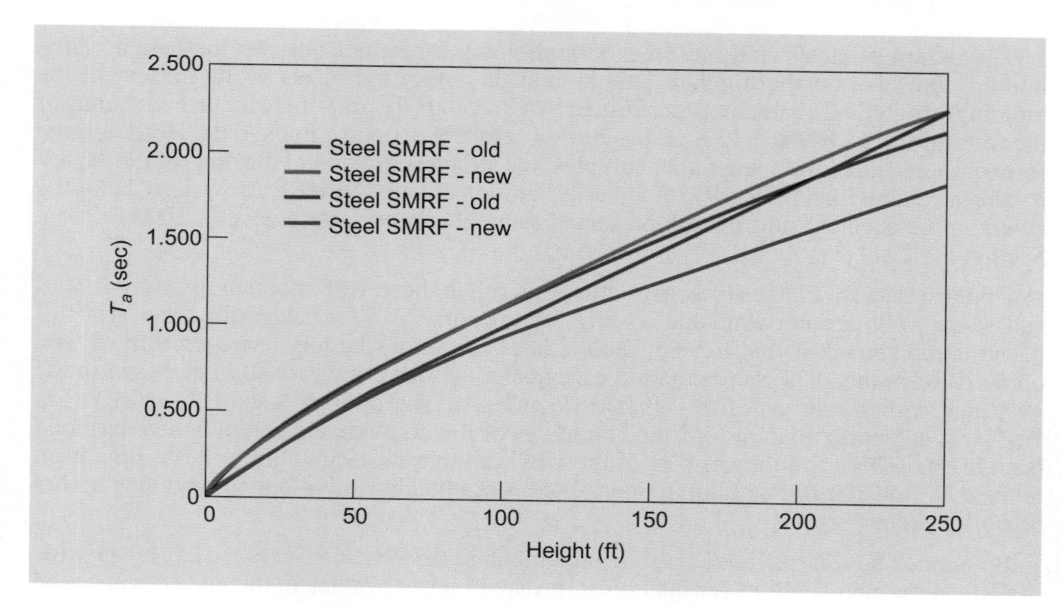

Figure 16-44 Comparison of approximate period formula from the 1997 and 2003 NEHRP *Provisions*

For buildings with other than moment-resisting frames, the approximate period was determined in editions of the UBC through 1985 from the following equation:

$$T_a = 0.05h_n / \sqrt{D}$$

D being the plan dimension of the building in the direction of analysis.

The above formula was retained in ATC 3-06, and in all the NEHRP *Provisions* through the 1991 edition. Periods for nine reinforced concrete shear wall buildings computed from accelerograph records during the 1971 San Fernando earthquake were compared against predictions by the above formula, and the above formula was seen to provide a reasonable lower bound to the measured data.

The above formula was changed to $T_a = 0.02h_n^{3/4}$ in the 1988 UBC and subsequently in the 1994 NEHRP *Provisions*. Goel and Chopra[45] have provided a comparison of predictions by this formula against the periods of 16 reinforced concrete shear wall buildings (27 data points) identified from their motions recorded during earthquakes. It was found that, for a majority of buildings, the code formula gave periods longer than the corresponding measured values.

The 1988 UBC introduced the following alternative to the 0.02 coefficient for structures with concrete or masonry shear walls;

$$C_t = 0.1 / \sqrt{A_c}$$

$$A_c = A_e\,[0.2+(D_e/h_n)2]$$

Where A_e = the minimum cross-sectional shear area in any horizontal plane in the first story, in square feet, of a shear wall.

And D_e = the length, in feet, of a shear wall in the first story in the direction parallel to the applied forces.

The above formula remained in the UBC through its 1997 edition. The development of the UBC formulas was described in detail in previous Blue Books (Seismology 1999, Section C105.2.2)

Goel and Chopra compared the above formula against some of the measured periods and concluded that the alternate formula almost always gives a value of the period that is much shorter than the corresponding measured period. In their opinion, ". . . the degree of conservatism seems excessive for most buildings considered in this investigation."

The period of shear wall buildings is highly dependent not only on the height of the building but also on the number, lengths and thicknesses of shear walls present in the building. Analytical evaluations performed by Goel and Chopra[45] indicate that equations of the form of Eq. 12.8-9 and 12.8-10 provide a reasonable good fit to the data. However, the form of these equations being rather complex, the simpler equation of the form of Eq. 12.8-7 has been retained from the 1997 and earlier editions of the NEHRP *Provisions*, with the newer, more accurate information presented as an alternative (Reference 2, 2000 Edition, Section 5.4.2).

Updated data for classes of construction other than those with moment resisting frames and shear walls are not available so far. As a result, C_t and x values for other types of construction shown in ASCE 7-05 Table 12.8-2 are values largely based on limited data obtained from the 1971 San Fernando earthquake that have traditionally been used in the 1997 and earlier editions of the NEHRP *Provisions*. The optional use of $T_a = 0.1N$ (Eq. 12.8-8) is an approximation for frame buildings of low to moderate height, which has long been in use. That formula was part of the UBC through its 1985 edition. It has now been revived by ASCE 7-05, with limitations on the story height (lower-bound) and the number of stories (upper-bound).

It ought to be recognized that all the equations in ASCE 7-05 Section 12.8.2.1 provide period estimates that are lower than most measured period values in the elastic range, and definitely much lower than nearly all measured values in the cracked section state for concrete buildings and the partially yielded state for steel buildings. However, these estimated period values, when used in the design base shear equations (Eqs. 12.8-3 and 12.8-4), provide design values that are judged to be appropriate and consistent with past design practice. For the usual case of a descending spectrum, the decrease in demand that is due to the increase in period as the structure deforms into the inelastic range is already taken into account by the R-value of a given structural system. Therefore, period formulas should provide the period of a structure in its elastic state.

Strictly speaking, the formulas presented for approximate period are appropriate only for structures with rigid diaphragms. Structures, especially low-rise structures with flexible diaphragms, will generally have periods related to the stiffness of the diaphragm rather than the stiffness of the vertical resisting elements. There are, at present, no proposals for considering this phenomenon in determining the approximate periods of such structures. ASCE 7, the IBC and the NEHRP *Provisions* do not prohibit the use of the approximate formulas in determining the periods of structures with flexible diaphragms.

Vertical distribution of seismic forces. It is generally well understood, and it has been succinctly pointed out in the 2003 NEHRP Commentary, that the distribution of lateral forces over the height of a structure is typically quite complex because these forces are the result of superposition of a number of natural modes of vibration. The relative contribution of these vibration modes to the overall distribution of lateral forces over the height of the building depends on a number of factors including the shape of the earthquake response spectrum, the natural period of vibration of the structure, and the characteristic shapes of the natural vibration modes that, in turn, depend on the magnitude and distribution of mass and stiffness over the height of the structure.

ATC 3-06, in 1978, changed the vertical distribution of lateral forces, which was in use in the UBC through its 1997 edition. In the UBC, for a structure having fundamental period, T, not exceeding 0.7 second, the entire design base shear, V, was distributed linearly along the height of the structure, with a zero value at the base and the maximum value at the top. This distribution was based on two assumptions: that the structure responds to earthquake ground motion in its fundamental mode, and that the fundamental mode shape is linear. For taller structures (having elastic fundamental period exceeding 0.7 second), an allowance was made to higher-mode response by concentrating a portion of the design base shear, $0.07TV$, at the top level. The remainder of the design base shear, $V - 0.07TV$, was then distributed linearly along the height of the structure, with a zero value at the base and a maximum value at the top.

The IBC/ASCE 7, following the 2003 and prior editions of the NEHRP *Provisions* and ATC 3-06, prescribes a triangular distribution of the entire design base shear for buildings having elastic fundamental period not exceeding 0.5 second (see ASCE 7-05 Section 12.8.3). This distribution is in view of the well-known fact that the influence of modes of vibration higher than the fundamental mode is small in the earthquake response of short-period structures and that, for regular structures, the fundamental mode shape departs little from a straight line.

The IBC/ASCE 7 prescribes a parabolic distribution of the entire design base shear, with a zero value at the base and a maximum value at the top, for structures having elastic fundamental period in excess of 2.5 seconds. As pointed out in the 2003 NEHRP *Provisions* Commentary, it has been demonstrated that although the earthquake response of long-period structures is primarily due to the fundamental natural mode of vibration, the influence of higher modes of vibration can be significant, and, in regular structures, the fundamental vibration mode lies approximately between a straight line and a parabola with the vertex at the base. The higher mode effects are adequately, indeed conservatively, accounted for when the entire design base shear, rather than only the first mode base shear, is distributed parabolically along the height of a structure.

For structures having periods between 0.5 second and 2.5 seconds a linear interpolation between a linear and a parabolic distribution of the design base shear along the height of a structure is allowed, or the entire design base shear may be distributed parabolically along the height, with a zero value at the base and the maximum value at the top.

Horizontal shear distribution and torsion. Before it branches into subsections, ASCE 7-05 Section 12.8.4 essentially defines the term seismic design story shear. The story shear in any story is the sum of the lateral forces acting at all levels above the story. Story x is the story immediately below level x.

The application of the requirements of this section depends on the diaphragm flexibility assumed as per Section 12.3.1. In most cases, the diaphragm may be modeled as fully rigid without in-plane deformability. However, as pointed out in the 1999 SEAOC *Blue Book* Commentary, there are structural configurations—such as vertical resisting elements that have large differences in stiffness or offset between stories, and diaphragms with irregular shapes and/or openings—where the effects of diaphragm deformability should be investigated. The use of the most critical results obtained from the fully rigid and the flexible diaphragm models should generally be acceptable.

Actual torsion, resulting from a noncoincidence of the center of mass and the center of rigidity at a particular floor level must be considered in design unless the diaphragm at that level is flexible. Flexible diaphragms, of course, cannot transmit torsion to the vertical lateral force-resisting elements.

The 1999 SEAOC *Blue Book* Commentary made a very important point that torsional shear should be subtracted from the direct load shear when the torsional shear is opposite to the direction of the seismic load. In such cases, the torsional shear must be due to the smaller torsion that is reduced by the accidental torsion requirements. Where the torsional shear is in the same sense and direction as the direct load shear, the more severe loading would be due to the larger torsion that is increased by the accidental torsion requirements and shear added to the direct load. The commentary goes on to explain in detail why the foregoing recommendations are justified and necessary.

According to ASCE 7-05 Section 12.8.4.2, in addition to the actual torsion, an accidental torsion must also be included in design, provided the diaphragm at a particular floor level is not flexible. In the fairly recent past, code-required practice used to include in design the actual torsion subject to a minimum of the accidental torsion. Where there is no actual torsion, obviously only the accidental torsion need be included.

The location and distribution of all of the weights that have an influence on seismic response cannot be determined with complete certainty. The requirement concerning displacement of the mass at each level was introduced to account for this deficiency.

The following additional reasons for requiring consideration of an accidental torsion in design are important:[2,22]

1. There may be unforeseeable differences between computed and actual values of stiffnesses, yield strengths and masses.
2. The real distribution of both the dead and the live loads may be nonuniform.
3. Nonstructural elements such as stairs and interior partitions may introduce eccentricities in the structural stiffness.
4. Differences in the seismic ground motion over the extent of the foundation may impose torsional loading on the structures.

ASCE 7-05 Section 12.8.4.2 makes the important point that where orthogonal forces are applied simultaneously on a structure, the required 5-percent eccentricity of the center of mass should be applied for only one of the orthogonal forces at a time. Application of the 5-percent displacement for both orthogonal directions concurrently is not consistent with the original intent of the accidental eccentricity provision.

Dynamic amplification of torsion applies to structures having torsional irregularity or extreme torsional irregularity (Horizontal Irregularities Type 1a or 1b of ASCE 7-05 Table 12.3-1) only, and does not apply to any structure assigned to SDC A or B.

The 1999 SEAOC *Blue Book* Commentary pointed out that the eccentricity amplification factor, A_x, is intended to represent the increased eccentricity caused by the yielding of perimeter elements. This factor provides a simple yet effective control on systems that might otherwise have excessive torsional yield in a given story. ASCE 7-05 Equation 12.8-14 is an empirical formula that was developed by the SEAOC Seismology Committee to encourage buildings with good torsional stiffness. The 1999 SEAOC *Blue Book* Commentary made the following important points:

1. In calculating A_x, the value of δ_{avg} should be found from the displacement resulting from the same torsional moment used to evaluate δ_{max}.
2. Where large eccentricity and low torsional rigidity result in a negative displacement on one side of a given level under static lateral forces, the value of δ_{avg} should be calculated as the algebraic average.
3. For dynamic analysis, the algebraic average needs to be found for each mode and then properly combined to produce the total modal response value for δ_{avg}.
4. Torsional analysis using the amplified accidental torsion need not be repeated beyond the analysis with the initial value of A_x. An iterative process was not intended.

Items 2 and 3 above also apply to the calculation of average story drift for the determination of Horizontal Irregularities Type 1a and 1b of Table 12.3-1.

It is important to note that in the 1994 edition of the NEHRP *Provisions* and the 1997 UBC, the factor, A_x, was used to amplify the accidental torsional moment, M_{ta}, only. In the 1997, 2000 and 2003 NEHRP *Provisions* and the 2000 IBC, the factor, A_x, amplifies the sum of the actual torsional moment, M_t, and the accidental torsional moment, M_{ta}. This is obviously a significantly more severe requirement intended to discourage the design and construction of torsionally irregular buildings. However, in ASCE 7-02 and -05, the factor, A_x, is used to amplify the accidental torsional moment, M_{ta}, only.

Overturning. ASCE 7-05 Section 12.8.5 requires that a structure be designed to resist overturning moments statically consistent with the design story shears determined in Section 12.8.3. At any level, the incremental changes of the design overturning moment are to be distributed to the various resisting elements in the same proportion as the distribution of the horizontal shears to these elements.

In the 1997 and earlier editions of the NEHRP *Provisions*, the overturning moment was modified by a factor, τ, to account for the effects of higher mode response in taller structures. In the 2000 NEHRP *Provisions*, the equivalent lateral force procedure was limited in application to structures that do not have significant higher mode participation. Thus, it was no longer necessary to include the factor τ in overturning calculations.

Many older building codes and design recommendations used to allow more drastic reduction of overturning moments relative to their value statically consistent with the design story shears. These reductions appeared to be excessive in light of structural damage observed during the 1967 Caracas, Venezuela, earthquake where a number of column failures were due primarily to the effect of overturning moments. The 1974 and subsequent editions of the SEAOC *Blue Book*, and also the 1976 and subsequent editions of the UBC, did not permit any reduction in overturning moments statically consistent with the story shears.

Drift determination and *P*-delta effects. As pointed out later on under Section 12.12 of ASCE 7-05, interstory drift caused by the design earthquake must be controlled. The design story drift is determined by ASCE 7-05 Section 12.8.6. The reasons for controlling drift are discussed under Section 12.12.

Determination of the design story drift involves the following steps (ASCE 7-05 Section 12.8.6):

1. Determine the lateral deflections at the various floor levels of a building by elastic analysis of a mathematical model of the building under the design base shear of Section 12.8.1, distributed along the height of the building as required by Section 12.8.3. Although ASCE 7-05, unlike the 1997 UBC, does not prescribe a set of detailed modeling requirements for static analysis, the modeling requirements of Section 12.7.3 must be followed. The lateral deflection at floor level x, obtained from this analysis, is δ_{xe}.
2. Amplify δ_{xe} by the deflection amplification factor, C_d, given in Table 12.12-1. The quantity, $C_d\delta_{xe}$, is an estimated design earthquake displacement at floor level x. This quantity is divided by the importance factor, I, because the forces under which the δ_{xe} displacement is computed are already amplified by I. Because the prescribed drift limits are tighter for buildings in higher occupancy categories, this division by I is important. Without it, there would be a two-fold tightening of drift limitations for buildings with seismic importance factors greater than one. The quantity, $C_d\delta_{xe}/I$, at floor level x is δ_x, the adjusted design earthquake displacement.
3. Calculate the design story drift, Δ_x, for story x (the story below floor level x) by deducting the adjusted design earthquake displacement at bottom of story x (floor level $x - 1$) from the adjusted design earthquake displacement at the top of story x (floor level x):

$$\Delta_x = \delta_x - (\delta_{x-1})$$

Figure 16-45 shows the schematic of story drift determination.

The Δ_x (or simply Δ) values must be kept within tolerable limits as given in Table 12.12-1. Two items are worth noting specifically:

1. Even in allowable stress design, the same strength-level seismic forces as given in Section 12.8.1 are to be used without any reduction.
2. For determining compliance with the story drift limitation of Section 12.12, the deflections, δ_x, may be calculated as indicated above for the seismic force-resisting system, using design forces corresponding to the fundamental period of the structure, T, calculated without the limit, $T < C_uT_a$, specified in Section 12.8.2. The same model of the seismic force-resisting system used in determining the deflections must be used for determining T. The waiver does not pertain to the calculation of drifts for determining P-delta effects on member forces, overturning moments, etc. If the P-delta effects determined in Section 12.8.7 are significant, the design story drift must be increased by the resulting incremental factor.

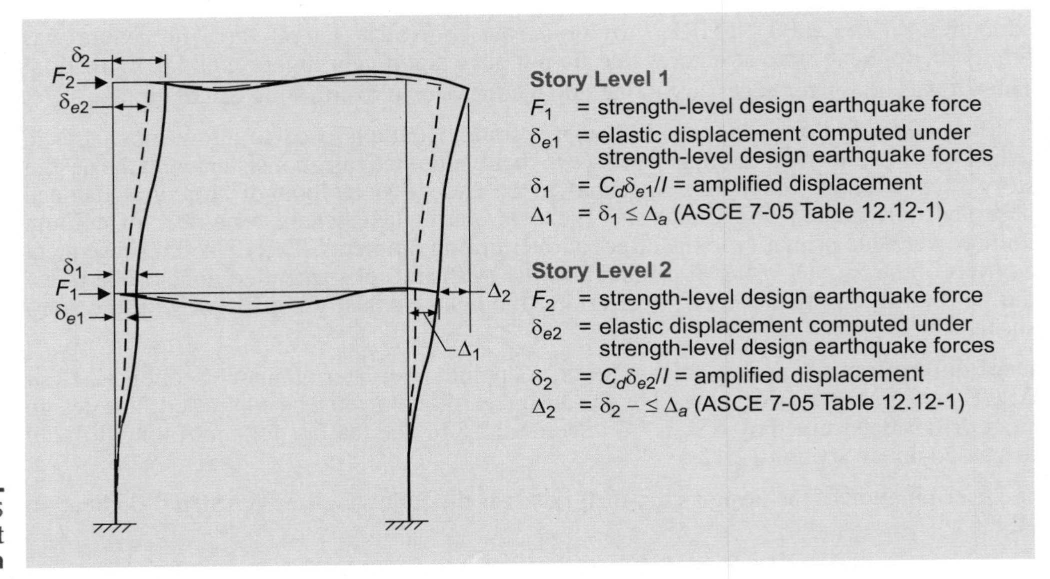

Figure 16-45 Story drift determination

***P*-delta effects.** The *P*-delta effects in a given story are due to the horizontal offset or eccentricity of the gravity load above the story. Referring to Figure 16-46, if the design story drift due that is to the lateral forces prescribed in Section 12.8.3 were Δ, the bending moments in the story would be augmented by an amount equal to Δ times the gravity load above the story. The ratio of the *P*-delta or secondary moment to the lateral-force story moment (primary moment, equal to story shear times story height) is designated as the stability coefficient as given by Equation 12.8.16. If the stability coefficient, θ, is less than 0.10 for every story, then the *P*-delta effects on story shears and moments and member forces may be ignored. If, however, the stability coefficient, θ, exceeds 0.10 for any story, then the P-delta effects on story drifts, shears, member forces, etc., for the whole building must be determined by a rational analysis.

Any of a number of rational analyses may be used. Many proprietary computer programs take *P*-delta effects into account. A mathematical description of the method employed by several popular programs is given by Wilson and Habibullah.[46] However, it must be noted that if a computer analysis is carried out that includes *P*-delta effect for computing Δ, the value of θ computed using Equation 12.8-16 needs to be divided by $(1+\theta)$ to negate the secondary effect that is already included in the calculated Δ.

An acceptable *P*-delta analysis, when required, may follow the procedure given in older editions of the SEAOC *Blue Book* Commentary, also reproduced in the 2003 NEHRP *Provisions* Commentary:

1. Compute the *P*-delta amplification factor, $a_d = \theta/(1-\theta)$, for each story. In this case, a_d accounts for the multiplier effect that is due to the initial story drift leading to another increment of drift, which would lead to yet another increment, etc. Thus both the effective shear in the story and the computed eccentricity would be augmented by a factor $1 + \theta + \theta^2 + \theta^3 \cdots$, which is $1/(1-\theta)$ or $(1 + a_d)$.
2. Multiply the story shear, V_x, in each story by the factor $(1 + a_d)$ for that story and recompute overturning moments and other seismic force effects corresponding to these augmented story shears.

The columns of moment frames that are designed with *P*-delta effects included need not be subject to the bending stress amplification factor for steel columns or the moment magnification factor for concrete columns, because these factors are intended to account for *P*-delta effects.

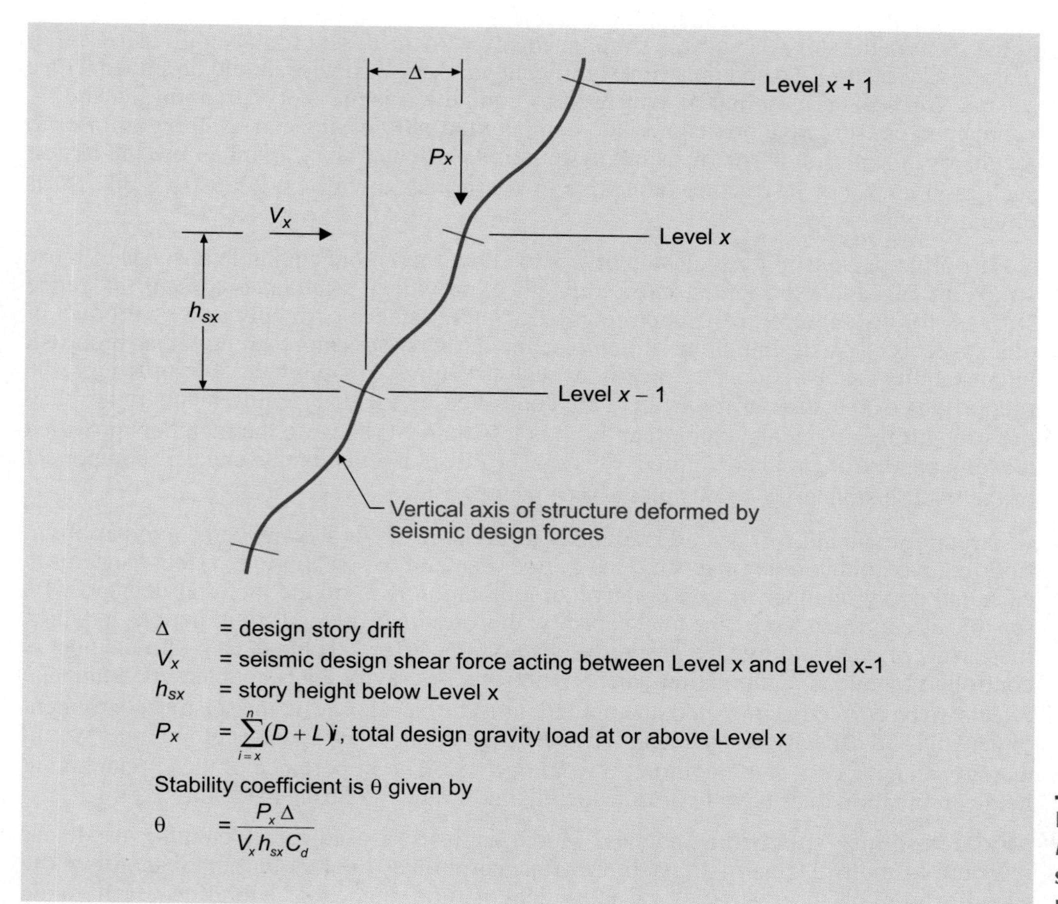

Figure 16-46 ***P*-delta symbols and notation**

It may be noted that Equation 12.8-16 uses Δ/C_d, rather than Δ, to compute the secondary moments. The 2003 NEHRP Commentary provides an elaborate explanation of this particular important aspect.

P-delta effects are potentially much more significant in buildings assigned to low seismic design categories than in buildings assigned to high seismic design categories. This is because the lateral stiffness of lateral force-resisting systems is required to be greater in the higher seismic design categories.

ASCE 7-05 has adopted a requirement, first introduced in the 1991 edition of the NEHRP *Provisions*, that the computed stability coefficient, θ, not exceed 0.25 or $0.5/\beta C_d$ where β is the ratio of shear demand to shear capacity for a particular story. The factor, βC_d, is an adjusted ductility demand considering that the seismic strength demand may be somewhat less than the code strength supplied. The adjusted ductility demand is not intended to incorporate overstrength beyond that computed by the means available in IBC Chapters 19, 21, 22, 23 and ASCE 7-05 Chapters 15 and 17.

According to the 2000 NEHRP Commentary, the purpose of this requirement is to protect structures from the possibility of stability failures triggered by post-earthquake residual deformation. The danger of such failures is real and may not be eliminated by apparently available overstrength. This is particularly true of structures assigned to low seismic design categories.

The computation of θ_{max} which, in turn, is based on βC_d, requires the computation of story strength supply and story strength demand. Story strength demand is simply the seismic design shear for the story under consideration. The story strength supply may be computed as the shear in the story that occurs simultaneously with the onset of first significant yield of the overall structure. To compute first significant yield, the structure should be loaded with a seismic force pattern similar to that used to compute seismic story strength demand. A simple and conservative procedure is to compute the ratio of demand to strength for each member of the seismic force-resisting system in a particular story and then use the largest such ratio as β. For a structure otherwise in conformance with the 2006 IBC, $\beta = 1.0$ is obviously conservative.

The principal reason for inclusion of β is to allow for a more equitable analysis of those structures in which substantial extra strength is provided, whether as a result of added stiffness for drift control, from code-required wind resistance or simply as a by-product of other aspects of the design. Some structures inherently possess more strength than required, but instability is not typically a concern for such structures. For many flexible structures, the proportions of the structural members are controlled by the drift requirements rather than the strength requirements; consequently, β is less than 1.0 because the members provided are larger and stronger than required. This has the effect of reducing the inelastic component of the total seismic drift, so β is placed as a factor on C_d.

Accurate evaluation of β would require consideration of all relevant load combinations to find the maximum value of seismic load effect demand to seismic load effect capacity in each and every member. A conservative simplification is to divide the total demand with seismic effects included by the total capacity; this considers all load combinations in which the effects of dead and live loads add to seismic force effects. If the design of a member is controlled by a load combination where the effects of gravity loads counteract earthquake effects, to be correctly computed, the ratio, β, must be based only on the seismic component of demand and capacity, not the total. The vertical load, P, in the P-delta computation would be less in such a case and, therefore, θ would be less. The importance of the counteracting load combination does have to be considered, but it rarely controls instability.

12.9 Modal response spectrum analysis. Modal analysis is used for calculating the linear response of multi-degree-of-freedom systems and utilizes the fact that the response is the superposition of the responses of individual natural modes of vibration, each mode responding with its own particular deformed shape, its own frequency and with its own modal damping. The response of the structure can, therefore, be obtained from the responses of a number of single-degree-of-freedom systems with properties chosen to be representative of the modes and the degree to which the modes are excited by the earthquake motion. For certain types of damping, the representation is mathematically exact. Numerous full-scale tests and analyses of earthquake response of structures have shown that the use of modal analysis, with viscously damped single-degree-of-freedom oscillators describing the responses of the structural modes, produces an accurate approximation of linear response.[2]

Modal analysis is useful in design. The equivalent lateral force procedure described above is simply a first mode application of this technique, which assumes all the mass of the structure to be active in the first mode. The purpose of modal analysis is to obtain the maximum response of the structure in each of its important modes, which are then summed in an appropriate manner.

This section contains the provisions for the modal analysis procedure. A mathematical model of the structure is to be constructed, in the same way as stipulated in Section 12.7.3, that represents the spatial distribution of mass and stiffness throughout the structure. In the model, cracked section properties of concrete and masonry structures must be considered, and the contribution of panel zone deformations to overall story drift must be included for steel moment frame systems.

Two-dimensional models can be used to represent independent orthogonal seismic force-resisting systems in regular structures. In structures that are irregular or possess orthogonal seismic force-resisting systems that are not independent, a three-dimensional model incorporating a minimum of three dynamic degrees of freedom (translation in two orthogonal plan directions and torsional rotation about the vertical axis) must be used.

Additional degrees of freedom are required in the model to account for the participation of nonrigid diaphragms in a structure's overall dynamic response.

Sections 12.9.1 through 12.9.3 outline the requirements regarding the minimum number of modes to be considered, calculations of base shear, story drift and member forces, and combination of these modal responses for the actual structure. Section 12.9.4 contains the provisions with regard to scaling of the base shear. These requirements are taken from the 2003 NEHRP *Provisions*. In particular, when the base shear obtained from a dynamic analysis is less than 85 percent of the base shear obtained by the equivalent lateral force procedure (where, in the equivalent lateral force procedure, the fundamental period, T, shall not exceed the coefficient, C_u, times the approximate period T_a), the dynamic analysis results must be scaled to no less than 85 percent of the equivalent lateral force procedure values. This upper limit on the period is imposed primarily to ensure that the design forces are not under-estimated through the use of a structural model that is excessively flexible. The 85-percent rule is felt to be a direct and less then complex way of providing an incentive for performing a dynamic analysis.

When the response spectrum analysis produce results that are larger than the equivalent lateral force procedure values, no scaling is permitted. The deletion of the scale-down feature in the 2003 NEHRP *Provisions* and in ASCE 7-05 is justified by pointing out that the equivalent lateral force procedure method may result in an underprediction of response for structures with significant higher mode participation. However, with the deletion, there will be no scale-down to static force levels even when a site-specific response spectrum is used in dynamic analysis. This probably places too much confidence in the geotechnical input. The confidence is now felt justified in view of the controls placed on the geotechnical input by the provisions in ASCE 7-05 Chapter 21.

ASCE 7-05 Sections 12.9.5 through 12.9.7 contain additional requirements for horizontal shear distribution of seismic forces, *P*-delta effect, and possible reduction of response for soil-structure interaction effect, respectively.

Instead of providing further commentary, Example 16-10, Part 2 of this chapter is introduced to illustrate in detail the modal analysis procedure. The step-by-step solution is probably the best help that can be provided to the reader, who may want to consult one of the standard references[47,48] for background, if needed.

12.10

Diaphragms, chords and collectors. Diaphragms are required to be designed to resist design forces from lateral floor loads as well as the transfer forces coming from discontinuities in the diaphragm or horizontal off-sets in the lateral force-resisting system. Equation 12.10-1 gives the diaphragm design forces as a function of story shear and the ratio of the weight of the diaphragm to the weight of all the diaphragms resisting the story shear. An upper-bound (nonmandatory) value and a lower-bound (mandatory) value are also given as $0.4S_{DS}Iw_{px}$ and $0.2S_{DS}Iw_{px}$. Transfer forces, when they are present, are to be added to this design force. For structures assigned to SDC D, E or F, the redundancy factor, ρ, as applicable to the whole structure, is to be applied to the part that represents the transfer forces. See Examples 16-11 and 16-12 for calculation of diaphragm forces in SDC D and higher.

Diaphragm design has attracted considerable attention in recent times, following the less than satisfactory performance of certain diaphragms in the Northridge earthquake of 1994.[49,50] There are indications[50] that if elastic diaphragm response in the design earthquake is desired, diaphragm design forces may have to be significantly larger than those given in Section 12.10.

ASCE 7-05 Section 12.10.2.1 requires collector elements, splices and their connections to resisting elements in buildings assigned to SDC C through F to have the design strength to resist the load combinations with overstrength of ASCE 7-05 Section 12.4.3.2 (see Example 16-12, which is specifically for a structure assigned to SDC D). As an exception, ASCE 7-05 allows such elements in structures or portions thereof braced entirely by light-frame shear walls to resist diaphragm forces in accordance with ASCE 7-05 Section 12.10.1.1. The collector (or drag strut) is part of the connection between the floor or roof diaphragm and the vertical lateral force-resisting elements. It is important that these connections be designed to prevent localized slip failure (or rupture), which would in turn prevent inelastic energy dissipation from developing in the lateral force-resisting system, as assumed. Collector elements and their connections must be designed to resist Ω_0 times the specified seismic design forces. The purpose is to ensure that inelastic energy dissipation occurs in the ductile lateral force-resisting elements (frame members, braces, walls), rather than in the collectors and connections.

12.11 Structural walls and their anchorage. One of the major hazards from buildings during an earthquake is the separation of walls, especially heavy concrete or masonry walls, from floors or roofs. To reduce this hazard, minimum connection forces are given. ASCE 7-05 Section 12.11.1 requires exterior and interior bearing walls and their anchorages to be designed for an out-of-plane force that is the larger of $0.1W_p$ or $0.40S_{DS}IW_p$, where W_p is the weight of the wall. To ensure that the walls and supporting framing systems interact properly, the interconnection of dependent wall elements and connections to the framing system must have sufficient ductility or rotational capacity, or strength, to stay as a unit. Large shrinkage or settlement cracks can significantly affect the desired interaction.

For concrete and masonry walls, the anchorage must provide a direct connection between the walls and the roof or floor construction capable of resisting the greater of the forces: (1) $0.40S_{DS}I\,W_p$, (2) $400S_{DS}I$ pounds per linear foot of wall or (3) 280 pounds per linear foot of wall. A minimum anchorage to bearing walls of diaphragms or other resisting elements is necessary. The wall anchorage requirements are illustrated in Figure 16-47. This figure can be compared with Figure 16-1, which gives comparable requirements for structures assigned to SDC A. Concrete and masonry walls must be designed to resist bending between anchors where the anchor spacing exceeds 4 feet.

For buildings assigned to SDC C through F with flexible diaphragms, the anchorage force is increased to $0.8S_{DS}IW_p$, where W_p is the weight of the wall tributary to the anchor (Figure 16-47). The provisions concerning anchorage to a flexible diaphragm have evolved over the years based largely on observation of performance of tilt-up concrete buildings in earthquakes. Tilt-up buildings have at times proved vulnerable to earthquake damage. The first indication of this was in the 1964 Anchorage, Alaska, earthquake, which caused several tilt-up buildings to collapse at an Air Force base. The vulnerability was not fully appreciated and understood until the 1971 San Fernando earthquake, which caused the collapse of many tilt-up buildings. The UBC provisions concerning anchorage of concrete and masonry walls

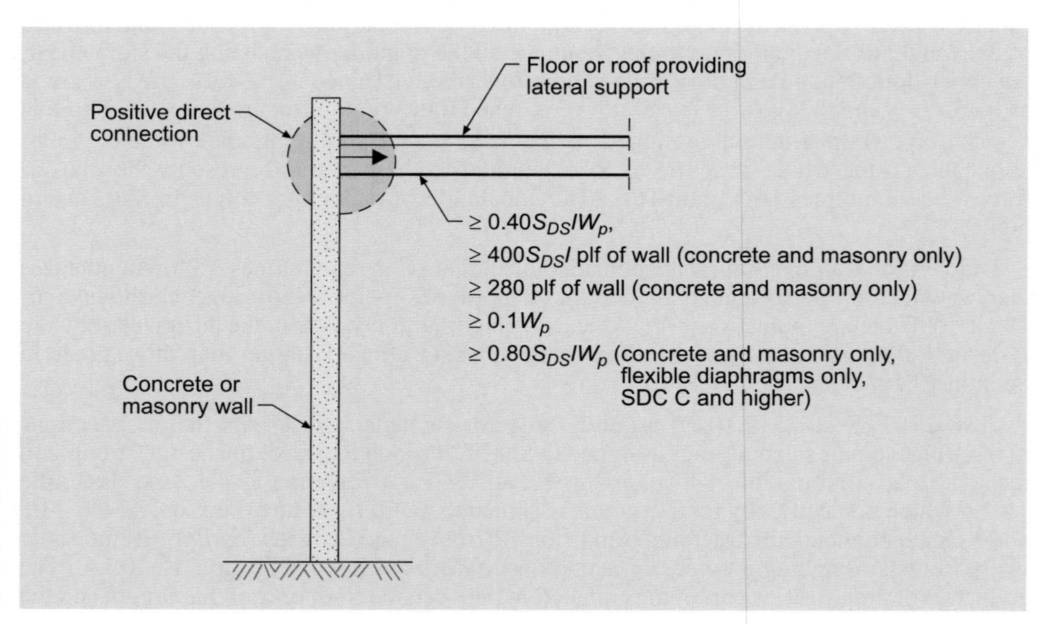

Figure 16-47 Minimum anchorage of walls to diaphragms or other elements providing lateral support

to flexible diaphragms were upgraded in almost every edition since the 1973 edition of that code. The Whittier Narrows earthquake of 1987[51] and the Northridge earthquake of 1994[52] have led to better understanding of the deficiencies in code provisions and to significant changes. The UBC changes, including post-Northridge changes, have been chronicled by Brooks.[53] The post-Northridge changes of the 1997 edition of the UBC are reflected in the ASCE 7-05 provisions of Section 12.11.2.1.

To summarize the information presented in Figure 16-47, ASCE 7-05 has several criteria for determining the force to be used to design anchors attaching structural (bearing and/or shear) walls to elements providing lateral support in buildings assigned to SDC B, C, D, E or F. These criteria are found in Section 12.11.1. They are:

$$F_p = 0.4S_{DS}IW_p \quad (1)$$

But not less than

$$F_p = 0.1W_p \quad (2)$$

where F_p is the design force in the anchor and W_p is the weight of the wall tributary to the anchor.

Section 12.11.2 has two additional criteria that apply to concrete and masonry structural walls only. These are:

$$F_p = 400S_{DS}I \text{ pounds per liner foot of wall} \quad (3)$$

and

$$F_p = 280 \text{ pounds per linear foot of wall} \quad (4)$$

In addition, where a concrete or masonry wall is anchored to a flexible diaphragm in a building assigned to SDC C, D, E or F, an additional criteria must be evaluated, which is:

$$F_p = 0.8S_{DS}IW_p \quad (5)$$

Because these criteria apply to seismic design, one would expect that the design force in the anchor would have some relationship to the mass of the wall tributary to the anchor and the design ground motion. However, two of the five criteria do not consider the mass (Eqs. 3 and 4) and two of the five criteria do not consider the ground motion (Eqs. 2 and 4). The criteria that do not consider mass and/or ground motion would appear to be arbitrary in nature—having, perhaps, only historical basis. This does not mean that it is not a good idea to have a lower-bound value of the design force. Eq. 2 serves this purpose rather well and is applicable to structural walls irrespective of the material of construction.

The arbitrary nature of Eq. 4, which considers neither mass nor ground motion, imposes a penalty on walls that are short and/or relatively thin (associated with less mass), as compared to walls that are taller and/or thicker (associated with more mass). Shorter, thinner walls are typically used in low-rise residential buildings, especially homes where 4-inch thick walls are used. Eq. 3 appears to be almost as arbitrary as Eq. 4, because it only considers ground motion. Comparing the two equations, Eq. 4 controls at values of S_{DS} less than 0.7, and Eq. 3 controls for values of S_{DS} greater than 0.7, assuming the seismic importance factor, I, to be equal to 1.0. One would logically expect Eq. 5 to be the governing equation as it accounts for mass and ground motion; however, Eq. 3 governs over Eq. 5 for tributary weights of less than 500 pounds per anchor.

To illustrate the impact of Eqs. 3 and 4, consider a 6-inch normal-weight concrete wall (weighing 75 psf), that carries an additional weight of 15 pounds per square foot to account for interior and exterior finishes added to the wall. If the wall encloses a 10-foot high story and the anchorage at the top of the wall is being designed, the height tributary to the anchor is 5 feet. Therefore, the weight tributary to the anchor is 450 pounds per linear foot of wall [(75 + 15) × 5 = 450]. For this wall, Figure 16-48a shows that Eq. 4 governs over all other criteria for values of S_{DS} less than 0.70. According to ASCE 7-05 Table 11.6-1, S_{DS} is less than 0.70 for all buildings assigned to SDC B and C (except, perhaps, some essential or hazardous facilities) and many buildings assigned to SDC D. For values of S_{DS} greater than 0.7, Eq. 3 governs, and this equation does not consider the mass of the wall. In fact, Eq. 3

governs over Eq. 5 in all situations where the weight tributary to the anchor is less than 500 pounds.

If the height of the wall is increased to 20 feet (10-foot tributary height), Figure 16-48b shows that Eq. 4 governs if the value of S_{DS} is less than approximately 0.38, assuming that the diaphragm is flexible (Eq. 5). If the diaphragm is rigid, the anchor design force is governed by Eq. 4 for values of S_{DS} less than approximately 0.71. For larger values of S_{DS}, Eq. 3 controls. It is interesting to note that Eq. 1, which recognizes mass and ground motion, never controls.

Upon superficial examination, the impact of requiring higher design forces than otherwise required by Eq. 1 or Eq. 5 appears to be inconsequential. However, when the higher forces are combined with the conservative requirements of Appendix D (Anchoring to Concrete) of ACI 318, it becomes extremely difficult to design anchors at the top of a thin concrete wall. 2006 IBC Section 1911 requires that all anchors carrying seismic forces in a structure assigned to SDC C, D, E or F must be designed by 2006 IBC Section 1912, which adopts Appendix D of ACI 318-05 by reference. The Section 1911 requirement has many other connotations and repercussions that are beyond the scope of discussion here.

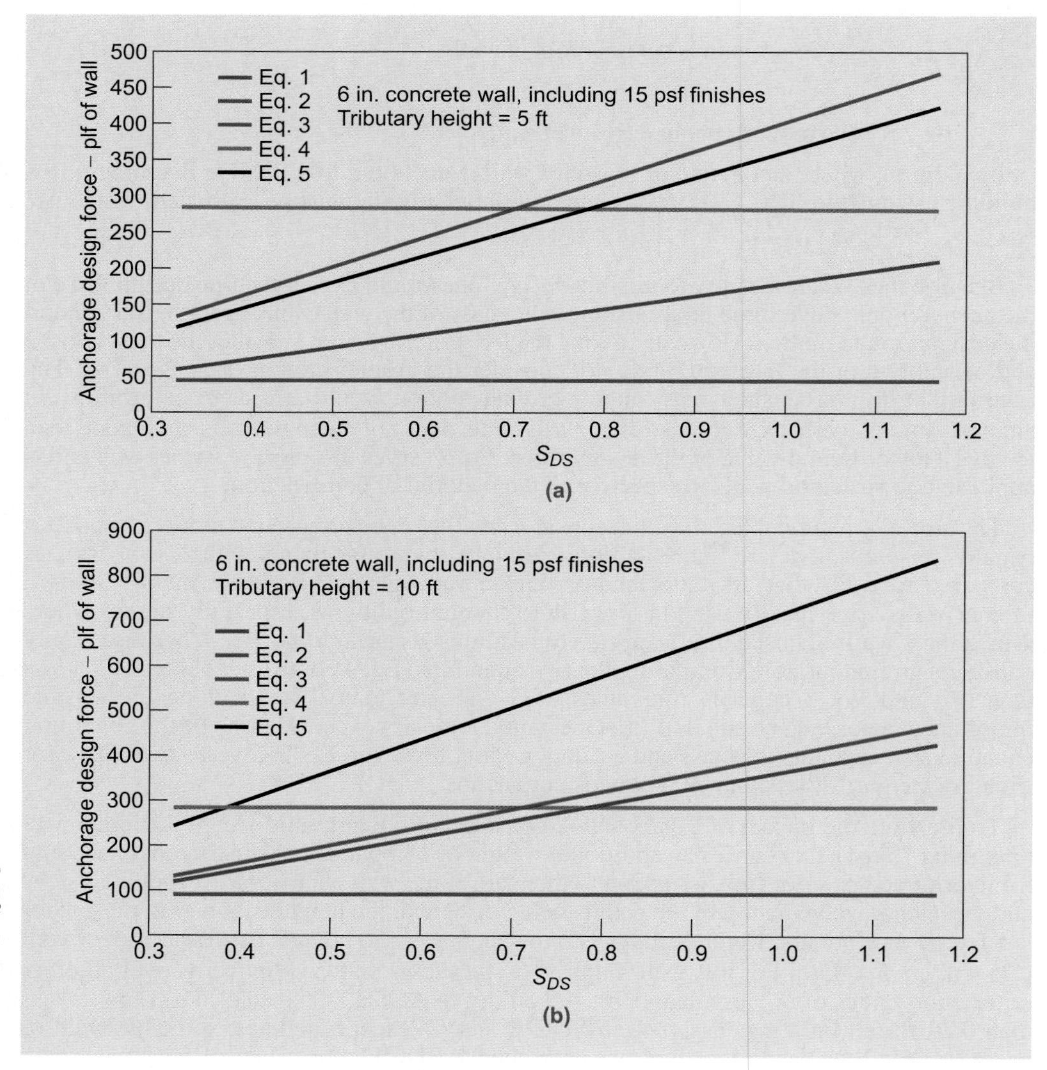

Figure 16-48 Comparison of various criteria for out-of-plane anchorage force for concrete and masonry wall

ASCE 7-05 Section 12.11.2.2 also contains additional detailed requirements for buildings assigned to SDC C through F concerning: (1) continuous ties or struts between diaphragm chords in Section 12.11.2.2.1, (2) steel elements of the wall anchorage system in Section 12.11.2.2.2, (3) wood diaphragms in Section 12.11.2.2.3, (4) metal deck diaphragms in Section 12.11.2.2.4, (5) diaphragm to wall anchorage using embedded straps in Section 12.11.2.2.5, (6) eccentrically loaded elements of the wall anchorage system in Section 12.11.2.2.6, and (7) anchorage force for pilasters in Section 12.11.2.2.7.

Drift and deformation. This section sets the limiting drifts for structures, except for those designed using the simplified analysis procedure in Section 12.14. The commentary to the 2003 NEHRP *Provisions* gives several reasons for controlling drift. **12.12**

One reason is "to control member inelastic strain. Although use of drift limitations is an imprecise and highly variable way of controlling strain, this is balanced by the current state of knowledge of what the strain limitations should be."

A stability problem is resolved by limiting the drift on the vertical load-carrying elements and the resulting secondary moments (*P*-delta effects). Large deformations with heavy vertical loads can lead to significant secondary moments from the *P*-delta effects. The drift limits indirectly provide upper bounds on these effects.

Buildings subjected to earthquakes need drift control to restrict damage to partitions, shaft and stair enclosures, glass and other fragile nonstructural elements and, more important, "to minimize differential movement demand on the seismic safety elements. Because general damage control for economic reasons is not the goal . . . and because the state of the art is not well developed in this area, the drift limits have been established without regard to considerations such as present worth of future repairs versus additional structural costs to limit drift. These are matters for building owners and designers to examine." To the extent that life safety might be excessively threatened, damage to nonstructural and seismic safety elements is a drift limit consideration.

The design story drift limits of ASCE 7-05 Table 12.12-1 reflect consensus judgment taking into account the goals of drift control outlined above. The prescribed limits on story drift depend upon the occupancy category, and generally become more restrictive for the higher occupancy categories, to provide a higher level of performance. The limits also depend on the type of structure.

The 1997 UBC drift limitations, given in Section 1630.10.3 of that code, did not depend upon the occupancy of a structure. They were 2.5 percent of story height for $T = 0.7$ sec and 2.0 percent of story height for $T > 0.7$ sec. Although there was no explicit consideration of occupancy in the UBC drift control approach, drift was implicitly considered by not dividing Δ_s by I in computing drift; higher drifts were calculated as a result for essential and hazardous facilities. However, this consideration was rather minimal, considering that the highest I-value of the 1997 UBC was 1.25.

Moment frames in structures assigned to Seismic Design Category D through F. A new drift restriction is imposed on structures with $\rho = 1.3$ by requiring that the story drift in an SDC D, E or F structure consisting only of moment-resisting frames for seismic resistance be limited to Δ_a/ρ, where Δ_a is the allowable story drift from ASCE 7-05 Table 12.12-1. Thus, the limiting story drift is reduced by a factor of 1.3, when compared to that of a redundant moment frame system. It was felt that because redundancy helps a structure, especially a moment frame system, to attain a comparatively large deflection without significant strength loss, the penalty for not having redundancy should not be confined to design strength only, but should also extend to the drift allowances. **12.12.1.1**

Diaphragm deflection. This section requiring that deflection in the plane of the diaphragm shall not exceed the permissible deflection of the attached elements is only sensible and virtually identical with 1997 UBC Section 1633.2.9 Diaphragms, Item 1. **12.12.2**

Building separation. 1997 UBC Section 1633.2.11 required that all structures must be separated from adjoining structures. Adjacent buildings on the same property must be separated by the square root of the sum of the squares of the Δ_M deflections of the two **12.12.3**

buildings. When a structure adjoined a property line not common to a public way, that structure was also required to be set back from the property line by at least the displacement Δ_M of that structure. An exception stated that smaller separation or property line setbacks might be permitted when justified by rational analysis based on maximum expected ground motions. The statement in ASCE 7-05 Section 12.12.3 is identical to the provisions in the 2000 IBC Section 1617.3 and ASCE 7-02 Section 9.5.2.8 (referenced in 2003 IBC Section 1617.3) and is much less explicit. Also, it talks about portions of the same structure, rather than separation between adjacent structures or property line setback. Note that the total deflection δ_x under which damaging contact between portions of a structure is to be avoided is $C_d\delta_{xe}/I$, which is the expected design earthquake deflection excluding consideration of the importance of a structure's occupancy.

12.12.4 Deformation compatibility for Seismic Design Categories D through F. This section requires deformation compatibility between structural members that are not part of the lateral force-resisting system and the lateral force-resisting system under design story drift Δ, which excludes consideration of the importance factor, I. 1997 UBC Section 1633.2.4 required deformation compatibility to be maintained under drift based on Δ_M or 0.0025 times the story height, whichever was greater. In other words, for very stiff systems, there was a lower-bound drift under which deformation compatibility had to be checked. Also, the 1997 UBC had a couple of requirements not included in ASCE 7-05. For concrete and masonry elements forming part of the lateral force-resisting system, the assumed flexural and shear stiffness properties were required not to exceed one-half of the gross section properties. Also, additional deformations that might result from foundation flexibility and diaphragm deflections were required to be considered. The 1997 UBC stated: "For concrete elements not part of the lateral force-resisting system, see Section 1921.7." ASCE 7-05 states, in an exception to Section 12.12.4, "Reinforced concrete frame members not designed as part of the seismic force-resisting system shall comply with Section 21.9 of ACI 318." The cited ACI 318 section number is erroneous. Reference should be made to Section 21.11 of ACI 318-05, which is the direct counterpart of Section 21.7 of ACI 318-95.

12.13 Foundation design. This section sets forth new criteria for modeling foundation stiffness for those analysis methods that permit the modeling of the load-deformation characteristics of the foundation-soil system (foundation stiffness), rather than making the assumption of a fixed base. Further information providing guidance on modeling the load-deformation characteristics of the foundation-soil system is provided in the Appendix to Chapter 7 of the 2003 NEHRP *Provisions* and Commentary (Section A7.2.3).

12.13.4 Reduction of foundation overturning. Because of the expected deviation from the results of the equivalent lateral force analysis of Section 12.8, which assumes a fixed-base of the building, Section 12.13.4 permits a 25 percent reduction in the overturning moments in the foundation when the structure is not an inverted pendulum or a cantilever column type structure. However, the allowable reduction is only 10 percent when the overturning moments are obtained from a modal analysis procedure, on account of the higher degree of accuracy in this procedure. Also, as mentioned before in the discussion of 2006 IBC Section 1605.3.2, this reduction is not allowed when using the alternative basic load combination for allowable stress design, because of a lower safety level offered by this load combination against overturning. Comparing the load combination equations that govern overturning:

2006 IBC Eq. 16-7 for strength design: $0.9D + 1.0E + 1.6H$

2006 IBC Eq. 16-15 for ASD: $0.6D + 0.7E + H$

2006 IBC Eq. 16-21 for alternative ASD: $0.9D + 0.7E$

and neglecting the term H, it is apparent that the first two equations are compatible with each other, with the former representing strength-level forces and the latter representing the corresponding service-level forces. The third equation, however, includes the earthquake load at service level ($0.7E$) while still having the dead load, which counteracts the overturning effect of lateral earthquake forces, at strength level ($0.9D$). As a result, the alternative basic ASD load combination provides a lesser margin of safety against overturning, compared to those provided by the basic load combinations for strength design and allowable stress design.

Requirements for structures assigned to Seismic Design Category C. This section gives design requirements for pole-type structural elements (Section 12.13.5.1), foundation ties (Section 12.13.5.2), and pile anchorage (Section 12.13.5.3). The foundation tie requirements are also found in Section 1808.2.23.1 of the 2006 IBC. The pile anchorage requirements are also found in Section 1808.2.23.1.1 of the 2006 IBC. The foundation tie requirement was in Section 1807.2 of the 1997 UBC. The minimum tie force was a fixed 10 percent of the larger column vertical load, while in ASCE 7-05, it is a function of S_{DS}. **12.13.5**

Requirements for structures assigned to Seismic Design Categories D through F. This section gives additional requirements for pole-type structural elements (Section 12.13.6.1), foundation ties (Section 12.13.6.2), general pile design requirements (Section 12.13.6.3), batter piles (Section 12.13.6.4), pile anchorage (Section 12.13.6.5), splices of pile segments (Section 12.13.6.6), pile-soil interaction (Section 12.13.6.7) and pile-group effects (Section 12.13.6.8). Under Section 12.13.6.2, ties are required to interconnect individual spread footings found on Site Class E or F, not just individual pile caps, drilled piers or caissons, as in Section 12.13.5.2. The minimum tie force does not change from Section 12.13.5.2 to 12.13.6.2. The general pile design requirement of Section 12.13.6.3 is very similar to the requirement represented by the first three sentences of 2006 IBC Section 1808.2.23.2.1. Batter piles are covered in Section 1808.2.23.2.3 of the 2006 IBC. The IBC requirement that the connection between batter piles and grade beams or pile caps be designed to resist the nominal strength of the pile acting as a short column is not found in ASCE 7-05. The ASCE 7-05 requirement concerning vertical and batter piles acting jointly to resist foundation forces as a group is not found in the 2006 IBC. The requirement that batter piles and their connections be capable of resisting forces and moments from the load combination including overstrength is common between ASCE 7-05 and the 2006 IBC. The anchorage requirements of Section 12.13.6.5 are found in 2006 IBC Section 1808.2.23.2.2. The splice requirements of Section 12.13.6.6 are found in 2006 IBC Section 1808.2.23.1.1, which is applicable in SDC C and above. The pile group effect provisions of Section 12.13.6.8 are found in 2006 IBC Section 1808.2.23.1.2, which is applicable in SDC C and above. **12.13.6**

Simplified alternative structural design criteria for simple bearing wall or building frame system. In recent years, engineers and building officials have become increasingly concerned that building codes in general, and the seismic design provisions of those codes in particular, have become increasingly complex, difficult to understand and to implement. The basic driving force behind this increasing complexity is the desire to provide design guidelines that assure reliable performance of structures. Because the response of buildings to earthquake ground shaking is by nature very complex, realistic accounting for these effects leads to increasingly complex provisions. However, it has also been recognized that for buildings to be reliably constructed to resist earthquakes, it is necessary that designers have sufficient understanding of the design provisions to be able to properly implement them. In recognition of this, the SEAOC Seismology Committee developed, for inclusion in the 1997 UBC, a conservative, simple method of determining design forces for certain simple buildings. This procedure, in a slightly modified form, was adopted into the 2000 IBC and appeared with some slight modifications in the 2003 IBC. **12.14**

The simplified design procedure of the 1997 UBC, the 2000 IBC and the 2003 IBC is not found in the 2006 or ASCE 7-05. Instead, a new complete set of simplified seismic design requirements is placed in stand-alone Section 12.14 of ASCE 7-05. The applicability of the simplified procedure is clearly defined through a list of limitations. The simplified procedure is limited in its applicability to simple and redundant structures falling under Occupancy Categories I and II, not exceeding three stories in height, where the seismic force-resisting elements are arranged in a torsion-resistant, regular layout. Furthermore, only bearing wall and building frame systems are allowed to be designed by this procedure. Moment frames are excluded from the simplified procedure because the simplified procedure does not require a drift check as part of the design, and drift is a major concern for moment frames. The simplified design procedure allows interstory drift to be taken as 1.0

percent of story height when an estimation of drift is required for determining structural separations or to meet specific design requirements. For easier and faster navigation, parts of ASCE 7-05 Table 12.2-1 are reproduced as Table 12.14-1, to provide the *R*-values for only the structural systems that are permitted to be designed by the simplified method. However, Table 12.14-1 does not include values for the System Overstrength Factor, Ω_o, and the Deflection Amplification Factor, C_d, because the simplified analysis procedure uses a blanket 2.5 value for Ω_o for all permitted structures, while not requiring any drift calculation using C_d. Furthermore, the load combination equations are reproduced in this section, with the modification that the Overstrength Factor, Ω_o, in the equations is taken as 2.5, while ρ is taken as 1.0. Also, because only simple torsion resistant structures are permitted, seismic loads in each building direction can be applied separately and orthogonal effects need not be considered.

The seismic base shear of simplified analysis varies (ASCE 7-05 Equation 12.14-11) with the building height and is 20 percent higher than the upper-bound design base shear of the equivalent lateral force procedure for a three-story building. The importance factor, *I*, is equal to 1.0, insofar as only Occupancy I and II structures may be designed by this procedure. Also, because only short-period structures may be designed by this procedure, the upper-bound design base shear governs. As noted previously, the response modification factor, *R*, to be used when utilizing this method is given in Table 12.14-1.

The vertical distribution of the design base shear is simpler than that given by ASCE 7-05 Section 12.8. The lateral force at a floor level is proportional to the weight at that floor level. When all the floors in a building support equal weight, the distribution of *V* along the height of the building is uniform.

For horizontal distribution of the lateral loads, the simplified analysis permits diaphragms constructed of untopped steel decking, wood structural panels or similar light-framed construction to be considered as flexible. This essentially eliminates the need to calculate diaphragm deflections for these types of systems.

As mentioned before, no drift limitation is imposed and no drift calculation is necessary, because only short and stiff structures are permitted to be designed by the simplified procedure. However, a design drift, where needed to estimate the necessary building separation, etc., is allowed to be taken as 1 percent of the building height.

Application of the simplified design procedure is illustrated in Example 16-8, Part 2 of this chapter.

It may be instructive to compare the simplified seismic design procedures of the 1997 UBC, particularly because the 1997 UBC procedure, with not very substantive modifications, was retained through the 2003 IBC. The slight differences between the 1997 UBC procedure and the 2000 and 2003 IBC procedure are pointed out at the end of the comparison. Limitations on the use of the respective procedures are and were given in ASCE Section 12.14.1.1 and UBC Section 1629.8.2. The limitations were fewer and less stringent in the UBC. The simplified design base shear Equation 12.14-11 of ASCE 7-05 and Formula (30-11) of the 1997 UBC are the same, if one uses the conversion: $S_{DS} = 2.5C_a$, except that the constant 1.2 of the UBC is replaced by the variable *F* in ASCE 7-05, which is 1.2 for three-story buildings, but lesser for shorter buildings. ASCE 7 allows the short-period site coefficient F_a (comparable to C_a/Z of the UBC) to be taken equal to 1.0 for rock sites and 1.4 for soil sites, rock sites and soil sites being defined in Section 12.14.8.1. The UBC did not have this feature. The uniform distribution of the design base shear along the height (for the same weight at the various floor levels) is also the same in the two codes. ASCE 7 allows and the UBC also allowed the design story drift, where needed, to be taken as one-percent of the story height. There are significant differences, however, between the ASCE 7 and the 1997 UBC provisions. The UBC required the use of a default soil profile S_E, rather than the usual S_D, in the simplified procedure; ASCE 7 does not. The UBC allowed the near-source factor, N_a, to be taken as no larger than 1.3 if irregularities of certain types did not exist in a structure; ASCE 7, of course, does not use near-source factors. On the other hand, ASCE 7 allows S_s to be taken no larger than 1.5*g*. The UBC specifically listed certain code sections that were not applicable when the simplified procedure was used (the redundancy factor ρ, however, was not allowed to be taken equal to unity without computations justifying that value); ASCE 7 does not give such a specific list. It does not need to, because simplified design provisions are placed in a self-contained section. Also, ρ is always equal to 1 in ASCE 7 simplified design. Finally, the UBC gave a modified Formula (33-1) for diaphragm design forces when the simplified procedure was used. Although ASCE 7 does not give a similar modification of the corresponding ASCE 7 Equation 12.10-1, it does, in effect, have a modified diaphragm design force equation for the simplified method in that Section 12.14.7.4 refers to Section 12.14.8.2 for determination of diaphragm design forces and not Equation 12.10-1. The simplified procedure in ASCE 7-05 thus permits a diaphragm to simply be designed to resist the seismic forces tributary to the diaphragm.

From the 1997 UBC to the 2000 and 2003 IBC, the conditions of applicability of simplified design changed in only one respect. The 2000 and 2003 IBC required the diaphragms in a building, not of light-frame construction, to be flexible for simplified seismic design to be applicable. This restriction was not in the 1997 UBC. There were a few differences in the design provisions themselves. The 1997 UBC required the use of a default soil profile type S_E, rather than the usual S_D, in the simplified procedure; the 2000 and the 2003 IBC did not. The UBC allowed the near-source factor, N_a, to be taken as no larger than 1.3 if irregularities of certain types did not exist in a structure; the 2000 and the 2003 IBC, of course, did not use near-source factors. The UBC specifically listed certain code sections that were not applicable when the simplified procedure was used; the 2000 and the 2003 IBC did not contain such a list. Finally, the UBC gave a modified Formula (33-1) for diaphragm design forces when the simplified procedure was used; the 2000 and the 2003 IBC did not give a similar modification of the corresponding IBC/ASCE 7 formula.

Chapter 13 *Seismic Design Requirements for Nonstructural Components*

General. This chapter stipulates the general requirements governing the design of all nonstructural components, including the applicability of the provisions, exemptions and the use of reference documents. **13.1**

All architectural, mechanical and electrical components in structures must be designed and constructed to resist the equivalent static forces and displacements determined in accordance with Chapter 13 of ASCE 7-05. The design and evaluation of support structures and attachments of architectural components and equipment must consider their flexibility as well as their strength.

This chapter is patterned after Chapter 6 of the 2003 NEHRP *Provisions* but is reformatted for user-friendliness and streamlined so that the information is presented in a logical sequence with redundancies eliminated.

General provisions applicable to nonstructural components are presented, followed by provisions specific to architectural components, followed by provisions specific to mechanical and electrical components.

Chapter 13 is divided up into the following six sections:

Section 13.1 General

Section 13.2 General Design Requirements

Section 13.3 Seismic Demands on Nonstructural Components

Section 13.4 Nonstructural Component Anchorage

Section 13.5 Architectural Components

Section 13.6 Mechanical and Electrical Components

The component design requirements establish minimum design criteria for architectural, mechanical and electrical components permanently attached to a structure, and for the attachments and supports of such components. The design criteria establish minimum equivalent static force levels and relative displacement demands for the design of components and their attachments to the structure, recognizing ground motion and structural amplification, component toughness and weight, and performance expectations. Five exemptions are made to the applicability of ASCE 7-05 Chapter 13 in ASCE 7-05 Section 13.1.4.

For the purposes of this chapter, components must be considered to have the same seismic design category as that of the structure they occupy or to which they are attached. Seismic design categories for structures are defined in ASCE 7-05 Section 11.6. In addition, all components are assigned a component importance factor, I_p, given in ASCE 7-05 Section 13.1.3.

The default value of I_p is 1.0 for typical components in normal service. Higher values of I_p are assigned for components that contain hazardous substances, must have a higher level of assurance of function, or otherwise require additional attention because of their life-safety characteristics. Component importance factors are either 1.0 or 1.5. In ASCE 7-05 Section 13.1.3, the value of I_p is based on:

1. Requirements of the component to function after a design earthquake, or
2. Occupancy category of the structure or facility, or
3. Whether the component contains hazardous materials.

In ASCE 7-05, nonstructural components that are assigned an $I_p = 1.5$ become part of a Designated Seismic System (see definition in ASCE 7-05 Section 11.2).

No specific reference document is stipulated by ASCE 7-05 for earthquake-resistant design of nonstructural components. Instead, Section 13.1.6 permits any reference document for this purpose as long as it is approved by the authority having jurisdiction and the minimum requirements in terms of force, drift and so forth that are stipulated in ASCE 7-05 Chapter 13 are met. Section 13.1.7 specifically addresses the use of reference documents that define acceptance criteria in terms of allowable stresses rather than strengths. Although the design of primary structural systems is strength-based, many nonstructural systems and components are proportioned using allowable stress design. There are a large number of groups that prepare standards for the design and installation of nonstructural systems and components. The use of these standards is now specifically permitted, provided that the force and displacement requirements of ASCE 7-05 are met.

13.2 General design requirements. Submission (to the authority having jurisdiction) of the manufacturer's certificate of compliance with ASCE 7-05 force requirements must be made when required by the contract documents or when required by the regulatory agency. Evaluation of the components needs to be carried out not only in terms of strength, but also in terms of flexibility.

ASCE 7-05 Section 13.2.5 permits equipment testing as an acceptable method to determine seismic capacity, in lieu of the analysis methods described in the preceding sections. Adaptation of a nationally recognized standard for qualification by testing is an acceptable alternative, provided that it is acceptable to the authority having jurisdiction and that the equipment seismic capacity at least meets the force and displacement provisions of ASCE 7-05 Section 13.3.

ASCE 7-05, in Section 13.2.6, also allows experience data based on nationally accepted procedures that are accepted by the authority having jurisdiction as an acceptable means of evaluating component performance.

ASCE 7-05 Section 13.2.7 requires that construction documents be prepared to comply with the requirements summarized in ASCE 7-05 Table 13.2-1. Additionally, construction documents need to include any quality assurance plan stipulated in Appendix 11A. These quality assurance provisions are meant to provide a reasonable level of assurance that the construction and installation of the aforementioned components be consistent with the basis of the supporting seismic design. However, 2006 IBC Section 1613.1 requires users to refer to Chapter 17 of the 2006 IBC for quality assurance provisions instead of using ASCE 7-05 Appendix 11A. According to the 2003 NEHRP *Commentary*, it is important that a registered design professional prepare construction documents for use by multiple trades and suppliers in the course of construction.

Special certification requirements for Designated Seismic Systems. This section requires seismic certification for: **13.2.2**

1. Active mechanical and electrical equipment that is required to function following a design earthquake, and
2. Components containing hazardous contents.

Certification demonstrating functionality after being subject to a design earthquake must be established by:

1. Approved shake table testing in compliance with ICC-ES AC 156[54] (ASCE 7-05 Section 13.2.5), or
2. Experience data (ASCE 7-05 Section 13.2.6), or
3. In the case of components with hazardous contents only, analysis.

The supplier is required to provide evidence demonstrating compliance with this requirement.

Testing alternative for seismic capacity determination. ASCE 7-05 Section 13.2.5 makes reference to ICC-ES AC 156, which is *ICC Evaluation Service Acceptance Criteria 156, Seismic Qualification by Shake-table Testing of Nonstructural Components and Systems*, 2004. ASCE 7-02 previously referenced IEEE-344,[55] a nuclear standard for shake table testing of electrical equipment that is very industry-specific. AC 156 is a more generally applicable procedure for shake table testing that is rapidly gaining wide acceptance. **13.2.5**

Seismic demands on nonstructural components. The seismic design force for any component is to be applied at the center of gravity of the component and should be assumed to act in any horizontal direction. **13.3**

Seismic design force. The design seismic force given in ASCE 7-05 Section 13.3.1 is dependent upon the weight of the system or component, the component amplification factor, the component acceleration at point of attachment to the structure, the component importance factor and the component response modification factor. **13.3.1**

The component amplification factor, a_p, represents the dynamic amplification of the component relative to the fundamental period of the structure, T. At the time the components are designed or selected, the structural fundamental period is not always defined or readily available. Also, the component fundamental period, T_p, is usually only accurately obtained by expensive shake-table or pull-back tests. A listing is provided in ASCE 7-05 Table 13.5-1 of a_p values ranging between 1.0 and 2.5 for architectural components and in ASCE 7-05 Table 13.6-1 for mechanical and electrical components, based on the expectation that the component will usually behave in either a rigid or a flexible manner. In general, if the fundamental period of the component is less than 0.06 second, no dynamic amplification is expected. It is not the intention of the ASCE 7-05 requirements to preclude more accurate determination of the component amplification factor when reasonably accurate values of both the structure and the component fundamental periods are available (see ASCE 7-05 Table 13.5-1, Footnote a).

The component response modification factor, R_p, represents the energy absorption capability of the component's structure and attachments. Conceptually, the R_p value considers both the overstrength and the deformability of the component's structure and attachments. The R_p values ranging between 1.0 and 12.0, given in ASCE 7-05 Table 13.5-1 for architectural components and ASCE 7-05 Table 13.6-1 for mechanical and electrical components, are carefully determined using the collective wisdom and experience of the responsible committee. In general, the following benchmark values were used:

$R_p = 1.25$ for elements of low deformability.

$R_p = 2.5$ for elements of limited deformability.

$R_p = 3.5$ for elements of high deformability.

ASCE 7-05 Equation 13.3-1 represents a trapezoidal distribution of floor accelerations within the structure, linearly varying from the acceleration at the ground, $0.4S_{DS}$, to the acceleration at the roof, $1.2S_{DS}$. The ground acceleration, $0.4S_{DS}$, is intended to be the same acceleration used as design input for the structure itself, including site effects.

Examination of recorded in-structure acceleration data in response to large California earthquakes reveals that a reasonable maximum value for the roof acceleration is four times the input ground acceleration to the structure. This finding is actually implemented with the 1997 UBC formula corresponding to ASCE 7-05 Equation 13.3-1. However, a close examination of recently recorded strong motion data at sites with peak ground accelerations in excess of 0.1g indicated that an amplification factor of 3 is more appropriate. A lower limit for F_p is set to assure a minimal seismic design force. The minimum value for F_p is determined by setting the quantity $(a_p / R_p)(1 + 2z/h) = 0.7$, which is equivalent to the minimum used in current practice.

To meet the need for a simple formulation, a conservative maximum value of F_p is also set. The maximum value of F_p is determined by setting the quantity $(a_p/R_p)(1 + 2z/h)$ equal to 4.0.

The provision for vertical seismic forces is similar to that for lateral force-resisting structural components by Equation 12.4-4, and is given by $0.2S_{DS}W_p$. The dead load effect D of Equation 12.4-4 is replaced by the component operating weight W_p.

Accelerations at any level may be determined by the modal analysis procedures of ASCE 7-05 Section 12.9, taking $R = 1$, in lieu of the forces determined by ASCE 7-05 Equation 13.3-1. In such cases, the seismic forces are determined by ASCE 7-05 Equation 13.3-4, using the accelerations obtained from the modal analysis. The upper and lower limits on F_p given by ASCE 7-05 Equations 13.3-2 and 13.3-3, respectively, are also applicable in this case.

13.3.2 Seismic relative displacements. Equations for seismic relative displacements, which are given in ASCE 7-05 Section 13.3.2, are needed for the design of cladding, stairwells, windows, piping systems, sprinkler components and other components that are connected to the structure(s) at multiple levels or points of connection. Two equations are given for each situation. Equations 13.3-5 and 13.3-7 yield *real* structural displacements as determined by elastic analysis under the code-prescribed seismic forces, with the deflection amplification factor, C_d, applied. Recognizing that elastic displacements are not always defined or available at the time a component is designed or procured, default Equations 13.3-6 and 13.3-8 also are provided and allow the use of structural drift limitations. Use of these default equations are intended to balance the need for a timely component design/procurement with the possible conservatism of their use. The lesser of the paired equations are intended to be acceptable for use. The effects of seismic relative displacements must be considered in combination with displacements caused by other loads.

ASCE 7-05 also allows the seismic relative displacement to be determined using modal analysis procedures.

13.4 Nonstructural component anchorage. Components must be attached to the building structure so as to be able to transfer the component forces. Such attachments must be bolted, welded or otherwise positively fastened, and frictional resistance produced by the effects of gravity loads must not be counted on. Details of the attachments are to be provided in the construction documents with sufficient information to verify compliance with the requirements in this section.

New provisions for cast-in-place anchors in concrete have been developed and these are incorporated into ASCE 7-05 Section 13.4.2. They include specific consideration of ductile and nonductile anchorage. In addition, new cyclic test procedures have been developed for post-installed anchors. Now, if post-installed anchors are prequalified for seismic applications in accordance with ACI 355.2, *Qualification of Post-Installed Mechanical Anchors in Concrete*,[56] the R_p-penalty on low-ductility post-installed anchorage may be waived.

As pointed out in the 2003 NEHRP *Commentary*, the requirements of ASCE 7-05 Section 13.4 reflect the fact that, in general, it is not desirable to rely upon anchors for energy dissipation. Inasmuch as the anchor represents the transfer of load from a relatively deformable material (steel) to a low-deformability material (concrete or masonry), the boundary conditions for ensuring deformable, energy-absorbing behavior in the anchor itself are, at best, difficult to achieve. On the other hand, the concept of providing a fuse, or deformable link, in the load path to the anchor is encouraged. This approach allows the designer to provide the necessary level of ductility and overstrength in the connection while at the same time protecting the anchor from overload. Generally, power-driven fasteners in concrete tend to exhibit variations in strength that are somewhat larger than those for post-drilled anchors and do not provide the same levels of reliability. As such, qualification under a simulated seismic test program must be demonstrated prior to use.

Architectural components. This section provides for the design of not only architectural components, but also their attachments and supports. **13.5**

Attachments are means by which components are secured or restrained to the lateral force-resisting system of a structure. Such attachments and restraints may include anchor bolts, welded connections and other fasteners.

Component supports are those members or assemblies of members, including braces, frames, struts and attachments, that transmit all loads and forces between the component and the building structure. Component supports also provide structural stability to the component to which they connect.

This section contains specific requirements for: (1) Exterior nonstructural wall elements and connections (Section 13.5.3), (2) Suspended ceilings (Section 13.5.6)—relevant ASTM standards and Recommendations of the Ceilings and Interior Systems Construction Association (CISCA) are referenced, (3) Glazing (Section 13.5.9)—the drift causing glass fallout from a curtain wall, storefront, or partition is to be determined by the recommended dynamic test method[57] of the American Architectural Manufacturers Association (AAMA), (4) Access floors including special access floors (Section 13.5.7) and (5) Partitions (Section 13.5.8)—independent bracing is required for partitions that are tall or are tied to the ceiling.

Forces and displacements. ASCE 7-05 Section 13.5.2 requires that architectural components be designed using the forces given in Section 13.3.1, with a_p and R_p values given in Table 13.5-1. **13.5.2**

Exterior nonstructural wall elements and connections. The requirements of ASCE 7-05 Section 13.5.3 are specifically for nonstructural wall panels or elements that are attached to or enclose the structure. Movements of the structure resulting from temperature changes must be accommodated. Elements addressed in this section must be supported by means of positive and direct supports or mechanical connections and fasteners in accordance with a number of requirements that are given. The requirements of ASCE 7-05 Section 13.5.9 are to be used for the design and installation of glass in glazed curtain walls and storefronts. **13.5.3**

Out-of-plane bending. Most walls are subject to out-of-plane forces when a building is subjected to earthquake ground motion. These forces and the resulting bending must be considered in the design of wall panels, nonstructural walls and partitions. Conventional limits based on deflections as a proportion of the span may be used with the applied force derived from Section 13.5.2. **13.5.5**

13.5.6 Suspended ceilings. Suspended ceilings, like other architectural components, must be designed to resist a seismic force F_P that is calculated from Equation 13.3-1. In calculating F_P, the weight of the ceiling W_P needs to include ceiling grid and panels, any light fixtures present on the ceiling, and other components that are laterally supported by the ceiling. W_P should not be taken less than 4 pounds per square foot. In addition, suspended ceilings must be designed to meet the requirements of either industry standard construction as modified in Section 13.5.6.2 or integral construction as set forth in Section 13.5.6.3.

13.5.7 Access floors. In addition to meeting the force requirements of ASCE 7-05 Section 13.3.1, access floors must meet the requirements of ASCE 7-05 Section 13.5.7.

13.5.8 Partitions. Bracing requirements for partitions are given in ASCE 7-05 Section 13.5.8.

13.5.9 Glass in glazed curtain walls, glazed storefronts and glazed partitions. The provisions in ASCE 7-05 Section 13.5.9 are taken from Section 6.3.7 of the 2003 NEHRP *Provisions*. Prior to the 2000 NEHRP *Provisions*, seismic design requirements for glass were nonexistent.

All glass in glazed curtain walls, glazed storefronts and glazed partitions are required to meet the relative displacement requirements of ASCE 7-05 Equation 13.5-1 or 0.5 inch, whichever is greater. This equation is meant to provide a life-safety level of performance. The drift causing glass fallout is to be determined in accordance with the accepted standard "Recommended Dynamic Test Method for Determining the Seismic Drift Causing Glass Fallout from a Wall System," by the American Architectural Manufacturers Association or by engineering analysis. The relative seismic displacement, D_p, that the glass component must be designed to accommodate is determined from ASCE 7-05 Equation 13.3-5 (please note: the first edition of ASCE 7-05 has misprinted this equation number as 13.3-2). This relative displacement is to be applied over the height of the glass component. The 1.25 factor on the right-hand side of Equation 13.5-1 is meant to reflect the uncertainties associated with calculated inelastic seismic displacements in buildings.

Three exceptions to meeting the relative displacement requirement are also given in this section. Glass with sufficient clearances from its frame can be exempted, provided Equation 13.5-2 is satisfied. The quantity D_{clear} in this equation depends on the geometry of the rectangular glass and the gaps between the vertical and horizontal glass edges and the frame. Both this equation and Equation 13.5-1 are based on the principle that a rectangular window frame becomes a parallelogram as a result of interstory drift, and that glass-to-frame contact occurs when the length of the shorter diagonal of the parallelogram is equal to the diagonal of the glass panel itself. The other exemptions in this section are for tempered monolithic glass and annealed or heat-strengthened laminated glass under certain conditions.

2003 NEHRP *Commentary* Section 6.3.7 gives additional background information on the seismic design requirements for glass, including a summary of the test programs that were conducted for seismic drift limits for glass components.

13.6 Mechanical and electrical component design. Requirements are given for mechanical and electrical components in ASCE 7-05 Section 13.6. The attachments of the mechanical and electrical components and their supports to the structure need to comply with Section 13.4.

This section contains specific requirements for: (1) Lighting fixtures (Section 13.6.1 Exception), (2) Utility and service lines (Section 13.6.6), (3) HVAC ductwork (Section 13.6.7), (4) Piping systems (Section 13.6.8), (5) Fire protection sprinkler systems (Sections 13.6.8.2 and 13.6.8.3), (6) Boilers and pressure vessels (Section 13.6.9), and (7) Elevators and escalators (Section 13.6.10). This section also contains specific prescriptive details for mechanical components (Section 13.6.3), electrical components (Section 13.6.4) and component supports (Section 13.6.5).

The force and seismic relative displacement requirements of ASCE 7-05 Sections 13.3.1 and 13.3.2, respectively, with a_p and R_p values given in ASCE 7-05 Table 13.6-1, must be met.

ASCE 7-05 Table 13.6-1 reflects the extensive treatment given to nonstructural components in ASCE 7-05. It also reflects the fact that the diversity and complexity of mechanical and electrical components installed in buildings has increased dramatically over the years. This table has been revised extensively even from the 2002 edition of ASCE 7.

General. The exception allowing lateral bracing of suspended components to be omitted is clarified with respect to the type of nonstructural components covered by the exception, as well as limits on the acceptable consequences of interaction between components. Certain nonstructural components that could represent a fire hazard following an earthquake are not eligible for the exception. **13.6.1**

Component period. Determination of the fundamental period of a piece of mechanical or electrical equipment can be by ASCE 7-05 Equation 13.6-1, if the component and attachment can be reasonably represented analytically by a simple spring and mass single-degree-of-freedom system. Otherwise, period must be determined from experimental test data or by analysis. **13.6.2**

Mechanical components. Requirements specific to mechanical components are given in ASCE 7-05 Section 13.6.3. Other than stipulating the separate provisions for different types of mechanical components through reference, this section also sets a general rule that any component with I_p greater than 1.0 needs to be designed for the seismic forces and relative displacements of Sections 13.3.1 and 13.3.2, respectively, in addition to meeting three other requirements. **13.6.3**

Electrical components. Requirements specific to electrical components are given in ASCE 7-05 Section 13.6.4. This section stipulates that any component with I_p greater than 1.0 needs to be designed for the seismic forces and relative displacements of Sections 13.3.1 and 13.3.2, respectively, in addition to meeting eight other requirements. **13.6.4**

Component supports. Component supports are differentiated here from component attachments to emphasize that the supports themselves—the structural members, braces, frames, skirts, legs, saddles, pedestals, cables, guys, stays, snubbers and tethers—even if fabricated by the mechanical or electrical component manufacturer, must be designed for seismic forces and displacements. **13.6.5**

Utility and service lines. Adequate flexibility must be provided for utilities at the interface of adjacent and independent structures to accommodate anticipated differential displacement. Consideration must be given for possible interruption of utility services for Occupancy Category IV structures. Vulnerability of underground utility lines at the interfaces between the structure and the ground should also be accounted for if the site class is E or F and where S_{DS} is equal to or greater than $0.33g$. **13.6.6**

HVAC ductwork. Attachments and supports of HVAC ductwork, in addition to meeting the force and displacement requirements of Sections 13.3.1 and 13.3.2, respectively, must meet the additional requirements in ASCE 7-05 Section 13.6.7. **13.6.7**

Piping systems. Attachments and supports for piping systems, in addition to meeting the force and displacement requirements of Sections 13.3.1 and 13.3.2, respectively, must meet the additional requirements of ASCE 7-05 Section 13.6.8. **13.6.8**

In previous editions of ASCE 7, fire sprinkler systems installed in accordance with NPFA 13, *Standard for the Installation of Sprinkler Systems*,[58] were deemed to comply with the seismic bracing requirements. Changes to nonstructural design procedures beginning with the 1997 UBC have increased seismic design demands on such systems in areas of high seismicity to levels greater than those provided for in NFPA 13. ASCE 7-05 permits the seismic provisions of NFPA 13 to be used, while maintaining conformity with the general seismic design requirements of ASCE 7-05. Fire sprinkler piping in SDCs D, E and F, designed in accordance with NFPA 13, is now subject to the general requirements for piping systems with an importance factor, $I_p = 1.5$.

Boilers and pressure vessels. Attachments and supports for boilers and pressure vessels, in addition to meeting the force and displacement requirements of Sections 13.3.1 and 13.3.2, **13.6.9**

respectively, must be designed to meet the additional requirements of ASCE 7-05 Section 13.6.9.

13.6.10 Elevator and escalator design requirements. Elevators, escalators, hoistway structural systems, elevator equipment and controller supports and attachments are required to meet the force and displacement provisions of ASCE 7-05 Sections 13.3.1 and 13.3.2. The seismic force requirements for elevators and escalators can also be met by designing in accordance with the ASME *Safety Code for Elevators and Escalators*,[59] which has adopted many requirements to improve the seismic response of elevators. Additional requirements are given in Sections 13.6.10.3 and 13.6.10.4.

Chapter 14 *Material-Specific Seismic Design and Detailing Requirements*

All of the material-specific requirements are collected and organized according to material type and included in this chapter. Note, however, that 2006 IBC Section 1613 exempts structures from having to comply with the requirements in ASCE 7-05 Chapter 14 because the material-specific seismic design requirements are covered in the materials chapters of the 2006 IBC (Chapters 19 through 23). Thus, this chapter is not discussed further.

Note that although ASCE 7-05 Chapter 14 itself is not adopted by the 2006 IBC, ASCE 7-05 Table 12.2-1, which is definitely adopted, makes repeated references to ASCE 7-05 Chapter 14 in the column entitled: "ASCE 7 Section where Detailing Requirements are Specified."

Chapter 15 *Seismic Design Requirements for Nonbuilding Structures*

15.1 General. The provisions of this chapter are adopted from the 2003 NEHRP *Provisions*. The design of nonbuilding structures must provide sufficient strength, stiffness and ductility, consistent with the requirements specified for buildings, to resist the effects of seismic ground motions as represented by the following:

1. Applicable strength and other design criteria must be obtained from the seismic requirements specified in other portions of ASCE 7-05 and codes and standards referenced from those portions.
2. Where applicable strength and other design criteria are not contained in or referenced by any other portion of ASCE 7-05, such criteria are to be obtained from reference documents. Where reference documents define acceptance criteria in terms of allowable stresses as opposed to strength, the design seismic forces are to be obtained from this chapter and used in combination with other loads as specified in ASCE 7-05 Section 2.4 and used directly with allowable stresses specified in the reference documents. Detailing must be in accordance with the reference documents.

As per ASCE 7-05 Section 11.2, a "nonbuilding structure similar to a building" (such as a pipe rack) is defined as "a nonbuilding structure that is designed and constructed in a manner similar to buildings, will respond to strong ground motion in a fashion similar to buildings, and have basic lateral and vertical seismic force-resisting system conforming to one of the types indicated in Tables 12.2-1 or 15.4-1."

For nonbuilding structures that are similar to buildings, the structural analysis procedure needs to be selected following the provisions in ASCE 7-05 Section 12.6, which are based on the seismic design category of the structure, its height and whether it is structurally regular or irregular. For nonbuilding structures that are not similar to buildings (such as flat-bottom tanks), however, any of the analysis methods approved by ASCE 7-05, i.e., equivalent lateral force procedure, modal analysis procedure, linear or nonlinear time history analysis or any other procedure prescribed in the reference documents listed in Chapter 23 of ASCE 7-05 may be used.

Reference documents. Reference documents are listed in Chapter 23 of ASCE 7-05. **15.2**

Nonbuilding structures supported by other structures. Figure 16-49 explains the provisions. **15.3**

ASCE 7-05 Chapter 13 governs the design of architectural, mechanical and electrical components supported by nonbuilding structures.

Structural design requirements. The provisions of this section govern the seismic design of nonbuilding structures included in the scope of Chapter 15. **15.4**

Where approved standards are available for nonbuilding structures, such standards are to be used in the seismic design of such structures, subject to the amendments given in this

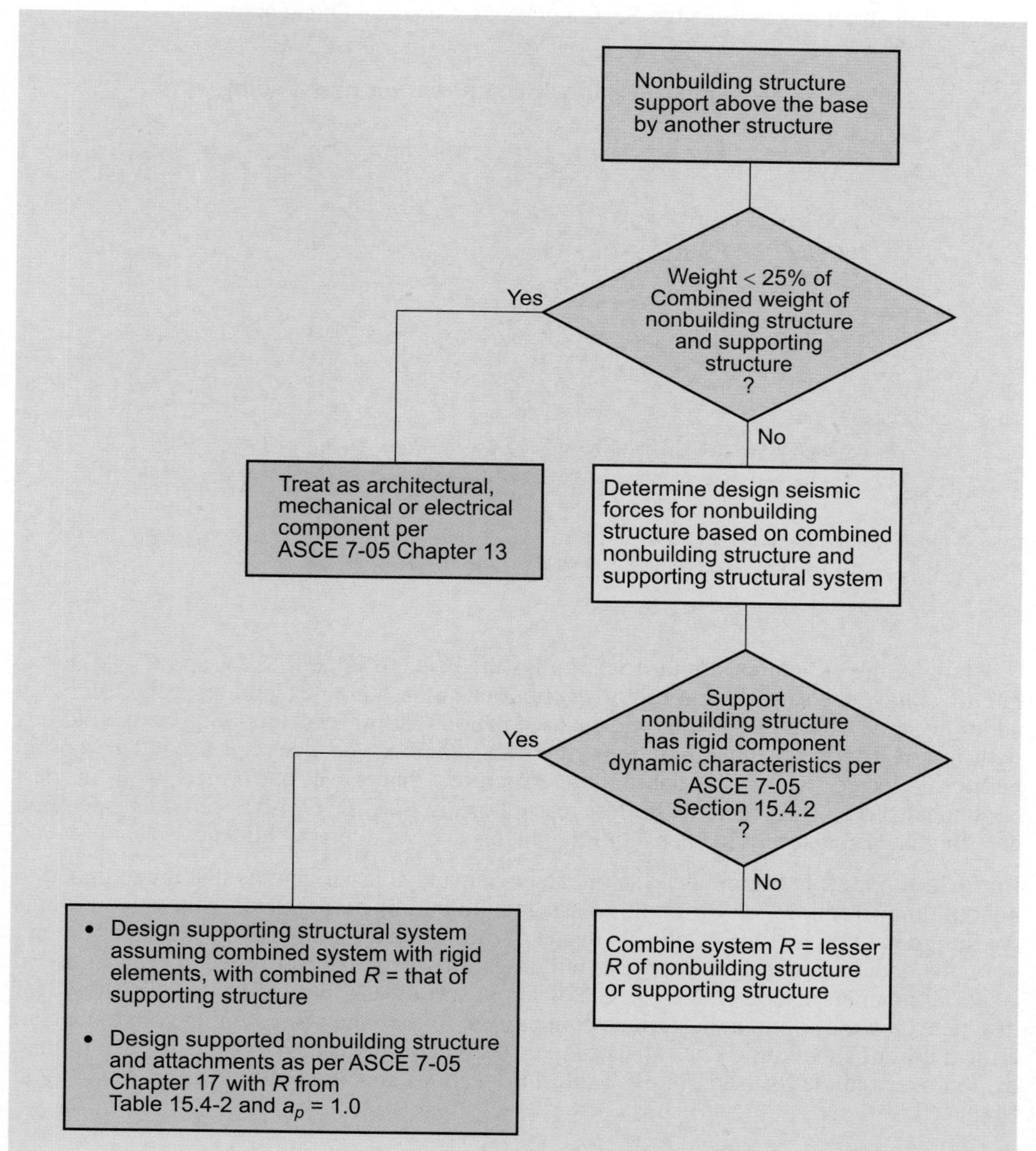

Figure 16-49
Code requirements for nonbuilding structures supported by other structures

section. Also, the requirements of Sections 15.5 (nonbuilding structures similar to buildings) and 15.6 (nonbuilding structures not similar to buildings) must be satisfied when the structure is subject to a seismic lateral force not less than that obtained from the equivalent lateral force procedure of Section 12.8. Note that nine additions and exceptions are made to those provisions for nonbuilding structures, as outlined in Section 15.4.1. The provisions of this section are summarized in Table 16-34.

ASCE 7-05 Section 12.3.4.1 permits ρ to be taken equal to 1.0 in the design of nonbuilding structures that are not similar to buildings.

ASCE 7-05 Section 15.4 contains provisions that were in the previous edition of ASCE 7; however, many substantive revisions are included. In the updated section, the seismic design coefficients (R, Ω_0, and C_d) table is split into two tables—one for structures similar to buildings (Table 15.4-1 Seismic Coefficients for Nonbuilding Structures Similar to Buildings), and the other for structures not similar to buildings (Table 15.4-2, Seismic Coefficients for Nonbuilding Structures NOT Similar to Buildings). Also, references to the applicable design and detailing requirements in other sections and reference documents are added to the two tables. The tables also prescribe height limits for structural systems of nonbuilding structures assigned to Seismic Design Category D and above.

Table 16-34. Minimum seismic design forces for nonbuilding structures

$$V = C_s W$$

$$= \frac{S_{D1} T_L L}{T^2 (R / I)} \leq \frac{S_{D1} W}{T (R / I)} \leq \frac{S_{DS} W}{R / I}$$

$$\geq 0.03 \text{ instead of } 0.01$$

$$\geq \frac{0.8 S_1 I W}{R} \text{ instead of } \frac{0.5 S_1 W}{R}$$

where:

R = Value gives in ASCE 7-05 Table 15.4-1 or 15.4-2, as applicable

I = From reference documents/ASCE 7-05 Table 11.5-1/Chapter 15

Important points:

1. Distribute seismic lateral forces vertically per reference document or ASCE 7-05 Section 12.8.3 or 12.9
2. Apply height limitations of ASCE 7-05 Table 15.4-1 or 15.4-2, as applicable
3. Note exceptions

Nonbuilding structures similar to buildings that use the same R-, Ω_0-, and C_d- values as buildings have the same height limits, restrictions and footnote exceptions as buildings. In addition, some structural systems may be used in nonbuilding structures similar to buildings with less restrictive height limitations, if lower specified R-values are used. This option permits selected types of nonbuilding structures that have performed well in past earthquakes to be constructed with less restrictions in SDC D, E and F, provided specified detailing is used and provided design force levels are considerably higher.

15.4.1.1 Importance factor. Importance factors and occupancy classifications that are assigned to nonbuilding structures vary from those assigned to building structures. Although buildings are designed to protect the occupants inside of the structure, nonbuilding structures are not normally occupied in the same sense as buildings. However, they still need to be designed in a special manner, insofar as they pose a different sort of risk in regard to public safety (for instance, they may contain hazardous compounds, or they may be essential components in critical lifeline functions). The value of I to be used in design is the largest value determined by approved standards, ASCE 7-05 Table 11.5-1, or as specified elsewhere in ASCE 7-05 Chapter 15.

Rigid nonbuilding structures. Nonbuilding structures that have a fundamental period, *T*, less than 0.06 second are defined as rigid nonbuilding structures. For such structures, ASCE 7-05 Equation 15.4-5 is to be used to determine the seismic base shear, *V*. The base shear is distributed over the height of the structure in accordance with the equivalent lateral force procedure (ASCE 7-05 Section 12.8.3). **15.4.2**

Loads. In addition to the dead load defined in ASCE 7-05 Section 12.7.2, normal operating contents of items such as tanks, vessels, bins, hoppers and the contents of piping must be included in the weight, *W*, for the purpose of calculating the design seismic forces. Snow and ice loads are to be included when they constitute 25 percent or more of *W*. **15.4.3**

Fundamental period. The rational methods for period determination contained in ASCE 7-05 Section 12.8.2 were developed for building structures. If a nonbuilding structure has dynamic characteristics similar to those of a building, the methods are directly applicable. However, Equations 12.8.7 through 12.8.10, which are specified for calculating the fundamental period of building structures, are not to be used for nonbuilding structures, because those equations are not relevant for commonly encountered nonbuilding structures. Thus, Equation 15.4-6 is provided specifically for nonbuilding structures as an alternative to carrying out a rational analysis. The equation is derived from Rayleigh's formula and involves application of a lateral force distribution based on the structural type and calculation of deflections at different levels. Accuracy of Rayleigh's formula depends on the assumed distribution of the lateral forces. However, as a property of Rayleigh's formula, any error in the assumption would always make the calculated time period lower than the actual, making the design conservative. **15.4.4**

If a nonbuilding structure is not similar to a building structure, other techniques for period calculation may be required. Some of the references for specific types of nonbuilding structures contain more accurate methods.

Drift limitations. The drift limits of ASCE 7-05 Section 12.12.1 are allowed to be exceeded when it can be shown by rational analysis, including consideration of *P*-delta effects, that they may be exceeded without adversely affecting structural stability. **15.4.5**

Material requirements. This section stipulates that the seismic requirements of ASCE 7-05 Chapter 14 apply, unless specifically exempted in Chapter 15. However, as mentioned before, 2006 IBC Section 1613 exempts structures from having to comply with the requirements in ASCE 7-05 Chapter 14 because the material-specific seismic design requirements are covered in the materials chapters of the 2006 IBC. **15.4.6**

Deflection limits and structure separation. Unless specifically amended, deflection limits and separation are to be determined in accordance with the applicable sections in ASCE 7-05. **15.4.7**

Site-specific response spectra. Site-specific seismic characteristics need to be taken into account when required by a reference document or the authority having jurisdiction. If a recurrence interval longer than 2,500 years, as considered by ASCE 7-05, is required by the reference document, then the longer interval should be considered.

Nonbuilding structures similar to buildings. In this section, specific guidance is provided for: Pipe Racks (Section 15.5.2), Steel Storage Racks (Section 15.5.3), Electrical Power Generating Facilities (Section 15.5.4), Structural Towers for Tanks and Vessels (Section 15.5.5), and Piers and Wharves (only those accessible to the public—Section 15.5.6). A typical example of a nonbuilding structure similar to buildings can be seen in Figure 16-50. **15.5**

Nonbuilding structures not similar to buildings. This section provides specific guidance for: Earth-Retaining Structures (Section 15.6.1), Stacks and Chimneys (Section 15.6.2), Amusement Structures (Section 15.6.3), Special Hydraulic Structures (Section 15.6.4), Secondary Containment Systems (Section 15.6.5), and Telecommunication Towers (Section 15.6.6). A typical example of a nonbuilding structure not similar to buildings can be seen in Figure 16-51. **15.6**

Figure 16-50 Nonbuilding structure similar to buildings

Figure 16-51 Nonbuilding structure not similar to buildings

15.7 Tanks and vessels. This section provides specific requirements for all tanks, vessels, bins and silos, and similar containers storing liquids, gases and granular solids supported at the base. Specific guidance is provided for Ground-Supported Storage Tanks and Liquids (Section 15.7.6), Water Storage and Water Treatment Tanks and Vessels (Section 15.7.7), Petrochemical and Industrial Tanks and Vessels Storing Liquids (Section 15.7.8), Ground-Supported Storage Tanks for Granular Materials (Section 15.7.9), Elevated Tanks and Vessels for Liquids and Granular Materials (Section 15.7.10), Boilers and Pressure Vessels (Section 15.7.11), Liquid and Gas Spheres (Section 15.7.12), Refrigerated Gas Liquid Storage Tanks and Vessels (Section 15.7.13), and Horizontal, Saddle Supported Vessels for Liquid or Vapor Storage (Section 15.7.14).

Chapter 16 Seismic Response History Procedures

This chapter includes the less commonly used dynamic analysis procedures: (1) linear response history procedure and (2) nonlinear response history procedure. They appear later in the standard because they are not used as often as other provisions. Although these procedures can be used for any structure, ASCE 7-05 Table 12.6-1 specifies when they are required to be used.

16.1

Linear response history analysis. This section provides the basic guidelines on conducting a linear response history analysis in terms of modeling criteria, selection of ground motions and scaling of the analytical results for design purposes.

For the mathematical modeling of a building, the requirements are the same as those given in ASCE 7-05 Section 12.7.

Several criteria are specified for the selection of ground motion acceleration time histories:

1. Selected ground motion time histories need to be consistent with the nature of maximum considered earthquake expected at the building location. It is preferable to derive the ground motion time histories from actual recorded time histories. However, quite often, an adequate number of recorded time histories may not be available, and synthetic records are permitted to be used.
2. At least three ground motion time histories must be selected. For a 3-D analysis, three pairs of orthogonal ground motions need to be selected. As explained in the Commentary to the 2003 NEHRP *Provisions*, the response history analyses are highly sensitive to the characteristics of the selected ground motions. Thus, in order to capture a range of structural response characteristics, the use of multiple ground motion time histories is essential. Because of this, in addition to specifying a minimum number of three, this chapter stipulates design incentives when seven or more ground motions are used. This will be discussed later.
3. Ground motions need to be scaled to the design level. To determine the scaling factors, 5-percent damped response spectra of all selected ground motion are constructed, and an average spectral value is computed at each period. The ground motions need to be scaled such that the average spectral values are not less than the corresponding values from the design response spectrum at the site (as given in ASCE 7-05 Section 11.4.5) within periods of $0.2T$ and $1.2T$, where T is the fundamental period of the structure in the direction of the analysis. For a 3-D analysis, 5-percent damped spectra from the orthogonal ground motions in each pair of selected ground motion are first combined by taking the square root of the sum of the squares (SRSS) of the spectral values at each period, and the combined values are then averaged over all the selected ground motion pairs at each period. Ground motions need to be scaled such that this average is not smaller than 1.3 times the corresponding value from the design response spectrum obtained from ASCE 7-05 Section 11.4.5 or 21.2 by more than 10 percent in the period range between $0.2T$ and $1.5T$. Also, the same scale factor is to be used for the two orthogonal components of any ground motion pair.

No specific requirements on how to perform a time history analysis are given in this chapter. However, a very clear guideline on how to apply the two orthogonal components of a ground motion pair in a 3-D analysis is given in ASCE 7-05 Section 12.5.

Once the analyses are complete and the structural responses to the selected ground motions are obtained, the design response parameters are to be scaled based on the occupancy of the structure as well as the type of seismic force-resisting system used. Thus, base shear, member forces and story drifts obtained from each ground motion analysis are multiplied by the quantity I/R, where I is the importance factor of the building (ASCE 7-05

Table 11.5-1) and R is the response modification factor of the seismic force-resisting system used (ASCE 7-05 Table 12.2-1). (Note that, although ASCE 7-05 Section 16.1.4 states precisely what is stated in the previous sentence, it does not make sense to multiply story drifts by I/R. We believe the intent and the proper thing to do is to multiply story drifts by C_d/R. See Section 12.9.4). Also, if the base shear calculated for a particular ground motion is less than the minimum design base shear as given by ASCE 7-05 Eq. 12.8-5 or Eq. 12.8-6, whichever is applicable, member forces corresponding to that ground motion need to be further scaled up by the ratio of the minimum base shear and the base shear calculated for that particular ground motion. This scaling is not required for story drifts.

Finally, where it is required (ASCE 7-05 Sections 12.3.3.3, 12.10.2.1, 12.14.7.3, 1808.2.23.2.1, 1808.2.23.2.3 and 1908.1.12) to use the load combinations with overstrength factor of ASCE 7-05 Section 12.4.3.2, $\Omega_0 Q_E$ need not exceed the maximum of the unscaled values of Q_E corresponding to the various ground motion (or ground motion pairs) considered. Unscaled in this context means not scaled by I/R or by the ratio of minimum base shear to the base shear calculated for a particular ground motion.

It is stipulated that the maximum member forces and story drifts from all analyses are to be used for design purposes when the number of ground motions (or the number of ground motion pairs, for 3-D analysis) is less then seven. However, when the number of ground motions (or ground motion pairs) is seven or more, average member forces as well as story drifts can be used for design. This is to encourage designers to use as many ground motions as possible so that a wide range of structural responses can be taken into account.

16.2 Nonlinear response history procedure. This is the most complex and computationally exhaustive procedure that, when used properly, provides the most accurate estimate of the structural response during a design ground motion. Whereas all other analysis methods included in the code try extrapolate the inelastic response of a building from its elastic response, this procedure directly accounts for the inelastic hysteretic behavior of structural members, thus predicting the structural response more realistically. A more detailed discussion on the strengths and shortcomings of this analysis method can be found in the 2003 NEHRP *Provisions* Commentary.

The same requirements regarding mathematical modeling of the structure as those for linear time history analysis apply for nonlinear analysis as well. However, additional requirements are given for modeling the post-yield behavior of the members. Hysteretic behavior of the elements needs to incorporate appropriate strength and stiffness degradations, and hysteretic pinching that are consistent with laboratory test data. Material overstrength and strain-hardening need to be taken into account as well. Foundations can be assumed fixed or can be modeled considering appropriate soil properties. When diaphragms are not rigid as per Section 12.3.1, they are also required to be modeled.

This section gives detailed requirements concerning when a 3-D analysis is required and when a 2-D analysis is permitted. This is in addition to requirements found in ASCE 7-05 Section 12.5.

Requirements concerning the selection of ground motions are the same as those for the linear time history procedure.

Unlike the linear analysis procedure, no scaling of the response parameters is required in the nonlinear procedure. An average can be taken of the maximum member forces, member inelastic deformations and story drifts from all individual analyses when seven or more ground motions are used. Otherwise, the maximum values of those response parameters from all analyses are to be used for design purposes. Member inelastic deformations need to be compared against the acceptable limits, which are also specified in this section. Maximum story drifts need to be smaller than 125 percent of the story drift limits specified in ASCE 7-05 Section 12.12.1. Because a nonlinear time history analysis gives the most realistic estimate of the structural response, the limit on story drift is less stringent than that for other analysis procedures.

As in the case of a linear time history procedure, nonlinear time history procedures are also very sensitive to the various modeling assumptions involved in the analysis. However, nonlinear procedures involve numerous modeling techniques to represent the different and complex characteristics of the hysteretic behavior of various structural components, and those models are still not standardized in a code format. As a result, there is scope for a wide variation in structural response an account of improper uses of those inelastic models. For this reason, there is a special provision for nonlinear time history analysis, which requires a review of the design and analysis by an independent team of registered design professionals who are experienced in the theory and methods involved in a nonlinear seismic analysis. The minimum review requirements are given in Section 16.2.5.

Chapter 17 *Seismic Design Requirements for Seismically Isolated Structures*

Background

Many modern buildings contain sensitive and expensive equipment that is vital in business, commerce, education and health care. The contents frequently are more valuable than the buildings themselves. Also, hospitals, communication and energy centers, and police and fire stations must be operational during and following an earthquake.

Conventional construction can cause very high floor accelerations in stiff buildings and large interstory drifts in flexible structures. These two factors cause difficulties in ensuring the safety of the building components and contents.

Mounting a building on an isolation system (base isolation) may prevent most of the horizontal movement of the ground from being transmitted to the building. This results in a significant reduction in floor accelerations and interstory drifts, thereby providing protection to the building contents and components.[60]

Base isolation also has significant benefits in seismic retrofit where, because of a desire to preserve the historical character of a building or to avoid disruption, conventional seismic strengthening may not be practical. The concept of base isolation was implemented in Japan and New Zealand before it was first used in the United States to isolate buildings from earthquake ground motions in 1986. Base-isolated structures have by now been built in at least 17 countries throughout the world. The total number of base-isolated structures built or under construction as of 1994 was in excess of 300.[61] Most of these were in Japan, New Zealand and the United States. Kelly[62] has provided an excellent history, and Buckle[63] an excellent world overview. Excellent state-of-the-art information is available in Reference 60. The EERI has devoted an entire issue of Earthquake Spectra to the theme of "Seismic Isolation: From Idea to Reality."[64] Perhaps the most lucid presentation of the concept has been provided by Bachman.[61] The material presented here is drawn largely from References 60 and 61.

Base isolation is founded on fundamental principles of earthquake engineering and accepted assumptions regarding earthquake ground motions. Most low-rise structures (less than 10 stories in height) have a fixed-base fundamental period of 1 second or less. It has been observed in past earthquakes that these structures typically respond to earthquake motions by amplifying their acceleration response by two to four times. However, for structural systems with periods of 2 to 3 seconds, the acceleration responses are no longer amplified and are in fact generally reduced. Base isolation takes advantage of this characteristic of structural response to earthquake ground motion.

In Figure 16-52, a comparison of the response of a conventional fixed-base structure and a base-isolated structure is illustrated. In a base-isolated structure, the columns of the structure are typically supported on individual base isolation bearings. These bearings usually have relatively low horizontal stiffness and relatively high vertical stiffness. The relatively low horizontal stiffness shifts the response characteristics of a typical lower-rise building system from maximum roof accelerations of $1.0g$ to $1.5g$ to maximum roof

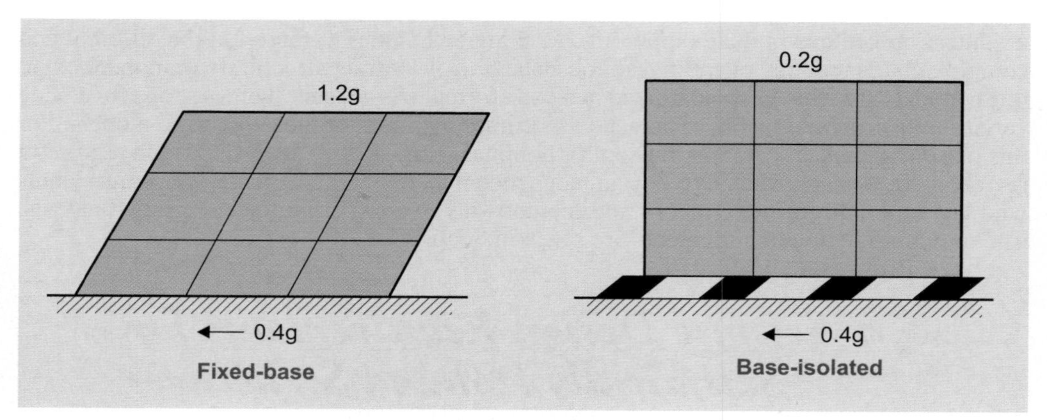

Figure 16-52
Seismic response of a conventional structure and a base-isolated structure

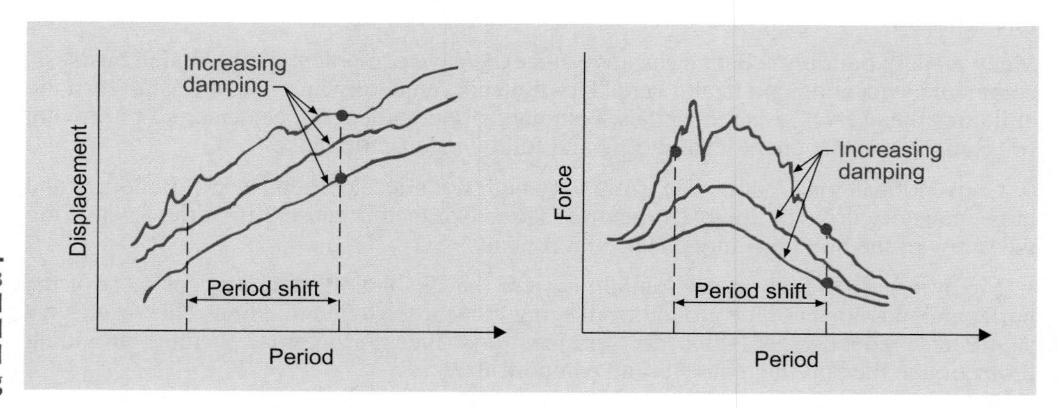

Figure 16-53
Effects of period shift and increased damping

accelerations of 0.2*g* to 0.3*g*. This reduction is further enhanced if the bearings have inherent increased damping (Figure 16-53).

Typically, bearings behave nonlinearly with equivalent viscous damping of 10 to 20 percent of critical or more. This compares with approximately 5 percent for conventional construction. It should be noted that, as the period shifts to larger values, the earthquake response displacements increase. This increase can also be tempered by an increase in damping (Figure 16-53). Because with base isolation the greatest portion of the response occurs across the base isolation bearings, almost all of the response displacement occurs between the base of the structure and the ground. This means that, in high earthquake zones, relative displacements of 6 to 24 inches must be accommodated across the isolator interface by utility lines and entrance ways during maximum design earthquake ground motions. The vertical stiffness of bearings is typically 50 to 1,000 times the horizontal stiffness so that rocking and vertical amplifications of response are minimized.

Types of bearings

The four types of bearings commonly available include: laminated, high-damping rubber bearings; laminated rubber bearings with lead core; steel sliding bearings using dished surfaces; and coil spring systems. Laminated rubber bearings (high-damping and lead core) are most often used. They consist of layered sheets of alternating rubber and steel plates vulcanized together. Typically square or round in plan, these bearings range in width from 14 to 45 inches and some contain a lead core to provide additional damping (Figure 16-54).

Basic elements of seismic isolation systems

There are three basic elements in any practical seismic isolation system. These are:

1. A flexible mounting so that the period of vibration of the total system is lengthened sufficiently to reduce the force response.

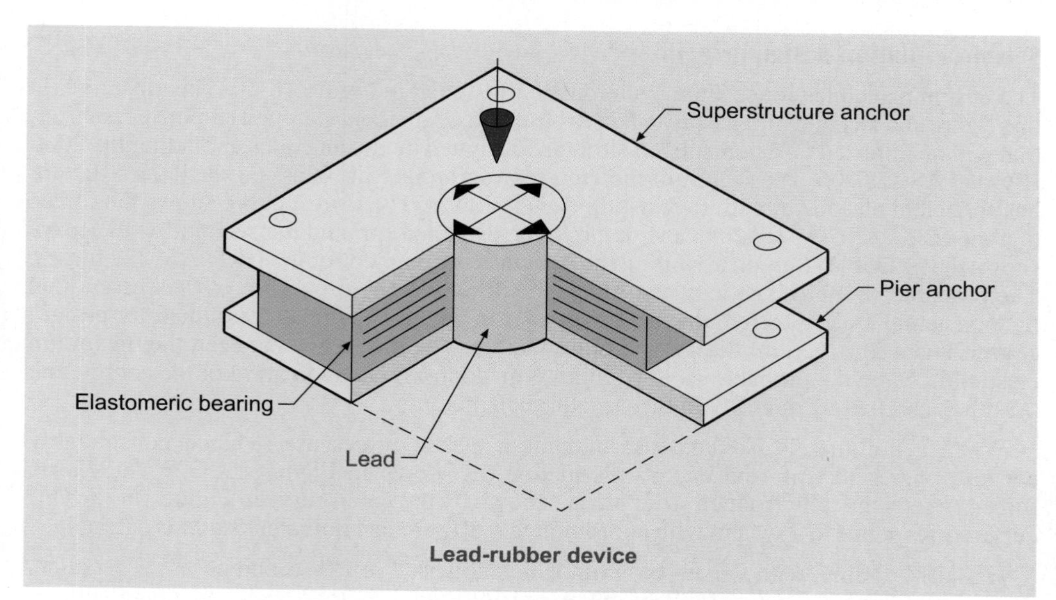

Figure 16-54
A mechanical energy dissipator

2. A damper or energy dissipator so that the relative deflections between building and ground can be controlled to a practical design level.
3. A means of providing rigidity under low (service) levels of excitation such as caused by wind and minor earthquakes.

Although lateral flexibility is highly desirable for high seismic loads, it is clearly undesirable to have a structural system that will vibrate perceptibly under frequently occurring excitations such as those caused by minor earthquakes or wind. Lead-rubber bearings (and other mechanical energy dissipators) provide the desired initial rigidity by virtue of their high elastic stiffness. Some other seismic isolation systems require a wind-restraint device for this purpose—typically a rigid component designed to fail under a given level of lateral load. This can result in a shock loading being transferred to the structure because of the sudden loss of load in the restraint. Nonsymmetrical failure of such devices can also introduce undesirable torsional effects in a building. Further, such devices need to be replaced after each failure.

Practicability

As pointed out by Mayes,[60] five developments over the last two decades are together responsible for the ever-increasing application of the seismic isolation concept:

1. The design and manufacture of high-quality elastomeric (rubber) pads, frequently called bearings, that are used to support the weight of the structure but at the same time protect it from earthquake-induced forces.
2. The design and manufacture of mechanical energy dissipators (absorbers) and high-damping elastomers that are used to reduce the movement across the bearings to practical and acceptable levels and to resist wind loads.
3. The development and acceptance of computer software for the analysis of seismically isolated structures, which includes nonlinear material properties and the time-varying nature of the earthquake loads.
4. The ability to perform shaking-table tests using real recorded earthquake ground motions to evaluate the performance of structures and provide results to validate computer modeling techniques.
5. The development and acceptance of procedures for estimating site-specific earthquake ground motions for different return periods.

Seismic isolation design principles[60]

The design principles for seismic isolation are illustrated in Figure 16-50. The top curve on this figure shows the realistic elastic forces based on a 5-percent-damped response spectrum that will be imposed on a nonisolated structure designed by recent codes, including the 2006 IBC and ASCE 7-05. The spectrum shown is for a soft rock (Site Class B) site if the structure has sufficient elastic strength to resist this level of load. The lowest curve shows the forces that the IBC / ASCE 7 requires a structure to be designed for, and the second-lowest curve shows the probable strength assuming the structure is designed for the IBC / ASCE 7 forces. The probable strength is at least two times the design strength because of the design load factors, actual material strengths that are greater in practice than those assumed for design, conservatism in structural design and other factors. The difference between the maximum elastic force and the probable yield strength is an approximate indication of the energy that must be absorbed by ductility in the structural elements.

When a building is isolated, the maximum elastic forces are reduced considerably because of period shift and energy dissipation, as shown in Figure 16-53.[64] The elastic forces on a seismically isolated structure are shown by the green curve in Figure 16-55. This curve corresponds to a system with approximately 30 percent equivalent viscous damping.

If a stiff building, with a fixed-base fundamental period of 1.0 second or less, is isolated, then its fundamental period will be increased into the 1.5- to 2.5-second range (Figure 16-55). This results in a reduced design force but, more important, in the 1.5- to 2.5-second range, the probable yield strength of the isolated building is approximately the same as the maximum forces to which it will be subjected. Therefore, there will be little or no ductility demand on the structural system, and the lateral design forces are reduced by approximately 50 percent.

Performance in earthquakes[61]

During the Northridge and Hyogo-Ken Nanbu (Kobe) Earthquakes, base-isolated buildings were for the first time subjected to strong ground motion associated with Richter Magnitude 6+ earthquakes. During the Northridge Earthquake, the Los Angeles County Fire Command and Control Facility and the University of Southern California (USC) Hospital experienced

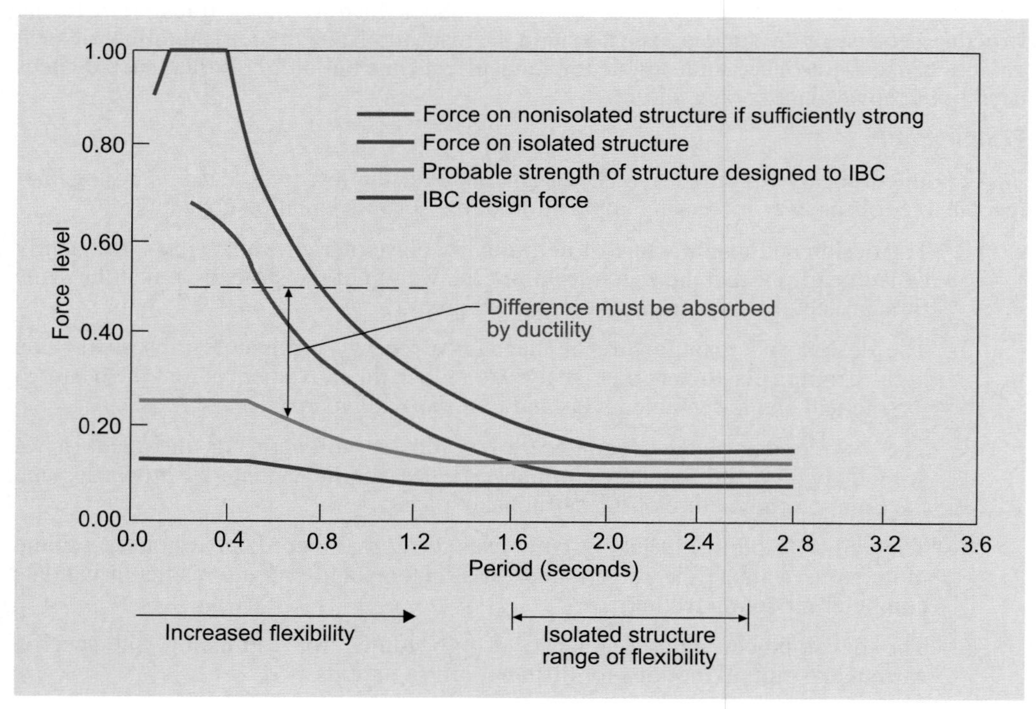

Figure 16-55 **Design principles of seismic isolation**

ground accelerations of 0.32g and 0.49g, respectively. The Los Angeles County Fire Command and Control Facility, supported by laminated high-damping rubber bearings, performed well and continued to function fully during and following the earthquake. Some intentionally sacrificial architectural tile was damaged at a seismic joint at the main entrance, as planned. However, the tile grout across the joint was thicker and stronger than anticipated and caused some localized *spike* type accelerations in one direction because of impact at the building entrance. This detail has since been replaced with a *pop-out* tile detail that will eliminate this impact in future earthquakes.

The USC Hospital, supported by laminated rubber bearings with lead cores, also performed exceptionally well with peak horizontal roof accelerations of 0.21 g.

In Kobe, two buildings located about 20 kilometers from the zone of maximum ground motion were base-isolated. One building, a large five-story postal computer center on lead rubber bearings, experienced ground accelerations in the range of 0.25g to 0.30g. Maximum roof accelerations were 0.11g with maximum isolator displacements of 100 to 110 mm. The other Kobe building, smaller and built on high-damping rubber bearings, performed well with no damage.

In all cases, the base-isolated buildings continued to function exactly as planned while nearby fixed-base buildings were significantly damaged and had their functions interrupted.

Cost issues[61]

When both fixed-base and base-isolated structures are evaluated based on current code design requirements, base-isolated structures initially cost typically 4 to 10 percent more to construct than fixed-base structures. The additional costs result from: the seismic isolation system, an additional first-floor beam-slab system and sub-basement, special mechanical and electrical details to accommodate isolator displacement, optional weather protection for the sub-basement and additional engineering costs. There may be some off-setting cost reduction of the base-isolated structural system that is due to the reduced structural system design forces.

If a fixed-base building and its contents are designed, strengthened and tested to have the same expected fully functional performance as a base-isolated building during design earthquake-level ground motions, the base-isolated building has been shown to be less expensive. In the case of the Los Angeles County Fire Command and Control Facility, the savings were estimated as approximately 6 percent.

Force-deflection characteristics of isolation systems

As pointed out in the commentary to Chapter 13 of the 2003 NEHRP *Provisions*, a general concern has long existed regarding the applicability of different types of isolation systems. Rather than addressing a specific method of base isolation, ASCE 7-05 provides general design requirements applicable to a wide range of possible seismic isolation systems. Although remaining general, the design requirements rely on mandatory testing of isolation-system hardware to confirm the engineering parameters used in the design and to verify the overall adequacy of the isolation system. Some systems may not be capable of demonstrating acceptability by test and, consequently, would not be permitted. In general, acceptable systems will: (1) remain stable for required design displacements, (2) provide increasing resistance with increasing displacement, (3) not degrade under repeated cyclic load, and (4) have quantifiable engineering parameters (e.g., force-deflection characteristics and damping).

Conceptually, there are four basic types of isolation system force-deflection relationships.[2,60] These idealized relationships are shown in Figure 16-56 with each idealized curve having the same design displacement, D_D, for the design earthquake. A linear isolation system is represented by Curve A and has the same isolated period for all earthquake load levels. In addition, the force generated in the superstructure is directly proportional to the displacement across the isolation system.

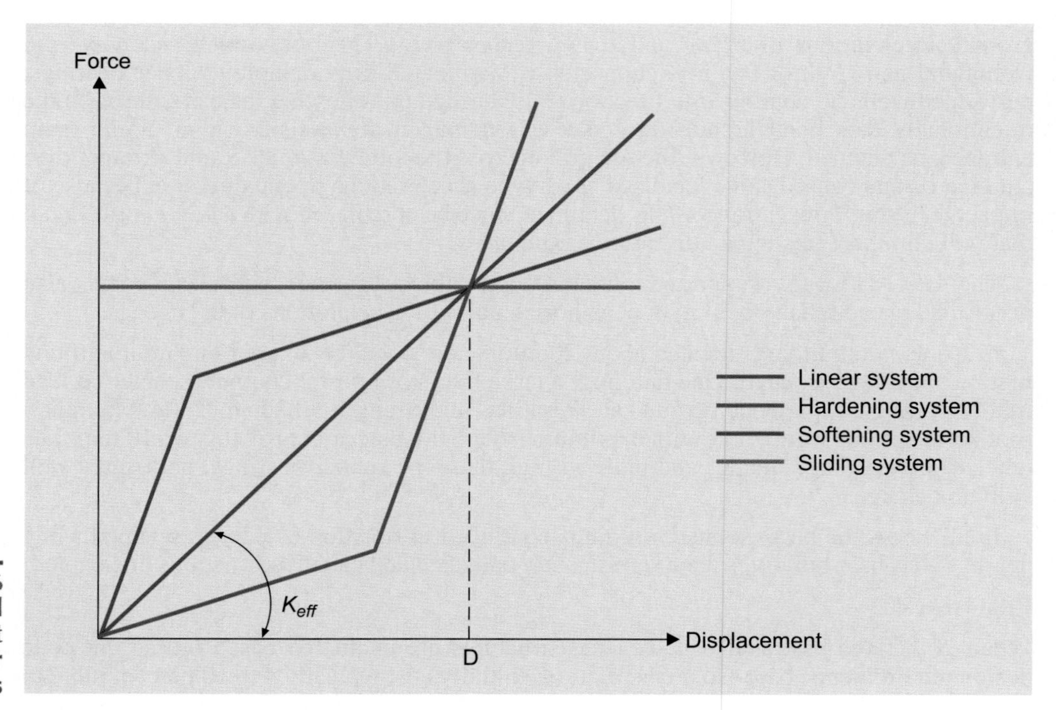

Figure 16-56 Idealized force-displacement relationship for isolation systems

A hardening isolation system is represented by Curve B. This system is soft initially (long effective period) and then stiffens (effective period shortens) as the earthquake load level increases. When the earthquake load level induces displacements in excess of the design displacement in a hardening system, the superstructure is subjected to higher forces and the isolation system to lower displacements, than a comparable linear system.

A softening isolation system is represented by Curve C. This system is stiff initially (short effective period) and softens (effective period lengthens) as the earthquake load level increases. When the earthquake load level induces displacements in excess of the design displacement in a softening system, the superstructure is subjected to lower forces, and the isolation system to higher displacements, than a comparable linear system.

A sliding isolation system is represented by Curve D. This system is governed by the friction force of the isolation system. Like the softening system, the effective period lengthens as the earthquake load level increases and loads on the superstructure remain constant. The displacement of the sliding isolation system after repeated earthquake cycles is highly dependent on the vibratory characteristics of the ground motion and may exceed the design displacement, D_D. Consequently, minimum design requirements do not adequately define peak seismic displacement for seismic isolation systems governed solely by friction forces.

Code provisions for base isolation

In the late 1980s, the leadership of the Structural Engineers Association of California realized the significant benefits of base isolation in seismic design. They also recognized that base isolation would only be widely accepted if code requirements existed for its design and implementation. Accordingly, in late 1989, the SEAOC Seismology Committee adopted an "Appendix to Chapter 2" of the SEAOC *Blue Book*[22] entitled, "General Requirements for the Design and Construction of Seismic-Isolated Structures." These requirements were adopted as an appendix to the 1991 UBC. The isolation appendix of the UBC used to be updated on an annual basis and the last version of these regulations may be found in the 1997 UBC Appendix Chapter 16, Division IV.

During development of the 1994 NEHRP *Provisions*, it was decided to use the then-latest version (1993 approved changes) of the SEAOC/UBC provisions as a basis for the development of the requirements included in the NEHRP *Provisions*. The only significant changes involved an appropriate conversion to strength design and making the requirements applicable on a national basis. For the 1997 NEHRP *Provisions*, it was decided to incorporate the 1997 UBC provisions. Because the 1997 UBC was based on strength design, Appendix Chapter 16, Division IV of the 1997 UBC and Chapter 13 of the 1997 NEHRP *Provisions* were almost identical, except for seismic criteria. The seismic criteria of the 1997 NEHRP *Provisions* were based on the then new national earthquake maps and the associated ground motion parameters.

Chapter 13 of the 2000 and 2003 NEHRP *Provisions* is essentially the same as that in the 1997 edition with a few exceptions. Section 13.2.3 (Section 13.2.1 in the 2003 NEHRP *Provisions*) was revised to correct an oversight concerning the importance factor. Also, Section 13.4.4.1 (2003 NEHRP *Provisions* Section 13.2.3.1) was revised to reflect that Site Class E is now covered by the seismic maps. Section 13.6.2.3 (2003 NEHRP *Provisions* Section 13.2.5.3) has been revised to provide additional wording to clarify the intent of the *Provisions*, whereas Section 13.9.2.1 (2003 NEHRP *Provisions* Section 13.6.1) has been changed to allow prototype bearings used for testing to be used in the construction of the structure if the registered design professional permits their use. The most significant change is the replacement of the brief Appendix to Chapter 13 with an extensive appendix (2003 NEHRP *Provisions* Chapter 15) with a complete set of design provisions for structures with damping systems. Chapter 17 of ASCE 7-05 is substantially the same as Chapter 13 of the 2003 NEHRP *Provisions*.

17.2 General design requirements.

Importance factor. Base isolated structures are to be assigned an occupancy category following the same provisions that are used for the fixed base building structures (ASCE 7-05 Table 1-1). However, unlike the fixed base structures, all base isolated structures are to be assigned an importance factor of 1.0, regardless of occupancy category.

MCE spectral response acceleration parameters. Maximum spectral accelerations at 0.2-second and 1-second time periods are to be obtained from spectral maps given in ASCE 7-05 Figures 22-1 through 22-10.

Configuration. The structural configuration of a building above the base determines its irregularity or lack thereof.

Isolation system. Various requirements of the isolation system are given in this section.

In addition to the requirements of gravity and lateral loads that are due to wind and earthquakes, other environmental conditions that must be considered in the design of the isolation system include: aging effects, creep, fatigue, operating temperature, and exposure to moisture or damaging substances.

Lateral displacement that is due to wind forces in the isolation system is required to be limited to a value similar to that required for other story heights.

If a base-isolated building is exposed to fire, the isolation system should be capable of supporting the weight of the building, as required for other vertical-load-carrying elements of the structure, but may end up having diminished resistance to lateral loads. [2]

The isolation system must be configured so as to develop a restoring force sufficient to avoid significant residual displacement in an earthquake. This is necessary to ensure that the isolated structure will not have a stability problem and that it will be in a condition to withstand aftershocks and future earthquakes.

Where a displacement restraint is used, explicit analysis of the isolated structure for the maximum considered earthquake is required to account for the effects of engaging the restraint.

The stability of the vertical-load-carrying elements of the isolation system must be checked by analysis and test, as required, for lateral seismic displacements equal to the total

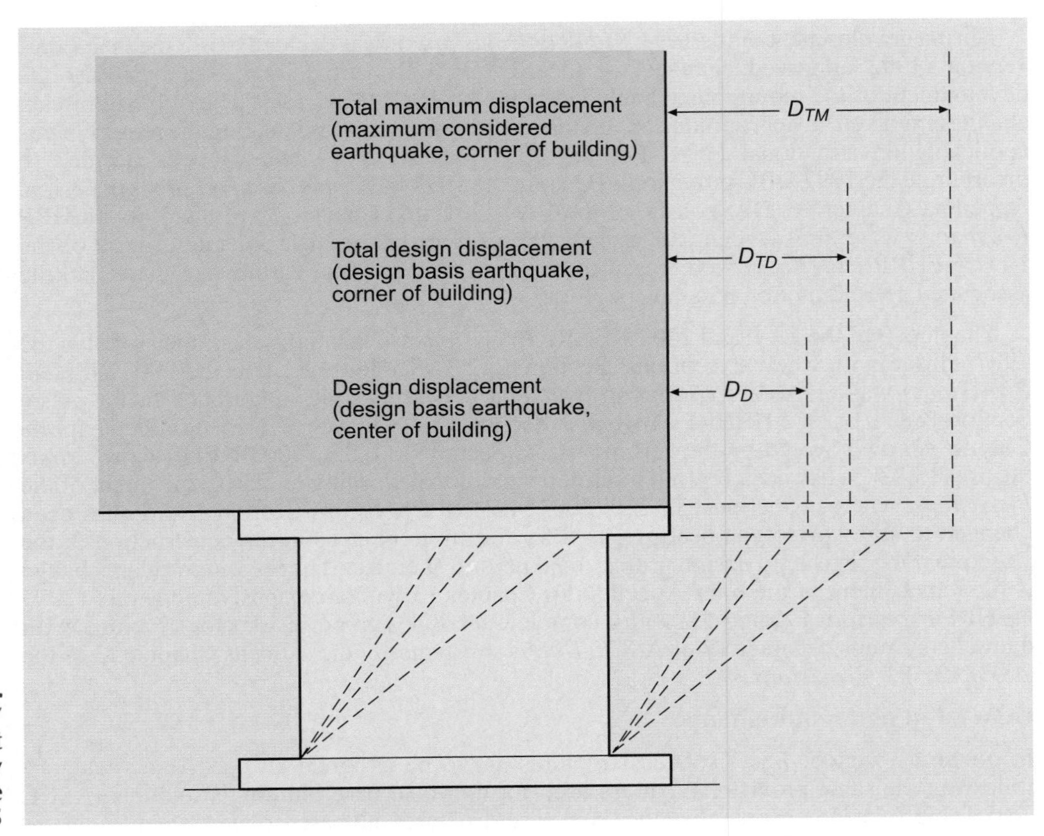

Figure 16-57 Displacement terminology used in the ASCE 7-05

maximum displacement (see Figure 16-57 for terminology). The vertical load to be used in checking the stability of an isolator must be calculated using both the maximum and minimum effects of dead load and live load and the peak axial force induced by the maximum considered earthquake ground motion instead of the design earthquake ground motion. The lateral forces are also to be calculated from the maximum considered ground motion instead of the design earthquake ground motion for checking the stability.

The intent of the overturning requirement is to prevent global structural overturning and overstress of elements that is due to local uplift.

Although most isolation systems are unlikely to require replacement of major components following an earthquake, ASCE 7-05 Section 17.2.4.8 provides for access for inspection and replacement as well as for certain specific inspections, including a periodic monitoring, inspection and maintenance program for the isolator system.

A test and inspection program is necessary for both fabrication and installation of the isolator system.

Structural system. The requirements of ASCE 7-05 Section 17.2.5 apply to the structural system above the isolation interface.

The requirement in ASCE 7-05 Section 17.2.5.1 emphasizes the need to provide for transfer of lateral forces from the isolation system across the isolation interface to the structural system above.

A minimum separation between an isolated structure and a rigid obstruction is required to be provided.

Base-isolated nonbuilding structures must be designed by ASCE 7-05 Chapter 15, using the design forces and design displacements calculated by the equivalent lateral force procedure of Section 17.5 or the dynamic lateral response procedure of Section 17.6.

Elements of structures and nonstructural components. This section applies to parts and portions of an isolated structure, permanent nonstructural components and the attachments to them, and the attachments for permanent equipment supported by a structure (collectively referred to here as components). For components at or above the isolation interface, the exception to ASCE 7-05 Section 17.2.6.1 provides a straightforward and usually conservative design procedure. However, according to the SEAOC *Blue Book* Commentary,[22] to obtain the full benefits of a seismically isolated building, it is necessary to base the design of these components on the dynamic response of the isolated structure, through the generation of floor spectra for the building. For critical buildings designed to remain operational after an earthquake, this approach to the design of nonstructural components and equipment may be necessary. ASCE 7-05 Section 17.2.6.2 addresses components crossing the isolation interface. To accommodate the differential movement between the isolated building and the ground, provisions for flexible utility connections must be made. In addition, rigid components crossing the interface (stairs, elevator shafts and walls) must have details to accommodate the differential movement at the isolator level without sustaining damage of such magnitude as may endanger life safety. Components below the isolation interface must be designed to ensure complete load path, per ASCE 7-05 Sections 12.1 and 13.

17.3

Ground motion for isolated systems. When the isolation system is designed using a design spectrum, it should be obtained in the same way as that for structural systems from design ground motions in accordance with ASCE 7-05 Section 11.4.5. When the structure is located on a near-fault site, site-specific ground motion hazard analysis needs to be performed as per ASCE 7-05 Section 21.2 to generate the design spectrum. A design spectrum from the maximum considered earthquake also needs to be generated for the purpose of determining maximum displacement of the isolation system, and it should not be less than 1.5 times the design spectrum obtained from the design earthquake.

When the response history procedure is used, pairs of orthogonal ground motion time histories need to be derived from real life recorded events by appropriate scaling. The same scale factor is to be applied to both the components of a pair. The source events should be selected considering the similar magnitude, fault distance and source mechanism as those for the site considered. As specified in ASCE 7-05 Section 17.6.3.4, at least three pairs of ground motions need to be used. Synthetically generated ground motions can also be used when an adequate number of recorded ground motion time histories is not available. The selected pairs of ground motions are also required to meet a minimum criterion that is described step-by-step below:

1) 5 percent-damped response spectra are to be obtained for all the selected ground motions.
2) Spectral values from the two orthogonal ground motions from a given pair are to be combined by SRSS (square root of the sum of the squares).
3) Average of the SRSS spectral values from all the selected ground motion pairs at different time periods are to be computed.
4) The effective period of the isolated structure at the design displacement, T_D, and the effective period of the isolated structure at maximum displacement, T_M, are calculated to be as per ASCE 7-05 Sections 17.5.3.2 and 17.5.3.4, respectively.
5) The average spectra computed as above between the time periods of $0.5T_D$ and $1.25T_M$ should not fall below 1.17 times the design spectra obtained in Section 17.3.1.

17.4

Analysis procedure selection. Three procedures are included in this section: equivalent lateral force procedure, response-spectrum analysis and time-history analysis.

Rather severe limitations are placed on the applicability of the equivalent lateral force procedure. The limitations cover the site location with respect to active faults; soil characteristics at the site, the height, regularity and stiffness characteristics of the building; and the characteristics of the isolation system.

The equivalent lateral force procedure is based on a spectral shape that varies with the inverse of the elastic fundamental period, $1/T$, consistent with the constant velocity domain. The procedure may not be used when the spectral demands may not be adequately characterized using the assumed shape, namely, for:[22]

1. Isolated buildings located in the near-field.
2. Isolated buildings on soft soil sites.
3. Long-period isolated buildings (beyond the constant velocity domain).

Further, the equivalent lateral force procedure may not be used for other than regular superstructures or for highly nonlinear isolation systems. The restrictions placed on the use of equivalent lateral force design procedures effectively require dynamic analysis for virtually all isolated structures. However, lower-bound limits on isolation system design displacements and structural design forces are specified by ASCE 7-05 in Section 17.6 as a percentage of the values obtained from the equivalent lateral force design formulas, even when dynamic analysis is used as the basis for design. These lower-bound limits on key design parameters ensure consistency in the design of isolated structures and serve as a safety net against gross under-design.[2] Table C13.2-2 of the 2003 NEHRP Commentary provides a summary of the lower-bound limits on parameters obtained from dynamic analysis.

Response-spectrum analysis is permitted for the design of all isolated buildings except: (1) buildings located on very soft soil sites (for which site-specific spectra must be established), or (2) buildings supported by highly nonlinear isolation systems for which the assumptions implicit in the definitions of effective stiffness and damping break down.

Time-history analysis is the *last resort* analysis procedure that must be used when the conditions for the applicability of the static and the response spectrum analyses cannot be met. It may be used for the analysis of any isolated building. The SEAOC *Blue Book* rightly cautions that the results of time-history analysis must be carefully reviewed to avoid any gross design errors or over-design.

17.5 Equivalent lateral force procedure. The lateral displacement given by ASCE 7-05 Equation 17.5-1 approximates peak design earthquake displacement of a single-degree-of-freedom, linear elastic system of period, T_D, and equivalent viscous damping, β_D. The lateral displacement given by Equation 17.5-3 approximates peak maximum considered earthquake displacement of a single-degree-of-freedom, linear elastic system of period, T_M and equivalent viscous damping, β_M.

Deformation characteristics of the isolation system. The deformation characteristics of the isolation system, used to determine the minimum lateral earthquake design displacements and forces on seismically isolated structures, are required to be based on properly substantiated tests conducted in accordance with ASCE 7-05 Section 17.8.

Minimum lateral displacements. Equation 17.5-1 gives an estimate of peak displacement expected in the isolation system in the design earthquake. In this equation, the spectral acceleration, S_{D1}, is the same as that required for the design of a conventional fixed-base structure of period, T_D, as given by Equation 17.5-2. A damping term, β_D, is used to increase or decrease the computed displacement when the equivalent damping coefficient of the isolation system is smaller or larger, respectively, than 5 percent of critical damping. Values of the coefficient, β_D, are given in ASCE 7-05 Table 17.5-1 for different values of isolation system damping, β_D.

Equation 17.5-3 gives an estimate of peak displacement expected in the isolation system in the maximum considered earthquake. In this equation, the spectral acceleration, S_{M1}, is the same as that required for the design of a conventional fixed-base structure of period, T_M, as given by Equation 17.5-4. The damping term, β_M, increases or decreases the computed displacement when the equivalent damping coefficient of the isolation system is smaller or larger, respectively, than 5 percent of critical damping. Values of the coefficient, β_M, are given in ASCE Table 17.5-1 for different values of isolation system damping, β_M.

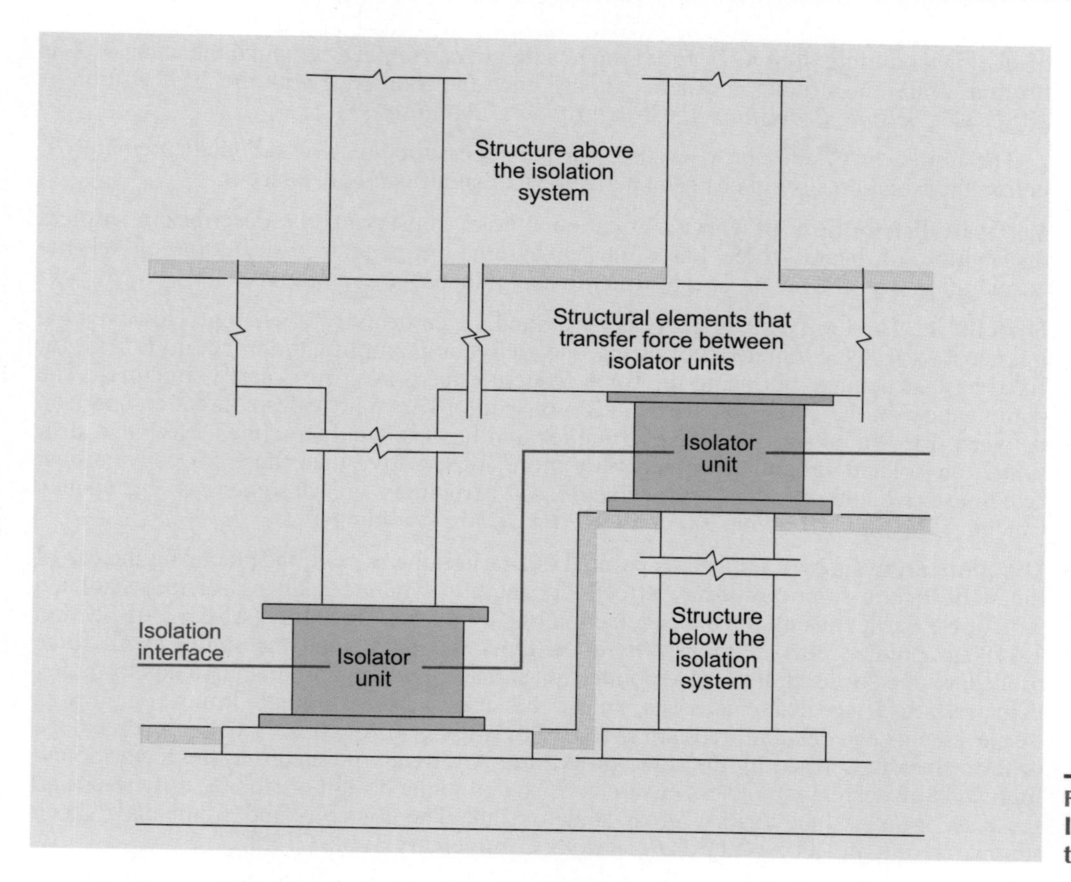

Figure 16-58
Isolation system terminology

As in the case of conventional structures, accidental eccentricity in both horizontal directions must be considered. Figure 16-56 defines the displacement terminology used in ASCE 7-05. Equation 17.5-5 (or Equation 17.5-6 for the maximum considered earthquake) provides a simplified formula for estimating the response including the effects of torsion in the absence of more refined analysis.

Minimum lateral forces. Figure 16-58 defines the terminology below and above the isolation system. Equation 17.5-7 gives the design base shear for all structural components at or below the seismic interface, without any reduction for ductile response. Equation 17.5-8 gives the design base shear for structural systems above the seismic interface. For structures with appreciable inelastic deformability, this equation includes an effective reduction factor of up to 2 for response beyond the strength design level. As explained in Reference 2, a factor of at least 2 is assumed to exist between the strength-design force level and the true yield level of the structural system. Thus, even with a reduction factor of 2, the structure is expected to remain essentially elastic under the design earthquake.

The 2006 IBC Section 1613.6 adds an exception to ASCE 7-05 Section 17.5.4.2, whereby restrictions placed on the use of Ordinary Steel Concentrically Braced Frames (Steel OCBFs) and Ordinary Steel Moment Frames (OMFs of steel) in a seismically isolated SDC D, E or F structure are relaxed significantly. Because of their poor performance during a large earthquake, ASCE 7-05 Table 12.2-1 severely limits the use of these systems in SDC D, E or F structures. However, although these limitations are valid for typical buildings with fixed bases where large ductility demands are expected, seismically isolated structures deserve a different approach, as the ductility demand on them is much smaller, owing to the much smaller value of the response reduction factor R. Thus, a significant cost reduction is possible by not requiring the implementation of special detailing schemes associated with special moment frames and braced frames. This added exception permits OCBFs and OMFs of steel in a building of SDC D, E or F up to a height of 160 feet, provided the value of R_I is taken as 1 and the structural systems are designed in accordance with the 2005 edition of AISC 341, *Seismic Provisions for Structural Steel Buildings*.[24]

The limits on V_s attempt to ensure that the superstructure will not yield prematurely before the isolation system has been activated and significantly displaced.

Vertical distribution of force. Equation 17.5-9 conservatively describes a vertical distribution of lateral force based on an assumed triangular distribution of seismic acceleration over the height of the structure above the isolation interface.

Drift limits. The limit on interstory drift is a fixed 1.5 percent of story height. However, the interstory drift is computed not using the deflection amplification factor, C_d, as for fixed-base structures, but using the force reduction factor, R_I, for isolated structures. The commentary on the 2003 NEHRP *Provisions* Section 13.3.5 provides a direct comparison between drift limits for fixed-base structures and base-isolated structures. Note that drift limits for isolated structures are generally more conservative than those for conventional fixed-base structures, even when fixed-based structures are designed as Occupancy Category IV (SUG III in the 2003 NEHRP *Provisions*) buildings.

17.6 Dynamic analysis procedure. As pointed out earlier, the restrictions placed on the use of the static lateral force procedure effectively require dynamic analysis for most isolated structures. Earlier discussion under selection of lateral force procedure (ASCE 7-05 Section 17.4) has pointed out situations where the dynamic rather than the static lateral force procedure must be used. It has also pointed out situations where time-history analysis, rather than response spectrum analysis, must be used. Lower-bound limits on design displacements and design forces are specified in ASCE 7-05 Section 17.6.4 as a percentage of the values prescribed by the static procedure. As already pointed out, the lower-bound limits on key design parameters ensure consistency in the design of isolated structures and serve as a safety net against gross underdesign. The lower bound limits have been summarized in the 2003 NEHRP *Provisions* Commentary Table C13.2-2.

Mathematical modeling of the isolation system needs to be done based on the deformation characteristics obtained by proper test procedures, and needs to include the location and forces on each individual isolation unit and the horizontal translation as well as the maximum possible torsional displacement of the structure above the isolation interface.

Seismic forces induced on the seismic force-resisting-system above the isolation interface depend of the relative flexibility of the isolation system. The stiffer the isolation system, the more the force induced on the structural elements. As a result, elastic design of the isolated structure is only permitted when the forces in the structural elements are within the inelastic limits, even when the effective stiffness of the nonlinear isolation system is assumed to be at its maximum. However, this condition can be assumed to be satisfied when the design base shear obtained from the response spectrum or the response history analysis for the structural system is at least 100 percent of V_s given by Equation 17.5-8 for irregular structures and 80 percent of V_s given by Equation 17.5-8 for regular structures, because Equation 17.5-8 derives the base shear from the maximum effective stiffness and the maximum displacement of the isolation system in the design earthquake.

Description of procedures. For both the response spectrum analysis and the response history analysis, a design ground motion is to be used for calculating the design forces and the design displacements of the isolation system as well as the isolated structure. Additionally, maximum considered earthquake ground motions are to be used for calculating the maximum displacement of the isolation system only.

For the response spectrum analysis, an upper bound is imposed on the assumed damping of the structure in its fundamental mode as 30 percent of the critical or the effective damping of the isolation system, whichever is less. Modal damping for the higher modes is to be selected in the same way it is done for a conventional fixed-base structure. As specified in ASCE 7-05 Section 12.9.1, an adequate number of modes needs to be included in the analysis to obtain a combined modal mass participation of at least 90 percent of the total seismic mass in the direction of analysis. For calculating the total design displacement and the total maximum displacement that include the effects of orthogonal horizontal translations as well as torsion, two horizontal ground motions need to be applied simultaneously: 100 percent of the ground motion in the critical direction and 30 percent of the ground motion in the orthogonal direction. It may be recalled that Section 17.6.2.1 requires that the centroid of the eccentric mass needs to be assumed at the most disadvantageous location in the modeling of the structural system to maximize the torsional effects.

As mentioned before, at least three sets of ground motions need to be included in a response history analysis. When more than seven pairs of ground motions are used, an average value of the response parameters can be used. For less than seven ground motion pairs, however, the maximum values of the response parameters need to be used.

Minimum lateral displacements and forces. Elastic forces and displacements for the isolation system are to be obtained directly from the dynamic analysis without reduction. However, elastic forces on the structural system above the isolation interface is to be reduced by dividing by the factor R_I as determined in accordance with ASCE 7-05 Section 17.5.4.2. Minimum values are also specified as percentage of those from the equivalent lateral force analysis. If the factored lateral forces on the structure, as obtained from the dynamic analysis, are less than the minimum required values, all response parameters need to be scaled up proportionally.

Inelastic story drift of the isolated structure should be calculated using ASCE 7-05 Equation 12.8-15, i.e., $\delta_x = C_d\,\delta_{xe}/I$, where δ_{xe} is the story drift of the isolated structure from the elastic dynamic analysis using design earthquake. ASCE 7-05 Section 17.6.4.4 specifies that factor C_d be taken equal to R_I. Also, it may be recalled that the importance factor I for isolated structures is always taken equal to 1, as per ASCE 7-05 Section 17.2.1.

Calculated inelastic story drift should be within the limiting drift of 1.5 percent of the story height for response spectrum analysis and 2 percent of the story height for response history analysis. However, if the design story drift exceeds $0.01h_{sx}/R_I$ (i.e. elastic drift exceeding $0.01h_{sx}$), the P-Δ effects need to be considered in the maximum displacement analysis using the maximum considered earthquake.

Design review. Review of design and analysis of the isolation system as well as of the isolator testing program is mandated in this section for reasons given in References 22 and 32. **17.7**

Testing. Calculation of the design displacements and forces per the provisions of ASCE 7-05 Chapter 17 requires the determination of the deformation characteristics and damping values of the isolation system. It is required that prototype systems be designed during the early phases of design to provide the basis for such determination. **17.8**

Chapter 18 *Seismic Design Requirements for Structures with Damping Systems*

This chapter is new in its entirety and contains damping-related definitions and notation within the chapter rather than in ASCE 7-05 Chapter 11, where the other definitions and notation are located. This chapter is based on Chapter 15 of the 2003 NEHRP *Provisions*. It was decided to include Chapter 18 in ASCE 7-05 because the techniques and design criteria have been determined to be well established and an increasing number of buildings are being designed with damping systems.

Chapter 15, Structures with Damping Systems, appears for the first time in the body of the 2003 NEHRP *Provisions*, having first appeared as an appendix (to Chapter 13) in the 2000 NEHRP *Provisions*. The primary resource document for the design of structures with dampers was the NEHRP *Guidelines and Commentary for Seismic Rehabilitation of Buildings*.[26] Although suitable for performance-based design, the terms, methods of analysis and response limits of the NEHRP *Guidelines* for existing building did not match those of the NEHRP *Provisions* for new structures. Therefore, new analysis methods were developed for structures with dampers based on nonlinear *pushover* characterization of the structure and calculation of peak response using effective (secant) stiffness and effective damping properties of the first (pushover) mode in the direction of interest. These are the same concepts as used in the 2003 NEHRP *Provisions* Chapter 13 or ASCE 7-05 Chapter 17 to characterize the force-deflection properties of isolation systems, modified to explicitly incorporate the effects of ductility demand (post-yield response) and higher-mode response of structures with dampers. In contrast to isolated structures, structures with dampers are in general expected to yield during strong ground shaking (similar to conventional structures), and their performance can be significantly influenced by the responses of higher modes.

The approach taken in developing ASCE 7-05 Chapter 18 for structures with damping systems is based on the following concepts:

1. The chapter is applicable to all types of damping systems, including both displacement-dependent damping devices of hysteretic or friction systems and velocity-dependent damping devices of viscous or visco-elastic systems.
2. The chapter provides minimum design criteria with performance objectives comparable to those for a structure with a conventional seismic force-resisting system (but also permits design criteria that will achieve higher performance levels).
3. The chapter requires structures with a damping system to have a seismic force-resisting system that provides a complete load path. The seismic force-resisting system must comply with the requirements of ASCE 7-05, except that the damping system may be used to meet drift limits.
4. The chapter requires design of damping devices and prototype testing of damper units for displacements, velocities and forces corresponding to those of the maximum considered earthquake (same approach as that used for structures with an isolation system).
5. The chapter provides linear static and response spectrum analysis methods for design of most structures that meet certain configuration and other limitation criteria (for example, at least two damping devices at each story configured to resist torsion). The chapter requires additional nonlinear response history analysis to confirm peak response for structures not meeting the criteria for linear analysis (and for structures close to major faults).

ASCE 7-05 Chapter 18 defines a damping system as:

> The collection of structural elements that includes all individual damping devices, all structural elements or bracing required to transfer forces from damping devices to the base of the structure, and all structural elements required to transfer forces from damping devices to the seismic force-resisting system.

The damping system is defined separately from the seismic force-resisting system, although the two systems may have common elements. The damping system may be external or internal to the structure and may have no shared elements, some shared elements or all elements in common with the seismic force-resisting system. Elements common to the damping system and the seismic force resisting system must be designed for a combination of the two loads of the two systems.

The seismic force-resisting system may be thought of as a collection of lateral force-resisting elements of the structure if the damping system was not functional (as if damping devices were disconnected). This system is required to be designed for not less than 75 percent of the base shear of a conventional structure (not less than 100 percent, if the structure is highly irregular), using an *R*-factor as defined in ASCE 7-05 Table 12.2-1. This system provides both a safety net against damping system malfunction as well as the stiffness and strength necessary for the balanced lateral displacement of the damped structure.

The chapter requires the damping system to be designed for the actual (nonreduced) earthquake forces (such as peak forces occurring in damping devices). For certain elements of the damping system, other than damping devices, limited yielding is permitted, provided such behavior does not affect damping system function or exceed the amount permitted by ASCE 7-05 for elements of conventional structures.

Reference should be made to 2003 NEHRP *Provisions* Chapter 15 Commentary for in-depth information on this chapter.

Chapter 19 *Soil Structure Interaction for Seismic Design*

This chapter adopts its requirements from Section 5.6 of the 2003 NEHRP *Provisions*. Detailed background to the provisions is available in the 2003 NEHRP *Provisions* Commentary. This section of the NEHRP *Provisions* dates back to ATC 3-06[32] and has undergone little significant change since then. The following is adapted from the 2003 NEHRP *Provisions* Commentary on Section 5.6.

Fundamental to the design requirements presented in ASCE 7-05 Sections 12.8 and 12.9 is the assumption that the motion experienced by the base of a structure during an earthquake is the same as the *free-field* ground motion, a term that refers to the motion that would occur at the level of the foundation if no structure were present. This assumption implies that the foundation-soil system underlying the structure is rigid and, hence, represents a *fixed-base* condition. Strictly speaking, this assumption never holds in practice. For structures supported on a deformable soil, the foundation motion generally is different from the free-field motion and may include an important rocking component in addition to a lateral or translational component. The rocking component, and soil-structure interaction effects in general, tend to be most significant for laterally stiff structures such as buildings with shear walls, particularly those located on soft soils. For convenience, the response of a structure supported on a deformable foundation-soil system may be referred to as the *flexible-base* response.

A flexibly supported structure also differs from a rigidly supported structure in that a substantial part of its vibrational energy may be dissipated into the supporting medium by radiation of waves and by hysteretic action in the soil. The importance of the latter factor increases with increasing intensity of ground shaking. There is, of course, no counterpart of this effect of energy dissipation in a rigidly supported structure.

The effects of soil-structure interaction accounted for in ASCE 7-05 Chapter 19 represent the difference in the flexible-base and fixed-base responses of the structure. This difference depends on the properties of the structure and the supporting medium, as well as the characteristics of the free-field ground motion.

The interaction effects accounted for in ASCE 7-05 Chapter 19 should not be confused with site effects, which relate to the fact that the characteristics of the free-field ground motion reduced by a dynamic event at a given site are functions of the properties and geological features of the subsurface soil and rock. The interaction effects, on the other hand, refer to the fact that the dynamic response of a structure built on the site depends, in addition, on the interrelationship of the structural characteristics and the properties of the local underlying soil deposits. The site effects are reflected in the values of the seismic coefficients employed in ASCE 7-05 Sections 12.8 and 12.9 and are accounted for only implicitly in ASCE 7-05 Chapter 19.

Two different approaches may be used to assess the effects of soil-structure interaction. The first involves modifying the stipulated free-field design ground motion, evaluating the response of the given structure to the modified motion of the foundation and solving simultaneously with additional equations that define the motion of the coupled system, whereas the second involves modifying the dynamic properties of the structure and evaluating the response of the modified structure to the prescribed free-field ground motion. When properly implemented, both approaches lead to equivalent results. However, the second approach, involving the use of the free-field ground motion, is more convenient for design purposes and provides the basis of the requirements presented in ASCE 7-05 Chapter 19.

The interaction effects in the approach used in ASCE 7-05 Chapter 19 are expressed by an increase in the fundamental natural period of the structure and a change (usually an increase) in its effective damping.

The provisions in ASCE 7-05 Chapter 19 were previously found in ASCE 7-02 Section 9.5.9. There is only one substantive change to the provisions, which is found in Section 19.1. The substantive change limits the use of the soil-structure interaction provisions to models that do not directly incorporate the effects of foundation flexibility. This is because if a flexible-base foundation, rather than a fixed-base foundation, is modeled in the analysis of a structure, then the effects of period lengthening on demand would already be included in the analysis. Consequently, the provisions of Section 19.1 should not be used for these cases, because the effects of period lengthening should not be counted twice.

Chapter 20 *Site Classification Procedure for Seismic Design*

This chapter isolates the site classification provisions in one chapter for the convenience of the user. The provisions are reformatted for clarity and understandability. Note that the same provisions are to be found in Section 1613.5.5 of the 2006 IBC.

ASCE 7-05 Section 11.4.2 states that each site is required to be assigned a site class; however, in order not to interrupt the flow of the seismic design requirements, it was decided to include the site classification procedure in a subsequent chapter. ASCE 7-05 Section 11.4.2 does include a statement, repeated in ASCE 7-05 Section 20.1, that Site Class D may be used unless the authority having jurisdiction or geotechnical data determine Site Class E or F soils to be present at the site.

20.1 Site classification. This section is very similar to 1997 UBC Section 1629.3. It should be noted that Soil Profile Types S_A, S_B, S_C, S_D, S_E and S_F of the 1997 UBC are the same as Site Classes A, B, C, D, E and F of the 2006 IBC, respectively. A substantive difference between the ASCE 7-05 and the 1997 UBC sections being discussed is that the following ASCE 7-05 provision is not to be found in the 1997 UBC: "Where site-specific data are not available to a depth of 100 ft, appropriate soil properties are permitted to be estimated by the registered design professional preparing the soil investigation report based on known geologic conditions." Also not to be found in either 1997 UBC Section 1629.3 or 2006 IBC Section 1613.5.2 is the ASCE 7-05 stipulation: "Site Classes A and B shall not be assigned to a site if there is more than 10 ft of soil between the rock surface and the bottom of the spread footing or mat foundation."

Table 20.3-1 of ASCE 7-05 and Table 16-J of the 1997 UBC are the same, except that the footnote to Table 16-J is to be found in ASCE 7-05 Section 20.3.2.

20.3.1 Site Class F. The provisions are the same as in 1997 UBC Section 1629.3.1; however, the exception to Item 1 is not to be found in the 1997 UBC. The reason for this exception, first added in the 2003 IBC, is that any amplifying effect of soil liquefaction on ground motion response spectra is expected to be in the long-period range and short-period response spectra will tend to be deamplified or remain unchanged by liquefaction.

20.3.2 Soft clay Site Class E. As noted above, this was a footnote to 1997 UBC Table 16-J.

Site Classes C, D and E. The same provisions are to be found in 1997 UBC Section 1636.2.5. **20.3.3**

Shear wave velocity for Site Class B. The same requirement is to be found in 1997 UBC Section 1636.2.6. **20.3.4**

Shear wave velocity for Site Class A. The same requirement is to be found in 1997 UBC Section 1636.2.6. **20.3.5**

$\bar{v}_S$, average shear wave velocity. This section is identical to Section 1636.2.1 of the 1997 UBC. **20.4.1**

$\bar{N}$, average field standard penetration resistance, and $\bar{N}_{ch}$ average standard penetration for cohesionless soil layers. The corresponding 1997 UBC section is 1636.2.2. In ASCE 7-05, "N_i is the standard penetration resistance (ASTM D 1586) not to exceed 100 blows per foot as directly measured in the field without corrections." This definition, unlike the corresponding 1997 UBC definition, is quite specific. The specified upper limit on N_i of 100 blows/foot for individual soil layers does not provide for a scenario where penetration refusal is met because of the presence of a rock layer. In ASCE 7-05, it is now stipulated that the N_i-value for rock layers is to be taken as 100 blows/foot as well. **20.4.2**

$\bar{S}_u$, average undrained shear strength. This section is essentially the same as Section 1636.2.3 of the 1997 UBC. However, in ASCE 7-05, "s_{ui} is the undrained shear strength in psf, not to exceed 5000 psf as determined in accordance with ASTM D 2166 or ASTM D 2850." This definition, unlike the corresponding 1997 UBC definition, is quite specific. **20.4.3**

Chapter 21 *Site-Specific Ground Motion Procedures for Seismic Design*

This chapter sets forth the requirements for the following two types of site-specific ground motion procedures: (1) site response analysis and (2) ground motion hazard analysis. The triggers requiring such analyses are found in Chapter 11; rather than using space in Chapter 11 to include the requirements, they are placed in Chapter 21. The subject is treated in a peripheral way in 1997 UBC Section 1631.2(2).

Although the site-specific ground motion procedures, now set forth in ASCE 7-05 Chapter 21, can be used to determine ground motions for any structure, ASCE 7-02 required "site-specific" analysis to be carried out in two very different situations: (1) for structures located on Site Class F soils and (2) for seismically isolated structures on sites that are *near-source* ($S_1 = 0.6g$). In order to clarify what is required for these two different situations, the site-specific procedure section in ASCE 7-05 is divided into two parts: (1) Section 21.1 Site Response Analysis (required for Site Class F soils, unless the exception to Section 20.3.1 is applicable) and (2) Section 21.2 Ground Motion Hazard Analysis (required for seismically isolated structures and structures with damping systems on sites with $S_1 = 0.6g$). The site response analysis is completely new, whereas the ground motion hazard analysis is based on the site-specific procedure in ASCE 7-02 Section 9.4.1.3. Revisions have been made to the latter to reflect language used in the 2003 NEHRP *Provisions*.

This new site response analysis of Section 21.1 determines the maximum considered earthquake (MCE) response spectrum of the building site by first starting with a bedrock response spectrum and then working upward to convert it to a response spectrum on the ground surface using the properties of the soil layer above the bedrock.

Design response spectrum. There is a new requirement in Section 21.3 (which was ASCE 7-02 Section 9.4.1.3.4) that reads: **21.3**

> For sites classified as Site Class F requiring site response analysis in accordance with Section 11.4.7, the design spectral response acceleration at any period shall not be taken less than 80 percent of S_a determined for Site Class E in accordance with Section 11.4.5.

This revision came about because of the observation that for certain Site Class F soils, such as those prone to liquefaction, large reductions in short- and intermediate-period ground motions could be computed, depending on the type of site response analysis and the material properties and constitutive models selected for the soils. Because of the lack of ground motion data recorded on Site Class F soils to corroborate the results from these analyses, the decision was made to limit the amount of reduction to 20 percent, which is the same as the maximum reduction ASCE 7-05 allows for site-specific ground motion procedures for other site classes (Reference 20), when compared to spectral acceleration values obtained using the ground motion maps and site coefficients of ASCE 7.

Chapter 22 *Seismic Ground Motion and Long-Period Transition Period Maps*

Chapter 22 contains the subject maps that are approximately 23 pages in length. The same maps are to be found in Chapter 3 of the NEHRP *Provisions*.

Maps showing the contours of 5-percent-damped 0.2-second and 1-second spectral response acceleration values for the Maximum Considered Earthquake (MCE) ground motion can be seen in Figures 22-1 through 22-14. The important features of these maps are discussed below.

Forty-eight conterminous states: The maps of the 48 conterminous states are based on the 2002 USGS probabilistic maps that incorporate improved earthquake data in terms of updated fault parameters (such as slip rates, recurrence time and magnitude), additional attenuation relations (also referred to as ground motion prediction equations) and data calculated at smaller grid spacing as compared to the 1996 USGS probabilistic maps that formed the basis of the 2003 IBC and ASCE 7-02. The grid spacing used to contour the 2002 USGS probabilistic maps of the 48 states was 0.05 degree. For the MCE maps used in the 2006 IBC, five smaller regions were also contoured using a grid spacing of 0.01 degree. The small grid spacing makes interpolation in regions with numerous faults more accurate. The changes found in the 2006 IBC maps are discussed for the western U.S. along with its regions and the central and eastern U.S. along with its regions, as some of the changes apply to one but not the other.

Western U.S.

- The ground motion values in the western U.S. are most affected by the attenuation equations, fault recurrence parameters and fault geometry of the Cascadia subduction zone (located in Oregon and Washington). Five attenuation equations for shallow crustal faults were used, including one equation for faults within an area of extensional tectonics. The ground motions from the Cascadia subduction zone are attenuated using two equations.
- The Pacific Northwest region includes primarily Washington and Oregon, and although a regional map was not considered necessary, the region is contoured using a grid spacing of 0.01 degree. This region is bounded by latitudes of 41° N to 49° N and longitudes of 123° W to 125° W. An areal source zone is included in the Puget lowland area around Seattle, which smoothes the ground motion values around Seattle and increases the values of S_s and S_1 on the order of 5 to 10 percent.

Central and Eastern U.S.

- Five attenuation equations are used for the 2006 maps. Region 3 and Region 4 are located in the central and eastern U.S.

Region 3 [Figure 22-7 and 22-8]: In the region primarily covering the New Madrid seismic zone (bounded by latitudes of 34° N to 39° N and longitudes of 87° W to 92° W), a regional map was considered necessary because of steep changes in the ground motion. A deterministic area is included in the New Madrid seismic zone, marked by the shaded area in Figures 22-7 and 22-8, after a review of the New Madrid ground motions. Because MCE rules state that deterministic ground motions only govern over probabilistic ground motions if the deterministic values are lower, the inclusion of this deterministic area somewhat reduces the size of the high-ground-motion region near the New Madrid fault.

Region 4 [Figure 22-9]: This regional map covers primarily South Carolina. The region is bounded by latitudes of 31° N to 35° N and longitudes of 77° W to 83° W. The ground motion values are calculated for a grid spacing of 0.05 degree instead of the smaller 0.01 degree. However a regional map was considered necessary because of the high gradient in the ground motion.

ALASKA: There are no differences in the maps between the 2003 and 2006 IBC for the state of Alaska.

HAWAII: There are no differences in the maps between the 2003 and 2006 IBC for the state of Hawaii.

PUERTO RICO AND THE U.S. VIRGIN ISLANDS: MCE ground motion maps [Figure 22-13] now provide contours of varying spectral values in Puerto Rico and the Virgin Islands, similar to those for the rest of the country, as opposed to the older practice of assigning a single value to the entire region. As a result of this update, most parts of Puerto Rico and the Virgin Islands have reductions in the mapped spectral values, except for the western one-third of Puerto Rico on the 0.2-second map and the extreme southwestern part of Puerto Rico on the 1-second map. The differences can be seen in Figures 16-10 and 16-11.

GUAM AND TUTUILLA: There are no differences in the maps between the 2003 and 2006 IBC for the territories of Guam and Tutuilla.

Chapter 23 *Seismic Design Reference Documents*

Chapter 23 includes a list of the reference documents referred to in the seismic provisions (ASCE 7-05 Chapters 11 through 22). This list is organized by the promulgating organization and includes helpful information about how to obtain each document and which edition is being referred to. In fact, this chapter is the only place a reader can find which particular edition of a standard is being referenced by ASCE 7-05. In ASCE 7-02, the reference documents were included within the provisions themselves, rather than collectively listed in one place. Also, in ASCE 7-02, the reference standards were assigned numbers such as 9.6-14 and 9.6-15. ASCE 7-05 refers to each reference document by its common name, such as ACI 318; this is a much more user-friendly approach.

REFERENCES

[1.]American Society of Civil Engineers, ASCE 7 Standard *Minimum Design Loads for Buildings and Other Structures*, ASCE 7-93, ASCE 7-95, ASCE 7-98, ASCE 7-02, ASCE 7-05, New York, NY, 1993, 1995, 1998, 2003 and 2005, respectively.

[2]Federal Emergency Management Agency, NEHRP (National Earthquake Hazards Reduction Program) *Recommended Provisions for Seismic Regulations for New Buildings and Other Structures*, Washington, DC, 1985, 1988, 1991, 1994, 1997, 2000, 2003.

[3]International Conference of Building Officials, *Uniform Building Code*, Whittier, CA, 1997, copyright held by International Code Council.

[4]Building Officials and Code Administrators International, *The BOCA National Building Code*, Country Club Hills, IL, 1993, 1996, 1999, copyright held by International Code Council.

[5]Jirsa, J.O., "Behavior of Elements and Subassemblages – R.C. Frames," Proceedings of a Workshop on Earthquake-Resistant Reinforced Concrete Building Construction, Berkeley, CA, July 1977, Vol. III, pp. 1196 – 1214.

[6]International Code Council, *Standard for Bleachers, Folding and Telescopic Seating and Grandstands*, Washington, DC, 2007.

[7]Peir, J.C., and Cornell, C.A., "Spatial and Temporal Variability of Live Loads," *Journal of the Structural Division*, ASCE, Proceedings Vol. 99, No. ST5, May 1973, pp. 903 – 922.

[8]McGuire, R.K., and Cornell, C.A., "Live Load Effects in Office Buildings," *Journal of the Structural Division*, ASCE, Proceedings Vol. 100, No. ST7, July 1974, pp. 1351 – 1366.

[9]Culver, C.G., "Survey Results of Fire Loads and Live Loads in Office Buildings", *NBS Building Science Series 85*, National Bureau of Standards, Washington, DC, May 1976.

[10]Ellingwood, B.R., and Culver, C.G., "Analyses of Live Loads in Office Buildings," *Journal of the Structural Division*, ASCE, Proceedings Vol. 103, No. ST8, August 1997, pp. 1551 – 1560.

[11]International Conference of Building Officials, *Handbook to the Uniform Building Code: An Illustrative Commentary*, Whittier, CA, 1998, copyright held by International Code Council.

[12]American Association of State Highway and Transportation Officials, *LRFD Bridge Design Specifications*, Fourth edition, 2007.

[13]International Code Council/American National Standards Institute, Inc., *Accessible and Usable Buildings and Facilities*, A117.1-03, Washington, DC, VA, 2004.

[14]Southern Building Code Congress International, *Standard Building Code*, Birmingham, AL, 1994, 1997, 2000, copyright held by International Code Council.

[15]The National Bureau of Standards, *Live Loads on Floors and Buildings*, Building Materials and Structures Publication No. 133, Washington, DC, 1952.

[16]O'Rourke, M., and Downey, C., "Rain-on-Snow Surcharge for Roof Design," *Journal of Structural Engineering*, ASCE, Vol. 127, No. 1, 2001, pp. 74 – 79.

[17]International Code Council, *Standard for Hurricane Resistant Residential Construction*, SSTD 10-99, Washington, DC, 1999.

[18]American Forest and Paper Association, *Wood Frame Construction Manual for One- and Two-Family Dwellings*, ANSI/AF&PA WFCM-2001, 2001.

[19]Durst, C.S., "Wind Speeds over Short Periods of Time," *Meteorological Magazine*, Vol. 89, 1960, pp. 181 – 187.

[20]American Society for Testing and Materials, *Specification for Performance of exterior Windows, Glazed Curtain Walls, Doors and Storm Shutters Impacted by Windborne Debris in Hurricanes*, ASTM E 1996-01, West Conshohocken, PA, 2001.

[21]American Society of Civil Engineers, *Flood Resistant Design and Construction Standard*, Reston, VA, 1998.

[22]Seismology Committee, Structural Engineers Association of California, *Recommended Lateral Force Requirements and Commentary*, San Francisco (later Sacramento), CA, 1974, 1988, 1996, 1999.

[23]Joyner, W.B., Fumal, T.E., and Glassmoyer, G., "Empirical Spectral Response Ratios for Strong Motion Data from the 1989 Loma Prieta, California Earthquake," Proceedings of the NCEER/SEAOC/BSSC Workshop on Site Response during Earthquakes and Seismic Code Provisions, November 18 – 20, 1992, University of Southern California, Los Angeles, Martin, G.M., Ed., 1994.

[24]American Institute of Steel Construction, *Seismic Provisions for Structural Steel Buildings*, AISC 341-05, Chicago, IL, 2005.

[25]International Code Council, *International Residential Code for One- and Two-Family Dwellings*, Washington, DC, 2006.

[26]Federal Emergency Management Agency, *Prestandard and Commentary for the Seismic Rehabilitation of Buildings*, FEMA 356, Washington, DC, November, 2000.

[27]American Society for Testing and Materials, *Standard Test Methods for Liquid Limit, Plastic Limit, and Plasticity Index of Soils*, ASTM D 4318-05, West Conshohocken, PA, 2005.

[28]American Society for Testing and Materials, *Standard Test Method for Unconfined Compressive Strength of Cohesive Soil*, ASTM D 2166-06, West Conshohocken, PA, 2006.

[29]American Society for Testing and Materials, *Standard Test Method for Unconsolidated-Undrained Triaxial Compression Test on Cohesive Soils*, ASTM D 2850-03, West Conshohocken, PA, 2003.

[30]American Society for Testing and Materials, *Standard Test Methods for Laboratory Determination of Water (Moisture) Content of Soil and Rock by Mass*, ASTM D 2216-05, West Conshohocken, PA, 2005.

[31]Newmark, N.M., and Hall, W.J., "Seismic Design Criteria for Nuclear Reactor Facilities," Proceedings, Fourth World Conference on Earthquake Engineering, Santiago, Chile, 1969, Vol. 2, pp. B-4 39 – 50.

[32]Applied Technology Council, *Tentative Provisions for the Development of Seismic Regulations for Buildings*, ATC Publication ATC 3-06, NBS Special Publication 510, NSF Publication 78-8, U.S. Government Printing Office, Washington, DC, 1978.

[33]Ghosh, S.K., Dowty, S., and Fanella, D.A., *2003 Analysis of Revisions to the IBC – Structural Provisions*, International Code Council, Washington, DC, 2003.

[34]Tsai, K.C., Weng, Y.T., Lin, M.L., Chen, C.H., Lai, J.W., and Hsiao, P.C., "Pseudo Dynamic Tests of a Full Scale CFT/BRB Composite Frame: Displacement Based Seismic Design and Response Evaluation." Proceedings of the International Workshop on Steel and Concrete Composite Construction, National Center for Research on Earthquake Engineering, Taiwan, pp. 165 – 176, 2003.

[35]Sabelli, R., and Bruneau, M., *AISC Design Guide 20: Steel Plate Shear Walls*, American Institute of Steel Construction, Chicago, IL, 2006.

[36]American Institute of Steel Construction, *Specification for Structural Steel Buildings*, AISC 360, Chicago, IL, 2005.

[37]American Concrete Institute, *Building Code Requirements for Structural Concrete and Commentary*, ACI 318-95, ACI 318-05, Detroit, MI, 1995, 200, respectively.

[38]Itani, A., and Goel, S.C., *Earthquake Resistant Design of Open Web Framing Systems*, Research Report No. UMCE 91-21, University of Michigan, Ann Arbor, MI, 1991.

[39]Newmark, N.M., and Hall, W., *Earthquake Spectra and Design*, Monograph, Earthquake Engineering Research Institute, Oakland, CA, 1982, 103 pp.

[40]Uang, C.M., "Establishing R (or R_N) and C_d Factors for Building Seismic Provisions," *Journal of Structural Engineering*, ASCE, Vo. 117, No. 1, 1991, pp. 19 – 28.

[41]Applied Technology Council, *Structural Response Modification Factors*, ATC 19, Redwood City, CA, 1995, 64 pp.

[42]Structural Engineers Association of California, *Seismic Design Manual, Volume I, Code Application Examples*, Sacramento, CA, April 1999.

[43]Cobeen, K., Dolan, J.D., and Russell, J., *Recommendations for Earthquake Resistance in the Design and Construction of Woodframe Buildings*, CUREE W-30, Richmond, CA, 2004.

[44]Goel, R.K. and Chopra, A.K., "Period Formulas for Moment-Resisting Frame Buildings," *Journal of Structural Engineering*, ASCE, Vol. 123, No.11, November 1997, pp.1454 – 1461.

[45]Goel, R.K., and Chopra, A.K., "Period Formulas for Concrete Shear Wall Buildings," *Journal of Structural Engineering*, ASCE, Vol. 124, No. 4, April 1998, pp. 426 – 433.

[46]Wilson, E.L., and Habibullah, A., "A Static and Dynamic Analysis of Multistory Buildings Including P-delta Effects," *Earthquake Spectra*, Earthquake Engineering Research Institute, Oakland, CA, Vol. 3, No. 2, 1987.

[47]Clough, R.W., and Penzien, J., *Dynamics of Structures,* 2nd Edition, McGraw-Hill, Inc., New York, 1993.

[48]Chopra, A.K., *Dynamics of Structures*, 2nd Edition, Prentice Hall, Englewood Cliffs, NJ, 2000.

[49]Iverson, J.K., and Hawkins, N.M., "Performance of Precast/Prestressed Concrete Building Structures During the Northridge Earthquake," *PCI Journal*, Vol. 39, No. 2, March – April 1994, pp. 38 – 55.

[50]Fleischman, R.B., Sause, R., Pessiki, S., and Rhodes, A.B., "Seismic Behavior of Precast Parking Structure Diaphragms," *PCI Journal*, Vol. 43, No. 1, January – February 1998, pp. 38 – 53.

[51]Hamburger, R.O., McCormick, D.L., and Holm, S., "The Whittier Narrows, California Earthquake of October 1, 1987—Performance of Tilt-up Buildings," *Earthquake Spectra*, Vol. 4, No. 2, 1988, pp. 219 – 253.

[52]Hamburger, R.O., and McCormick, D.L., "Implications of the January 17, 1994 Northridge Earthquake on Tiltup and Masonry Buildings with Wood Roofs," *Seminar Papers*, Structural Engineers Association of Northern California, San Francisco, CA, 1994.

[53]Brooks, H., "Building Code Provisions Affecting Tilt-Up Including Proposed Post-Northridge Code Changes," Proceedings of the Third National Concrete & Masonry Engineering Conference, San Francisco, CA, June 1995, pp. 173 – 191.

[54]ICC-ES AC 156, *Acceptance Criteria for Seismic Qualification Testing of Nonstructural Components*, International Code Council Evaluation Service, Whittier, CA, 2004.

[55]IEEE 344, *Recommended Practice for Seismic Qualification of Class 1E Equipment for Nuclear Power Generating Stations*, Institute of Electrical and Electronics Engineers, Inc., Piscataway, NJ, 1987.

[56]ACI 355.2/355.2R, *Qualification of Post-Installed Mechanical Anchors in Concrete*, American Concrete Institute, Farmington Hills, MI, 2004.

[57]American Architectural Manufacturers Association, "Recommended Dynamic Test Method for Determining the Seismic Drift Causing Glass Fallout from a Wall System," Publication No. AAMA 501.6-2001, 2001.

[58]NPFA 13, *Standard for the Installation of Sprinkler Systems*, National Fire Protection Association, Quincy, MA, 2002.

[59]American Society of Mechanical Engineers, *Safety Code for Elevators and Escalators*, ASME A17.1, New York, NY, 1996.

[60]Mayes, R.L. "Design of Structures with Seismic Isolation," Chapter 13, *The Seismic Design Handbook*, F. Naeim ed., Van Nostrand Reinhold, New York, NY, 1989, pp. 413 – 438.

[61]Bachman, R.E., "Base Isolation: A New Concept for Earthquake Resistant Design," *Structure*, National Council for Structural Engineering Associations, Spring 1995.

[62]Kelly, J.M. "A Seismic Base Isolation: Its History and Prospects," *Joint Sealing and Bearing Systems for Concrete*, Publication SP-70, American Concrete Institute, Detroit, MI, 1982.

[63]Buckle, I.G. "Development and Application of Base Isolation and Passive Energy Dissipation: A World Overview," Publication ATC 17, Applied Technology Council, Palo Alto, CA, March 1986.

[64]Theme Issue – Seismic Isolation: From Idea to Reality, *Earthquake Spectra*, Earthquake Engineering Research Institute, Vol. 6, No. 2, May 1990.

CHAPTER

16

Part 2 STRUCTURAL DESIGN EXAMPLES

Example 1. Design Axial Force, Shear Force and Bending Moment for Shear Wall Due to Lateral and Gravity Loads (Strength Design)
Example 2. Design Axial Force, Shear Force and Bending Moment for Shear Wall Due to Lateral and Gravity Loads
Example 3. Design Axial Force, Shear Force and Bending Moment for Shear Wall Due to Lateral and Gravity Loads
Example 4. Calculation of Live Load Reduction
Example 5. Design of a 20-Story Reinforced Concrete Building for Wind Forces
Example 6. Calculation of Wind Pressures for a Low-Rise Building
Example 7. Redundancy (ρ) and Concrete Shear Walls
Example 8. Simplified Design Procedure
Example 9. Design of Multistory Reinforced Concrete Building Subjected to Earthquake Forces
Example 10. Dynamic Analysis Procedure Response Spectrum Analysis
Example 11. Calculation of Diaphragm Design Forces
Example 12. Partial Diaphragm Design
Example 13 Calculation of Collector Strength
Example 14. Lateral Force on Elements of Structures

Example 1 *Design Axial Force, Shear Force and Bending Moment for Shear Wall Due to Lateral and Gravity Loads (Strength Design)*

Load Effect	Symbol	Axial Force (kips)	Shear Force (kips)	Bending Moment (ft-kips)
Dead load effect	D	5381	0	0
Live load effect	L	837	0	0
Wind load effect	W	0	289	18,711
Effect of (horizontal) design earthquake forces	Q_E	0	1571	118,596

Code Formula	Combination	Axial Force (kips)	Shear Force (kips)	Bending Moment (ft-kips)
(16-1)	$1.4D$ [(1)]	7533	0	0
(16-2)	$1.2D + 1.6L$ [(2)]	7796	0	0
(16-4)	$1.2D + 1.6W + 0.5L$ [(3)]	6876	462	29,938
(16-5)	$1.2D + (\rho Q_E + 0.2S_{DS}D) + 0.5L$ [(4)]	7952	1571	118,596
(16-6)	$0.9D + 1.6W$ [(5)]	4843	–462	–29,938
(16-7)	$0.9D + (\rho Q_E + 0.2S_{DS}D)$ [(6)]	3767	–1571	–118,596

Notes:

(1) The effect of fluid load (F) not considered here.

(2) The effect of fluid load (F), lateral earth pressure (H), self-straining force (T), roof live load (L_r), snow load (S) or rain load (R) not considered here.

(3) The live load was assumed to be less than 100 psf; the effect of roof live load (L_r), snow load (S) or rain load (R) *not* considered here.

(4) The live load was assumed to be less than 100 psf; $\rho = 1.0$ and $S_{DS} = 1.0$; the effect of snow load (S) not considered here.

(5) The effect of lateral earth pressure (H) not considered here.

(6) The effect of lateral earth pressure (H) not considered here; $\rho = 1.0$ and $S_{DS} = 1.0$.

Example 2 *Design Axial Force, Shear Force and Bending Moment for Shear Wall Due to Lateral and Gravity Loads (Allowable Stress Design Using Basic Load Combinations)*

Load Effect	Symbol	Axial Force (kips)	Shear Force (kips)	Bending Moment (ft-kips)
Dead load effect	D	5381	0	0
Live load effect	L	837	0	0
Wind load effect	W	0	289	18,711
Effect of (horizontal) design earthquake forces	Q_E	0	1571	118,596

Code Formula	Combination	Axial Force (kips)	Shear Force (kips)	Bending Moment (ft-kips)
(16-8)	D [(1)]	5381	0	0
(16-9)	$D + L$ [(2)]	6218	0	0
(16-13)	$D + 0.75W + 0.75L$ [(3)]	6009	217	14,033
(16-13)	$D + 0.75(0.7)(\rho Q_E + 0.2S_{DS}D) + 0.75L$ [(4)]	6574	825	62,263
(16-14)	$0.6D + W$ [(5)]	3229	–289	–18,711
(16-15)	$0.6D + 0.7(\rho Q_E + 0.2S_{DS}D)$ [(5)]	2475	–1100	–83,017

Notes:

(1) The effect of fluid load (F) not considered here.

(2) The effect of fluid load (F), lateral earth pressure (H), and self-straining force (T) not considered here.

(3) The effect of fluid load (F), lateral earth pressure (H), roof live load (L_r), snow load (S) or rain load (R) not considered here.

(4) The effect of fluid load (F), lateral earth pressure (H), roof live load (L_r), snow load (S) or rain load (R) not considered here; $\rho = 1.0$ and $S_{DS} = 1.0$.

(5) The effect of lateral earth pressure (H) not considered here.

Example 3 *Design Axial Force, Shear Force and Bending Moment for Shear Wall Due to Lateral and Gravity Loads (Allowable Stress Design using Alternate Basic Load Combinations)*

Load Effect	Symbol	Axial Force (kips)	Shear Force (kips)	Bending Moment (ft-kips)
Dead load effect	D	5381	0	0
Live load effect	L	837	0	0
Wind load effect	W	0	289	18,711
Effect of (horizontal) design earthquake forces	Q_E	0	1571	118,596

Code Formula	Combination	Axial Force (kips)	Shear Force (kips)	Bending Moment (ft-kips)
(16-16)	$D + L$ (1)	6218	0	0
(16-17)	$D + L + \omega W$ (2)	6218	376	24,324
(16-17)(3)	$0.67D + L + \omega W$	4424	–289	–18,711
(16-19)	$D + L + \omega W/2$ (4)	6218	188	12,162
(16-20)	$D + L + \rho Q_E + 0.2S_{DS}D)/1.4$ (4)	6987	1122	84,678
(16-21)	$0.9D + (\rho Q_E + 0.2S_{DS}D)/1.4$ (4)(5)	4073	–1122	–84,678

Notes:

(1) The effect of roof live load, snow load or rain load not considered here.

(2) ω = 1.3 as wind forces are calculated per ASCE 7-05.

(3) In 2006 IBC Section 1605.3.2, there is a statement that reads: "For load combinations that include the counteracting effects of dead and wind loads, only two-thirds of the minimum dead load likely to be in place during a design wind event shall be used." This is obviously the same as multiplying the dead load by 0.67, which is not very different from 0.6 (IBC Equation 16-14, see Example 2). The live load effect will help in counteracting the wind load effect when using the "alternative basic" ASD load combinations, and as a result, they are somewhat less conservative than the "basic" load combinations.

(4) The effect of snow load not considered here; ρ = 1.0 and S_{DS} = 1.0.

(5) In seismic design, the alternative basic load combinations are likely to result in a more economical structure, because 90 percent, rather than 60 percent, of the design dead load effects can be counted upon to counteract service-level earthquake effects.

Example 4 *Calculation of Live Load Reduction*

The computation of live load reduction for roof and floor per Sections 1607.9 and 1607.11 is illustrated in the following example.

Given: Plan dimensions: (as shown)

Story height		= 12 ft
Live load:	Floors	= 50 psf (for a typical office building, see Table 1607.1)
Roof		= 20 psf (assumed)
Dead load:	Floors	= 120 psf (assumed)
Roof		= 90 psf (assumed)

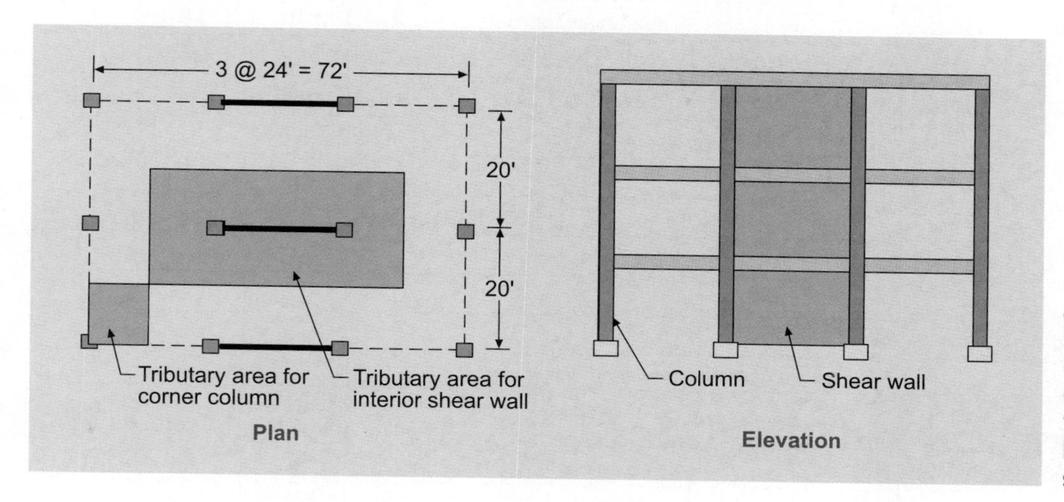

Figure E4-1 Plan and elevation of the example building studied

Example 4—(continued)

Live Load Reduction for Corner Column

Level	A_T (ft²)	D (psf)	D (kips)	L_0	K_{LL}	$K_{LL} \times A_T$	R (%)	R (max) (%)	Reduction (%)	L (psf)	L (kips)
3	120	90	10.8	20	see below for roofs				0	20	2.4
2	240	120	14.4	50	4	960	27	40	27	36.5	8.8
1	360	120	14.4	50	4	1440	35	40	35	32.5	11.7
							$R = 1 - L/L_0$				

Live Load Reduction for Corner Column (Alternate Method)

Level	A_T (ft²)	D (psf)	D (kips)	L_0	r	R	R (max)	R (max)*	Reduction (%)	L (psf)	L (kips)
3	120	90	10.8	20	see below for roofs				0	20	2.4
2	240	120	14.4	50	0.08	7.2	60	78.5	7.2	46.4	11.1
1	360	120	14.4	50	0.08	16.8	60	78.5	16.8	41.6	15.0
Roof		R_1 =		1.00							
		R_2 =		1.00							
		L_r =		20	2006 IBC Equation 16-27						
		L_r (max) =		20							
		L_r (min) =		12							

* as given by the second equation in Section 1607.9.2

Live Load Reduction for Interior Shear Wall

Level	A_T (ft²)	D (psf)	D (kips)	L_0	K_{LL}	$K_{LL} \times A_T$	R	R (max)*	Reduction (%)	L (psf)	L (kips)
3	960	90	86.4	20	see below for roofs				40	12	11.5
2	1920	120	115.2	50	1	1920	41	60	41	29.5	56.6
1	2880	120	115.2	50	1	2880	47	60	47	26.5	76.3
							$R = 1 - L/L_0$				

Live Load Reduction for Interior Shear Wall (Alternate Method)

Level	A_T (ft²)	D (psf)	D (kips)	L_0	r	R	R (max)	R (max)*	Reduction (%)	L (psf)	L (kips)
3	960	90	86.4	20	see below for roofs				40	12	11.5
2	1920	120	115.2	50	0.08	141.6	60	78.5	60.0	20	38.4
1	2880	120	115.2	50	0.08	218.4	60	78.5	60.0	20	57.6
Roof		R_1 =		0.60							
		R_2 =		1.00							
		L_r =		12	2006 IBC Equation 16-27						
		L_r (max) =		20							
		L_r (min) =		12							

Example 5 *Design of a 20-Story Reinforced Concrete Building for Wind Forces*

General

The main wind force-resisting system of a 20-story reinforced concrete office building is designed following the requirements of the 2006 IBC concerning wind forces.

According to Section 1609.1.1 (Determination of wind loads), wind loads on every building or structure shall be determined in accordance with Chapter 6 of ASCE 7-05 (*Minimum Design Loads for Buildings and Other Structures*). The 20-story example building does not fall under any of the four exceptions provided in Section 1609.1.1.

Design Criteria

Wind Design Data:

Location of building: Los Angeles, California

Plan dimensions of the building: 130 ft × 130 ft

Height of the building: 258 ft = (255 ft + 3-ft parapet)

Building is partially enclosed per definition under ASCE 7-05 Section 6.2

Building is regular in shape (as shown in Figure E5.1)

Assume Exposure B (per ASCE 7-05 Section 6.5.6.3)

Wind importance factor for the building I = 1.0 (Category II, ASCE 7-05 Table 6-1)

Calculation of Flexibility of Structure

A structure is considered flexible per definition under ASCE 7-05 Section 6.2 if it has a fundamental natural frequency of less than 1 Hz (i.e., a fundamental period $T > 1$ sec). Using the approximate frequency formula of ASCE 7-05 Equation C6-17, the frequency $n_1 = 100/H$ (H is the building height) = 100 / 255 = 0.4 Hz., which indicates that the building is flexible.

Allowed Procedures

The design for wind loads of buildings and other structures, including the main wind force-resisting system (MWFRS) and components and cladding elements thereof, needs to be based on one of the following procedures:

(1) Method 1 - The Simplified Procedure of ASCE 7-05 Section 6.4, applicable to buildings meeting the requirements specified in that section. This method is not applicable to the example building, insofar as all the conditions under Section 6.4.1 are not met (for example, the height of the building is not less than 60 feet).

(2) Method 2 - The Analytical Procedure of ASCE 7-05 Section 6.5 for buildings meeting the requirements specified in that section. This method is applicable to the example building as it satisfies the two conditions set forth in Section 6.5.1 (given below):

1. The building is regular in shape as defined in ASCE 7-05 Section 6.2.
2. The building does not have response characteristics making it subject to across wind loading, vortex shedding, instability that is due to galloping or flutter; or does not have a site location for which channeling effects or buffeting in the wake of upwind obstructions warrant special consideration.

(3) Method 3 - The Wind Tunnel Procedure of ASCE 7-05 Section 6.6. This method is not necessary for the example building.

Design Procedure

The design load that is due to wind can be calculated after determining the following quantities (ASCE 7-05 Section 6.5.3, as modified by Section 1609):

1. The basic wind speed, V, and wind directionality factor, K_d, per ASCE 7-05 Section 6.5.4 or 2006 IBC Section 1609.3.
2. An importance factor, I, per ASCE 7-05 Section 6.5.5.
3. An exposure category and velocity pressure exposure coefficient, K_z or K_h, as applicable, per ASCE 7-05 Section 6.5.6.

Example 5—(continued)

4. A topographic effect factor, K_{zt}, per ASCE 7-05 Section 6.5.7.
5. A gust effect factor, G or G_f, as applicable, per ASCE 7-05 Section 6.5.8.
6. An enclosure classification per ASCE 7-05 Section 6.5.9 (or Section 1609.2).
7. Internal pressure coefficient, GC_{pi}, per ASCE 7-05 Section 6.5.11.1.
8. External pressure coefficients, C_p or GC_{pf}, or force coefficients, C_f, as applicable, per ASCE 7-05 Section 6.5.11.2 or 6.5.11.3.
9. Velocity pressure, q_z or q_h, as applicable, per ASCE 7-05 Section 6.5.10.
10. Design wind pressure, p, or load, F, per ASCE 7-05 Section 6.5.12, 6.5.13, 6.5.14 or 6.5.15, as applicable.

Basic Wind Speed (IBC Section 1609.3, ASCE 7-05 Section 6.5.4)

The basic wind speed, V, used in the determination of design wind loads on buildings and other structures is as given in IBC Figure 1609 or ASCE 7-05 Figure 6-1, except as provided in ASCE 7-05 Section 6.5.4.1 or 6.5.4.2. The wind must be assumed to come from any direction. According to ASCE 7-05 Section 6.5.4.4, the wind directionality factor, K_d, must be determined from ASCE 7-05 Table 6-4. This factor is to be applied in conjunction with the design load combinations of 2006 IBC Section 1605.2.1.

V = 85 mph (for Los Angeles from ASCE 7-05 Figure 6-1)

K_d = 0.85 (from ASCE 7-05 Table 6-4 for main wind force-resisting system)

Importance Factor (ASCE 7-05 Section 6.2)

An importance factor, I, for the building needs to be determined from ASCE 7-05 Table 6-1, based on the building and structure categories listed in IBC Table 1604.5 (or ASCE 7-05 Table 1-1).

The example office building can be placed under Category II.

Importance factor I = 1.00

Exposure Categories (IBC Section 1609.4)

For each wind direction considered, an exposure category that adequately reflects the characteristics of ground surface irregularities needs to be determined for the site at which the building is to be constructed. For urban and suburban areas, wooded areas or other terrain with numerous closely spaced obstructions having the size of single-family dwellings or larger, Surface Roughness B is applied. Exposure B shall apply where the ground surface roughness condition, as defined by Surface Roughness B, prevails in the upwind direction for a distance of at least 2600 feet or 20 times the height of the building, whichever is greater.

Assume Exposure Category B for the example building in both orthogonal directions.

Based on the exposure category, a velocity pressure exposure coefficient, K_z or K_h, as applicable, needs to be determined from ASCE 7-05 Table 6-5.

For MWFRS of building, Case 2 of ASCE 7-05 Table 6-3 is applicable.

K_z = $2.01\ (z/z_g)^{2/\alpha}$ for 15 ft $\leq z \leq z_g$

Where: z = height above ground level, ft

α = 3-second gust speed power law exponent

z_g = nominal height of the atmospheric boundary layer, ft

From ASCE 7-05 Table 6-2 (for Exposure B) α = 7.0

z_g = 1200 ft

The values of K_z at different height levels and K_h are shown in Table E5.1.

Example 5—(continued)

Topographic Effects (ASCE 7-05 Section 6.5.7)

Wind speed-up effects at isolated hills, ridges and escarpments, constituting abrupt changes in the general topography, located in any exposure category, need to be included in the design when the conditions set forth in ASCE 7-05 Section 6.5.7.1 are met. This effect is given by a factor K_{zt}:

$$K_{zt} = (1 + K_1 K_2 K_3)^2 \quad \text{(ASCE 7-05 Eq. 6-3)}$$

Where: K_1, K_2 and K_3 are given in ASCE 7-05 Fig. 6-4. For the example building, assuming it to be situated on level ground (i.e., with *H*, as shown in ASCE 7-05 Figure 6-4, equal to zero), the K_{zt} factor may be taken equal to 1.

Gust Effect Factor (ASCE 7-05 Section 6.5.8)

The structure is flexible as defined in ASCE 7-05 Section 6.2. So the gust effect factor must be calculated by (ASCE 7-05 Section 6.5.8.2):

$$G_f = 0.925\left[\frac{1+1.7I_{\bar{z}}\sqrt{g_Q^2Q^2+g_R^2R^2}}{1+1.7g_vI_{\bar{z}}}\right] \quad \text{(ASCE 7-05 Eq. 6-8)}$$

where: $g_Q = 3.4$

$g_v = 3.4$

$$g_R = \sqrt{2\ln(3600n_1)} + \frac{0.577}{\sqrt{2\ln(3600n_1)}} \quad \text{(ASCE 7-05 Eq. 6-9)}$$

$$I_{\bar{z}} = c\left(\frac{33}{\bar{z}}\right)^{1/6} \quad \text{(ASCE 7-05 Eq. 6-5)}$$

$$Q = 1/\sqrt{1+0.63\{(B+h)/L_{\bar{z}}\}^{0.63}} \quad \text{(ASCE 7-05 Eq. 6-6)}$$

$$L_{\bar{z}} = \ell\left(\frac{\bar{z}}{33}\right)^{\bar{\varepsilon}} \quad \text{(ASCE 7-05 Eq. 6-7)}$$

From ASCE 7-05 Table 6-2 (Exposure B):

$\hat{\alpha} = 0.25$

$\bar{b} = 0.45$

$c = 0.3$

$\bar{\varepsilon} = 0.33$

$\ell = 320$ ft

$z_{min} = 30$ft

So: $\bar{z} = 0.6h = 0.6 \times 255 = 153$ ft

$\geq z_{min} = 30$ ft Use $\bar{z} = 153$ ft

$B = 130$ ft (plan dimension of building normal to wind direction)

$L = 130$ ft (plan dimension of building parallel to wind direction)

$I_{\bar{z}} = 0.3\ (33/153)^{1/6} = 0.231$

$L_{\bar{z}} = 320\ (153/33)^{0.33} = 533.6$ ft

$$Q = 1/\sqrt{1+0.63\{(130+255)/533.6\}^{0.63}} = 0.813$$

$n_1 = 0.4$ *Hz* (building natural frequency)

$$g_R = \sqrt{2\ln(3600n_1)} + \frac{0.577}{\sqrt{2\ln(3600n_1)}} = 3.96 \quad \text{(ASCE 7-05 Eq. 6-9)}$$

Example 5—(continued)

$R = \sqrt{\frac{1}{\beta} R_n R_h R_B (0.53 + 0.47 R_L)}$ (ASCE 7-05 Eq. 6-10)

β = 0.02 (ASCE 7-05 Section C6.5.8 states that a damping ratio of 0.02 can be used for concrete buildings)

$R_n = 7.47\, N_1/(1+10.3\, N_1)^{5/3}$ (ASCE 7-05 Eq. 6-11)

$N_1 = \dfrac{n_1 L_{\bar{z}}}{\bar{V}_{\bar{z}}}$ (ASCE 7-05 Eq. 6-12)

$R_\ell = 1/\eta - (1-e^{-2\eta})/2\eta^2$ for $\eta > 0$ (ASCE 7-05 Eq. 6-13a)

$R_\ell = 1$ for $\eta = 0$ (ASCE 7-05 Eq. 6-13b)

where the subscript ℓ in ASCE 7-05 Eq. 6-13 shall be taken as h, B and L, respectively.

$R_\ell = R_h$ setting $\eta = 4.6 n_1 h / \bar{V}_{\bar{z}}$

$R_\ell = R_B$ setting $\eta = 4.6 n_1 B / \bar{V}_{\bar{z}}$

$R_\ell = R_L$ setting $\eta = 15.4 n_1 L / \bar{V}_{\bar{z}}$

$\bar{V}_{\bar{z}} = \bar{b}\left(\dfrac{\bar{z}}{33}\right)^{\bar{\alpha}} V\left(\dfrac{88}{60}\right)$ (ASCE 7-05 Eq. 6-14)

The above quantities can be computed as follows:

$\bar{V}_{\bar{z}} = 0.45 \times (153/33)^{0.25}\, 85\, (88/60)$ = 82.3 ft/s

$\eta_h = 4.6 \times 0.4 \times 255 / 82.3$ = 5.70

corresponding R or $R_h = 1/5.70 - (1-e^{-2 \times 5.70})/2 \times 5.70^2$ = 0.16

$\eta_B = 4.6 \times 0.4 \times 130 / 82.3$ = 2.91

corresponding R or $R_B = 1/2.91 - (1-e^{-2 \times 2.91})/2 \times 2.91^2$ = 0.29

$\eta_L = 15.4 \times 0.4 \times 130 / 82.3$ = 9.73

corresponding R or $R_L = 1/9.73 - (1-e^{-2 \times 9.73})/2 \times 9.73^2$ = 0.10

$N_1 = 0.4 \times 533.6 / 82.3$ = 2.59

$R_n = 7.47 \times 2.59 / (1+10.3 \times 2.59)^{5/3}$ = 0.076

$R = \sqrt{(1/0.02) \times 0.076 \times 0.16 \times 0.29 \times (0.53 + 0.47 \times 0.10)}$ = 0.32

$G_f = 0.925\left\{1 + 1.7 \times 0.231\sqrt{3.4^2 \times 0.813^2 + 3.96^2 \times 0.32^2}\right\}/\{1 + 1.7 \times 3.4 \times 0.231\} = 0.868$

Enclosure Classifications (ASCE 7-05 Section 6.5.9 or 2006 IBC Section 1609.2)

For the purpose of determining internal pressure coefficients, all buildings must be classified as enclosed, partially enclosed or open, as defined in Section 6.2 of ASCE 7-05 or 2006 IBC Section 1609.2.

Based on the conditions set forth in ASCE 7-05 Section 6.2 or 2006 IBC Section 1609.2, the building is assumed to be enclosed.

Internal Pressure Coefficients (ASCE 7-05 Section 6.5.11.1)

Internal pressure coefficients, GC_{pi}, are determined from ASCE 7-05 Figure 6-5 based on building enclosure classifications established using ASCE 7-05 Section 6.5.9.

$GC_{pi} = +0.18$ (ASCE 7-05 Figure 6-5)

-0.18

The reduction factor for large volume buildings is not applicable for the example building, assuming proper partitioning of each floor area.

External Pressure Coefficients (ASCE 7-05 Section 6.5.11.2)

For MWFRS, C_p coefficients are given in ASCE 7-05 Figure 6-6.

Example 5—(continued)

Wall: L/B = 130/130 = 1

Windward C_p = 0.8 To be used with q_z

Leeward C_p = – 0.5 To be used with q_h

Side: C_p = – 0.7 To be used with q_h

Roof: h/L = 255/130 = $1.96 \geq 1.0$

θ = $0 < 10^{\circ}$ (assuming a flat roof)

Horizontal distance from windward edge

0 to $h/2$: C_p = – 1.3 To be used with q_h

> $h/2$: C_p = – 0.7

The above values for roof can be reduced depending on the area

Roof Area = 130 × 130 = 16,900 sq ft $\geq$ 1000 sq ft

Reduction factor = 0.8 (Figure 6-6 of ASCE 7-05)

Velocity Pressure (ASCE 7-05 Section 6.5.10)

Velocity pressure, q_z, evaluated at height z shall be calculated by the following equation:

q_z = $0.00256\, K_z\, K_{zt}\, K_d\, V^2 I$ (psf) (ASCE 7-05 Eq. 6-15)

K_d is wind directionality factor defined in ASCE 7-05 Table 6-4; K_z is the velocity pressure exposure coefficient defined in ASCE 7-05 Table 6-3; K_{zt} is the topographic effect factor defined in ASCE 7-05 Equation (6-3); q_h is the velocity pressure calculated using the above equation at mean roof height h.

Table E5.1 shows the computation of velocity pressure using the above expression.

Design Wind Force p (ASCE 7-05 Section 6.5.12)

ASCE 7-05 Section 6.5.12 is applicable here because the building is considered enclosed. In particular, Section 6.5.12.2.3 is applicable, because the building is flexible and because the design force is being computed for the main wind force-resisting system. The design wind pressure shall be determined from the following equation:

p = $q\, G_f C_p - q_i\, (GC_{pi})$ (ASCE 7-05 Eq. 6-19)

Where: q = q_z for windward walls, evaluated at height z from the ground

= q_h for leeward walls, side walls and roofs, evaluated at height h

q_i = q_h for windward walls, side walls, leeward walls and roofs of enclosed buildings

The distribution of the design wind pressures and loads along the height of the building is shown in Tables E5.2 and E5.3.

Design Wind Pressure p_p for Parapets

(ASCE 7-05 Section 6.5.12.2.4)

Effects of parapets on MWFRS loads:

p_p = $q_p GC_{pn}$ (ASCE 7-05 Eq. 6-20)

q_p = velocity pressure evaluated at the top of the parapet

GC_{pn} = combined net pressure coefficient

= +1.5 for windward parapet

= -1.0 for leeward parapet

At top of parapet, h = 255 + 3 = 258 ft, $K_z = 2.01(z/z_g)^{2/\alpha} = 2.01(258/1200)^{2/7} = 1.296$ (α = 7, z_g = 1200 ft for exposure B from ASCE 7-05 Table 6-2 and Table 6-3):

$q_p = 0.00256\, K_z\, K_{zt}\, K_d\, V^2\, I = 0.00256 \times 1.296 \times 1 \times 0.85 \times 85^2 \times 1 = 20.38$ psf

- For windward parapet:

  $p_p = 20.38 \times 1.5 = 30.57$ psf Force = 30.57 × 3 × 130 / 1000 = 12.0 kips

- For leeward parapet:

Example 5—(continued)

$p_p = 20.38 \times (-1.0) = -20.38$ psf Force = 20.38 × 3 × 130 / 1000 = 7.9 kips

Thus, at the roof level, 12.0 + 7.9 = 19.9 kips is added to the design wind force for MWFRS computed earlier (shown in Tables E5.2 and E5.3).

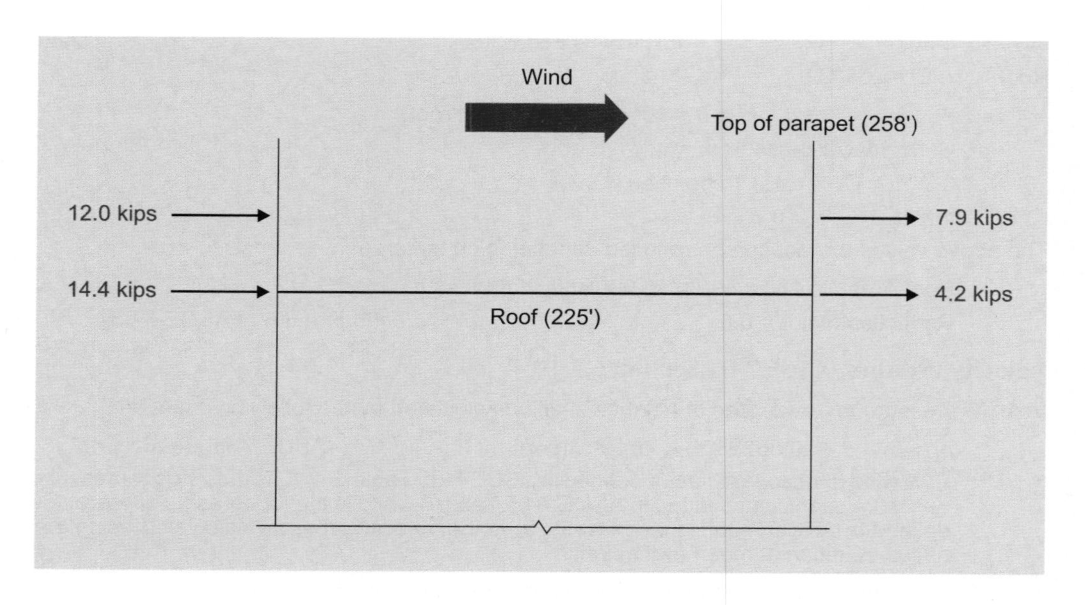

Lateral Analysis

According to ASCE 7-05 Section 6.5.12.3, the main wind-force-resisting systems of buildings of all heights, whose wind loads have been determined according to ASCE 7-05 Section 6.5.12.2.1 and Section 6.5.12.2.3, must be designed for the full and partial wind load cases of ASCE 7-05 Figure 6-9 (Cases 1 through 4). These four cases need to be considered in three-dimensional analyses. A three-dimensional elastic analysis of the structure was performed under the lateral forces shown in Table E5.3 using the SAP 2000 computer program. In the model, rigid diaphragms were assigned at each level, and rigid-end offsets were defined at the ends of each member so that results were automatically obtained at the faces of each support. The mathematical model must consider cracked section properties.

The stiffnesses of members used in the analyses were as follows:

For columns and shear walls $I_{eff} = I_g$

For beams $I_{eff} = 0.5I_g$ (not considering slab contribution)

P-delta effects were considered in the lateral analysis.

Results of Analysis

Based on the results of SAP 2000 analysis, Case 1 shown in ASCE 7-05 Figure 6-9 controls the wind design. For example, the maximum shear force and bending moment at the base of each shear wall under Case 1 wind loads were found to be 284 kips, and 10,571 ft-kips, respectively. The maximum shear force and bending moment at the base of each shear wall under Case 2 wind loads were found to be 264 kips, and 9695 ft-kips, respectively. Because of the location of the shear walls within the plan of the building, the wind-induced axial force in each shear wall is equal to zero.

Example 5—(continued)

Table E5-1. **Computation of K_z, K_h and q_z for the example building**

Height, z, ft	K_z	K_{zt}	K_d	V mph	I	q_z psf
0	0.575	1	0.85	85	1	9.04
15	0.575	1	0.85	85	1	9.04
17.5	0.601	1	0.85	85	1	9.44
30	0.701	1	0.85	85	1	11.01
42.5	0.774	1	0.85	85	1	12.17
55	0.833	1	0.85	85	1	13.10
67.5	0.883	1	0.85	85	1	13.89
80	0.927	1	0.85	85	1	14.58
92.5	0.966	1	0.85	85	1	15.19
105	1.002	1	0.85	85	1	15.75
117.5	1.035	1	0.85	85	1	16.27
130	1.065	1	0.85	85	1	16.75
142.5	1.093	1	0.85	85	1	17.19
155	1.120	1	0.85	85	1	17.61
167.5	1.145	1	0.85	85	1	18.00
180	1.169	1	0.85	85	1	18.38
192.5	1.192	1	0.85	85	1	18.73
205	1.213	1	0.85	85	1	19.07
217.5	1.234	1	0.85	85	1	19.40
230	1.254	1	0.85	85	1	19.71
242.5	1.273	1	0.85	85	1	20.01
255	1.291	1	0.85	85	1	20.30

$\alpha = 7$

z_g = 1200 ft (for exposure B from ASCE 7-05 Table 6-2)

The velocity pressure exposure coefficient K_z may be determined from the following formula: (ASCE 7-05 Table 6-3)

for 15 ft $\leq z \leq z_g$, $K_z = 2.01(z/z_g)^{2/\alpha}$

for $z < 15$ ft, $K_z = 2.01(15/z_g)^{2/\alpha}$

K_h = 1.291 i.e., K_z at a height of 255 ft

q_h = 20.30 psf i.e., q_z at a height of 255 ft

Example 5—(continued)

Table E5-2. Calculation of design wind pressure for the example building

Location	Height Z ft	External Pressure				Internal Pressure			Combine Pressure	
		q_z psf	C_p	G_f	$q_zG_fC_p$ psf	q_i	GC_{pi}	q_iGC_{pi} psf (+)	$q_zG_fC_p$ + q_iGC_{pi} (psf)	$q_zG_fC_p$ - q_iGC_{pi} (psf)
Windward	0	9.04	0.8	0.868	6.28	20.30	0.18	3.65	9.93	2.63
	15	9.04	0.8	0.868	6.28	20.30	0.18	3.65	9.93	2.63
	17.5	9.44	0.8	0.868	6.56	20.30	0.18	3.65	10.21	2.91
	30	11.01	0.8	0.868	7.65	20.30	0.18	3.65	11.30	4.00
	42.5	12.17	0.8	0.868	8.45	20.30	0.18	3.65	12.10	4.80
	55	13.10	0.8	0.868	9.10	20.30	0.18	3.65	12.75	5.45
	67.5	13.89	0.8	0.868	9.65	20.30	0.18	3.65	13.30	6.00
	80	14.85	0.8	0.868	10.12	20.30	0.18	3.65	13.77	6.47
	92.5	15.19	0.8	0.868	10.55	20.30	0.18	3.65	14.20	6.90
	105	15.75	0.8	0.868	10.94	20.30	0.18	3.65	14.59	7.29
	117.5	16.27	0.8	0.868	11.30	20.30	0.18	3.65	14.95	7.65
	130	16.75	0.8	0.868	11.63	20.30	0.18	3.65	15.28	7.98
	142.5	17.19	0.8	0.868	11.94	20.30	0.18	3.65	15.59	8.29
	155	17.61	0.8	0.868	12.23	20.30	0.18	3.65	15.88	8.58
	167.5	18.00	0.8	0.868	12.50	20.30	0.18	3.65	16.15	8.85
	180	18.38	0.8	0.868	12.76	20.30	0.18	3.65	16.41	9.11
	192.5	18.73	0.8	0.868	13.01	20.30	0.18	3.65	16.66	9.36
	205	19.07	0.8	0.868	13.24	20.30	0.18	3.65	16.89	9.56
	217.5	19.40	0.8	0.868	13.47	20.30	0.18	3.65	17.12	9.82
	230	19.71	0.8	0.868	13.69	20.30	0.18	3.65	17.34	10.04
	242.5	20.01	0.8	0.868	13.89	20.30	0.18	3.65	17.54	10.24
	255	20.30	0.8	0.868	14.10	20.30	0.18	3.65	17.75	10.45
Leeward	ALL	20.30	–0.5	0.868	–8.81	20.30	0.18	3.65	–5.16	–12.46
Side Walls	ALL	20.37	–0.7	0.868	–12.38	20.30	0.18	3.65	–8.73	–16.03
Roof	ALL	20.37	–1.04	0.868	–18.39	20.30	0.18	3.65	–14.74	–22.04

Note that the internal pressures do not come into play in the overall design of the building. One can go from this table to Table E5.3 by using just the external pressures on the windward and leeward faces.

Example 5—(continued)

Table E5-3. Calculation of design wind loads for lateral analysis

Height z ft	Tributary Height ft	Building Dimension ft	Windward Face		Leeward Face		Combined Design Wind Load kips
			Design Wind Pressure psf	Design Wind Load kips	Design Wind Pressure psf	Design Wind Load kips	
(Col. 1)	*(Col. 2)*	*(Col. 3)*	*(Col. 4)*	*(Col. 5)*	*(Col. 6)*	*(Col. 7)*	*(Col. 8)*
17.5	15	130	10.21	19.9	–5.16	–10.1	30.0
30	12.5	130	11.30	18.4	–5.16	–8.4	26.7
42.5	12.5	130	12.10	19.7	–5.16	–8.4	28.0
55	12.5	130	12.75	20.7	–5.16	–8.4	29.1
67.5	12.5	130	13.30	21.6	–5.16	–8.4	30.0
80	12.5	130	13.77	22.4	–5.16	–8.4	30.8
92.5	12.5	130	14.20	23.1	–5.16	–8.4	31.5
105	12.5	130	14.59	23.7	–5.16	–8.4	32.1
117.5	12.5	130	14.59	24.3	–5.16	–8.4	32.7
130	12.5	130	15.28	24.8	–5.16	–8.4	33.2
142.5	12.5	130	15.59	25.3	–5.16	–8.4	33.7
155	12.5	130	15.88	25.8	–5.16	–8.4	34.2
167.5	12.5	130	16.15	26.2	–5.16	–8.4	34.6
180	12.5	130	16.41	26.7	–5.16	–8.4	35.1
192.5	12.5	130	16.66	27.1	–5.16	–8.4	35.5
205	12.5	130	16.89	27.4	–5.16	–8.4	35.8
217.5	12.5	130	17.12	27.8	–5.16	–8.4	36.2
230	12.5	130	17.34	28.2	–5.16	–8.4	36.6
242.5	12.5	130	17.54	28.5	–5.16	–8.4	36.9
255	6.25	130	17.75	26.4*	–5.16	–12.1**	38.5

Plus and minus signs signify pressures acting toward and away from the surfaces, respectively. Thus, the forces listed in Column 5 and Column 7 are acting in the same direction. The values in Column 8 are the sums of the absolute values in Column 5 and Column 7.

*Total windward wind force at roof = force from one-half the top story + force on parapet = 14.4 + 12.0 = 26.4 kips

**Total leeward wind force at roof = force from one-half the top story + force on parapet = 4.2 + 7.9 = 12.1 kips

Example 5—(continued)

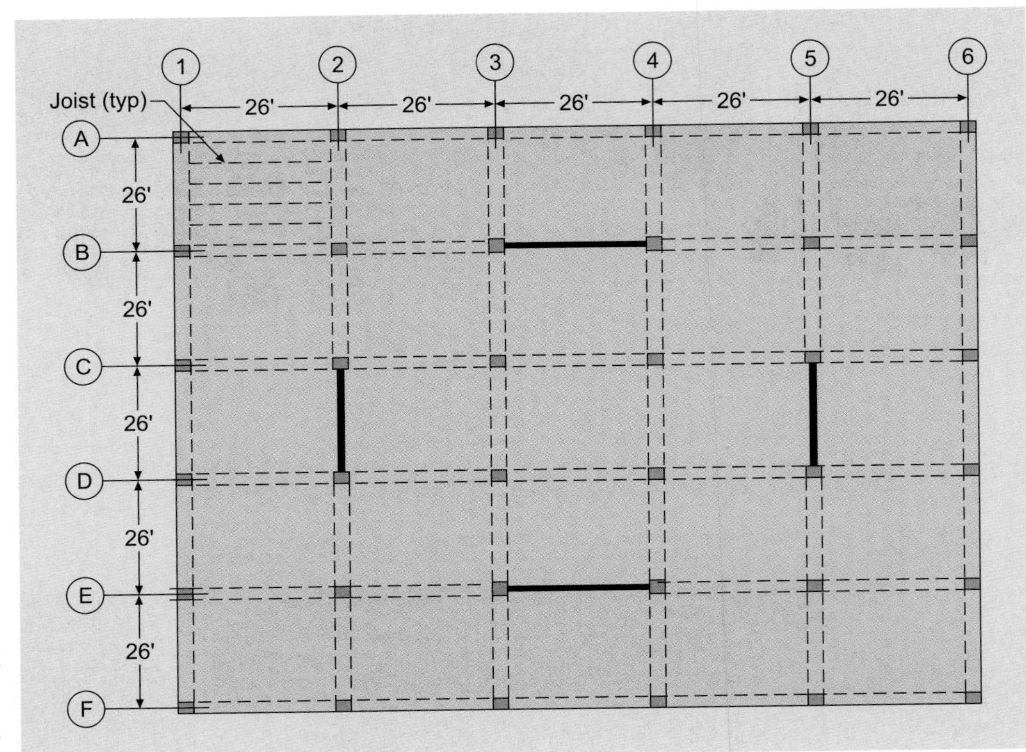

Figure E5-1 Plan of example building considered

Example 5—(continued)

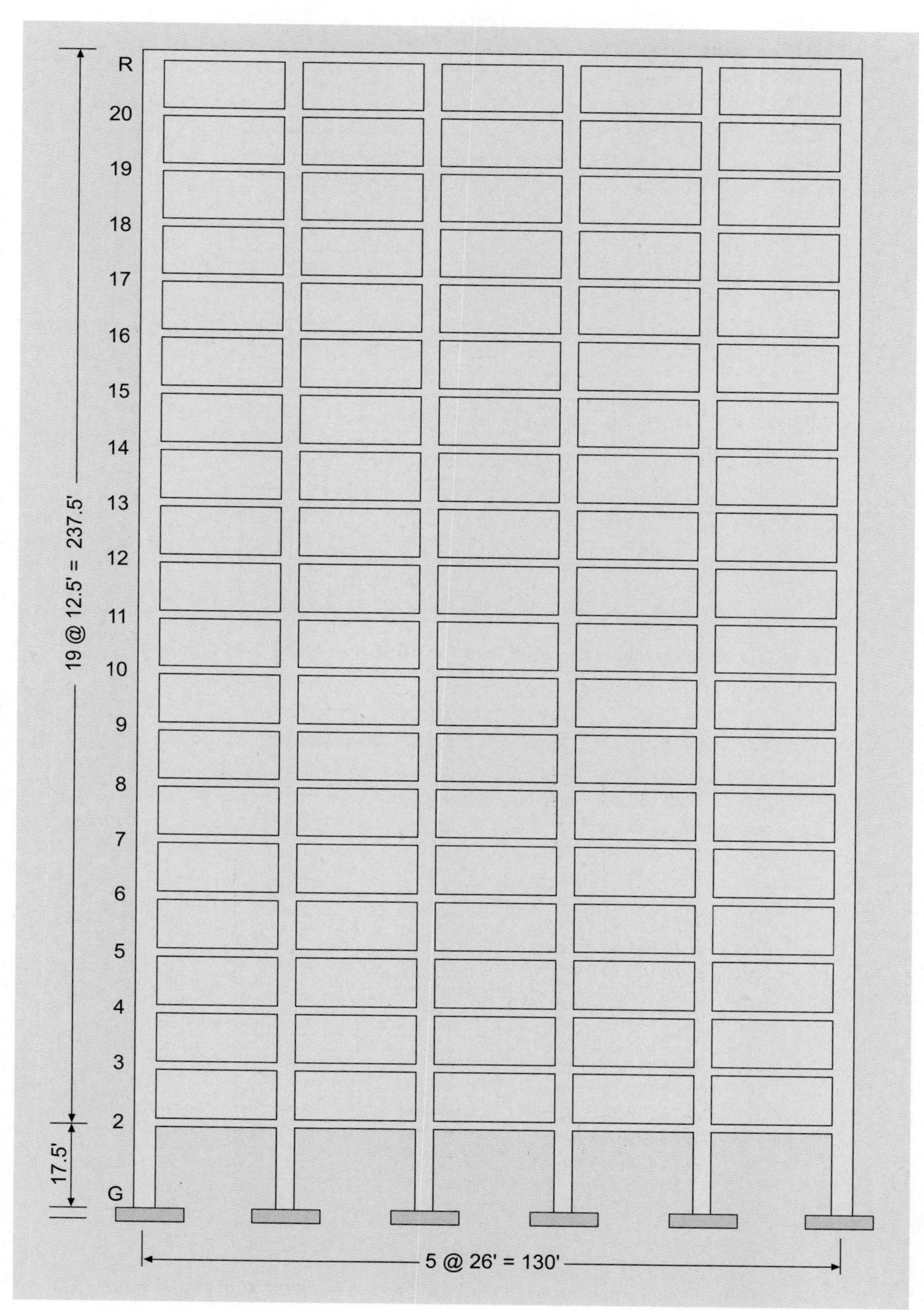

Figure E5-2 Elevation of example building considered

Example 5—(continued)

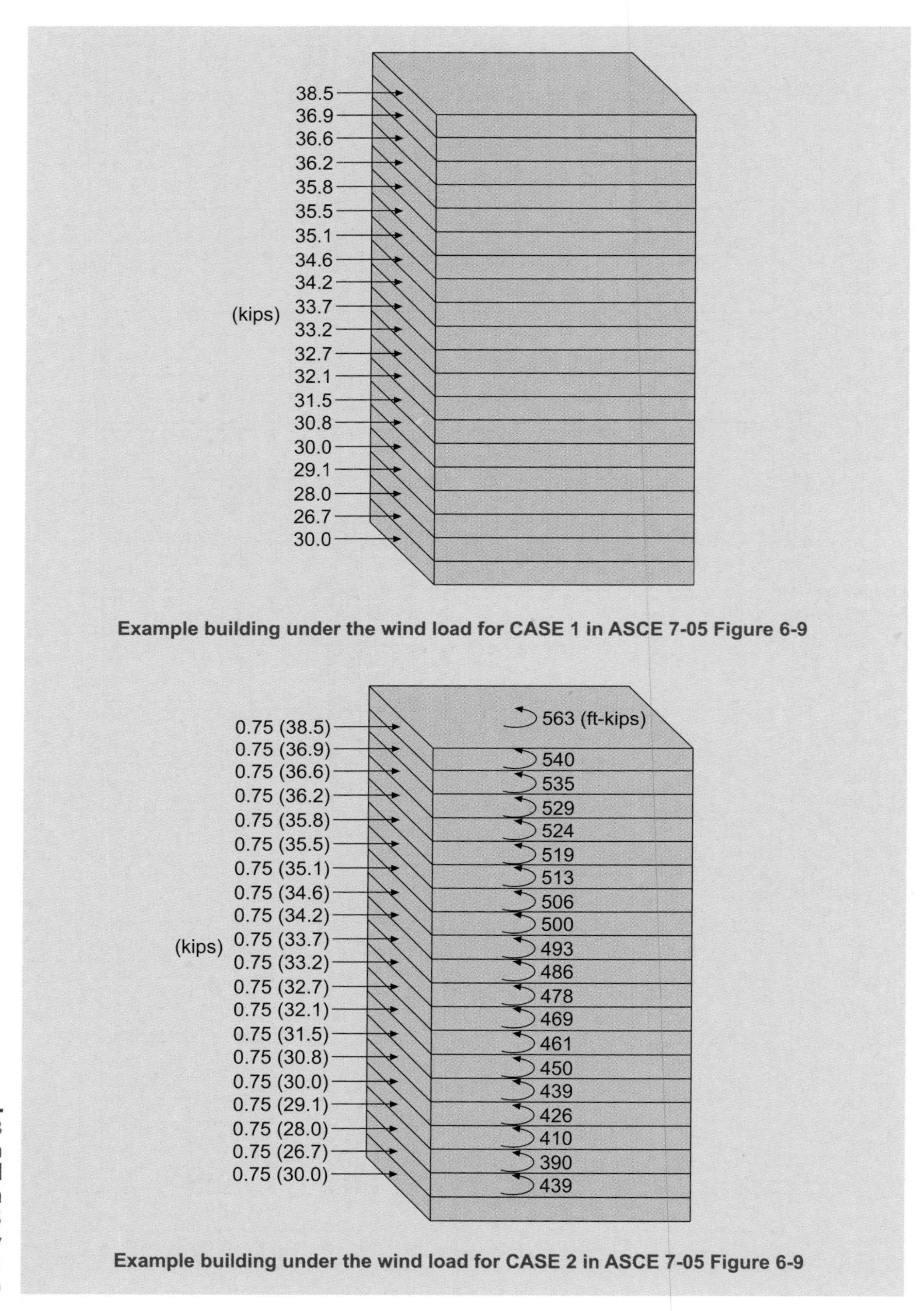

Figure E5-3 **Lateral forces on MWFRS for wind (adapted from ASCE 7-05 Figure 6-9, only showing CASE 1 and 2)**

Example 6 *Calculation of Wind Pressures for a Low-Rise Building*

General

This example illustrates the use of the simplified procedure of ASCE 7-05 to determine design pressures for the MWFRS (Main Wind Force Resisting System of a low-rise building). The building data are as follows:

Location: Memphis, Tennessee

Terrain: Flat farmland

Dimensions: 200 ft × 250 ft in plan (Figure E7.1)

Eave height of 20 ft

Roof Slope 4:12 (18.4 degrees)

Framing: Moment frame spans the 200-ft direction

Moment frame spacing is 25 ft

Cross bracing in 250-ft direction

Openings uniformly distributed

Low-Rise Building

Per ASCE 7-05 Section 6.4, the simplified provisions are applicable only to low-rise buildings (subject to other restrictions). A low-rise building is defined as follows (ASCE 7-05 Section 6.2):

1. The mean roof height h must be less than or equal to 60 ft (18 m).
2. The mean roof height h must not exceed the least horizontal dimension.

The mean roof height for the example building = 20 + 100 × 4/(12 × 2) = 36.7 ft

Least horizontal dimension = 200 ft > 60 ft 60 ft governs

Mean roof height, h = 36.7 ft < 60 ft Condition (2) in ASCE 7-05 Section 6.4.1.1 is met

The example building is a not simple diaphragm building as defined in ASCE 7-05 Section 6.2 [A building in which both windward and leeward wind loads are transmitted through floor and roof diaphragms to the same vertical MWFRS (e.g., no structural separations)]. Thus Condition (1) in ASCE 7-05 Section 6.4.1.1 is not met.

The simplified design procedures still applied here to the building because the design of this example building has been featured in ASCE's *Guide to the Use of Wind Load Provisions of ASCE 7-95* (E6.1) since the first edition of the publication. This provides a rare opportunity for comparison with the past.

The example building is enclosed and a regular-shaped building as defined in ASCE 7-05 Section 6.2 and it is not located in a wind-borne debris region. Thus, Conditions (3) and (4) in Section 6.4.1.1 are both met.

The example building is one-story and the fundamental natural frequency is much greater than 1 Hz according to ASCE 7-05 Section C6.5.8. Thus, Condition (5) in Section 6.4.1.1 is met. The example building does not have response characteristics making it subject to across wind loading, vortex shedding, instability that is due to galloping or flutter; and does not have a site location for which channeling effects or buffeting in the wake of upwind obstructions warrant special consideration. Thus, Condition (6) in Section 6.4.1.1 is also met.

The example building is symmetrical with a gable roof with $\theta \leq 45°$ (18.4°). Thus, Condition (7) in Section 6.4.1.1 is met.

The example building has a flexible diaphragm and is exempted from torsional load cases as indicated in Note 5 of ASCE 7-05 Figure 6-10. Thus Condition (8) in Section 6.4.1.1 is met.

Therefore, but for the violation of Condition 1, Method 1, simplified procedure in ASCE 7-05 Section 6.4 can be applied.

Exposure, Building Classification and Basic Wind Speed

The building is located in flat farmland; therefore, use Exposure C (Section 1609.4).

Building Category is II, per definitions of ASCE 7-05 Table 6.1 - Wind importance factor I = 1.0

Example 6—(continued)

The basic wind speed can be taken as 90 mph (Figure 1609).

Main Wind Force Resisting System (MWFRS)

Simplified design wind pressures, p_s, for the MWFRSs of low-rise simple diaphragm buildings represent the net pressures (sum of internal and external) to be applied to the horizontal and vertical projections of building surfaces as shown in ASCE 7-05 Figure 6-2. For the horizontal pressures (Zones A, B, C, D), p_s is the combination of the windward and leeward net pressures. Pressures shown in ASCE 7-05 Figure 6-2 are applied to the horizontal and vertical projections and are for exposure B, h = 30 ft, I = 1.0, and K_{zt} = 1.0. Adjust to other conditions using ASCE 7-05 Equation 6-1.

Edge strip width: = 0.1 B = 0.1 × 200 = 20 ft

= 0.4 × mean roof height = 0.4 × 36.7 = 14.7 ft . . . governs

= 0.04 B = 0.04 × 200 = 8 ft

= 3 ft

End Zone: 2 times the edge strip width = 2 × 14.7 = 29.4 ft

Table E6.1 shows the detailed calculation of the wind pressures on different zones (surface locations). Figures E6.2 and E6.3 show the direction of wind pressure on different surface locations of the low-rise building.

Reference:

E6.1. Mehta, K.C., and Delahay, J.M., *Guide to the Use of Wind Load Provisions of ASCE 7-02,* American Society of Civil Engineers, New York, NY, 2004.

Example 6—(continued)

Table E6-1. Calculation of wind pressure[1, 2, 3]

Basic Wind Pressure

			Horizontal Loads				Vertical Loads				Overhangs	
			End Zone		Interior Zone		End Zone		Interior Zone			
	Load Direction	Roof Angle	Wall	Roof	Wall	Roof	Wind-ward Roof	Leeward Roof	Wind-ward Roof	Leeward Roof	End Zone	Interior Zone
ASCE 7-05 Figure 6-2	Trans erse	0 to 5	12.8	–6.7	8.5	–4	–15.4	–8.8	–10.7	–6.8	–21.6	–16.9
		15	16.1	–5.4	10.7	–3.0	–15.4	–10.1	–10.7	–7.7	-21.6	-16.9
		20	17.8	–4.7	11.9	–2.6	–15.4	–10.7	–10.7	–8.1	-21.6	-16.9
Transverse*		18.4	17.3	–4.9	11.5	–2.7	–15.4	–10.5	–10.7	–8.0	-21.6	-16.9
Longitudinal		All	12.8	–6.7	8.5	–4	–15.4	–8.8	–10.7	–6.8	-21.6	-16.9

Revised for Exposure and Height

Loading in Transverse Direction (Fig. E6.2)										
Wind Pressure	25.2	–7.2	16.8	–4.0	–22.5	–15.3	–15.6	–11.6	–31.5	–24.7
Building Surface Location	A	B	C	D	E	F	G	H	NA	NA

Loading in Longtitudinal Direction (Fig. E6.2)										
Wind Pressure	18.7	–9.8	12.4	–5.8	–22.5	–12.8	–15.6	–9.9	–31.5	–24.7
Building Surface Location	A		B	C	D	E	F	G	H	NA

Adjustment factor for Building Height and Exposure = 1.46 (ASCE 7-05 Figure 6-2)

Notes: 1. Pressures are applied in accordance with the loading diagrams shown in ASCE 7-05 Figure 6-2, as shown in Fig. E6.3

2. The basic wind pressures are obtained from ASCE 7-05 Figure 6-2 (for wind speed of 90 mph and roof angle of 18.4 degrees).

The mean roof height and exposure for this figure are assumed to be 30 ft and B, respectively.

These wind pressures are modified for mean roof height of 36.7 ft and Exposure C (the adjustment factor is 1.46 for this case), as shown.

3. If roof pressure under horizontal loads is less than zero, use zero.

* Interpolated for appropriate roof angle.

Example 6—(continued)

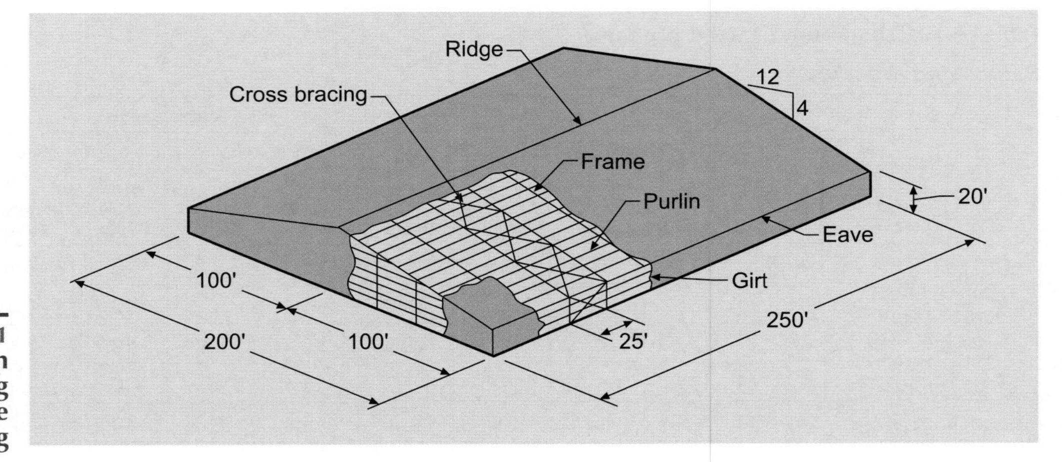

Figure E6-1
Dimension and framing of low-rise building

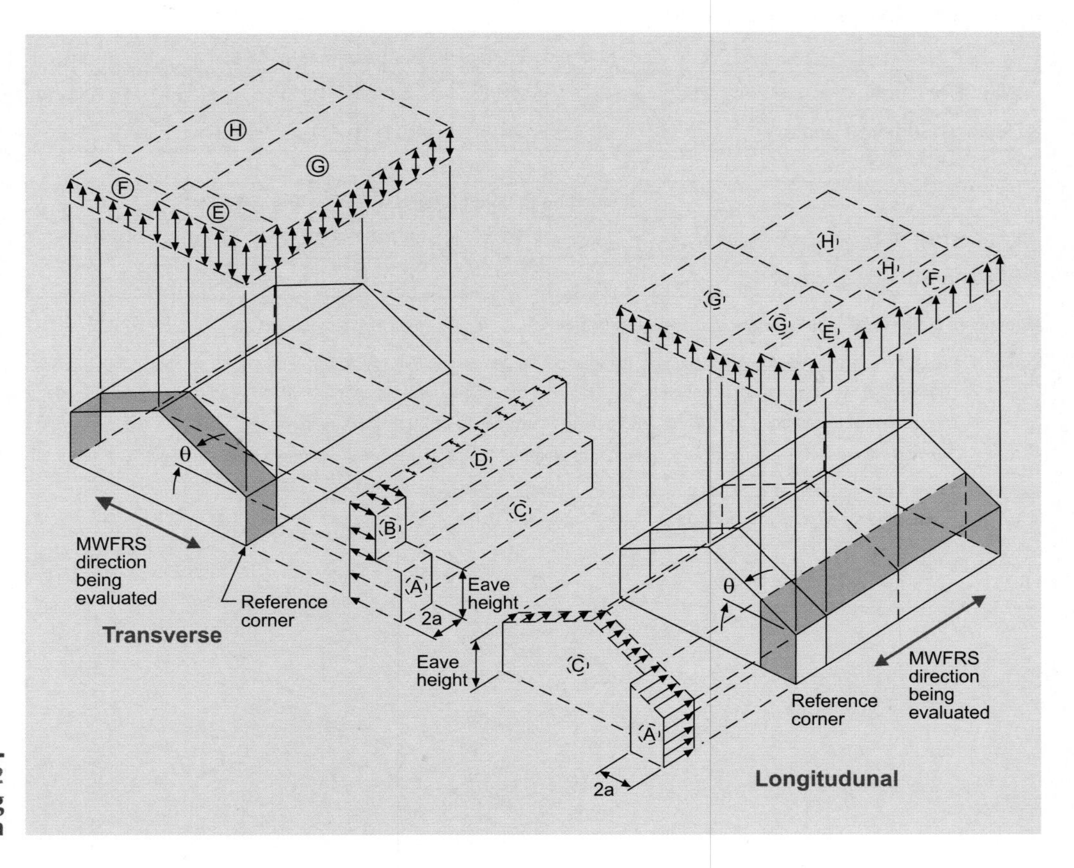

Figure E6-2
MWF loading diagram

Example 6—(continued)

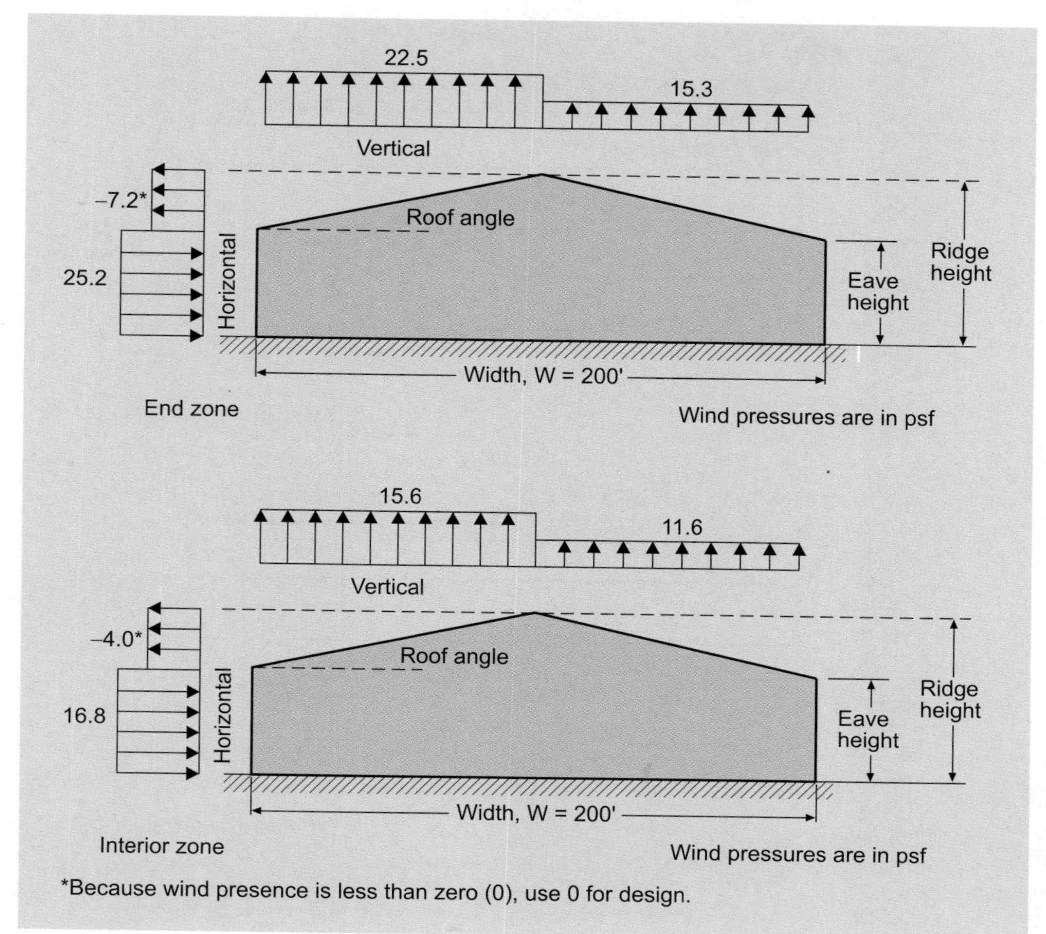

Figure E6-3 (a) **Application of MWFRS loads—loading in transverse direction**

Example 6—(continued)

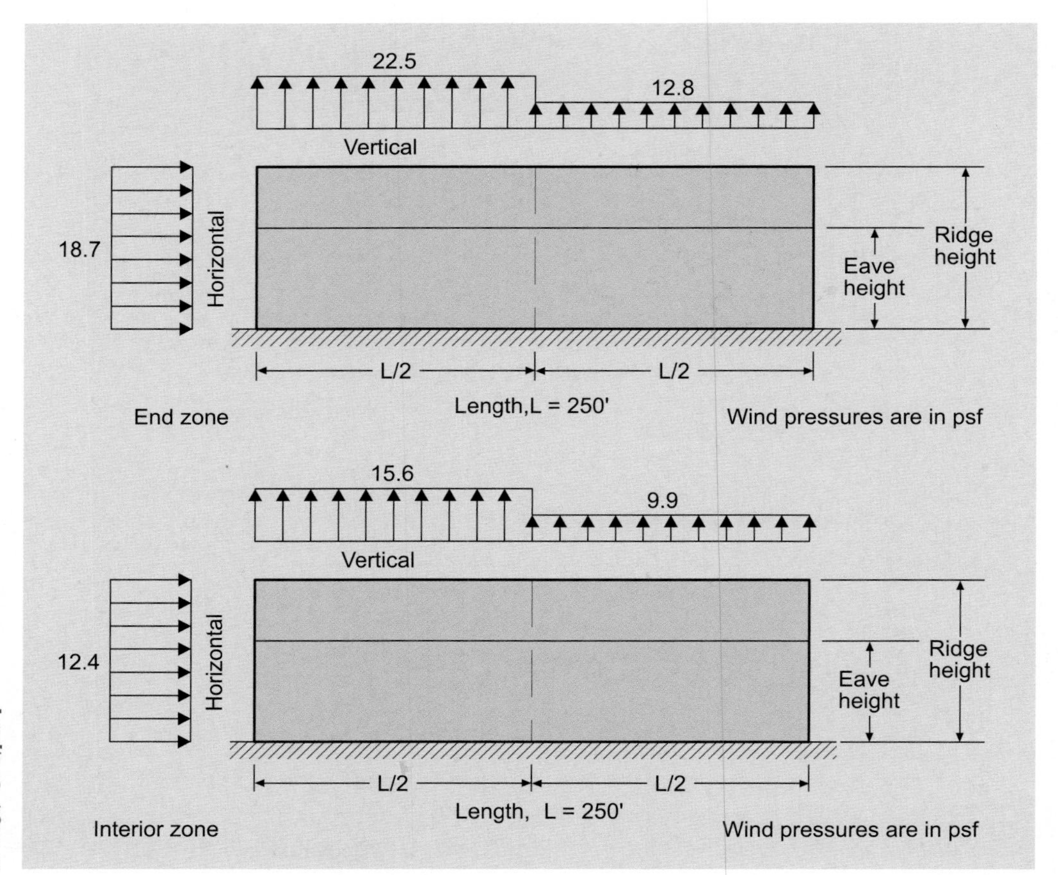

Figure E6-3 (b) Application of MWFRS loads—loading in longitudial direction

Example 7 *Redundancy (ρ) and Concrete Shear Walls*

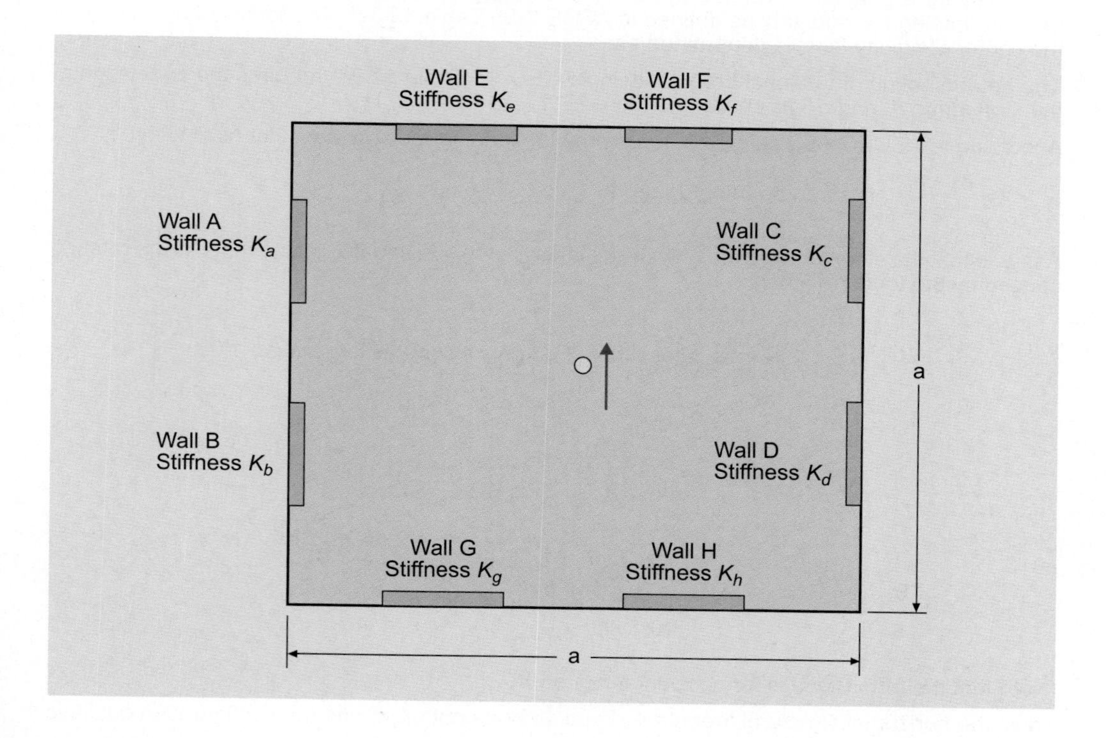

Given information:

SDC D, one story, concrete shear wall building.

$K_a = K_b = K_c = K_d = K_e = K_f = K_g = K_h = \mathbf{K}$

All walls have same nominal shear strength, V_n.

The story height is 18 feet.

The length of each shear wall is 15 feet. Let *a* denote the horizontal dimension of this building.

The redundancy factor ρ must be determined and used in ASCE 7-05 Equation 12.4-3 to determine the horizontal seismic load effect. None of the conditions listed in Section 12.3.4.1 applies, and thus Section 12.3.4.2 must be used to determine whether ρ is 1.0 or 1.3.

Because there are two shear walls on each of the perimeter lines of resistance and the building is completely regular, ASCE 7-05 Section 12.3.4.2(b) might allow a factor ρ of 1.0. However, the length of each shear wall is less than the story height, the number of bays as defined by ASCE 7-05 Section 12.3.4.2(b) is less than two, and thus the configuration does not automatically qualify for a redundancy factor of 1.0 described in ASCE 7-05 Section 12.3.4.2(b). The configuration will therefore be analyzed using the method outlined in ASCE 7-05 Section 12.3.4.2(a); namely, by removing a wall and assessing the effect on story shear strength and on building torsion. In this example, Wall C will be removed. Because of the symmetry of the system, the removal of one wall covers the cases of the removal of each of the other walls.

The story shear strength before removal of a wall is the sum of the strengths of the 4 walls resisting the shear force in the direction under consideration (provided that the orthogonal walls have sufficient strength to resist the torsion, which in this case is only the accidental torsion). If one wall is removed, the story shear strength is the sum of the strengths of the three remaining walls resisting the seismic force in the direction under consideration (again, the orthogonal walls need to have the strength to resist the forces resulting from building torsion, which in this case may be substantial).

Thus the reduction in strength is only 25 percent.

Example 7—(continued)

To qualify for a ρ factor of 1.0, the system with one wall removed must also be checked for an extreme torsional irregularity as defined in ASCE 7-05 Table 12.3-1. The torsional rigidity about the center of rigidity (CR) is determined as:

The determination of torsional irregularity in ASCE 7-05 Table 12.3-1 requires the evaluation of the story drifts d_a and d_b, as shown in Figure E7.1.

According to ASCE 7-05 Table 12.3-1, extreme torsional irregularity does not exist when

$\delta_b, < 1.4\left(\frac{\delta_a + \delta_b}{2}\right)$, this can be transformed to $\frac{\delta_b}{\delta_a} < 2.33$

Assume that the story drift caused only by the lateral force V is equal to d, and that θ is the rotation caused by the torsion T, then

$$\frac{\delta_b}{\delta_a} = \frac{\delta + \frac{2}{3}a\theta}{\delta - \frac{1}{3}a\theta} = \frac{\frac{\delta}{a\theta} + \frac{2}{3}}{\frac{\delta}{a\theta} - \frac{1}{3}}$$

This ratio is less than 2.33 only if δ/(aθ) is larger than 1.08.

$$\frac{\delta}{a\theta} = \frac{\frac{\frac{1}{3}V}{K}}{a\frac{T}{J}} = \frac{\frac{\frac{1}{3}V}{K}}{a\frac{(\frac{1}{6}a + 0.05a)V}{J}} = \frac{\frac{\frac{1}{3}V}{K}}{a\frac{(\frac{1}{6} + 0.05)Va}{\frac{5}{3}Ka^2}} = \frac{\frac{\frac{1}{3}V}{K}}{\frac{(\frac{1}{6} + 0.05)}{\frac{5}{3}}\frac{V}{K}} = 2.56 > 1.08$$

(Note that the term 0.05a is for accidental torsion.)

Thus, the horizontal structural irregularity Type 1b does not exist, and the configuration qualifies for a ρ factor of 1.0.

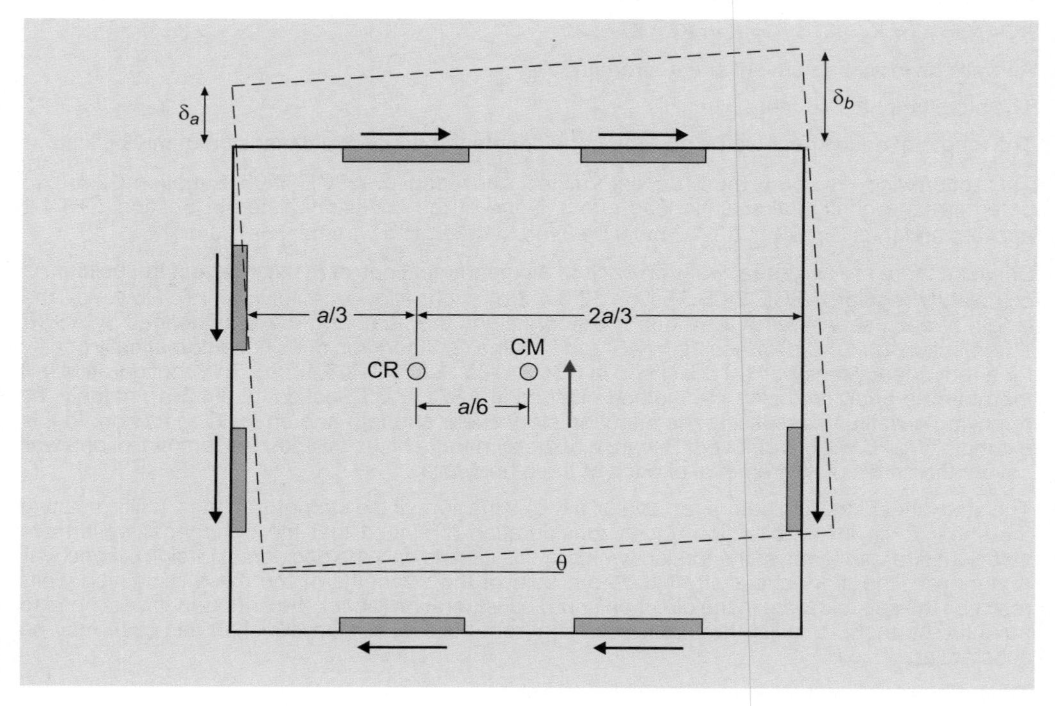

Figure E7-1 **Shear wall resistance**

Example 8 *Simplified Design Procedure*

A two-story building, for which the lateral force-resisting system in the direction considered consists of a building frame system featuring special reinforced concrete shear walls, is considered for the example. The following parameters are assumed for the building:

Weight at each floor, w_x = 390 kips

Response modification factor, R = 6

Design spectral response acceleration at short periods, S_{DS} = 0.75 at the location of the building

1. Determine seismic base shear (ASCE 7-05 Section 12.14.8.1)

The effective seismic weight, $W = 2 \times 390 = 780$ kips

$$V = \frac{FS_{DS}}{R}W \quad \text{(ASCE 7-05 Eq. 12.14-11)}$$

$$= 1.1 \times 0.75 \times 780/6 = 107 \text{ kips}$$

2. Determine vertical distribution (ASCE 7-05 Section 12.14.18.2)

$$F_x = \frac{w_x}{W}V \quad \text{(ASCE 7-05 Eq. 12.14-12)}$$

$$= 390 \times 107/780 = 53.5 \text{ kips at each level}$$

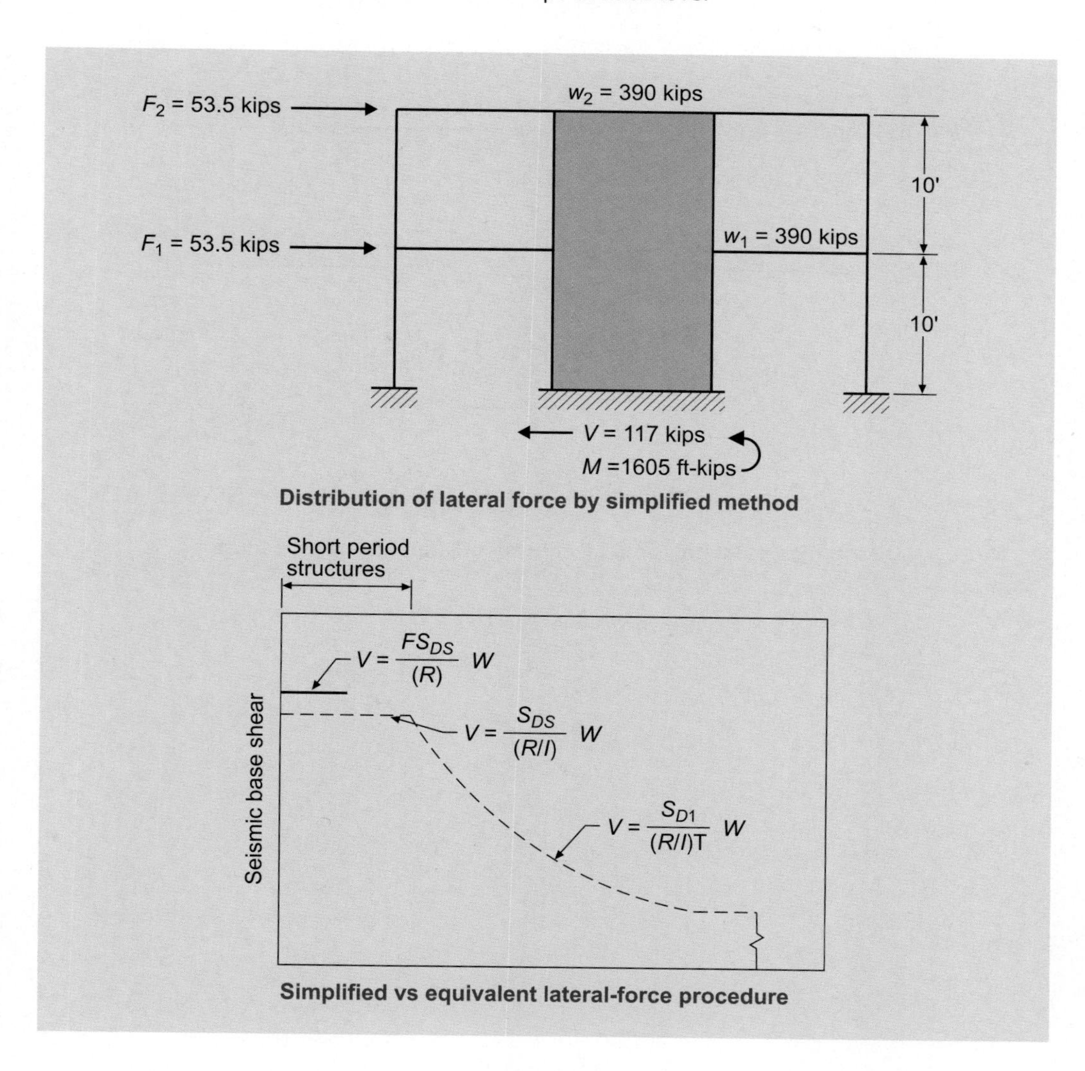

Distribution of lateral force by simplified method

Simplified vs equivalent lateral-force procedure

Example 8—(continued)

3. For comparison with equivalent lateral-force procedure seismic base shear (ASCE 7-05 Section 12.8.1)

By the equivalent lateral-force procedure,

$$V = \frac{S_{DS}}{R/I}W = 1.0 \times 1.0 \times 780 / 8 = 97.5 \text{ kips}$$

Vertical distribution (ASCE 7-05 Section 12.8.3)

$$F_x = \frac{w_x h_x}{\sum_1^2 w_i h_i} V$$

$$F_1 = \frac{390 \times 10}{390 \times 10 + 390 \times 20} \times 97.5 = 32.5 \text{ kips}$$

$$F_2 = \frac{390 \times 20}{390 \times 10 + 390 \times 20} \times 97.5 = 65.0 \text{ kips}$$

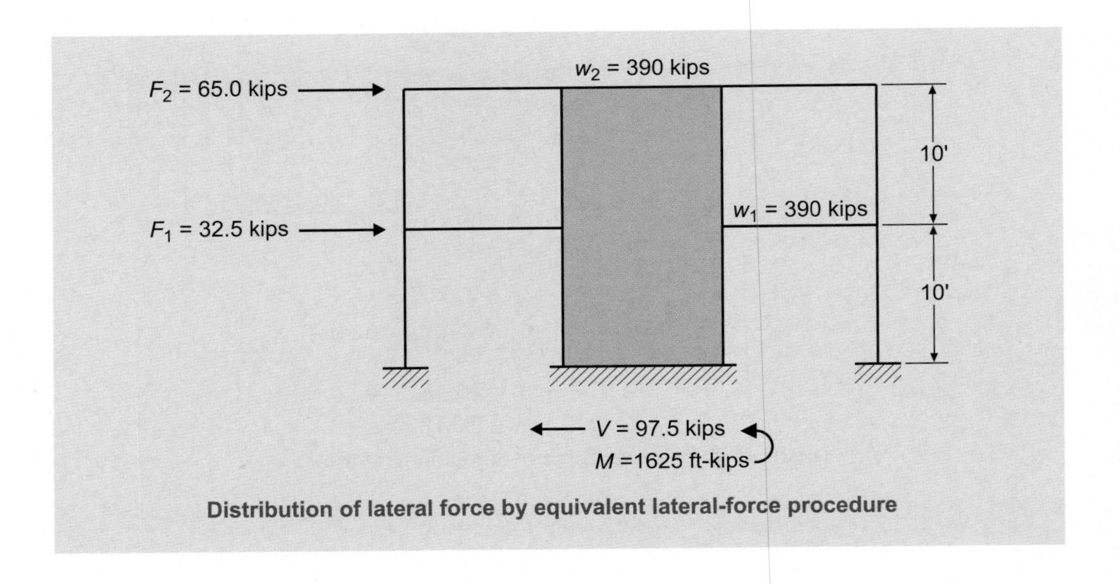

Distribution of lateral force by equivalent lateral-force procedure

Example 9 *Design of Multistory Reinforced Concrete Building Subjected to Earthquake Forces*

General

A 20-story reinforced concrete building is designed following the requirements of the 2006 IBC. Design wind forces for the same building were calculated in Example 5. The building is located in Los Angeles (on Site Class D). The static lateral-force procedure is used as the basis of design.

The building is symmetrical about both principal plan axes. Along each axis a dual structural system is utilized for resistance to lateral forces.

A dual system is defined as a structural system with the following features (ASCE 7-05 12.2.5.1):

1. The moment frames shall be capable of resisting at least 25 percent of the design seismic forces.
2. The total seismic force resistance is to be provided by the combination of the moment frames and the shear walls or braced frames in proportion to their rigidities.

Design Criteria

A typical plan and elevation of the building are shown in Figures E9.1 and E9.2, respectively. The member sizes for the structure are chosen as follows:

Spandrel beams: 34 × 24 in. (width = 34 in.)

Interior beams: 34 × 24 in.

Columns: 34 × 34 in.

Shear walls:

- Grade to 9th floor: 16 in. thick
- 10th floor to 16th floor: 14 in. thick
- 17th floor to roof: 12 in. thick

Shear wall boundary elements: 34 × 34 in.

Material properties: Concrete (all members): $f_c' = 4000$ psi

All members are constructed of normal weight concrete ($w_c = 145$ pcf)

Reinforcement: $f_y = 60{,}000$ psi

Service loads:
Loads on Floors:

Superimposed dead load (SDL): 20 psf—partition and equipment

Live load (LL): 80 psf—per practice, minimum 50 psf (IBC Table 1607.1)

Loads on Roof: 10 psf SDL – roofing
+ 200 kips for penthouse and equipment

20 psf LL (IBC Section 1607.11.2.1)

Joists and Topping: 86 psf

Cladding: 8 psf

Seismic Design Data:

The maximum considered earthquake spectral response acceleration at short period, $S_s = 1.5$g, and that at 1.0 sec period, $S_1 = 0.6$g.

Assume Occupancy Category *II* (IBC Table 1604.5), which gives seismic importance factor, $I = 1.0$ (ASCE 7-05 Table 11.5-1)

Use default soil type = S_D

Site coefficient $F_a = 1.0$ (2006 IBC Table 1613.5.3 (1))
(ASCE 7-05 Table 11.4-1)

Site coefficient $F_v = 1.5$ (2006 IBC Table 1613.5.3(2)
(ASCE 7-05 Table 11.4-2)

Example 9—(continued)

$S_{MS} = F_a S_s = 1.0 \times 1.5 = 1.5g$ (2006 IBC Eq. 16-37) (ASCE 7-05 Eq. 11.4-1)

$S_{M1} = F_v S_1 = 1.5 \times 0.6 = 0.9g$ (2006 IBC Eq. 16-38) (ASCE 7-05 Eq. 11.4-2)

Design Spectral Response Acceleration Parameters (at 5% damping):

At short periods: $S_{DS} = 2/3\ S_{MS}/g = 2/3 \times 1.5 = 1.0$ (2006 IBC Eq. 16-39) (ASCE 7-05 Eq. 11.4-3)

At 1.0 Section period: $S_{D1} = 2/3\ S_{M1}/g = 2/3 \times 0.9 = 0.6$ (2006 IBC Eq. 16-40) (ASCE 7-05 Eq. 11.4-4)

Dual system (special RC shear walls with SMRF)
indicates $R = 7$; $C_d = 5.5$ (ASCE 7-05 Table 12.2-1)

where: R and C_d are the response modification factor and the deflection amplification factor, respectively.

Seismic Design Category: based on both S_{DS} (2006 IBC Table 1613.5.6(1)) and S_{D1} (2006 IBC Table 1613.5.6(2)), the seismic design category for the example building is D.

Design Basis

Calculation of the design base shear and distribution of that shear along the height of the building using the equivalent lateral-force procedure (which is used in a majority of designs) is appropriate and is allowed by the 2006 IBC for buildings with $T < 3.5T_s$ in SDC D (ASCE 7-05 Table 12.6-1). T and T_s are calculated in the following section, and T is less than $3.5T_s$. Therefore the equivalent lateral-force procedure (ASCE 7-05 Section 12.8) is used here. These calculations are required even when design is based on dynamic analysis.

Weights at Each Floor Level

Table E9.1 shows the weights (self-weight + SDL) at each floor level. The weights are calculated as follows:

Total weight of the building: $W = \sum_{i=1}^{20} W_i = 67{,}246$ kips

EQUIVALENT LATERAL FORCE PROCEDURE (ASCE 7-05 Section 12.8)

Design Base Shear (ASCE 7-05 Section 12.8.1)

$V = C_s W$ (ASCE 7-05 Eq. 12.8-1)

where: $C_s = \dfrac{S_{D1} I}{RT} (T \le T_L), \dfrac{S_{D1} T_L I}{RT^2} (T > T_L)$ (ASCE 7-05 Eq. 12.8-3)

$\le \dfrac{S_{DS} I}{R}$ (ASCE 7-05 Eq. 12.8-2)

≥ 0.01 (ASCE 7-05 Eq. 12.8-5)

$\ge 0.5 \dfrac{S_1 I}{R}$ (ASCE 7-05 Eq. 12.8-6) {where $S_1 > 0.6g$}

For the example building considered

$S_{DS} = 1.0$

$S_{D1} = 0.6$

$S_1 = 0.6$

$R = 7$

$I = 1.0$

Approximate fundamental period T (ASCE 7-05 Section 12.8.2.1)

$T = C_T (h_n)^{3/4}$ (ASCE Eq. 12.8-7)

Example 9—(continued)

C_T = 0.02 for a dual system

h_n = total height = 255 feet

T = $0.02 \times (255)^{3/4} = 1.28$ sec < T_L=12 sec (T_L = long-period transition period obtained from ASCE 7-05 Figure 22-16)

$$\frac{S_{D1}I}{RT}W = \frac{0.6 \times 67{,}246}{7 \times 1.28} = 4503 \text{ kips . . . governs}$$

$$\frac{S_{DS}I}{R}W = \frac{1.0 \times 67{,}246}{7} = 9607 \text{ kips}$$

$$\frac{0.5 \times S_1 I}{R}W = \frac{0.5 \times 0.6 \times 67{,}246}{7} = 2882 \text{ kips} \qquad \text{(ASCE Eq. 12.8-6)}$$

$0.01\ S_{DS}I\ W = 0.044 \times 67{,}246 = 2959$ kips

Use V = 4503 kips

Vertical Distribution of Base Shear (ASCE 7-05 Section 12.8.3)

Distribute the base shear as follows:

$$F_x = C_{vx} V \qquad \text{(ASCE 7-05 Eq. 12.8-11)}$$

$$C_{vx} = \frac{w_x h_x^k}{\Sigma w_i h_i^k} \ (i = 1 \text{ to } 20) \qquad \text{(ASCE 7-05 Eq. 12.8-12)}$$

$$T = 1.28 \text{ sec} \qquad T_s = \frac{S_{D1}}{S_{DS}} = \frac{0.6}{1.0} = 0.6 \text{ sec} \qquad T < 3.5T_s = 2.1 \text{ sec}$$

$$k = 1 + 0.5\,(T - 0.5) \le 2 = 1.39 \qquad \text{(ASCE 7-05 Section 12.8.3)}$$

The distribution of the design base shear along the height of the building is shown in Table E9.1.

Lateral Analysis

Three-dimensional analysis of the structure was performed under the lateral forces shown in Table E9.1 using the SAP 2000 computer program. To account for accidental torsion, the mass at each level was assumed to be displaced from the center of mass by a distance equal to 5 percent of the building dimension perpendicular to the direction of force (ASCE 7-05 Section 12.8.4.2). In the model, rigid diaphragms were assigned at each level, and rigid-end offsets were defined at the ends of each member so that results were automatically obtained at the faces of each support.

The stiffnesses of members used in the analyses were as follows:

For columns and shear walls $I_{eff} = I_g$

For beams $I_{eff} = 0.5I_g$ (considering slab contribution)

P-Δ effects were considered in the lateral analysis. Note that this effect is allowed to be neglected in many situations as explained later (ASCE 7-05 Section 12.8.7).

The lateral displacements of the example building, computed elastically under the distributed lateral forces of Table E9.1, are shown in Table E9.2.

Modification of Approximate Period

Often the use of period by the approximate method (ASCE 7-05 Section 12.8.2) results in a conservative design. It is appropriate to use a more rational method for computation of period to reduce the design forces. However, the modified period must not exceed the approximate period by more than a factor (referred to as coefficient C_u) shown in ASCE 7-05 Table 12.8-1. The following expression is used for a rational method:

$$T = 2\pi\sqrt{\Sigma w_i \delta_i^2 / g\Sigma F_i \delta_i}$$

where: the values of F_i represent any lateral force distributed approximately; the elastic deflections, δ_i, must be calculated using the applied lateral forces, F_i.

Example 9—(continued)

Table E9.2 shows the F_i and corresponding δ_i based on approximate period. The modified period can be found as follows:

$$T = 2\pi\sqrt{\frac{\Sigma w_i \delta_i^2}{g \Sigma F_i \delta_i}}$$

$$= 2\pi\sqrt{\frac{1{,}769{,}653}{386 \times 27{,}785}} \quad \text{(Table E9.2)}$$

= 2.55 sec

$\leq$ 1.4 × T from Approximate Method (ASCE 7-05 Table 12.8-1 for S_{D1} > 0.4)

$\leq$ 1.4 × 1.28 = 1.792 sec . . . governs

Revised Design Base Shear

Using the modified period of T = 1.792 sec, the design base shear is recalculated:

$$\frac{S_{D1}I}{RT}W = \frac{0.6 \times 67{,}246}{7 \times 1.792} = 3216 \text{ kips . . . governs} \quad \text{(ASCE 7-05 Eq. 12.8-3)}$$

$$\frac{S_{DS}I}{R}W = \frac{1.0 \times 67{,}246}{7} = 9607 \text{ kips} \quad \text{(ASCE 7-05 Eq. 12.8-2)}$$

$$\frac{0.5 \times S_1 I}{R}W = \frac{0.5 \times 0.6 \times 67{,}246}{7} = 2882 \text{ kips} \quad \text{(ASCE 7-05 Eq. 12.8-6)}$$

Use V = 3216 kips

Figure E9.3 shows the graphical representation of the above three expressions in nondimensionalized form.

Distribute the base shear as follows:

$$F_x = C_{vx} V \quad \text{(ASCE 7-05 Eq. 12.8-11)}$$

$$C_{vx} = \frac{w_x h_x}{\Sigma w_i h_i^k} (i = 1 \text{ to } 20) \quad \text{(ASCE 7-05 Eq. 12.8-12)}$$

T = 1.792 sec

k = 1 + 0.5 (T – 0.5) = 2 = 1.65 (ASCE 7-05 Section 12.8.3)

The distribution of the design base shear along the height of the building is shown in Table E9.3.

Results of Analysis

The maximum shear force and bending moment at the base of the shear wall were found to be 1571 kips and 118,596 ft-kips, respectively. Because of the location of the shear walls within the plan of the building, the earthquake-induced axial force in each shear wall is equal to zero.

The lateral displacement at every floor level (δ_{xe}) are shown in Table E9.4. The maximum inelastic response displacements (δ_x) and story drifts are computed and shown in Table E9.4.

δ_x is calculated per ASCE 7-05 Section 12.8.6:

$$\delta_x = \frac{C_d \delta_{xe}}{I} \quad \text{(ASCE Eq. 12.8-15)}$$

Story Drift Limitation

According to ASCE 7-05 Section 12.12.1, the calculated story drift, δ, as shown in Table E9.4, shall not exceed 0.020 times the story height (ASCE 7-05 Table 12.12-1: for all other buildings, Occupancy Category I or II).

Example 9—(continued)

Floor	Maximum allowable drift	Largest drift: (Table E9.4)
1st	0.02 × 17.5 ft = 4.2 inches > 0.96 in.	O.K.
Others	0.02 × 12.5 ft = 3.0 inches > 2.87 in.	O.K.

P-Δ Effects

According to ASCE 7-05 Section 12.8.7, *P*- Δ effects on story shears and moments, the resulting member forces and moments, and story drifts induced by these effects need not be considered when the stability coefficient, θ, as detemined by the following formula, is equal to or less than 0.1:

$$\theta = \frac{P_x \Delta}{V_x h_{sx} C_d}$$ (ASCE 7-05 Eq. 12.8-16)

where: P_x = the total unfactored vertical force

Δ = the design story drift

V_x = the seismic shear force acting between levels x and x-1

h_{sx} = the story height below level x

C_d = the deflection amplification factor

In the lateral analysis performed using SAP 2000, the *P*-Δ effects were included. However, for illustration purposes, the stability coefficient is calculated as shown in Table E9.5. As the maximum stability coefficient θ (= 0.053) is less than 0.1, the *P*-Δ effect need not be considered.

Redundancy factor, ρ (ASCE 7-05 Section 12.3.4)

Use ρ = 1 (The configuration of this building qualifies for a ρ factor of 1.0, according to ASCE 7-05 Section 12.3.4.2).

Example 9—(continued)

Table E9-1. Lateral forces by Equivalent Lateral-Force Procedure using Approximate Period

V = 4503 kips, T = 1.28 sec, k = 1.390					
Floor Level x	Weight w_x, kips	Height h_x, ft	$w_x h_x^k$, ft^k-kips	Lateral Force F_x, kips	Story Shear V_x, kips
1	2	3	4	5	6
21	2987	255.0	6,611,897	448	448
20	3338	242.5	6,890,261	467	916
19	3338	230.0	6,401,592	434	1350
18	3338	217.5	5,923,177	402	1751
17	3352	205.0	5,478,250	371	2123
16	3366	192.5	5,040,491	342	2465
15	3366	180.0	4,591,376	311	2776
14	3366	167.5	4,154,270	282	3058
13	3366	155.0	3,729,711	253	3311
12	3366	142.5	3,318,309	225	3536
11	3366	130.0	2,920,757	198	3734
10	3380	117.5	2,548,410	173	3907
9	3394	105.0	2,188,593	148	4055
8	3394	92.5	1,835,054	124	4179
7	3394	80.0	1,499,709	102	4281
6	3394	67.5	1,184,252	80	4361
5	3394	55.0	890,873	60	4422
4	3394	42.5	622,547	42	4464
3	3394	30.0	383,628	26	4490
2	3559	17.5	190,174	13	4503
Σ	67,246	Σ	66,403,331	4503	

Example 9—(continued)

Table E9-2. Calculation of period by Rational Analysis (Equivalent Lateral-Force Procedure)

Floor Level x	Weight w_x, kips	Lateral Force F_x, kips	Displacement δ_x, in.	$w_x\delta_x^2$, kip-in.^{2}	$F_x\delta_x$, kip-in.
1	2	3	4	5	6
21	2987	448	8.73	227,883	3,913
20	3338	467	8.37	233,850	3,909
19	3338	434	7.98	212,485	3,463
18	3338	402	7.57	191,461	3,045
17	3352	371	7.13	170,310	2,644
16	3366	342	6.67	149,705	2,281
15	3366	311	6.20	129,243	1,927
14	3366	282	5.70	109,246	1,607
13	3366	253	5.18	90,457	1,312
12	3366	225	4.66	73,016	1,048
11	3366	198	4.12	57,067	815
10	3380	173	3.58	43,259	619
9	3394	148	3.04	31,314	450
8	3394	124	2.52	21,630	313
7	3394	102	2.03	13,918	207
6	3394	80	1.55	8,180	124
5	3394	60	1.11	4,159	66
4	3394	42	0.73	1,804	31
3	3394	26	0.41	557	11
2	3559	13	0.18	110	2
Σ	67,246	4503	Σ	1,769,653	27,785

Example 9—(continued)

Table E9-3. Lateral forces by Equivalent Lateral-Force Procedure using Period from Rational Analysis

V = 3216 kips, T = 1.792 (Section) k = 1.65					
Floor Level x	**Weight** w_x, **kips**	**Height** h_x, **ft**	$w_x h_x^k$, **ft^k-kips**	**Lateral Force** F_x, **kips**	**Story Shear** V_x, **kips**
1	2	3	4	5	6
21	2987	255.0	13,438,884	353	353
20	3338	242.5	13,914,857	363	716
19	3338	230.0	12,840,714	333	1049
18	3338	217.5	11,796,400	304	1352
17	3352	205.0	10,827,953	276	1629
16	3366	192.5	9,882,799	250	1879
15	3366	180.0	8,925,196	224	2103
14	3366	167.5	8,001,448	199	2302
13	3366	155.0	7,112,751	175	2478
12	3366	142.5	6,260,443	153	2630
11	3366	130.0	5,446,031	131	2761
10	3380	117.5	4,690,662	112	2873
9	3394	105.0	3,970,793	93	2966
8	3394	92.5	3,275,782	76	3041
7	3394	80.0	2,627,861	59	3101
6	3394	67.5	2,030,462	45	3146
5	3394	55.0	1,487,930	32	3178
4	3394	42.5	1,006,019	21	3199
3	3394	30.0	592,901	12	3211
2	3559	17.5	274,321	5	3216
Σ	67,246	Σ	128,404,210	3216	

Example 9—(continued)

Table E9-4. **Lateral displacements and drifts of example building by Equivalent Lateral-Force Procedure (in.) (along outer frame line F)**

Floor Level	δ_{xe}	C_d	d_x	Drift
1	2	3	4	5
21	6.47	5.50	35.59	1.49
20	6.20	5.50	34.10	1.60
19	5.91	5.50	32.51	1.65
18	5.61	5.50	30.86	1.82
17	5.28	5.50	29.04	1.87
16	4.94	5.50	27.17	1.93
15	4.59	5.50	25.25	2.04
14	4.22	5.50	23.21	2.09
13	3.84	5.50	21.12	2.15
12	3.45	5.50	18.98	2.20
11	3.05	5.50	16.78	2.20
10	2.65	5.50	14.58	2.20
9	2.25	5.50	12.38	2.09
8	1.87	5.50	10.29	2.04
7	1.50	5.50	8.25	1.93
6	1.15	5.50	6.33	1.82
5	0.82	5.50	4.51	1.54
4	0.54	5.50	2.97	1.32
3	0.30	5.50	1.65	0.94
2	0.13	5.50	0.72	0.72
1	0.00	0.00	0.00	0.00

Example 9—(continued)

Table E9-5. **Calculation of Stability Coefficient**

Story Level	*DL* psf	*LL* psf	Area sq ft	P_x kips	V_x kips	h_{sx} ft	Drift, Δ in.	θ
1	2	3	4	5	6	7	8	9
21	177	12	16,900	3194	353	12.5	1.49	0.016
20	198	32	33,800	7774	716	12.5	1.60	0.021
19	198	32	50,700	11,661	1049	12.5	1.65	0.022
18	198	32	67,600	15,548	1352	12.5	1.82	0.025
17	198	32	84,500	19,435	1629	12.5	1.87	0.027
16	199	32	101,400	23,423	1879	12.5	1.93	0.029
15	199	32	118,300	27,327	2103	12.5	2.04	0.032
14	199	32	135,200	31,231	2302	12.5	2.09	0.034
13	199	32	152,100	35,135	2478	12.5	.15	0.037
12	199	32	169,000	39,039	2630	12.5	2.20	0.040
11	199	32	185,900	42,943	2761	12.5	2.20	0.041
10	200	32	202,800	47,050	2873	12.5	2.20	0.044
9	201	32	219,700	51,190	2966	12.5	2.09	0.044
8	201	32	236,600	55,128	3041	12.5	2.04	**0.045**
7	201	32	253,500	59,066	3101	12.5	1.93	0.044
6	201	32	270,400	63,003	3146	12.5	1.82	0.044
5	201	32	287,300	66,941	3178	12.5	1.54	0.039
4	201	32	304,200	70,879	3199	12.5	1.32	0.035
3	201	32	321,100	74,816	3211	12.5	0.94	0.026
2	210	32	354,900	85,886	3216	17.5	0.72	0.017

Example 9—(continued)

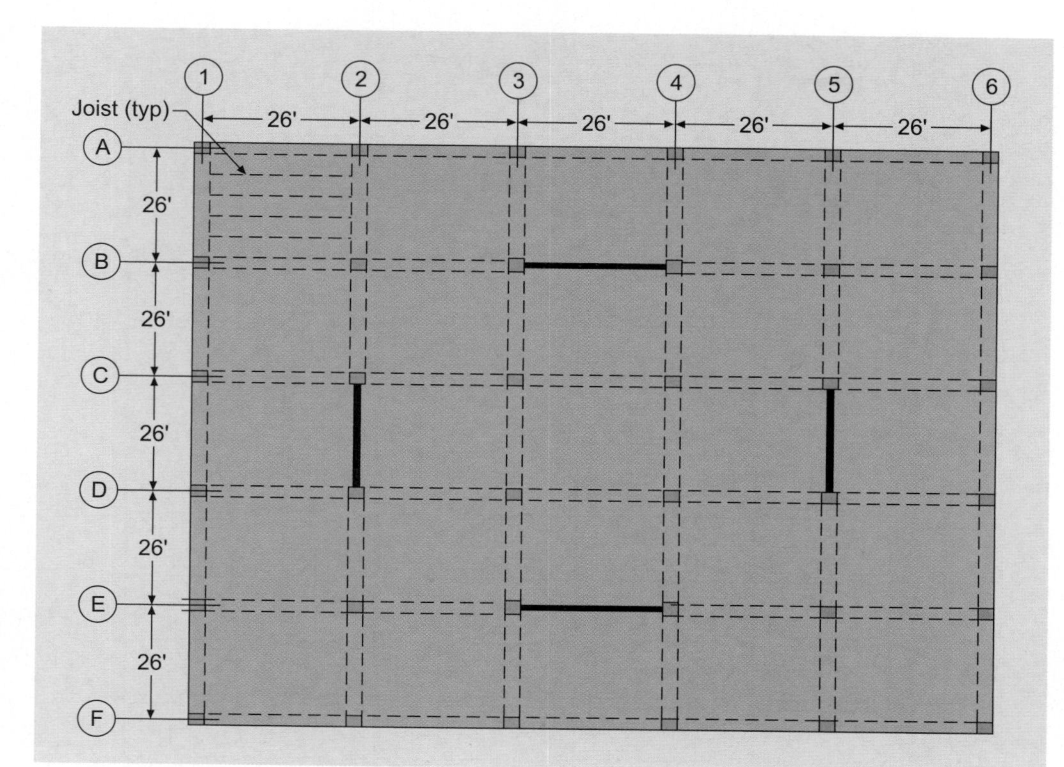

Figure E9-1
Plan of example office building

Example 9—(continued)

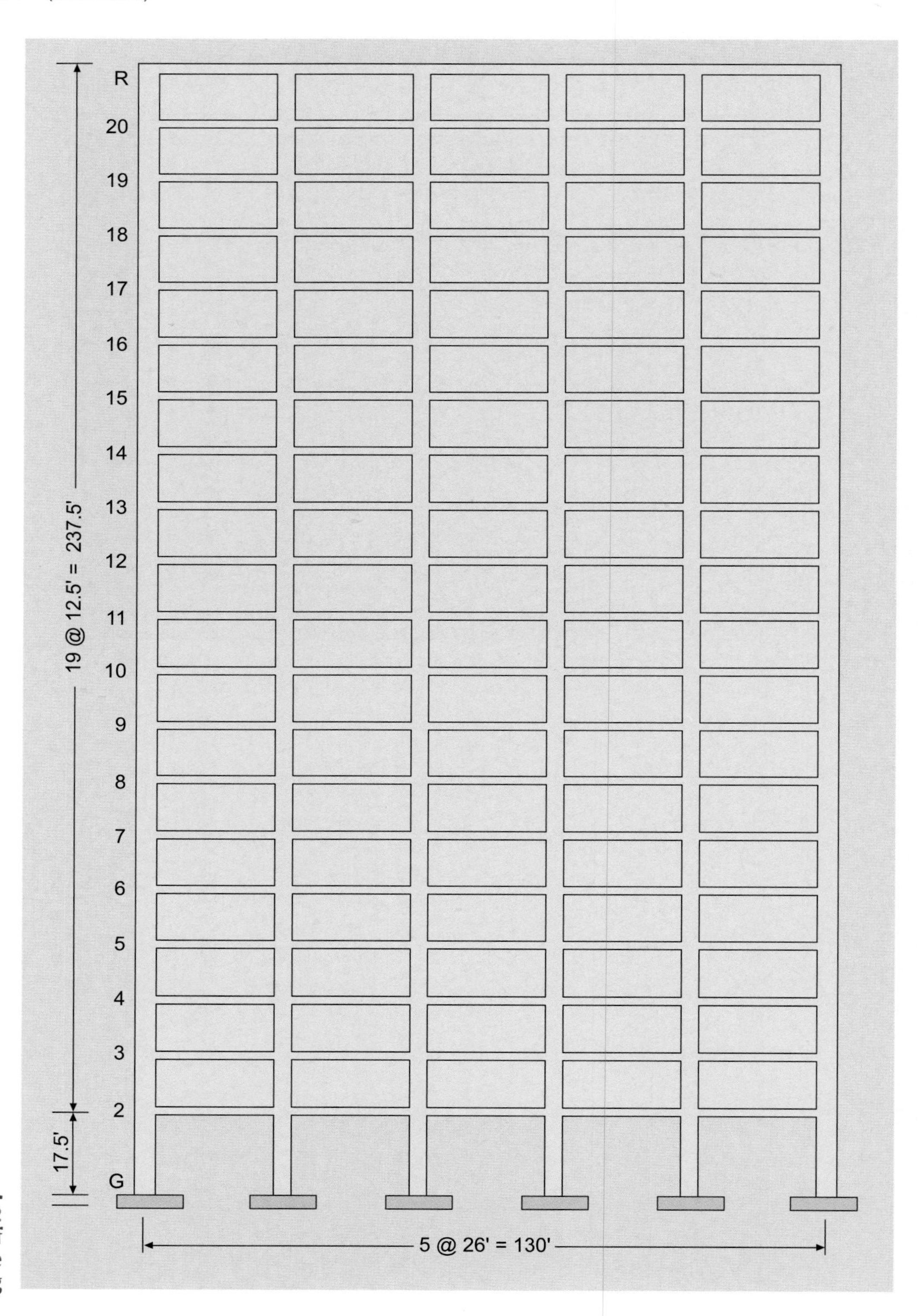

Figure E9-2 Elevation of example office building

Example 9—(continued)

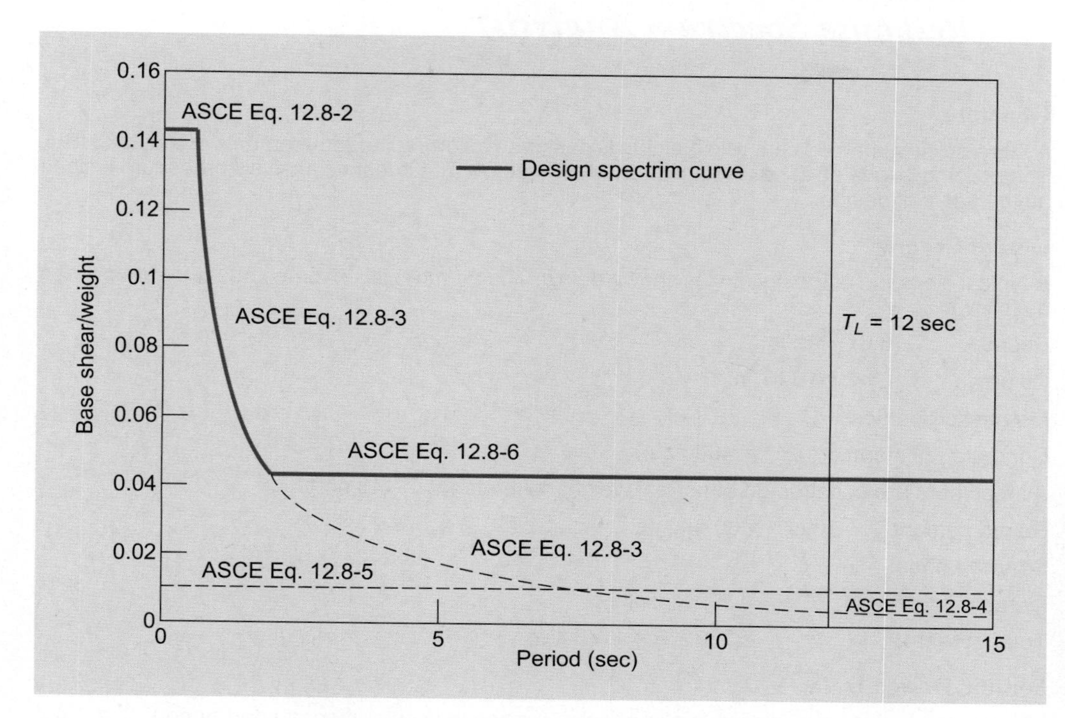

Figure E9-3
Design response spectrum (for structure located where S_1 is equal to or greater than 0.6g)

Example 10 *Dynamic Analysis Procedure (Response Spectrum Analysis)*

General

A three-story reinforced concrete building is designed following the requirements of the 2006 IBC. The building is located in Los Angeles (on Site Class D). The dynamic analysis procedure is used as the basis of design.

Design Criteria

A typical elevation of the building is shown below. The member sizes for the structure are chosen as follows:

Beams 16.67 × 12 in.

Columns 16.67 × 16.67 in.

Material properties:

Concrete (all members): f_c' = 4000 psi

All members are constructed of normal weight concrete (w_c = 145 pcf)

Reinforcement: f_y = 60,000 psi

Service Loads:

Assumed floor load = 390 kips/floor

Total weight W = 390 × 3 = 1170 kips

Seismic Design Data:

The maximum considered earthquake spectral response acceleration at short period, $S_s = 1.5g$, and that at 1-sec period, $S_1 = 0.6g$.

Assume standard Occupancy Category II and seismic importance factor, I = 1.0 (ASCE 7-05 Table 11.5-1)

Use default soil type = S_D

Site coefficient F_a = 1.0 [IBC Table 1613.5.3(1), ASCE 7-05 Table 11.4-1]

Site coefficient F_v = 1.5 [IBC Table 1613.5.3(2), ASCE 7-05 Table 11.4-2]

Adjusted S_s $= S_{MS} = F_a S_s$ (ASCE 7-05 Eq. 11.4-1)

$= 1.0 \times 1.5 = 1.5g$

Adjusted S_1 $= S_{M1} = F_v S_1$ (ASCE 7-05 Eq. 11.4-2)

$= 1.5 \times 0.6 = 0.9g$

Design Spectral Response Acceleration Parameters (at 5% damping):

At short periods: $S_{DS} = 2/3\ S_{MS}/g$ (ASCE 7-05 Eq. 11.4-3)

$= 2/3 \times 1.5 = 1.0$

At 1 sec period: $S_{D1} = 2/3\ S_{M1}/g$ (ASCE 7-05 Eq. 11.4-4)

$= 2/3 \times 0.9 = 0.6$

For an SMRF system, $R = 8$; $C_d = 6.5$ (ASCE 7-05 Table 12.2-1)

where: R and C_d are the response modification factor and the deflection amplification factor, respectively.

Seismic Design Category: based on both S_{DS} [IBC Table 1613.5.6(1), ASCE 7-05 Table 11.6-1] and S_{D1} [IBC Table 1613.5.6(2), ASCE 7-05 Table 11.6-2], the seismic design category for the example building is *D*.

Example 10—(continued)

Design Basis

Calculation of the design base shear and distribution of that shear along the height of the building using the equivalent lateral-force procedure (which is used in a majority of designs) is not appropriate and is not allowed by the *International Building Code* for regular buildings with $T > 3.5\ T_S$ in SDC D (ASCE 7-05 Table 12.6-1). In these cases, a dynamic analysis based procedure (ASCE 7-05 Section 12.9 or Chapter 21) must be used. In order to determine the appropriate analytical procedure that can be used in this example (ASCE 7-05 Table 12.6-1), it is also necessary to check if $T > 3.5\ T_S = 3.5 \times 0.6 = 2.10$ sec, where $T_S = S_{D1}/S_{DS} = 0.6/1{,}0 = 0.6$ sec (ASCE 7-05 Section 11.4.5). The fundamental period T is determined in accordance with ASCE 7-05 Section 12.8.2. In lieu of a more exact analysis, an approximate fundamental period is computed by ASCE 7-05 Equation 12.8-7.

Approximate period parameter $c_t = 0.016$ (ASCE 7-05 Table 12.8-2)

Approximate period parameter $x = 0.9$ (ASCE 7-05 Table 12.8-2)

Period $T_a = C_t h_n^x = 0.016 \times (30)^{0.9} = 0.342 \text{ sec} < 2.10 \text{ sec}$

Thus, the equivalent lateral-force procedure can be used. However, for illustration purposes, the dynamic analysis procedure (ASCE 7-05 Section 12.9) has been used.

- **Given:**

h_s = 10 feet

w = 390 kips/floor

E = 4000 ksi

I_{col} = 4599 in.4 each column (taken equal to $0.7Ig$)

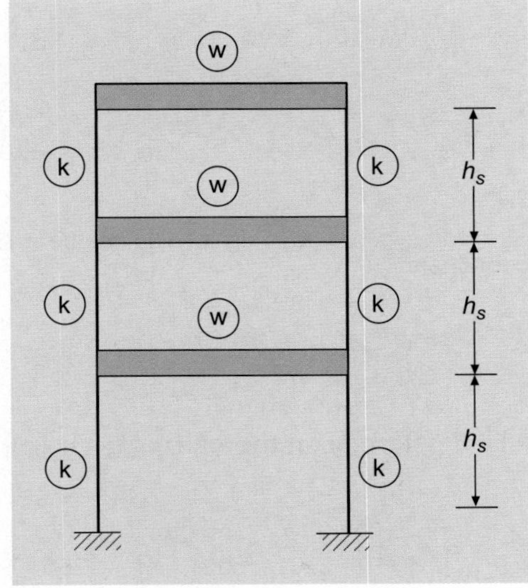

- **Determined mass matrix:**

$m = w/g = 390/386.4$

$= 1.0 \text{ kip-sec}^2/\text{in.}$

$$[m] = \begin{bmatrix} 1 & 0 & 0 \\ 0 & 1 & 0 \\ 0 & 0 & 1 \end{bmatrix}$$

Example 10—(continued)

- **Determined stiffness matrix:**

$12EI / h_s^3 = 12 \times 4000 \times 9000/(12 \times 10)^3 = 250$ kips/in.

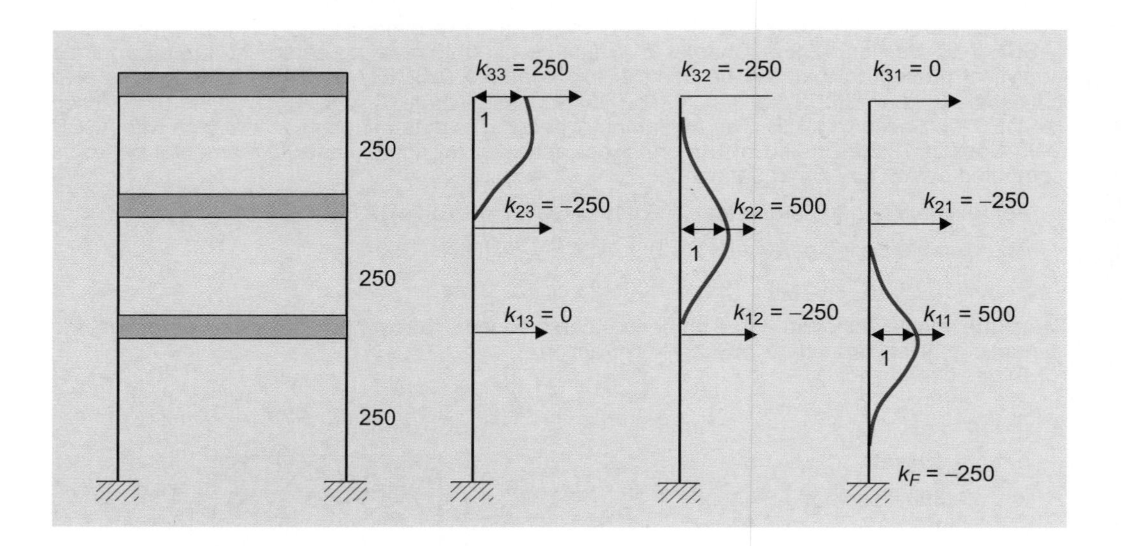

K_y = force corresponding to displacement of coordinate i testing from a unit displacement of coordinate j

$$[k] = 250\begin{bmatrix} 2 & -1 & 0 \\ -1 & 2 & -1 \\ 0 & -1 & 1 \end{bmatrix}$$

- **Find determinant for matrix [*k*] - ϖ^2[*m*]**

$$[k] - \omega^2[m] = \begin{bmatrix} 500-\omega^2 & -250 & 0 \\ -250 & 500-\omega^2 & -250 \\ 0 & -250 & 250-\omega^2 \end{bmatrix}$$

Setting the determinant of the above matrix equal to zero yields the following frequencies:

ω_1 = 7.036 radians/sec

ω_2 = 19.685 radians/sec

ω_3 = 28.491 radians/sec

The period is equal to $2\pi/\omega$:

T_1 = 0.893 sec

T_2 = 0.319 sec

T_3 = 0.221 sec

First mode:

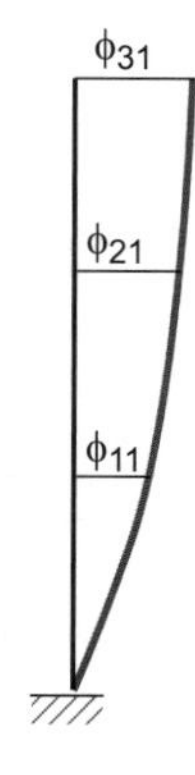

$$\begin{pmatrix} 500-7.036^2 & -250 & 0 \\ -250 & 500-7.036^2 & -250 \\ 0 & -250 & 250-7.036^2 \end{pmatrix}\begin{bmatrix} \phi_{31} \\ \phi_{21} \\ \phi_{11} \end{bmatrix} = \begin{pmatrix} 0 \\ 0 \\ 0 \end{pmatrix}$$

$\phi_{31} = 1.0$ $\phi_{21} = 0.802$ $\phi_{11} = 0.445$

Example 10—(continued)

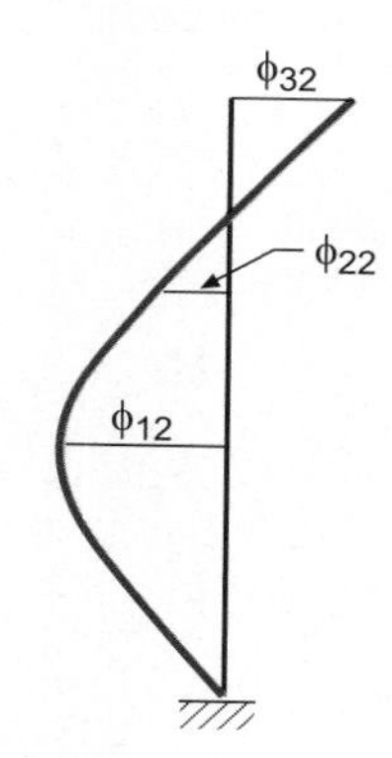

Second mode:

$$\begin{pmatrix} 500-19.685^2 & -250 & 0 \\ -250 & 500-19.685^2 & -250 \\ 0 & -250 & 250-19.685^2 \end{pmatrix} \begin{bmatrix} \phi_{32} \\ \phi_{22} \\ \phi_{12} \end{bmatrix} = \begin{pmatrix} 0 \\ 0 \\ 0 \end{pmatrix}$$

$\phi_{32} = 1.0 \quad \phi_{22} = -0.55 \quad \phi_{12} = -1.22$

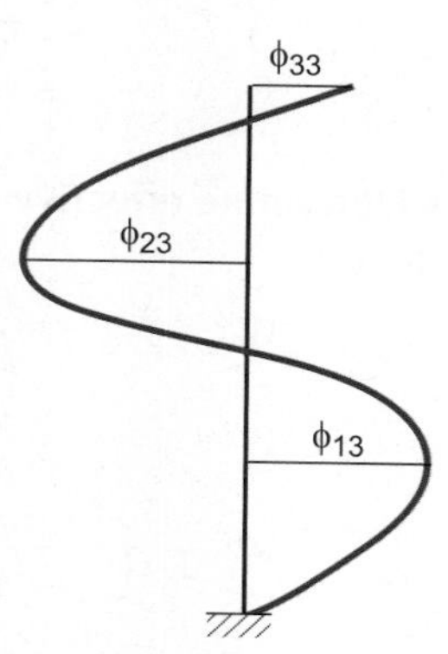

Third mode:

$$\begin{pmatrix} 500-28.491^2 & -250 & 0 \\ -250 & 500-28.491^2 & -250 \\ 0 & -250 & 250-28.491^2 \end{pmatrix} \begin{bmatrix} \phi_{33} \\ \phi_{23} \\ \phi_{13} \end{bmatrix} = \begin{pmatrix} 0 \\ 0 \\ 0 \end{pmatrix}$$

$\phi_{33} = 1.0 \quad \phi_{23} = -2.25 \quad \phi_{13} = 1.802$

The portion of base shear contributed by the mth mode, V_m, shall be determined from the following formulas:

$$V_m = C_{sm}\overline{W}_m$$

$$\overline{W}_m = L_m^{\,2} / M_m$$

$$L_m = \sum_{i=1}^{n} w_i \phi_{im}$$

$$M_m = \sum_{i=1}^{n} w_i \phi_{im}^2$$

Where: C_{sm} = the modal seismic response coefficient

$\overline{W}_m$ = the effective modal gravity load

w_i = the portion of total gravity load, W, of the building at Level i

ϕ_{im} = the displacement amplitude at the ith level of the building when vibrating in its mth mode.

- **Determine modal mass and participation factor for each mode**

$$L_1 = \sum_{i=1}^{3} \frac{\omega_i \phi_{i1}}{g} = 1.0 \text{ kip-sec}^2\text{/in.} (\phi_{11} + \phi_{21} + \phi_{31})$$

$= 1.0\ (0.445 + 0.802 + 1.0) = 2.247$ kip-sec²/in.

$$M_1 = \sum_{i=1}^{3} \frac{\omega_i \phi_{i1}}{g} = 1.0 \text{ kip-sec}^2\text{/in.} (\phi^2{}_{11} + \phi^2{}_{21} + \phi^2{}_{31})$$

$= 1.0\ (0.445^2 + 0.802^2 + 1.0^2) = 1.0\ (0.198 + 0.643 + 1)$

$= 1.841$ kip-sec²/in.

Example 10—(continued)

$L_2 = \sum_{i=1}^{3} \frac{\omega_i \phi_{i2}}{g}$ = 1.0 kip-sec²/in. $(\phi_{12} + \phi_{22} + \phi_{32})$

= 1.0 (– 1.22 – 0.55 + 1.0) = – 0.77 kip-sec²/in.

$M_2 = \sum_{i=1}^{3} \frac{\omega_i \phi_{i2}^2}{g}$ = 1.0 kip-sec²/in. $(\phi^2_{12} + \phi^2_{22} + \phi^2_{32})$

= 1.0 (1.488 + 0.303 + 1) = 2.791 kip-sec²/in.

$L_3 = \sum_{i=1}^{3} \frac{\omega_i \phi_{i3}}{g}$ = 1.0 kip-sec²/in. $(\phi_{13} + \phi_{23} + \phi_{33})$

= 1.0 (1.802 – 2.25 + 1.0) = 0.552 kip-sec²/in.

$M_3 = \sum_{i=1}^{3} \frac{\omega_i \phi_{i3}^2}{g}$ = 1.0 kip-sec²/in. $(\phi^2_{13} + \phi^2_{23} + \phi_{33})$

= 1.0 (3.247 + 5.063 + 1) = 9.310 kip-sec²/in.

- **Determine effective weight and participating mass (pm) for each mode**

$$W_1 = \frac{L_1^2}{M_1} g = \frac{2.247^2}{1.841} \times 386.4 \times \frac{\text{kip}-\text{sec}^2}{\text{in.}} \times \frac{\text{in.}}{\text{sec}^2} = 1059.58 \text{ kips}$$

$$W_2 = \frac{L_2^2}{M_2} g = \frac{(-0.77)^2}{2.791} \times 386.4 = 82.09 \text{ kips}$$

$$W_3 = \frac{L_3^2}{M_3} g = \frac{(0.552)^2}{9.310} \times 386.4 = 12.65 \text{ kips}$$

$$\Sigma W_i = 1154.32 \text{ kips}$$

$$PM_1 = \frac{W_1}{W} = \frac{1059.58}{3 \times 386.4} = 0.914$$

$$PM^2 = \frac{W_2}{W} = \frac{82.09}{3 \times 386.4} = 0.071$$

$$PM_3 = \frac{W_3}{W} = \frac{12.65}{3 \times 386.4} = 0.011$$

$$\Sigma PM = 0.996 \approx 1.0$$

Therefore, consideration of the above three modes (Modes 1, 2, 3) is sufficient per ASCE 7-05 Section 12.9.1. Indeed, the consideration of just the first mode would have been sufficient (as $PM_1 \geq 0.90$).

- **Modal seismic design coefficients, C_{sm}**

$$C_{sm} = \frac{S_{am}}{R/I}$$

where: S_{am} = the modal design spectral response acceleration at period T_m determined from the design response spectrum.

In the example considered here, the procedure of ASCE 7-05 Section 11.4 will be followed. Under this procedure, the spectral response acceleration, S_a, can be expressed by the following equations (ASCE 7-05 Figure 11.4-1):

For $T_s < T < T_L$ $\quad S_a = \frac{S_{D1}}{T}$

$T_0 < T < T_s$ $\quad S_a = S_{DS}$

$T < T_0$ $\quad S_a = 0.6 S_{DS} \frac{T}{T_0} + 0.4 S_{DS}$

Example 10—(continued)

where: $T_s = S_{D1}/S_{DS}$ and $T_o = 0.2\ T_s$; $T_L = 12$ sec (from ASCE 7-05 Section 11.4.5 and Figure 22-15).

For the example building, $T_s = 0.6/1.0 = 0.60$ sec, $T_o = (0.2)\ (0.60) = 0.12$ sec

Mode 1: $T_1 = 0.893$ sec $C_{s1} = \dfrac{S_{D1}}{0.893 \times 8/1} = 0.084g$

> 0.60 sec

< 12 sec

Mode 2: $T_2 = 0.319$ sec $C_{s2} = \dfrac{S_{DS}}{8/1} = 0.125g$

> 0.12 sec

< 0.60 sec

Mode 3: $T_3 = 0.221$ sec $C_{s3} = \dfrac{S_{DS}}{8/1} = 0.125g$

> 0.12 sec

< 0.60 sec

- **Base Shear Using Modal Analysis**

$$V_m = C_{sm}\overline{W}_m = \frac{L_m^2}{M_m} C_{sm}$$

Mode 1: $V_1 = 0.084 \times 1059.58 = 89.3$ kips

Mode 2: $V_2 = 0.125 \times 82.09 = 10.3$ kips

Mode 3: $V_3 = 0.125 \times 12.65 = 1.6$ kips

The modal base shears are combined by the SRSS method to give the resultant base shear

$V_d = [89.3^2 + 10.3^2 + 1.6^2]^{1/2} = 90$ kips

- **Design Base Shear Using Equivalent Lateral-Force Procedure (ASCE 7-05 Section 12.8)**

Design Base Shear (ASCE 7-05 Section 12.8.1)

$V = C_s W$ (ASCE 7-05 Eq. 12.8-1)

where: $C_s = \dfrac{S_{D1} I}{RT}$ (ASCE 7-05 Eq. 12.8-3)

$\leq \dfrac{S_{DS} I}{R}$ (ASCE 7-05 Eq. 12.8-2)

≥ 0.01 (ASCE 7-05 Eq. 12.8-5)

(Note that Supplement No.2 to ASCE 7-05 has changed this minimum coefficient to 0.044 $S_{DS}I$)

and $C_s \geq \dfrac{0.5 S_1 I}{R}$ (ASCE 7-05 Eq. 12.8-6) {where $S_1 = 0.6$g}

For the example building considered

$S_{DS} = 1.0$

$S_{D1} = 0.6$

$S_1 = 0.6$

$R = 8$

$I = 1.0$

Approximate fundamental period T (ASCE 7-05 Section 12.8.2.1)

$T_a = C_T (h_n)^{0.9}$ (ASCE 7-05 Eq. 12.8-7)

Example 10—(continued)

C_T = 0.016 for an SMRF system

h_n = total height = 30 ft

T_a = $0.016 \times (30)^{0.9} = 0.342$ sec

T from rational analysis = 0.893 sec, which should not exceed the approximate period by more than a factor of C_u (obtained from ASCE 7-05 Table 12.8-1).

C_u = 1.4

T = min. of (0.893, 1.4 × 0.342 = 0.479) = 0.479 sec

$$\frac{S_{D1}I}{RT}W = \frac{0.6 \times 1 \times 1170}{8 \times 0.479}\ \frac{0.6 \times 1 \times 1170}{8 \times 0.479} = 183 \text{ kips}$$

$$\frac{S_{DS}I}{R}W = \frac{1.0 \times 1 \times 1170}{8} = 146 \text{ kips . . . governs}$$

$0.01W$ = 0.01× 1170 = 11.7 kips

(Note that Supplement No.2 to ASCE 7-05 has changed this minimum design base shear to 0.044 $S_{DS}IW$=0.044 × 1.0 × 1 × 1170 = 51.5 kips, which still does not govern)

$$\frac{0.5S_1I}{R}W = \frac{0.5 \times 0.6 \times 1 \times 1170}{8} = 43.9 \text{ kips}$$

Use V = 146 kips

• Scaling of Elastic Response Parameters for Design

According to ASCE 7-05 Section 12.9.4, a base shear V shall be calculated in each of the two orthogonal horizontal directions using the calculated fundamental period of the structure T in the relevant direction and the procedures of ASCE 7-05 Section 12.8, except where the calculated fundamental period exceeds C_uT_a, in which case C_uT_a shall be used in lieu of T in that direction. Where the statistical sum of the modal base shears V_t is less than 85 percent of the calculated base shear V using the equivalent lateral force procedure, the forces, but not the drifts, shall be multiplied by $0.85V/V_t$.

Use V = 146 kips

0.85 × 146 kips (85 percent of the equivalent lateral force base shear) > 90 kips (modal base shear)

So the modal forces must be scaled up.

Scale factor = 0.85 × 146/90 = 1.379

The modified modal base shears are as follows:

V_1 = 1.379 × 89.3 = 123.4 kips V_2 = 1.379 × 10.3 = 14.2 kips

V_3 = 1.379 × 1.6 = 2.2 kips $V = [123.4^2 + 14.2^2 + 2.2^2]^{1/2}$ = 124 kips

• Distribution of Modal Base Shears

Lateral force at Level x (1 to 3) for mode m (1 to 3) is to be calculated as:

$$F_{xm} = C_{vxm} V_m$$

$$C_{vxm} = \frac{w_x \phi_{xm}}{\sum_{i=1}^{n} w_i \phi_{im}}$$

where: C_{vxm} = The vertical distribution factor in the mth mode,

V_m = The total design lateral force or shear at the base in the mth mode.

w_i, w_x = The portion of the total gravity load of the building, W, located at or assigned to Level i or x.

ϕ_{im} = The displacement amplitude at the ith level of the building when vibrating in its mth mode.

Example 10—(continued)

ϕ_{xm} = The displacement amplitude at the xth level of the building when vibrating in its mth mode.

The distribution of the modal base shear for each mode is shown in the table below and also in the figure below.

Mode = 1		V_m =	123.4 kips	
Level, *i*	**Weight, w_i**	ϕ_{i1}	$w_i\,\phi_{i1}$	F_{i1}
3	390	1	390.0	54.9
2	390	0.802	312.8	44.1
1	390	0.445	173.6	24.4
		Σ =	**876.3**	**123.4**
Mode = 1		V_m =	**123.4 kips**	
Level, *i*	**Weight, w_i**	ϕ_{i2}	$w_i\,\phi_{i2}$	F_{i2}
3	390	1	390.0	–17.7
2	390	–0.55	–214.5	9.8
1	390	–1.22	–475.8	22.1
		Σ =	**–300.3**	**14.2**
Mode = 1		V_m =	**123.4 kips**	
Level, *i*	**Weight, w_i**	ϕ_{i3}	$w_i\,\phi_{i3}$	F_{i3}
3	390	1	390.0	3.9
2	390	–2.25	–877.5	–8.9
1	390	1.802	702.8	7.2
		Σ =	**215.3**	**2.2**

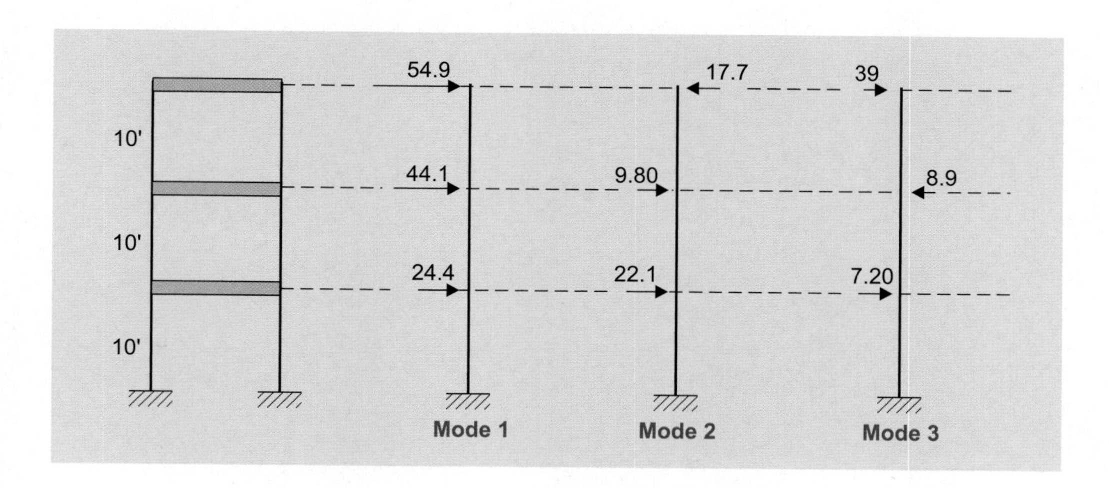

• Lateral Analysis

The stiffnesses of members used in the analyses were as follows:

For columns $I_{eff} = 0.7I_g$

For beams $I_{eff} = 0.5I_g$

Example 10—(continued)

The bending moments and shear forces at beam ends at various levels for different modes are obtained from lateral analysis and are shown in the table below:

Design Negative Bending Moments at Beam Ends (ft-kips)				
Level, *i*	**Mode 1**	**Mode 2**	**Mode 3**	**Resultant, *M***
3	216	–40	4.6	220
2	285	–37	0.17	287
1	269	–8	–2.1	269
Design Shear Forces at Beam Ends (kips)				
Level, *i*	**Mode 1**	**Mode 2**	**Mode 3**	**Resultant, *M***
3	24	–4.5	0.5	24.4
2	31	–4.1	0	31.3
1	30	–0.9	–0.24	30.0

Example 11 *Calculation of Diaphragm Design Forces*

Diaphragm design forces are calculated for the multistory building of Example 9.

The last column gives the final diaphragm design forces calculated by ASCE 7-05 Equation 12.10-1 and the provisions in Section 12.10.1.1.

V = 4503 kips, T = 1.28 (sec) k = 1.390										
Floor Level x	Weight w_x, kips	Height h_x, ft	$w_x h_x^k$, ft^k-kips	Lateral Force F_x, kips	$\sum_{i=x}^{n} F_i$	$\sum_{i=x}^{n} w_i$	F_{px} Equation 12.10-1)	Max Force $0.4S_{DS}I$ w_{px}	Min Force $0.2S_{DS}I$ w_{px}	Design force F_{px}
1	2	3	4	5	6	7	8	9	10	11
21	2987	255.0	6,611,897	448	448	2987	448	1195	597	597
20	3338	242.5	6,890,261	467	916	6325	483	1335	668	668
19	3338	2300	6,401,592	434	1350	9663	466	1335	668	668
18	3338	217.5	5,923,177	402	1751	13,001	450	1335	668	668
17	3352	205.0	5,478,250	371	2123	16,353	435	1341	670	670
16	3366	192.5	5,040,491	342	2465	19,719	421	1346	673	673
15	3366	180.0	4,591,376	311	2776	23,085	405	1346	673	673
14	3366	167.5	4,154,270	282	3058	26,451	389	1346	673	673
13	3366	155.0	3,729,711	253	3311	29,817	374	1346	673	673
12	3366	142.5	3,318,309	225	3536	33,183	359	1346	673	673
11	3366	130.0	2,920,757	198	3734	36,549	344	1346	673	673
10	3380	117.5	2,548,410	173	3907	39,929	331	1352	676	676
9	3394	105.0	2,188,593	148	4055	43,323	318	1358	679	679
8	3394	92.5	1,835,054	124	4179	46,717	304	1358	679	679
7	3394	80.0	1,499,709	102	4281	50,111	290	1358	679	679
6	3394	67.5	1,184,252	80	4361	53,505	277	1358	679	679
5	3394	55.0	890,873	60	4422	56,899	264	1358	679	679
4	3394	42.5	622,547	42	4464	60,293	251	1358	679	679
3	3394	30.0	383,628	26	4490	63,687	239	1358	679	679
2	3559	17.5	190,174	13	4503	67,246	238	1424	712	712
Σ	67,246		66,403,331	4503						

Example 12 *Partial Diaphragm Design*

This example illustrates the design of the roof diaphragm of the three-story building shown in Figure E12.1. The same building was analyzed in Example 10. The lateral forces shown are those obtained from the Mode 1 analysis in that example.

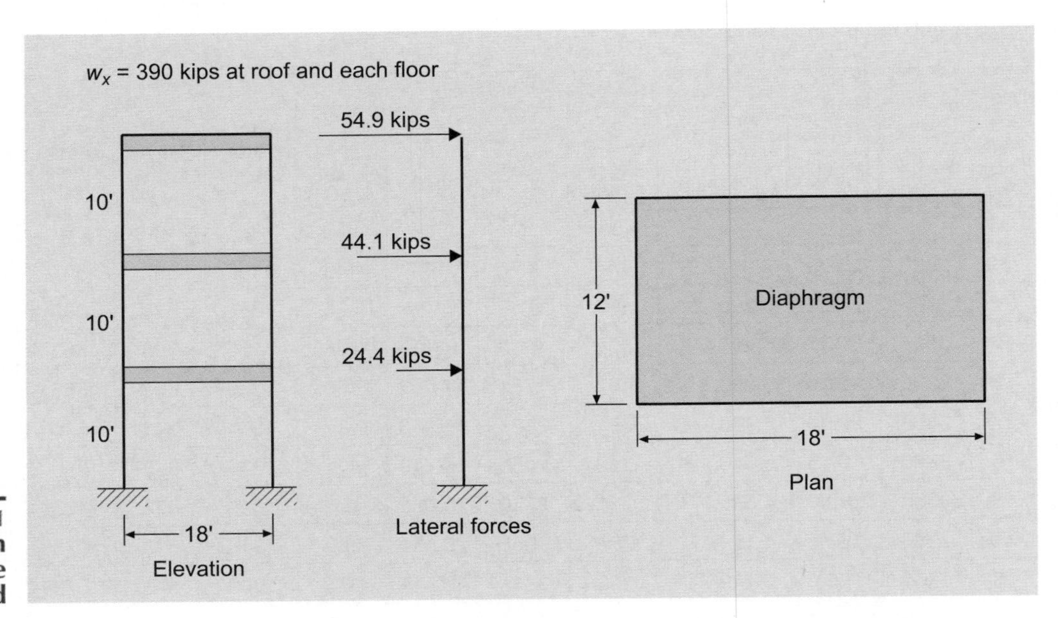

Figure E12.1 Plan and elevation of example building studied

Calculation of diaphragm forces for structure in Seismic Design Category D (based on ASCE 7-05 Equation 12.10-1)

Level, i	F_i	ΣF_i ($i = x$ to n)	w_{px}	$\Sigma \omega_i$ ($i = x$ to n)	F_{px}	Max $0.4 S_{DS} I w_{px}$	Min $0.2 S_{DS} I w_{px}$	Design F_{px}
roof	54.9	54.9	390	390	54.9	156	78	78
2	44.1	99.0	390	780	49.5	156	78	78
1	24.4	123.4	390	1170	41.1	156	78	78

All values are in kips

Number of levels, n = 3 Importance Factor, I = 1.0

S_{DS} = 1.0 SDC (based on S_{DS}) = D (ASCE 7-05 Table 11.6-1)

S_{D1} = 0.6 SDC (based on S_{D1}) = D (ASCE 7-05 Table 11.6-2)

Input: Diaphragm force, F_{px} = 78 kips (at roof level)

Span length, L = 18 ft

Span width, B = 12 ft

Approx. uniformly distributed load, $w = F_{px}/L$ = 4.33 klf

Example 12—(continued)

Design of Chord Steel along Exterior:

Max. bending moment, M = $w L^2/8$ = 175 ft-kips (conservatively)

Chord Tension, $T_c = M/B$ = 14.6 kips

Area of steel required, A_s = $T_c / (\phi f_y)$ = 0.27 sq in. (using Gr. 60 steel & $\phi = 0.9$)

Use 2 #4 bars; Area of 1 bar = 0.2 sq in.

Total steel area provided = 0.4 sq in. > 0.27 sq in.

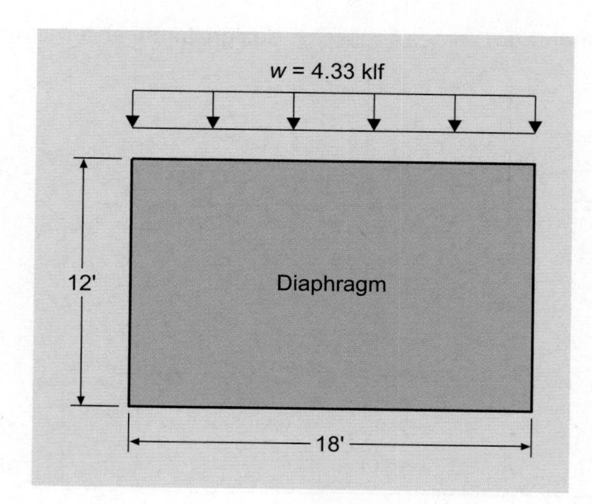

Example 13 *Calculation of Collector Strength*

(Adapted from Example 44 of SEAOC Seismic Design Manual Vol. 1[E13.1]) In this example, the collector strength is calculated in SDC B per ASCE 7-05 Section 12.10.

Given: A tilt-up building with a panelized wood roof has a partial shear wall along the collector element line (as shown below)

S_{DS}	= 0.24
S_{D1}	= 0.13
Importance factor, I	= 1.0
Roof dead load	= 20 psf
Wall weight	= 100 psf

Calculate the roof diaphragm forces, collector design force at tie to wall, and required collector strength.

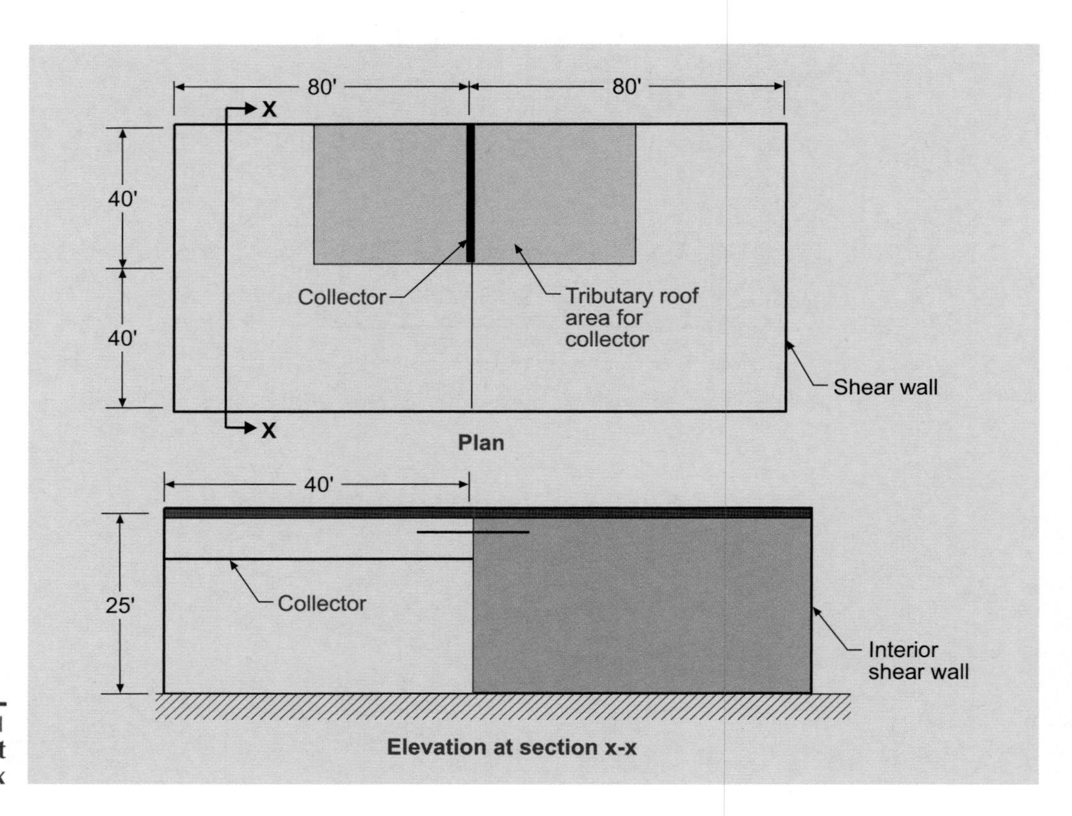

Figure E13-1 Elevation at section x-x

SDC (based on S_{DS}) = B (ASCE 7-05 Table 11.6-1)

SDC (based on S_{D1}) = B (ASCE 7-05 Table 11.6-2)

$T_0 = 0.2 \times S_{D1} / S_{DS} = 0.2 \times 0.13 / 0.24 = 0.108$ sec

$T_S = S_{D1} / S_{DS} = 0.13 / 0.24 = 0.54$ sec

Approximate fundamental period T (ASCE 7-05 Section 12.8.2.1)

$T = C_T (h_n)^{3/4}$ (ASCE Eq. 12.8-7)

C_T = 0.02 for the example building

h_n = total height = 25 feet

$T = 0.02 \times (25)^{3/4} = 0.22$ sec

$T_0 < T < T_S$ Thus, ASCE Eq. 12.8-2 will control the design.

Example 13—(continued)

$$F = \frac{S_{DS}I}{R}W = \frac{1.0 \times S_{DS} \times W}{2} = 0.5\ S_{DS}W$$

(R = 2.0, for bearing wall systems — light-framed walls with shear panels in ASCE 7-05 Table 12.2-1)

Use ASCE 7-05 Equation 12.10-1 for diaphragm forces in SDC B

Weight of the diaphragm and other elements of the structure attached to it

w_p = 20 × 160 × 80 + 100 × (25/2) × (2 × 80 + 2 × 160 + 40) = 906 kips

Diaphragm design force = 0.5 × 0.24 × 906 = 109 kips (ASCE Equation 12.10-1)

In designing the collector element, it is necessary to modify the weight, w_p, calculated above.

The tributary roof weight and out-of-plane wall weight is

w_p = 20 × 80 × 40 + 100 × (25/2) × (80) = 164 kips

Collector design force = 0.5 × 0.24 × 164 = 20 kips

ASCE 7-05 Section 12.10.2.1 states that in structures assigned to SDC C, D, E or F, collector elements shall resist the load combinations with overstrength of Section 12.4.3.2. In this example (SDC B), this does not apply.

Reference:

E13.1. Structural Engineers Association of California, *2006 IBC Structural/Seismic Design Manual*, Volume 1, Sacramento, CA, 2006.

Example 14 *Lateral Force on Elements of Structures*

Determine lateral seismic forces for design of concrete tilt-up wall panel shown below. Embedded plates to be provided at 5-ft centers to match joist spacing, with ledger angle welded to each embedded plate at each joist bearing. Wall to be designed as structural component according to ASCE 7-05 Section 12.11.

Design data

Seismic Design Category D

S_{DS} = 1.30 (based on geotechnical report)

Component Importance Factor, I_p = 1.0 (ASCE 7-0 5 Section 13.1.3)

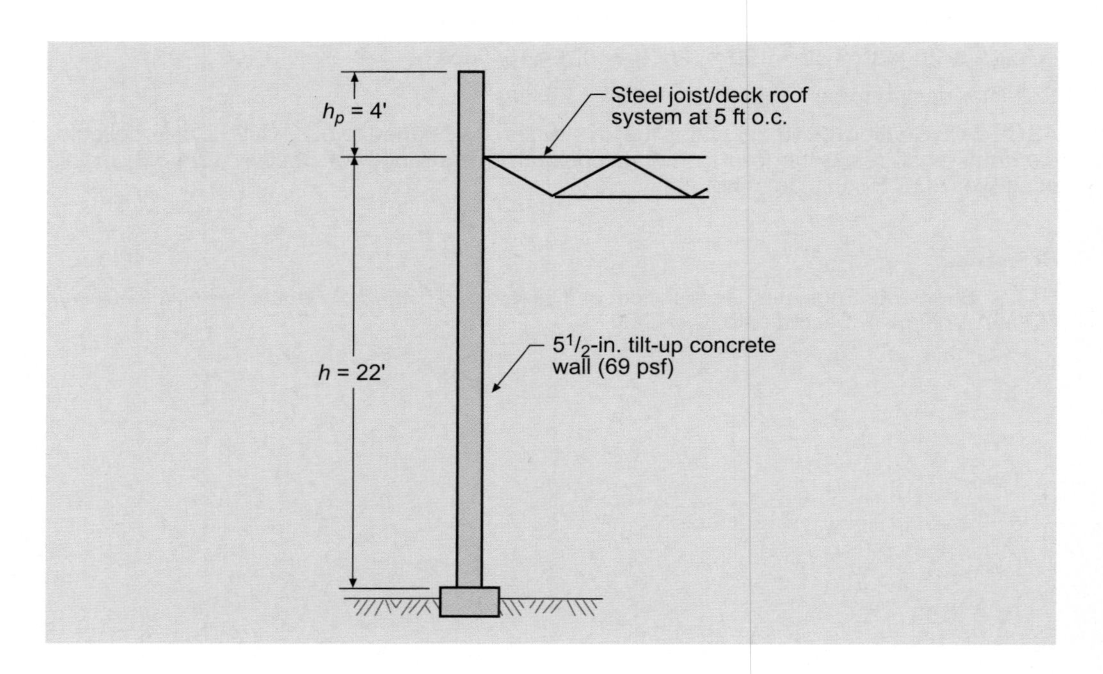

PARAPET PORTION (ASCE 7-05 Section 13.3.1)

Seismic lateral forces for the nonstructural portion of the wall panel

$$F_p = \frac{0.4a_p S_{DS} I_p}{R_p}\left[1+\frac{2z}{h}\right]W_p \qquad \text{(ASCE 7-05 Eq. 13.3-1)}$$

need not be more than . . . $1.6S_{DS}I_pW_p$ (ASCE 7-05 Eq. 13.3-2)

must not be less than . . . $0.3S_{DS}I_pW_p$ (ASCE 7-05 Eq. 13.3-3)

a_p = 2.5 ASCE 7-05 Table 13.5-1

R_p = 2.5

S_{DS} = 1.30

z = h = 22 ft

F_p = $[0.4(2.5)(1.30)(1.0)/2.5]\ (1 + 2 \times 22/22)W_p \geq 0.3(1.30)(1.0)W_p$

F_p = $1.56W_p \geq 0.39W_p$

F_p $\leq 1.6(1.30)(1.0)W_p = 2.08W_p$

F_p = 1.56(69) = 107.6 psf (plf/ft width of wall)

Example 14—(continued)

Note that F_p shall be assumed to act in any horizontal direction (the figure in this example only shows the out-of-plane distributed loading) and this wall portion shall be designed for a concurrent vertical force $\pm 0.2S_{DS}W_p$.

WALL PORTION (ASCE 7-05 Section 12.11.1)

Design for out-of-plane forces of this structural portion:

$F_p = 0.40IS_{DS}w_w \geq 0.10w_w$

$F_p = [0.40(1.0)(1.30)]\ w_w \geq 0.10w_w$

$F_p = 0.52w_w \geq 0.10w_w$

$F_p = 0.52(69) = 35.9$ psf (plf/ft width of wall)

Note that structural wall shall be designed to resist bending between anchors where the anchor spacing exceeds 4 ft (in this example, it is 5 ft).

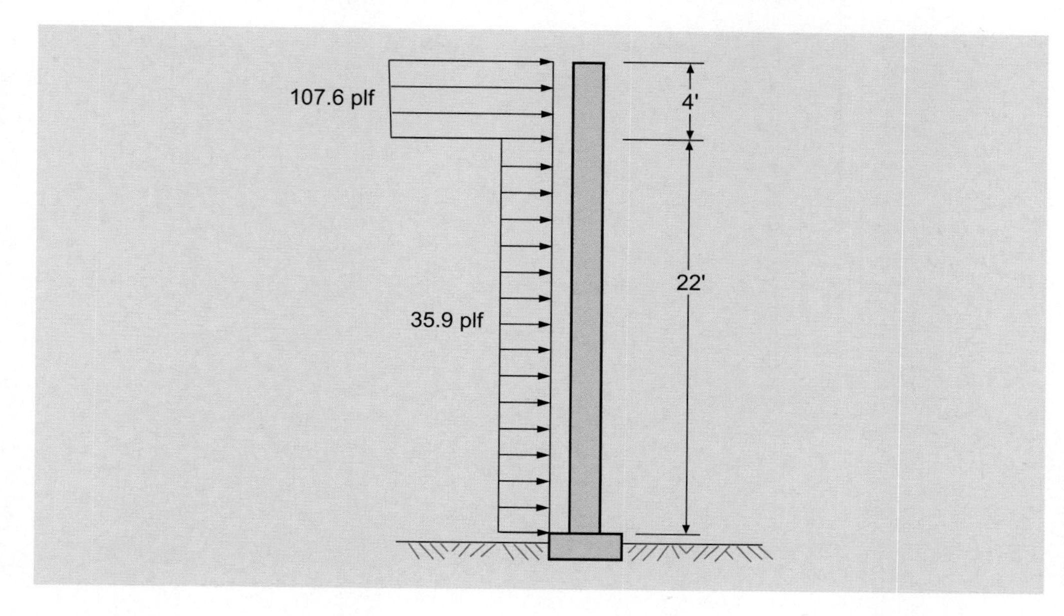

Figure E14-1 Seismic forces for wall design (per ft width)

Seismic design force for wall anchorage at roof joists

In SDC C and above, wall anchorage to diaphragms is governed by the provisions of ASCE 7-05 Section 12.11.2.

Elements of wall anchorage system must be designed for the out-of-plane force specified in ASCE 7-05 Equation 12.11-1 if the diaphragm is flexible as defined in ASCE 7-05 Section 12.3.1.1. (This section states that diaphragms constructed of untopped steel decking are permitted to be idealized as flexible in structures in which the vertical elements are concrete shear wall. Thus, this example building is a flexible-diaphragm building.)

$F_p = 0.8S_{DS}Iw_w$ (ASCE 7-05 Eq. 12.11-1)

$F_p = [0.8(1.30)(1.0)](69) = 71.8$ psf

ASCE 7-05 Section 12.11.2 states that F_p used for design of the elements of the wall anchorage system must not be less than ($400S_{DS}I_E$) plf of wall or 280 lb/linear ft of wall. Using a joist spacing of 5 ft

$5(71.8) = 358.8$ plf

Wall anchorage force . . .

$358.8 \times 26^2/2 = 22R$

Example 14—(continued)

$$R = 5.51 \text{ kips} > 400(1.3)(1.0) \times 5 = 2.60 \text{ kips}$$
$$> 280 \times 5 = 1400 \text{ lb} = 1.4 \text{ kips}$$

ASCE 7-05 Section 12.11.2.2 gives the additional requirements for diaphragms in structures assigned to SDC C through F. The strength design forces for steel elements of the structure wall anchorage system, with the exception of anchor bolts and reinforcing steel, shall be increased by 1.4 times the forces otherwise required by this section. All steel elements of wall anchorage system (plates/angles/welds) must be designed for a strength-level force of 5.51 × 1.4 = 7.71 kips.

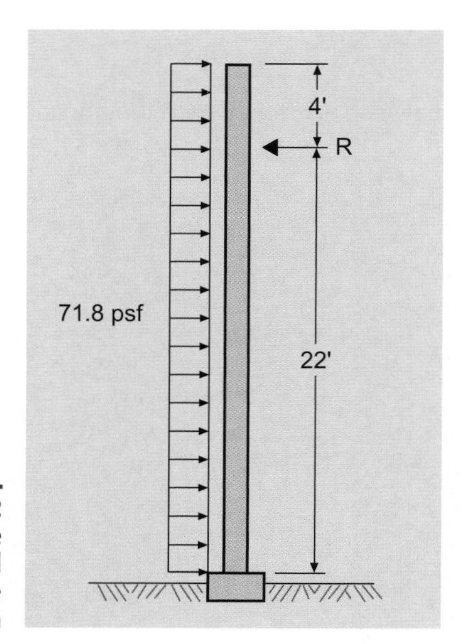

Figure E14-2 Seismic force for wall anchorage design

CHAPTER

17

STRUCTURAL TESTS AND SPECIAL INSPECTIONS

The primary goal of Chapter 17 of the *International Building Code* (IBC) is to improve the quality and workmanship of certain structural systems by requiring structural testing, special inspection and structural observation. Many of the requirements are specifically intended to improve the quality of the lateral force resisting system when buildings are subjected to the design wind or seismic event. To accomplish these goals, the chapter sets forth provisions for quality of materials, workmanship, testing and labeling of materials incorporated into the construction of buildings or structures regulated by the code. Generally, all materials used must conform to the requirements in the code or the applicable standards referenced by the code. Specific tests and standards are referenced in other parts of the code; e.g., Chapter 35 has reference standards for material requirements.

This chapter also provides the requirements for special inspection of the construction at various stages, special testing for seismic resistance, structural observation by the registered design professional and alternative methods to establish test procedures for products that do not have applicable standards.

Several successful code changes to Chapter 17 of the 2006 IBC reorganized and clarified the provisions to make them more useable. The most significant of these changes are as follows:

- The most significant change to Chapter 17 of the 2006 IBC reorganized and clarified the special inspection and quality assurance provisions to more clearly convey their intent. The term *statement of special inspections* is consistently used throughout the 2006 IBC instead of the terms *special inspection program* and *quality assurance plan*. There has been some confusion over the difference between statement of special inspections, or *special inspection program*, as it has also been referred to, and the quality assurance plans for wind and seismic. The quality assurance plans for wind and seismic were intended to be an extension of the special inspection program. The provisions that were formerly in the *quality assurance plan* requirements of Sections 1705 and 1706 in the 2003 IBC are now included as part of the statement of special inspections when required by the seismic or wind criteria.
- The statement of contractor responsibility was intended to be separate from the statement of special inspections prepared by a registered design professional. Including the statement of contractor responsibility within the quality assurance plan requirements for wind and seismic resistance has also been a source of confusion. Therefore, the statement of contractor responsibility requirements has been consolidated and put in Section 1706, eliminating the redundant language of the 2000 and 2003 IBC, where it appeared in both the wind and seismic quality assurance plans.
- Sections 1704.7, 1704.8 and 1704.9 provide specific requirements for special inspection of existing site soils, fill placement, load bearing requirements and installation and testing of pile and and pier foundations. Three new tables are provided that identify the specific inspection tasks. Although the provisions in the 2000 and 2003 editions of the IBC require special inspection for site preparation, placement of fill and in-place density (compaction) testing, the code relied more on the geotechnical engineer's soils report for specific details. The changes to Sections 1704.7, 1704.8 and 1704.9 in the 2006 IBC provide more concise guidance for the special inspection and testing tasks pertaining to site preparation, controlled fill placement, and determination of soil bearing capacity, as well as installation of pile and pier foundations. Three new tables clarify the requirements and identify the specific inspection tasks, and indicate whether continuous or periodic special inspection is required. The new tables are similar to other special inspection tables in Chapter 17.

- Several code changes clarified the provisions related to the inspection and testing of mechanical and electrical components for seismic resistance. The mechanical and electrical components in Section 1707.7 of the 2003 IBC that require special inspection based on seismic design category were reformatted into a list of itemized requirements and relocated to Section 1707.8, where they are now more easily understood. The additional item 5 was added to the list for consistency with ASCE 7 and the NEHRP *Provisions.*[1] Footnote b of ASCE 7-05 Table 13.6-1 provides an option to reduce the design seismic force on vibration isolated components and systems where a clearance of 0.25 inches or less is provided between the equipment support frame and the restraint. To verify that the design intent is achieved in the field, these equipment installations are subject to special inspection.
- Sections 1707.7.1, 1707.7.2 and Section 1707.7.3 of the 2003 IBC were deleted because they were based on the 1991 NEHRP, which has changed with subsequent editions. They did not fit well within the rest of the special inspection requirements because they duplicated other sections and incorrectly included requirements for component testing and contractor quality control. The last sentence in 2003 IBC Section 1707.7.2, however, does pertain to special inspection of designated seismic systems requiring seismic certification, and this is reflected in 2006 IBC Section 1707.9. The statement was retained but revised to be consistent with the seismic certification requirements.
- A duplicate statement on special inspection requirements located in 2003 IBC Section 1708.5 (which duplicated the last sentence in 2003 IBC Section 1707.7.2) was deleted, providing further clarification of the provisions. The first sentence in 2006 IBC Section 1708.5 was added to require the registered design professional to specify the applicable seismic qualification requirements on the construction drawings for designated seismic systems.

The above code changes to the special inspection provisions of the 2006 IBC make the requirements much easier to understand and enforce.

Section 1701 *General*

This section sets forth the scope for Chapter 17 and the general requirements for both new and used materials.

1701.2

New materials. Testing is required for all materials that are not specifically provided for in the code. For example, a composite wood material that is not listed in Chapter 23 would be required to follow the procedures set forth in this chapter. A similar provision for acceptance of alternative materials, systems or methods for which the standards are not adopted in the code is set forth in Section 104.11. This section restates that alternative or new materials and methods may be used if it can be established by tests or other means that the performance of the new material or method will equal that required by the code for the replaced product.

1701.3

Used materials. Materials may be reused, provided that they meet *all* the code requirements for new materials. Note, however, that Section 104.9.1 specifically restricts the use of used equipment and devices unless specifically approved by the building official. One should always exercise caution in approving reuse of materials. The applicable material or design standards must be consulted to determine if reuse of materials is allowed or prohibited. For example, reuse of high-strength A490 structural bolts is prohibited by the AISC specifications. Even a piece of used structural steel should be carefully checked for conformance to the design specifications, applicable standards and code requirements.

Section 1702 *Definitions*

1702.1 General. Definitions of various terms help in the understanding and application of code requirements. The definitions are provided in this chapter so that the reader does not constantly need to refer back to Chapter 2. These definitions are also cross-referenced in Chapter 2.

APPROVED AGENCY. The definition of this term is needed in order to effect the requirements of Section 1703.1. The word *approved* means "acceptable to the building official or authority having jurisdiction" (see the definition of *approved* in IBC Section 202). The basis for approval of an agency for a particular activity by the building official includes the competence or technical capability to perform the work in accordance with Section 1703.

CERTIFICATE OF COMPLIANCE. An *approved* fabricator is required to submit a Certificate of Compliance for work performed without special inspection. See Section 1704.2.2 for discussion of fabricator approval.

DESIGNATED SEISMIC SYSTEM. The designated seismic system consists of those architectural, electrical and mechanical systems and their components that require design in accordance with Chapter 13 of ASCE 7 for which the component importance factor, I_p, is greater than 1 as prescribed in Section 13.1.3 of ASCE 7. Section 13.1.3 of ASCE 7 lists three components that have an importance factor of 1.5, which include components required to function after an earthquake, such as fire sprinkler systems, components containing hazardous materials and components in Occupancy Category IV structures necessary for continued operation of the facility. Occupancy Category IV structures are described in IBC Table 1604.5. Section 13.1.4 of ASCE 7 lists those nonstructural components that are exempt from the requirements of Chapter 13.

FABRICATED ITEM. Fabricated items are materials assembled prior to installation in a building and are referred to in Section 1704.2. The definition is provided to clarify the intent of the code, as the term *fabricated items* could easily be interpreted to mean items for which special inspection is not intended by the code. An example of a fabricated item for which special inspection is required is a roof truss.

INSPECTION CERTIFICATE. An inspection certificate is an identification applied to a product indicating that the individual product has been inspected by an approved agency. The inspection certificate is obligatory for components subject to special inspection requirements for seismic resistance (see Section 1707). Note that the requirements for an inspection certificate differ from the requirements for a label. An inspection certificate is issued for the specific piece of a product when it is inspected and is an ongoing process, whereas a label requires only that a representative sample of a product be periodically tested.

LABEL. A label is an identification for a product as set forth in Section 1703.5. A label identifies the manufacturer, the function of the product and the quality control agency that allows the use of its label based on periodic audits and inspections of the manufacturer's facilities.

MANUFACTURER'S DESIGNATION. The manufacturer's self-certification that a product complies with a given standard. There is no independent certification.

MAIN WIND-FORCE-RESISTING SYSTEM. The Main Wind-Force-Resisting System (MWFRS) is one of the systems designed to resist wind loads. The other system designed to resist wind loads is Components and Cladding. The MWFRS comprises those structural elements that provide lateral support and stability for the overall structure. In general, the MWFRS receives wind loading from more than one surface. In contrast, Components and Cladding are on the exterior envelope of the building and receive loading directly from the wind pressure. Examples of MWFRS are diaphragms, chords, collectors, shear walls, etc.

MARK. A manufacturer's identification of a product stating who made the product and the intended function. It includes neither certification that the product meets any given standard nor third party verification.

SPECIAL INSPECTION. That category of inspection for which special knowledge, special attention or both are required. For example, inspection of complete penetration welds requires both special knowledge and special attention to ensure that the requirements of the code are met. Note that the special inspector does not have the same authority as that of the jurisdiction inspector. The role of the special inspector is to report discrepancies to the contractor and building official.

SPECIAL INSPECTION, CONTINUOUS. Continuous full-time inspection is required where compliance of the work or product cannot be determined after incorporation into the building or structure. For example, one cannot determine whether a multipass fillet weld is in compliance with the code requirements unless each pass of the weld is inspected during the welding process.

SPECIAL INSPECTION, PERIODIC. Intermittent or part-time inspection, which may be allowed when the compliance of the work or product can be determined after being incorporated into the structure. For example, compliance with the design nailing requirements of a wood shear wall can be determined after completion of the wall (but before closure); hence, verification by periodic special inspection is adequate.

STRUCTURAL OBSERVATION. Note that structural observation by the registered design professional does not replace any of the requisite special inspections or jurisdiction inspections required by Section 109. Structural observation is for determining general conformance with the design intent, and not for specific conformance with the design documents and the code.

Section 1703 *Approvals*

1703.1

Approved agency. The word *approved* means "acceptable to the building official or the authority having jurisdiction" (see the definition of *approved* in IBC Section 202.1). The basis for approval of an agency for a particular activity by the code or building official may include the capacity or technical capability to perform the work in accordance with Section 1703. For example, if an agency wishes to be approved for the special inspection of structural welds, the agency should submit evidence that its welding inspector is certified in accordance with applicable American Welding Society (AWS) or American Society of Nondestructive Testing (ASNT) requirements.

1703.1.1

Independent. The agency should have objectivity as well as competence. Objectivity can be measured by the agency's financial and fiduciary independence. The agency should have no financial ties to the organization it inspects. For example, a testing laboratory checking concrete strength should have no financial ties to either the contractor, its subcontractors or the concrete supplier.

1703.1.2

Equipment. The building official should evaluate any testing laboratory for compliance with the requirements of applicable standards. To be approved, the laboratory should have its equipment calibrated at least annually. The laboratory must have the necessary calibrated equipment. For example, the laboratory must have the requisite temperature and humidity controlled storage for concrete cylinders in order to certify concrete cylinder strength.

1703.1.3

Personnel. The building official should evaluate both the experience and qualifications of personnel. An agency may have personnel with the appropriate certifications but not the necessary experience. Supervisory and inspection personnel should have certifications as well as the requisite experience and/or education. For example, a concrete technician may be certified in accordance with the ACI technician program.

If the services being provided by an inspection or testing agency come under the purview of the professional registration laws of the state of jurisdiction, the building official should request evidence that personnel are qualified to perform the work in accordance with the requisite professional registration. It should be noted that the structural observer does not have the same authority as the jurisdiction inspector. The role of the structural observer is to report discrepancies to the contractor and building official.

1703.2 Written approval. A written approval by the building official is required for all material, appliance, equipment or system incorporated into the work in order to have a documented record of approval and the basis for approval.

1703.3 Approved record. Records must be kept for all approvals, including conditions and limitations of approval. The approvals must be kept on file and available for public inspection. The records must demonstrate compliance with the code requirements of any material, appliance, equipment or system incorporated into the structure.

1703.4 Performance. When conformance with the code is predicated on the performance and quality of materials or products, the building official must require the submission of test reports from an approved agency establishing this conformance. In the absence of such reports, the building official should require specific data that show compliance with the intent of the applicable code requirements in accordance with Chapter 16 (see IBC Sections 1604.6 and 1604.7). For example, core tests of in-situ concrete could be used to determine compliance with design strength if the sample cylinders required in Section 1905.6 were destroyed.

Materials and products must be subjected to various levels of quality control and identification in order to determine that the material complies with the requirements of the code. The degree of quality control and identification to which a material must be subjected is based on its relative importance to the structure's performance and function. By use of the terms *mark*, *manufacturer's designation*, *label* and *inspection certificate*, the code establishes a hierarchy of quality control and identification as follows:

Level 1: Manufacturer identifies the material or product with the name of the material or product, the manufacturer's name and the intended usage (see *Mark*).

Level 2: Manufacturer identifies the material or product as in Level 1 and also certifies compliance with a given standard or set of rules (see *Manufacturer's designation*).

Level 3: Manufacturer identifies the material or product as in Level 1. An approved quality control agency performs periodic audits of the manufacturer's facilities and QA/QC procedures.

Level 4: Each batch of material or individual product is inspected by an approved quality control agency (see *Inspection Certificate*).

1703.4.1 Research and investigation. This section is intended to implement the requirements of Section 104.11 for use of innovative or alternative materials, design and methods of construction. For example, an innovative prestressed concrete system that did not emulate the performance of cast-in-place concrete could be evaluated in accordance with this section.

1703.4.2 Research reports. Evaluation reports prepared by approved agencies, such as those published by the organizations affiliated with the model code groups, e. g., ICC Evaluation Service, may be accepted as part of the data needed by the building official to form the basis of approval of a material or product. Such reports supplement the resources of the building official and eliminate the need for the official to conduct detailed analysis on every new product. The building official is not mandated to approve the evaluation reports issued by the model code affiliated organizations, as these reports are advisory only and are intended for technical reference. Technically, such reports are approved under alternative materials and methods of construction.

Labeling. When materials or assemblies are required to be labeled by the code, such as plywood, fire doors, etc., the labeling must be in accordance with the procedures outlined in this section and its subsections. Labeling of materials or assemblies is an indication that the materials or assemblies have been subjected to testing, inspection and/or other operations by the labeling agency. The presence of a label does not necessarily indicate compliance with code requirements. For example, use of plywood sheathing rated C-C, although labeled, would not comply with the code if Structural I sheathing was required to resist the design lateral forces. **1703.5**

The installation of labeled products must comply with the specific requirements and limitations of the labeled product. For example, a fire-rated door is labeled with the hardware requirements specified on the label. The building inspector must ensure that the hardware used in the installation of the door meets the labeling requirements to ensure that the door complies with the code.

Testing. For a material or product to be labeled, the labeling agency is required to perform testing on representative samples of the material or product in accordance with standards referenced by the code. An example is factory built fireplace assemblies that must be tested in accordance with UL 127 (see Chapter 35). **1703.5.1**

Inspection and identification. The approved agency whose label is applied to a material or product must perform periodic inspections of the manufacture of the material or product to determine that the manufacturer is indeed producing the same material or product as tested and labeled. For example, if the labeling agency had tested $^1/_2$-inch C-C plywood sheathing made with five plies but the manufacturer was now making the plywood with only three plies, the agency would need to withdraw the use of its label and listing. **1703.5.2**

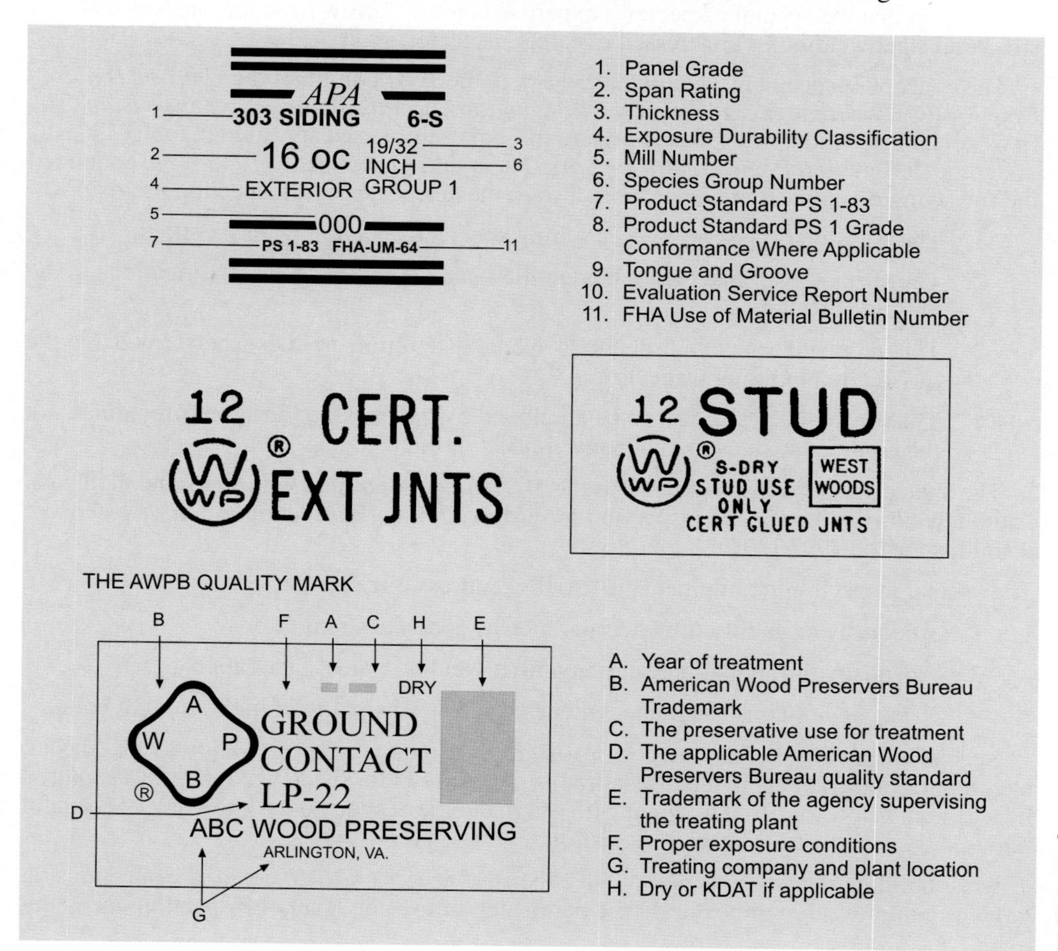

Figure 17-1 Examples of lumber grade labels

1703.5.3 Label information. This specifies the minimum information necessary on a label for the building inspector to determine that the installed material conforms with the approved plans. See Figure 1 as an example of typical lumber grade labels.

1703.6 Heretofore approved materials. If a material was approved under a previous edition of the code, it may continue to be used, provided its use is not detrimental to the health and safety of the building occupants or the general public. The code is not retroactive.

1703.7 Evaluation and follow-up inspections. This applies where the structural component cannot be inspected after completion of a prefabricated assembly. An example might be a prefabricated shear wall consisting of plywood over a welded steel frame; the welding must be inspected prior to the installation of the plywood and the entire assembly inspected after application of the plywood. The testing and inspection should follow the concepts set forth above for Section 1703.

Section 1704 *Special Inspections*

The oldest mechanism for providing quality assurance in construction is the process known as *special inspection*. The purpose of special inspection is to ensure proper fabrication, installation and placement of components or materials that require special knowledge or expertise, such as welding of structural steel or placement of grouted masonry. The knowledge and duties of a special inspector differ from that of the jurisdiction building inspector in that the special inspector's expertise is more narrow in scope, such as that of a structural steel welding or prestressed concrete inspector.

The concept of special inspection dates back to the 1927 edition of the *Uniform Building Code* (UBC), where it existed under the designation *Special Engineering Supervision*. The first special inspection provisions similar to those presently used appeared in the 1943 UBC under the designation *Registered Inspectors*. The requirements in the 1943 UBC contained the following essential elements that are also in the current special inspection provisions:

1. The particular types of work requiring special inspection were specified.
2. The special inspector had to be qualified and demonstrate his qualifications to the building official.
3. The requisite special inspections were in addition to those performed by the jurisdiction building inspector.
4. The special inspector was to be employed by the owner or design professional, not the contractor, so as to avoid any conflict of interest.

The special inspection provisions in the 1997 UBC continued with the same elements. Although similar, the special inspection provisions in this IBC section are more extensive than those in the 1997 UBC.

The special inspection requirements in the IBC address three areas:

1. Adequacy or quality of materials, such as concrete strength
2. Adequacy of fabrication, such as a fillet weld on a steel fabrication
3. Adequacy of construction techniques, such as tensioning of high-strength bolts

Special inspection is that category of inspection requiring special knowledge, special attention or both. The knowledge required is generally more specialized than that required by a general inspector. An individual with a high degree of specialized knowledge is usually required; hence, the term special inspection.

Most building departments do not have the staff necessary to do detailed inspections on large or complex structures, nor do the permit fees allow the level of inspection necessary for the types of construction where extra care in quality control must be exercised to assure compliance with the code. Hence, the need for special inspection.

ICC offers a Model Program For Special Inspection[2] that provides the building official with guidance on the administration and implementation of the special inspection requirements of the *International Building Code* (IBC). The guidance is based on recommended practices and the consensus of building officials and design professionals, as well as inspection and testing agencies. The duties and responsibilities of the building official, special inspector, project owner, engineer or architect of record, contractor and building official are covered in the guide. Suggested forms are also included that can be easily adapted to the specific needs of the jurisdiction.

A good example of a checklist document for special inspection is *Guidelines for Special Inspection in Construction* by the California Council of Testing and Inspection Agencies.[3] Copies of this document are also available from any of the member organizations of the council, which include most of the major testing laboratories in California.

Note that there are additional special inspection requirements in Section 1707 for seismic resistance.

General. Special inspection is required for various types of structural elements and systems as prescribed in Section 1704. These include structural systems constructed of steel (Section 1704.3), concrete (Section 1704.4), masonry (Section 1704.5), wood systems (Section 1704.6), site soils (Section 1704.7), pile foundations (Section 1704.8) and pier foundations (Section 1704.9). In addition, special inspection is required for alternate materials and unusual designs when required by the building official as prescribed in Section 1704.13. **1704.1**

The owner is responsible for the employment of special inspectors meeting the approval of the building official. The owner is also responsible for all costs associated with the employment of special inspectors. Note that the special inspectors must be employed by the owner, or the responsible registered design professional acting as the owner's agent, not by the contractor or builder. This ensures independence of the special inspector and avoids any conflict of interest that could occur if the special inspector were employed by the contractor or builder. Note also that the special inspections required are in addition to the jurisdiction inspections required by Section 109.

Exceptions to the requirement for special inspections are only for minor work or work not required to be designed or sealed by a registered design professional. The exception for R-3 occupancies and U occupancies accessory to a residential occupancy are for those structures typically not required to be designed by a registered design professional or those designed and constructed in accordance with the *International Residential Code.* Exception 1 does not necessarily mean that the inspections listed are not required, only that they are not required to be made by a special inspector. Also, Exception 1 refers to "conditions in the jurisdiction" as a possible exception. The primary *conditions* envisioned by the code refer either to the jurisdiction having the resources and skill level necessary to perform the requisite special inspection tasks, thus obviating the need for a special inspector, or the work being of a minor nature in the opinion of the building official. Note that this exception for special inspection cannot be invoked by the owner. One purpose of the exception is to allow jurisdictions to perform special inspections if the jurisdiction so desires.

Statement of special inspections. The permit applicant must submit a detailed statement outlining the required special inspections and designate those who will perform the special inspections. The responsible registered design professional is required to prepare the statement of special inspections because the special inspections relate directly to the design and construction documents required by Sections 106.1 and 1603. The statement of special inspections must conform to the detailed requirements described in Section 1705. **1704.1.1**

A statement of special inspections is not required for conventional wood frame buildings constructed in accordance with prescriptive provisions in Section 2308. It should be noted, however, that Sections 2308.1.1 and 2308.4 permit portions and elements of an otherwise conventional building to be designed in accordance with the engineering provisions of the code, and therefore the engineered elements and portions may require special inspection.

The statement of special inspections need not be prepared by the registered design professional if prepared by a qualified person approved by the building official for construction that is not designed by or required to be designed by a registered design professional.

1704.1.2 Report requirement. Records of each inspection must be kept and submitted to the building official so as to document compliance with the code. The records must include all inspections made and compliance with the code requirements, as well as all violations and discrepancies. A final report must show that all required special inspections have been made and that discrepancies have been resolved before a certificate of occupancy can be issued. It is the responsibility of the special inspector to document and submit inspection records to the building official and to the registered design professional in responsible charge. The final special inspection report documenting resolution of discrepancies must be submitted at a time agreed upon by the permit applicant and the building official prior to the commencement of work.

1704.2 Inspection of fabricators. This section should be used in conjunction with Section 1703.7 relating to evaluation and follow up reports. The use of a special inspector does not relieve the fabricator from his own required quality control procedures and personnel.

1704.2.1 Fabrication and implementation procedures. The special inspector acts in the quality assurance role in this instance by verifying that the quality control procedures that the fabricator has in place will ensure compliance and that the fabricator follows the requisite quality control procedures. For example, welding of structural steel is to be done in accordance with the requirements of AWS D1.1.[4] AWS D1.1 requires that the fabricator have its own quality control procedures and personnel. The fabricator's QC personnel are responsible for determining conformance of the welds with the requirements of AWS D1.1 by applicable visual and nondestructive testing (NDT) methods. Note that NDT is supplementary to, not in place of, visual inspection. The special inspector acts as the owner's agent for verification by auditing the fabricator's quality control (QC) program. The fabricator's welding inspector is responsible for: 1) making sure that all welders are qualified in accordance with D1.1; 2) making sure that weld procedure specifications (WPS) are in place for all welds (both prequalified and unqualified welds); 3) visually inspecting fit-up of the weld; 4) verifying that the machine settings, electrodes and other parameters match those set forth in the WPS; 5) observing the welding operation; 6) performing any required NDT; and 7) keeping documentation for each weld made in the fabrication shop. The special inspector, on the other hand, is responsible for reviewing the qualification records of the welders, for determining that the WPS were suitable for the specified weld and were properly qualified, for reviewing NDT procedures and records, and for observing a representative number of welds in order to ensure that the fabricator's QC program is adequate and being followed.

1704.2.2 Fabricator approval. Special inspection is not required where a fabricator has been specifically preapproved by the building official. This preapproval is based on a review of the fabricator's written procedures and quality assurance/quality control program. The fabricator should be periodically audited by an independent, approved special inspection agency. This section is intended to apply to programs such as the Quality Certification Program for steel fabricators by AISC, the Plant Certification Program for steel joists by the Steel Joist Institute or the Plant Certification Program for precast concrete products by the Precast/Prestressed Concrete Institute.[5]

1704.3 Steel construction. This section sets forth the special inspection requirements for the fabrication and erection of structural steel elements. Detailed requirements are in Table 1704.3 and the subsections of Section 1704.3. Table 1704.3 sets forth the verification and inspection requirements, whether the frequency of inspection is to be continuous or periodic, the appropriate referenced standard and the applicable IBC code section.

Exception 1 eliminates the need for special inspection in certain cases. Special inspection is not required if the fabricator does not alter the properties of the parent material by welding, thermal cutting or heating operations. For example, if the members being fabricated were cut by mechanical means, such as a band saw, and punched or drilled for bolt holes, with no application of heat, special inspection would not be required. But if the same members were cut with an oxy-acetylene torch, special inspection *would* be required. Even if special inspection is not required, the fabricator must provide evidence that his tracking procedures are adequate to verify that the material used in any member meets the required specification, is of the proper grade and has an associated mill test report.

Exception 2 does not exempt the work from special inspection. The exception relaxes the amount of special inspection for welding from continuous to periodic when the quality of the weld can be readily determined after completion of the weld. For example, welded studs can be given the *hammer* test, and small, single pass fillet welds can be readily inspected by visual inspection and dye penetrant methods for conformance with AWS D1.1 requirements after the weld is completed.

Welding. Welding inspection must be in accordance with AWS D1.1. The weld inspector should be (but is not required to be) an AWS Certified Weld Inspector. Any nondestructive testing such as dye penetrant or ultrasonic testing should be performed by an inspector qualified under the American Society for Nondestructive Testing (ASNT) and should be certified as an ASNT Level II NDT inspector for that process. **1704.3.1**

Welding inspection relies on *visual inspection* as the primary method used to evaluate conformance of welds with the applicable quality requirements. Visual inspection is *supplemented* by various nondestructive test methods such as ultrasonic and dye penetrant methods. Visual inspection means that inspection is performed not just after welding, as is often practiced, but prior to, during *and* after welding. A Welding Procedure Specification should be prepared and approved for each weld prior to welding (including prequalified welds). Weld metal and shielding gases should be checked for conformance with the WPS. Joints should be examined prior to welding to ensure that fit-up, bevel preparation requirements, alignment, preheat and other variables match the WPS requirements. Prior to and during welding, the welding equipment settings should be verified to be within the correct amperage and voltage range per the WPS. The welding techniques of the welder should be observed during welding, particularly observation of the weld pool and electrode travel rates. After each pass of weld is completed, visual inspection should verify that the width, depth, convexity, undercut and other requirements of D1.1 are met. Use of ultrasonic testing alone after completion of the weld without *visual inspection* as noted above, does *not* meet the requirements of AWS D1.1.

Details. The special inspector must verify that all framing elements conform to the requirements of the approved construction documents, e.g., the beam-column joint must have the correct geometry, the correct number and type of fasteners, etc. **1704.3.2**

High-strength bolts. This section references the AISC specification for high-strength bolting[6]. In general, the use of high-strength bolts is required to conform to the provisions of the *Specification for Structural Joints Using ASTM A325 or A490 Bolts* as approved by the Research Council on Structural Connections.[7] Hence, the special inspection is to be done in accordance with the applicable RCSC requirements. **1704.3.3**

Note that Section 1704.3.3.3 requires continuous inspection for pretensioning bolts by the calibrated wrench or turn-of-nut methods.

General. High-strength bolts are used in two types of connections: 1) friction connections where the clamping force provided by pretensioning of the bolts is necessary for proper performance of the connection, and 2) bearing connections where friction, and hence pretensioning, is not critical to the performance of the connection. Bearing connections require only snug-tight bolts that bring the layers of steel into contact, whereas friction connections require that the clamping force is obtained fully by pretensioning the bolts to a specified minimum tensile force. **1704.3.3.1**

There are three types of high-strength bolts, and the type specified depends on the usage. Matching nuts and washers must be used with high-strength bolts. Consult the appropriate RCSC *Specification* for correct type of bolt, matching nuts and washer requirements. Note that all A325 and A490 bolts are quenched and tempered (Q and T). Welding, thermal cutting or sustained high temperatures will change the metallurgy and strength of the bolt and must be avoided.

Occasionally there is a need for high-strength bolts of diameter or length exceeding what is available for ASTM A325 and ASTM A490 bolts. The specification permits ASTM A449 bolts and ASTM A354 Grade BC and BD threaded rods to be used in joints requiring diameters in excess of $1^1/_2$ inch or lengths in excess of 8 inches. It should be noted that it is generally preferable to specify anchor rods as ASTM F1554 material. Refer to the Commentary to ANSI/AISC 360 and the RCSC for more detailed information.

The *Structural Bolting Handbook,*[8] published by Steel Structures Technology Center, Inc., is a good resource on installation, tightening and inspection of high-strength bolts.

1704.3.3.2 **Periodic monitoring.** Periodic monitoring is permitted in two cases: 1) in a friction connection when it can be determined after tensioning that the required tension has been induced in the bolt, as with a calibrated direct tension indicator, or 2) in a bearing connection using snug-tight bolts.

1704.3.3.3 **Continuous monitoring.** Continuous monitoring is required when it is necessary to observe the tightening of the bolt in order to determine that the minimum pretensioning force is induced in the bolt. Use of a calibrated torque wrench or turn-of-nut method without match marking both require continuous monitoring.

Table 1704.3 Required Verification and Inspection of Steel Construction. This table presents the requirements for special inspection in a concise format along with the inspection frequency, reference standard and applicable code section. Some of the items in the table are also required in the preceding code text. For Item 2, inspection of high-strength bolting, see the discussion in Section 1704.3.3. Items 1, 3 and 4 are verification that the correct material and grade are being used. Item 5, inspection of welding, only summarizes the welding inspection requirements and the required frequency. Refer to documents AWS D1.1, D1.3 and D1.4 for more complete inspection requirements. Note that the requirements for welding of reinforcing steel are in this table rather than in the concrete or masonry section. These requirements are placed in this table because the inspection requirements are similar to structural steel welding and the personnel and equipment involved are often the same as those inspecting structural steel welding. See analysis of Section 1704.3.1 for a discussion of the term visual inspection for welding. See also the commentary in AWS D1.1, D1.3 and D1.4.

1704.4 **Concrete construction.** This section presents the special inspection requirements for the construction of concrete foundations, structures or elements. Detailed requirements are covered in Table 1704.4, which sets forth the verification and inspection requirements, whether the frequency of inspection is to be continuous or periodic, the reference standard and the IBC code section. Exceptions for special inspection are concrete elements that are nonstructural such as sidewalks, lightly loaded elements such as slabs on grade, concrete foundations that are not heavily loaded or plain concrete basement or foundation walls.

1704.4.1 **Materials.** Constituent materials for concrete such as aggregate, cement, admixtures and water must conform to the requirements set forth in Section 1903 and the standards of Chapter 3 of ACI 318. When sufficient documentation is not available to verify that constituent materials conform to these requirements, the building official should require testing of the materials.

Table 1704.4. Required Verification and Inspection of Concrete Construction. This table presents the requirements for special inspection of concrete in a concise format along with the inspection frequency and reference standard. The table summarizes the required inspections and test samples necessary to verify that the in-place concrete meets the code requirements. Refer to ACI 318[9] and the applicable ASTM standards for more details of the constituent material tests, sampling of fresh concrete, testing of slump or air content, casting of test specimens and other test requirements.

Item 1: Inspections per Item 1 should check that the reinforcement is of the correct size and grade, as required by the approved drawings and specifications, and is properly placed prior to placement of concrete. Proper placement of reinforcement has a significant impact on the integrity and strength of reinforced concrete. The reinforcement should be placed within the tolerances set forth in ACI 318 Section 7.5 and ACI 117, *Standard Tolerances for Concrete Construction and Materials*. Additional requirements of ACI 318, such as surface conditions of reinforcement (Section 7.4), spacing limitations (Section 7.6) and concrete protection for reinforcement or cover (Section 7.7), must also be checked.

Item 2: Note that Item 2, inspection of reinforcing steel welding, is referred to in Table 1704.3, Item 5b, for reinforcing steel.

Item 3: Item 3 comes into effect when anchor bolts are designed for the higher load allowed by special inspection (see Section 1911.5). Most jobs that require special inspection for concrete will be designed to take advantage of the higher allowable bolt loads with special inspection. Hence, if special inspection is required for the concrete, most likely it will also be required for the anchor bolts. Proper placement and embedment of anchor bolts is of extreme importance. If an anchor bolt is set too low for proper thread engagement, there are few satisfactory methods to remedy the situation. The common practice of placing a puddle weld in the nut is of questionable value, as neither the bolt nor the nut may be weldable material. The chemistry of nuts is not controlled by the ASTM requirements, and the chemistry allowed for an A307 bolt is such that it may or may not be weldable. In either case, the weld is not a prequalified weld in accordance with AWS D1.1 and a Procedure Qualification Record must be developed to qualify the welding procedure. Note also that if the anchor bolt is a high-strength bolt such as an ASTM A 325 or ASTM A 449, the bolt is quenched and tempered and application of heat by welding may destroy the strength of the bolt and make the bolt brittle.

Item 4: Item 4 is a particularly important verification on larger jobs that may have many required mix designs with differing strength requirements and aggregate sizes.

Item 5: Sampling of fresh concrete for making specimens for strength tests is extremely important for proper quality control. Properly sampled and prepared specimens are necessary to determine that the concrete will meet or exceed the design strength. The frequency of sampling should be in accordance with IBC Section 1905.6.2 and ACI 318 Section 5.6.2—one set of specimens for each class of concrete not less than once per day, once per 150 cubic yards or once per 5,000 square feet of slab or wall. Sampling should be done in accordance with ASTM C172 to ensure representative samples for determining compressive strength. The tests specified in Item 5 may be supplemented by other tests such as unit weight or air content.

Item 6: Observation of the actual placement is important to determine that the fresh concrete is properly handled so that it does not segregate during placement and that the concrete is properly consolidated by vibration. The mixing requirements (see IBC Section 1905.8), conveying requirements (see IBC Section 1905.9) and depositing requirements (see IBC Section 1905.10) should be strictly enforced to ensure proper placement with adequate consolidation and without segregation. Concrete voids or *rock pockets* can adversely affect design strength and are unattractive.

Item 7: Maintenance of proper cure is essential to obtaining quality concrete that will reach the design strength. Concrete that is not properly cured often will be below design strength and may suffer degradation at the surface from use or from environmental effects much earlier than properly cured concrete.

Item 8: The inspections required by Item 8 are of extreme importance as the strength of a prestressed member is highly dependent on proper prestressing. When checking the application of prestressing force, both the force applied to the tendon and the tendon elongation should be checked simultaneously to ensure that the tendon has not been hung up in the tendon sheath.

Item 9: Criteria for the erection procedures of precast concrete must be provided on the design drawing by the design engineer. The drawings should identify each panel to be cast and should specify: dimensions and thickness of panels, reinforcement grade, size and location, location of inserts, and minimum concrete strength at lifting and in-service. The special inspector should ensure that the erection process complies with the approved procedures.

Item 10: Concrete strength must be verified prior to stressing tendons used in post-tensioned concrete. Form supports for prestressed concrete should not be removed until sufficient prestressing has been applied.

Item 11: Inspection of concrete forms for proper dimensions and location is essential for adequate performance on concrete members such as beams, columns, walls and structural slabs.

1704.5 Masonry construction. This section sets forth the special inspection requirements for the construction of masonry foundations, structures or elements. Detailed requirements are in Tables 1704.5.1 and 1704.5.3. These tables set forth the verification and inspection requirements, whether the frequency of inspection is to be continuous or periodic, the applicable section of the reference standard and the relevant IBC code section. Table 1704.5.1, Level 1 Special Inspection, is required for engineered masonry in structures that are not classified as essential facilities (Occupancy Category I, II or III in Table 1604.5). Table 1704.5.3, Level 2 Special Inspection, with its higher requirements is required for engineered masonry in structures designated as essential facilities (Occupancy Category IV in Table 1604.5). Exempted from special inspection are masonry elements that are nonstructural such as glass unit masonry and masonry veneers when used in nonessential structures; empirically designed masonry structures; plain masonry foundation walls constructed in accordance with the prescriptive empirical requirements of Chapter 18, Tables 1805.5 (1) through 1805.5 (4); and masonry fireplaces, heaters and chimneys constructed in accordance with Sections 2111, 2112 or 2113.

Masonry structures and elements designed using allowable stress design are covered in IBC Section 2107 and the Masonry Standards Joint Committee (MSJC) standards ACI 530/ASCE 5/TMS 402[10] and ACI 530.1/ASCE 6/TMS 602.[11] Masonry structures and elements designed using strength design are covered in IBC Section 2108 and the ACI 530/ASCE 5/TMS 402 standard. Structures subjected to seismic load effects must comply with Section 2106. See also the testing requirements for seismic resistance in the analysis of Section 1708.

1704.5.1 Empirically designed masonry, glass unit masonry and masonry veneer in occupancy category IV. Masonry elements of these types of construction in structures designated as essential facilities must be inspected to Level 1 requirements, which are periodic inspections except for grouting and welding of reinforcing steel. Note that empirically designed masonry is not permitted in the following cases: buildings assigned to Seismic Design Category D, E or F; seismic force resisting systems in buildings in Seismic Design Category B and C; buildings sited where the basic 3-second gust wind speed exceeds 110 mph. Where empirical design is not permitted, allowable stress design in accordance with Section 2107 or strength design in accordance with Section 2108 must be provided. Note the other restrictions on the use of empirically designed masonry outlined in Section 2109.1.1.

1704.5.2 Engineered masonry in occupancy category I, II or III. Engineered masonry elements designed using either allowable stress or strength design in nonessential facilities must be inspected to Level 1 requirements, which are periodic inspections except for grouting, preparation of grout or mortar specimens, and welding of reinforcing steel, which are continuous inspections.

1704.5.3 Engineered masonry in occupancy category IV. Engineered masonry elements designed using either allowable stress or strength design in facilities not designated as essential must be inspected to Level 2 requirements. Level 2 inspection requires continuous inspection for grout space prior to grouting, placement of grout, placement of anchors, preparation of grout or mortar specimens and welding of reinforcing steel.

Table 1704.5.1 Level 1 Special Inspection. This table presents the requirements for Level 1 special inspection of masonry in a concise format along with the inspection frequency and reference standard. Level 1 special inspection is used for engineered masonry in nonessential facilities and empirically designed masonry, glass unit masonry and masonry veneer in essential facilities. Note that all inspections are periodic except for welding of reinforcing bars, placement of grout and preparation of specimens, which are continuous. The table summarizes the required inspections and test samples necessary to verify that the in-place masonry meets the code requirements. Refer to ACI 530/ASCE 5/TMS 402 and ACI 530.1/ASCE 6/TMS 602 and the applicable ASTM standards for additional details of the constituent material tests, sampling of mortar or grout, preparation of prisms and other test requirements.

Item 1: Inspections under Item 1 should check that the mortar is being properly batched and mixed. The influence of the mortar on the strength of the finished masonry is directly related to the degree of grouting. The mortar has little influence on the strength of a fully grouted wall, but it has a significant influence on a plain masonry wall. If the mortar deviates too much from the specifications, the mason will often reject the mortar because it has poor workability; e.g., it is too wet or too harsh (oversanded). The inspector should also look for fully mortared head and bed joints, particularly in plain masonry walls. In addition, the reinforcing steel must be of the correct size, type and grade, and placed and held in the correct position. Proper placement of reinforcement has a significant impact on the integrity and strength of reinforced masonry. The reinforcement should be placed within the tolerances set forth in ACI 530.1/ASCE 6/TMS 602, Section 3.4.

Item 2: Most of the inspections for Item 2 are verifications that materials are of the correct type and grade and located in accordance with the approved plans. Although welding of reinforcing bars is not referred back to Table 1704.3, Steel, ACI 530/ASCE 5/TMS 402 requires use of AWS D1.4.

Item 3: In grouted masonry, cleanliness of the cells prior to grouting is essential to ensure continuity and proper bonding. Cells to be grouted should be clean and free from debris and mortar projections greater than $^1/_2$ inch.

Item 4: Grouting must be in accordance with ACI 530/ASCE 5/TMS 402, Section 3.5. Grout should be initially consolidated by vibration and reconsolidated after initial water loss and settlement while the grout is still plastic.

Item 5: Sampling of fresh grout or mortar for the making of specimens for strength tests is extremely important to ensure quality control. Properly sampled and prepared specimens are necessary to determine that the grout or mortar will meet or exceed the minimum design strength. The frequency of sampling should be not less than one set of specimens for each 5,000 square feet of wall. Sampling and testing for grout should be done per ASTM C 1019, for mortar per ASTM C 270 and for prisms per ASTM C 1314.

Table 1704.5.3 Special Inspection of Masonry Construction—Level 2. This table presents the requirements for Level 2 special inspection of masonry along with the inspection frequency and applicable sections from the code and referenced standard. Level 2 special inspection is used for engineered masonry in buildings designated as essential facilities. Note that some inspections are periodic, but inspections for welding of reinforcing bars, placement of grout, inspection of grout space prior to grouting and preparation of specimens must be continuous. The table summarizes the required inspections and test samples necessary to verify that the in-place masonry meets the requirements of the code and the referenced standard. Refer to ACI 530/ASCE 5/TMS 402 and ACI 530.1/ASCE 6/TMS 602 and the applicable ASTM standards for additional details of the constituent material tests, sampling of mortar or grout, preparation of prisms and other testing requirements.

Item 1: Inspections under Item 1 should check that the mortar is being properly batched and mixed. The influence of the mortar on the strength of the finished masonry is directly related to the degree of grouting. The mortar has little influence on the strength of a fully grouted wall but has a significant influence on a plain masonry wall. If the mortar deviates from the specifications, the mason will often reject the mortar because it has poor workability, meaning that it is too wet or too harsh (oversanded). The inspector should also look for fully mortared head and bed joints, particularly in plain masonry walls. In addition to the reinforcing steel being the correct size, type and grade, proper placement of reinforcement has a significant effect on the structural integrity and strength of reinforced masonry. The reinforcement should be placed within the tolerances set forth in ACI 530.1/ASCE 6/TMS 602, Section 3.4. Cleanliness is next to godliness in grouted masonry. Cells to be grouted should be clean and free from debris and mortar projections greater than $^{1}/_{2}$ inch. Grouting should be in accordance with ACI 530/ASCE 5/TMS 402, Section 3.5. Grout should be initially consolidated by vibration and reconsolidated after initial water loss and settlement while the grout is still plastic.

Item 2: Most of the inspections for Item 2 verify that materials of the correct type and grade are in the location shown on the approved plans. Although welding of reinforcing bars is not specifically referred back to Table 1704.3 for steel, the table does refer to ACI 530/ASCE 5/TMS 402, which requires compliance with AWS D1.4.

Item 3: Sampling of fresh grout or mortar for the making of specimens for strength tests is extremely important to maintain adequate quality control. Properly sampled and prepared specimens are necessary to determine that the grout or mortar will meet or exceed the minimum design strength. The frequency of sampling should be not less than one set of specimens for each 5,000 square feet of wall. Sampling and testing for grout should be done per ASTM C1019; for mortar per ASTM C270; and for prisms per ASTM C1314.

1704.6 Wood construction. This section sets forth the special inspection requirements for wood construction. The requirement is for inspection of prefabricated elements such as wood trusses. (See IBC Section 1704.2.) For portions of wood structures designated as the seismic lateral force resisting system, see also the special inspection requirements for seismic resistance discussed in the analysis of Section 1707. Special inspection is also required for site-built assemblies. High-load diaphragms designed in accordance with Table 2306.3.2 require special inspection. The special inspector is required to inspect the diaphragm for proper sheathing grade and thickness; proper size, species and grade of framing members; and proper fastener type, size, spacing and edge distance from sheathing and framing members.

1704.7 Soils. This section covers special inspection requirements for site preparation, engineered fills and load-bearing requirements. The load-bearing capacity of the site soil and any fill has a significant impact on the structural integrity of a building supported by the fill. For example, settlements in an improperly compacted fill can cause significant structural distress. Differential settlements of $^{1}/_{4}$ inch in a 20-foot grade beam can induce stresses that exceed the yield limits. Hence, fills should be engineered and compaction carefully controlled. The special inspection tasks outlined in Table 1704.7 must be performed to verify compliance with the approved foundation and soils report required by Section 1802.2. Table 1704.7 is a new table that was added to the 2006 IBC to clarify the specific tasks required for special inspection of soils. The special inspector is required to (1) verify materials below footings are adequate to achieve the design bearing capacity; (2) verify that excavations are extended to the proper depth and have reached proper bearing material; (3) perform classification and testing of controlled fill materials; (4) verify use of proper materials, densities and lift thicknesses during placement and compaction of controlled fill material and (5) verify that the site and subgrade have been properly prepared prior to placement of controlled fill.

The exception exempts placement of controlled fills 12 inches or less in depth from having to comply with the special inspection requirements.

1704.8 Pile foundations. Special inspection is required for installation and testing of pile foundations. The special inspection tasks outlined in Table 1704.8 are required to verify compliance with the approved foundation and soils report required by Section 1802.2. Table

1704.8 requires continuous special inspection to (1) verify pile materials, sizes and lengths comply with the requirements; (2) determine capacities of test piles and conduct additional load tests; (3) observe driving operations and maintain complete and accurate records for each pile; and (4) verify placement locations and plumbness, confirm type and size of hammer, record number of blows per foot of penetration, determine required penetrations to achieve design capacity, record tip and butt elevations and document any pile damage.

For driven piles, records should, at a minimum, contain for each pile a driving log showing the number of hammer blows for each foot (and the energy) and the resistance at final penetration in blows per inch or blows per foot as applicable. The inspector should verify that the hammer throttle is set correctly to give the desired energy. The driving log should also give information on the duration and cause of any delays, the depth of any pre-excavation, and the elevation of the tip and the cutoff. Any known or suspected pile damage, as well as any observation of pile drift or heave, should be noted.

For each drilled pile, data should include a drilling log showing the types of soils encountered for each foot and the material stratum at the required tip elevation. The inspector should verify that the soil at the required tip elevation is the correct soil. The drilling log should also give information on the duration and cause of any delays, data on the rebar cage and concreting procedures, casing or other procedures necessary to prevent intrusion of ground water, and results of any concrete strength tests.

For specialty piles, the special inspector must perform additional inspections in accordance with the registered design professional's recommendations.

Pier foundations. Special inspection is required for installation and testing of pier foundations. The special inspection tasks outlined in Table 1704.9 are required to verify compliance with the approved foundation and soils report required by Section 1802.2. Table 1704.9 requires the special inspector to observe drilling operations and maintain complete and accurate records for each pier and verify placement locations and plumbness, confirm pier diameters, bell diameters (if applicable), lengths, embedment into bedrock (if applicable) and adequate end-bearing strata capacity. **1704.9**

Sprayed fire-resistant materials. This presents the requirements for the special inspection of spray applied fire-resistant materials (SFRM). For an SFRM to perform as intended, its application must be within the proper range for certain parameters as determined by the system manufacturer. These include temperature at time of application, substrate conditions, thickness, bond strength and density. These parameters must be checked prior to and during installation of the SFRM. **1704.10**

Structural member surface conditions. The integrity of an SFRM system primarily depends on the condition of the surface of the steel member to which it is applied. For proper performance, the system must be fully adhered to the surface. **1704.10.1**

Application. During the application and curing of SFRMs, several parameters must be controlled, including ambient temperature, temperature of the substrate and the temperature of the applied SFRM. Control of these temperatures is necessary so that the chemical reactions needed to make the SFRM bond to the steel substrate properly occur. The minimum or maximum temperatures needed for proper bond and cure depend on the particular system. **1704.10.2**

Thickness. The SFRM must be applied at the correct thickness for the system to provide the required fire resistance. The sampling and thickness determinations are based on ASTM E605. This standard also provides testing methods commonly used by industry. See Reference 11 for additional information on inspection of SFRM. **1704.10.3**

Floor, roof and wall assemblies. The frequency of thickness measurements for membrane assemblies, i.e., walls, floors and roofs, is based on square footage. **1704.10.3.1**

Structural framing members. The frequency of thickness measurements for structural framing members such as columns is based on the number of members with a minimum of 25 percent of the structural members on each floor. **1704.10.3.2**

1704.10.4 Density. The SFRM must be applied at the correct density in order for the system to provide the required fire resistance. The sampling and density determinations are based on ASTM E 605.

1704.10.5 Bond strength. The bond of the SFRM to the steel substrate is essential for the SFRM to provide the required fire resistance. The sampling and bond strength determinations are based on ASTM E 736. The minimum cohesive/adhesive bond strength of 150 pounds per square foot is based on the American Institute of Architects (AIA) Master Specifications. The sampling rates for bond strength, as set forth in IBC Sections 1704.11.5.1 and 1704.11.5.2, are lower than those for thickness and density, as the determinations for thickness and density give an indirect indication of bond.

Special inspection is required for mastic and intumescent fire-resistant coatings applied to structural elements and decks in accordance with the Technical Manual AWCI 12-B, *Standard Practice for the Testing and Inspection of Field Applied Film Intumescent Fire-resistive Materials.*[12] Like spray and other applied fire proofing, the need for special inspection of mastic and intumescent fire-resistant coatings is important because proper application of the materials requires special expertise to ensure that they are installed in accordance with the manufacturer's instructions and their listings. If not installed and applied properly, they may not perform as expected under design fire conditions. The special inspection is to be based on the degree of fire-resistance designated in the approved construction documents. AWCI 12-B is published by The Association of the Wall and Ceiling Industries and was added to the referenced standards listed in Chapter 35 of the 2006 IBC.

1704.11 Mastic and intumescent fire-resistant coatings. The special inspector verifies that coatings are applied to structural elements and decks in accordance with AWCI 12-B based on the required fire-resistance as shown in the approved construction documents.

1704.12 Exterior insulation and finish systems (EIFS). Special inspection for EIFS systems should be based on the manufacturer's installation instructions. Critical areas necessary for adequate EIFS performance are proper installation of the waterproofing membrane and installation of flashings at windows, doors, joints, eaves, corners and penetrations.

1704.13 Special cases. This section requires special inspections for proposed work that is unique or unusual and not specifically addressed in the code or in standards referenced by the code.

1704.14 Special inspection for smoke control. This section, although related to mechanical systems rather than structural or architectural systems, is required by the code because the mechanical ductwork and signaling devices are likely to be concealed during the building construction and the ductwork needs to be leakage tested prior to concealment.

Section 1705 *Statement of Special Inspections*

There were several code changes to the 2006 IBC that reorganized and clarified the special inspection provisions of Chapter 17 to more clearly convey the intent. In the 2006 IBC, the term *statement of special inspections* is used instead of the terms *special inspection program* and *quality assurance plan*. Provisions that were formerly covered in the quality assurance plan requirements of Sections 1705 and 1706 in the 2003 IBC are now included as part of the statement of special inspections when required by the seismic or wind criteria. In the 2006 IBC, the statement of contractor responsibility requirements have been consolidated and put in a separate Section 1706, eliminating the redundant language of the 2000 and 2003 IBC, where they appeared in both the wind and seismic quality assurance plans. These code changes make the special inspection provisions in the 2006 IBC much easier to understand and enforce.

General. When special inspection, special inspection for seismic resistance, or structural testing for seismic resistance is required by Sections 1704, 1707 or 1708, the registered design professional is required to prepare a detailed statement of special inspections. The statement of special inspections must be submitted by the permit applicant and is a condition of permit issuance as prescribed in Section 1704.1.1. The statement of special inspection must identify ordinary special inspections required by Section 1704, special inspection requirements for seismic resistance prescribed in Section 1705.3, special inspection requirements for wind resistance prescribed in Section 1705.4 and the seismic special inspection and testing requirements covered by Sections 1707 and 1708. **1705.1**

Content of statement of special inspections. The statement of special inspections must identify the work requiring special inspection or testing. It should include the type and extent of each special inspection or test and indicate whether the inspections are to be continuous or periodic. The statement also includes the additional special inspection or testing requirements for seismic or wind resistance as prescribed by Sections 1705.3, 1705.4, 1707 or 1708. **1705.2**

Seismic resistance. Section 1705.3 requires the statement of special inspections to include elements of the seismic force resisting system, the designated seismic system and additional systems and components listed in the section. The designated seismic system consists of those architectural, electrical and mechanical systems and components that require design in accordance with Chapter 13 of ASCE 7 for which the component importance factor, I_p, is greater than 1.0 as prescribed in Section 13.1.3 of ASCE 7. Section 13.1.3 of ASCE 7 lists three classes of components that have a component importance factor of 1.5. These are components required to function after an earthquake, including fire sprinkler systems, components containing hazardous materials and components in Occupancy Category IV structures that are deemed to be necessary for continued operation of the facility. **1705.3**

The items required to be identified in the statement of special inspections pertaining to seismic resistance are triggered by the seismic design category assigned to the building. Special inspection for seismic resistance is required for the seismic force resisting system in buildings assigned to Seismic Design Categories C, D, E or F; for the designated seismic system in Seismic Design Categories D, E or F; and for the additional architectural, mechanical and electrical components listed under items 3, 4 and 5 of Section 1705.3.

The statement of special inspections need not include seismic resistance for structures in any of the following three cases: (1) structures of light-frame construction where the design spectral response acceleration at short periods, S_{DS}, does not exceed 0.5g and the height of the structure does not exceed 35 feet; or (2) structures of reinforced concrete or masonry structural system where the design spectral response acceleration at short periods, S_{DS}, does not exceed 0.5g and the height of the structure does not exceed 25 feet and (3) detached one- or two-family dwellings not exceeding two stories in height, provided the structure does not have plan or vertical irregularities.

Seismic requirements in the statement of special inspections. Depending on the seismic design category assigned to the structure, Section 1705.3 requires specific seismic requirements to be included in the statement of special inspections. The elements of the seismic force resisting system, the designated seismic system and the additional systems and components listed in the section are also to be included in the statement. **1705.3.1**

In addition, seismic special inspections required by Section 1707 and seismic structural testing required by Section 1708 must be included in the statement of special inspections. Seismic special inspection and seismic structural testing are required in areas of relatively high seismicity, and are triggered by the seismic design category assigned to the building. Special inspection for seismic resistance and seismic structural testing are required for the seismic force resisting system in buildings assigned to Seismic Design Categories C, D, E or F; for the designated seismic system in Seismic Design Categories D, E or F; and for architectural, mechanical and electrical components in Seismic Design Category C, D, E and F as described in Sections 1707.7 and 1707.8.

1705.4 Wind resistance. Depending on the basic wind speed and exposure category of the building site, Section 1705.4 requires the statement of special inspections to identify the elements of the main wind-force-resisting system and components and cladding that are subject to special inspection. The specifics are detailed in Section 1705.4.2.

1705.4.1 Wind requirements in the statement of special inspections. The statement of special inspections must identify the elements of the main wind-force-resisting system and components and cladding that are subject to special inspection where the basic wind speed is greater than or equal to 120 mph in wind Exposure Category B, or where the basic wind speed is greater than or equal to 110 mph in wind Exposure Category C or D. See Section 1609.4 for descriptions of exposure categories B, C and D.

1705.4.2 Detailed requirements. As a minimum, the code lists six items that are subject to special inspection: (1) roof cladding and roof framing connections; (2) wall connections to roof and floor diaphragms and framing; (3) roof and floor diaphragm systems; (3) roof and floor diaphragm systems, including collectors, drag struts and boundary elements; (4) vertical wind-force-resisting systems, including braced frames, moment frames and shear walls; (5) wind-force-resisting system connections to the foundation and (6) fabrication and installation of systems or components in wind-borne debris regions required to meet the impact-resistance requirements in accordance with Section 1609.1.2. Note that the exception allows the installation of manufactured systems or components that have an approved label showing compliance with the wind-load and impact-resistance requirements without requiring special inspection. The exception in this case is similar to the exception to special inspection for items that are manufactured in an approved fabricators shop. See Section 1704.2.

Section 1706 *Contractor Responsibility*

The statement of contractor responsibility was never intended to be part of the statement of special inspections prepared by a registered design professional. Including the requirements for the statement of contractor responsibility within the quality assurance plan requirements for wind and seismic resistance was a source of confusion. In the 2006 IBC, the statement of contractor responsibility requirements are consolidated and relocated in a separate Section 1706, eliminating the redundant language of the 2000 and 2003 IBC where it appeared in both the wind and seismic quality assurance plans.

Section 1706 requires the contractor responsible for construction of the main wind-force-resisting system, the seismic force resisting system, the designated seismic system and those wind resisting components listed in the statement of special inspections to submit a statement of responsibility to the owner and building official. The contractor's statement of responsibility is to be submitted prior to commencement of work on a particular structural system or component. The statement is required to acknowledge awareness of special requirements contained in the statement of special inspections, acknowledge that control will be exercised to achieve conformance with construction documents, outline the procedures for exercising quality control, specify the methods and frequency of reporting, and identify and provide qualifications of key personnel responsible for implementing the requirements contained in the statement of special inspections.

Section 1707 *Special Inspections for Seismic Resistance*

Section 1707 requires special inspection for seismic resistance in addition to the general special inspection requirements covered in Section 1704. The requirements are triggered by the seismic design category of the building. Special inspection for elements of the seismic resistance is required for the seismic force resisting system in buildings assigned to Seismic Design Categories C, D, E or F; for the designated seismic system in Seismic Design Categories D, E or F; and for architectural, mechanical and electrical components in Seismic Design Category C, D, E and F as described in Sections 1707.7 and 1707.8, respectively.

The seismic force resisting system consists of those structural elements and systems that provide lateral stability of the structure and are specifically designed to provide resistance to the anticipated seismic forces. The designated seismic system consists of those architectural, electrical and mechanical systems and components that require design in accordance with Chapter 13 of ASCE 7 for which the component importance factor, I_p, is greater than 1.0 as prescribed in Section 13.1.3 of ASCE 7. Section 13.1.3 of ASCE 7 lists three classes of components that have a component importance factor of 1.5. These are components required to function after an earthquake including fire sprinkler systems, components containing hazardous materials, and components in Occupancy Category IV structures that are deemed to be necessary for continued operation of the facility.

1707.1

Special inspection for seismic resistance. Special inspection for seismic resistance is only required for structures in SDC C, D, E or F. Special inspection for designated systems is a function of the Component Importance Factor, I_P, as defined in Section 1702, which references ASCE 7.

1707.2

Steel. The requirements are the same as those set forth in Table 1704.3, Item 5, continuous inspection of welding in accordance with AISC 341.

1707.3

Structural wood. The inspections are to ensure continuity of load path within the seismic lateral-force-resisting system. The walls will transfer their inertial loads to the diaphragms, which will in turn transmit the inertial loads to the lateral-force-resisting system and thence to the foundation. Particular care should be given to the nailing of diaphragms and shear walls. Common nails are often specified in the design, but smaller diameter sinkers or power driven (gun) nails are often substituted in the field because the smaller diameter nails have a lower lateral resistance. For example, the lateral resistance of a 0.131-inch diameter power driven nail used as a replacement for a 10d common nail in Douglas Fir-Larch is only 76 pounds, whereas the value for the 10d common nail is 90 pounds. Of additional importance is the connection of drag struts or collectors to shear walls and the proper installation and tightening of hold-down bolts in shear walls. The exception provides that special inspection is not required for diaphragm and shear panel construction where the fastener spacing is more than 4 inches on center. Where the fastener spacing is greater than 4 inches there is lower demand and less potential for splitting; therefore, special inspection is less critical.

1707.4

Cold-formed steel framing. The inspections are to ensure continuity of load path within the seismic lateral-force-resisting system; i.e., the walls will transfer their inertial loads to the diaphragms, which will in turn transmit the inertial loads to the lateral-force-resisting system and thence to the foundation. Particular care should be given to connections of braces and hold-downs.

1707.5

Pier foundations. Pier foundations in buildings assigned to Seismic Design Category C, D, E or F require periodic special inspection during placement of reinforcement and continuous special inspection during concrete placement. Pier foundations are similar to piles in that their load-carrying capacity is achieved by skin friction or end bearing or a combination of both. However, piers are relatively short in comparison to their width, having lengths less than or equal to 12 times the least horizontal dimension.

1707.6 Storage racks and access floors. Proper anchorage is critical to keep racks from overturning. This is especially important for racks 8 feet or more in height located in areas of relatively high seismicity such as Seismic Design Category D, E or F.

1707.7 Architectural components. Exterior cladding and veneers can be a serious safety hazard if detached from the structure during seismic shaking as well as potentially blocking exit paths. The code requires periodic special inspection for exterior cladding, nonbearing walls and partitions and veneer in Seismic Design Category D, E or F. The exception provides that no special inspection is required for these elements in buildings 30 feet or less in height, for light-weight cladding and veneer weighing 5 psf or less and for light-weight partitions weighing 15 psf or less.

1707.8 Mechanical and electrical components. Inspection is necessary for components that must function in postearthquake conditions, such as emergency electrical systems, or for anchorage of mechanical equipment, piping and ducting using or carrying flammable or hazardous materials.

A code change to the 2006 IBC clarified provisions related to the inspection and testing of mechanical and electrical components for seismic resistance. Periodic special inspection is required for the following items in structures in Seismic Design Category C, D, E or F: (1) during the anchorage of electrical equipment for emergency or standby power systems; (2) during installation of piping systems intended to carry flammable, combustible or highly toxic contents and their associated mechanical units; (3) during the installation of HVAC ductwork that contains hazardous materials and (4) during installation of vibration isolation systems in structures where the construction documents require a nominal clearance of 0.25 inches or less between the equipment support frame and restraint. Periodic special inspection is required for the installation of the anchorage system of all other electrical equipment in structures in Seismic Design Category E or F.

1707.9 Designated seismic system verifications. The special inspector must verify that anchorage or mounting systems conform to the certificate of compliance.

1707.8 Seismic isolation system. The performance of seismic isolators is critical to the performance of isolation systems. See the analysis of Section 1623.5 for the inspection requirements.

Section 1708 *Structural Testing for Seismic Resistance*

Section 1708 sets forth the required structural testing for seismic resistance. The structural testing for seismic resistance requirements are triggered by the seismic design category of the building. Section 1708.2 requires special seismic testing for elements of the seismic force resisting system in buildings assigned to Seismic Design Category C, D, E or F; for the designated seismic system in Seismic Design Category D, E or F; and for architectural, mechanical and electrical components in Seismic Design category C, D, E or F. Testing is required for reinforcement and prestressing steel per Section 1708.3, for structural steel per Section 1708.4, for mechanical and electrical equipment in the designated seismic system per Section 1708.5 and for seismically isolated structures per ASCE 7 Section 17.8. Components in Category IV structures that are necessary for continued operation of the facility must comply with the seismic qualification requirements in Section 1708.5.

1708.1 Masonry. Certificates of compliance for materials used is required for all categories. Testing is required only as set forth below. In actuality, the testing is for verification that actual compressive strength, f'_m, using the proposed constituent masonry, mortar and grout, meets or exceeds that specified prior to construction. Compressive strength, f'_m, can be determined by testing the strength of the constituent materials or the strength of masonry prisms.

Empirically designed masonry and glass unit masonry in occupancy category I, II or III. Certificates of compliance for the materials used are required in accordance with Table 1708.1.1. **1708.1.1**

Empirically designed masonry and glass unit masonry in occupancy category IV. Certificates of compliance for the materials used are required, and verification of compressive strength, f'_m, prior to construction, is required in accordance with Table 1708.1.2. **1708.1.2**

Engineered masonry in Occupancy Category I, II or III. Certificates of compliance for the materials used are required, and verification of compressive strength, f'_m, prior to construction is required in accordance with Table 1708.1.2. **1708.1.3**

Engineered masonry in Occupancy Category IV. Certificates of compliance for the materials used are required, and verification of compressive strength, f'_m, prior to construction, is required. In addition, verification of the proportions in the mortar and grout as delivered to the jobsite is required. See Table 1708.1.4. **1708.1.4**

Testing for seismic resistance. Structural testing for seismic resistance is required only for structures in Seismic Design Category C, D, E or F. Structural testing for designated systems is a function of the Component Importance Factor, I_P, as defined in Section 1702. See also the analysis of Section 1708.5. **1708.2**

Reinforcing and prestressing steel. Certified mill reports are used in lieu of testing to establish conformance to code requirements. However, when reinforcing steel conforming to ASTM A615 is used in special moment-resisting frames or in the boundary elements of shear walls in Seismic Design Category D, E or F, the steel yield and ultimate strengths must be specifically tested. **1708.3**

Additionally, if A615 reinforcing steel is to be welded, the chemical composition must be determined. The weldability of the steel is determined by its chemical content. AWS D1.4, *Structural Welding Code—Reinforcing Steel*, bases the welding requirements, including preheat and interpass temperature, on the carbon equivalent, which is based on the steel chemistry and the bar size. See the analysis of Section 1704.3.1 concerning requirements for welding procedures.

Structural steel. The required testing is as set forth in AISC 341[13], supplemented by the following requirements: **1708.5**

1. All complete joint penetration and partial joint penetration groove welds subject to net tension forces must be tested by nondestructive methods.
2. The acceptance criteria for nondestructive testing are either those for static loading or dynamic loading as designated by the registered design professional.
3. Base metal thicker than $1^1/_2$ inches subject to through thickness weld shrinkage strains. For example, a T-joint or the column flange in a typical beam-column connection must be tested for discontinuities such as lamellar tearing after testing. See Figure 2 for an example of a weld susceptible to weld shrinkage strains.

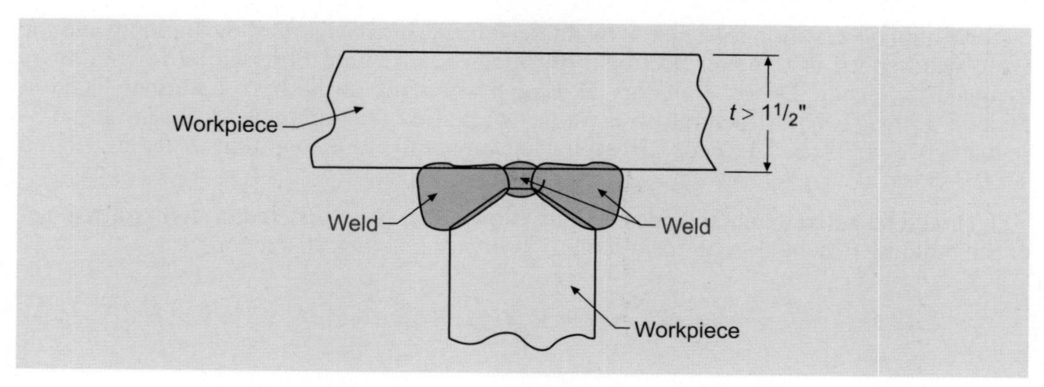

Figure 17-2 Weld susceptable to shrinkage stresses

1708.5 Seismic qualification of mechanical and electrical equipment. In the 2006 IBC, Section 1708.5 was retitled "Seismic qualification of mechanical and electrical equipment" to correctly describe the subject matter of the provision. The section was revised to require the registered design professional to provide on the construction drawings the applicable seismic qualification requirements for designated seismic systems. This helps ensure that the seismic qualification requirements are provided to the manufacturer, who must submit the certificate of compliance for the designated seismic system. The last sentence of 2003 IBC Section 1708.5 pertaining to labeling and anchorage was relocated to a new Section 1707.9 in the 2006 IBC.

The section requires the registered design professional to identify the applicable seismic qualification requirements for the designated seismic systems on the construction documents. In addition, the manufacturers of the designated seismic system components are required to test or analyze the component and its mounting system or anchorage and submit a certificate of compliance to the registered design professional and the building official. The qualification must be by one of the following methods: shake table testing, three-dimensional shock tests, a dynamic analytical method, or historical data demonstrating acceptable seismic performance; or by a more rigorous analysis that provides equivalent safety. The qualification method and certificate of compliance is to be accepted by the registered design professional and approved by the building official.

1708.6 Seismically isolated structures. Testing requirements for seismically isolated structures are covered by Section 17.8 of ASCE 7.

Section 1709 *Structural Observations*

1709.1 Structural observation. Structural observation requirements first appeared in the 1988 UBC and were applicable to buildings in high seismic risk areas. The purpose of structural observation is to ensure that critical elements of the lateral-force-resisting system is constructed in general conformance with the design as shown in the approved structural drawings and specifications. Because the registered design professional is most familiar with the design and the details of the lateral-force-resisting system, he or she is the most appropriate person to execute the requisite observation. Note that structural observation is in addition to, not in place of, special inspection, and it does not replace or waive any of the required inspections by the jurisdiction inspector.

Structural observation consists of visual observation of the structural systems by the registered design professional for general conformance with the approved construction documents at significant construction stages and at the completion of the structural system. The owner is required to employ the registered design professional to perform those structural observations described in Section 1702. Structural observation does not include or waive the inspections performed by the jurisdiction as required by Section 109 or the special inspections required by Section 1704 or 1707.

Structural observation is triggered by the seismic design category of the building and the basic wind speed at the site. Section 1709 requires structural observation for buildings assigned to Seismic Design Category D, E or F when any of the five conditions listed in Section 1709.2 exist. Structural observation is required for buildings where the basic 3-second gust wind speed exceeds 110 mph when any of the four conditions listed in Section 1709.3 exist.

It should be noted that both code sections allow either the registered design professional or the building official to require structural observation at their discretion.

Overview of Chapter 17

Chapter 17 is designed to improve construction quality in structural systems through the following requirements:

- Special inspection is required for various types of structural materials, elements and systems as prescribed in Section 1704.
- The registered design professional must provide a detailed statement of special inspections in accordance with Section 1705.
- The contractor responsible for constructing the wind or seismic force resisting system must provide a statement of responsibility in accordance with Section 1706.
- Additional special inspections for seismic resistance may be required in accordance with Section 1707.
- Specific structural testing for seismic resistance may be required in accordance with Section 1708.
- Structural observation by the registered design professional may be required in accordance with Section 1709.

Special inspection is required for various types of structural elements and systems as prescribed in Section 1704. These include structural systems constructed of steel (Section 1704.3), concrete (Section 1704.4), masonry (Section 1704.5), wood systems (Section 1704.6), site soils (Section 1704.7), pile foundations (Section 1704.8) and pier foundations (Section 1704.9). In addition, special inspection is required for alternate materials and unusual designs when required by the building official as prescribed in Section 1704.13.

The requirements for special seismic inspection and seismic testing are triggered by the seismic design category of the building. The requirements for structural observation are triggered by the seismic design category of the building and the basic wind speed at the site.

Example Problem

A new fire station building is proposed in Grass Valley, California. The building will be constructed of fully grouted reinforced concrete masonry unit (CMU) walls with a steel truss roof system. What are the specific quality assurance requirements for the masonry construction according to Chapter 17 of the 2006 IBC?

Solution: Because many of the special inspection, structural observation and structural testing requirements of the IBC are based on seismic design category, the first step is to determine the seismic design category of the proposed building.

Seismic Design Category. Determine the occupancy category of the structure from IBC Table 1604.5. The building is a fire station and therefore assigned to Occupancy Category IV.

The latitude and longitude of the building site are determined from the Grass Valley USGS quadrangle map as follows:

Latitude = 39.219 degrees, Longitude = – 121.060 degrees.

From the USGS Seismic Design Parameters program, the mapped short and long period mapped spectral accelerations are:

The mapped spectral acceleration for short periods, $S_S = 0.515$

The mapped spectral acceleration for 1-second period, $S_1 = 0.198$

These mapped spectral acceleration values are for Site Class B soils. Assuming Site Class D without a geotechnical report that establishes the site class is permitted under Section 1613.5.2. Because the existing soil at the site is firm, rocky soil, the seismic design category will be determined assuming Site Class D as the default soil profile. From Tables 1613.5.3(1) and 1613.5.3(2), the soil factors F_a and F_v for Site Class D and S_S and S_1 are found:

$F_a = 1.39$

$F_v = 2.01$

The design spectral accelerations S_{DS} and S_{D1} are $^2/_3$ of the soil modified spectral acceleration values as follows:

$S_{DS} = 2/3\ (F_a S_S) = 2/3 \times 1.39 \times 0.515 = 0.477$

$S_{D1} = 2/3\ (F_v S_1) = 2/3 \times 2.01 \times 0.198 = 0.265$

Refer to the 2003 NEHRP Commentary (FEMA 450) for a discussion of the $^2/_3$ factor.

Entering Table 1613.5.6(1) with $S_{DS} = 0.477$ and Occupancy Category IV, we find the building is assigned to Seismic Design Category D. Entering Table 1613.5.6(2) with $S_{D1} = 0.265$ and Occupancy Category IV, we find the building is assigned to Seismic Design Category D. The building is therefore assigned to Seismic Design Category D.

Basic wind speed. Figure 1609 indicates that the basic 3 second gust wind speed for Grass Valley, California, is 85 mph. There are no special requirements in Chapter 17 based on wind speed.

Permissible design methods. Section 2109 specifically prohibits the use of the empirical design method for buildings in Seismic Design Category D, E or F. Therefore, the proposed fire station building must be designed by the engineering provisions prescribed by the either allowable stress design procedure of Section 2107 or the strength design procedure of Section 2108.

Special inspection. Because the proposed building is classified as "Engineered masonry in Occupancy Category IV," Section 1704.5.3 requires Level 2 special inspection in accordance with Table 1704.5.3.

Structural testing for seismic resistance. Because the proposed building is classified as "Engineered masonry in Occupancy Category IV," Section 1708.1.4 requires Level 2 quality assurance in accordance with Table 1708.1.4. The table requires certificates of compliance for masonry construction materials, verification of f'_m prior to construction and every 5,000 square feet during construction, and verification of proper proportions of mortar and grout materials delivered to the job site. Certified mill test reports are required for reinforcing steel used in the boundary elements of special reinforced masonry shear walls. Where ASTM A615 reinforcing is used in boundary elements of shear walls, testing in accordance with ACI 318 is required. Components in Category IV structures that are necessary for continued operation of the facility must comply with the seismic qualification requirements in Section 1708.5.

Structural observation. Because the building is classified in Occupancy Category IV and assigned to Seismic Design Category D, Section 1709.2 requires structural observation in accordance with Section 1709.

Summary of requirements:

- The proposed fire station is in Occupancy Category IV in accordance with Table 1604.5.
- The Seismic Design Category of the building was determined to be Seismic Design Category D based on assumed Site Class D soil conditions.
- The concrete masonry structural system must be designed in accordance with the engineering provisions prescribed by the allowable stress design procedure of Section 2107 or the strength design procedure of Section 2108.
- As a condition of permit issuance, the registered design professional must provide a statement of special inspections to be submitted by the permit applicant.
- Level 2 special inspection must be provided for the masonry construction in accordance with Table 1704.5.3.
- Additional special inspection for seismic resistance is required in accordance with Section 1707.
- Level 2 quality assurance is required to be provided for the masonry construction in accordance with Table 1708.1.4.
- Structural testing for seismic resistance is required for the seismic force resisting system, the designated seismic system and for architectural, mechanical and electrical components in accordance with Sections 1708.2 through 1708.5. Components that are necessary for continued operation of the facility must comply with the seismic qualification requirements in Section 1708.5.
- Structural observation for seismic resistance in accordance with Section 1709 is required.

The contractor responsible for construction of the main wind and seismic force resisting systems and the designated seismic system or wind resisting components must submit a statement of responsibility to the owner and building official prior to commencement of work on a particular structural system or component.

Section 1710 *Design Strengths of Materials*

This section requires structural materials to conform to the applicable design standards or accepted engineering practice in the absence of such standards.

Conformance to standards. Structural materials must conform to design standards, approved rules or accepted engineering principles and practice. **1710.1**

New materials. Materials not explicitly covered by the code are allowed, subject to testing that demonstrates adequate performance (see Section 1711). **1710.2**

Section 1711 *Alternative Test Procedures*

Test reports from approved agencies may be used as the basis for approval of materials not covered by the code or any approved rules. New materials are mentioned in Section 1701.2. Note that this section references the use of new, innovative or alternative materials from Section 104.11. The building official has the authority to accept such materials based on reports from an approved agency that is independent from the material supplier.

Section 1712 *Test Safe Load*

Testing to determine a safe load is required when a structure or component cannot be designed or analyzed in accordance with accepted engineering principles and practices, or where the construction design method does not fully comply with the respective material design standards found in Chapter 35.

Section 1713 *In-Site Load Tests*

This section covers load testing for existing structures or portions thereof where the load-bearing capacity or stability is in doubt. The requirements should not be confused with the test procedures set forth in Section 1714 for materials, components and assemblies prior to construction.

1713.1 **General.** When there is doubt as to the structural integrity, load-carrying capacity or stability of an existing structure or portion thereof, the building official has the option of requiring a structural analysis or a load test or both. If the structural analysis shows that the structure is not capable of safely carrying the code-design loads, the structure must be load-tested.

1713.2 **Test standards.** When load test procedures are given in a referenced standard, those procedures should be followed. An example would be the test procedures in the SJI standard specification.[14] When the referenced standard lacks a load test procedure or there is no reference standard for the structure or portions thereof, a registered design professional should develop a test procedure and test protocol that simulate the actual loading conditions, both loads and displacements, that the structure is expected to experience.

1713.3 **In-situ load tests.** In-situ tests fall into two categories: procedures specified and procedures not specified (see below). The test must be performed under the supervision of the registered design professional.

1713.3.1 **Load test procedure specified.** When the applicable standard has a test procedure and acceptance criteria specified, the standard shall apply.

1713.3.2 **Load test procedures not specified.** When there is no applicable load test procedure in a referenced standard, the structure or portion thereof should be tested with a loading protocol that simulates the actual loads and deformations, both lateral and vertical, that the structure is expected to receive. For the gravity system, the vertical loads shall be twice the unfactored design loads. Dead loads should include expected partition loads.

Section 1714 *Preconstruction Load Tests*

General. When the load carrying capacity and the physical properties of materials or methods of construction are not amenable to analysis by accepted engineering methods, or the material or method does not comply with applicable standards, the structural load capacity and physical properties must be determined by tests. This section is applicable to components, assemblies and elements of structures; e.g., windows. This section is not applicable to load testing of existing structures (see analysis of Section 1713). **1714.1**

Load test procedures specified. When an applicable standard has a test procedure and acceptance criteria, the standard shall apply. **1714.2**

Load test procedures not specified. When there is no applicable load test procedure in a referenced standard, the structure or portion thereof should be tested with a loading protocol that simulates the actual loads and deformations, both lateral and vertical, that the structure is expected to receive. For components, assemblies and elements that are not part of the lateral force resisting system, the loading and acceptance criteria are set forth in Section 1714.3.1. **1714.3**

Test procedure. The loading procedure and acceptance criteria use commonly accepted engineering practices to test the adequacy of the component or assembly to resist structural failure at the design loads. This procedure should be used only for loads for which the upper bounds can be established with a reasonable degree of accuracy such as dead and live loads. This procedure should not be used for earthquake loads and should be used with care for wind loads. **1714.3.1**

Deflection. Deflection under design load is limited to those set forth in Section 1604.3. **1714.3.2**

Wall and partition assemblies. The assembly must be tested for simultaneous vertical and lateral loads, both with and without window framing. **1714.4**

Exterior window and door assemblies. Doors and window assemblies are generally qualified by tests. The allowable wind pressures for smaller units made of identical components, including glass thickness, may be higher as determined by analysis, provided that an additional test is done on the assembly with the highest pressure to validate the analysis. **1714.5**

Exterior windows and doors. These window and door assemblies must be tested and labeled to indicate conformance with the referenced American Architectural Manufacturers Association standard. The products tested and labeled in accordance with the standard are not subject to the analysis and testing requirements of Section 2403.2 and the deflection limitations specified in Section 2403.3. **1714.5.1**

Exterior windows and door assemblies not provided for in Section 1714.5. Exterior window and door assemblies not covered by Section 1714.5.1 must be tested in accordance with ASTM E 330. The test load is equal to 1.5 times the design pressure, as determined per Chapter 16, applied for 10 seconds. Exterior window and door assemblies covered by this section and containing glass are required to conform to the requirements of Section 2403. **1714.5.2**

Test specimens. Test specimens should be representative. Tests must be conducted or witnessed by an independent approved agency. **1714.6**

Section 1715 Material and Test Standards

This section sets forth the test standards and acceptance criteria for joist hangers and concrete and clay roof tiles.

1715.1 Test standards for joist hangers and connectors.

1715.1.1 Test standards for joist hangers. Testing is typically performed using specimens of Douglas Fir-Larch (specific gravity 0.50) or Southern Pine (specific gravity 0.55) per ASTM D 1761.

1715.1.2 Vertical load capacity for joist hangers. This section sets forth the test and acceptance criteria for testing joist hangers. Note that items 1 and 2 are the strength limit states of the hanger, item 3 is a deflection limit state, and items 4 and 5 are design load limit states for the fasteners and wood, respectively.

1715.1.3 Torsional moment capacity for joist hangers. The allowable torsional moment capacity is based on the lateral movement of the top or bottom of the joist.

1715.1.4 Design value modifications for joist hangers. Only the design value based on the allowable design load for the fasteners or wood member (items 4 or 5 in Section 1715.1.2) are allowed to be increased for duration of load factors specified in the NDS, but may not exceed the capacity determined by items 1, 2 or 3. No duration of load increase is allowed for capacities determined by ultimate load tests or limited by deflection.

REFERENCES

[1]NEHRP, *NEHRP (National Earthquake Hazard Reduction Program) Recommended Provisions for New Buildings and Other Structures (FEMA 450)*, Building Seismic Safety Council, Washington, DC, 2003.

[2]*Model Program For Special Inspection,* International Code Council, Washington, DC.

[3]CCTIA, *Guidelines for Special Inspection in Construction*, California Council of Testing and Inspection Agencies, c/o Testing Engineers, 2811 Adeline Street, Oakland, CA 94608, 1996.

[4]AWS D1.1-04, *Structural Welding Code—Steel*, American Welding Society, Miami, FL, 2004.

[5]PCI MNL 116-85, *Manual for Quality Control for Plants and Production of Precast Concrete Products*, Precast/Prestressed Concrete Institute, Chicago, IL, 1987.

[6]AISC 360-05, *Specification for Structural Steel Buildings*, American Institute for Steel Construction, Inc., Chicago, IL, 2005.

[7]RCSC-04, *Specification for Structural Joints Using A325 or A490 Bolts* (June 30, 2004), Research Council on Structural Connections, Chicago, IL, 2004.

[8]SSTC, *Structural Bolting Handbook 2006, (based on June 30, 2004 edition of RCSC Specification)*, Steel Structures Technology Center, Inc., Novi, MI, 2006.

[9]ACI 318-05, *Building Code Requirements for Structural Concrete*, American Concrete Institute, Farmington Hills, MI, 2005.

[10]ACI 530/ASCE 5/TMS 402, *Building Code Requirements for Masonry Structures*, American Concrete Institute, Farmington Hills, MI, 2005.

[11]ACI 530.1/ASCE 6/TMS 602, *Specification for Masonry Structures*, American Concrete Institute, Farmington Hills, MI, 2005.

[12]Technical Manual 12-B, First Edition; *Standard Practice for the Testing and Inspection of Field Applied Thin Film Itumescent Fire-resistive Materials; an Annotated Guide*, The Association of the Wall and Ceiling Industries International, Falls Church, VA, 1998.

[13]AISC 341, *Seismic Provisions for Structural Steel Buildings,* including Supplement No. 1, American Institute of Steel Construction, Chicago, IL, 2005.

[14] SJI, *Standard Specification for Joist Girders, Open Web Steel Joists (K-Series), Longspan Steel Joists (LH Series) and Deep Longspan Steel Joists (DLH Series)*, Steel Joist Institute, Myrtle Beach, SC, 2005.

BIBLIOGRAPHY

AISC 341-05, *Seismic Provisions for Structural Steel Buildings,* including Supplement No. 1, 2006, American Institute for Steel Construction, Inc., Chicago, IL, 2005.

AWS D1.4-98, *Structural Welding Code—Reinforcing Steel*, American Welding Society, Miami, FL, 1998.

ICC, *The BOCA National Building Code—Commentary Volume 2*, International Code Council, Washington, DC, 1999.

ICC, *Handbook to the Uniform Building Code: An illustrative commentary*, International Code Council, Washington, DC, 1998.

ICC, *Proposed Changes to the Final Draft of the International Building Code*, International Code Council, Washington, DC, January 1999.

ICC, *1998 Proposed Changes to the First Draft of the International Building Code*, International Code Council, Washington, DC, April 1998.

UBC-IBC Structural Comparison & Cross Reference, International Code Council, Washington, DC, 2000.

CHAPTER

18

SOILS AND FOUNDATIONS

Section 1604.4 requires that all structural systems have a continuous load path from the point of origin to the resisting element. The resisting element for buildings and other structures is the foundation that ultimately transfers the loads to the supporting soil. Therefore, the satisfactory performance of the foundation system is critical to the satisfactory performance of the overall structure. Most building structures are designed assuming a fixed unyielding base that is not subjected to large total or differential settlements or displacements. Shallow foundations on firm soils will generally perform satisfactorily if the requirements of these provisions are followed.

Foundation design, however, becomes a significant factor for large structures, embedded structures—such as a tall building constructed over a multilevel basement garage—structures on soft soils, structures supporting rotating or reciprocating equipment, and structures sensitive to differential displacements. It is important to have a good knowledge of the behavior of the various foundation types, including their limitations. In addition, for structures subject to high wind forces or seismic ground motion, special consideration must be given to the lateral load path, and, in the case of deep foundation pile-supported structures, the ability of the deep foundation to survive the displacements and curvatures imposed on the pile by seismic ground motion.

Sufficient understanding of the behavior and limitations of the various deep and shallow foundations is necessary to determine that the foundation and the supported structure will provide the intended serviceability. It is important to determine whether the estimated total and differential settlements of the foundation are compatible with the selected structure type. For example, a stiff bearing wall structure with openings may be more sensitive to differential settlements than a more flexible light-frame structure. When considering seismic ground motion, sufficient knowledge of ground shaking effects on the foundation is important, particularly in soft soils.

Section 1801 *General*

1801.1 **Scope.** The provisions apply to foundations not subject to scour or water pressure by wind or wave action. For example, these provisions would not be applicable to a beach-front structure that would be subjected to wave runup and inundation during hurricanes. See Section 1612 for flood loads.

1801.2 **Design.** Bearing pressures, stresses and lateral pressures used in this chapter are allowable pressures or stresses, not strength level values. These allowable foundation pressures are to be used with the allowable stress design load combinations set forth in Section 1605.3 unless noted otherwise.

1801.2.1 **Foundation design for seismic overturning.** This section was added to clarify that when strength design loads are used to proportion the foundations, the seismic overturning effects are permitted to be reduced in accordance with ASCE 7 Section 12.13.4. This maximum recognizes that the seismic forces determined in accordance with the ASCE 7 standard are based on strength design, not allowable stress design (ASD). Foundations proportioned in accordance with ASD procedures have historically performed satisfactorily. Because of expected deviation from the results from the equivalent lateral force method, which assumes a fixed base of the building, overturning effects at the foundation are permitted to be reduced 25 percent for structures other than inverted pendulum or cantilever column systems when designed by the equivalent lateral force procedure. Overturning effects at the foundation are permitted to be reduced 10 percent for structures designed by the modal analysis method because of the higher degree of accuracy of the procedure. Note that these reductions cannot be used with the alternative basic ASD load combinations of Section 1605.3.2

Section 1802 *Foundation and Soils Investigations*

The reorganization of this section consolidates all the conditions for which a soils investigation and report is necessary into Section 1802.2. The information necessary to be included in a soils report is in Section 1802.6.

General. A foundation and soils investigation must be conducted when required by the building official. In general, the investigation should be required unless the foundation is designed and constructed in accordance with the presumptive allowable foundation pressures and lateral bearing pressures set forth in Section 1804.2. Some minimal knowledge of soil classification at the bearing elevation of the foundation is required by Section 1804.2. A registered design professional should be used as required by the professional practice laws of the state in which the jurisdiction lies. The practice of geotechnical or soils engineering is a branch of civil engineering and is generally regulated by the various states. **1802.1**

Where required. A foundation and soils investigation is required for any of the adverse subsurface conditions listed in Sections 1802.2.1 through 1802.2.7. **1802.2**

In Seismic Design categories A and B, the exception allows the building official some flexibility in requiring a foundation and soils investigation when the soil conditions of the site are already known from other soils reports. For example, if the site is located in an area where there are reasonably uniform and horizontal soil strata and a soils report is available for the adjacent parcels, then a new soils report should not be necessary.

Questionable soil. Where the classification, strength or compressibility is uncertain, or where bearing capacity of soil in excess of the presumptive value is claimed, the code requires the building official to obtain a geotechnical report. The phrase, *safe sustaining power* was revised in the 2006 IBC to be "classification, strength or compressibility" because the term *sustaining power* has no real meaning. Two cases trigger a foundation soils investigation and report: **1802.2.1**

1. Where the design load bearing value is greater than the presumptive allowable foundation pressures and lateral bearing pressures set forth in Section 1804.2.
2. Where the type of soil, the bearing capacity or the stiffness of the soil is questionable, such as in areas subject to liquefaction from strong ground shaking, in areas containing soft or sensitive clays such as bay muds, or in areas with unconsolidated or improperly consolidated fills.

Generally, the first case will be triggered for commercial buildings other than light-frame construction, whereas the second case could be triggered for all buildings.

Expansive soils. Expansive soils are those that shrink and swell appreciably because of changes in soil moisture content. See Section 1802.3.2 for classification of expansive soils, and Section 1805.8 for mitigation methods for expansive soils. Frost heave is not considered in this section. **1802.2.2**

Groundwater table. This section may cause significant changes to the conventional approach to foundation design and construction for light-frame buildings with subsurface floors, either a basement or a hillside building with a floor cut into the hillside. A foundation and soils investigation is required to show that the ground-water table is at least five feet below the elevation of the lowest floor level unless waterproofing is provided in accordance with Section 1807. **1802.2.3**

1802.2.4 Pile and pier foundations. It is generally not economical or feasible to use pier or pile foundations without a soils investigation and report.

1802.2.5 Rock strata. If a rock stratum is being used for bearing, the characteristics of the layer must be known in sufficient detail to classify the rock per Section 1802.3.

1802.2.6 Seismic Design Category C. For all structures in Seismic Design Category (SDC) C, a foundation and soils investigation is required to evaluate liquefaction, slope stability and surface rupture caused by faulting or lateral spreading. Significant ground motion can occur even in areas with moderate seismic risk. Liquefaction typically occurs at sites with loose sands and high water tables. Surface rupture generally occurs with large magnitude earthquake events. Lateral spreading generally occurs adjacent to waterways where a saturated soil has a free edge.

1802.2.7 Seismic Design Categories D, E and F. In addition to the investigation required for SDC C by Section 1802.2.6, the investigation must evaluate the additional lateral pressures on basement or retaining walls from ground shaking, the detrimental effects of liquefaction or soil strength loss, and mitigation methods. The potential for liquefaction and soil strength loss must be evaluated using the peak acceleration based on a site specific study and including the effects of soil amplification.

Liquefaction causes loss of bearing capacity with resulting large differential or total settlements. Structures with high height to width ratios that have liquefaction occur under a portion of the structure are subject to overturning. Many of the structures damaged in various Japanese earthquakes suffered the consequences of liquefaction.

Soil strength loss is associated with *sensitive* or *quick* clays that are sensitive to remolding effects. These soft clays typically occur in marine (or former marine) environments and are often called *bay muds*. These clays lose significant strength when remolded. Remolding and subsequent strength loss occurs when a pile foundation through these clays is subjected to strong ground shaking. As a consequence, lateral support for the pile is lost.

Mitigation methods for liquefaction include:

1. In-situ densification of the loose sands subject to liquefaction.
2. Use of pile foundations penetrating through the liquefying layers. The pile capacity must be developed through bearing or skin friction in the soils below the liquefying layers.
3. Use of rigid raft foundations that can minimize the effects of settlement caused by loss of bearing capacity.

Mitigation methods for soils susceptible to strength loss such as *sensitive* or *quick* clays include:

1. Replacement of the soft clay.
2. In-situ consolidation of the soft clays by preloading or water removal.
3. Use of pile foundations penetrating through the soft clay layers. The pile capacity must be developed through bearing or skin friction in the soils below the soft clay layers.

If using pile foundations, the effects of the layers of liquefaction or soils susceptible to strength loss on the curvature (and hence, moment) demands on the pile must be investigated. Often these soil-induced curvatures place a much higher moment demand on the piles than would be determined from conventional lateral force *P-y* analyses. The curvatures are significantly increased at the interface between the soft soils and stiffer soils. The curvature demands increase as the ratio of stiffness of the stiff layer to the soft layer increases.

Soil classification. **1802.3**

General. Soils are classified in accordance with the Unified Soil Classification System as set forth in ASTM D 2487. **1802.3.1**

Expansive soils. Expansive soils are those that shrink and swell appreciably in accordance with changes in soil moisture content. Expansive soils are present or prevalent in all 50 states. Soils must meet all four of the criteria to be classified as expansive, not just a high Plasticity or Expansion Index. See Section 1805.8 for design methods to mitigate the effects of expansive soils. **1802.3.2**

The section allows two different ways to identify expansive soils. The first option involves meeting four criteria: 1) Plasticity Index (PI) of 15 or greater determined by ASTM D 4318; 2) more than 10 percent of the soil particles pass a No. 200 sieve determined by ASTM D 422; 3) more than 10 percent of the soil particles are less than 5 micrometers in size as determined by ASTM D 422; and 4) an Expansion Index greater than 20 according to ASTM D 4892. The second option is to determine if the Expansion Index (according to ASTM D 4892) is greater than 20. If the Expansion Index is determined, the first three tests need not be conducted. In other words, Expansion Index > 20 is both a necessary and sufficient condition by itself.

Investigation. The investigation must consider all the evaluations set forth in this section as applicable. Slope stability is a new variable. **1802.4**

Exploratory boring. The investigation must be under the supervision of a registered design professional experienced in soils exploration. **1802.4.1**

Soil boring and sampling. Whenever the allowable bearing capacity is in doubt or a foundation and soils investigation is necessary, exploratory borings, conducted in accordance with generally accepted practices, are necessary to determine the soil characteristics and determine the load-bearing capacity. All borings and sampling should be done under the supervision of the registered design professional. **1802.5**

Reports. The objective of a foundation and soils investigation is to produce all the necessary information for design and construction of the structure foundations. To ensure that the report of the foundation and soils investigation meets this objective and provides enough data to ensure compliance with the code and a safe foundation, the report must contain at a minimum the enumerated items. See also report requirements for compacted fill in Section 1803.4. Other data such as results of consolidation tests, compaction curves, sieve analyses, Attaberg Limits tests, Plasticity and Expansion Index tests, etc., should be included in the report where available. **1802.6**

Section 1803 *Excavation, Grading and Fill*

Some of the requirements relating to excavation, grading and placement of fill were taken from the 1997 UBC, Chapter 33. See also the special inspection requirements in the analysis of Section 1704.7.

Excavations near footings or foundations. The intent of this section is clear—do not remove lateral and subjacent support. For example, the area shown in Figure 18-1 should not be excavated without providing support for the foundation. **1803.1**

Placement of backfill. Backfill should be performed in accordance with an approved soils report. If no soils report is needed, the backfill should be placed in maximum 6-inch layers free from any rocks or cobbles larger than 4 inches and compacted to a minimum 90-percent relative density (Modified Proctor per ASTM D 1557) using the appropriate compaction equipment. The 2003 IBC introduced the use of controlled low-strength material (CLSM) as **1803.2**

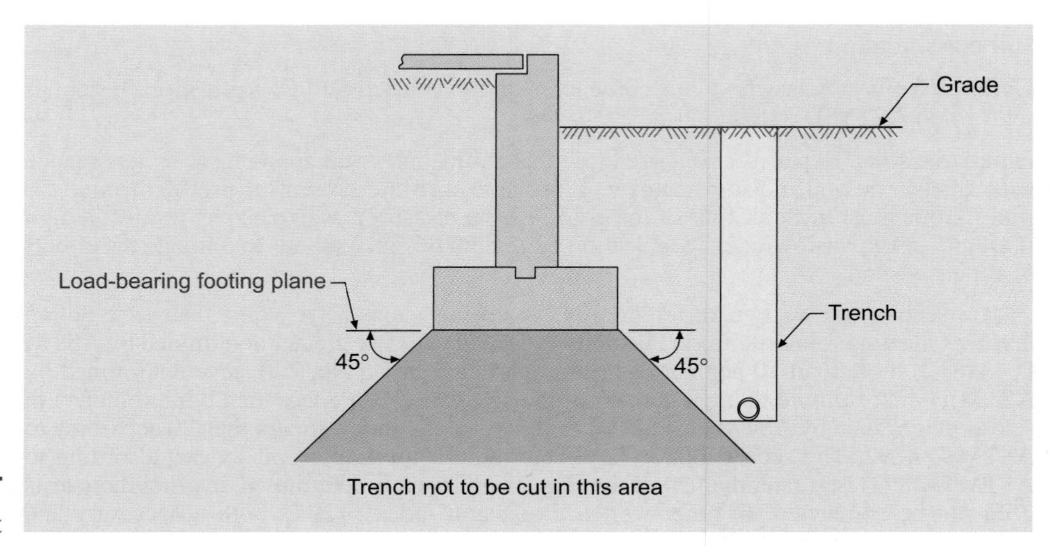

Figure 18-1
Lateral support

an acceptable backfill material that need not be compacted. See further discussion under Section 1803.6.

1803.3 Site grading. The general requirement is that surface water must drain away from foundations. Minimum slope is 5 percent for a distance of 10 feet. A change in the 2006 IBC permits an alternate method if physical obstructions or lot lines prohibit the 5 percent slope for a minimum of 10 feet horizontally. In this case, swales or impervious surfaces must have a minimum 2 percent slope where located within 10 feet of the building foundation.

1803.4 Grading and fill-in flood hazard areas. In general, grading is not permitted in designated flood hazard areas. Changes in the configuration or shape of floodways by grading or fill can divert erosive flows and increase wave energies that could increase forces and adversely affect adjacent buildings and structures. The restrictions in the code are consistent with provisions related to fill in ASCE 24, Flood Resistant Design and Construction. The exceptions permit grading in flood hazard areas, provided an engineering analysis demonstrates that the proposed grading will not increase flood levels or otherwise adversely affect the design flow. The last exception allows grading in flood hazard areas that are not designated floodways, provided the overall effect of the encroachment does not increase the design flood elevation by more than 1 foot at any point.

1803.5 Compacted fill material. When compacted fill is used for foundation support, the foundation and soils investigation report required in Section 1802.6 must also contain the items listed in this section. See also the special inspection requirements in analysis of Section 1704.7.

1803.6 Controlled low-strength material (CLSM). CLSM was introduced into the 2003 IBC as an acceptable backfill material that need not be compacted. Prior to the 2003 IBC, CLSM would need to be approved under the alternative materials, design and methods of construction provisions in Section 104.11. It is a self-compacted, cementitious material used as backfill instead of compacted backfill. It has also been referred to as *flowable fill* and *lean mix backfill*. CLSM has compressive strength of 1200 psi or less. Most CLSM applications require unconfined compressive strengths of 200 psi or less to allow for future excavation of CLSM. CLSM is composed of water, portland cement, aggregate and fly ash. It is a fluid material with typical slumps of 10 inches or more and has a consistency similar to a milk shake. Although there is no referenced standard for CLSM in the IBC, there is a national report promulgated by the ACI Committee 229, entitled "ACI 229R-99 Controlled Low-Strength Materials."

Section 1804 *Allowable Load-bearing Values of Soils*

Design. The presumptive bearing values in the code are allowable stress values, not strength level values. **1804.1**

Presumptive load-bearing values. The presumptive allowable bearing and lateral pressures must be used unless a foundation and soils investigation substantiates higher values. The term *unprepared fill* refers to fill that was not placed and compacted in accordance with an approved soils report. **1804.2**

The format of IBC Table 1804.2 comes from the 1997 UBC. However, some of the unwieldy footnotes from the UBC version have been moved into the text of the code. Each of the model codes used a different approach to generate their allowable values. Hence, there was a disparity between the values shown in the model codes. The tabular values for allowable foundation pressure cannot be increased for width or depth as was allowed by the UBC. Jurisdictions that used the UBC may see larger footings under the IBC 2000 provisions. Note, however, that the allowable foundation pressures are permitted to be increased by one third with the alternative ASD load combinations of Section 1605.3.2 for combinations including wind or earthquake so as to be consistent with previous editions of the model codes.

The classifications in Table 1804.2 are from the Unified Soil Classification System. Most foundation-bearing strata can be classified into one of the classifications in the table. The allowable bearing pressures and lateral bearing values are based on long experience with the behavior of these materials. However, the presumptive value for CL and CH may not be conservative for *soft* clays, depending on the degree of consolidation of the clay. The selection of an allowable bearing pressure should take into account the strength of weaker underlying soil strata so that the pressure in any weaker stratum does not exceed the allowable pressure, particularly in clay soils. Because of this, it is important to know the soil profile and classifications of the different strata.

Lateral sliding resistance. The classifications for lateral bearing in Table 1804.2 are from the Unified Soil Classification System. The lateral bearing values are based on long experience with the behavior of these materials. The limitation on frictional resistance for silts and clays is intended to provide structural stability and improve serviceability. **1804.3**

Increases in allowable lateral sliding resistance. The formulae for lateral bearings (found in IBC Section 1805.7) were originally for outdoor advertising structures. For these structures, deflections of $^1/_2$ inch at the surface do not affect serviceability. Thus, the allowance of two times the tabular value for these structures is permitted. **1804.3.1**

Section 1805 *Footings and Foundations*

This section contains design requirements for footings and foundations, including minimum footing size and depth, distance of footings from slopes, prescriptive foundation (basement) wall designs, foundations employing lateral bearing (pole foundations), and design procedures for expansive soils.

General. In order to minimize differential settlement, footings must be constructed on undisturbed native soil, compacted fill material or controlled low-strength material (CLSM). Where constructed on fill, the material must be properly placed and compacted to achieve adequate density in accordance with Section 1803.5. CLSM must be placed and tested in accordance with Section 1803.6. **1805.1**

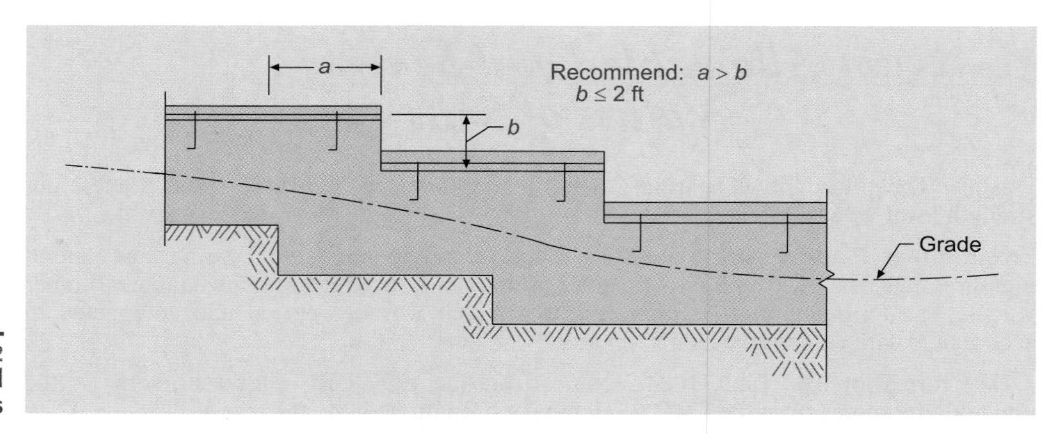

Figure 18-2 Stepped foundations

Footings are required to be stepped when the slope of the bearing surface exceeds 1 in 10. No recommendations or restrictions are provided. Figure 18-2 schematically represents a satisfactory stepped foundation. The figure shows a recommended horizontal overlap of the top of the foundation wall beyond the step in the foundation to be larger than the vertical step in the foundation wall at that point. This is recommended to keep any crack propagation approximately at a 45-degree angle. To keep this cracking to a minimum, it is also recommended that the height of each step not exceed 1 to 2 feet. Other measures to protect against cracking, such as special reinforcing details, may also be needed.

1805.2 Depth of footings. Footings should always be placed a minimum of 12 inches deep on either firm undisturbed earth or properly compacted fill.

1805.2.1 Frost protection. To prevent frost heave during winter and subsequent settlement upon thawing, footings should be placed on a stratum with adequate load-bearing resistance that is below the frost line. Frost heave occurs because of the increased soil volume from the freezing of pore water in the soil. Clay soils, particularly saturated clays, are most susceptible to frost heave. Well-drained sands and gravels will not be susceptible to significant movement. If the foundations are built on soils that can freeze, the resulting frost heave, which is rarely uniform, can cause serious damage from differential settlements.

The frost line is defined as the lowest depth below the ground surface to which a temperature of 32°F extends. The factors governing the depth of the frost line are air temperature, the length of time the air temperature is below freezing (32°F), and the soil's thermal conductivity. Frost lines vary significantly throughout the country from no penetration in southern Florida to 100 inches in the northern regions of Michigan and Maine. Data on frost penetration is available from the U.S. Department of Commerce Weather Bureau. See Figure 18-3.

The code offers three options for ensuring adequate frost protection for foundations. The first option—the most common and simplest to accomplish—is to construct the bottom of the footing below the frost line for the particular locality. The second option is to construct the footing in accordance with the referenced standard, ASCE 32, *Design and Construction of Frost Protected Shallow Foundations*. The third option, which is often encountered in areas where bedrock is prevalent, is to construct the footing on solid rock.

Note the exception where frost protected foundation is not required: free-standing buildings classified in Occupancy Category I, floor area of 600 square feet or less for light-frame construction or 400 square feet or less for other than light-frame construction, and eave height of 10 feet or less. Note that the term *light-frame construction* is defined in Section 201 as a system that uses repetitive wood or light-gage steel framing members.

The code prohibits footings from bearing directly on frozen soil unless the soil is permanently frozen. Permafrost may not meet this condition insofar as permafrost is considered soil that remains in a frozen state for more than two years in a row.

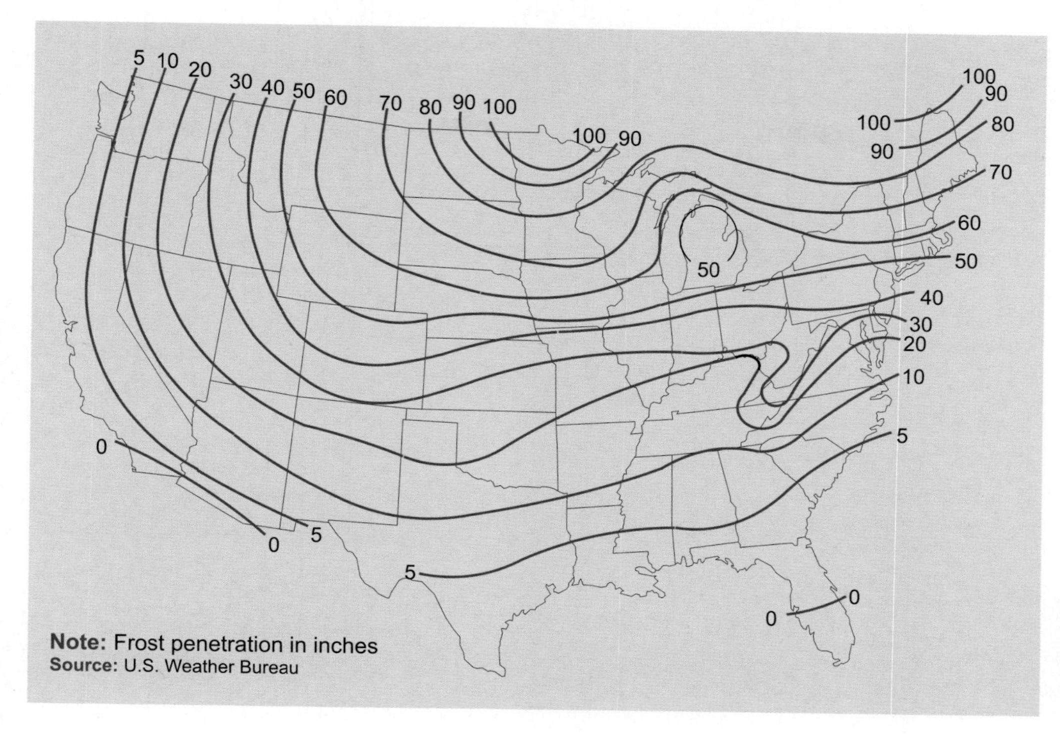

Figure 18-3 **Frost penetration depths**

Isolated footings. This restriction is intended to minimize the influence of vertical and lateral loads from footings at a higher elevation on footings at a lower elevation. See Figure 18-4. **1805.2.2**

Shifting or moving soils. For example, loose sands. **1805.2.3**

Footings on or adjacent to slopes. The provisions of this section apply only to buildings placed on or adjacent to slopes steeper than 1 vertical to 3 horizontal. **1805.3**

Building clearance from ascending slopes. This setback requirement is intended to provide protection to the structure from shallow slope failure (sloughing) and protection for erosion and slope drainage. The setback space also provides access around the structure and helps to create a light and open environment. IBC Figure 1805.3.1 depicts the criterion for the setback or clearance. Figure 18-5 also depicts the criteria set forth in this item for determination of the toe of the slope when the slope exceeds 1:1. **1805.3.1**

Footing setback from descending slope surface. The setback requirement at the top of slopes is intended to provide vertical and lateral support for the foundations and minimize the possibility of shallow bearing failure of the foundation because of lack of lateral support. The setback also provides space for drainage away from the slope without creating too steep a drainage profile, which could cause erosion problems. The setback space also provides access around the structure and helps to create a light and open environment. IBC Figure 1805.3.1 depicts the criterion for the setback or clearance. Figure 18-6 herein depicts the criteria set forth in this item for determination of the toe of the slope when the slope exceeds 1:1. **1805.3.2**

It is possible to locate a structure closer to the slope than indicated in IBC Figure 1805.3.1. The footing of the structure may be located on the slope itself, provided that the depth of embedment of the footing is such that the face of the footing at the bearing plane is set back from the edge of the slope at least H/3.

Pools. Figure 18-7 depicts the criteria for the design of swimming pool walls near the top of a descending slope. The wall must be sufficient to resist the hydrostatic water pressure without support from the soil to protect against failure of the pool wall should localized **1805.3.3**

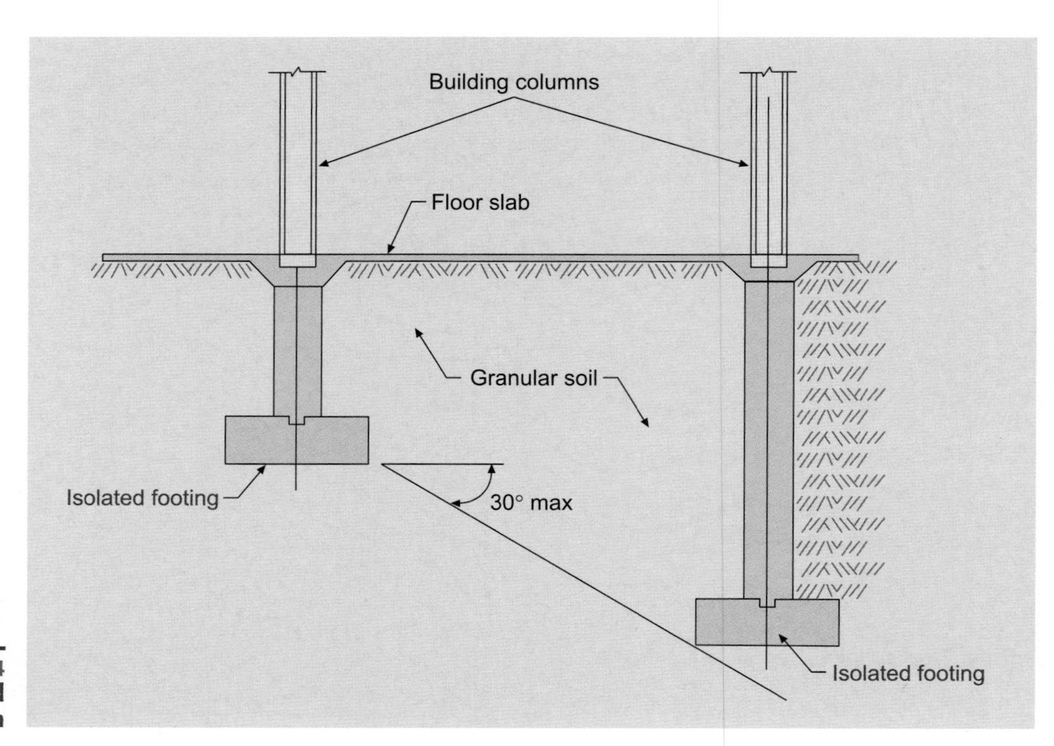

Figure 18-4
Isolated foundation

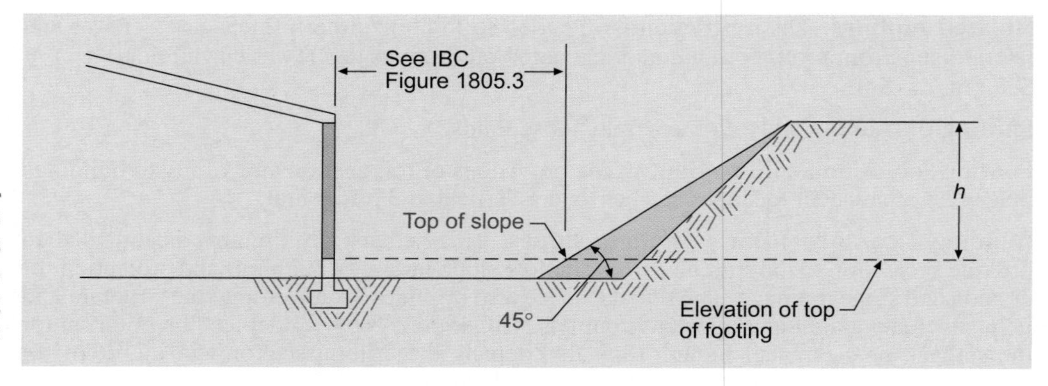

Figure 18-5
Buildings adjacent to ascending slope exceeding 1:1

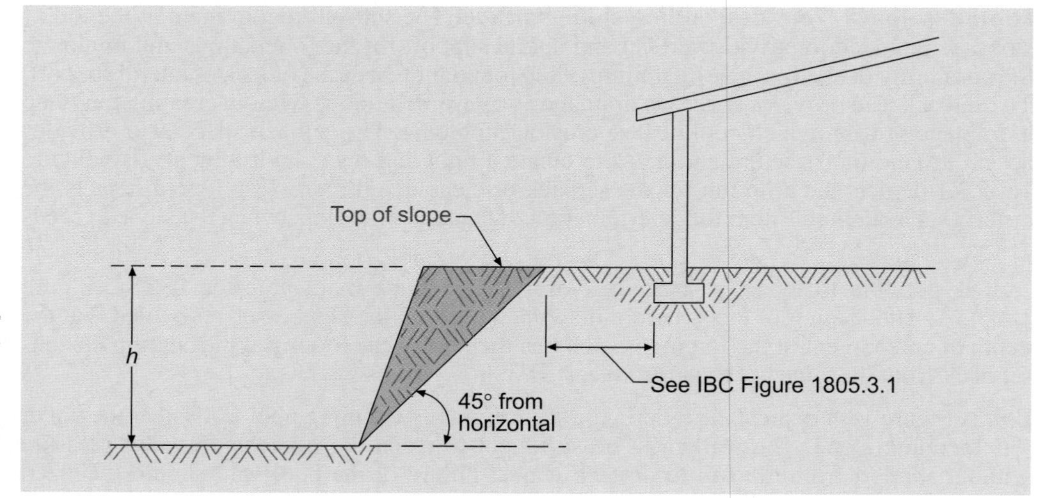

Figure 18-6
Buildings adjacent to descending slope exceeding 1:1

minor slope movement or sloughing occur. The pool setback should be established as one-half of the setback required by IBC Figure1805.3.1.

Foundation elevation. Figure 18-8 depicts the criteria from this section for the elevation of the exterior foundations relative to the street, gutter or point of inlet of a drainage device. The elevation of the street or gutter shown is that point at which drainage from the site reaches the street or gutter. **1805.3.4**

This requirement is intended to protect the structure from water encroachment in the case of heavy or unprecedented rains. This requirement may be modified if the building official finds that positive drainage slopes are provided to drain water away from the building and that the drainage pattern is not subject to temporary flooding from clogged drains, landscaping or other impediments.

Alternate setback and clearance. This alternate procedure allows the building official to approve alternate setbacks and clearances from slopes, provided that the intent of Section 1805.3 is met. This section gives the building official the authority to require a foundation and soils investigation by a qualified geotechnical engineer to establish that the intent of Section 1805.3 is met and specifies the minimum parameters to be investigated. **1805.3.5**

Footings. This section contains the design requirements for footings and foundations, including minimum footing size and depth. **1805.4**

Design. Footings should be designed for approximately equal settlements to minimize differential settlements. For footings on sands, this may require unequal footing pressures to affect equal settlements. For example, see Terzaghi, et al.[1] **1805.4.1**

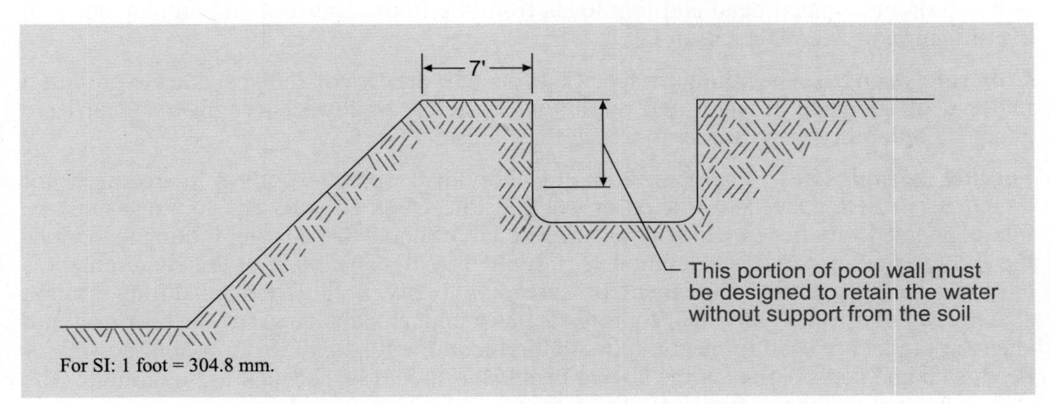

Figure 18-7 Swimming pool adjacent to descending slope

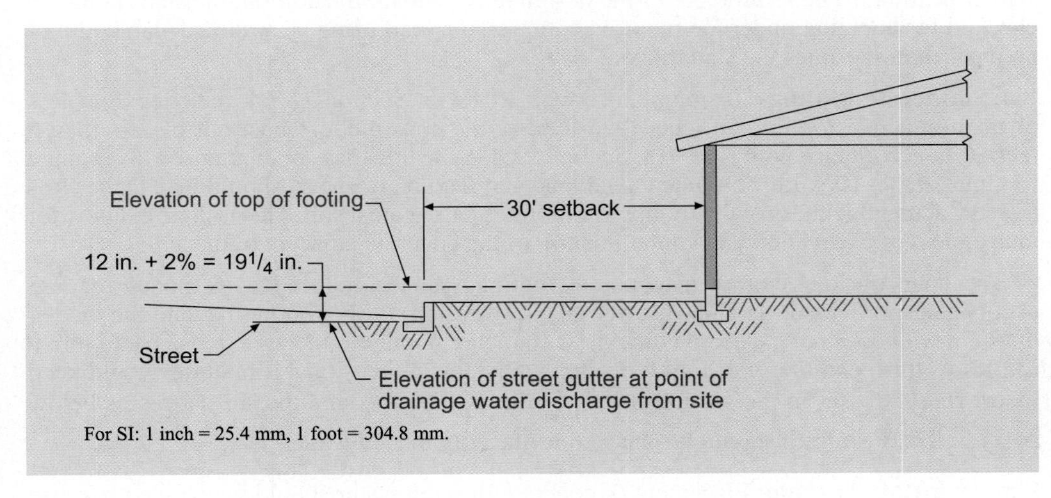

Figure 18-8 Footing elevation on graded sites

1805.4.1.1 Design loads. Footings are to be designed using full dead load (including overlying fill materials), floor and roof live loads, snow load, wind or seismic forces, and any other loads required by Section 1605 that will produce the most severe loading. Live loads acting at the foundation may be reduced based on the reduced probability of simultaneous occurrence of maximum live loads. The topic of reduction of live loads for footing design has not been previously addressed in some model codes. However, this section specifically permits live load reduction as specified in Sections 1607.9 and 1607.11 for the foundation design.

1805.4.1.2 Vibratory loads. Footings supporting equipment should be designed to minimize the transmission of vibratory loads to the soils. The dynamic interaction of the footing, equipment and soil mass should be analyzed, and the footing *tuned* to minimize the transmission. As a rough rule of thumb, footings for rotating or reciprocating equipment should have a mass that is at least four times the mass of the equipment.

Vibratory loads from equipment foundations that are transmitted to the soil can cause significant and damaging settlements. The transmitted vibration will cause densification of granular materials, particularly loose or medium dense sands. The reduction in volume can cause large settlements depending on the initial density of the sands. In saturated granular materials, such as loose or medium dense sands with a high water table, the transmitted vibrations can cause a buildup or pore pressure and liquefaction of the sands, with resulting loss of bearing capacity and settlements. In saturated clays, the vibrations can enhance the drainage of water from the pores and increase long-term settlements.

1805.4.2 Concrete footings. Footings may be designed, or the requirements of Table 1805.4.2 may be used, for structures with light-framed walls of conventional construction, where frost heave or expansive soils are not a problem. Table 1805.4.2, which originated with the UBC, is based on anticipated dead and live loads from the floors and roof and an assumed soil classification of ML, MH, CL or CH.

1805.4.2.1 Concrete strength. The minimum specified concrete strength of 2500 psi is set to provide a material of adequate strength and durability. Concrete of lower strength may not have adequate durability, particularly in freeze-thaw areas.

1805.4.2.2 Footing seismic ties. Interconnection of individual spread footings is required for structures in SDC D, E and F sited on soils in Site Class E or F. The footings must be interconnected with ties capable of transmitting a force equal to the larger footing load times the short period response acceleration, S_{DS}, divided by 10. The intent of this requirement is to minimize differential movement or spreading between the footings during ground shaking, and have the individual footings act as a unit. If slabs on grade, or beams within slabs on grade, are used to meet the tie requirement, the load path from footing to slab or beam/slab and across joints in the slab or beam/slab should be checked for continuity. The slab or beam must be reinforced for the design tension load. In addition, the slab should be checked for buckling under the required compression load using an assumed slab width of no more than six times the slab thickness.

1805.4.2.3 Plain concrete footings. In compliance with ACI 318, Section 22.7.4, the edge thickness of plain concrete footings in other than light-frame construction must not be less than 8 inches. In accordance with ACI 318, Section 22.4.8, the thickness or depth used to compute footing stresses (flexure, combined axial load and flexure, or shear) should be 2 inches less than the actual thickness of the footing for footings cast against soil. This is done to allow for unevenness of excavation and contamination of the concrete adjacent to the soil.

The edge thickness of plain concrete footings can be reduced to 6 inches for R-3 occupancies, provided that the edge distance (projection) of the footing beyond the face of the stem wall does not exceed the thickness (6 inches depth = 6 inches extension). Figure 9 illustrates this condition. For lightly loaded walls, this dimensional limitation should keep the flexural stresses in the footing below the limit of $5\phi\sqrt{f'_c}$ and the shear stresses below $2\phi\sqrt{f'_c}$. These stresses should be checked for heavily loaded walls.

1805.4.2.4 Placement of concrete. Placement of concrete through water should be avoided because of the increased risk of segregation and dilution of the concrete paste. When concrete is placed under water, by tremie or other approved method, the mix must be different from the standard mix used for ordinary concrete foundations. The mix must be proportioned so that it is plastic with high workability and will flow without segregation. The desired consistency can be obtained by using rounded aggregates, high sand contents, entrained air, and superplasticizers. Higher cement contents are necessary to compensate for the increase in the water/cementitious materials ratio caused by dilution from placement through water. Minimum cement content should be 600 pounds per yard.

Protection of concrete. Concrete footings should not be placed during rain, sleet, snow or freezing weather without protection against freezing or without protection against increase in water content at the surface from rain while plastic. See ACI 306R and ACI 306.1 for cold weather concrete operations. **1805.4.2.5**

Forming of concrete. The soil should have sufficient strength and cohesion that the shape, dimensions and vertical sides of the excavation can be maintained without sloughing prior to and during the concreting operations. Excavations in loose granular materials must be formed. **1805.4.2.6**

Masonry unit footing. Masonry footings were widely used until the middle of the last century when they were replaced by steel or wood grillage footings, which in turn were replaced by more economical plain or reinforced concrete footings. Masonry footings were often built of stone cut to a specific size or rubble masonry of random size stones bonded with mortar. Although seldom used, masonry footings may be constructed of hard-burned brick set in cement mortar to support light-weight buildings. **1805.4.3**

Dimensions. Type M mortar is suitable for unreinforced masonry below grade or with earth contact. Type S should be used for reinforced footings. Projections of the footing beyond a wall or pier should not exceed one half of the footing depth to keep the shear and flexural stresses in the footing within safe limits. For example, a footing with a 12-inch depth should project no more than 6 inches beyond the face of the wall. **1805.4.3.1**

Offsets. The stepping back, or racking, of successive courses of the foundation wall supported by a masonry footing must not exceed $1^1/_2$ inches for a single course or 3 inches for a double course. Where wide footings are necessary for bearing, the wall must be stepped back to keep the footing projection within the limits of Section1805.4.3.1. See Figure 18-9. **1805.4.3.2**

Steel grillage footings. Steel grillage footings were used extensively during the latter half of the last century, but the development of reinforced concrete made grillage footings obsolete, except for underpining work. **1805.4.4**

A typical grillage consists of two or more tiers of steel beams, with each tier placed at right angles to the tier below. The beams in each tier are usually held together by a system of

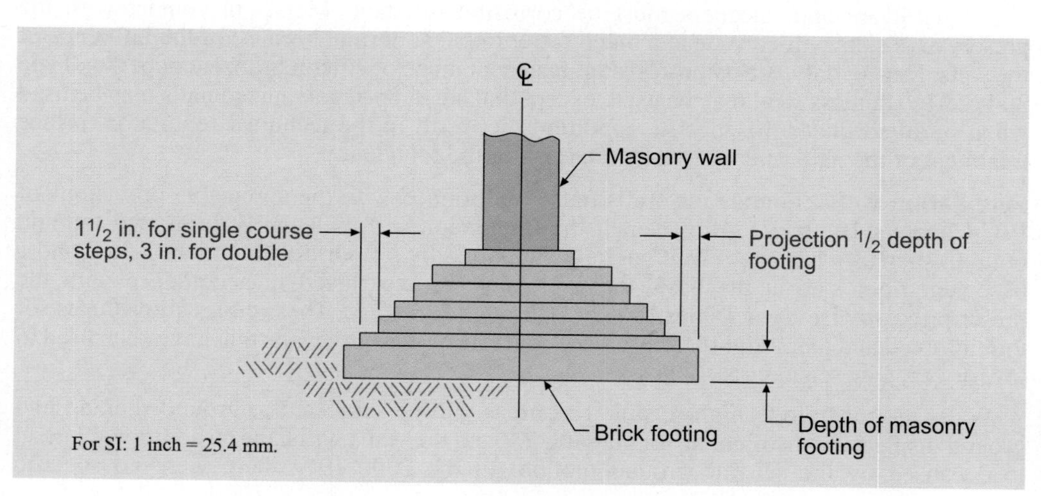

Figure 18-9 Brick footing wall offsets

bolts and pipe spacers. For construction of new grillage footings, the beams should be clean and unpainted and the entire grillage system filled with and encased in concrete with at least three inches of cover. The grillage should be placed on a concrete pad at least 6 inches thick to distribute the load evenly to the soil.

1805.4.5 Timber footings. Use of timber footings is allowed only for Type V structures. Timber must be pressure treated to American Wood Preservers Association (AWPA) U1 (Commodity Specification A, Use Category 4B) standard, except for foundations permanently below the groundwater table. The pressure preservative treatment protects the timber from decay, fungi and harmful insects. The AWPA Use Category System is based on the end use hazard, similar to other international standards for wood treatment. The Use Category System (UCS) is used to specify the wood treatment based on the desired wood species and the environment of the intended end use. There are six Use Categories, which describe the exposure conditions that wood may be subject to in service. ICC in partnership with AWPA publishes all 24 AWPA standards that are referenced in the 2006 IBC. Stronger preservatives are necessary to prevent marine borers when timber foundations are used in coastal brackish or marine environments.

Preservative treatment by the pressure process within the limitations of the AWPA standards should not significantly affect the strength of the wood. Part of the process, however, involves the conditioning of the wood prior to treatment by steaming or boiling under vacuum. This conditioning can cause reduction in strength. This strength loss is recognized in the AFPA *National Design Specification for Wood Construction* by use of the untreated factor, C_u, which provides an increase in the tabular design values for untreated timber poles and piles. Note that the NDS states that load duration factors greater than 1.6 are not allowed for structural members that are pressure treated with water-borne preservatives. This restriction would apply to impact loads that have a duration factor of 2.0.

Untreated timber may be used when the footings are completely embedded in soil below the groundwater table. Long experience has shown that timber permanently confined in water will stay sound and durable indefinitely. Wood submerged in fresh water cannot decay, because the necessary air is excluded. Because ground-water levels can sometimes change appreciably, untreated timber should only be used at depths sufficiently below the water table so that small drops in the water level will not expose the timbers to air.

1805.4.6 Wood foundations. The requirements set forth in AFPA Technical Report No. 7 must be rigidly followed. The wood foundation system is an assembly similar to a fire assembly—no substitution of materials or methods is allowed. All lumber and plywood must be treated in accordance with AWPA U1 (Commodity Specification A, Use Category 4B, Section 5.2) and must be identified and labeled in accordance with Section 2303.1.8.1. ICC in partnership with AWPA publishes all 24 AWPA standards that are referenced in the 2006 IBC.

All hardware and fasteners must be corrosion resistant. Metals in contact with the preservative salts will corrode at a much faster rate that normal because of the influence of the salts. Hence, only corrosion-resistant fasteners made of silicon bronze, copper, or Type 304 or 316 stainless steel may be used, except that hot-dipped galvanized nails may be used when installed under the specific conditions set forth in the technical report for surface treatment of the nails and moisture protection of the foundation.

1805.5.5 Foundation walls. Foundation walls must be designed with the applicable provisions of IBC Chapters 19 and 21. However, if the foundation wall is laterally supported top and bottom, such as a basement wall laterally supported by a floor diaphragm at the top and a basement floor slab at the base, the wall may be constructed in accordance with the prescriptive provisions of Tables 1805.5(1) through 1805.5(5). These tables allow the use of unreinforced and plain (lightly reinforced) concrete or masonry walls that have been used in low or very low seismic risk areas.

Walls in moderate to high seismic risk areas will be subjected to ground shaking and ground displacements of unknown magnitudes, and the walls will have an additional lateral load caused by the seismic ground motion. In the 2000 IBC, there were no specific requirements based on regional seismic considerations. The tables did not take into consideration additional lateral seismic soil pressures or the additional lateral inertial loads from the floor diaphragm and superstructure. Hence, these prescriptive provisions were thought by some to be inadequate for use in structures classified in SDC C, D, E and F. The prescriptive foundation wall tables were of particular concern in the western states where earthquakes are relatively frequent and destructive. Code changes were made to the 2003 IBC that were designed to address these concerns by the addition of a new Section 1805.5.5, which covers seismic requirements based on seismic design category. Specific seismic requirements for concrete and masonry foundation walls are covered in Sections 1805.5.5.1 and 1805.5.5.2, based on the seismic design category of the building.

Additionally, if the prescriptive provisions are used, sufficient soil investigation should be done to properly classify the retained soils as indicated in the tables in accordance with the Unified Soil Classification Method (see IBC Section 1802.3). The construction materials must at a minimum comply with IBC Section 1805.5.2.

1805.5.1

Foundation wall thickness. The minimum wall thicknesses are specified in the appropriate sections, based on the thickness of the supported wall, soil loads, unbalanced backfill height and overall height of the wall. Rubblestone walls cannot be less than 16 inches thick where permitted.

1805.5.1.1

Thickness based on walls supported. These minimum thickness provisions are to facilitate support of the wall above. These thickness provisions are empirical and have been used successfully in low or very low seismic risk areas. Additional seismic requirements for concrete and masonry foundation walls are covered in Sections 1805.5.5.1 and 1805.5.5.2, based on the seismic design category of the building. See Section 2104.2 if corbeling is necessary or desired to match the width of a masonry cavity wall above the foundation wall.

1805.5.1.2

Thickness based on soil loads, unbalanced backfill height and wall height. Wall thickness must comply with the requirements of Tables 1805.5(1) through 1805.5(5). Note the requirement for Type M or S mortar, and the definition of unbalanced fill. Note that masonry units must be laid in running bond.

1805.5.1.3

Rubble stone. Because rubble masonry uses rough stones of irregular shape and size, a larger thickness, as compared to hollow unit masonry or concrete, is required for adequate bonding of the stone and mortar. Rubble stone foundation walls are not permitted in SDC C, D, E or F.

1805.5.2

Foundation wall materials. This section specifies the material requirements for walls constructed in accordance with the prescriptive tables. Note the effective depth, *d*, in the title of the table. Placement of reinforcing at the prescribed *d* is critical to develop adequate flexural strength necessary to resist the combined vertical and lateral soil loads.

In the 2000 IBC the foundation wall material requirements consisted of five items that applied to both concrete and masonry. In the 2006 IBC the section was subdivided into Section 1805.5.2.1 with six specific requirements for concrete walls, and Section 1805.5.2.2 with seven specific requirements for masonry walls. Concern has been expressed that the prescriptive foundation wall provisions do not impose a limitation on the maximum axial loads that the walls should support. To resolve this concern, a conservative maximum unfactored axial load of $1.2tf'_c$ for concrete and $1.2tf'_c$ for masonry were added. The maximum unfactored axial load is based on a compressive stress on the outside face of the wall that is due to the axial load and bending moment induced by the backfill that is well below that permitted by ACI 318 or ACI 530/ASCE 5/TMS 402. Although this axial load limitation has merit, it requires a calculation to determine actual maximum axial load acting on a given wall. Tables 1 and 2 show the maximum unfactored allowable axial load for the typical construction materials.

Table 18-1. Maximum permissible axial load for concrete walls based on $1.2tf'_c$ in pounds per foot of wall

Wall Thickness (inches)	f'_c = 2000 psi	f'_c = 3000 psi
7.5	22,500	27,000
9.5	28,000	34,200
11.5	34,500	41,400

Table 18-2. Maximum permissible axial load for masonry walls based on $1.2tf'_c$ in pounds per foot of wall

Wall Thickness (inches)	f'_m = 1500 psi	f'_m = 2000 psi
7.625	13,725	18,300
9.625	17,325	23,100
10.625	19,125	25,500
11.625	20,925	27,900

Note that in the 2006 IBC the prescriptive foundation wall tables were revised so that the first four tables cover masonry foundation walls and the last table covers concrete foundation walls.

1805.5.3 Alternative foundation wall reinforcement. The code permits equivalent cross section of reinforcing, provided the spacing does not exceed 72 inches and the bar size does not exceed #11. However, a critical consideration for basement walls is crack control. If the reinforcing is spaced too widely, crack widths will be larger than if the reinforcing is more closely spaced. See ACI 318, Section 10.6, for a discussion of reinforcing spacing for crack control for one-way slabs. If alternative reinforcement is used, it is preferable to reduce bar size and spacing rather than increase bar size and spacing. There is the additional consideration of development of the reinforcing. The bar size must be small enough that the reinforcing, and any splices, can be adequately developed. Development refers to the embedment of the reinforcing to adequately develop the bond between the reinforcing and concrete. A good rule of thumb to prevent splitting of concrete masonry is that the bar size number should not exceed t-1 where t is the nominal thickness of the wall in inches.

1805.5.4 Hollow masonry walls. The masonry is required to be solid in order to distribute the concentrated force into the hollow unit masonry.

1805.5.5 Seismic requirements. As indicated above, the 2000 IBC had no specific seismic requirements for the prescriptive foundation wall provisions. This was of particular concern in the western states where earthquakes are relatively frequent and destructive. In the 2003 IBC, these concerns were addressed by adding specific seismic related requirements in a new Section 1805.5.5, which covers seismic requirements based on seismic design category. Specific seismic requirements for concrete and masonry foundation walls are covered in Sections 1805.5.5.1 and 1805.5.5.2, based on the seismic design category of the building. These requirements are summarized below and in Table 3.

1. Seismic requirements for concrete foundation walls constructed in accordance with Table 1805.5(5) are as follows:

   1.1. Seismic Design Category A and B – Two #5 bars are required around window and door openings, which must extend at least 24 inches beyond the corners of the openings.

   1.2. Seismic Design Category C, D, E and F – The prescriptive tables are not allowed to be used except as permitted for plain concrete members in accordance with Section 1908.1.5, which modifies ACI 318, Section 22.10. The modification states that structural plain concrete members are not permitted in Seismic Design Category C, D, E or F except for structural plain concrete basement, foundation or other walls below the base in detached one- and two-family dwellings three stories or less in height constructed with stud-bearing walls. Additional restrictions apply to dwellings in Seismic Design Category D or E, where the walls cannot exceed 8 feet in height, cannot be less than 7.5 inches thick, and can retain no more than 4 feet of unbalanced fill. The last requirement states that the walls must be reinforced in accordance with ACI Section 22.6.6.5.

2. Seismic requirements for masonry foundation walls constructed in accordance with Tables 1805.5(1) through 1805.5(4) are as follows:

2.1. Seismic Design Category A and B – no additional requirements apply.

2.2. Seismic Design Category C – additional requirements are outlined in Section 2106.4, which covers discontinuous members that are part of the lateral-force-resisting system, such as columns, pilasters and beams that support reactions from walls or frames, but no specific requirements for foundation walls. Refer to Section 1.14.5 of ACI 530/ASCE 5/TMS 402 for other requirements.

2.3. Seismic Design Category D – must conform to the requirements of Seismic Design Category C, as well as Section 2106.5. If the masonry foundation wall also serves as a shear wall that resists seismic forces, the requirements of Section 2106.5 must be complied with.

2.4. Seismic Design Category E and F – must conform to the requirements of Seismic Design Category C and D, as well as Section 2106.6, which requires conformance with Section 1.14.7 of ACI 530/ASCE 5/TMS 402.

Table 18-3. Seismic requirements for masonry foundation walls

Seismic Design Category	IBC Section	ACI 530/ASCE 5/TMS 402
C	2106.4	1.14.5
D	2106.5	1.14.6
E, F	2106.6	1.14.7

Foundation wall drainage. If a drainage system is not placed behind the wall to drain the ground water away from the wall, hydrostatic pressures, which can easily exceed the lateral pressures from the retained soil, will occur. For example, the active lateral pressure from a well-graded granular soil may be in the range of 30 to 35 pounds per square foot, whereas the hydrostatic pressure is 62.4 pounds per cubic foot. Hence, the equivalent fluid pressures on a wall that was designed as drained, but constructed without an effective drainage system, could be subjected to pressures approximately three times the design pressure. **1805.5.6**

Pier and curtain wall foundations. Pier foundations may not be used to support structures assigned to SDC E and F. **1805.5.7**

Foundation plate or sill bolting. The requirements for the bolting of foundation sill plates were moved to Chapter 23, Wood. This was done to clarify that the prescriptive anchor bolt provisions are applicable only to structures constructed using wood conventional construction provisions (see IBC Section 2308). Anchor bolt designs must be engineered when the conventional construction provisions are not used. **1805.6**

Designs employing lateral bearing. The design criteria for the use of poles embedded in the ground, or in concrete footings in the ground and unconstrained at the ground surface, were developed for the Outdoor Advertising Association of America, Inc. The research was conducted at Purdue University from 1938 to 1940, and continued in 1947 at the University of Notre Dame. The results of this research were used by the association for the design of outdoor advertising structures, which had previously used trussed A-frame supporting systems. Charts and a nomograph were developed, which the association used for the design of poles as cantilever uprights for support of its outdoor advertising structures. These data **1805.7**

were subsequently submitted through the International Conference of Building Officials' (ICBO) code change process and were incorporated into the 1964 edition of the *Uniform Building Code* (UBC).

The criteria relate to lateral bearing and apply to a vertical pole considered a column embedded in either earth or in a concrete footing in the earth and used to resist lateral loads. In order for the pole to meet the conditions of research that resulted in the code formula, the code requires that the backfill in the annular space around a column that is not embedded in a concrete footing be either of 2,000 psi (13.78 Mpa) concrete or of clean sand thoroughly compacted by tamping in layers not more than 8 inches in depth.

The original design criteria established for the Outdoor Advertising Association of America, Inc. resulted in a $^1/_2$-inch lateral pole deformation at the surface of the ground. These criteria were also based on field tests conducted in a range of sandy and gravelly soils and of silts and clays.

The IBC employs allowable lateral bearing stresses in IBC Table 1804.2, which are considerably lower than those developed for the Outdoor Advertising Association of America. Consequently, Section 1804.3.1 permits a doubling of the lateral bearing values for isolated poles and poles supporting structures that can safely tolerate the $^1/_2$-inch movement at the ground surface.

1805.7.1 Limitations. The limitations imposed by this section are intended for both structural stability and serviceability. The limitation of the frictional resistance for silts and clays is consistent with the UBC, which also limits the sliding resistance to one-half the dead load.

The limitations on the types of construction that use the lateral support of poles are based on the brittle nature of the materials. To prevent excessive distortions that would cause the cracking of these brittle materials, the code limits the use of the poles unless some type of rigid cross bracing is provided to limit the deflections to those that can be tolerated by the materials.

Wood poles must be treated in accordance with AWPA U1. Sawn timber posts are Commodity Specification A, Use Category 4B, and round timber posts are Commodity Specification B, Use Category 4B.

1805.7.2 Design criteria. See IBC Section 1804.3 for allowable values of lateral bearing. Note that the two-time increase allowed per Section 1804.3.1 may only be used for structures where deflection of $^1/_2$ inch at the surface is tolerable, e.g., signs, flagpoles and light poles.

1805.7.2.1 Nonconstrained. See Section 1805.7 for the empirical basis of the formula. This formula should be used only for minor foundations of moderate size, which will fit within the constraints of the data from which the formula was developed. For large size piers, i.e., more than 2 feet in diameter, a more appropriate method should be used. See Winterkorn, et al.[2]

1805.7.2.2 Constrained. The term *pavement* means a rigid pavement such as reinforced concrete that will form a fulcrum for the column. Columns in flexible pavements such as asphalt concrete must use the formula in Section 1805.7.2.1 for unconstrained conditions.

1805.7.2.3 Vertical load. There is no requirement to consider a combined lateral and vertical load. The vertical loads for which the formulae were derived were less than 0.1 $F_c A_g$. If there are vertical loads greater than 0.1 $F_c A_g$, these formulae should not be used.

1805.7.3 Backfill. Backfill in accordance with the requirements is necessary to achieve the strength predicted by the formulae. The required backfill was used as part of the research conducted to develop the formulae. Note that the sand should be compacted to a relative density of at least 85 percent.

1805.8 Design for expansive soils. The requirements to mitigate the effects of expansive soils are set forth. In addition to mitigation by foundation design, the effects of expansive soils may also be mitigated by removal of the expansive soils or stabilization by chemical means, presaturation or dewatering. Expansive soils are cohesive soils, typically high plasticity clays, with a high Plasticity Index and a high Swell Index.

Foundations. The large volume changes in expansive soils caused by changes in the soils' water content can cause significant differential deflections in a building if not uniform. In a typical building on expansive soils, the soils at the perimeter of the building will have seasonal moisture changes, whereas the soils at the interior of the building will remain at a fairly constant moisture content. The perimeter foundations will rise and fall with the seasonal volume changes in moisture content, whereas the soil at the interior footings or slab will not have any volume changes, because of a constant moisture content. The resulting differential displacements between the interior and exterior footings can cause significant structural distress. Hence, the requirements that the foundation be designed to resist the differential volume changes and to minimize racking or differential displacements in the structure. **1805.8.1**

Slab-on-ground foundations. The slab-on-ground or raft foundation design methods cited in the section result in a raft that has sufficient stiffness to bridge differential displacements caused by the volume changes in the supporting soil. **1805.8.2**

This section was revised in the 2006 IBC to add clarification to the requirements. Design moments, shears and deflections are to be determined in accordance with WRI/CRSI Design of Slab-on-Ground Foundations or PTI *Standard Requirements for Analysis of Shallow Concrete Foundations on Expansive Soils*. Once the design moments, shears and deflections are determined from the applicable standard, then conventionally reinforced (nonprestressed) foundations on expansive soils must be designed in accordance with WRI/CRSI *Design of Slab-on-Ground Foundations*, and post-tensioned foundations on expansive soils must be designed in accordance with PTI *Standard Requirements for Design of Shallow Post-Tensioned Concrete Foundations on Expansive Soils*.

The exception that exempted engineering analysis for slab-on-ground systems that have performed adequately in similar soil conditions when permitted by the building official was deleted from the 2006 IBC. Instead of the exception, the code permits alternate methods of analysis, provided the methodology is rational and the basis for the analysis and design parameters are available for peer review.

Removal of expansive soil. Removal of the expansive soil is an acceptable mitigation method and is the preferred method if the stratum of expansive soil is near the surface and reasonably thin. This method may also be the least expensive method if the expansive soil is at the surface. **1805.8.3**

Stabilization. Expansive soils may be stabilized so that the moisture content does not change; hence, there will be no volume changes to cause differential displacements. Stabilization can be by chemical methods, by presaturating the soils to a maximum swell and capping the expansive layer to keep the moisture content constant, or by dewatering to a minimum shrinkage and providing drainage to keep the moisture content constant. **1805.8.4**

Seismic requirements. Specific requirements for foundations in Seismic Design Categories C, D, E or F are contained in Section 1908, which contains the modifications to ACI 318. In Seismic Design Categories C, D, E or F, concrete must have a specified compressive strength of not less than 3,000 psi. However, 2,500 psi concrete strength is permitted in Group R or U occupancies of light-framed construction two stories or less in height. **1805.9**

Buildings in Seismic Design Category D, E and F are required to comply with ACI 318 Sections 21.10.1 through 21.10.3 except for detached one- and two-family dwellings of light-frame construction two stories or less in height. ACI 318 Section 21.10 covers foundation requirements in general, and Sections 21.10.2 and 21.10.3 cover requirements for footings, mat foundations, pile caps, grade beams and slabs on grade.

Note that plain concrete is either unreinforced or lightly reinforced concrete that contains less reinforcing than required to meet the minimum reinforcement requirements set forth in ACI 318, Section 10.5.

Section 1806 *Retaining Walls*

In the 2003 IBC, retaining wall requirements were relocated from Chapter 16 to Chapter 18. Although the 1999 BOCA/NBC, 1999 SBC and ASCE 7-98 contained some requirements for retaining walls, they were very limited in scope. The 1997 UBC has more detailed requirements for retaining walls, which are essentially the same as the provisions in the IBC. The IBC requires retaining walls to be designed to resist overturning, sliding and excessive foundation bearing pressure with a safety factor of at least 1.5 against lateral sliding and overturning using allowable stress design loads. See discussion under Section 1610 and Table 1610.1 for soil lateral loads.

Section 1807 *Dampproofing and Waterproofing*

This covers the requirements for waterproofing and dampproofing those parts of substructure construction that need to be provided with moisture protection. Sections 1807.1 through 1807.4 identify the locations where moisture barriers are required and specify the materials to be used and the methods of application. The provisions also deal with subsurface water conditions, drainage systems and other protection requirements.

Dampproofing requirements are outlined in Section 1807.2, and waterproofing requirements are covered in Section 1807.3. Although both terms are intended to apply to the installation and the use of moisture barriers, dampproofing does not furnish the same degree of moisture protection as does waterproofing.

Dampproofing generally refers to the application of one or more coatings of a compound or other materials that are impervious to water, which are used to prevent the passage of water vapor through walls or other building components, and which restrict the flow of water under slight hydrostatic pressure. Waterproofing, on the other hand, refers to the application of coatings and sealing materials to walls or other building components to prevent moisture from penetrating in either a vapor or liquid form, even under conditions of significant hydrostatic pressure. Hydrostatic pressure is created by the presence of water under pressure. This pressure can occur when the ground-water table rises above the bottom of the foundation wall, or the soil next to the foundation wall becomes saturated with water caused by uncontrolled storm water runoff.

1807.1 **Where required.** This section is an overall requirement specifying that waterproofing and dampproofing applications are to be made to horizontal and vertical surfaces of those below-ground spaces where the occupancy would normally be adversely affected by the intrusion of water or moisture. Moisture or water in a floor below grade can cause damage to structural members such as columns, posts or load-bearing walls, as well as pose a health hazard by promoting growth of bacteria or fungi. Moisture can adversely affect any mechanical and electrical appliances that may be located at that level. It can also cause a great deal of damage to goods that may be located or stored in that lower level. These vertical and horizontal surfaces include foundation walls, retaining walls, underfloor spaces and floor slabs. Waterproofing and dampproofing are not required in locations other than residential and institutional occupancies where the omission of moisture barriers would not adversely affect the use of the spaces. An example of a location where waterproofing or dampproofing would not be required is in an open parking structure, provided the structural components are individually protected against the effects of water. Waterproofing and dampproofing are not permitted to be omitted from residential and institutional occupancies where people may be sleeping or services are provided on the floor below grade. A person waking in a flooded basement may find themselves in a very hazardous situation, particularly if the possibility exits of an electrical charge in the water caused by electrical service at that level.

Section 1807.1.1 addresses the type of problem faced when a portion of a story is above grade, whereas Section 1807.1.2 limits any infiltration of water into crawl spaces so as to protect this area from potential water damage and prevent ponding of water in the crawl space. These sections reference other applicable sections of Chapter 18.

1807.1.1

Story above grade. The provisions of this section require that where a basement is deemed to be a *story above grade*, the section of the basement floor that occurs below the exterior ground level and the walls that bound that part of the floor are to be dampproofed in accordance with the requirements of Section 1807.2.

The use of dampproofing, rather than waterproofing, is permitted here because high hydrostatic pressure will not tend to develop against the walls if the basement is a story above grade and the ground level adjacent to the basement wall is below the basement floor elevation for not less than 25 percent of the basement perimeter.

Any water pressure that may occur against the walls below ground or under the basement floor would be relieved by the water drainage system required in this section. The drainage system would be installed at the base of the wall construction in accordance with Section 1807.4.2 for a minimum distance along those portions of the wall perimeter where the basement floor is below ground level.

Because of the relationship of grade to the basement floor and the inclusion of foundations drains, the potential for hydrostatic pressure buildup is not significant. Therefore, a ground-water table investigation, waterproofing and a basement floor gravel base course is not required.

The objective of this section is to prevent moisture migration in basement spaces. In story-above-grade construction that meets the requirements of this section, the basement floor would be only partly below ground level (sometimes a small part) and the need for section-required moisture protection is unnecessary. Dampproofing of the floor slab would be required, however, in accordance with Section 1807.2.1.

1807.1.2

Underfloor space. Essentially, the requirements of this section are designed to prevent any ponding of water in underfloor areas such as crawl spaces. Crawl spaces are particularly susceptible to ponding of water insofar as they are usually uninhabitable spaces that are infrequently observed. Water can build up in these spaces and remain for an extended period of time without being noticed by the building occupants. Stagnant water collected under a building can result in a serious health concern. Water buildup in a crawl space can also damage the structural integrity of the building. Wood exposed to water can deteriorate and rot, and concrete and masonry exposed to water can deteriorate with a loss of strength.

Steel exposed to water or high humidity can eventually rust to the extent that effective structural capability is jeopardized. Water buildup in a crawl space can also damage mechanical or electrical appliances located in the space by causing corrosion of electrical parts or metal skins and deterioration to insulation used to protect heating elements.

Where it is known that the water table can rise to within 6 inches of the outside ground level, or where there is evidence that surface water cannot readily drain from the site, then the finished ground surface in underfloor spaces is to be set at an elevation equal to the outside ground level around the perimeter of the building unless an approved drainage system is provided. For the drainage system to be approved, it must be demonstrated that the system will be adequate to prevent the infiltration of water into the underfloor space. This is done by determining the maximum possible flow of water near the foundation wall and footing and designing the drainage system to remove that flow of water as it occurs, thereby preventing the buildup of water at the foundation wall.

To prevent the ponding of water in the underfloor space from a rise in the ground-water table, or from storm water runoff, the finished ground level of an underfloor space is not to be located below the bottom of the foundation footings.

Dampproofing, waterproofing or providing subsoil drainage is not necessary if the ground level of the underfloor space is as high as the ground level at the outside of the building perimeter, as the foundation walls do not enclose an interior space below grade. Compliance with Sections 1807.2, 1807.3 and 1807.4 would be required where the finished ground surface of the underfloor space is below the outside ground level.

1807.1.2.1 Flood hazard areas. The requirement is to prevent water from ponding under the structure if it is flooded. Under-floor spaces of Group R-3 buildings that meet the requirements of FEMA/FIA-TB-11 need not comply with the requirements in this section.

1807.1.3 Ground-water control. After completion of building construction, it is necessary to maintain the water table at a level that is at least 6 inches below the bottom of the lower floor to prevent the flow or seepage of water into the basement. Where the site consists of well-draining soil and the highest point of the water table occurs naturally at or lower than the required level stated above, then there is no need to provide any kind of a site drainage system specifically designated to control the ground-water level. Where the soil characteristics and the site topography are such that the water table can rise to a level that will produce a hydrostatic pressure against the basement structure, a site drainage system may be installed to reduce the water level. When ground-water control in accordance with this section is provided, waterproofing in accordance with Section 1807.3 is not required.

There are many types of site drainage systems that can be employed to control ground-water levels. The most commonly used systems may involve the installation of drainage ditches or trenches filled with pervious materials, sump pits and discharge pumps, well point systems, drainage wells with deep-well pumps, sand-drain installations, etc. This section requires that all such systems be designed and constructed using accepted engineering principles and practices based on considerations that include the permeability of the soil, amount and rate at which water enters the system, pump capacity, capacity of the disposal area and other such factors that are necessary for the complete design of an effective drainage system.

1807.2 Dampproofing required. Where a ground-water table investigation has established that the high water table will occur at such a level that the building substructure will not be subjected to significant hydrostatic pressure, then dampproofing in accordance with this section and a subsoil drain in accordance with Section 1807.4 are sufficient to control moisture in the floor below grade. Because the wall will not be subject to high hydrostatic pressure, the more restrictive provisions of waterproofing, as provided for in Section 1807.3, are not required. Wood foundation systems specified in Section 1805.4.6 are to be dampproofed as required by AFPA TR7.

1807.2.1 Floors. Floors requiring dampproofing in accordance with Section 1807.2 are to employ materials specified in Section 1807.2.1. The dampproofing materials must be placed between the floor construction and the supporting gravel or stone base as shown in Figure 10. Even if a floor base is not required, dampproofing should be placed under the slab.

The installation is intended to provide a moisture barrier against the passage of water vapor or seepage into below-ground spaces.

The dampproofing material most commonly used for underslab installations consists of a polyethylene film not less than 6 mil in thickness, which is applied over the gravel or stone base required in Section 1807.4.1. Care must be used in the installation of the material over the rough surface of the base and during the concreting operation so as not to puncture the polyethylene. Joints must be lapped at least 6 inches.

Dampproofing materials can also be applied on top of the base concrete slab if a separate floor is provided above the base slab, because the dampproofing is provided to prevent moisture infiltration of the interior space, and not the concrete slab.

Materials commonly used for dampproofing floors are listed in Table 18-4.

1807.2.2 Walls. Walls requiring dampproofing in accordance with Section 1807.2 are first to be prepared as required in Section 1807.2.2.1, then coated with any of the bituminous materials

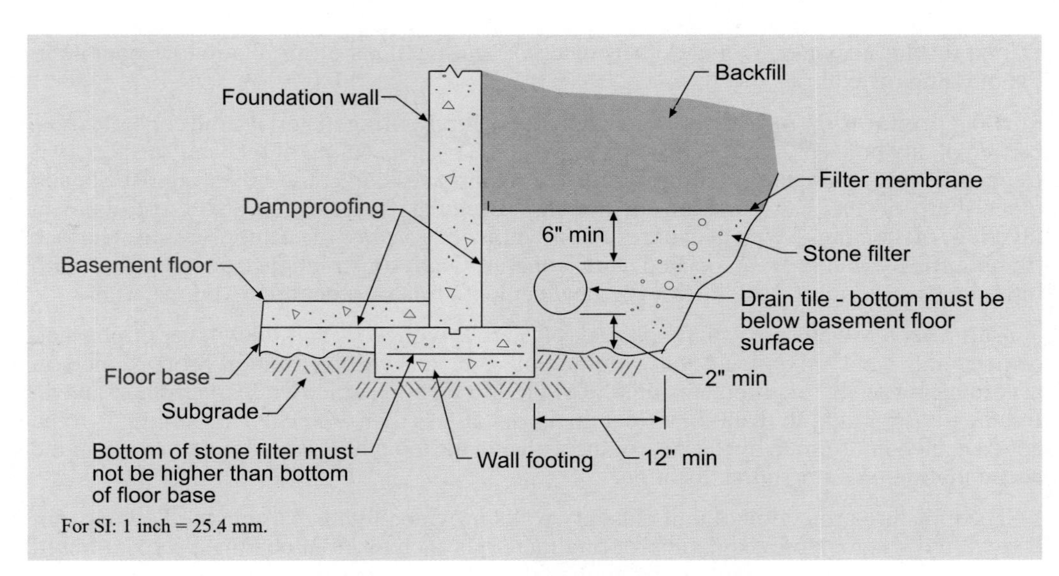

Figure 18-10 A foundation drainage system

listed in Table 18-1 or by other approved materials and methods of application. Approved materials are those that will prevent moisture from penetrating the foundation wall.

Coatings are applied to cover prepared exterior wall surfaces extending from the top of the wall footings to slightly above ground level so that the entire wall that contacts the ground is protected. Surfaces are usually primed to provide a bond coat and then dampproofed with a protection coat of asphalt or tar pitch.

Dampproofing materials for walls may be any of the materials specified in Section 1807.3.2 for waterproofing. Table 18-4 gives a list of bituminous materials that can be used, including the applicable standards that may be used as the basis of acceptance of such materials. Included in Table 18-4 is ASTM D 1668 for glass fabric that is treated with asphalt (Type I), coal-tar pitch (Type II) or organic resin (Type III).

Table 18-4. **Dampproofing materials**

Material	Specification
Asphalt	ASTM D 449
Asphalt primer	ASTM D 41
Coal-tar	ASTM D 450
Concrete and masonry oil primer (for coal-tar applications only)	ASTM D 43
Treated glass fabric	ASTM D 1668

Surface-bonding mortar complying with ASTM C 887 may be utilized. This specification covers the materials, properties and packaging of dry, combined materials for use as surface-bonding mortar with concrete masonry units that have not been prefaced, coated or painted. Because this specification does not address design or application, the manufacturer's recommendations should be followed. This standard covers proportioning, physical requirements, sampling and testing. The minimum thickness of the coating is $^{1}/_{8}$ inch.

Acrylic-modified cement coatings may be utilized at the rate of 3 pounds per square foot. These types of materials have been used successfully as dampproofing materials for foundation walls. Surface-bonding mortar and acrylic-modified cement are limited in use to dampproofing. The ability of these two types of products to bridge nonstructural cracks, as required in Section 1807.3.2 for waterproofing materials, is not known. Therefore, their use is limited to dampproofing and they are not permitted to be used as waterproofing.

Dampproofing may also include other materials and methods of installation acceptable to the building official.

1807.2.2.1 Surface preparation of walls. Before applying dampproofing materials, the concrete must be free of any holes or recesses that could affect the proper sealing of the wall surfaces. Air trapped beneath the dampproofing coating or membranes can cause blistering. Rocks and other sharp objects can puncture membranes. Irregular surfaces can also create uneven layering of coatings, which can result in vulnerable areas of dampproofing. Surface irregularities commonly associated with concrete wall construction can be sealed with bituminous materials or filled with portland cement grout or other approved materials.

Unit masonry walls are usually parged (plastered) with a $^1/_2$-inch-thick layer of portland cement and sand mix (1:2$^1/_2$ by volume) or with a Type M mortar proportioned in accordance with the requirements of ASTM C 270 and applied in two $^1/_4$-inch-thick layers. In no case is parging to result in a final thickness of less than $^3/_8$ inch. The parging is to be coved at the joint formed by the base of the wall and the top of the wall footing to prevent the accumulation of water at that location.

The moisture protection of unit masonry walls provided by the parging method may not be required where approved dampproofing materials such as grout coatings, cement-based paints or bituminous coatings can be applied directly to the masonry surfaces.

1807.3 Waterproofing required. Waterproofing installations are intended to provide moisture barriers against water seepage that may be forced into below-ground spaces by hydrostatic pressure.

Where a ground-water table investigation has established that the high-water table will occur at such a level that the building substructure will be subjected to hydrostatic pressure, and where the water table is not lowered by a water control system, as described in the commentary to Section 1807.4.2, all floors and walls below ground level are to be waterproofed in accordance with Sections 1807.3.1 and 1807.3.2.

1807.3.1 Floors. Because floors required to be waterproofed are subjected to hydrostatic uplift pressures, such floors must, for all practical purposes, be made of concrete and designed and constructed to resist the maximum hydrostatic pressures possible. It is particularly important that the floor slab be properly designed, as severe cracking or movement of the concrete would allow water seepage into below-ground spaces. The ability of the waterproofing materials to bridge cracks is limited. Concrete floor construction is to comply with the applicable provisions of Chapter 19.

Materials used for waterproofing below-ground floors are to conform to the requirements of Section 1807.3.1.

Below-ground floors subjected to hydrostatic uplift pressures are to be waterproofed with membrane materials placed as underslab or split-slab installations, including such materials as rubberized asphalt, butyl rubber and neoprene or with polyvinyl chloride or polybutylene films not less than 6 mil in thickness, lapped at least 6 inches. There are many proprietary membrane products available that are specifically made for waterproofing floors and walls (i.e., polyethylene sheets sandwiched between layers of asphalt), which may be used when approved by the building official.

All membrane joints are to be lapped and sealed in accordance with the manufacturer's instructions to form a continuous, impermeable waterproof barrier.

1807.3.2 Walls. Walls that are required to be waterproofed in accordance with Section 1807.3 must first be prepared as required in this section and then waterproofed with the required membrane-type installations.

The walls must be designed to resist the hydrostatic pressure anticipated at the site, as well as any other lateral loads to which the wall will be subjected, such as soil pressures or seismic loads. As with the floors required to be waterproofed, it is particularly important that the walls required to be waterproofed be properly designed to resist all anticipated loads, as cracking and other damages would allow water seepage into below-ground spaces.

Water seepage can lead to the deterioration of the foundation as wood rots, concrete and masonry erode and steel rusts. Such deterioration can cause structural failure of the foundation. More important, failure of the foundation wall can lead to structural failure of the building because the foundation supports the building structure. Concrete or masonry construction must comply with the applicable provisions of Chapters 19 and 21, respectively.

Table 18-5 lists materials commonly used for the installation of moisture barriers in wall construction and the related standards that may be used as a basis for acceptance of such materials.

Table 18-5. Waterproofing materials

Material	Specification
Asphalt-saturated asbestos felt	ASTM D 250
Asphalt-saturated burlap fabric	ASTM D 1327
Asphalt-saturated cotton fabric	ASTM D 173
Asphalt-saturated organic felt	ASTM D 226
Coal-tar-saturated burlap fabric	ASTM D 1327
Coal-tar-saturated cotton fabric	ASTM D 173
Coal-tar-saturated organic felt	ASTM D 227

Asphalt and coal-tar products are not compatible and should not be used together.

Waterproofing installations are to extend from the bottom of the wall to a height not less than 12 inches above the maximum elevation of the groundwater table. The remainder of the wall below ground level (if the height is small) may be either waterproofed as a continuation of the installation or must be dampproofed in accordance with the requirements of Section 1807.2.

This section requires that waterproofing must consist of two-ply hot-mopped felts. The practice of the waterproofing industry is to select the number of plies of membrane material based on the hydrostatic head (height of water pressure against the wall). As a general practice, if the head of water is between 1 and 3 feet, two plies of felt or fabric membrane are used; between 4 and 10 feet, three-ply construction is needed; and between 11 and 25 feet, four-ply construction is necessary.

Waterproofing installations may also use polyvinyl chloride materials of not less than 6 mil thick, 40 mil polymer-modified asphalt, or 6 mil polyethylene. These materials have been widely recognized for their effectiveness in bridging nonstructural cracks. Other approved materials and methods may be used, provided that the same performance standards are met. All membrane joints must be lapped and sealed in accordance with the manufacturer's instructions.

Surface preparation of walls. Before applying waterproofing materials to concrete or masonry walls, the wall surfaces must be prepared in accordance with the requirements of Section 1807.2.2.1, which requires the sealing of all holes and recesses. Surfaces to be waterproofed must also be free of any projections that might puncture or tear membrane materials that are applied over the surfaces. **1807.3.2.1**

Joints and penetrations. This section requires that all joints occurring in floors and walls and at locations where floors and walls meet, as well as all penetrations in floors and walls, be made watertight by approved methods. Sealing the joints and penetrations in the waterproofing is of primary importance to ensure the effectiveness of the waterproofing. If the joints or penetrations are not sealed properly, they can develop leaks, which become a passageway for water to enter the building. Because the remainder of the foundation is wrapped in waterproofing, moisture can actually become trapped in the foundation walls or floor slab, and serious damage to these structural components can occur. Such methods may involve the use of construction keys between the base of the wall and the top of the footing, or, if there is a hydrostatic pressure, floor and wall joints may require the use of **1807.3.3**

manufactured waterstops made of metal, rubber, plastic or mastic materials. Floor edges along the walls and floor expansion joints may employ any of a number of preformed expansion joint materials, such as asphalt, polyurethane, sponge rubber, self-expanding cork, cellular fibers bonded with bituminous materials, etc., which all comply with applicable ASTM standards or other approved specifications. A variety of sealants may be used together with the preformed joint materials. Gaskets made of neoprene and other materials are also available for use in concrete and masonry joints. The National Roofing Contractor's NRCA Roofing and Waterproofing Manual provides details for the reinforcement of membrane terminations, corners, intersections of slabs and walls, through-wall and slab penetrations and other locations.

Penetrations in walls and floors may be made watertight with grout or manufactured fill materials and sealants made for the purpose.

1807.4 **Subsoil drainage system.** Subsoil drainage systems are required to drain the area adjacent to basement walls to eliminate hydrostatic loads.

This section covers subsoil drainage systems in conjunction with dampproofing (see Section 1807.2) to protect below-ground spaces from water seepage. Such systems are not used where basements or other below-ground spaces are subject to hydrostatic pressure, because they would not be effective in disposing of the amount of water anticipated if hydrostatic pressure conditions exist. Ground-water tables may be reduced to acceptable levels by methods described in the commentary to Section 1807.1.3.

The details of subsoil drainage systems are covered in the requirements of Sections 1807.4.2 through 1807.4.3.

1807.4.1 **Floor base course.** This section requires that floors of basements, except for story-above-grade construction, must be placed on a gravel or stone base not less than 4 inches thick. Not more than 10 percent of the material is to pass a No. 4 sieve to provide a porous installation and provide a capillary break. Material that passes a No. 4 sieve would be silt or clay that does not permit the free movement of water through the floor base, but allows for upward migration by capillary action.

This requirement serves three purposes. The first is to provide an adjustment to the irregularities of a compacted subgrade so as to produce a level surface upon which to cast a concrete slab. The second purpose is to provide a capillary break so that moisture from the soil below will not rise to the underside of the floor. Finally, where required, the porous base can act as a drainage system to expel underslab water by means of gravity, or the use of a sump pump or other approved methods.

The exception allows for the omission of the floor base when the natural soils beneath the floor slab consist of well-draining granular materials such as sand, stone or mixtures of these materials. Some caution, however, is justified in the use of this exception. If the granular soils contain an excessive percentage of fine materials, the porosity and the ability of the soil to provide a capillary break may be considerably diminished. The exception should be applied only if the natural base is equivalent to the floor base otherwise required by this section.

1807.4.2 **Foundation drain.** This describes in considerable detail the materials and features of construction required for the installation of foundation drain systems.

This type of drain system is suitable where the water table occurs at such elevation that there is minimal hydrostatic pressure exerted against the basement floor and walls and where the amount of seepage from the surrounding soil is so small that the water can be readily discharged by gravity or by mechanical means into sewers or ditches. The objective is to combine the protection afforded by the dampproofing of walls and floors (see Section 1807.2) and that given by the perimeter drains to maintain below-ground spaces in a dry condition.

A foundation drain system usually consists of the installation of drain tiles made of clay or concrete or of drain pipes of corrugated metal or nonmetallic pipes surrounded by crushed stone or gravel and a filter membrane material (filter fabric). The foundation drain is set adjacent to the wall footing and extends around the perimeter of the building. Drain tiles are placed end to end with open joints to permit water to enter the system. Metallic and nonmetallic drains are made with perforations at the invert (bottom) section of the pipe and are installed with connected ends. Where drain tile or perforated drain pipe is used, the invert must not be set higher than the basement floor line so that water conveyed by the drain does not seep into the filter material and then create a hydrostatic pressure condition against the foundation wall and footing. The inverts should not be placed below the bottom of the adjacent wall footings to avoid carrying away fine soil particles whose loss, in time, could possibly undermine the footing and cause settlement of the foundation walls.

Tile joints or pipe perforations should be covered with an approved filter membrane material to prevent them from becoming clogged and to prevent fine particles that may be contained in the surrounding soil from entering the system and being carried away by water.

The filter material around the drain tiles or pipes (not to be confused with filter membrane material) should consist of selected gravel and crushed stone containing not more than 10 percent of material that passes a No. 4 sieve. The filter materials should be selected to prevent the movement of particles from the protected soil surrounding the drain installation into the drain. Filter material is to be placed in the excavation so that it will extend out from the edge of the wall footing a distance of at least 12 inches, with the bottom of the fill being no higher than the bottom of the base under the floor (see Section 1807.4.1) and the top of the footing.

Requiring the bottom of the foundation drain to be no higher than the bottom of the floor base is necessary so that if the water table rises into the floor base, it will also be able to rise unobstructed into the foundation drain. The foundation drain will then drain the water away from the building, as required by Section 1807.4.3. The top of the filter fill material must be covered with an approved filter membrane to allow water to pass through to the perimeter drain tile or pipes without allowing fine soil materials to enter the drainage system.

Drain tiles or pipes are to be installed in the filter bed and should be seated on at least 2 inches of filter material and covered with at least 6 inches of filter material to maintain good water flow into the drain tile of pipe.

1807.4.3

Drainage discharge. This section references the *International Plumbing Code* (IPC) for requirements for installing piping systems for the disposal of water from the floor base and the foundation drains. Chapter 11 of the IBC considers the piping materials, applicable standards and methods of installation of subsurface storm drains to facilitate water discharge either by gravity or mechanical means.

Where the soil at the site consists of well-drained granular materials such as gravel or sand-gravel mixtures to prevent the occurrence of hydrostatic pressure against the foundation walls and under the floor slab, the use of a dedicated drainage system as prescribed in the IPC is not required, because the site soils would permit natural drainage.

Section 1808 *Pier and Pile Foundations*

These sections cover the requirements for deep foundation systems (piles and piers). New definitions have been added to clarify the terms used in Section 1808, which also contains the requirements that are common for all types of piles and piers. Sections 1809, 1810 and 1811 cover the more specific requirements for driven piles, cast-in-place piles and composite piles, respectively. Section 1812 covers the requirements that are specific to pier foundations. For the most part, the requirements in these sections are based on the Board for Coordination of the Model Codes (BCMC) report on Pile Foundation Systems published in July, 1979, as adopted by the other model codes. The exception is the seismic requirements, which are based on the NEHRP *Provisions*. The provisions in the 2000 IBC cover more deep foundation types and are more detailed than the requirements in the UBC. Required material properties (e.g., allowable concrete or steel stresses) were also added to types of piles where none were specified in any of the model codes, and some requirements were modified as compared to the other model codes. The intent was to coordinate all the material property requirements, not just the new ones, to eliminate conflicts or inconsistencies among the model codes.

1808.1 Definitions. One major improvement in the IBC over the legacy model codes was to clearly define piles and piers, and to standardize the definitions of other terms that are used inconsistently in the design, enforcement and scientific communities. As originally defined in the 2000 IBC, piles are relatively slender (long in comparison to their cross-sectional dimension), whereas piers are relatively stocky (short in comparison to their cross-sectional dimension). The definitions were arrived at by balancing the definitions found in various foundation engineering references[1, 2, 3, 4, 5, 6] with the code language already in use.

Piles and piers are defined based on the ratio of the length to least dimension. A more accurate method to classify piers and piles is by considering the stiffness of the member relative to the surrounding soil similar to the relative stiffness parameter, β, for the beam-on-an-elastic foundation problem. Piles tend to be flexible relative to soil stiffness whereas piers are stiff relative to the soil stiffness. See, for example, Winterkorn, et al.[2]

Some changes occurred to the definitions in the 2003 IBC and 2006 IBC. In the 2003 IBC the definition of flexural length, steel cased pile and timber pile were added. Most significant however, is the new type of piles known as micropiles, which was added to the 2006 IBC. Micropiles are 12-inch-diameter or less bored, grouted-in-place piles incorporating steel pipe casing and/or steel reinforcement. See discussion of micropiles under Section 1810.8.

1808.2 Piers and piles—general requirements.

1808.2.1 Design. A clarification was made to the 2003 IBC allowing piles to be designed as piers where either of the following conditions exists: 1) Group R-3 and U occupancies not exceeding two stories of light-frame construction, or 2) where the surrounding foundation materials furnish adequate lateral support for the pile. This allowance is subject to the approval of the building official. As indicated, the term *light-frame construction* is defined in Section 201 as a system that uses repetitive wood or light-gage steel framing members. The change to Section 1808.2.1 replaced the exceptions in Section 1811.2, which were in conflict with the definition of piers.

1808.2.2 General. The purpose of a foundation investigation is to define the general subsurface stratifications of soil and rock materials, determine the soil and rock profiles and locate the groundwater table. Such information will help in selecting the type of pile foundation and in estimating pile lengths. Furthermore, a foundation investigation is often required to render data on specific soil properties, such as shear strength, relative density, compressibility of the soil and other such findings that will help in analyzing subsurface conditions for determining design loads, type of pile foundations, driving criteria, suitable bearing strata, probable durability of pile materials relative to the particular soil conditions found at the construction site, etc.

Subsurface information is normally obtained by means of test borings that yield suitable samples of soils and rock and give the depths from which they are obtained. Sometimes, certain in-situ tests are also conducted, but these should not be made without first making test borings.

This section outlines the kinds of information to be derived from a foundation investigation and report. In addition to the items listed in the code provision, the investigation could include other valuable and applicable data that would help in the evaluation, such as:

- Information on existing construction at the site or at adjacent sites, including the type and condition of these structures, age, types of foundations used, performance data, etc.

- Information on the existence of deleterious substances in the soils or other conditions that could seriously affect the durability and structural performance of the piles.
- Information on the geologic conditions at the site, which could include such items as the existence of mines, earth cavities, underground streams or other adverse water conditions, history of seismic activity, etc.

The 2000 IBC included a section (1807.2.6) covering pile spacing that required the spacing of piers and piles to be in accordance with the soils report. This section was deleted in the 2003 IBC because pile spacing was added to the general requirements for piers and piles under Section 1808.2.2. Determination of the proper spacings of a pile group in relation to the type of pile foundation employed and the soil conditions encountered is a matter of design. The spacing of piles must be such that the loads transferred to the load-bearing strata do not exceed the safe load-bearing values of the supporting strata as determined by test borings, field load tests or other approved methods.

Special types of piles. Pile types are basically classified in accordance with the structural material used, such as concrete, steel or wood. They can also be categorized in accordance with the method of construction or installation. There are many variations of pile types used in the construction of deep foundations, including some special or proprietary types beyond the scope of the code. This section, as well as other sections of Chapter 18 that deal with pile foundations, generally include only those basic pile types commonly used today. **1808.2.3**

Special pile types that are not specifically included in the provisions of the code are not precluded from use, provided that adequate information covering test data, calculations, structural properties, load capacity and installation procedures is submitted and accepted by the building official.

Pile caps. Pile caps are to be of reinforced concrete and designed in accordance with the requirements of ACI 318. For footings (pile caps) on piles, computations for moments and shears may be based on the assumption that the load reaction from any pile is concentrated at the pile center. See ACI 318 for loads and reactions of footings on piles. **1808.2.4**

The soil immediately under the pile cap should not be considered to provide any support for vertical loads. For a more detailed explanation of this requirement, see the commentary to Section 1808.2.9.3.

The heads of all piles are to be embedded not less than 3 inches into pile caps, and the edges of the pile caps are to extend at least 4 inches beyond the closest sides of all piles. The degree of fixity between a pile head and the concrete cap depends on the method of connection required to satisfy design considerations.

Stability. A group of piles designed to support a common load or to resist horizontal forces must be braced or rigidly tied together to act as a single structural unit that will provide lateral stability in all directions. Piles connected by a rigid, reinforced concrete pile cap are deemed to be sufficiently braced to meet the intent of this provision. **1808.2.5**

Three or more piles are generally used to support a building column load or other isolated, concentrated load. In a three-pile group, lateral stability is assured by requiring that the piles are located such that they will not be less than 120 degrees apart as measured from the centroid of the group in a radial direction.

For stability of a pile group supporting a wall structure, the piles are braced by a continuous, rigid footing and are alternately staggered in two lines at least 1 foot apart and symmetrically located on each side of the center of gravity of the wall. Other approved pile arrangements may be used to support walls, provided the piles are adequately braced and lateral stability of the foundation construction is assured.

This section clarifies that for pile or pier groups to be considered to provide lateral stability, they must meet the radial spacing requirements defined herein. With the exception of smaller one- and two-family dwellings, foundation walls are required to be supported by two lines of piles or piers. For lightweight construction, such as R-3 buildings not exceeding two stories or 35 feet in height, a single row of piles located within the width of the wall is permitted. These specifications should provide the building official with better guidelines as to what is acceptable.

1808.2.6 Structural integrity. Piles are generally installed by either driving, vibration, jacking, jetting, direct weight or a combination of such methods. Most types of piles are exposed to some degree of damage during placement. However, with knowledge of soil conditions and the proper selection of equipment, installation methods and techniques, damage may be prevented or minimized.

Due care must be exercised during pile installation to avoid interference with adjacent piles or other structures so as to leave their strength and load capacity unimpaired. If any pile is damaged during installation so as to affect its structural integrity, the damage must be satisfactorily repaired or the pile rejected.

Many of the code provisions within Section 1808 are to ensure pile structural integrity by adhering to proper installation procedures. The code cannot cover all possibilities, however, thus establishing the need for the general nature of the section.

1808.2.7 Splices. This section specifies the requirements for splicing of piles. One provision that raised questions during the drafting process is the second sentence of the section, which requires that splices shall develop not less than 50 percent of the least capacity of the pier or pile in bending. Although this may appear to be allowing the pile or pier to be underdesigned in bending, the preceding sentence does require the splice to be able to transmit the driving and service loads acting on the pile or pier at that location. The 50 percent requirement actually provides more strength where the bending moments are low, insofar as it is based on the capacity of the pile or pier, not the design loads.

Although it is physically and economically better to drive piles in one piece, site conditions sometimes necessitate that piles be driven in spliced sections. For example, when the soil or rock-bearing stratum is located so deep below the ground that the leads on the driving equipment will not receive full-length piles, it becomes necessary to install the piles sectionally or, where possible, to take up the extra length by setting the tip in a pre-excavated hole (see commentary, Section 1808.2.13). When piles are installed in areas such as existing buildings with restricted headroom, they are also required to be placed in spliced sections. There are a number of other reasons for field-splicing piles, such as restrictions on shipping lengths or the use of composite piles.

This provision requires that splices be constructed to provide and maintain true alignment and position of the pile sections during installation. Splices must be of sufficient strength to transmit the vertical and lateral loads on the piles, as well as to resist the bending stresses that may occur at splice locations during the driving operations and under long-term service loads. Splices are to develop at least 50 percent of the value of the pile in bending. Consideration should be given to the design of splices at locations where the piles may be subject to tension. Splices that occur in the upper 10 feet of pile embedment are to be designed to resist the bending moments and shears at the allowable stress levels of the pile material, based on an assumed pile load eccentricity of 3 inches, unless the pile is properly braced. Proper bracing of a spliced pile is deemed to exist if stability of the pile group is furnished in accordance with the provisions of Section 1808.2.4, provided that other piles in the group do not have splices in the upper 10 feet of their embedded length.

There are different methods employed in splicing piles depending upon the materials used in pile construction. For example, timber piles are spliced by one of two commonly used methods. The first method uses a pipe sleeve with a length of about four to five times the diameter of the pile. The butting ends of the pile are sawn square for full contact of the two pile sections, and the spliced portions of the timber pile are trimmed smoothly around their periphery to fit tightly into the pipe sleeve. The second splicing method involves the use of steel straps and bolts. The butting ends of the pile sections are sawn square for full contact and proper alignment, and the four sides are planed flat to receive the splicing straps. This type of splicing can resist some uplift forces.

Splicing of precast concrete piles usually occurs at the head portions of the piles. After the piles are driven to their required depth, pile heads are cut off or spliced to the desired elevation for proper embedment in the concrete pile caps. Any portion of the pile that is cracked or shattered by the driving operations or cutting off of pile heads should be removed and spliced with fresh concrete. To cut off a precast concrete pile section, a deep groove is chiseled around the pile exposing the reinforcing bars, which are then cut off (by torch) to desired heights or extensions. The pile section above the groove is snapped off (by crane) and a new pile section is freshly cast to tie in with the precast pile.

Steel H-piles are spliced in the same manner as steel columns, usually by welding the sections together. Welded splices may be welded-plate or bar splices, butt-welded splices, special welded splice fittings or a combination of these. Spliced materials should be kept on the inner faces of the H-pile sections to avoid forcing a hole in the ground larger than the pile, causing at least a temporary loss in frictional value and lateral support that might result in excessive bending stresses.

Steel pipe piles may be spliced by butt welding, sometimes using straps to guide the sections and to provide more strength to the welded joint. Another method is to use inside sleeves having a driving fit, with a flange extending between the pipe sections. By applying bituminous cement or compound on the outside of the ring before driving, a water-tight joint is obtained.

Allowable pier or pile loads. **1808.2.8**

Determination of allowable loads. The IBC specifies that the determination of allowable loads shall be based on three methods: **1808.2.8.1**

1. An approved driving formula
2. Load tests
3. Foundation analysis

In most cases, the allowable loads will be determined by a combination of Items 2 and 3. However, there may be circumstances where the soil conditions, such as granular soils, and the types of piles selected are such that the use of an approved dynamic pile-driving formula can be an aid to a qualified practitioner in establishing reasonable but safe allowable loads for the foundation system. Nevertheless, some literature indicates that "the use of a complicated formula is not recommended since such formulas have no greater claim to accuracy than the more simple ones."[7]

The dynamic pile-driving formula included in the 1970 and earlier editions of the UBC was dropped from the code because of its unreliability for cohesive soils. It is interesting to note that the earliest editions of the UBC utilized the so-called Engineering News formula, $R = 12WH/S+c$, which is the most simple of the dynamic pile-driving formulas. In 1937, the Pacific Coast formula was adopted into the UBC until its deletion prior to the 1973 edition of the code. This was one of the more complex dynamic pile-driving formulas and was based on a dynamic pile-driving formula developed by Terzaghi. However, as stated previously, in the hand of a qualified practitioner, a dynamic pile-driving formula does have some utility even though the UBC no longer provides such a formula.

There are two general considerations for determining pile capacity as required for the design and installation of pile foundations. The first consideration involves the determination of the underlying soil or rock characteristics. The second is the application of approved driving formulas, load tests or accepted methods of analysis to determine the pile capacities required to resist the applied axial and lateral loads, as well as to provide the basis for the proper selection of pile-driving equipment.

Driving criteria. For many decades, it has been the practice to try to predict the capacity of a pile from its resistance to driving. The usual procedure has been to make such determinations by the application of pile-driving formulas, none of which have been completely dependable. The singular premise used in the development of these formulas is simple and is best expressed by R.B. Peck as follows: The greater the resistance of a pile to driving, the greater the pile's capacity to support load. With complex engineering problems, **1808.2.8.2**

however, occasionally there are special circumstances under which there will be exceptions to general statements of this kind.

There are many pile-driving formulas. The simplest and most widely used formula in the United States is the Engineering News formula. This particular expression, and other formulas in common use today, have all generally shown poor correlations with load test results. Such comparisons are considerably better, however, when they are applied to the determination of pile capacity in soils consisting of free-draining, coarse-grained materials such as sand and gravel. In soils such as silt, clay and fine sand the water cannot escape fast enough during driving operations to not have an adverse influence on the frictional resistance of the piles. As a consequence, information may be unreliable.

This section limits the allowable compression load on a pile as established by an approved driving formula to a maximum of 40 tons. Generally, the use of pile-driving formulas to determine pile capacity should be avoided except, perhaps, in cases involving small jobs where the piles are to be driven in well-drained granular soils and load testing cannot be economically justified.

1808.2.8.3 **Load tests.** This section specifies the standards to be used to load test piles or piers, where higher compressive loads than allowed in other sections of the code are exceeded. See the discussion under Section 1808.2.8.3.1 for test evaluation methods.

Questions were raised at the public hearings as to whether or not ASTM D 4945, which is a dynamic test, is sufficient by itself to verify the pile capacity. Many standards, including ASTM D 4945 indicate that a dynamic test may not be sufficient without a static test (ASTM D 1143) to calibrate the results, but leave it up to the registered design professional to decide if the dynamic test is sufficient. Other standards require a static load test to calibrate the dynamic test.

The safest method for determining pile capacity is by load test. A load test should be conducted wherever feasible and used where the pile capacity is intended to exceed 40 tons per pile (see Section 1808.2.8.3). Test piles are to be of the same type and size as intended for use in the permanent foundation and installed with the same equipment, by the same procedure, and in the same soils intended or specified for the work.

Load tests are to be conducted in accordance with the requirements of ASTM D 1143, which covers procedures for testing vertical or batter foundation piles, individually or in groups, to determine the ultimate pile load (pile capacity) and whether the pile or pile group is capable of supporting the loads without excessive or continuous settlement. Recognition, however, must be given to the fact that load-settlement characteristics and pile capacity determinations are based on data derived at the time and under conditions of the test. The long-term performance of a pile or group of piles supporting actual loads may produce behaviors that are different than those indicated by load test results. Judgment based on experience must be used to predict pile capacity and expected behavior.

The load-bearing capacity of all piles, except those seated on rock, does not reach the ultimate load until after a period of rest. The results of load tests cannot be deemed accurate or reliable unless there is an allowance for a period of adjustment. For piles driven in permeable soils such as coarse-grained sand and gravel, the waiting period may be as little as two or three days. For test piles driven in silt, clay or fine sand, the waiting period may be 30 days or longer. The waiting period may be determined by testing (i.e., by redriving piles) or from previous experience.

This section also requires that at least one pile be tested in each area of uniform subsoil conditions. The statement should not be misconstrued to mean that the area of test is to have only one uniform stratum of subsurface material, but rather that the soil profile, which may consist of several layers (strata) of different materials, must represent a substantially unchanging cross section in each area to be tested.

The allowable pile load to be used for design purposes is not to be more than one-half of the ultimate pile capacity, as determined by the load test in which the net settlement of the test pile is not to exceed 0.01 inch per ton or more than a total of $^{3}/_{4}$ inch. The rate of penetration of permanent foundation piles must be equal to or less than that of the test pile(s).

All production piles should be of the same type, size and approximate length as the prototype test pile, as well as installed with the same or comparable equipment and methods. They should also be installed in soils similar to those for the test pile.

1808.2.8.3.1

Load test evaluation. This section was added to the 2003 IBC which gives three specific methods that are acceptable for performing pile load tests. Other methods are permitted at the discretion of the building official.

1808.2.8.4

Allowable frictional resistance. Resistance that is due to skin friction is limited to a maximum of 500 psf unless a greater value is permitted by the building official based on recommendations of an approved soil investigation or a greater value is substantiated by load test methods described in Section 1808.2.8.3.

1808.2.8.5

Uplift capacity. This section gives both the designer and building official needed guidance on criteria to use for design of piles or piers for uplift. The 1997 UBC was silent on the requirements, but the 2000 IBC now requires a test or approved method of analysis, with a safety factor of 3. The capacity of pile or pier groups is also limited to two-thirds of the weight of the group and the soil contained in the group. This is consistent with requirements in other sections in the code on uplift and overturning, where the dead load resistance is limited to two-thirds of the weight. The maximum allowable uplift load cannot exceed the ultimate load capacity determined by the methods described in Section 1808.2.8.3 divided by a factor of safety of two.

When piles are designed to withstand uplift forces, they act in tension and are actually friction piles. The amount of tension that can be developed not only depends on the strength properties of the pile, but also on the frictional or cohesive properties of the soil. The uplift or tensile resistance of a pile is not necessarily a function of its load-bearing capacity under compressive load. For example, the tensile resistance of a friction pile in clay will usually be about the same value as its load-bearing capacity, as the skin friction developed in such soils is very large. In contrast, a friction pile in sand or in other granular materials will develop a tensile resistance considerably less than its load-bearing capacity.

Where the properties of the soil are known, the ultimate uplift resistance value of a pile can be determined by approved analytical methods. This section requires that where the ultimate tensile value is determined by analysis, a safety factor of 3 must be applied to establish the allowable uplift load of the pile.

The best way to determine the response of a vertical or batter pile to a static tensile load (uplift force) applied axially to the pile is by applying an extraction test in accordance with the requirements of ASTM D 3689. The maximum allowable uplift load is not to be more than one-half of the total test load. This section of the code gives a limitation on the upward movement of the pile in compliance with the provisions of the ASTM D 3689 test method. The measurements of pile movement in the standard test procedure, however, are time-dependent incremental measurements and should be adhered to in determining allowable pile load.

The allowable uplift load on a group of piles is to be reduced from the value obtained on a single pile as briefly described in this section of the code and in compliance with comprehensive analytical methods.

To be effective in resisting uplift forces as tension members of a foundation system, piles must be well-anchored into the pile cap by adequate connection devices. In turn, the cap must be designed for the uplift stresses. Piles must also be designed to take the tensile stresses imposed by the uplift forces. For example, concrete piles must be reinforced with longitudinal steel to take the full net uplift. Special consideration needs to be given in the design of pile splices that are intended to act in tension.

1808.2.8.6

Load-bearing capacity. The load-bearing capacity of a pile foundation is determined as a pile-soil system. For example, the load-bearing capacity of a single pile is the function of either the structural strength of the pile or the supporting strength of the soil. The load-bearing capacity of the pile is controlled by the smaller value obtained in the two considerations. The load-bearing capacity of a pile group may be greater than, equal to or less than the capacity of a single pile multiplied by the number of piles in the group, depending on pile spacing and soil conditions.

Because the supporting strength of the soil generally controls the load-bearing capacity of a pile foundation, this section requires that the load-bearing capacity of an individual pile or a group of piles be not more than one-half of the ultimate load capacities of the piles in terms of the supporting load-bearing strata.

Sometimes, weaker layers of soil underlie the soil load-bearing strata supporting a pile foundation and may cause damaging settlements. Under such subsurface conditions, it must be determined by an approved method of analysis that the safety factor has not been reduced to a figure less than 2. Otherwise the piles are to be driven to deeper load-bearing soils to obtain adequate and safe support, or the design capacity is to be reduced and the number of piles increased.

1808.2.8.7 **Bent piers or piles.** Piles that are discovered to have sharp or sweeping bends because of obstructions encountered during the driving operations or for any other cause are to be analyzed by an approved method, or a representative pile is to be load tested to determine its load-carrying capacity. Otherwise, the piles could be used at some reduced capacity as determined by test or analysis; or, if necessary, the piles can be abandoned and replaced.

1808.2.8.8 **Overloads on piers or piles.** Because of subsurface obstructions or other reasons, it is sometimes necessary to offset piles a small distance from their intended locations so that they are not driven out of position. In such cases, the load distribution in a group of piles may be changed from the design requirements and cause some of the piles to be overloaded. This section requires that the maximum compressive load on any pile caused by mislocation not exceed 110 percent of the allowable design load. Piles exceeding this limitation must be extracted and redriven in the proper location or other approved remedies applied, such as installing additional piles to balance the group.

1808.2.9 **Lateral support.** This section provides needed guidance to the designer and building official on what constitutes adequate lateral support for piles and piers. Section 1808.2.9.1 specifies that any soil other than fluid soil is allowed to be considered to provide lateral support to prevent buckling of piles or piers, and Section 1808.2.9.3 establishes the acceptance criteria for tests that establish the allowable lateral load.

1808.2.9.1 **General.** Liquefaction causes loss of lateral bearing capacity with resulting loss of support for piles. Loss of lateral support can also occur from the soil strength loss associated with sensitive or quick clays that are sensitive to remolding effects. These soft clays typically occur in marine or former marine environments and are often called *bay muds*. These clays lose significant strength when remolded as might occur when a pile foundation is moved through muds by seismic induced displacements.

1808.2.9.2 **Unbraced piles.** (Column Action.) The code requires that piles standing unbraced in air, water or material not capable of providing adequate lateral support shall be designed in accordance with the column formulas of the code. Obviously, water and air do not provide lateral support. On the other hand, most soils do provide lateral support, although exceptionally loose and unconsolidated fills, liquified sands, and remolded clays are inadequate to provide lateral support.

1808.2.9.3 **Allowable lateral load.** Because of wind loads, unbalanced building loads, earth pressures and seismic loads, it is inevitable that individual piles or groups of vertical piles supporting buildings or other structures will be subjected to lateral forces. The distribution of these lateral forces to the piles largely depends on how the loads are carried down through the structural framing system and transferred through the supporting foundation to the piles. The amount of lateral load that can be taken by the pile foundation is a function of 1) the type of pile used; 2) the soil characteristics, particularly in the upper 10 to 30 feet of the piles; 3) the embedment of the pile head (fixity); 4) the magnitude of the axial compressive load on the piles; 5) the nature of the lateral forces; and 6) the amount of horizontal pile movement deemed acceptable.

The degree of fixity of the pile head is an important design consideration under very high lateral loading unless some other method, such as the use of batter piles, is employed to resist lateral loads. The fixing of the pile head against rotation reduces the lateral deflection. In general, pile butts are embedded 3 to 4 inches into the pile cap (see Section 1808.2.3) with no ties to the cap. These pile heads are neither fixed nor free, but somewhere in between. Such construction is satisfactory for many loading conditions, but not for high seismic loads.

The magnitude of friction developed between the surfaces of two structural elements in contact with each other is a function of the weight or load applied. The larger the weight, the greater the frictional resistance developed. In the design of pile foundations, frictional resistance between the soil and the bottom of the pile caps (footings) should not be relied on to provide lateral restraint, because the vertical loads are transmitted through the piles to the supporting soil below and to the ground immediately under the pile caps. Only the weights of the pile caps can supply some frictional resistance insofar as such footings are constructed by placing fresh concrete on the soil, thus providing a positive contact. The weights of the pile caps in comparison to the magnitude of loads and lateral forces transmitted to the piles is nominal and not significant from a structural design standpoint. Also, in rare occurrences, soils have been known to settle under pile caps, leaving open spaces and thus eliminating the development of any frictional restraint.

Where vertical piles are subjected to lateral forces exceeding acceptable limitations, the use of batter piles may be required. Lateral forces on many structures are also resisted by the embedded foundation walls and the sides of the pile caps.

The allowable lateral-load capacity of a single pile or group of piles is to be determined either by approved analytical methods or load tests. Load tests are to be conducted to produce lateral forces that are twice the proposed design load; however, in no case is the allowable pile load to exceed one-half of the test load, which produces a gross lateral pile movement of 1 inch as measured at the ground surface.

1808.2.10

Use of higher allowable pier or pile stresses. In those sections of Chapter 18 that specifically deal with the several types of piles most commonly used in the construction of pile foundations, there are limitations placed on the stresses that can be used in the pile design. In most cases, the allowable stresses are stated as a percentage of some limiting strength property of the pile material. For example, in the case of piles made of steel materials, the allowable stresses are prescribed as a percentage of the yield strengths of the several grades of steel that can be used for pile construction. For concrete, the allowable stress is stated as a percentage of the 28-day specified compressive strength of the material. The allowable stresses for timber piles are based on tabulations of already reduced stresses. The reduced stresses are based on the strength values of different species of wood and reductions in strength caused by preservative treatment.

The allowable design stresses designated in Chapter 18 for each of the different types of piles are intended to provide a factor of safety against the dynamic forces of pile driving that may cause damage to the pile, and to avoid overstresses in the pile under the design loads and other loads that may be induced by subsoil conditions.

This section allows the use of higher allowable stresses when evidence supporting the values is submitted and approved by the building official. The data submitted to the building official should include analytical evaluations and findings from a foundation investigation as specified in Section 1808.2.1, and the results of load tests performed in accordance with the requirements of Section 1808.2.8.3. The technical data and the recommendation for the use of higher stress values must come from a registered design professional who is knowledgeable in soil mechanics and experienced in the design of pile foundations. This registered design professional must supervise the pile design work and witness the installation of the pile foundation so as to certify to the building official that the construction satisfies the design criteria. In any case, the use of greater design stresses should not result in design loads that are larger than one-half of the test loads (see Section 1808.2.8.3).

1808.2.11 Piles in subsiding areas. Where piles are driven through subsiding soils and derive their support from underlying firmer materials, the subsiding soils cause an additional load to the piles through so-called negative friction. This negative friction is actually a downward friction force on the piles, which increases the axial load on the piles. The code permits an increase in the allowable stress on the piles if an analysis of the foundation investigation indicates that the increase is justified.

1808.2.12 Settlement analysis. The purpose of a settlement analysis is to provide the data needed to design a pile foundation system that will maintain the stability and structural integrity of the supported building or structure.

The load-bearing stratum of every soil must support the loads transferred through the pile system, as well as the weight of all soil above. The capability of the strata underlying the pile tips to support additional loads without detrimental settlement can often be determined by analytical procedures. Serious settlements in a pile foundation system, particularly differential settlements, can cause great structural damage to the supported structure and the foundation itself.

Although the settlement analysis of an individual pile is complex, the analysis of a group of piles is significantly more complicated because of the overlapping soil stresses caused by closely spaced piles. Analytical procedures vary with the type of piles and especially with the soil conditions. Settlement analysis would generally include cases involving point-bearing piles on rock, and in granular soils and hard clay. It would also involve friction piles in sand and gravel soils, and in clay materials.

Load tests are often used to aid in the analysis. In the case of pile foundations in clay soils, however, there are no practical ways to determine long-term settlement from load tests, and therefore only approximations of settlement may be derived from laboratory tests.

1808.2.13 Pre-excavation. There are several important reasons for the use of pre-excavation to facilitate the installation of foundation piles. Some of these purposes are:

- To install piles through upper strata of hard soil.
- To penetrate through subsurface obstructions, such as timbers, boulders, rip rap, thin stone strata and the like.
- To reduce or eliminate the possibility of ground heave that could lift adjacent structures or piles already driven.
- To reduce ground pressures resulting from soil displacement during driving and to prevent the lateral movement of adjacent piles or structures.
- To reduce the amount of driving required to seat the piles in their proper load-bearing strata.
- To reduce the possibility of damaging vibrations or jarring of adjacent structures, as well as reduce the amount of noise, all of which are associated with pile driving operations.
- To accommodate the placement of piles that may be somewhat longer than the leads of the pile driving equipment.

The two most common methods employed in pre-excavation in operations are prejetting and predrilling. Jetting is usually effective in most types of soils, except very coarse and loose gravel and highly cohesive soils. Jetting is most effective in granular materials. Generally, jetting in cohesive soils is not very practical or especially useful and should be avoided in soils containing very coarse gravel, cobbles or small boulders. These stones cannot be removed by the jet and tend to collect at the bottom of the hole, preventing pile penetration below that depth.

Jetting operations must be carefully controlled to avoid excessive loss of soil, which could affect the load-bearing capacity of piles already installed or the stability of adjacent structures.

Piles should be driven below the depth of the jetted hole until the required resistance or penetration is obtained. Before this pre-excavation method is used, consideration should be given to the possibility that jetting, unless strictly controlled, can adversely affect load transfer, particularly as it involves the placement of nontapered piles.

Predrilling or coring before driving is effective in most types of soils and is a more controllable method of pre-excavation than jetting. The risk of adversely affecting the structural integrity of adjacent piles or structures or the frictional capacity of piles is considerably less than jetting.

Predrilling can be performed as a dry operation or as a wet rotary process. Dry drilling can be done by using a continuous-flight auger or a short-flight auger attached to the end of a drill stem or kelly bar. Wet drilling requires a hollow-stem continuous-flight auger or a hollow drill stem employing the use of spade bits. When the wet rotary process of predrilling is used, bentonite slurry or plain water is circulated to keep the hole open. As in the case of jetting, piles should be driven with tips below the predrilled hole. This is necessary to prevent any voids or very loose or soft soils from occurring below the pile tip.

There are other methods used for pre-excavation purposes, such as the dry tube method and spudding, but such procedures are seldom used. In any case, the methods to be employed for pre-excavation are subject to the approval of the building official.

Installation sequence. As displacement piles are driven within a group, progressive compaction of the surrounding soil occurs, particularly where it involves closely spaced piles. This can cause piles to be deflected off-line because of the buildup of unequal soil pressures around the piles. Soil compaction during driving operations can cause extreme variations in pile lengths within a group, with some piles failing to reach specified load-bearing material. Ground heave is another effect of soil compaction (see Commentary, Section 1808.2.19). **1808.2.14**

To prevent or reduce significantly the problems associated with soil compaction, the driving sequence of pile installations becomes an important consideration. For example, if the outer piles of a group are driven first so that the inner piles, because of soil compaction, fetch up to specified sets (hammer blows) at much higher elevations than the outer piles, the total load-bearing value of the group will be adversely affected. As another example, starting pile driving at the edge of a group makes the piles progressively more difficult to drive and results in a one-sided bearing group. The general driving practice is to work from the center of a group outward. For large groups consisting of rows of widely spaced piles, driving can be done progressively from one side to the other.

Use of vibratory drivers. The use of vibratory drivers for the installation of piles is not applicable to all types of soil conditions. They are effective in granular soils with the use of nondisplacement piles, such as steel H-piles and pipe piles driven open ended. Vibratory drivers are also used for extracting piles or temporary casings employed in the construction of cast-in-place concrete piles. **1808.2.15**

Vibratory drivers, either low or high frequency, cause the pile to penetrate the soil by longitudinal vibrations. Although this type of pile driver can produce good results in the installation of nondisplacement piles under favorable soil conditions, the greatest difficulty is the lack of a reliable method of estimating the load-bearing capacity. After the pile has been installed with a vibratory driver, pile capacity can be determined by using an impact-type hammer to set the pile in its final position.

One method to determine pile capacity is to calibrate the power consumption in relation to the rate of penetration. Nonetheless, the use of a vibratory driver is only permitted where the pile load capacity is established by load tests in accordance with the requirements of Section 1808.2.8.3.

1808.2.16 Pile driveability. Piles must be of a size, strength and stiffness capable of resisting without damage:

- crushing caused by impact forces during driving
- bending stresses during handling
- tension from uplift forces or from rebound during driving
- bending stresses caused by horizontal forces during driving
- bending stresses caused by pile curvatures

Additionally, the pile must be capable of transmitting dynamic driving forces to mobilize the required ultimate pile capacity within the soil without severe elastic energy losses. Pile driveability depends on the pile stiffness, which is a function of pile length, cross-sectional area and modulus of elasticity. Yield strength does not affect stiffness. Thus, caution should be observed in the use of high-yield-strength steels for high loads on smaller cross sections requiring high dynamic driving energy. A wave equation analysis would reflect pile stiffness or driveability. The selection of pile types and dimensional requirements for driveability is a function of soil characteristics.

1808.2.17 Protection of pile materials. Unless properly protected, piles may deteriorate because of biological, chemical or physical actions caused by particular conditions that exist or that may later develop at the site. The durability of piles will be long lasting if care is taken in the selection and protection of pile materials.

Some of the problems associated with pile durability are as follows:

- Untreated timber piles may be successfully used if they are entirely embedded in earth and their butts (cutoffs) are below the lowest ground-water level or are submerged in fresh water. The risk, however, is in situations where unexpected lowering of the water table occurs and exposes the upper parts of piles to decay and insect attack. Such conditions may especially occur in urban areas where the water table may be significantly lowered by pumping or deep drainage. There is also the remote possibility that wood piles will be damaged by the percolation of ground water heavily charged with alkali or acids.
- Wood piles extending above the water table or exposed to air or saltwater are subject to decay, attack by insects and marine borers, and need to be pressure treated with preservatives conforming to the requirements of AWPA C1 and C2.
- Steel piles that are driven and embedded entirely in undisturbed soil are generally not significantly affected by corrosion caused by oxidation, regardless of soil types or soil properties. The reason for this is that undisturbed soil is so deficient in oxygen at levels only a few feet below the ground line or below the water table that progressive corrosion is inhibited. However, where upper portions of steel piles protrude above ground into the air, where piles are placed in corrosive soils, or where ground water contains deleterious substances from sources such as coal piles, alkali soils, active cinder fills, chemical wastes from manufacturing operations or other sources of pollutants, the steel may be subject to corrosive action. Under such conditions, steel piles may be protected by a concrete encasement or a suitable coating, extending from a level slightly above ground to a depth below the layer of disturbed earth. For piles above ground level that are exposed to air and subject to rusting, the steel should be protected by being painted, as any other type of structural steel construction would be protected.
- Steel piles installed in saltwater or exposed to a saltwater environment are subject to corrosion, and therefore should be protected with approved coatings or encased in concrete.
- Concrete piles, plain or reinforced, that are entirely embedded in undisturbed earth are generally considered permanent installations. The level of the water table does

not normally affect the durability of concrete piles. Ground water that readily flows through granular materials or through disturbed soil and contains deleterious substances can have deteriorating effects. Concrete piles embedded in impervious clay materials will not generally suffer from ground water containing harmful substances. The primary deleterious substances that attack concrete are acids and sulphates. In the case of acids, it is best to use an alternative pile material if the acid attack is potentially destructive as coating piles may be ineffective because of soil abrasion during driving. Concrete can be attacked, however, by exposure to soils with high sulfate content. In the case of high alkaline soils with sulphate salts, Type V portland cement may be used. Where the exposures are only moderate, a Type II portland cement will usually be adequate. If the piles are in a marine environment, Type V or Type II portland cement is also indicated to provide the necessary sulphate resistance.

The conditions of the underground environment should be ascertained so as to protect piles against possible corrosion of either the concrete or exposed load-bearing steel. For example, corrosion can be caused by chemical attack from the corrosive soil, industrial wastes, organic fills, oxidation or even by electrolytic action. Corrosion by oxidation is generally very minor and often disregarded. Corrosion caused by electrolytic action or by destructive chemicals on load-bearing steel can be protected by suitable coating, concrete encasement and cathodic protection. This also applies to bare steel piles previously discussed. Concrete can be protected from chemical attack by the use of special cements, dense concrete mixtures and special coatings.

Piles installed in saltwater, such as for buildings or other structures in waterfront construction, are subject to chemical action on concrete materials coming from polluted waters, frost action on porous concrete, spalling of concrete and rusting of steel reinforcement. Spalling may become particularly serious under tidal conditions where alternate wetting and drying occurs coupled with cycles of freezing and thawing. Spalling can be minimized or prevented by providing additional cover over the reinforcement; by the use of rich, dense concrete; air entrainment and suitable concrete admixtures; and, in the case of precast piles, by careful handling to minimize stresses and avoid cracking during placement. See analysis of Chapter 19 for more detailed information on concrete quality and concrete materials.

1808.2.18

Use of existing piers or piles. This section allows the reuse of existing piers or piles where sufficient information is submitted to the building official to demonstrate they are adequate. This introduces flexibility for both the building designer and the building owner to make use of existing materials where it makes sense to do so.

Piles remaining after structures are demolished should not be used for the support of new loads, unless evidence shows them to be adequate. This is because of the lack of soil data or detailed information on the piling material, and because of the unavailability of the pile driving records made during the construction of these older buildings or structures. As such, the true condition of the piles is unknown and, over time, the piles may have deteriorated, or their load capacity may have been reduced. Such piles may be used, however, if they are load tested or retracted and redriven to verify their capacities. Only the lowest allowable load capacity as determined by test data or redriving information should be used in the design.

1808.2.19

Heaved piles. Piles that are driven into saturated plastic clay materials can often displace a volume of soil equal to that of the piles themselves. When this happens, the soil displacement sometimes occurs as ground heave and may lift adjacent piles already driven. Under such conditions, heaved piles may no longer be properly seated and a loss of pile capacity occurs. Heaved piles must be redriven to firm bearing to again develop the required capacity and penetration. If heaved piles are not redriven, their capacity must be verified by load tests made in accordance with the requirements of Section 1808.2.8.3.

This section applies only to piles that can be safely redriven after installation. Heaved uncased cast-in-place concrete piles or sectional piles with joints that cannot take tension should be abandoned and replaced. When redriving heaved piles, a comparable driving system or the same as that of the initial driving should be employed. It should be noted that in redriving concrete-filled pipe piles, the driving characteristics of the pile have been altered and the pile is substantially stiffer than when the empty pipe was initially driven. In such cases, the required driving resistance would be less than originally required.

One method used to prevent or reduce objectionable soil displacement is to remove some of the soil in the spaces to be occupied by the piles. This is done by predrilling the pile holes (see commentary, Section 1808.2.13).

1808.2.20 Identification. All pile materials must be identified for conformity to the code requirements and construction specifications. Information such as strength (species and grade for timber piles), dimensions and other pertinent information are required. Such identifications must be provided for all piles, whether they are taken from manufacturer's stock or made for a particular project. Identifications are to be maintained from the point of manufacture through the shipment, on-site handling, storage and installation of the piles. Manufacturers, upon request, usually furnish certificates of compliance with construction specifications. In the absence of adequate data, piles must to be tested to demonstrate conformity to the specified grade.

In addition to mill certificates (steel piles), identification is made through plant manufacturing or inspection reports (precast concrete and timber piles) and delivery tickets (concrete). Timber piles are stamped (labeled) with information such as producer, species, treatment and length.

Identification is essential when high-yield-strength steel is specified. Frequently, pile cutoff lengths are reused and pile material may come from a jobber, a contractor's yard or a material supplier. In such cases, mill certificates are not available and the steel should be tested to see if it complies with the code requirements and the project specifications.

1808.2.21 Pier or pile location plan. A plan clearly showing the designation of all piles on a project by an identification system is to be filed with the building official before the installation of piling is started. The pile inspector (see analysis of Section 1808.2.22) must keep piling logs and other records and submit written reports based on this identification system. The use of such a system becomes particularly important at sites where the variations in soil profiles are so extensive that it becomes necessary to manufacture piles of different lengths to satisfy bearing conditions.

The building official should also be furnished copies of all modifications to the original pile location plan that may be necessary as the work proceeds (as-built drawings). This would show piles added, deleted or relocated. Such records would facilitate the use of existing piles in the future (see Section 1808.2.18) if the structure is altered or another structure is built on the site.

1808.2.22 Special inspection. See analysis of Sections 1704.8 and 1704.9.

1808.2.23 Seismic design of piers or piles. This section specifies the requirements for pile or pier designs in SDC C through F. Requirements for interconnection of pile caps, anchorage of the piles to pile caps, transverse reinforcement in the pile, strength of pile splices, consideration of group effects, and a consideration of nonlinear interaction of the pile and the soil begin in SDC C. This reflects the CRDC's concern that significant ground motions can occur in SDC C, despite the approximate correspondence to Zone 2 in the 1997 UBC. Additional requirements are imposed on piles and piers in SDC D through F, including a requirement that the upper portion of piles be detailed as special moment-resisting-frame columns, to prevent failure of the piles under severe ground motions.

The purpose of this section is to include the pile bending and curvatures resulting from horizontal ground movement during an earthquake in the structural design. The reinforcement in the pile, required to resist the pile curvature effect, increases ductility of the foundation such that bending or shear failure is precluded.

1808.2.23.1 Seismic Design Category C. Interconnection of piles and caissons is necessary to prevent differential movement of the components of the foundation during an earthquake. It is well known that a building must be thoroughly tied together if it is to successfully resist earthquake ground motion.

Individual piles, piers or pile caps required for structures in SDC C must be interconnected with ties capable of transmitting a force equal to the larger pile cap or column load times the short period response acceleration, S_{DS}, divided by 10. The intent of this requirement is to minimize differential movement or spreading between the footings during ground shaking. If slabs on grade, or beams within slabs on grade, are used to meet the tie requirement, the load path from footing to slab or beam/slab and across joints in the slab or beam/slab should be checked for continuity. The slab or beam must be reinforced for the design tension load. In addition, the slab should be checked for buckling under the required compression load using an assumed slab of no more than six times the slab thickness.

Piles in structures subject to seismic ground shaking are likely to be subjected to uplift (tension) forces, either by design or because of insufficient resistance to overturning forces by gravity loads. Hence, concrete piles and concrete-filled steel pipe piles must be able to develop the strength of the pile (in tension) in the connection to the cap. This is accomplished by the requirement that the pile be embedded in the cap by a distance equal to the development length. The development length *may not be reduced* by the ratio $A_{required}$ / $A_{supplied}$. Alternative means, such as increasing concrete confinement, may be used to reduce the development length.

Similarly, the various types of steel piles are required to develop the strength in tension and to transmit this strength to the cap by positive means other than bond to the bare steel; e.g., welded studs or welded reinforcement must be used.

Splices must develop the full strength of the pile, both tension and compression, for all pile types.

Moments, shears and deflections must be based on nonlinear soil-pile interaction. If using pile foundations in soils with lenses of soft clays, lenses subject to liquefaction, or soils susceptible to strength loss from remolding, the effects of the these layers on the curvature and, hence, moment demands on the pile should be investigated. Often these soil-induced curvatures place a much higher moment demand on the piles than would be determined from conventional lateral force *P-y* analyses.

Note that Section 1808.2.23.1.2 states that if the ratio of the depth of embedment of the pile to the pile diameter or width is less than or equal to six, the pile may be assumed to be rigid. Note that for $L/D \leq 12$, Section 1808 defines it as a pier, not a pile. See Winterkorn, et al[2] for methods of analysis of rigid piers.

1808.2.23.2

Seismic Design Category D, E or F. In addition to the requirements for SDC C, moments, shears and deflections must be based on nonlinear soil-pile interaction. If using pile foundations in soils with layers of soft to medium stiff clays, layers subject to liquefaction, or soils susceptible to strength loss from remolding (see Site Class E and F), the effects of these layers on the curvature and, hence, moment demands on the pile, must be investigated. Often these soil-induced curvatures place a much higher moment demand on the piles than would be determined from conventional lateral force *P-y* analyses. In addition, at the interfaces of the layers described above and stiffer layers, plastic hinging may occur. Hence, confinement reinforcement per the concrete special-moment frame provisions must be provided at these interfaces and at the pile-to-cap connection for concrete piers and piles. See ASCE 7 Chapter 12.

Grade beams must be designed as ductile unless the beam is strong enough to resist the anticipated maximum earthquake force as set forth in the load combination of IBC Section 1605.4 and ASCE 7 Section 12.4. That is, grade beams must be either strong or ductile.

Anchorage of piles or piers into pile caps must consider the combined effects of uplift and pile fixity. For piles not subject to uplift by design or not required to provide rotational restraint, the anchorage must develop at least 25 percent of the strength of the pile in tension. For piles subject to uplift or required to provide rotational restraint, the anchorage must develop the lesser of the tensile capacity of the pile or 1.3 times the uplift capacity based on the soil strength. Because of the large variability in soils, it would be prudent to design these piles for the full tensile capacity rather than 1.3 times the uplift capacity.

Batter piles have performed poorly in past earthquakes. This is because the batter piles are laterally stiff relative to vertical piles and resist most of the seismic induced inertial forces. The piles are not usually designed to resist the actual forces, but are designed to resist an inertial force reduced by an assumed ductility. However, the batter piles are axially stiff and generally not detailed for ductility; hence the failures. To preclude this type of failure in batter piles, the piles and their connections must be designed to resist the anticipated maximum earthquake forces from the load combinations of IBC Section 1605.4.

Section 1809 *Driven Pile Foundations*

Except for steel H-piles, driven piles covered in this section are the displacement type. That is, as the pile is driven, a volume of soil is displaced by the pile volume, resulting in compaction of the surrounding soils.

1809.1 Timber piles. One of the new definitions added to Section 1808.1 of the 2006 IBC was for timber piles, which are round, tapered timbers with the small (tip) end embedded into the soil. The definition was added to be consistent with the timber pile definition in the AF&PA NDS Supplement for Timber Pole and Piles.

Timber piles, although not having the high load capacities of steel or concrete piles, are the most commonly used type of pile, mainly because of their availability and ease of handling. Timber piles are shaped from tree trunks and are tapered because of the natural taper of the trunk. Round timber piles are generally made from Southern Pine in lengths up to 80 feet or Pacific Coast Douglas Fir in lengths up to 120 feet. Other species that are used are red oak and red pine in lengths up to 60 feet.

Untreated timber piles that are embedded permanently below the ground-water level (fresh water only, not brackish or marine conditions) may last indefinitely. If embedded above the water table, the piles are subject to decay, and if above ground are also subject to insect attack. Hence, piles should be preservative treated.

1809.1.1 Materials. Timber piles may be either end-bearing or friction piles. ASTM D 25 sets forth the minimum circumference at the butt and the tip based on pile taper, as well as quality of the wood and tolerances on straightness, knots, twist of grain and other requirements.

1809.1.2 Preservative treatment. Piles must be pressure treated to prevent decay and insect attack. The 2006 IBC references AWPA U1 (Commodity Specification E, Use Category 4C) for round timber piles and AWPA U1 (Commodity Specification A, Use Category 4B) for sawn timber piles. See also discussion in Section 1805.4.5 for timber footings.

1809.1.3 Defective piles. Damage to the pile, including breakage, should be suspected when there is a sudden drop in penetration resistance while driving that cannot be explained by the soil profile. The pile should be withdrawn for examination. If penetration resistance should suddenly increase, driving should be stopped to avoid possible damage. A significant problem encountered during installation of timber piles is damage from overdriving. Overdriving can cause failure by bending, brooming of the tip, crushing, brooming at the butt end, or splitting or breaking along the pile section. See Figure 18-11.

1809.1.4 Allowable stresses. Allowable stress for timber piles and poles are in accordance with the provisions of the referenced standard, AF&PA NDS.

1809.2 Precast concrete piles.

1809.2.1 General. Precast concrete piles are manufactured as conventionally reinforced concrete or as prestressed concrete. Both types can be formed by bed casting, centrifugal casting,

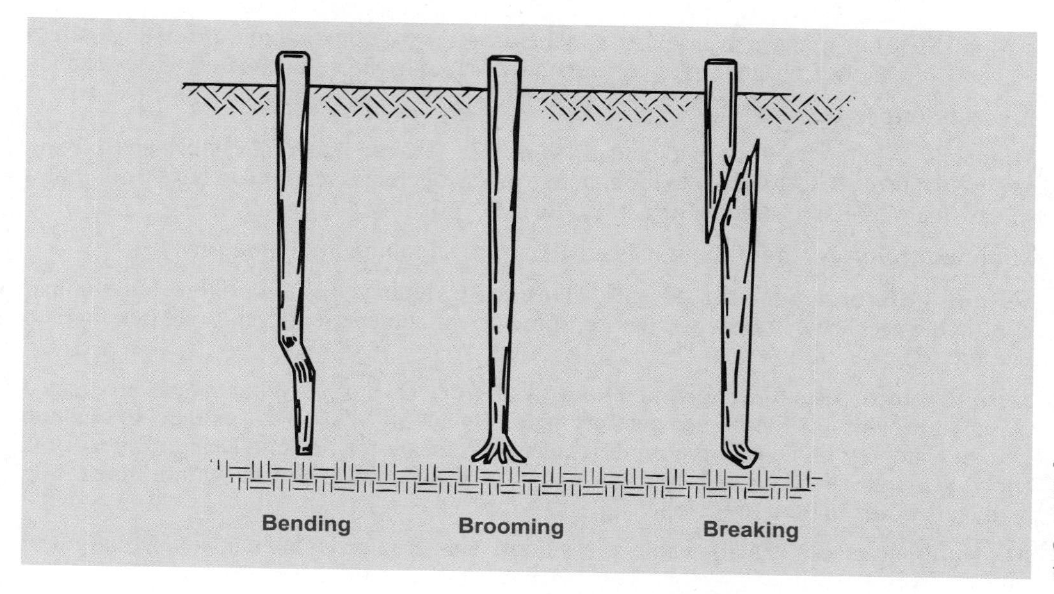

Figure 18-11
Effect of overdriving timber piles

slipforming or extrusion methods. Piles are usually square, octagonal or round, and either solid or hollow. Precast piles, which may be either friction or end-bearing piles, are of the displacement type and are driven into place.

1809.2.1.1

Design and manufacture. Piles must be designed to resist driving and handling stresses in addition to anticipated service loads. In long piles, tensile stresses resulting from driving may govern the design. In shorter piles, the handling loads may dominate.

1809.2.1.2

Minimum dimension. This is the minimum practical dimension to accommodate reinforcement.

1809.2.1.3

Reinforcement. The closely spaced spirals or ties at the ends are to accommodate radial tensile principle stresses from driving.

1809.2.1.4

Installation. Care must be used during installation to prevent damage during handling and driving. The proper cushion must be used at the driving end. Recommendations for design, manufacture and installation are given in ACI 543R. Damage to piles can be classified into four types:

1. Spalling at the butt or head (driving end) caused by high or irregular compressive stress concentrations. The spalling may be caused by insufficient cushioning, pile butt not square with the pile longitudinal axis, hammer and pile not aligned, reinforcing steel not flush or below the top of the pile allowing the hammer force to be transmitted through the steel, or insufficient transverse reinforcement.
2. Spalling at the tip, which is usually caused by an extremely high driving resistance such as when the tip is bearing on a rock.
3. Breaking or transverse cracking. This is caused by the rarefaction wave reflected from the tip. When the hammer strikes the cushion or head, a compression wave is produced that travels down the pile. The wave can be reflected from the tip as a rarefaction (tension) wave or a compression wave depending on the soil stiffness. Rarefaction waves usually occur when the soil at the tip is soft with very little resistance to penetration, causing tension waves that can cause significant tensile damage. This phenomenon usually occurs only in long piles exceeding 50 feet. Prestressed piles have more resistance to rarefaction damage than do conventionally reinforced piles. The hammer energy should be reduced when driving long piles through soft soils.

4. Spiral or transverse cracking may be caused by a combination of torsional stress and rarefaction stress. Torsion is usually caused by excessive restraint in the leads.

1809.2.2 Precast nonprestressed piles.

1809.2.2.1 Materials. Although piles are excluded from ACI 318, the material, quality and mixing requirements of ACI 318 apply to the concrete mix. Concrete is required to have a minimum specified compressive strength of not less than 3,000 psi.

1809.2.2.2 Minimum reinforcement. Four bars are the practical minimum for placement.

1809.2.2.2.1 Seismic reinforcement in Seismic Design Category C. Minimum longitudinal reinforcing steel and transverse ties or spirals confinement reinforcing are required to provide some ductility.

1809.2.2.2.2 Seismic reinforcement in Seismic Design Category D, E or F. Spirals or ties are spaced closer to provide for the higher ductility requirements in SDC D and higher. In Seismic Design Category D, E or F, the reinforcing requirements for Seismic Design Category C apply as well as additional transverse confinement reinforcement within three pile diameters of the bottom of the pile cap.

1809.2.2.3 Allowable stresses. The allowable stress limits are set to provide an adequate margin of safety.

1809.2.2.4 Installation. Handling and driving forces will likely govern pile strength requirements. Through the use of steam cure and Type III cements, 75 percent of specified strength can be achieved relatively quickly.

1809.2.2.5 Concrete cover. Cover requirements are set to provide protection and minimize steel corrosion.

1809.2.3 Precast prestressed piles. The purpose of prestressing piles is to place the concrete under a compressive stress so that hairline cracks caused by any subsequent tensile stress that may occur from handling, driving, superimposed loads or seismic imposed curvatures, and which are larger than the prestressed compression stress, will close when the tensile stresses are removed. This is easily achievable for handling and driving loads, but may not be feasible for seismic imposed curvatures.

Prestressed piles can be either pretensioned or post-tensioned. Pretensioned piles are generally cast full length in a casting bed at a manufacturing plant and often contain only prestressing steel reinforcement. Post-tensioned piles may be plant cast or site cast and generally contain mild steel reinforcing to resist handling stresses.

1809.2.3.1 Materials. Higher strength concrete is used in piles to reduce the volume changes, which reduces prestress losses, and to provide a more dense concrete to reduce cover requirements.

1809.2.3.2 Design. Minimum prestress is set to minimize cracking from handling and driving stresses.

1809.2.3.2.1 Design in Seismic Design Category C. Spiral transverse (confining) reinforcement is required to mitigate the effects of soil induced curvatures from seismic ground displacements. See discussion in commentary to Section 1809.2.23 and subsections. The volumetric requirement is the same as for columns in ductile frames. (See ACI 318, Section 21.4.2.) When using prestressing strand to anchor the pile in the cap, the development length should be sufficient to develop the strength of the strand.

1809.2.3.2.2 Design in Seismic Design Category D, E or F. These transverse reinforcing requirements are based on testing of prestressed piles in New Zealand and the subsequent recommendations by the Prestressed Concrete Institute (PCI). The requirements result in prestressed piles with good ductility without creating construction problems from reinforcement congestion. In Item 2, the "distance from the underside of the pile cap to the point of zero curvature" is determined as in Section 1810.1.2.1 below. See Sheppard[8] and Joen, et al.[9]

1809.2.3.2.3 Allowable stresses. The term $0.33f_c'$ is the same as conventionally reinforced piles. Because prestressing places additional compressive stresses on the pile, this stress must be subtracted from the allowable compressive stress; hence, the subtractive term $-0.27f_{pc}$. The term f_{pc} is the effective prestress on the gross area, which is the prestressing force remaining after losses have occurred.

Installation. Because of the precompression, less care is needed in the handling and driving than for conventionally reinforced piles, and prestressed piles are, in general, more durable than conventionally reinforced precast concrete piles. **1809.2.3.4**

Concrete cover. Cover requirements are set to minimize steel corrosion. **1809.2.3.5**

Structural steel piles. Structural steel piles are characterized by axial high-load capacity. Piles may be H-piles or steel pipe piles. **1809.3**

H-piles are usually used as deep end-bearing piles because they are essentially nondisplacement that can readily penetrate solid strata to reach rock or other suitable hard bearing strata such as dense gravels. Ideally, steel H-piles are driven to hard or medium hard rock.

Driven pipe piles are displacement type piles if driven closed; nondisplacement if driven open. Pipe piles are made of seamless or welded pipes and are frequently filled with concrete after driving. Pipe piles conforming to ASTM A 252 are used in both friction and end-bearing applications. Pipe piles may be driven open ended or closed ended. Open ended pipe piles are generally used when the soils investigation shows rock or a suitable end bearing stratum close to the ground surface, especially if the loads to be supported are large. The pipe is driven to bearing, the soils forced into the pipe during driving are cleaned out, and the pipe filled with concrete. Closed end piles are generally used as friction piles when a suitable bearing stratum is not available at suitable depths. There are several proprietary closed end pipe piles available.

Materials. H-piles are typically available in ASTM A 36 and A 572 steel. Pipe piles are fabricated from ASTM A 252 or A 283 plate. ASTM A 252 is a specification specifically for welded pipe piles. Piles using ASTM A 690 and A 992 are used in common practice and were added to the 2006 IBC. The Pile Driving Contractors Association (PDCA) Recommended Design Specifications For Driven Bearing Piles (PDCA 2001-9) contains both these material specifications for steel piles. **1809.3.1**

Allowable stresses. The stresses allowed consider stability. Tests have shown that for H-piles driven to refusal in rock through soils that provide full lateral support, the stresses at failure can approach the yield stress of the material. Hence, higher stresses are allowed if a soils investigation and load tests are performed. **1809.3.2**

Dimensions of H-piles. H-piles are proportioned to withstand the impact stresses from hard driving. The flange and web thicknesses are usually equal. The flange widths are proportioned such that the section modulus, S_y, in the weak axis is approximately one third of S_x. **1809.3.3**

Dimensions of steel pipe piles. This requirement differs from previous model code requirements in that it replaces an empirical requirement for a minimum thickness of $^1/_4$ inch with a more rational requirement that relates the minimum thickness to diameter and hammer energy by requiring a minimum area per kip-foot of energy. This requirement equates to a wall thickness of 0.27 inch for a 10-inch pipe driven with a hammer energy of 25 kip-feet. Note that if the wall thickness is less than 0.179 inch, a driving shoe is required to prevent local buckling at the tip from hard driving, regardless of diameter or hammer energy. Note that this minimum wall thickness of $^3/_{16}$ was changed to 0.179 inch in the 2003 IBC to be consistent with the most common minimum thickness for closed end pipe piles. ASCE 20-96 "Standard Guidelines for the Design and Installation of Pile Foundations" as well as the recommendations of the Driven Pile Committee of the Deep Foundations Institute list 0.179 inches as the minimum wall thickness. **1809.3.4**

Section 1809.3.5 Design in Seismic Design Category D, E or F, which imposed limitations on flange width to thickness or diameter to wall thickness ratio, was deleted in the 2003 IBC because it effectively prohibited the use of H piles in these seismic design categories. The limitations were essentially the same as beams and columns that are expected to undergo significant inelastic deformation in order to resist local buckling. For 50 kis steel, the criteria restricted the $b_f/2t_f$ ratio for flanges of HP sections to 7.1, although the $b_f/2t_f$ ratio of HP sections range from 9.0 to 14.4.

Section 1810 *Cast-in-Place Concrete Pile Foundations*

1810.1 General. Cast-in-place (CIP) concrete piles are installed by placing concrete into holes preformed by drilling or by driving a temporary or permanent casing to the required bearing depth. Drilled or augered piles are also known as cast-in-drill-hole or CIDH piles. CIP piles may be either cased or uncased. Uncased piles are difficult to construct when below the ground-water table. Except for enlarged base piles, the concrete in CIP piles is not subjected to driving forces, only the forces imposed by the service loads and downdrag from settlement. One advantage of drilled CIP piles is that the tip elevation can easily be adjusted to have the tip on the correct bearing stratum. Reinforcement is installed during the concreting operation. CIP drilled piles are of the nondisplacement type, i.e., the soil is not displaced or compacted by the drilling operation. CIP piles constructed by first driving a closed end shell are displacement piles, where the soil surrounding the shell is displaced and compacted during the driving operation. CIP piles constructed by driving an open-ended casing without a mandrel or temporary tip closure are nominally a nondisplacement pile, although some compaction around the shell may occur in cohesive soils, and densification may occur from driving in granular soils.

1810.1.1 Materials. The slump requirements stated are for cased piles. Slump requirements must be adjusted for other conditions. For example, concrete placed in uncased drilled holes needs to be in the 6 – 8 inch range so that the concrete flows readily into the irregularities in the drilled hole. Use of superplasticizers will provide the desired slump while keeping the water/cementitious material ratio low.

1810.1.2 Reinforcement. With two exceptions, reinforcing steel must be assembled into a cage and placed in the hole or casing prior to concreting, not stabbed after concreting. One exception is dowels less than 5 feet in length, and the other exception is for auger injected piles, which are placed by injecting the concrete through a hollow stem auger. The last sentence referencing "augered uncased cast-in-place piles" applies only to these auger injected piles.

1810.1.2.1 Reinforcement in Seismic Design Category C. Minimum steel requirements are established to provide some ductility. The minimum reinforcement must be continued throughout the flexural length of the pile. The term *flexural length* was not defined in the 2000 IBC. However, the term as defined in the 1997 UBC (from which the requirements were derived) was added to the 2003 IBC as the "length of the pile from the first point of zero lateral deflection to the underside of the pile cap or grade beam." The point of zero lateral deflection can be determined from the *P-y* analysis. Other clarification changes that were made in the 2003 IBC based on the 2000 NEHRP provisions were: *augered* piles were specifically named, the tie diameter was changed from $^1/_4$ inch to $^3/_8$ inch, and the region where transverse confinement reinforcing occurs below the pile cap was changed from two times to three times the least dimension of the pile.

1810.1.2.2 Reinforcement in Seismic Design Category D, E or F. In addition to the requirements for SDC C, the minimum reinforcement ratio is higher and extends over a longer length to improve ductility. In addition, closed ties or spirals are required to provide confinement in regions of plastic hinging, i.e., at the pile-cap interface, at the interface of soft to stiff layers, and in liquefaction zones. Confinement reinforcing should also be used for bay muds and sensitive clays. See also Section 1802.2.7. Clarification changes that were made in the 2003 IBC based on the 2000 NEHRP provisions were: *augered* piles were specifically named, the length of the pile over which the minimum longitudinal reinforcement is required was clarified with the flexural length defined within the provision itself, and specific references to ACI 318 were added for transverse confinement reinforcing requirements.

Concrete placement. Holes should be free from debris, loose soils or water. Water should be baled from the hole prior to concreting because excess water will cause segregation and dilution of the cement paste. If the water cannot be removed, the concrete should be placed by tremie methods and the concrete mix enriched and adjusted accordingly. When not placed by tremie, concrete should always be placed through an elephant trunk (funnel hopper) to avoid segregation **1810.1.3**

Enlarged base piles. Enlarged base piles are intended to be end-bearing type piles that spread the bearing load over a larger area than a prismatic pile, thereby increasing capacity. Enlarged based piles may be either cased or uncased. Enlarged base piles are used only in granular soils, which, because of the voids between soil particles, allow densification of the soils around the pile tip without creating excessive pressures. One type is the compacted base type, which consists of a bulb-shaped footing formed after driving the shaft casing to its final depth. See Section 1810.2.3 for a discussion of forming the bulb. Another type is the concrete-pedestal type, in which a truncated cone or pyramid shaped precast concrete tip larger than the steel casing diameter is driven into the soil with the casing. See Figure 18-12. **1810.2**

Materials. Concrete used for the bulb type of pile must have a zero slump to be stiff enough to be compacted by the drop weight. **1810.2.1**

Allowable stresses. Pile capacity must be based on concrete strength alone without consideration of soil capacity, and the area of the internal cross section of the pile, i.e., casing inside diameter. **1810.2.2**

Installation. Installation must employ the same methods used to install the load test piles. The compacted base pile is usually installed by driving a steel casing. A zero slump concrete plug is placed at the tip of the casing and impacted with a heavy drop weight, thereby driving the casing and plug. Sometimes a gravel plug is used rather than zero slump concrete. When the casing and plug have been driven to the required depth, the plug is driven out and the bulb is formed by progressively compacting additional layers of zero slump concrete. A welded reinforcing steel cage is added where required. If the pile is to be uncased, the shaft is formed by compacting zero slump concrete in small lifts as the casing is withdrawn. If the pile is to be cased, as would be required for piles through peats or other organic soils, the shaft is formed by inserting a steel shell inside the drive casing after forming the bulb, withdrawing the drive casing, and filling the shell with conventional concrete. The precast base type pile is installed with the precast base placed at the tip of a mandrel driven steel shell. **1810.2.3**

A problem that occurs with the precast base type and the cased bulb type piles is that an annular space between the casing and the soil remains. Either the pile must be designed as a slender reinforced concrete column governed by buckling, or the annular space must be

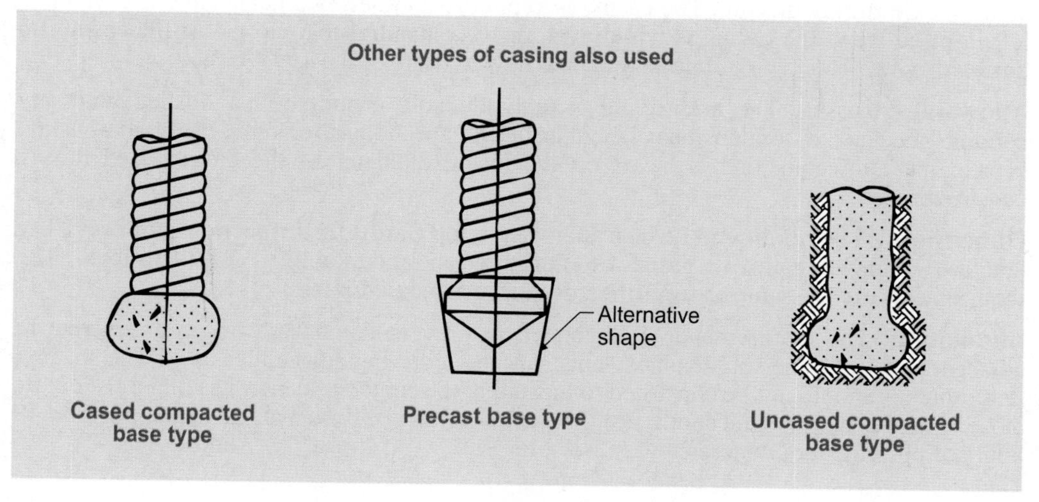

Figure 18-12 Enlarged base pile—cased or uncased shafts

filled to provide the requisite lateral support. The usual practice is to fill the annular space by pumping grout.

1810.2.4 Load-bearing capacity. The load-bearing capacity of enlarged base piles must be determined from load tests.

1810.2.5 Concrete cover. Concrete cover requirements are to provide protection and minimize corrosion of the steel in uncased piles.

1810.3 Drilled or augered uncased piles. This type of nondisplacement pile is installed by drilling or augering a hole and filling the uncased hole with concrete, either during or after withdrawing the auger.

1810.3.1 Allowable stresses. The allowable stresses for drilled uncased piles are the same as for cast-in-place piles for which holes are formed by machine drilling with auger or bucket type drills, with or without temporary casing. Concrete is placed by conventional methods including tremies or funnel hoppers. In the 2000 and 2003 IBC, the allowable stresses for augered piles are lower because of the problems of installation of augered piles. In this type of pile, the holes are formed by a hollow stem auger and concrete is injected through the hollow stem auger as the auger is withdrawn. Reinforcement, if placed without lateral ties, is also placed through ducts in the hollow stem auger. The provision in the 2006 IBC was changed to permit both concrete drilled or augered cast-in-place uncased piles to have an allowable stress limit of 33 percent of the 28-day specified compressive strength (f'_c), and the allowable compressive stress in the reinforcement was changed from 34 percent to 40 percent of the yield strength of the steel (or 25,500 psi). These changes were made for consistency with ASCE *Standard Guidelines for the Design and Installation of Pile Foundations* (ASCE 20-96).

1810.3.2 Dimensions. The minimum 12-inch diameter is for inspection purposes. The length-to-diameter ratio is based on construction and stability considerations.

1810.3.3 Installation. These requirements are intended to result in a satisfactory pile. The construction of drilled piles is fraught with problems: caving, ground water, etc. See BOCA *National Building Code*[10] Section 1820.3 for a discussion of problems.

1810.3.4 Reinforcement. The concrete must be quite fluid to allow for placement into the concrete. The concrete slump must exceed the limits of Section1810.1.1.

1810.3.5 Reinforcement in Seismic Design Category D, E or F. The reinforcement must meet minimum longitudinal reinforcement ratios and transverse reinforcement spacing to provide ductility.

1810.4 Driven uncased piles. This type of displacement pile is constructed by driving a temporary casing, removing the soils from the casing, and placing concrete in the hole as the casing is removed. The casing is driven with a closed end, thereby displacing and compacting adjacent soil during driving. The casing is kept closed either by a detachable tip, which is left in place when the casing is withdrawn, or by a mandrel that closes off the casing tip during driving. This type of pile will appear new to users of the 1997 UBC.

1810.4.1 Allowable stresses. The area of the actual pile hole cannot be established with any certainty because of dimensional irregularities of the hole after removal of the casing. Hence, the allowable load is based on the inside diameter of the casing used during construction.

1810.4.2 Dimensions. The minimum 12-inch diameter is primarily for inspection purposes. The length-to-diameter ratio is based on construction and stability considerations. The requirements are the same as for drilled or augered uncased piles.

1810.4.3 Installation. The requirements are intended to prevent damage to uncured concrete in adjacent piles from the soil displacements caused by driving adjacent casings. The spacing requirements should not be construed to mean that the center-to-center spacing must not be closer than one half the pile depth; just that casing cannot be driven at that spacing within 48 hours of placement of concrete.

Concrete cover. Concrete cover requirements are to provide protection and minimize corrosion of the steel. **1810.4.4**

Steel-cased piles. The steel-cased pile is the most widely used type of cast-in-place concrete pile. This pile type is characterized by a thin steel shell and is a displacement pile. This pile type consists of a closed-end light-gage steel shell or a thin-walled pipe driven into the soil and left permanently in place, reinforced when required for uplift, lateral bending or seismic induced curvatures, and filled with concrete. The shell or pipe is usually driven with a removable mandrel. The shell is either a constant section or a tapered shape. Steel encased piles are generally friction piles. **1810.5**

Materials. The steel shell must have sufficient strength to remain watertight and not collapse from ground pressure when the mandrel is removed. **1810.5.1**

Allowable stresses. The allowable stress is $0.33 f'_c$ as for other concrete piles. The steel shell is too thin to act as a composite pile but does act as confinement reinforcement. The allowable stress can be increased to $0.40 f'_c$ if the shell meets the confinement conditions in Section 1810.5.2.1 through 1810.5.2.4, in regard to thickness, strength and diameter. **1810.5.2**

Shell thickness. The shell thickness is not considered load carrying. **1810.5.2.1**

Shell type. The shell must be seamless or spirally welded. **1810.5.2.2**

Strength. The yield strength of steel normally used for the casings is 30 ksi. **1810.5.2.3**

Diameter. The maximum diameter is set so that the volumetric ratio of shell to concrete is sufficient to provide confinement for the concrete. **1810.5.2.4**

Installation. The restrictions on driving are primarily to avoid damage to uncured concrete in adjacent piles. **1810.5.3**

Reinforcement. Reinforcing is required when bending or tensile stresses are induced in the piles by imposed loads. **1810.5.4**

Seismic reinforcement. Reinforcement is required where seismic induced displacements cause bending in the pile. **1810.5.4.1**

Concrete-filled steel pipe and tube piles. Concrete-filled steel piles are either seamless or welded pipe, or closed end tubular piles with either straight or tapered sections that are driven into the soil. The piles may be installed as either friction or end-bearing piles. This pile type is characterized by a steel shell that is thicker than the thin shell used in the steel-cased pile of Section1810.5. Hence, both the concrete and steel shell are assumed to carry load compositely. If driven open ended, the earth core is removed from the shell prior to concreting. The shell may be driven with an internal mandrel. **1810.6**

Materials. The steel shell must be seamless or welded. **1810.6.1**

Allowable stresses. The allowable stress on the concrete is $0.33 f'_c$, as it is for other concrete piles. The steel shell acts compositely with the concrete. For mandrel driven pipe piles the walls are usually so thin that the strength of the shell is disregarded. **1810.6.2**

Minimum dimensions. The minimum diameter is primarily for inspection purposes. **1810.6.3**

Reinforcement. Reinforcing is required when bending or tensile stresses are induced in the piles by imposed loads. **1810.6.4**

Seismic reinforcement. For piles used in a structure assigned to SDC C, D, E or F, the minimum wall thickness is necessary to provide a minimum degree of concrete confinement, which is necessary for ductile concrete. Seismic induced displacements cause bending in the pile, thus requiring minimum reinforcement. **1810.6.4.1**

Placing concrete. A change was made to the 2006 IBC that permits concrete to be chuted directly into smooth sided pipes and tubes without a centering funnel hopper because many closed end pipes are filled without a hopper, and the pipe itself functions as a hopper. The American Association of State Highway and Transportation Official (AASHTO) **1810.6.5**

installation specifications do not require or even mention a hopper for pipe piles. The main purpose of the centering funnel for drilled piles is to prevent the concrete from encountering the soil at the perimeter of the hole, which is generally not a problem for pipes and tubes. The term *smooth sided* was included to prevent possible segregation from the ridges or corrugations if nonsmooth pipes or tubes are used.

1810.7 **Caisson piles.** The caisson pile, commonly known as a drilled-in caisson, is installed as a special type of high-load-capacity pile. Caissons are characterized by a structural steel core, an upper-cased section extending to bedrock, and a lower uncased tip that is socketed into rock.

1810.7.1 **Construction.** The caisson pile is a cased cast-in-place concrete pile that is formed by 1) driving a heavy-wall open-ended pipe down to bedrock, 2) cleaning out the soil materials within the pipe, 3) drilling an uncased socket into the bedrock, 4) inserting a structural steel core into the pipe and 5) filling the entire pipe and drilled socket with concrete.

The core material is usually made of hot-rolled structural steel wide flange or I-beam sections, or steel rails. This section specifies that the steel core is to extend full length from the base of the drilled socket to the top of the steel pipe or, as an alternative and depending on design requirements, the steel core may extend halfway up the pipe or as a stub core to a distance in the pipe at least equal to the depth of the socket.

1810.7.2 **Materials.** The pipe is a welded or seamless pipe conforming to ASTM A252 with a minimum thickness of $^{3}/_{8}$ inch.

1810.7.3 **Design.** The strength of the cassion is developed in combined friction and end-bearing of the rock socket.

1810.7.4 **Structural core.** The core material is usually made of hot-rolled structural steel wide flange or I-beam sections, or steel rails. This section specifies that the steel core is to extend full length from the base of the drilled socket to the top of the steel pipe or, as an alternative and depending on design requirements, the steel core may extend halfway up the pipe, or as a stub core to a distance in the pipe at least equal to the depth of the socket.

1810.7.5 **Allowable stresses.** The concrete and heavy walled shell are designed to act compositely.

1810.7.6 **Installation.** The section requires the rock socket and pile to be clean of foreign materials before filling with concrete to ensure adequate bond. Concrete should not be placed through water unless a tremie is used.

1810.7.8 **Micropiles.** A new definition and section on micropiles were added to the 2006 IBC. Micropiles are defined as 12-inch-diameter or less bored, grouted-in-place piles incorporating steel pipe (casing) and/or steel reinforcement. Prior to their inclusion in the code, the use of micropiles had to be approved under the alternative materials, design and methods of construction provisions in Section 104.11. The provisions are based on the recommendations of the ADSC/DFI (International Association of Foundation Drilling/ Deep Foundations Institute) Committee on Micropiles, and are intended to provide a uniform standard for micropiles, and eliminate inconsistencies in the design and installation of micropiles.

The new IBC provisions are based primarily on the *Massachusetts Building Code* (MBC). The MBC has been used on hundreds of micropile projects, including many outside of Massachusetts, since its inception in 1988. The IBC section is not a direct transcription of the MBC requirements but was modified to follow the format of the IBC as well as incorporate some technical changes:

1) The maximum allowable stress for grout was increased. The MBC limits the allowable stress on grout to $0.33f'_c$, but not to exceed 1600 psi. The IBC provision removes the limit of 1600 psi because of the use of high shear colloidal mixers that have become more readily available to micropile contractors over the past 5 years. The high shear action of the mixers provides more thorough hydration of the cement than is achieved with paddle mixers and allows lower water/cement ratios to be used, which results in significantly higher compressive strengths for the

conventional neat cement grouts. In addition, the 2000 FHWA publication, "Micropile Design and Construction Guidelines, Implementation Manual" (FHWA-SA-97-070), sets the allowable stress on grout at $0.4f'_c$, with no limit on f'_c.

2) The maximum allowable stress for steel was increased. The MBC limits the allowable stress on steel to 0.4 F_y, but not exceeding 24,000 psi. The IBC provision increases the limit to 32,000 psi because high-strength steel reinforcing bars and pipes have become readily available and are frequently used in micropiles at these design stresses. The 32 ksi is based on 0.4 x 80 ksi. ACI 318 permits a maximum usable strain at extreme concrete compression fiber of 0.003, which corresponds to a steel stress of 87 ksi. The 2004 DFI/ADSC publication "Guide to Drafting a Specification for High Capacity Drilled and Grouted Micropiles for Structural Support" sets the allowable stress on steel at $0.4F_y$ with no limit on F_y. DFI/ADSC limits the useable F_y to 80 ksi, with $F_c = 0.4\ F_y$. The 2000 FHWA aforementioned publication sets the allowable stress on steel at 0.47 F_y. The publication also has a discussion on the 80 ksi limit.

3) Load testing requirements, corrosion protection and quality control provisions are provided elsewhere in Chapter 18; therefore, they were not included in the IBC section on micropiles.

Section 1811 *Composite Piles*

Composite piles, as used herein, refer to piles placed in series, such as a cast-in-place concrete pile placed over a submerged wood pile. The maximum load is limited by the capacity of the weakest section.

Section 1812 *Pier Foundations*

Pier foundations are isolated masonry or cast-in-place concrete structural elements extending into firm materials. Piers are relatively short relative to their width, with lengths less than or equal to twelve times the least dimension of the pier. Hence, piers are rigid or semirigid. See Winterkorn, et al[2] for methods of analysis of rigid piers. The provisions for piers in this 2000 IBC section are primarily exceptions to the requirements of Section 1808.2 when used for one- or two-family residential structures and appurtenant U occupancies not exceeding two stories built with light-frame construction.

1812.1 **General.** The piers must meet all applicable provisions outlined in this section as well as the applicable requirements that apply to piers and piles in Section 1808.2.

1812.2 **Lateral dimensions and height.** The minimum dimension must be 2 feet, and the length must be less than or equal to twelve times the least dimension of the pier. If the the length exceeds 12 times the least horizontal dimension, it meets the definition of a pile.

The three exceptions in the 2000 IBC were deleted and replaced with a new Section 1808.2.1 that was inserted above the general provisions in Section 1808.2.2. This was done because there was confusion and some conflict between the exceptions and the general requirements for piers and piles. The new section in 2006 IBC essentially permits piles to be designed as piers in accordance with this section where either of the following conditions exists: 1) Group R-3 and U occupancies not exceeding two stories of light-frame construction, or 2) where the surrounding foundation materials furnish adequate lateral support for the pile. Note that the use of this exemption from the pile requirements is subject to the approval of the building official.

1812.3 Materials. Concrete with f'_c = 2,500 psi is allowed.

1812.4 Reinforcement. Minimum reinforcement is 0.25 percent in the top one third of the pier, but not less than 10 feet.

1812.7 Masonry. Masonry may be used subject to dimensional limitations.

1812.8 1812.8 Concrete. Plain (unreinforced) concrete may be used subject to dimensional limitations; otherwise, the pier must be reinforced per the column provisions of ACI 318.

REFERENCES

[1] Terzaghi, Karl and Peck, Ralph B., *Soil Mechanics in Engineering Practice*, John Wiley and Sons, 1967.

[2] Winterkorn, Hans F., and Fang, Hsai-Yang, *Foundation Engineering Handbook*, Van Nostrand Reinhold Company, Inc., New York, NY, 1975.

[3] Bowles, J. E., *Foundation Analysis and Design*, 5th Edition, McGraw Hill, Inc., 1996.

[4] Johnson, S. M., and Kavanagh, T. C., *The Design of Foundations for Buildings*, McGraw Hill, Inc.

[5] Liu, C., and Evett, J. B., *Soils and Foundations*, 2nd Edition, Prentice-Hall, Inc.

[6] Sowers, G. B., and Sowers, G. F., *Introductory Soil Mechanics in Engineering Practice*, Macmillan Publishing Co.

[7] Committee on the Bearing Value of Pile Driving Foundations, *Pile-driving Formulas—Progress Report*, Proceedings, May, American Society of Civil Engineers, New York, NY, 1941.

[8] Sheppard, David A., "Seismic Design of Prestressed Concrete Piling", *PCI Journal*, March-April, 1993.

[9] Joen, Pam Hoat, and Park, Robert, "Simulated Seismic Load Tests on Prestressed Concrete Piles and Pile-Cap Connections", *PCI Journal*, November-December, 1990.

[10] ICC, *The BOCA National Building Code—Commentary Volume 2*, International Code Council, Inc., Washington, DC, 1999.

BIBLIOGRAPHY

ACI 229R, *Controlled Low-Strength Materials*, American Concrete Institute, Farmington Hills, MI, 1999.

ACI 318-95, *Building Code Requirements for Structural Concrete*, American Concrete Institute, Farmington Hills, MI, 1995.

ACI 530/ASCE 5/TMS 402 , *Building Code Requirements for Masonry Structures*, American Concrete Institute, Farmington Hills, MI, 1995.

ACI 530.1/ASCE 6/TMS 602 , *Specification for Masonry Structures*, American Concrete Institute, Farmington Hills, MI, 1995.

ICC, *Handbook to the Uniform Building Code: An illustrative commentary*, International Code Council, Washington, DC, 1998.

Federal Highway Administration, *Micropile Design and Construction Guidelines, Implementation Manual*, FHWA Publication No. FHWA-SA-97-070, McLean, Virginia, June 2000.

ICC, *Proposed Changes to the Final Draft of the International Building Code*, International Code Council, Washington, DC, January 1999.

ICC, *1998 Proposed Changes to the First Draft of the International Building Code*, International Code Council, Washington, DC, April 1998.

Joint Micropile Committee of The Deep Foundations Institute (DFI) and The International Association of Foundation Drilling (ADSC), *Guide to Drafting a Specification for High Capacity Drilled and Grouted Micropiles for Structural Support*, Dallas, Texas, 2004.

NEHRP, *NEHRP (National Earthquake Hazard Reduction Program) Recommended Provisions for the Development of Seismic Regulations for New Buildings (and Other Structures)*, Building Seismic Safety Council, Washington, DC, 1994 (1997).

SEAOC, *Recommended Lateral Force Requirements and Commentary*, Structural Engineers Association of California, Sacramento, CA, 1996.

UBC-IBC Structural Comparison & Cross Reference, International Code Council, Washington, DC.

Zeevaert, L., *Foundation Engineering for Difficult Soil Conditions*, 2nd Edition, Van Nostrand Reinhold Company, Inc., New York, NY.

CHAPTER 19

Part 1 2006 IBC CONCRETE PROVISIONS

Of the three legacy codes, the *Uniform Building Code*™ (UBC)[1] adopted standards by transcription, while the BOCA *National Building Code* (BOCA/NBC)[2] and the *Standard Building Code* (SBC)[3] adopted standards by reference. The last edition of the UBC, dated 1997, transcribed the text of the 1995 edition of ACI 318 *Building Code Requirements for Structural Concrete* (ACI 318-95)[4] with amendments, which were printed in italics. The last editions of the BOCA/NBC and the SBC, dated 1999, adopted ACI 318-95 by reference, with Chapters 2 through 7 of that standard reproduced for the convenience of building inspectors and other building department personnel.

Chapter 19 of the 2003 *International Building Code* (IBC) adopted the 2002 edition of the ACI 318 Standard (ACI 318-02)[5] for concrete design and construction. Portions of Chapters 2 through 7 of ACI 318-02 were reproduced, with just a few amendments printed in italics, in IBC Sections 1902 through 1907. The remainder of ACI 318-02 was adopted by reference, subject to seven amendments listed in Section 1908. Section 1909, Structural Plain Concrete, reproduced parts of ACI 318-02 Chapter 22, referred to the rest of it, reformatted and rearranged reproduced text and made two deviations from ACI 318-02. Section 1910, Seismic Design Provisions, classified shear walls into plain concrete, detailed plain concrete, reinforced concrete and special reinforced concrete, and then prescribed seismic design provisions by the Seismic Design Category: one set for SDC B, a second set for SDC C and a third set for SDC D, E, and F. Section 1913 adopted Appendix D, Anchoring to Concrete, of ACI 318-02, and made one significant amendment to the Appendix D provisions. Chapter 19 of the 2003 IBC also included the following provisions not contained in ACI 318-02:

1911 – Minimum Slab Provisions

1912 – Anchorage to Concrete – ASD

1914 – Shotcrete

1915 – Reinforced Gypsum Concrete

1916 – Concrete-Filled Pipe Columns

The 2005 edition of the ACI 318 Standard (ACI 318-05)[6] is the primary reference document for concrete design and construction in the 2006 IBC. In a fairly significant change, Sections 1902 through 1907 no longer reproduce portions of Chapters 2 through 7 of ACI 318-05. The various subsections within those sections have been maintained. However, the subsections simply refer the user to the corresponding sections or subsections in Chapters 2 through 7 of ACI 318-05. This change was the result of a code change submitted by the American Concrete Institute (ACI). The following was cited as reason for the change: "After considerable analysis of the legal implications related to continued permission to allow ICC to reprint portions of this standard [ACI 318] in the *International Building Code*, ACI has determined that it is not prudent to allow ICC to continue to reprint the standard. The primary reason for this is that the ACI loses significant control over the use of our document and therefore compromises the integrity of the copyright."

In another significant change to the concrete chapter, Section 1910 has been eliminated, with the contents either incorporated in Section 1908 or deleted as being unnecessary. Many of the provisions in Section 1910 were repeated from what is in ACI 318, or from Section 1908. In addition, the section contained several modifications to ACI 318, which should have been in Section 1908. The expectation is that with the placement of all modifications to ACI 318 in Section 1908, ACI Committee 318 will consider the modifications to determine if they are appropriate for inclusion in the ACI 318 standard.

All sections subsequent to Section 1910 have been renumbered as a result of the elimination of Section 1910. For instance, the previous Section 1916 is now Section 1915.

Section 1912, Anchorage to Concrete-Strength Design, references Appendix D, Anchoring to Concrete, of ACI 318-05 for the strength design of cast-in and post-installed anchors embedded in concrete for purposes of transmitting structural loads from one connected element to the other. The provisions of Appendix D appeared for the first time in ACI 318-02, and are revised in ACI 318-05, as described in the appendix of this chapter. In Section 1912.1, one exception is made to these provisions, which pertains to the concrete breakout strength requirements of Section D.4.2.2 for single anchors exceeding 2 inches in diameter and/or 25 inches of embedment depth. Amendments are also made to ACI 318-05 Sections D.3.3.2 though D.3.3.5 in 2006 IBC Section 1908. A brief background on the development of Appendix D is given in the discussion of Section 1912 below.

Section 1901 *General*

1901.1

Scope. Chapter 19 provides minimum requirements governing the materials, quality control, design and construction of structural concrete elements of any structure erected under the requirements of the 2006 IBC.

1901.2

Plain and reinforced concrete. This section requires that structural concrete be designed and constructed in accordance with the requirements of Chapter 19 and ACI 318 as amended in Section 1908 of the 2006 IBC. Chapter 35 specifies that ACI 318-05 in particular must be used. This section states that Chapter 19 does not govern the design and construction of soil-supported slabs (i.e., slabs on grade), unless the slab transmits vertical loads or lateral forces from other portions of the structure to the soil, although Sections 1904, Durability Requirements, and 1910, Minimum Slab Provisions, do apply. A similar statement is to be found in Section 1.1.6 of ACI 318-05, although that statement does not make ACI 318 Chapter 4, Durability Requirements, applicable to slabs on grade. As mentioned, the equivalent of Section 1910 does not exist in ACI 318-05. Section 1.1.6 of ACI 318-95 mentioned gravity loads only; reference to lateral forces was added in the 1999 edition of ACI 318.

1901.3

Source and applicability. Sections 1902 through 1907 are formatted after the provisions for structural concrete in ACI 318-05. However, the actual provisions are adopted through reference only.

1901.4

Construction documents. This section reproduces from ACI 318-05, Section 1.2.1, the list of items to be shown on design drawings and specifications. Two items on the ACI list are omitted from the IBC. These are (1) the name and date of issue of the code and supplement to which the design conforms, and (2) live load and other loads used in design. See Section 1603 for additional items.

1901.5

Special inspection. Section 1704 of the 2006 IBC contains special inspection requirements for structural concrete. ACI 318-05, in general, does not contain provisions for special inspection of structural concrete, except that Section 1.3.5 requires continuous inspection of the placement of the reinforcement and concrete in special moment frames resisting seismic forces in regions of high seismic hazard, or in structures assigned to high seismic performance or design categories. Section 1901.5 provides a cross reference to the inspection requirements of IBC Chapter 17.

Section 1902 *Definitions*

This section simply references Chapter 2 of ACI 318-05 for the definitions of terms related to concrete construction used in 2006 IBC Chapter 19. No definition is actually contained in this section.

Section 1903 *Specifications for Tests and Materials*

1903.1 **General.** This section specifies that materials required to produce concrete and the testing of such materials must be in compliance with the provisions of Chapter 3 of ACI 318-05. The referenced provisions include those regarding cement, aggregates, water, steel reinforcement, admixtures and storage of materials. The corresponding ACI 318-05 provisions are discussed in Sections 19.4.3 through 19.4.8 below.

Chapter 17 of the 2006 IBC contains special inspection requirements that are not included in ACI 318-05, except that Section 1.3.5 does require continuous inspection of reinforcement and concrete placement in special moment frames. Chapter 17 also contains test requirements that are not included in ACI 318-05. Section 1903.1 makes these IBC Chapter 17 requirements applicable whenever required by the provisions of that chapter.

1903.2 **Glass fiber reinforced concrete.** This section in the 2006 IBC, which is not to be found in ACI 318-05, requires that glass fiber reinforced concrete (GFRC) and the materials used in such concrete conform to PCI MNL 128, *Recommended Practice for Glass Fiber Reinforced Concrete Panels*.[7] This publication contains the latest information on the planning, design, manufacture and installation of GFRC panels.

Cement. (ACI 318-05 Section 3.2) ACI 318 requires cements to conform to ASTM C 150, *Specification for Portland Cement*;[8] ASTM C 595, *Specification for Blended Hydraulic Cements*[9] (excluding Types S and SA, which are not intended as principal cementing constituents of structural concrete); ASTM C 845, *Specification for Expansive Hydraulic Cement*;[10] and ASTM C 1157, *Performance Specification for Hydraulic Cement*.[11] ASTM C 1157 was included in the 2000 IBC as a modification to ACI 318-99, since this ASTM standard had not been adopted into ACI 318-99. ACI 318-05 Section 3.2.2 requires that "cement used in the work shall correspond to that on which selection of concrete proportions was based." Refer to ACI 318-05 Commentary Section R3.2.2 for explanation of this requirement.

Aggregates. (ACI 318-05 Section 3.3) ACI 318-05 requires concrete aggregates to conform to either ASTM C 33, *Specification for Concrete Aggregates*,[12] or ASTM C 330, *Specification for Lightweight Aggregates for Structural Concrete*.[13] However, aggregates conforming to these ASTM specifications are not always economically available and, in many instances, noncomplying materials have a long history of satisfactory performance. Such nonconforming materials are permitted with the approval of the building official when acceptable evidence of satisfactory performance is provided.

ACI 318-05, Section 3.3.2 specifies size limitations on aggregates to ensure the proper encasement of reinforcement and to minimize honeycombing. These requirements are waived if, in the judgment of the engineer, workability and methods of construction are such that concrete can be placed without honeycombs or voids.

Water. (ACI 318-05 Section 3.4) Almost any water that is drinkable and has no pronounced taste or odor can be used as mixing water for making concrete. However, some waters that are not fit for drinking may still be suitable for mixing concrete.

Excessive impurities in mixing water may not only affect setting time and concrete strength but may also cause efflorescence, staining, corrosion of reinforcement, volume instability and reduced durability. Reference 14 advises that certain optional limits may be set on chlorides, sulfates, alkalis and solids in the mixing water, or that appropriate tests may be performed to determine the effect that the impurities have on various properties. Some impurities may have little effect on strength and setting time, yet they may adversely affect durability and other properties.

According to Reference 14, water containing no more than 2000 parts per million (ppm) of total dissolved solids can generally be used satisfactorily for making concrete. Water containing more than 2000 ppm of dissolved solids should be tested for its effects on strength and time of set.

Concern over high chloride content in mixing water is chiefly due to the possible adverse effect of chloride ions on the corrosion of reinforcing steel or prestressing strand. In fact, prestressing strand and aluminum embedments are singled out as areas of concern in ACI 318-05, Section 3.4.2. Chloride ions attack the protective oxide film formed on the steel or aluminum by the highly alkaline (pH>12.5) chemical environment present in concrete. Chlorides can be introduced into concrete with the separate mix ingredients—admixtures, aggregates, cement and mixing water—or through exposure to deicing salts, seawater or salt-laden air in coastal environments. Placing an acceptable limit on chloride content for any one ingredient, such as mixing water, is difficult considering the several possible sources of chloride ions in concrete. An acceptable limit depends primarily upon the type of structure and the environment to which it is exposed during its service life.[14] ACI 318-05, Section 4.4.1 is devoted to this topic.

Water of questionable suitability can be used for making concrete if mortar cubes (ASTM C 109)[15] made with it have 7-day and 28-day strengths equal to at least 90 percent of the corresponding strengths of companion specimens made with drinkable or distilled water. Reference 14 advises that ASTM C 191[16] tests should additionally be made to ensure that impurities in the mixing water do not adversely shorten or extend the setting time of the cement. Acceptable criteria for water to be used in concrete are given in ASTM C 94 *Specification for Ready-Mixed Concrete*[17] and the American Association of State Highway and Transportation Officials (AASHTO) T 26.[18]

Steel reinforcement. (ACI 318-05 Section 3.5) This section regulates reinforcement as well as welding of reinforcement to be placed in concrete.

Reinforcement is required to be deformed reinforcement, except that plain reinforcement is allowed for spirals or prestressing steel. Reinforcement consisting of structural steel, steel pipe or steel tubing is permitted as specified in ACI 318. Tables 19-1, 19-2 and 19-3 show the different reinforcement types recognized by ACI 318. Other metal elements such as inserts, anchor bolts or plain bars for dowels at isolation or contraction joints, are not normally considered to be reinforcement under the provisions of ACI 318.

Welding of reinforcing bars is required to conform to "*Structural Welding Code—Reinforcing Steel*," AWS D1.4 of the American Welding Society.[19] This document covers aspects of welding reinforcing bars, including criteria to qualify welding procedures.

The type and location of welded splices and other required welding of reinforcing bars must be indicated on the design drawings or in the project specifications. Much of this is also required under Section 1.2.1 of ACI 318-05.

Weldability of steel is based on its chemical composition or carbon equivalent. The AWS D1.4 welding code establishes preheat and interpass temperatures for a range of carbon equivalents and reinforcing bar sizes. Carbon equivalent is calculated from the chemical composition of the reinforcing bars. For bars other than ASTM A 706, *Specification for Low-Alloy Steel Deformed and Plain Bars for Concrete Reinforcement*,[20] the producer of the reinforcing bars does not routinely provide the chemical analysis required to calculate the carbon equivalent. For welding reinforcing bars other than ASTM A 706 bars, the design drawings or project specifications need to specifically require that results of the chemical analysis be furnished.

Table 19-1. Deformed reinforcement recognized in ACI 318-05

Product	ASTM Specification	Grade or Type		Minimum Yield Strength ksi	Minimum Tensile Strength ksi
Reinforcing bars	A615	40*		40	60
		60		60	90
		75*		75	100
	A706	60		60 (78 max)	80
	A 996**	Type R	50	50	80
			60	60	90
		Type R	40	40	70
			60	60	90
Bar mats[a]	A 184	-		-	-
Wire, deformed	A 496	-		75	85
Welded wire reinforcement, plan (W 1.2 and larger)	A 185	-		65	75
Welded wire reinforcement, deformed	A 497	-		70	80
Galvanized reinforcing bars[b]	A 767	-		-	-
Epoxy–coated reinforcing bars[b]	A 775 A 934	-		-	-
Epoxy–coated wires[c] and welded wire reinforcement[d]	A 884	-		-	-

*Grade 40 bars are furnished only in sizes 3 through 6. Grade 75 bars are furnished only in sizes 6 through 18.
** Bars are furnished only in sizes 3 through 8.
[a]Same as reinforcing bars, except only available with Grade 40 bars (ASTM A 615) and grade 60 bars (ASTM A 615 or ASTM A 706).
[b]Same as reinforcing bars.
[c]Same as wires.
[d]Same as welded wire reinforcement.

The ASTM A 706 specification covers low-alloy steel reinforcing bars intended for applications requiring controlled tensile properties or welding. Weldability is accomplished in the ASTM A 706 specification by limits or controls on chemical composition or carbon equivalent. The producer is required by the ASTM A 706 specification to report the chemical composition and carbon equivalent.

The AWS D1.4 welding code requires the contractor to prepare written welding procedure specifications conforming to the requirements of the welding code. Appendix A of the welding code contains a suggested form that shows the information required for such a specification.

It is often necessary to weld to existing reinforcing bars in a structure when no mill test report of the existing reinforcement is available. This condition is particularly common in alterations or additions to existing buildings. The AWS D1.4 welding code provides guidance concerning this situation.

Table 19-2. **Plain reinforcement and prestressing tendons recognized in ACI 318-05**

Product	ASTM Specification	Grade or Type	Minimum Yield Strength ksi	Minimum Tensile Strength ksi
Reinforcing bars	A615	40	40	60
		60	60	90
		75	75	100
	A706	60	60	80
Wire, plain	A 82	-	70	80
Prestressing wire	A 421	-	199.75-212.5	235-250
Prestressing wire, low-relaxation	A 421 + supplement	-	199.75-212.5	235-250
Prestressing strand, stress-relaxation	A416	250	212.5	250
		270	229.5	270
Prestressing strand, low-relaxation	A416	250	225	250
		270	243	270
Prestressing bar	A 722	Type I	127.5	150
		Type II	120	150

Table 19-3. **Structural steel, steel pipe or tubing recognized by ACI 318-05 for use in composite compression member**

Product	ASTM Specification
Structural Steel Used with Reinforcing Bars in Compression Members Meeting Requirements of Section 10.16.7 or 10.16.8	
Carbon structural steel	A 36
High-strength low-alloy structural steel	A 242
High-strength low-alloy columbium-vanadium structural steel	A 572
High-strength low-alloy structural steel with 50 ksi minimum yield strength up to 4 in. thick	A 588
Steel Pipe or Tubing for Composite Compression Members Composed of Steel Encased Concrete Core Meeting Requirements of Section 10.16.6	
Seamless and welded black and hot-dipped galvanized steel pipe	A 53 Grade B
Cold-formed welded and seamless carbon steel structural tubing in rounds and shapes	A 500
Hot-formed welded and seamless carbon steel structural tubing	A 501

AWS D1.4 does not cover welding of wire to wire, and of wire or welded wire reinforcement to reinforcing bars or structural steel elements. Some guidance on this subject is available in the ACI 318-05 Commentary, Section R3.5.2.

Note that ASTM A 616, *Specification for Rail-Steel Deformed and Plain Bars for Reinforcement*, and ASTM A 617, *Specification for Axle-Steel Deformed and Plain Bars for Concrete Reinforcement*, have been replaced by ASTM A 996, *Specification for Rail-Steel and Axle-Steel Deformed Bars for Concrete Reinforcement*.[21] Reinforcing bars conforming to ASTM A 996 are generally not available in all but a few areas of the U.S. ASTM 996 bars can be Type Rail Symbol (with a rail symbol stamped on the bars), Type R (with the letter R stamped on the bars) or Type A (with the letter A stamped on the bars). The first two types are rail steel, the third type is axle steel. ACI 318 requires rail-steel reinforcing bars to conform to the provisions for Type R bars, which are required to meet more restrictive provisions for bend tests.

Note that the term *welded wire fabric* has been changed consistently to *welded wire reinforcement* throughout ACI 318-05. This is for consistency with ASTM A 185 and A 497, in which the change had been made earlier.

Admixtures. (ACI 318-05 Section 3.6) ACI 318-05 Section 3.6.1 requires admixtures used in concrete to be subject to prior approval by the registered design professional.

The admixtures recognized for use in concrete in Section 3.6 of ACI 318-05 are listed in Table 19-4.

Table 19-4. Concrete admixtures recognized by ACI 318-05

Product	ASTM Specification	Desired Effect
Accelerators	C 494, Type C	Accelerate setting and develop early strength
Air-entraining admixtures	C 260	Improve durability in environments of freeze-thaw, deicers, sulfate, and alkali reactivity improved workability
Fly ash or other pozzolans	C 618	Pozzolanic activity improve workability, plasticity, sulfate resistance; reduce alkakli reactivity, permeability, heat of hydration partial cement replacement filler
Ground granulated blast furnace slag	C 989	Same as above
Retarders	C 494, Type B	Retard setting time
Silica fume	C 1240	Same as fly ash or other pozzolans
Superplasticizers	C 1017, Type 1	Reduce water-cement ratio Produce flowing concrete
Superplasticizer and retarder	C 1017, Type 2	Produce flowing concrete with retarded set Reduce water
Water reducer	C 494, Type A	Reduce water demand by at least 5%
Water reducer and accelerator	C 494, Type E	Reduce water (minimum 5%) and accelerate set
Water reducer and retarder	C 494, Type D	Reduce water (minimum 5%) and retard set
Water reducer—high range	C 494, Type F	Reduce water demand (minimum 12%)
Water reducer—high range and retarder	C 494, Type G	Reduce water demand (minimum 12%) and retard set

Storage of materials. (ACI 318-05 Section 3.7) This section requires proper storage of cementitious materials and aggregates to prevent contamination or deterioration, and prohibits the use of any material so affected in concrete.

Section 1904 *Durability Requirements*

Durability requirements are adopted through reference to Chapter 4 of ACI 318-05, with certain exceptions. The provisions of that chapter emphasize the importance of considering durability requirements before the designer selects f'_c and cover over reinforcing steel.

As pointed out in the commentary to Chapter 4 of ACI 318-05, maximum water-cementitious materials ratios of 0.40 to 0.50 that may be required for concrete exposed to freezing and thawing, sulfate in soils or waters, or for preventing corrosion of reinforcement, will typically be equivalent to requiring an f'_c of 5000 to 4000 pounds per square inch (psi), respectively. Generally, the required average concrete strengths, f'_{cr}, will be 500 to 700 psi higher than the specified compressive strength, f'_c. Because it is difficult to accurately determine the water-cementitious materials ratio of concrete during production, the f'_c specified should be reasonably consistent with the water-cementitious materials ratio required for durability. This will help ensure that the required water-cementitious materials ratio is actually obtained in the field. Because the usual emphasis in inspection is on strength, test results substantially higher than the specified strength may lead to a lack of concern for quality and production of concrete that exceeds the maximum water-cementitious materials ratio. Thus, an f'_c of 3000 psi and a maximum water-cementitious materials ratio of 0.45 should not be specified for a parking structure, if the structure will be exposed to deicing salts.

The commentary to Chapter 4 of ACI 318-05 also points out that the chapter does not include provisions for especially severe exposures, such as acids or high temperatures, and is not concerned with aesthetic considerations such as surface finishes. These items should be covered in project specifications.

1904.1

Water-cementitious materials ratio. This section simply references ACI 318-05 Section 4.1.

1904.2

Freezing and thawing exposures. This section mostly references ACI 318-05 Section 4.2. ACI 318 Section 4.2.1 requires normal-weight as well as lightweight concrete exposed to freezing and thawing or deicing chemicals to be entrained with air in amounts specified in ACI 318 Table 4.2.1, subject to a tolerance of ± 1.5 percent.

ACI 318 Section 4.2.2 requires that concrete subject to the exposures listed in Table 4.2.2 must not exceed the maximum water-cementitious materials ratios given in that table, and must not fall below the minimum specified compressive strength requirements of that table. Although the minimum strength requirements are given for normal-weight as well as for lightweight concrete, the maximum water-cementitious materials ratio requirements are given only for normal-weight concrete. This is because determination of absorption of lightweight aggregates is uncertain, making calculation of the water-cementitious materials ratio uncertain.

2006 IBC Section 1904.2.2 has added an exception to the ACI 318 provisions requiring that in buildings less than four stories in height, that house Group R (residential) Occupancies, normal-weight concrete subject to weathering (freezing and thawing), as determined from Figure 1904.2.2, or deicer chemicals, must comply with the requirements of Table 1904.2.2. Neither Figure 1904.2.2 nor Table 1904.2.2 is part of ACI 318-05. The exception, the table, and the figure are adopted into the IBC from the *International Residential Code*® (IRC®).[22] Table 1904.2.2 mandates a minimum specified compressive strength as a function of the concrete element and exposure, but no maximum water-cementitious materials ratio. Figure 1904.2.2 shows the geographic regions within the U.S. mainland that are classified as having negligible, moderate and severe weather exposures for the purposes of Table 1904.2.2. It is clearly indicated in a note to the figure that the boundary lines defining the weather regions are only approximate. Building officials in communities near a boundary line must establish applicable exposure classifications.

ACI 318 Section 4.2.3 provides Table 4.2.3, which establishes limitations on the amount of fly ash, other pozzolans, silica fume and slag that can be included in concrete exposed to deicing chemicals.

1904.3

Sulfate exposures. This section references Section 4.3 of ACI 318-05. ACI 318 Section 4.3.1 requires concrete exposed to injurious concentrations of sulfates from soil and water to be made with a sulfate-resisting cement. ACI 318 Table 4.3.1 lists the appropriate types of cement and the maximum water-cementitious materials ratios and minimum strengths for various exposure conditions.

ACI 318 Section 4.3.2 prohibits the use of calcium chloride as an admixture in concrete that is to be exposed to severe and very severe sulfate-containing solutions, as defined in

Table 4.3.1. Such use, because of the calcium content of the admixture, would adversely affect the performance of cement in resisting sulfate attack.

1904.4 Corrosion protection of reinforcement. This section references Section 4.4 of ACI 318-05. The maximum concentration of water soluble chloride ion permitted in hardened concrete from all sources is given in ACI 318 Table 4.4.1, Section 4.4.1. The soluble chlorides induce corrosion, whereas chlorides that are chemically combined with other ingredients have little or no corrosive effect. Additional information on the effects of chlorides on the corrosion of reinforcing steel is given in the *Guide to Durable Concrete* reported by ACI Committee 201[23] and the *Corrosion of Metals in Concrete* reported by ACI Committee 222.[24]

To provide the properties of concrete that will resist the harmful effects of exposure to chlorides from deicing chemicals, salts, saltwater, seawater, brackish water or spray from these sources, ACI 318 Section 4.4.2 requires concrete mixtures to conform to the maximum water-cementitious materials ratio (in the case of normal-weight aggregate concrete only) and the minimum specified compressive strength requirements of Table 4.2.2.

IBC Section 1907.7 provides the minimum concrete cover requirements for the protection of reinforcement through reference to ACI 318 Section 7.7. Section 1907.7.5 mandates that such requirements be suitably increased in corrosive environments or under other severe exposure conditions in accordance with ACI 318 Section 7.7.5. Section 18.16 of ACI 318-05 contains provisions for corrosion protection of unbonded prestressing tendons.

Section 1905 *Concrete Quality, Mixing and Placing*

As pointed out in the commentary to Chapter 5 of ACI 318-05, which is referenced by IBC Section 1905 for most of its provisions, the requirements for proportioning concrete mixes are based on the premise that concrete must provide both adequate durability (Section 1904) and strength. The criteria for acceptance of concrete are intended primarily to protect the safety of the public. Section 1905 describes procedures by which concrete of adequate strength can be obtained, and provides procedures for checking the quality of the concrete during and after its placement.

The provisions of Sections 1905.2, 1905.3 and 1905.4, through reference to ACI 318 Sections 5.2, 5.3 and 5.4, respectively, together with Section 1904, establish required mix proportions. The basis for determining the adequacy of concrete strength is in Section 1905.6.

Section 1905 also provides minimum criteria for preparation of equipment and place of deposit (Section 1905.7), mixing (Section 1905.8), conveying (Section 1905.9), depositing (Section 1905.10) and curing (Section 1905.11) of concrete. Requirements for cold weather concreting and hot weather concreting (Sections 1905.12 and 1905.13, respectively) are also given.

1905.1 General. The basic premises governing the designation and evaluation of concrete strength are presented in Section 1905.1.1. It is emphasized that the average strength of concrete produced should always exceed the specified value of f'_c used in the structural design calculations. This is based on probabilistic concepts and is intended to ensure that adequate concrete strength will be developed in the structure. Additionally, the durability requirements of Section 1904 are to be satisfied.

According to Section 1905.1.1, the minimum specified compressive strength of concrete is required to be 2500 psi for concrete designed and constructed in accordance with IBC Chapter 19. This is the same requirement as that given in ACI Section 5.1.1. However, Section 1905.1.1 goes on to state that no maximum specified compressive strength shall apply unless restricted by a specific provision of the IBC or ACI 318-05. Although this statement is not included in ACI Section 5.1.1, it is contained in ACI Section 1.1.1.

Selection of concrete proportions. This section simply references Section 5.2 of ACI 318-05. **1905.2**

Recommendations for selecting proportions for concrete are given in detail in *Standard Practice for Selecting Proportions for Normal, Heavyweight, and Mass Concrete*[25] reported by ACI Committee 211. It provides two methods for selecting and adjusting proportions for normal-weight concrete: the estimated weight and absolute volume methods. Example calculations are shown for both methods. Proportioning by the absolute volume method is also illustrated in Reference 14. Recommendations for lightweight concrete are given in *Standard Practice for Selecting Proportions for Structural Lightweight Concrete*[26] also reported by ACI Committee 211.

ACI Section 5.2.1 requires that the selected water-cementitious materials ratio be low enough, or in the case of lightweight concrete, the compressive strength be high enough, to satisfy both the strength criteria (Section 5.3 or 5.4) and the special exposure requirements (Chapter 4). At the same time, workability needed for proper placement must be provided.

The use of field experience or laboratory trial mixtures (Section 1905.3) is emphasized as the preferred method for selecting concrete mixture proportions. When no prior experience or trial mixture data are available, estimation of the water-cementitious materials ratio as prescribed in ACI 318-05 Section 5.4 is permitted, subject to approval by the registered design professional.

Proportioning on the basis of field experience and/or trial mixtures. This refers to ACI 318-02 Section 5.3. **1905.3**

Proportioning from field data. A previously used concrete mix design may be used for a new project if strength test data and standard deviations (ACI 318-05 Commentary Section R5.3.1) show that the mixture is acceptable. Durability requirements must also be met. The statistical data must essentially represent the same materials, proportions and concreting conditions to be used on the new project. The data used for proportioning must be from a concrete with an f'_c within 1000 psi of the strength required for the proposed work. Also, the data must represent at least 30 consecutive tests or two groups of consecutive tests totaling at least 30 tests (one test requires the average strength of two cylinders from the same sample). If only 15 to 29 consecutive tests are available, an adjusted standard deviation can be obtained by multiplying the standard deviation (s) for the 15 to 29 tests and a modification factor from ACI 318-05 Table 5.3.1.2. The data must represent 45 or more days of tests.

The modified standard deviation is then used in ACI 318-05 Equations 5-1 through 5-3, depending on the magnitude of the specified compressive strength of the concrete. The average compressive strength from the test record must equal or exceed the required average compressive strength, f'_{cr}, of ACI 318 in order for the concrete proportions to be acceptable. When the specified compressive strength f'_c is less than or equal to 5000 psi, the f'_{cr} for the selected mix proportions is equal to the larger of:

$$f'_{cr} = f'_c + 1.34s \quad \text{ACI 318-05 Equation (5-1)}$$

$$f'_{cr} = f'_c + 2.33s - 500 \quad \text{ACI 318-05 Equation (5-2)}$$

WHERE:

f'_{cr} = required average compressive strength of concrete used as the basis for selection of concrete proportions, psi

f'_c = specified compressive strength of concrete, psi

s = standard deviation, psi

When f'_c is greater than 5000 psi, the f'_{cr} for the selected mix proportions is equal to the larger of:

$f'_{cr} = f'_c + 1.34s$ ACI 318-05 Equation (5-1)

$f'_{cr} = 0.90f'_c + 2.33s$ ACI 318-05 Equation (5-3)

When field strength test records do not meet the previously discussed requirements, f'_{cr} can be obtained from ACI 318-05 Table 5.3.2.2. A field strength test record, several strength test records, or tests from trial mixtures must be used for documentation showing that the average strength of the mixture is equal to or greater than f'_{cr}.

If fewer than 30 but not fewer than 10 tests are available, the tests may be used for average strength documentation if the time period is not less than 45 days. Mix proportions may also be established by interpolating between two or more test records if each meets the above and project requirements. If a significant difference exists between the mixes used in the interpolation, a trial mix should be considered to check strength gain. If the test records meet the above requirements and limitations of ACI 318, the proportions for the mix may then be considered acceptable for the proposed work. For concrete strengths over 5000 psi, where the average strength documentation is based on laboratory trial mixtures, it may be appropriate to increase f'_{cr} calculated in Table 5.3.2.2 to allow for a reduction in strength from laboratory trials to actual concrete production.

Proportioning by trial mixtures. When field test records are not available or are insufficient for proportioning by field experience methods, the selection of concrete proportions may be based on trial mixes. The trial mixes must use the same materials proposed for the work. At least three mixes with three different water-cementitious materials ratios or cement contents must be made to produce a range of strengths that encompass f'_{cr}. The trial mixes must have slump and air content within ±0.75 in. and ±0.5 percent, respectively, of the maximum permitted. At least three cylinders per water-cementitious materials ratio must be made and cured according to ASTM C 192.[27] At 28 days or the designated test age, the compressive strength of the concrete must be determined by testing the cylinders in compression. A water-cementitious materials ratio

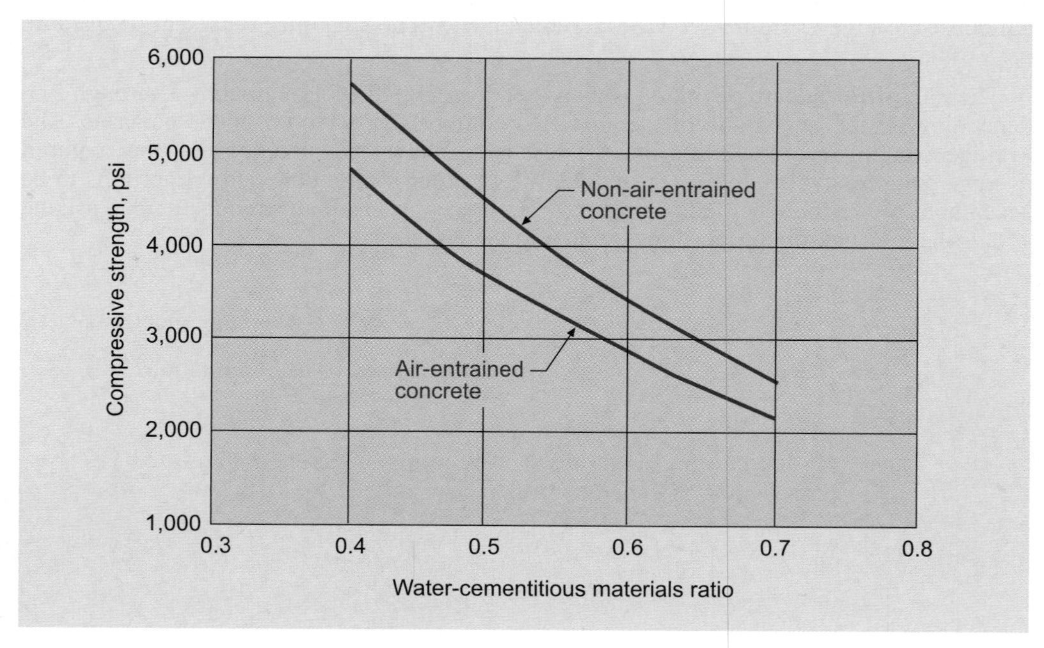

Figure 19-1 Typical trial mixture of field data strength curves[14]

versus strength curve (for example, see Figure 19-1) can then be plotted and the proportions interpolated from the data. It is also good practice to test the properties of the newly proportioned mix in a trial batch.

Proportioning without field experience or trial mixtures. This section refers to ACI 318-05 Section 5.4, which makes special provisions for mix proportioning based on experience where test data are not available and it is impractical to make trial batches. Permission of the registered design professional is a requirement of Section 5.4. Because combinations of different ingredients may vary considerably in strength level, this procedure is not permitted for specified compressive strengths greater than 5000 psi, and the required average strength f'_{cr} must exceed f'_c by 1200 psi. The purpose of this provision is to allow work to continue when there is an unexpected interruption in concrete supply and there is not sufficient time for tests and evaluation, or in small structures where the cost of trial mix data is not justified. **1905.4**

Average strength reduction. This section refers to Section 5.5 of ACI 318-05, which permits the amount by which f'_{cr} must exceed f'_c to be reduced, provided the three criteria stipulated in this section are met. **1905.5**

Evaluation and acceptance of concrete. The criteria for evaluation and acceptance of concrete follow Sections 5.6.1 through 5.6.5 of ACI 318-05. **1905.6**

Qualified technicians. This section specifically requires field testing of concrete to be done by qualified field testing technicians and laboratory testing of concrete to be done by qualified laboratory testing technicians. According to ACI 318-05 Commentary Section R5.6.1, a qualified field testing technician should be certified in accordance with the requirements of ACI Concrete Field Testing Technician—Grade I Certification Program, or the requirements of ASTM C 1077[28] (*Standard Practice for Laboratories Testing Concrete and Concrete Aggregates for Use in Construction and Criteria for Laboratory Evaluation*), or an equivalent program. A qualified laboratory testing technician should be certified in accordance with the requirements of ACI Concrete Laboratory Testing Technician, Concrete Strength Testing Technician, or the requirements of ASTM C 1077. **1905.6.1**

Frequency of testing. This specifies that strength tests for each class of concrete placed each day should be taken not less than once a day, nor less than once for each 150 cubic yards of concrete, nor less than once for each 5,000 square feet of surface area for slabs or walls. The average strength of two cylinders is required for each test. Additional specimens may be required when high-strength concrete is involved or where structural requirements are critical. The specimens should be laboratory-cured. Specifications may require that additional specimens be made and field-cured as nearly as practical in the same manner as the concrete in the structure. A 7-day test cylinder, along with the two 28-day test cylinders, is often made and tested to provide an early indication of strength development. As a rule of thumb, the 7-day strength is about 60 percent to 75 percent of the 28-day strength, depending upon the type and amount of cement, water-cementitious materials ratio, curing temperature and other variables. **1905.6.2**

Strength test specimens. This section refers to ACI 318 Section 5.6.3 for sampling and testing requirements for specimens prepared for acceptance testing of concrete in accordance with Section 1905.6.2, and for criteria for acceptance of test results. **1905.6.3**

Samples must be taken in accordance with ASTM C 172, *Standard Practice for Sampling Freshly Mixed Concrete*.[29]

Premolded specimens for strength testing must be made and cured in accordance with ASTM C 31, *Standard Practice for Making and Curing Concrete Test Specimens in the Field*.[30]

The standard test specimen for compressive strength of concrete with a maximum aggregate size of 2 inches or smaller is a cylinder 6 inches in diameter by 12 inches high. For larger aggregates, the diameter of the cylinder should be at least three times the maximum-size aggregate, and the height should be twice the diameter. Although rigid metal molds are preferred, paraffined cardboard, plastic or other types of disposable molds conforming to ASTM C 470[31] may be used. They should be placed on a smooth, level surface and filled carefully to avoid distortion of their shape. Unless required by the project specification, cylinders smaller than 6 inches by 12 inches must not be made in the field.

The cylinders, cured under standard laboratory conditions, must be tested in compression at a specified rate of loading at 28 days of age or at the designated test age, in accordance with the requirements of ASTM C 39, *Standard Test Method for Compressive Strength of Cylindrical Concrete Specimens*.[32]

The concrete strength is considered to be satisfactory as long as averages of any three consecutive strength tests equals or exceeds the specified f'_c, and no individual strength test falls below the specified f'_c by more than 500 psi when f'_{cr} is 5,000 psi or less, or by more than $0.10f'_c$ when f'_c is more than 5,000 psi. When concrete fails to meet either criterion, steps must be taken to increase the average of the subsequent concrete test results. See ACI 318-05 Section 5.6.5 for investigation of low-strength test results.

1905.6.4 Field-cured specimens. Strength tests of cylinders cured under field conditions (in conformance with the requirements of ASTM C 31) may be required by the building official to check the adequacy of curing and protection of concrete in the structure. These tests and the evaluation of results need to be carried out in accordance with ACI 318 Section 5.6.4.

Procedures for protecting and curing concrete are required to be improved when the strength of field-cured cylinders at the test age designated for determination of f'_c is less than 85 percent of that of companion laboratory-cured cylinders. However, results for the job-cured cylinders are considered satisfactory if the job-cured cylinders exceed the specified f'_c by more than 500 psi, even though they fail to reach 85 percent of the strength of the companion laboratory-cured cylinders.

1905.6.5 Low-strength test results. This section refers to ACI 318 Section 5.6.5 concerning the procedure to be followed when strength tests have failed to meet the specified acceptance criteria. If, after properly assessing the significance of low test results and determining whether they indicate need for concern, further investigation is deemed necessary, such investigation may include nondestructive tests, or in extreme cases, strength tests of cores taken from the structure. ACI Section 5.6.5.3, contains requirements for handling and testing cores.

For cores, if required, conservatively safe acceptance criteria are provided that should ensure structural adequacy of virtually any type of construction. Core tests yielding an average of 85 percent of the specified strength is entirely realistic. To expect core tests to yield 100 percent of f'_c is not realistic, insofar as differences in the size of specimens, conditions of obtaining samples and procedures of curing do not permit equal values to be obtained. When the core tests fail to provide assurance of structural adequacy, it may be practical, particularly in the case of floor or roof systems, for the building official to require a load test (Chapter 20 of ACI 318-05).

1905.7 Preparation of equipment and place of deposit. This section references Section 5.7 of ACI 318-05. Detailed recommendations for mixing, handling and transporting, and placing concrete are given in *Guide for Measuring, Mixing, Transporting and Placing Concrete* reported by ACI Committee 304.[33]

ACI Section 5.7 mandates the use of clean equipment and the cleaning of forms and reinforcement thoroughly before beginning to deposit concrete. In particular, sawdust, nails, wood pieces and other debris that may collect inside the forms must be removed. Reinforcement must be thoroughly cleaned of ice, dirt, loose rust, mill scale or other coatings. Water must be removed from the forms.

1905.8 Mixing. This section refers to ACI 318-05 Section 5.8.

All concrete must be mixed thoroughly until it is uniform in appearance, with all ingredients evenly distributed. Reference 14 points out that mixers should not be loaded above their rated capacities and must be operated at approximately the speeds for which they were designed. Increased output should be obtained by using a larger mixer or additional mixers, rather than by speeding up or overloading the equipment on hand. If the blades of the mixer become worn or coated with hardened concrete, the mixing action is going to be less efficient. Badly worn blades should be replaced and hardened concrete should be removed periodically.

If concrete has been adequately mixed, samples taken from different portions of a batch will have essentially the same unit weight, air content, slump and coarse aggregate content. Maximum allowable differences in test results within a batch of ready-mixed concrete are given in ASTM C 94.

The necessary time of mixing will depend on many factors, including batch size, stiffness of the batch, size and grading of the aggregate, and the efficiency of the mixer. Excessively long mixing times need to be avoided to guard against grinding of the aggregates.

1905.9

Conveying. This section references Section 5.9 of ACI 318-05.

Each step in the handling and transporting of concrete must be controlled to maintain uniformity within a batch and from batch to batch. Segregation of the coarse aggregate from the mortar or of water from the other ingredients must be avoided.

The equipment for handling and transporting concrete is required to be capable of supplying concrete to the place of deposit continuously and reliably under all conditions and for all methods of placement. Placing equipment commonly utilized are pumps, belt conveyors, pneumatic systems, wheelbarrows, buggies, crane buckets and tremies.

As pointed out in ACI 318 Commentary Section R5.9, serious loss in strength can result when concrete is pumped through pipe made of aluminum or aluminum alloy because of hydrogen generated by the reaction between the cement alkalis and the aluminum. It is, therefore, important not to use equipment made of aluminum or aluminum alloys for pump lines, tremies or chutes other than short ones.

1905.10

Depositing. This section refers to Section 5.10 of ACI 318-05 for provisions concerning depositing of concrete.

Concrete must be deposited continuously as near as possible to its final position. To avoid segregation, concrete should not be moved horizontally over too long a distance as it is being placed in forms or slabs.

In general, concrete should be placed in horizontal layers of uniform thickness, each layer having been thoroughly consolidated before the next is placed. The rate of placement should be rapid enough so that a layer of concrete has not yet set when a new layer is placed upon it. This avoids flow lines, seams and planes of weakness (cold joints) that result when freshly mixed concrete is placed on hardened concrete. Reference 14 recommends that layers be about 6 to 20 inches thick for reinforced members, and 15 to 20 inches thick for mass concrete, the thickness depending on the width between forms and the amount of reinforcement.

Rehandling concrete can cause segregation of the materials. Hence, caution is sounded against this practice.

Consolidation is the process of compacting fresh concrete to mold it within the forms and around embedded items and reinforcement and to eliminate stone pockets, honeycomb and entrapped air. It should not remove significant amounts of intentionally entrained air from air-entrained concrete. Proper mechanical consolidation makes possible the placement of stiff mixtures with the low water-cementitious materials ratios and high coarse aggregate contents associated with high-quality concrete, even in highly reinforced elements. Vibration, either internal or external, is the most widely used method for consolidation of concrete. Detailed recommendations for consolidation of concrete are given in *Guide to Consolidation of Concrete* reported by ACI Committee 309.[34]

ACI 318-05 Commentary Section R5.10 provides a recommended practice "when conditions make consolidation difficult, or where reinforcement is congested . . ."

Where construction joints are required, they must be in accordance with ACI 318-05 Section 6.4.

1905.11 Curing. This section references ACI 318-05 Section 5.11 curing provisions.

Curing is the maintenance of satisfactory moisture content and temperature in concrete during some definite period immediately following placement and finishing, so that the desired properties may develop. The need for adequate curing of concrete cannot be overemphasized. Curing has a strong influence on such properties of hardened concrete as durability, strength, water tightness, abrasion resistance, volume stability and resistance to freezing and thawing and deicer salts.[14] Recommendations for curing concrete are given in detail in *Standard Practice for Curing Concrete* reported by ACI Committee 308.[35]

Accelerated curing procedures require careful attention to obtain uniform and satisfactory results. Preventing moisture loss during the curing is essential. Guidance is available in *Accelerated Curing of Concrete at Atmospheric Pressure—State of the Art* reported by ACI Committee 517.[36]

Although accelerated curing produces early strength gain, the compressive strength of steam-cured concrete is not as high as that of similar concrete continuously cured under moist conditions at moderate temperatures. Also, the modulus of elasticity of steam-cured specimens may vary from that of specimens moist-cured at normal temperatures.

ACI 318-05 Commentary Section R5.11.3 suggests that when steam curing is used, it is advisable to base the concrete mix proportions on steam-cured test cylinders.

1905.12 Cold weather requirements. This section references ACI 318-05 Section 5.12. Detailed recommendations concerning cold weather concreting are given in *Cold Weather Concreting* reported by ACI Committee 306.[37]

1905.13 Hot weather requirements. ACI 318-05 Section 5.13 is referenced here. Detailed recommendations concerning hot weather concreting are given in *Hot Weather Concreting* reported by ACI Committee 305.[38]

Section 1906 *Formwork, Embedded Pipes and Construction Joints*

1906.1 Formwork. This section references Section 6.1, Design of formwork, from ACI 318-05.

As pointed out in ACI 318-05 Commentary Section R6.1, only minimum performance requirements for formwork, necessary to provide for public health and safety, are prescribed. Detailed information on formwork for concrete is given in *Guide to Formwork for Concrete*[39] reported by ACI Committee 347 and in *Formwork for Concrete*[40] prepared under the direction of ACI Committee 347. The former provides recommended practice for design and construction of formwork, including recommendations for loads and pressures. The latter is a manual that extensively describes systems and provides design procedures, design aids and examples. Valuable information on formwork is also contained in Reference 41.

1906.2 Removal of forms, shores and reshores. The removal of formwork for multistory construction must be part of a planned procedure considering the temporary support of the whole structure as well as that of each individual member. Such a procedure must be worked out prior to construction and, according to ACI 318-05 Commentary Section R6.2, should be based on structural analyses taking into account the following items, as a minimum:

1. The structural system that exists at the various stages of construction and the construction loads corresponding to those stages. The construction loads are frequently at least as large as the specified live loads.
2. The strength of the concrete at the various ages during construction.
3. The influence of deformations of the structure and shoring system on the distribution of dead loads and construction loads during the various stages of construction. The deformations themselves are of interest and must be limited. At

early ages, a structure may be adequate to support the applied loads, but may deflect sufficiently to cause permanent damage.

4. The strength and spacing of shores or shoring systems used, as well as the method of shoring, bracing, shore removal and reshoring, including the minimum time intervals between successive operations.
5. Any other loading or condition that affects the safety or serviceability of the structure during construction.

Further information on many of the above aspects may be found in Reference 42.

For multistory construction, the strength of the concrete in beams and slabs during the various stages of construction must be substantiated by tests of field-cured specimens or other methods approved by the building official.

Conduits and pipes embedded in concrete. This adopts Section 6.3 of ACI 318-05 by reference. **1906.3**

Conduits, pipes and sleeves made of any material that is not harmful to concrete is allowed to be embedded in the concrete, subject to the approval of the registered design professional, provided such embedment does not have an adverse impact on the strength of the structure. The contractor must not be permitted to install conduits, pipes, ducts or sleeves that are not shown on the plans or that are not approved by the registered design professional.

The use of aluminum in structural concrete is prohibited, unless it is effectively coated or covered. Aluminum reacts with concrete and, in the presence of chloride ions, may also react electrolytically with steel, causing cracking and/or spalling of the concrete. Aluminum electrical conduits present a special problem because stray electric current accelerates the adverse reaction.

Construction joint. This reproduces Section 6.4 of ACI 318-05. **1906.4**

Construction joints occur where concrete work is concluded for the day; they separate areas of concrete placed at different times. ACI 318-05 Commentary Section R6.4 points out that, for the integrity of the structure, it is important for all construction joints to be defined in construction documents and constructed as required. Any deviations should be approved by the registered design professional.

When freshly mixed concrete is placed on or against recently hardened concrete, certain precautions need to be taken to secure a well-bonded, watertight joint. The hardened concrete must be clean, sound, fairly level and reasonably rough with some coarse aggregate particles exposed. Any laitance, soft mortar, dirt, wood chips, form oil or other foreign materials should be removed because they would interfere with proper bonding of the subsequent placement. All construction joints are required to be wetted immediately before new concrete placement, and all standing water removed.

Construction joints are to be located where they cause the least weakness in the structure. Where shear that is due to gravity load is not significant, as is usually the case in the middle of the span of flexural members, a simple vertical joint may be adequate. The necessity to transfer the effects of lateral forces may require special design treatment of construction joints. Shear keys, intermittent shear keys, diagonal dowels or shear friction (ACI 318-05 Section 11.7) may be used whenever a force transfer is required.

The casting of beams, girders or slabs supported by columns or walls is required to be delayed until concrete in the vertical support members is no longer plastic. This is necessary to prevent cracking at the interface of the slab and supporting member caused by bleeding and settlement of plastic concrete in the supporting member.

Separate placement of slabs and beams, haunches and similar elements is permitted only when shown in design drawings or specifications and where provisions have been made for all necessary transfer of forces.

Section 1907 *Details of Reinforcement*

This section is based on Chapter 7 of ACI 318-05.

Good reinforcing details are vital to satisfactory performance of reinforced concrete structures. Standard practice for reinforcing details has evolved gradually. The ACI *Detailing Manual*[43] provides recommended methods and standards for preparing design drawings, typical details and drawings for fabrication and placing of reinforcing steel in reinforced concrete structures. As an aid to designers, "Recommended Industry Practice for Estimating, Detailing, Fabrication, and Field Erection of Reinforcing Materials" are included in the CRSI *Manual of Standard Practice*,[44] for direct reference in project drawings and specifications. The WRI *Structural Detailing Manual*[45] provides information on detailing welded wire fabric reinforcement systems. Chapter 3 of Reference 46 effectively summarizes much available information on reinforcement detailing.

1907.1 Hooks. This refers to Section 7.1, Standard hooks, of ACI 318-05.

The standard hooks defined in Section 7.1 of ACI 318-05 are illustrated in Figure 19-2.

1907.2 Minimum bend diameters. This section refers to ACI 318-05 Section 7.2.

Standard bends in reinforcing bars are described in terms of the inside diameter of bend, because this is easier to measure than the bend radius. The primary factors affecting the minimum bend diameter are the practicality of bending without breakage and the avoidance of crushing of concrete inside the bend.

Standard hooks for primary reinforcement

Bar size, No.	Minimum finished bend diameter*
3 through 8	$6d_b$
9, 10, 11	$8d_b$
14 and 18	$10d_b$

*measured on inside of bar

Standard hooks for stirrups and tie reinforcement

Bar size, No.	Minimum finished bend diameter*
3 through 5	$4d_b$
6 through 8	$6d_b$

*measured on inside of bar

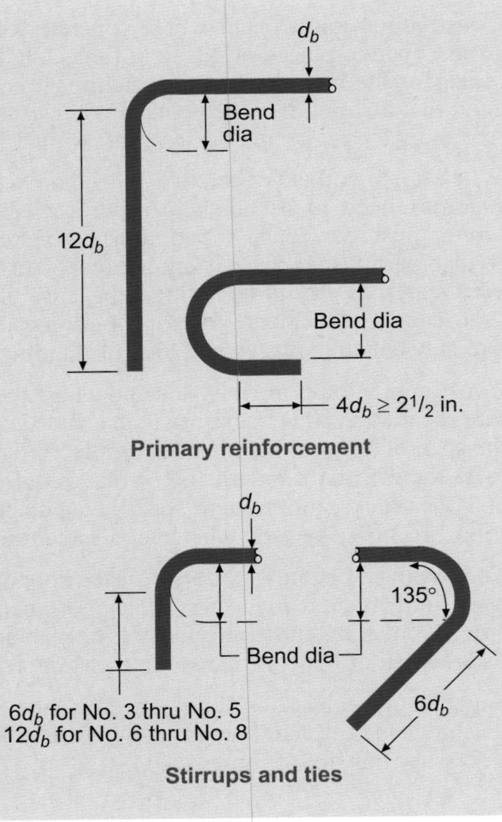

For SI: 1 inch = 25.4 mm, 1 degree = 0.01745 rad.

Figure 19-2 Standard hooks of ACI 318

The minimum bend diameters for bars used as primary reinforcement and as stirrup or tie reinforcement are shown in Figure 19-2. The minimum diameters of bend in welded wire reinforcement (plain or deformed) used for stirrups and ties are shown in Table 19-5.

Table 19-5. Minimum diameters of bend in welded wire reinforcement

Wire Size	Minimum Bend Diameter
D6 and smaller	$2d_b$
All other	$4d_b$

Note: Bend with inside diameter of less than 8 d_b must not be closer than 4 d_b to nearest welded intersection.

Bending. This section refers to Section 7.3 of ACI 318-05. **1907.3**

All reinforcement must be bent cold unless otherwise permitted by the registered design professional. For unusual bends, special fabrication including heating may be required and the registered design professional must give approval to the techniques used.

Reinforcing bars partly embedded in concrete are frequently subjected to bending and straightening in the field. Protruding bars often must be bent to provide clearance for construction operations. Field bending and straightening may also be required because of incorrect fabrication or accidental bending. Bars partly embedded in concrete must not be field-bent without authorization from the registered design professional. Tests using ASTM A 615 Grade 60 deformed bars provide guidelines for field bending and straightening, and heating if necessary, of such bars partly embedded in concrete. Reference 46 summarizes the recommendations of Reference 47 regarding field bending and straightening of reinforcing bars.

Surface conditions of reinforcement. At the time of concrete placement, reinforcement must be clean of any material that may interfere with bond between concrete and reinforcement, although epoxy coating in conformance with ASTM A 775,[48] A 934[49] or A 884[50] is specifically permitted. Guidance with regard to the effects of rust and mill scale on bond characteristics of deformed reinforcing bars is provided in Reference 51. Research has shown that a normal amount of rust increases bond. Guidance for evaluating the degree of rusting on prestressing steel is given in Reference 52. **1907.4**

Placing reinforcement. This references ACI 318-05 Section 7.5. **1907.5**

Support for reinforcement, including prestressing tendons, is required to adequately secure the reinforcement against displacement during concreting. The CRSI *Manual of Standard Practice* gives an in-depth treatise on types and typical sizes of supports for reinforcement.

ACI 318-05 Commentary Section R7.5.1 emphasizes the importance of rigidly supporting the beam stirrups, in addition to the main flexural reinforcement, directly to the formwork.

If concrete placement is by pumping, it is imperative that the pipelines and pipeline support system be supported above and independently of the chaired reinforcement. Otherwise, the surging action of the pipeline during pumping operations can dislodge the reinforcement.

The tolerances provided in ACI 318-05 Section 7.5.2 apply simultaneously to concrete cover and member effective depth, *d*. The dimension *d* is critical because any deviation in this dimension, especially for members of lesser depth, can have an adverse effect on the strength provided in the completed construction. The permitted variation in *d* takes this strength reduction into account. The critical dimensional tolerances for locating the longitudinal reinforcement are illustrated in Figure 19-3, adapted from Reference 46.

For ends of bars and longitudinal location of bends, the tolerance is ±2 in., except at discontinuous ends of brackets and corbels where the tolerance is $\pm^1/_2$ in. and at discontinuous ends of other members where the tolerance is ±1 in. The tolerance for minimum concrete cover, which is given in ACI 318-05 Section 7.5.2.1, also applies at

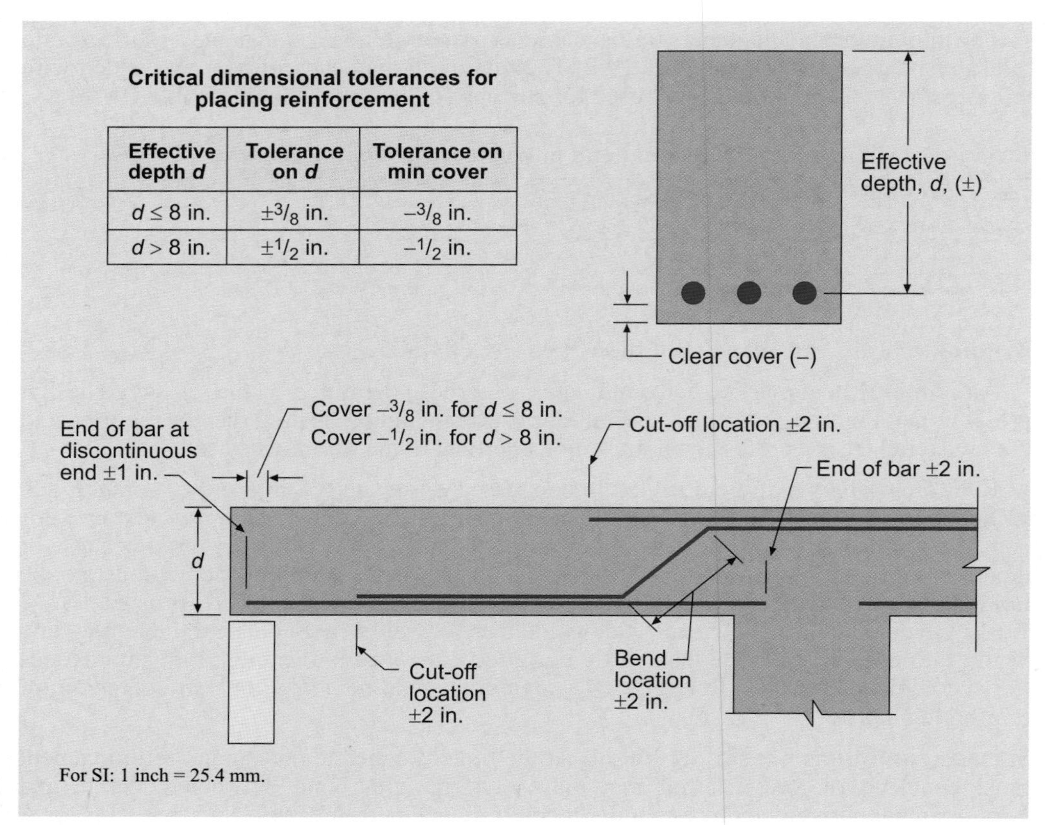

Critical dimensional tolerances for placing reinforcement

Effective depth d	Tolerance on d	Tolerance on min cover
$d \leq 8$ in.	$\pm 3/8$ in.	$-3/8$ in.
$d > 8$ in.	$\pm 1/2$ in.	$-1/2$ in.

Figure 19-3 Dimensional tolerances for placing reinforcement

discontinuous ends of members. These tolerances are also illustrated in Figure 19-3 for members other than brackets and corbels.

Note that welding of crossing bars (tack welding) for assembly of reinforcement is prohibited except as specifically authorized by the registered design professional.

1907.6 Spacing limits for reinforcement. This section refers to ACI 318-05 Section 7.6.

The minimum spacing limitations of ACI 318 Sections 7.6.1 through 7.6.3 and Section 7.6.7 are illustrated in Figure 19-4, adapted from Reference 46.

In walls and slabs other than concrete joist construction, the spacing of primary flexural reinforcement must not exceed three times the wall or slab thickness or 18 inches.

Bond research[53] has shown that bar cutoffs within bundles should be staggered. Bundled bars, limited to no more than four bars to a bundle (Figure 19-5), must be tied, wired or otherwise fastened together to ensure remaining in position, whether vertical or horizontal.

Spacing of pretensioning tendons is handled separately from that of reinforcing bars. Center-to-center spacing of pretensioning tendons at each end of a member is specified (which have been converted into clear spacing in Figure 19-4). This is because center-to-center spacing is what was used in the referenced research, and converting to clear spacing is unnecessary; and measurements for templates used in manufacture are always center-to-center. Closer vertical spacing or bundling of tendons is permitted in the middle portion of the span. Post-tensioning ducts may be bundled subject to certain preconditions.

A reduction, first introduced in ACI 318-99, is allowed in the minimum spacing requirement of prestressing strands for concrete strengths at the transfer of prestress of 4000 psi or greater, based on several research projects sponsored by the Federal Highway Administration.[54,55]

Clear distance between bars or bar bundles, or tendons

Reinforcement type	Member type	Clear distance
Deformed bars	Flexural members*	$d_b \geq 1$ in.
	Compression members, tied or spirally reinforced*	$1.5\ d_b$ $\geq 1^1/_2$ in.
Pretensioning tendons	Standards	$3\ d_b$
	Wires	$4\ d_b$

*Also see ACI Section 3.3.2

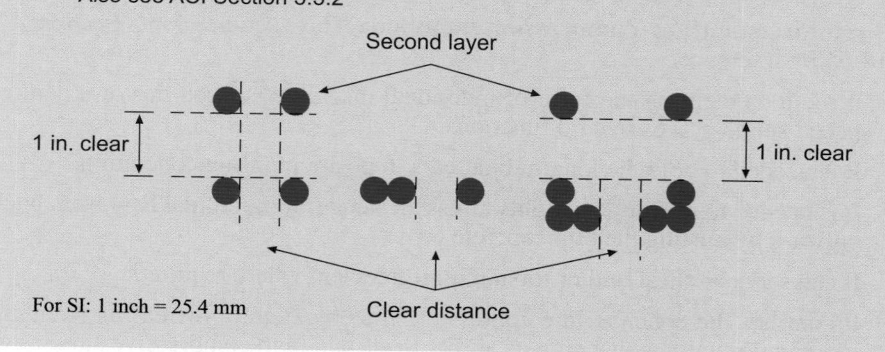

Figure 19-4
Clear distance between bars, bar bundles or tendons

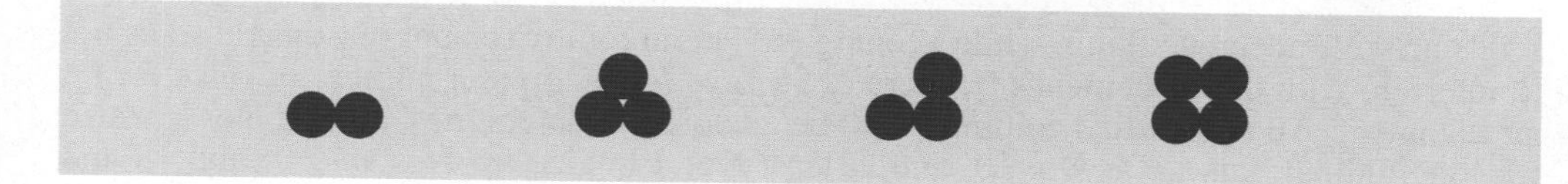

Figure 19-5
Bundling of reinforcing bars

1907.7

Concrete protection for reinforcement. This section references ACI 318-05 Section 7.7.1 through 7.7.5, and reproduces ACI 318-05 Section 7.7.6 and also 7.7.7 in a slightly modified form. Whereas Section 7.7.7 of ACI 318-05 refers to the general building code, the 2006 IBC Section 1907.7.7 refers to the IBC itself.

Concrete cover requirements are specified for nonprestressed cast-in-place concrete members cast against and permanently exposed to earth, exposed to earth or weather, and not exposed to weather or in contact with the ground (interior). Similar requirements are given for prestressed cast-in-place concrete and precast concrete (prestressed and nonprestressed) manufactured under plant-controlled conditions. The term *manufactured under plant controlled conditions* does not necessarily mean that precast members must be manufactured in a plant. Structural elements precast at the job site will also qualify under this section if the control of form dimensions, placing of reinforcement, quality of concrete and curing procedure are equivalent to those normally expected in a plant operation. For members with pretensioned strands, the minimum concrete cover specified in ACI 318-05 Section 7.7.3, which is intended to provide minimum protection against weather and other effects, may not be sufficient to transfer or develop the stress in the strand. In such cases, it may be necessary to increase the cover as required.

Larger diameter bars and bundled bars require slightly larger cover. Corrosive environments or fire protection may also warrant increased cover. Special note should be taken of the Commentary recommendations (ACI 318-05 Commentary Section R7.7.5) for increased cover when concrete will be exposed to external sources of chlorides in service, such as deicing salts and seawater. As noted in ACI 318-05 Commentary Section R7.7, alternative methods of protecting the reinforcement from weather may be used if they provide protection equivalent to the additional concrete cover required in ACI 318 Sections

7.7.1(b), 7.7.2(b) and 7.7.3(a), as compared to ACI 318 Sections 7.7.1(c), 7.7.2(c) and 7.7.3(b), respectively.

1907.8 Special reinforcement details for columns. This section adopts by reference Section 7.8 of ACI 318-05, which provides special detailing requirements for offset bent bars in columns (Figure 19-6) and requirements governing load transfer in structural steel cores of composite compression members.

1907.9 Connections. This refers to Section 7.9 of ACI 318-05, which requires that at connections of principal framing members such as beams and columns, enclosures be provided for splices of continuing reinforcement, and for end anchorage of reinforcement terminating in such connections. This confinement may be provided by surrounding concrete or internal closed ties, spirals or stirrups.

1907.10 Lateral reinforcement for compression members. This section adopts Section 7.10 of ACI 318-05 by reference.

Lateral reinforcement, in the form of individual discretely spaced ties or a continuous closely spaced spiral, serves several functions:

1. It is needed to hold the longitudinal bars in position during concreting.
2. It is needed to prevent the highly stressed, slender longitudinal bars from buckling outward by busting the thin concrete cover.
3. It can serve as shear and/or torsion reinforcement where required.
4. It confines the concrete in compression, thereby slightly increasing strength, but more importantly, enhancing its ability to sustain high compressive stresses over a significant range of deformations.

To achieve adequate tying, while keeping the lateral reinforcement reasonable so as not to interfere with the placement of concrete, Section 7.10.5 gives a number of rules for tie arrangement. All bars of tied columns shall be enclosed by lateral ties, at least No. 3 in size for longitudinal bars up to No. 10, and at least No. 4 in size for Nos. 11, 14 and 18 and bundled longitudinal bars. The spacing of the ties shall not exceed 16 diameters of longitudinal bars, 48 diameters of tie bars or the least dimension of the column. The ties shall be so arranged that every corner and alternate longitudinal bar shall have lateral support provided by the corner of a tie having an included angle of not more than 135 degrees, and no bar shall be farther than 6 inches clear on either side from such a laterally

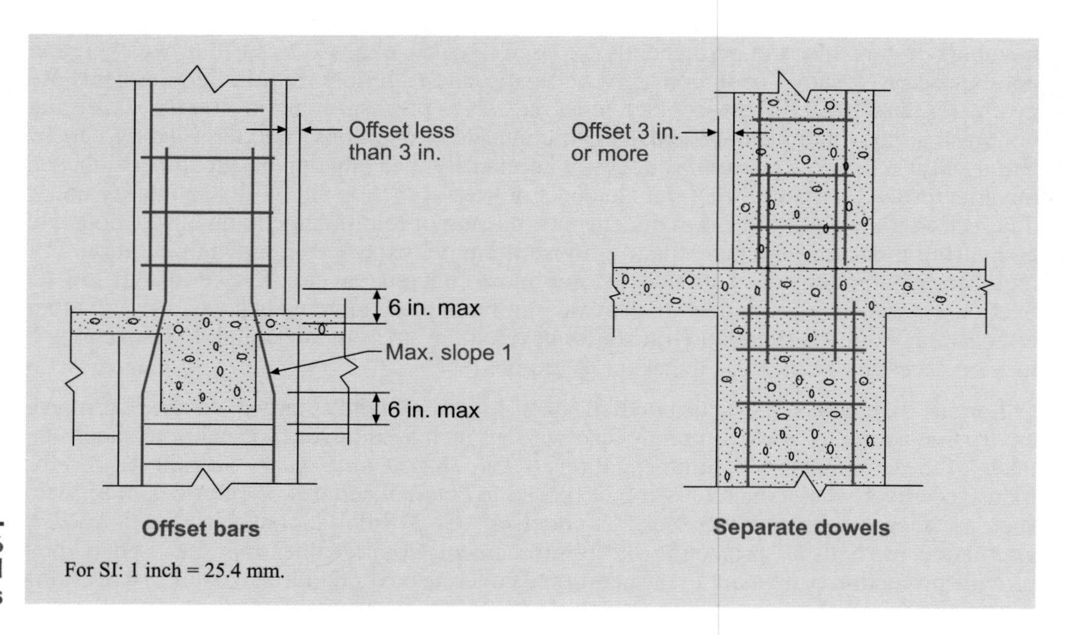

Figure 19-6 Special column details

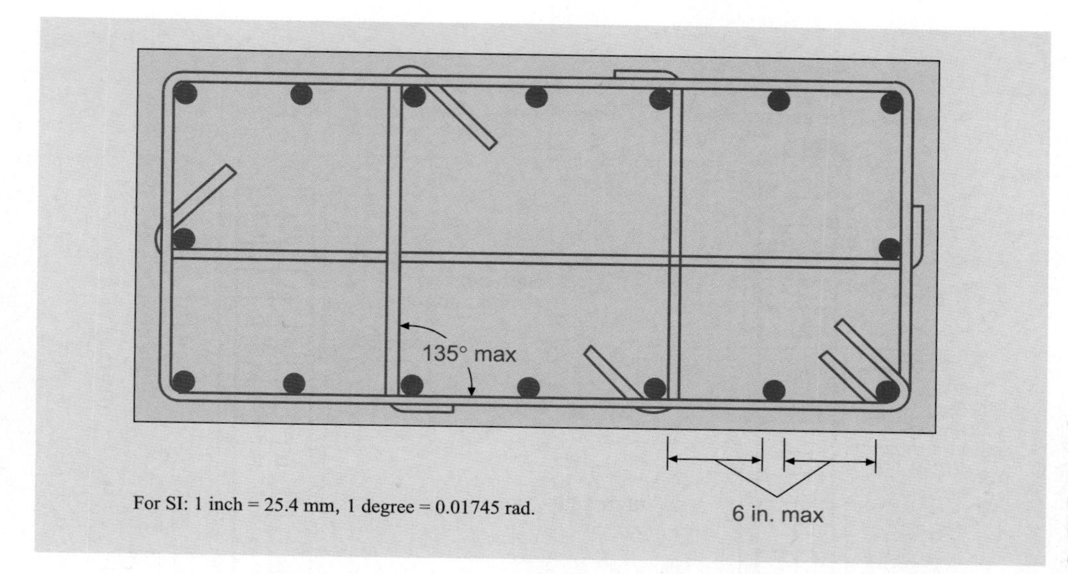

Figure 19-7 Lateral support of column bars by hoops and cross-ties

supported bar (Figure 19-7). Deformed wire or welded wire fabric of equivalent area may be used instead of ties. Where the bars are located around the periphery of a circle, complete circular ties may be used. Ties must be located at no more than one-half a tie spacing above the top of a floor slab in any story or the top of a footing, and no more than one-half a tie spacing below the lowest horizontal reinforcement in a slab or drop panel above. If beams or brackets from four directions frame into a column, ties may be terminated not more than 3 inches below the lowest horizontal reinforcement in the shallowest beams or brackets (Figure 19-8). Where anchor bolts are utilized in the tops of columns or pedestals, the bolts must be enclosed by lateral reinforcement that also surrounds at least four vertical bars of the column or pedestal. Lateral reinforcement consisting of at least two-No. 4 or three-No. 3 bars must be distributed within 5 inches of the top of the member.

For spirally reinforced columns, lateral reinforcement requirements are given in Section 7.10.4. Spirals must consist of a continuous bar or wire not less than $^3/_8$ inch in diameter, and the clear spacing between turns of the spiral must not exceed 3 inches nor be less than 1 inch Spiral reinforcement must extend from the top of footing or slab in any story to the level of the lowest horizontal reinforcement in slabs, drop panels or beams above. If beams or brackets do not frame into all sides of the column, ties must extend above the top of the spiral to the bottom of the slab or drop panel (Figure 19-9). In addition, to improve structural performance, a minimum ratio of spiral steel is imposed (Section 10.9.3).

1907.11

Lateral reinforcement for flexural members. This section refers to Section 7.11 of ACI 318-05.

Where compression reinforcement is used to increase the flexural strength of a member, or to control long-term deflection, Section 7.11.1 requires that ties or stirrups enclose such reinforcement. Requirements for size and spacing of the ties or stirrups are the same as for ties in tied columns. Welded wire reinforcement of equivalent area may be used. The ties or stirrups must extend throughout the distance where the compression reinforcement is required for flexural strength or deflection control. Section 7.11.1 is interpreted not to apply to reinforcement located in a compression zone to help assemble the reinforcing cage and hold the web reinforcement in place during concrete placement.

Enclosing reinforcement required by Section 7.11.1 is illustrated by the U-shaped stirrup in Figure 19-10(a); the continuous bottom portion of the stirrup satisfies the enclosure intent of Section 7.11.1 for the two bottom bars shown. A completely closed stirrup (Figures 19-10(b) and 19-10(c)) is ordinarily not necessary, except in cases of high moment reversal, where the reversal requires both top and bottom longitudinal reinforcement to be designed as compression reinforcement.

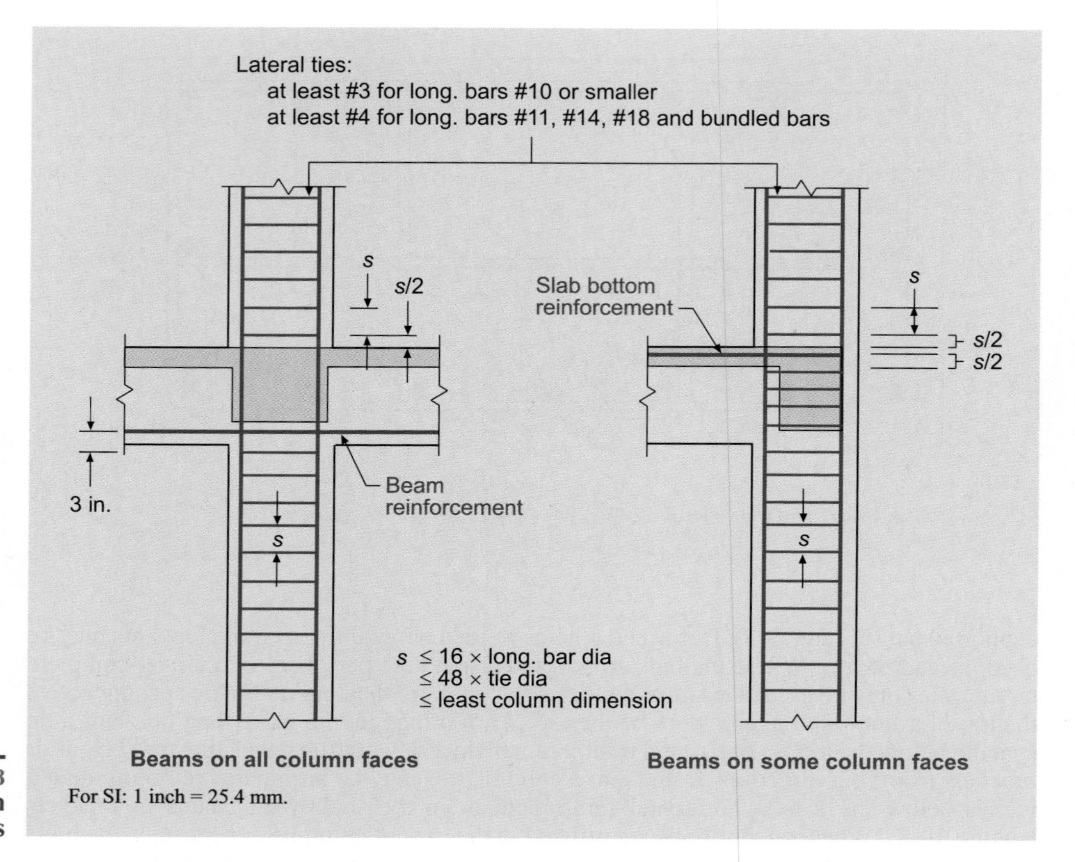

Figure 19-8 Termination of column ties

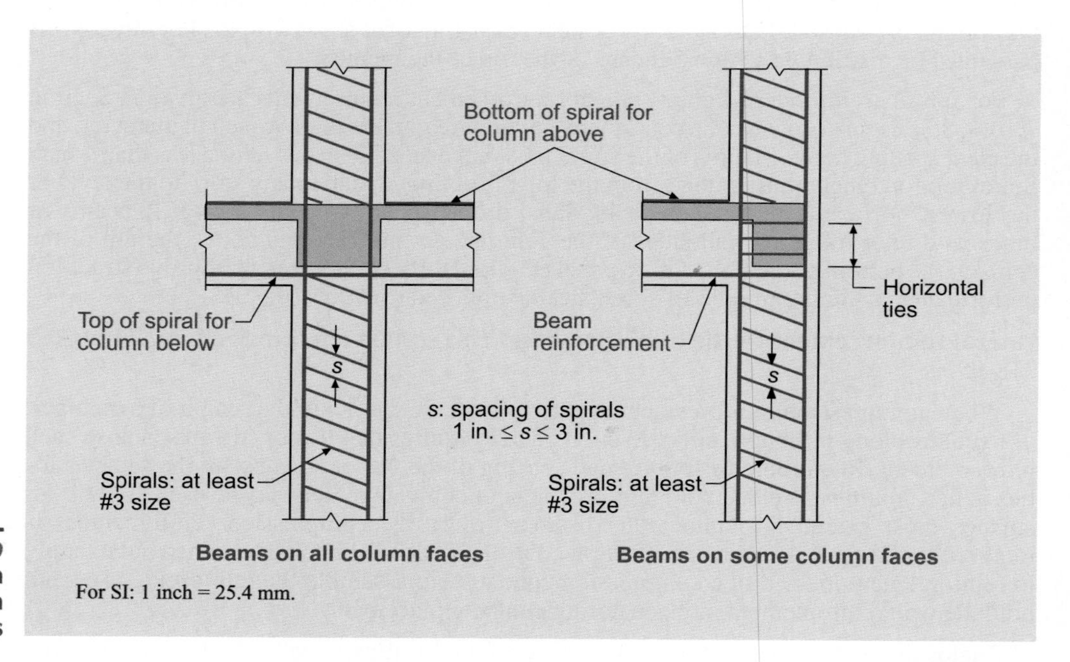

Figure 19-9 Termination of column spirals

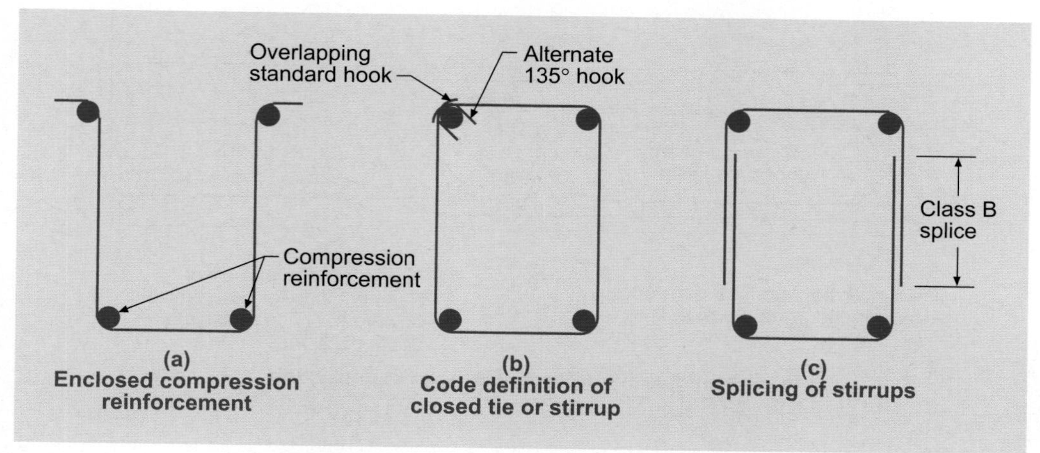

Figure 19-10
Closed tie or stirrup

Torsion reinforcement, where required, must consist of completely closed stirrups, closed ties or spirals, as required by Section 11.6.4. See Figures 3-10 and 3-9 of Reference 46 for closed stirrup details recommended and not recommended, respectively, for members subject to high torsion.

1907.12

Shrinkage and temperature reinforcement. This section adopts ACI 318-05 Section 7.12 by reference.

Minimum shrinkage and temperature reinforcement normal to primary flexural reinforcement is required for structural floor and roof slabs (not slabs on ground) where the flexural reinforcement extends in one direction only. Minimum steel ratios, based on the gross concrete area, are 0.0020 for Grade 40 and 50 deformed bars; 0.0018 for Grade 60 deformed bars or welded wire reinforcement (plain or deformed); and $0.0018 \times 60{,}000/f_y$, but not less than 0.0014, for reinforcement with a yield strength greater than 60,000 psi. Spacing of shrinkage and temperature reinforcement must not exceed 5 times the slab thickness or 18 inches. Splices and end anchorages of such reinforcement must be designed for the specified yield strength f_y in tension in accordance with Chapter 12.

Prestressing steel conforming to Section 3.5.5 may be used for shrinkage and temperature reinforcement in structural slabs (Section 7.12.3). The tendons must provide a minimum average compressive stress of 100 psi on the gross concrete area, based on effective prestress after losses. Spacing of tendons must not exceed 6 feet. When the spacing is greater than 54 inches, additional bonded reinforcement must be provided at slab edges.

1907.13

Requirements for structural integrity. ACI 318-05 Section 7.13 is adopted by reference in this section.

Experience has shown that the overall integrity of a structure can be substantially enhanced by minor changes in detailing of reinforcement. It is the intent of ACI 318-05 Section 7.13 to improve the redundancy and ductility in structures so that in the event of damage to a major supporting element or an abnormal loading, the damage may be confined to a relatively small area and the structure may have a better chance to maintain overall stability. The requirements of Section 7.13 are summarized in Figure 19-11.

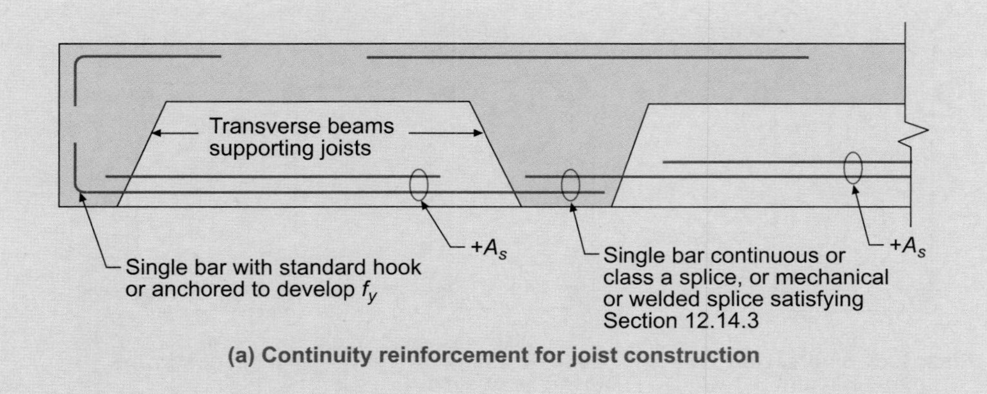

(a) Continuity reinforcement for joist construction

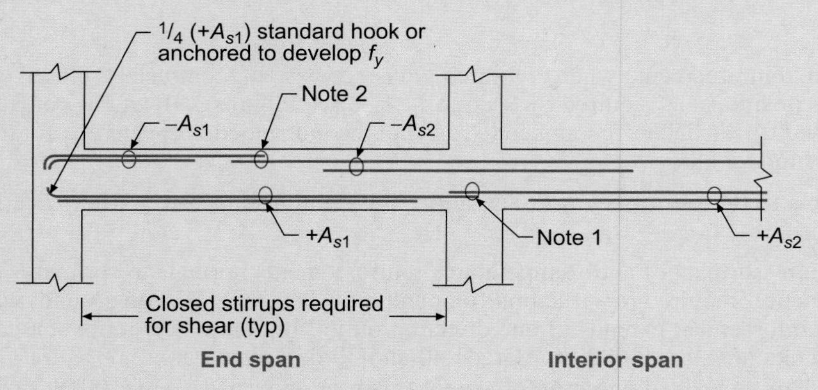

Note: (1) larger of $^1/_4$ $(+A_{s1})$ or $^1/_4$ $(+A_{s2})$ but not less than 2 bars, continuous or spliced with Class A splices, or mechanical or welded splice satisfying Section 12.14.3

(2) larger of $^1/_6$ $(-A_{s1})$ or $^1/_6$ $(-A_{s2})$ but not less than 2 bars, continuous or spliced with Class A splices, or mechanical or welded splice satisfying Section 12.14.3

(b) Continuity reinforcement for perimeter beams

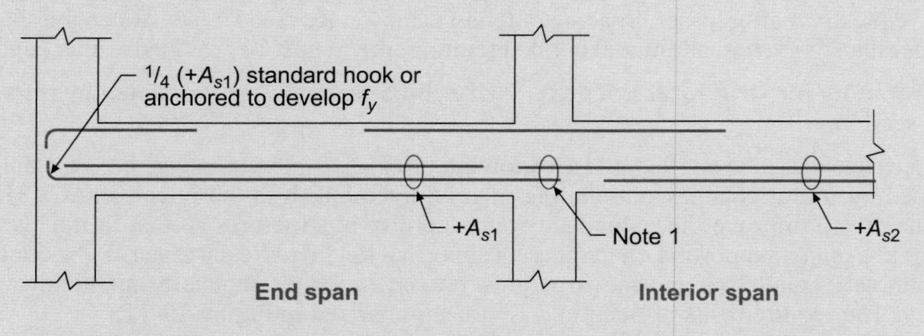

Note: (1) larger of $^1/_4$ $(+A_{s1})$ or $^1/_4$ $(+A_{s2})$ but not less than 2 bars, continuous or spliced with Class A splices, or mechanical or welded splice satisfying Section 12.14.3

(c) Continuity reinforcement for other than perimeter beams without closed stirrups

Figure 19-11 Requirements for structural integrity

Section 1908 *Modifications to ACI 318*

General. A number of modifications to the provisions of ACI 318-05 are made in this section. A summary of the modifications can be found in Table 19-6. More detailed description is provided in Sections 19.9.1.1 through 19.9.1.16. **1908.1**

Table 19-6. 2006 IBC modifications to ACI 318-05

Item	Subject	ACI 318-05 Section Modified
1	Beams in Ordinary Moment Frames assigned to SDC B	Section 10.5
2	Columns in Ordinary Moment Frames assigned to SDC B	Section 11.11
3	Additional definition of terms	Section 21.1
4	Clarifying ACI 318 description of regions of low, moderate and high seismic risk in terms of SDC A, B, C, D, E and F	Section 21.2.1
5	Prestressing steel in flexural members resisting seismic forces	Section 21.2.5
6	Anchorage for unbonded post-tensioning tendons resisting seismic forces in SDC C, D, E or F structures	Section 21.2
7	Prestressing steel in Special Moment Frames	Section 21.3
8	Transverse reinforcement in wall piers not part of Special Moment Frames	Section 21.7
9	Special precast structural walls	Section 21.8
10	Seismic design of concrete foundations	Section 21.10.1.1
11	Lap splices in members not part of lateral-force-resisting-system	Section 21.11
12	Columns supporting reactions from discontinuous stiff members in intermediate moment frames	Section 21.12.5
13	Requirements for Intermediate Precast Structural Walls	Section 21.13
14	Requirements for Detailed Plain Concrete Structural Walls	Section 22.6
15	Application of plain concrete in SDC C, D, E or F structures	Section 22.10
16	Anchors in SDC C, D, E or F structures	Section D.3.3

Additional detailing requirements, to be found in the 2003 NEHRP *Provisions*,[56] but not in the ACI 318-05, are imposed on the ordinary moment frame (OMF) of reinforced concrete assigned to SDC B. At least two longitudinal reinforcing bars must be provided continuously through the top and the bottom of flexural members, and developed within exterior columns or boundary elements. **1908.1.1**

Another additional detailing requirement from 2003 NEHRP *Provisions* is added to ACI 318 for OMFs assigned to SDC B. A column having a clear height to maximum plan dimension ratio of 5 or less must be designed for the maximum shear force that can develop in the column, computed per ACI 318-05 Section 21.12.3. This modification is intended to prevent shear failure preceding flexural failure in columns of ordinary moment frames that are part of the lateral-force-resisting system. **1908.1.2**

Modifications are made to existing definitions and new definitions are added to those in ACI 318-05 Section 21.1. The definition of *Design Displacement* is modified to reference Section 12.8.6 of ASCE 7-05.[57] Definitions of *Detailed Plain Concrete Structural Wall*, *Ordinary Precast Structural Wall*, *Ordinary Structural Plain Concrete Wall* and *Wall Pier* are added to Section 21.1. The definitions of *Ordinary Structural Plain Concrete Wall* and **1908.1.3**

Detailed Plain Concrete Structural Wall are, in fact, relocated from 2003 IBC Section 1910.2, after being rephrased to omit any part that was transcribed from ACI 318 and to only refer to that standard. Furthermore, the definition of *Ordinary Reinforced Concrete Shear Walls*, as found in 2003 IBC Section 1910.2, is split into two definitions, to make a distinction between an *Ordinary Precast Structural Wall* and a cast-in-place *Ordinary Reinforced Concrete Structural Wall*, and placed in this section. Making this distinction was deemed necessary, as the cast-in-place walls are assigned higher *R*-value for seismic design than the precast walls. However, because the existing ACI 318-05 definition of *Ordinary Reinforced Concrete Structural Wall* does not make this distinction, a modification of that definition was necessary. It should be noted that *Shear Walls* are now being referred to as *Structural Walls* to make the building code consistent with ACI 318, although the term *Shear Walls* continues to be used in Table 12.2-1 of ASCE 7-05.

1908.1.4 This IBC modification indicates that Seismic Design Category (SDC) A or B corresponds to low seismic risk or low seismic performance or design categories in ACI 318-05. SDC C corresponds to moderate seismic risk or intermediate seismic performance or design categories, and SDC D, E or F corresponds to high seismic risk or high seismic performance or design categories in ACI 318-05. This section now recognizes *Intermediate Precast Structural Walls* separately from *Ordinary Precast Structural Walls* because *Intermediate Precast Structural Walls* have now separate seismic design coefficients from those of *Ordinary Precast Structural Walls* and *Ordinary RC Stuctural Walls*.

1908.1.5
1908.1.6
1908.1.7 Section 21.2.5 of ACI 318-05 requires "reinforcement resisting earthquake-induced flexural and axial forces in frame members and in structural wall boundary elements" to comply with ASTM A 706. ASTM A 615 Grade 40 and 60 reinforcement is permitted subject to two supplementary requirements. This obviously precludes the use of prestressing tendons to resist earthquake-induced flexural and axial forces in frame members and in shear wall boundary elements. These modifications permit such usage, provided the conditions in new ACI Sections 21.2.5.2, 21.2.9 and 21.3.2.5 are satisfied.

The use of prestressing tendons, in conjunction with deformed reinforcing bars, in frames resisting earthquake-induced forces was first allowed in the 1994 NEHRP *Provisions*, based on research by Ishizuka and Hawkins.[58] The prestress f_{pc} calculated for an area equal to the member's shortest cross-sectional dimension multiplied by the perpendicular dimension (worded this way to specifically address T-beams, L-beams and other such shapes) was restricted in the 1994 NEHRP *Provisions* to the greater of 350 psi or $f_c'/12$. This restriction and the condition set forth in the added Section 21.2.5.2 came directly from the research reported in Reference 58. When the 1997 UBC adopted the same provision, the limit on f_{pc} was changed to the lesser of 350 psi or $f_c'/12$. The requirement to demonstrate the satisfactory seismic performance of anchorages for tendons was dropped, perhaps inadvertently. A third condition was added requiring that shear strength provided by prestressing tendons not be considered in design. In the 1997 NEHRP *Provisions*, the limit was increased to the lesser of 700 psi or $f_c'/6$, based on newer research.[59] The second condition remained unchanged from the 1994 NEHRP *Provisions*, including the requirement to demonstrate the satisfactory seismic performance of tendon anchorages. The third condition added by the 1997 UBC was not adopted. The 2000 NEHRP *Provisions* and the 2003 IBC have the same provision that is in the 1997 NEHRP *Provisions*. The 2003 NEHRP *Provisions* made significant organizational changes that are reflected in the 2006 IBC. The 2006 IBC has also included an unrelated change in Section 1908.1.5.

It was decided that ACI 318 Section 21.2.5.3, created by the 2003 IBC and dealing with prestressed beams, should be a modification to ACI 318 Section 21.3.2, rather than to Section 21.2.5. This modification clarifies that the requirements that prestressing steel not provide more than one-quarter of the member moment strength and be anchored at or beyond the exterior face of a joint apply only to flexural members. It is more appropriate to place these requirements within ACI 318 Section 21.3, which contains the provisions for "Flexural Members of Special Moment Frames." The phrase *for flexural members* is also added to the text for this purpose and the whole provision is placed in 2006 IBC Section 1908.1.7.

Similarly, the anchorage provision for unbonded posttensioning tendons resisting seismic forces in SDC C, D, E or F structures, previously found in ACI 318 Section 21.2.5.4, created by the 2003 IBC, is now a modification to ACI 318 Section 21.2.9, as was deemed appropriate, and is placed in 2006 IBC Section 1908.1.6.

The rest of the 2003 IBC modifications to ACI 318 Section 21.2.5 remain in 2006 IBC Section 1908.1.5, with an important further addition as described below.

As described in Part 2 of this chapter, ACI 318-05 adds a new sentence under Section 21.2.5 requiring that, "The value of f_{yt} for transverse reinforcement including spiral reinforcement shall not exceed 60,000 psi." Under 2006 IBC Section 1908.1.5, the applicability of this restriction is narrowed by adding: "For computing shear strength," in front of the requirement.

Two of the functions of transverse reinforcement in a reinforced concrete member are to confine the concrete and to act as shear reinforcement. There has been enough testing of columns[60,61,62] with high-strength confinement reinforcement (f_{yt} ranging up to 120 ksi and beyond) to show that there is no detriment to such use. The 2006 IBC, therefore, uses the ACI 318-05 upper limit on the yield strength of transverse reinforcement solely to limit the width of possible shear cracks to acceptable levels. This does not preclude the use of high-strength transverse reinforcement for confining the core of a concrete member.

1908.1.8

This section represents one of the modifications made by the 1997 UBC to the ACI 318 requirements, which is not part of ACI 318-05. The section specifies transverse reinforcement requirements for wall piers, defined in Section 1908.1.1 (Figure 19-12), when they are *not* designed as part of a special moment frame. The requirements were added to the 1991 edition of the UBC out of concern that thin column-like elements between openings in shear walls were being designed without proper transverse reinforcement. There are two important exemptions to the requirements of Section 1908.1.7. A wall pier satisfying the requirement of deformation compatibility with the lateral-force-resisting system need not be detailed by Section 21.7.10.2. Wall piers laterally supported by much stiffer shear walls along the same line within a story also need not be detailed by Section 21.7.10.2. In the 2006 IBC, ACI 318 Section 21.7.10.2 is revised to make it consistent with ASCE 7-05 Section 14.2.2.9.

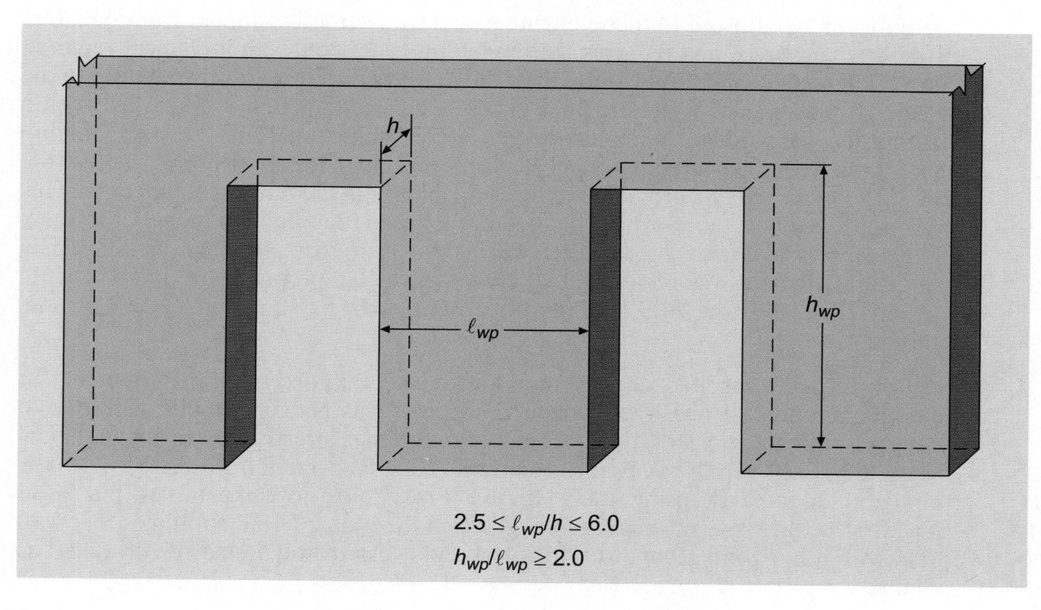

Figure 19-12
Wall piers and wall segments

1908.1.9 ACI 318-05 Section 21.8.1 requires that precast special structural walls satisfy all the requirements for cast-in-place special structural walls as specified in Section 21.7, as well as the requirements for precast intermediate structural walls specified in Sections 21.13.2 and 21.13.3. However, supplemental requirements are added for precast intermediate walls in ACI 318 Section 21.13 by 2006 IBC Section 1908.1.13 (discussed later). As a result of that modification, this section adds the supplemental ACI 318 Section 21.13.4 to the requirements for special precast structural walls.

1908.1.10 2006 IBC requires that the seismic design of foundations comply with relevant provisions in ACI 318-05, unless they are modified by Chapter 18, Soils and Foundations, of the 2006 IBC. In general, the provisions of 2006 IBC Chapter 18 are more stringent than those found in ACI 318-05 Section 21.10.

1908.1.11 This modification exempts the requirement that lap splices of column reinforcement be confined to the middle half of the column height in columns that are not part of the lateral-force-resisting system (gravity columns) of a building assigned to SDC D or above, where the column is going to remain elastic under the design earthquake displacements. In short, this amendment allows lap splices in such members to be located anywhere, including the column ends. This amendment is justified because gravity columns are required to sustain their gravity-load-carrying capacity under the design displacement, whereas columns that are part of the lateral-force-resisting system are required to sustain full gravity as well as lateral loads under the same displacement. Applicability of this exemption is restricted to "structures where the seismic-force-resisting system does not include special moment frames." This modification is now in 2006 IBC Section 1908.1.11.

1908.1.12 This modification requires columns supporting reactions from discontinuous stiff members in a building assigned to SDC C to be designed using the special seismic load combinations of Section 1605.4 and to be provided with closely spaced transverse reinforcement over their full height, extended above and below the column by a certain specified distance.

1908.1.13 This section modifies ACI 318 Section 21.13, Intermediate precast structural walls, by first renumbering Section 21.13.3 to Section 21.13.4, and then adding three new Sections 21.13.3, 21.13.5 and 21.13.6.

- Section 21.13.3 specifies a minimum ductility requirement for a steel element used in a connection between wall panels or between wall panels and the foundation, so that yielding can be allowed in that element. The new section is adopted from the 2003 NEHRP *Provisions*, and requires that a yielding steel connection element retain at least 80 percent of its design strength at the deformation level corresponding to the design displacement of the structure. Thus, a clear distinction is made between a ductile steel element and a nonductile one. ACI 318 Section 21.13.2 allows yielding in steel reinforcement as well in steel elements present in the connection, whereas the 2003 IBC modified this provision to restrict yielding in the reinforcement only. This was done because no formal distinction between ductile and nonductile steel elements was available at the time and it was felt that allowing yielding in a nonductile element was not desirable. However, with a proper definition of ductile elements in place, the 2003 IBC modification to ACI 318 Section 21.13.2 is now omitted.
- Section 21.13.5 specifies a minimum transverse reinforcement requirement for intermediate precast wall piers when not designed as part of a moment frame, to prevent a premature shear failure. Similar requirements already existed for wall piers meant to be used in SDC D, E or F structures (2006 IBC Section 1908.1.8). The new provision, adopted from the 2003 NEHRP *Provisions*, now expands the scope to SDC C structures as well. In a second addition, Section 21.13.6 stipulates that wall piers with a horizontal-length-to-thickness-ratio less than 2.5 shall be designed as columns.

1908.1.14 This modification adds new Sections 22.6.7 through 22.6.7.2 to specify the reinforcement requirement that detailed plain concrete structural walls need to conform to in addition to that required for ordinary plain concrete structural walls. Note that detailed plain concrete structural walls are not recognized in ACI 318-05.

1908.1.15

This modification replaces ACI 318 Section 22.10 in its entirety, while serving essentially the same purpose, i.e., to prohibit the use of plain concrete elements in structures assigned to Seismic Design Category C, D, E or F, with several specific exceptions that permit the use of plain concrete in foundations. Exceptions in the 2006 IBC are described in more detail than in ACI 318 Section 22.10. Also, the 2006 IBC clearly defines the scope of the provisions as extending to structures assigned to Seismic Design Category C, D, E or F, instead of "structures designed for earthquake induced forces in regions of high seismic risk or assigned to high seismic performance or design categories," used in ACI 318-05 Section 22.10.

1908.1.16

This modifies ACI 318-05 Sections D.3.3.2 through D.3.3.5, based on similar provisions found in the 2003 NEHRP *Provisions*. The modification clearly defines the scope of the provisions as covering structures assigned to SDC C, D, E or F. ACI 318-05 Section D.3.3.4, as modified by 2006 IBC Section 1908.1.16, requires that in structures assigned to Seismic Design Category C, D, E or F, anchors shall be designed to be governed by tensile or shear strength of a ductile steel element, unless ACI 318 Section D.3.3.5, as modified by 2006 IBC Section 1908.1.16, is satisfied. That modified ACI 318 section reads: "Instead of D.3.3.4, the attachment that the anchor is connecting to the structure shall be designed so that the attachment will undergo ductile yielding at a load level corresponding to anchor forces no greater than the design strength of anchors specified in D.3.3.3 [$0.75fN_n$ and $0.75fV_n$, where f is given in D.4.4 or in D.4.5, and N_n and V_n are determined in accordance with D.4.1], or the minimum design strength of the anchors shall be at least 2.5 times the factored forces transmitted by the attachment." The underlined words are added by 2006 IBC Section 1908.1.16 to ACI 318-05 Section D.3.3.5 and constitute a particularly important modification.

Section 1909 *Structural Plain Concrete*

This section reproduces parts of Chapter 22 of ACI 318-05, refers to the rest of it, reformats and rearranges the reproduced text (Table 19-7), but otherwise makes only two deviations from ACI 318-05.

Table 19-7. Correlation between Section 1909 (Structural Plain Concrete) of the 2006 IBC and Chapter 22 of ACI 318-05

2006 IBC Section Nos.	ACI 318-05 Section Nos.
1909.1 1909.1.1	22.1.2
1909.2	22.2.2 22.7.3
1909.3	22.3
1909.4	Refers to 22.4-22.8
1909.5	Refers to 22.9
1909.6	22.6.6.3
1909.6.1	22.1.1.1
1909.6.2 1909.6.3	22.6.6.2 22.6.6.5

First, under Section 1909.2, Limitations, the IBC directs the user to Section 1908.1.15 for additional limitations on the use of structural plain concrete.

Second, an exception is added under Section 1909.4, Design. For detached one- and two-family dwellings and other occupancies less than two stories in height of light-frame construction, the required edge thickness of ACI 318 (8 inches) is permitted to be reduced to 6 inches, provided the footing does not extend more than 4 inches on either side of the supported wall.

Section 1910 *Minimum Slab Provisions*

Section 1905.0 of the 1999 BOCA/NBC, Section 1900.4.4 of the 1997 UBC and Section 1909.1 of the 1999 SBC contained identical minimum thickness requirements for floor slabs supported directly on the ground. These are included in Section 1910 of the 2006 IBC. In addition, provisions aimed at retarding vapor transmission through the floor slab are included. These provisions are essentially the same as those found in the 1999 edition of the BOCA/NBC (Section 1905.1) and the 1999 edition of the SBC (Section 1909.2). BOCA and SBCCI have published explanations of their provisions,[63,64] which are used as source material for the following discussion.

It is specifically stated in ACI 318-05 Section 1.1.6 that it does not govern the design and construction of soil-supported slabs, unless the slab transmits vertical loads or lateral forces from other portions of the structure to the soil. This does not preclude the application of any ACI 318 provision that would assure the proper strength, durability and abrasion resistance of concrete slabs on ground. Slabs on ground may be plain concrete without any reinforcement or may contain shrinkage and temperature reinforcement in the form of reinforcing bars or wire mesh.

The minimum thickness of $3^1/_2$ inches is based on the fact that the largest aggregate size typically used in the construction of slabs on grade is 1 inch. ASTM C 33 allows 5 percent of 1-inch graded material to be larger than 1 inch in size, but smaller than $1^1/_2$ inch. Therefore, in accordance with the requirement of ACI 318-05 Section 3.3.2 that the nominal size of coarse aggregate must be no larger than one-third the depth of a slab, the minimum depth becomes $3^1/_2$ inch (3 times 1 inch plus a $^1/_2$-inch allowance for larger stones).

Although good quality concrete is practically impermeable to the passage of water (that is not under significant pressure), concrete is not impervious to the passage of water vapors. If the surface of the slab is not sealed, water vapor will pass through the slab. If a floor finish such as linoleum, vinyl tile, wood flooring or any type of covering is placed on top of the slab, the moisture is trapped in the slab. Any floor finish adhering to the concrete may eventually loosen or buckle or blister.

Many of the moisture problems associated with enclosed slabs on ground can be prevented or minimized by installing vapor retarders, such as polyethylene sheeting or other approved materials, between the slab and the ground. Such retarders are needed under slabs in habitable spaces. Where garages, utility areas and similar spaces are not to be heated or occupied, the use of vapor barriers is generally not necessary. If moisture migration is not expected to be a problem based on the occupancy of a building, Exception 3 permits the vapor retarder to be omitted. The vapor retarder provisions of Section 1910 should not change current practice in UBC territory, because typically the soil investigation report already required such vapor retarders.

Section 1911 *Anchorage to Concrete— Allowable Stress Design*

The 2006 IBC provides two distinct methods for designing anchorage to concrete. Section 1911 presents an allowable stress design (ASD) method of limited applicability. Section 1912 presents a strength design method of broader applicability.

1911.1

General. The ASD provisions of this section are adapted from the 1997 edition of the UBC. They are based on limited test data on headed bolts cast in normal-weight concrete subjected to static loading. It is, therefore, not applicable to hooked (J- or L-) bolts, lightweight aggregate concrete, post-installed anchors (meaning anchors installed in hardened concrete) or when load combinations include earthquake effects.

1911.2

Allowable service load. Table 1911.2 provides allowable service loads for headed bolts or stud anchors cast in normal-weight concrete that are subject to tension only or shear only. The table values are for a particular range of bolt sizes ($^1/_4$ to $1^1/_4$ in.) and concrete strengths (2500, 3000 and 4000 psi). When using the tabulated values, minimum edge distances and spacing between bolts, specified in the table, must be carefully maintained. Edge distances are measured from the face of the bolt to the sides and ends of concrete members. Section 1911.3 provides for decreases of up to 50 percent in edge distances and/or spacing, with corresponding decreases in allowable service loads.

For a headed bolt or stud anchor subjected to combined tension and shear loading, Section 1911.2 requires that the applied service-level load in tension and the applied service-level load in shear satisfy the interaction formula given in that section, which involves the allowable values given by Table 1911.2.

1911.3

Required edge distance and spacing. The purpose of this section is to expand the applicability of Table 1911.2 by providing for situations where bolt locations cannot meet the required minimum edge distances and/or the spacing requirements of the table. Both the minimum edge distance and the minimum spacing requirements of the table may be reduced by up to 50 percent for a corresponding 50-percent reduction in the allowable service loads. Where reduction in edge distance and spacing is less than 50 percent, the allowable service loads may be determined by linear interpolation.

1911.4

Increase in allowable load. This section allows a one-third increase in the tabulated values of allowable service loads of Table 1911.2 when service loads include those that are due to wind, in addition to dead loads, live loads and other loads such as snow. Wind loads are transitory in nature. It is considered highly unlikely that two or more variable loads such as wind and live loads will attain their design values simultaneously. This is the rationale behind the one-third increase in allowable stresses permitted when the alternate basic load combinations of Section 1605.3.2 are used in design, and load combinations include the effects of wind or seismic loads.

1911.5

Increase for special inspection. The provisions of this section have been taken from footnotes 5 and 6 to Table 19-D (Allowable Service Load on Embedded Bolts) of the 1997 UBC, which is essentially identical to Table 1911.2 of the 2006 IBC. These footnotes first appeared in the 1976 edition of the UBC. Footnote 5 applies to allowable service loads in tension, and states that the values shown are for work without special inspection; where special inspection is provided, values may be increased 100 percent. Footnote 6 applies to allowable service loads in shear, and states that the values shown are for work with or without special inspection.

See Example 4 in Part 3 of this chapter.

Section 1912 *Anchorage to Concrete—Strength Design*

IBC Section 1912 adopts the provisions of Appendix D, Anchoring of Concrete, of ACI 318-05 by reference. A brief discussion on the development of Appendix D, first introduced in the 2002 edition of ACI 318, follows.

As of the late 1990s, the primary sources of design information for connections to concrete using cast-in-place anchors were Appendix B of ACI 349[65] and the *PCI Design Handbook*.[66] The design of connections using post-installed anchors typically was performed using information from the anchor manufacturers.

ACI Committee 318, with the support of ACI Committee 355 (Anchorage to Concrete) and ACI Committee 349 (Concrete Nuclear Structures), took the lead in developing provisions for both cast-in-place and post-installed mechanical anchors. During the code cycle leading to ACI 318-99, ACI Committee 318 approved a proposed Appendix D to ACI 318 dealing with the design of anchorages to concrete. At the same time, ACI Committee 355 was in the process of developing a test method for evaluating the performance of post-installed anchors in concrete. Final adoption of Appendix D in ACI 318-99 depended on the approval of this test method.

The test method for evaluating the performance of post-installed anchors was not completed in time to meet the publication deadline of ACI 318-99. A subsequent attempt was made to process Appendix D with provisions for only cast-in-place anchors, but this move failed to garner support within ACI. Thus, ACI 318-99 was issued without any provisions for fastening to concrete.

The concrete industry then asked ACI for permission to submit the provisions of the proposed Appendix D on cast-in-place anchors to the IBC as a replacement for the strength design provisions of IBC Section 1912. ACI agreed to permit the submission of this copyright-protected material for inclusion in the IBC, provided it was understood that ACI retained exclusive copyright to the material.

During the code cycle leading to the 2003 IBC, a code change was submitted and subsequently approved to remove the IBC anchorage provisions and to reference the new provisions in Appendix D of ACI 318-02.

As noted previously, Appendix D of ACI 318-05 contains provisions for both cast-in-place and post-installed mechanical anchors. One exception is made to these provisions in Section 1912.1; this exception pertains to the concrete breakout strength requirements of Section D.4.2.2 for single anchors exceeding 2 inches in diameter and/or 25 inches of embedment depth. Anchors that are not within the scope of Appendix D are to be designed by an approved procedure.

Section 1913 *Shotcrete*

This section contains provisions very similar to those in Section 1911 of the BOCA/NBC, Section 1924 of the UBC and Section 1915 of the SBC (latest editions). BOCA and SBCCI published commentaries on their shotcrete provisions,[63,64] from which the material here is largely drawn.

1913.1 **General.** Shotcrete is pneumatically projected concrete or mortar. Other terms such as spraycrete, sprayed concrete and gunite are also associated with shotcrete construction. Shotcrete needs to conform to Chapter 19 requirements for plain or reinforced concrete, unless specifically exempted by Section 1913.

Proportions and materials. Proportions of shotcrete mixtures should be determined prior to the beginning of construction by trial applications on test specimens. The test specimens should be representative of the in-place application (flat, vertical, overhead) and the shotcrete should be applied using the same materials and equipment that will be used for construction. **1913.2**

Aggregate. For construction applications in which the shotcrete will be several inches thick, coarse aggregate may be used in the mixture. In those cases, the aggregate size is limited to $^3/_4$ inch, to minimize the effects of rebounding during placement and the creation of voids in the shotcrete. Rebound refers to shotcrete that ricochets off the receiving surface. See Section 1913.6. **1913.3**

Reinforcement. The size and the spacing of the reinforcement are required to be such as to minimize interference with the high-velocity placement of the shotcrete and to ensure that the reinforcement is completely covered. The clearance between the form and the reinforcement may vary depending on whether concrete or mortar is used for the shotcrete. The use of reinforcing bars or welded wire fabric can also affect the minimum clearance from the form. The exception allowing reduction of required clearances, subject to approval of the building official and preconstruction testing, should be noted. **1913.4**

Noncontact lap splices are preferred, to minimize the creation of weak sections in the shotcrete. Where possible, at least 2 inches should separate lapped bars. Welded wire reinforcement should be lapped by one square in all directions.[63] When adequate encasement can be shown, contact lap splices are permitted with the approval of the building official.

Preconstruction tests. The preconstruction tests provided for in this section are at the discretion of the building official. It was part of the UBC provisions, but not part of the BOCA/NBC or the SBC provisions. The requirements are quite explicit and self-explanatory. **1913.5**

Rebound. As mentioned earlier, rebound is shotcrete that ricochets off the receiving surface. The position of the work (flat, vertical or overhead), layer thickness, discharge pressure, cement content, water content, size and gradation of aggregate, and type and amount of reinforcement can affect the amount of rebounding that occurs.[63] Rebounded material may not be reused or worked back into the construction, and must be removed from the surface prior to placement of additional layers of shotcrete. **1913.6**

Joints. Construction joints are generally tapered to a thin edge over a width of approximately 12 inches. Square construction joints should be avoided, except as specifically permitted by this section. **1913.7**

Damage. After placement, any shotcrete that lacks uniformity or that exhibits segregation, honeycombing or delamination, or which contains dry patches, slugs, voids or sand pockets (porous areas low in cement content), or that sags or sloughs must be removed and replaced. **1913.8**

Curing. As in most construction involving cementitious materials, proper curing practices need to be followed from the time of completion of the shotcrete application, as outlined in this section. **1913.9**

Strength tests. The BOCA/NBC and the SBC required strength tests of shotcrete to be made in accordance with the quality assurance provisions of ACI 506.2.[67] Test specimens were required to be obtained from the in-place shotcrete or from a test panel that was representative of the work, and tested in accordance with ASTM C 42, *Test Method for Obtaining and Testing Drilled Cores and Sawed Beams of Concrete*.[68] The IBC, like the UBC, spells out strength test provisions, including sampling requirements and acceptance criteria. **1913.10**

Section 1914 Reinforced Gypsum Concrete

Provisions on reinforced gypsum concrete were included in Section 1925 of the 1997 UBC and Section 1914 of the 1999 SBC.

The specifications of Section 1914 cover poured-in-place reinforced gypsum concrete over permanent formboard. Gypsum concrete is used as a structural material in the construction of roof decks or slabs and as a nonstructural material in floor topping.

Gypsum concrete is required to conform to the specifications of ASTM C 317, *Specification for Gypsum Concrete*.[69] The design and application of reinforced gypsum concrete must be in accordance with the requirements of ASTM C 956, *Specification for Installation of Cast-in-Place Reinforced Gypsum Concrete*.[70]

The minimum thickness of reinforced gypsum concrete is required to be 2 inches, except that this may be reduced to no less than $1^1/_2$ inches if certain conditions given in Section 1914.2 are met.

Section 1915 Concrete-Filled Pipe Columns

These provisions are largely the same as those of Section 1912 of the 1999 edition of the BOCA/NBC. The BOCA Commentary[63] is used as a source for this discussion.

1915.1 **General.** According to Reference 60, steel pipe should be manufactured to the requirements of ASTM A 501[71] for hot-formed welded and seamless carbon steel of round, square or rectangular shape for general structural purposes. The steel must have a minimum tensile strength of 58,000 psi and minimum yield strength of 36,000 psi.

Cold-formed welded and seamless carbon steel of square or rectangular shape conforming to the Grade B requirements of ASTM A 500,[72] having a minimum tensile strength of 58,000 psi and a minimum yield strength of 36,000 psi, may also be used for concrete-filled pipe columns.

1915.2 **Design.** The load-carrying capacity of concrete-filled pipe columns may be computed in accordance with approved rules such as those in Section 10.16 of ACI 318-05, or may be determined by load tests.

1915.3 **Connections.** Conditions for making structural connections to concrete-filled pipe columns are prescribed. If connections require welding to the steel shell, it must be done before the core is filled with concrete, unless it is possible to demonstrate that the concrete will not be damaged from the heat of welding.

1915.4 **Reinforcement.** Reinforcement must comply with the requirements of this section as well as with the applicable provisions of ACI 318. A minimum clearance of 1 inch between any such reinforcement and the outer shell must be provided. This is to permit the concrete to flow around the reinforcement and bond to it, which is necessary for composite action.

1915.5 **Fire-resistance-rating protection.** Concrete-filled pipe columns are required to be fire-resistance rated in accordance with IBC Table 601. Irrespective of whether the protective cover is of concrete, concrete with a metal encasement or a metal cover that encases other fire-insulating materials, it must not be considered to contribute to the load-carrying capacity of the column.

1915.6 **Approvals.** Concrete-filled pipe columns, including their connection details and splices, which are shop-fabricated as pre-engineered items, are subject to certain inspection and approval requirements spelled out in this section.

REFERENCES

[1] *Uniform Building Code*, International Conference of Building Officials, Whittier, CA, 1997, copyright held by International Code Council.

[2] *The BOCA National Building Code*, Building Officials and Code Administrators International, Country Club Hills, IL, 1993, 1996, 1999, copyright held by International Code Council.

[3] *Standard Building Code*, Southern Building Code Congress International, Birmingham, AL, 1994, 1997, 1999, copyright held by International Code Council.

[4] ACI Committee 318, *Building Code Requirements for Structural Concrete (ACI 318–95) and Commentary (ACI 318R–95)*, American Concrete Institute, Farmington Hills, MI, 1995.

[5] ACI Committee 318, *Building Code Requirements for Structural Concrete (ACI 318–02) and Commentary (ACI 318R–02)*, American Concrete Institute, Farmington Hills, MI, 2002.

[6] ACI Committee 318, *Building Code Requirements for Structural Concrete (ACI 318–05) and Commentary (ACI 318R–05)*, American Concrete Institute, Farmington Hills, MI, 2005.

[7] *Recommended Practice for Glass Fiber Reinforced Concrete Panels*, PCI MNL 128, Precast and Prestressed Concrete Institute, Chicago, IL, 2001.

[8] *Specification for Portland Cement*, ASTM C 150, American Society for Testing and Materials, West Conshohocken, PA, 2005.

[9] *Specification for Blended Hydraulic Cements*, ASTM C 595, American Society for Testing and Materials, West Conshohocken, PA, 2005.

[10] *Specification for Expansive Hydraulic Cement*, ASTM C 845, American Society for Testing and Materials, West Conshohocken, PA, 2004.

[11] *Performance Specification for Hydraulic Cement*, ASTM C 1157, American Society for Testing and Materials, West Conshohocken, PA, 2003.

[12] *Specification for Concrete Aggregates*, ASTM C 33, American Society for Testing and Materials, West Conshohocken, PA, 2003.

[13] *Specification for Lightweight Aggregates for Structural Concrete*, ASTM C 330, American Society for Testing and Materials, West Conshohocken, PA, 2004.

[14] *Design and Control of Concrete Mixtures*, 14th Edition, Portland Cement Association, Skokie, IL, 2002.

[15] *Test Method for Compressive Strength of Hydraulic Cement Mortars (Using 2-in. or [50-mm] Cube Specimens)*, ASTM C 109, American Society for Testing and Materials, West Conshohocken, PA, 2005.

[16] *Test Method for Time of Setting of Hydraulic Cement by Vicat Needle*, ASTM C 191, American Society for Testing and Materials, West Conshohocken, PA, 2004.

[17] *Specification for Ready-Mixed Concrete*, ASTM C 94, American Society for Testing and Materials, West Conshohocken, PA, 2004.

[18] *Quality of Water to Be Used in Concrete*, T 26–96, American Association of State Highway and Transportation Officials, Washington, DC, 1996.

[19] *Structural Welding Code—Reinforcing Steel*, ANSI/AWS D1.4–98, American Welding Society, Miami, FL, 1998.

[20] *Specification for Low-Alloy Steel Deformed and Plain Bars for Concrete Reinforcement*, ASTM A 706, American Society for Testing and Materials, West Conshohocken, PA, 2005.

[21] *Specification for Rail-Steel and Axle-Steel Deformed Bars for Concrete Reinforcement*, ASTM A 996, American Society for Testing and Materials, West Conshohocken, PA, 2005.

[22] *International Residential Code*, International Code Council, Washington DC, VA, 2003.

[23] ACI Committee 201, *Guide to Durable Concrete*, ACI 201.2 R–92, American Concrete Institute, Farmington Hills, MI, 1992. Also, *ACI Manual of Concrete Practice.*

[24] ACI Committee 222, *Corrosion of Metals in Concrete*, ACI 222 R–96, American Concrete Institute, Farmington Hills, MI, 1996. Also, *ACI Manual of Concrete Practice.*

[25] ACI Committee 211, *Standard Practice for Selecting Proportions for Normal, Heavyweight, and Mass Concrete,* ACI 211.1–98, American Concrete Institute, Farmington Hills, MI, 1998. Also, *ACI Manual of Concrete Practice.*

[26]ACI Committee 211, *Standard Practice for Selecting Proportions for Structural Lightweight Concrete*, ACI 211.2–91, American Concrete Institute, Farmington Hills, MI, 1991. Also, *ACI Manual of Concrete Practice*.

[27]*Standard Practice for Making and Curing Concrete Test Specimens in the Laboratory*, ASTM C 192, American Society for Testing and Materials, West Conshohocken, PA, 2005.

[28]*Standard Practice for Laboratories Testing Concrete and Concrete Aggregates for Use in Construction and Criteria for Laboratory Evaluation*, ASTM C 1077, American Society for Testing and Materials, West Conshohocken, PA, 2005.

[29]*Standard Practice for Sampling Freshly Mixed Concrete*, ASTM C 172, American Society for Testing and Materials, West Conshohocken, PA, 2004.

[30]*Standard Practice for Making and Curing Concrete Test Specimens in the Field*, ASTM C 31, American Society for Testing and Materials, West Conshohocken, PA, 2003.

[31]*Specification for Molds for Forming Concrete Test Cylinders Vertically*, ASTM C 470, American Society for Testing and Materials, West Conshohocken, PA, 2002.

[32]*Standard Test Method for Compressive Strength of Cylindrical Concrete Specimens,* ASTM C 39, American Society for Testing and Materials, West Conshohocken, PA, 2005.

[33]ACI Committee 304, *Guide for Measuring, Mixing, Transporting and Placing Concrete*, ACI 304 R–00, American Concrete Institute, Farmington Hills, 2000. Also, *ACI Manual of Concrete Practice*.

[34]ACI Committee 309, *Guide for Consolidation of Concrete*, ACI 309 R–96, American Concrete Institute, Farmington Hills, MI, 1996. Also, *ACI Manual of Concrete Practice*.

[35]ACI Committee 308, *Standard Practice for Curing Concrete*, ACI 308–92, American Concrete Institute, Farmington Hills, MI, 1992. Also, *ACI Manual of Concrete Practice*.

[36]ACI Committee 517, *Accelerated Curing of Concrete at Atmospheric Pressure—State of the Art*, ACI 517.2R–92, American Concrete Institute, Farmington Hills, MI, 1992.

[37]ACI Committee 306, *Cold Weather Concreting*, ACI 306 R–88, American Concrete Institute, Farmington Hills, MI, 1988. Also, *ACI Manual of Concrete Practice*.

[38]ACI Committee 305, *Hot Weather Concreting*, ACI 305 R–99, American Concrete Institute, Farmington Hills, MI, 1999. Also, *ACI Manual of Concrete Practice*.

[39]ACI Committee 347, *Guide to Formwork for Concrete*, ACI 347 R–03, American Concrete Institute, Farmington Hills, MI, 2003. Also, *ACI Manual of Concrete Practice*.

[40]Hurd, M. K., *Formwork for Concrete*, SP–4, 7th Edition, American Concrete Institute, Farmington Hills, MI, 2005.

[41]Johnston, D. W., "Design and Construction of Concrete Formwork," Chapter 7, *Concrete Construction Engineering Handbook*, (E. G. Nawy, Editor–in–Chief), CRC Press, Boca Raton, FL, 1997, pp. 7.1–7.48.

[42]Ghosh, S. K, "Construction Loading in High-Rise Buildings," Chapter 8, *Concrete Construction Engineering Handbook*, (E. G. Nawy, Editor–in–Chief), CRC Press, Boca Raton, FL, 1997, pp. 8.1–8.60.

[43]ACI Committee 315, *ACI Detailing Manual*, SP–66, American Concrete Institute, Farmington Hills, MI, 2004.

[44]*Manual of Standard Practice*, 27th Edition, Concrete Reinforcing Steel Institute, Schaumburg, IL, 2001.

[45]*Structural Detailing Manual*, Wire Reinforcement Institute, Hartford, CT, 1994.

[46]Fanella, D. A., and Rabbat, B. G., Ed., *Notes on ACI 318–02*, Portland Cement Association, Skokie, IL, 2002.

[47]Babaei, K., and Hawkins, N. M., "Field Bending and Straightening of Reinforcing Steel," *Concrete International*, Vol. 14, No. 1, January 1992.

[48]*Specification for Epoxy-Coated Steel Reinforcing Bars*, ASTM A 775, American Society for Testing and Materials, West Conshohocken, PA, 2004.

[49]*Specification for Epoxy-Coated Prefabricated Steel Reinforcing Bars*, ASTM A 934, American Society for Testing and Materials, West Conshohocken, PA, 2004.

[50]*Specification for Epoxy-Coated Steel Wire and Welded Wire Reinforcement*, ASTM A 884, American Society for Testing and Materials, West Conshohocken, PA, 2004.

[51]Kemp, E. L., Brezny, F. S., and Unterspan, J. A., "Effect of Rust and Scale on the Bond Characteristics of Deformed Reinforcing Bars," *ACI Journal, Proceedings* Vol. 65, No. 9, September 1968, pp. 743–756.

[52]Sason, A. S., "Evaluation of Degree of Rusting on Prestressed Concrete Strand," *PCI Journal*, Vol. 37, No. 3, May–June 1992, pp. 25–30.

[53]ACI Committee 408, "*Bond Stress – The State of the Art*", *ACI Journal, Proceedings* Vol. 63, No. 11, November 1966, pp. 1161–1188.

[54]Deatherage, J. H., Burdette, E. G., and Chew, C. K., "Development Length and Lateral Spacing Requirements of Prestressing Strand for Prestressed Concrete Bridge Girders," *PCI Journal*, Vol. 39, No. 1, January–February 1994, pp. 70–83.

[55]Russell, B. W., and Burns, N. H., "Measured Transfer Lengths of 0.5 and 0.6 in. Strands in Pretensioned Concrete," *PCI Journal*, Vol. 41, No. 5, September–October 1996, pp. 44–65.

[56]*NEHRP (National Earthquake Hazards Reduction Program) Recommended Provisions for the Development of Seismic Regulations for New Buildings*, Building Seismic Safety Council, Washington, D. C., 1985, 1988, 1991, 1994, 1997, 2000, 2003.

[57]*Minimum Design Loads for Buildings and Other Structures*, ASCE 7-05, American Society of Civil Engineers, New York, 2005.

[58]Ishizuka, T., and Hawkins, N. M., *Effect of Bond Deterioration on the Seismic Response of Reinforced and Partially Prestressed Concrete and Ductile Moment Resistant Frames*, Report SM 87–2, Department of Civil Engineering, University of Washington, Seattle, 1987.

[59]Stanton, J., Stone, W. C., and Cheok, G. S., "A Hybrid Reinforced Precast Frame for Seismic Regions," *PCI Journal*, Vol. 42, No. 2, March–April 1997, pp. 20–32.

[60]Muguruma, H., and Watanabe, F. (1990), "Ductility Improvement of High-Strength Concrete Columns with Lateral Confinement," *Proceedings, Second International Symposium on High-Strength Concrete*, SP-121, American Concrete Institute, Detroit, MI, 47–60.

[61]Muguruma, H., Nishiyama, M., Watanabe, F., and Tanaka, H. (1991), "Ductile Behavior of High-Strength Concrete Columns Confined by High-Strength Transverse Reinforcement," *Evaluation and Rehabilitation of Concrete Structures and Innovations in Design*, SP-128, American Concrete Institute, Detroit, MI, 877–891.

[62]Sugano, S., Nagashima, T., Kimura, H., and Ichikawa, A. (1990), "Experimental Studies on Seismic Behavior of Reinforced Concrete Members of High-Strength Concrete," *Proceedings, Second International Symposium on High-Strength Concrete*, SP-121, American Concrete Institute, Detroit, MI, 61–87.

[63]Building Officials and Code Administrators International, *The BOCA National Building Code/1999 – Commentary.* Country Club Hills, IL, 1999, copyright held by International Code Council.

[64]Southern Building Code Congress International, *An Illustrated Commentary to the 1999 Edition of the Standard Building Code*, Birmingham, AL, 1999, copyright held by International Code Council.

[65]ACI Committee 349, "Code Requirements for Nuclear Safety Related Concrete Structures," ACI 349–97, *ACI Manual of Concrete Practice*, 1999 (Appendix B – Steel Embedments, pp. 349–76 , 349–82).

[66]*PCI Design Handbook*, 5th Edition, Precast/Prestressed Concrete Institute, Chicago, IL, 1999.

[67]ACI Committee 506, *Specifications for Shotcrete*, ACI 506.2–95, American Concrete Institute, Detroit, MI, 1995.

[68]*Test Method for Obtaining and Testing Drilled Cores and Sawed Beams of Concrete*, ASTM C 42, American Society for Testing and Materials, West Conshohocken, PA, 2004.

[69]*Specification for Gypsum Concrete*, ASTM C 317, American Society for Testing and Materials, West Conshohocken, PA, 2000.

[70]*Specification for Installation of Cast-in-Place Reinforced Gypsum Concrete*, ASTM C 956, American Society for Testing and Materials, West Conshohocken, PA, 2004.

[71]*Specification for Hot-Formed Welded and Seamless Carbon Steel Structural Tubing*, ASTM A 501, American Society for Testing and Materials, West Conshohocken, PA, 2001.

[72]*Specification for Cold-Formed Welded and Seamless Carbon Steel Structural Tubing in Rounds and Shapes*, ASTM A 500, American Society for Testing and Materials, West Conshohocken, PA, 2003.

Chapter 10 Flexure and Axial Loads

The axial load limit of $0.10 f'_c A_g$ in Section 10.3.5 is clarified to be a limit on factored axial compression load.

10.6.4

The maximum spacing of reinforcement closest to the tension face, for purposes of crack control, is now given by:

$$s = 15\left(\frac{40{,}000}{f_s}\right) - 2.5c_c \leq 12\left(\frac{40{,}000}{f_s}\right) \qquad \text{(19A-1)}$$

with f_s in psi, whereas in ACI 318-02 it was given by:

$$s = \frac{540}{f_s} - 2.5c_c \leq 12\left(\frac{36}{f_s}\right) \qquad \text{(19A-2)}$$

with f_s in ksi. f_s is the service-level stress in the tension reinforcement and c_c is the clear cover to the tension reinforcement. This change reflects the higher service level stresses that occur in flexural reinforcement with the use of the load combinations introduced in the 2002 edition of the standard. Instead of calculating, f_s may now be taken equal to $0.67f_y$ (40,000 psi for Grade 60 mild reinforcement), whereas it could be taken equal to $0.6f_y$ (36,000 psi for Grade 60 reinforcement) under ACI 318-02. It should be noted that, for the default value of the service-level stress, the maximum spacing limitation itself has not changed.

Section 10.6.7 on skin reinforcement in deep members has been modified as follows:

"~~If the effective depth d~~ Where h of a beam or a joist exceeds 36 in., longitudinal skin reinforcement shall be uniformly distributed along both side faces of the member ~~for a distance $d/2$ nearest the flexural tension reinforcement~~. Skin reinforcement shall extend for a distance $h/2$ from the tension face. The spacing ~~s_k between longitudinal bars or wires of the skin reinforcement shall not exceed the least of $d/6$, 12 in., and $1000A_b/(d-30)$~~ s shall be as provided in 10.6.4, where c_c is the least distance from the surface of the skin reinforcement or prestressing steel to the side face. It shall be permitted to include such reinforcement in strength computations if a strain compatibility analysis is made to determine stress in the individual bars or wires. ~~The total area of longitudinal skin reinforcement in both faces need not exceed one half of the required flexural tensile reinforcement.~~"

The changes in Section 10.6.7 are intended to simplify the crack control provisions for skin reinforcement and make those provisions consistent with those required for flexural tension reinforcement. The size of skin reinforcement is not specified; research[84] has indicated that the spacing rather than bar size is of primary importance.

As indicated earlier, Section 10.9.3 has been modified to permit the use of spiral reinforcement with specified yield strength of up to 100,000 psi. For spirals with f_{yt} greater than 60,000 psi, only mechanical or welded splices may be used.

Section 10.13.6 requires that in addition to load combinations involving lateral loads, the strength and stability of the structure as a whole under factored gravity loads must be considered. In Items (a) and (b) of that section, "1.4 dead load and 1.7 live load" of ACI 318-02 has been replaced by "factored dead and live loads" in ACI 318-05, thus supplying a much needed clarification.

Chapter 11 Shear and Torsion

11.6.7 In a significant change, an alternative design procedure for torsion design is introduced in Section 11.6.7, which more realistically addresses thin, deep spandrel beams, which are common in precast concrete construction. Design for torsion now must be in accordance with Sections 11.6.1 through 11.6.6, or 11.6.7. The design for torsion in Sections 11.6.1 through 11.6.6 is based on a thin-walled tube, space truss analogy.

Section 11.6.7, titled "Alternative design for torsion," states:

"For torsion design of solid sections within the scope of this code with an aspect ratio, h/b_t, [h = total depth of section, b_t = width of that part of cross section containing the closed stirrups resisting torsion] of three or greater, it shall be permitted to use another procedure, the adequacy of which has been shown by analysis and substantial agreement with results of comprehensive tests. Sections 11.6.4 (Details of torsional reinforcement) and 11.6.6 (Spacing of torsion reinforcement) shall apply."

Commentary Section R11.6.7 cites examples of such procedures. One cited procedure is an extension to prestressed concrete sections of the torsion procedures of pre-1995 editions of ACI 318. The fourth edition of the *PCI Design Handbook*[85] describes the procedure, which has also been republished in a recent issue of the *PCI Journal*.[86] The procedure was experimentally verified by the tests described in Reference 87. Another cited procedure is that of Reference 88.

Section 11.6.4.2, which requires transverse torsional reinforcement to be anchored in ways indicated by Items (a) or (b), has had Item (a) modified as follows: "(a) A 135-deg standard hook or seismic hook, as defined in 21.1, around a longitudinal bar."

Chapter 13 Two-Way Slab Systems

13.2.5 Dimensional requirements for drop panels are relocated to this new section from their earlier location in ACI 318-02 Sections 13.3.7.1 and 13.3.7.2. Also, it is clarified that drop panels are required to conform to these provisions only when they are used to reduce the amount of negative reinforcement over a column or minimum required slab thickness. A new Commentary Section R13.2.5 points out that drop panels with dimensions less than those specified in 13.2.5 may be used to increase slab shear strength.

Chapter 14 Walls

The f in Eq. (14-1), giving the design axial load strength of a wall eligible to be designed by the empirical design method, was 0.7 in ACI 318-02. Now the same f must correspond to compression-controlled sections in accordance with 9.3.2.2. This is for consistency with Chapter 9.

For similar reasons, under 14.8, Alternative design of slender walls, the previous requirement that the reinforcement ratio should not exceed $0.6\rho_{bal}$ is replaced by the requirement that the wall be tension-controlled, leading to approximately the same reinforcement ratio.

Chapter 15 Footings

An important clarification of 15.5.3 has been provided by replacing "Other pile caps shall satisfy one of 11.12, 15.5.4, or Appendix A" with "Other pile caps shall satisfy either Appendix A, or both 11.12 and 15.5.4." Section 15.5 deals with shear design of footings.

Chapter 18 Prestressed Concrete

Tendons of continuous post-tensioned beams and slabs are usually stressed at a point along the span where the tendon profile is at or near the centroid of the concrete cross section. Therefore, interior construction joints are usually located within the end thirds of the span, rather than the middle third of the span as required by 6.4.4. This has had no known detrimental effect on the performance of such beams. Thus, 6.4.4 is now excluded from application to prestressed concrete.

ACI 318-02 required prestressed two-way slab systems to be designed as Class U, which meant that f_t could be up to $7.5\sqrt{f_c'}$. ACI 318-05 restricts f_t in such states to $6\sqrt{f_c'}$, thus limiting the permissible flexural tensile stress in two-way prestressed slabs to the same value as in ACI 318-99 and prior codes. Section 18.4.4.4 has been modified as follows:

> "Where ~~If~~ h ~~the effective depth~~ of a beam exceeds 36 in., the area of longitudinal skin reinforcement consisting of reinforcement or bonded tendons shall be provided as required by 10.6.7."

In Commentary Section R18.10.3, the statement that for statistically indeterminate structures, the moments that are due to reactions induced by prestressing forces, referred to as secondary moments, are significant in both elastic and inelastic states is now supported by three added references.[89,90,91] The sentence, "When hinges and full redistribution of moments occur to create a statically determinate structure, secondary moments disappear" has been deleted. This removes an unnecessary and potentially confusing sentence.

Section 18.12.4 no longer refers to "normal live loads," because it is largely meaningless.

Chapter 21 Special Provisions for Seismic Design

A new term, design story drift ratio, is defined as the relative difference of design displacements between the top and the bottom of a story, divided by the story height. This is part of a change in 21.11 that is discussed later.

21.2.5

As mentioned earlier, Sections 9.4 and 10.9.3 are modified to allow the use of spiral reinforcement with specified yield strength of up to 100 ksi. However, a sentence added to Section 21.2.5 specifically prohibits such use in members resisting earthquake-induced forces in structures assigned to Seismic Design Category D, E or F. This is largely the result of some misgiving that high-strength spiral reinforcement may be less ductile than conventional mild reinforcement and that spiral failure has in fact been observed in earthquakes. There are fairly convincing arguments, however, against such specific prohibition. Spiral failure, primarily observed in bridge columns, has invariably been the result of insufficient spiral reinforcement, rather than the lack of ductility of the spiral reinforcement. Also, prestressing steel, which is primarily the high-strength steel available on the U.S. market, is at least as ductile as welded wire reinforcement, which is allowed to be used as transverse reinforcement. For this reason, the 2006 IBC has further modified this requirement in Section 1908.1.5. See earlier discussion of that IBC section in this chapter.

21.7.2.3

In a very significant and beneficial change, the requirements of Section 21.7.2.3 are revised to remove the reference to beam-column joints in Section 21.5.4, which modifies the development length requirements of Chapter 12 for longitudinal beam bars terminating at exterior beam-column joints of structures assigned to high seismic design categories. In ACI 318-02, Section 21.7.2.3 required that all continuous reinforcement in structural walls must be anchored or spliced in accordance with the provisions for reinforcement in tension in Section 21.5.4. This was very confusing to the user, because 21.5.4 is really not applicable to situations covered by this section. This problem existed with ACI 318 editions prior to 2002 as well.

Because actual forces in longitudinal reinforcement of structural walls may exceed calculated forces, it is now required that reinforcement in structural walls be developed or spliced for f_y in tension in accordance with Chapter 12. The effective depth of member referenced in Section 12.10.3 is permitted to be taken as $0.8\ell_w$ [ℓ_w= length of wall] for walls. Requirements of Sections 12.11, 12.12, and 12.13 need not be satisfied, because they address issues related to beams and do not apply to walls. At locations where yielding of longitudinal reinforcement is expected, $1.25f_y$ is required to be developed in tension, to account for the likelihood that the actual yield strength exceeds the specified yield strength, as well as the influence of strain-hardening and cyclic load reversals. Where transverse reinforcement is used, development lengths for straight and hooked bars may be reduced as permitted in Sections 12.2 and 12.5, respectively, because closely spaced transverse reinforcement improves the performance of splices and hooks subjected to repeated cycles of inelastic deformation.

21.9.5.3 The special confinement requirements, as specified in this section, for structural truss elements, struts, ties, diaphragm chords and collector elements are revised whereby the threshold compressive stresses are increased for cases where the design forces for those structural components are amplified by a factor Ω_0 to account for the overstrength in the vertical elements of the seismic-force-resisting system. For those cases, the special transverse reinforcement is now required when compressive stress at any section exceeds $0.5f'_c$, up from $0.2f'_c$ specified for general cases. Also, the special transverse reinforcement is required to be continued at least up to a section where the compressive stress falls below $0.4f'_c$, up from $0.15\sqrt{f'_c}$ specified for general cases. As done in general cases, stresses are calculated for factored forces using a linear elastic model and gross-section properties of the elements considered. In recent seismic codes and standards, collector elements of diaphragms are required to be designed for forces amplified by an overstrength factor Ω_0, which ranges between 2 and 3 for concrete structures, depending upon the document selected and on the type of seismic-force-resisting system. Thus, the threshold compressive stress values are also factored by approximately 2.5, the average of 2 and 3.

21.9.5.4 For the same reason as described under Section 21.7.2.3, Section 21.9.5.4 is also revised to remove any reference to Section 12.5.4 for the purpose of development length calculation for continuous reinforcement in diaphragms, trusses, ties, chords and collector elements. Section 21.9.5.4 now requires that all continuous reinforcement in diaphragms, trusses, struts, ties, chords and collector elements be developed or spliced for f_y in tension.

21.11.5 In a very significant change, provisions for shear reinforcement at slab-column joints have been added in a new section 21.11.5 to reduce the likelihood of punching shear failure in two-way slabs without beams. A prescribed amount and detailing of shear reinforcement is required unless either 21.11.5(a) or (b) is satisfied.

Section 21.11.5(a) requires calculation of shear stress that is due to the factored shear force and induced moment according to 11.12.6.2. The induced moment is the moment that is calculated to occur at the slab-column joint where subjected to the design displacement defined in Section 21.1. Section 13.5.1.2 and the accompanying commentary provide guidance on selection of the slab stiffness for the purpose of this calculation.

Section 21.11.5(b) does not require the calculation of induced moments and is based on research[90] that identifies the likelihood of punching shear failure considering interstory drift and shear that is due to gravity loads. The requirement is illustrated in the newly added Figure R21.11.5, reproduced here as Figure 19A-2. The requirement can be satisfied in several ways: adding slab shear reinforcement (Figure 19A-3), increasing slab thickness, designing a structure with more lateral stiffness to decrease interstory drift or a combination of two or more of these factors.

If column capitals, drop panels or other changes in slab thickness are used, the requirements of Section 21.11.5 must be evaluated at all potential critical sections.

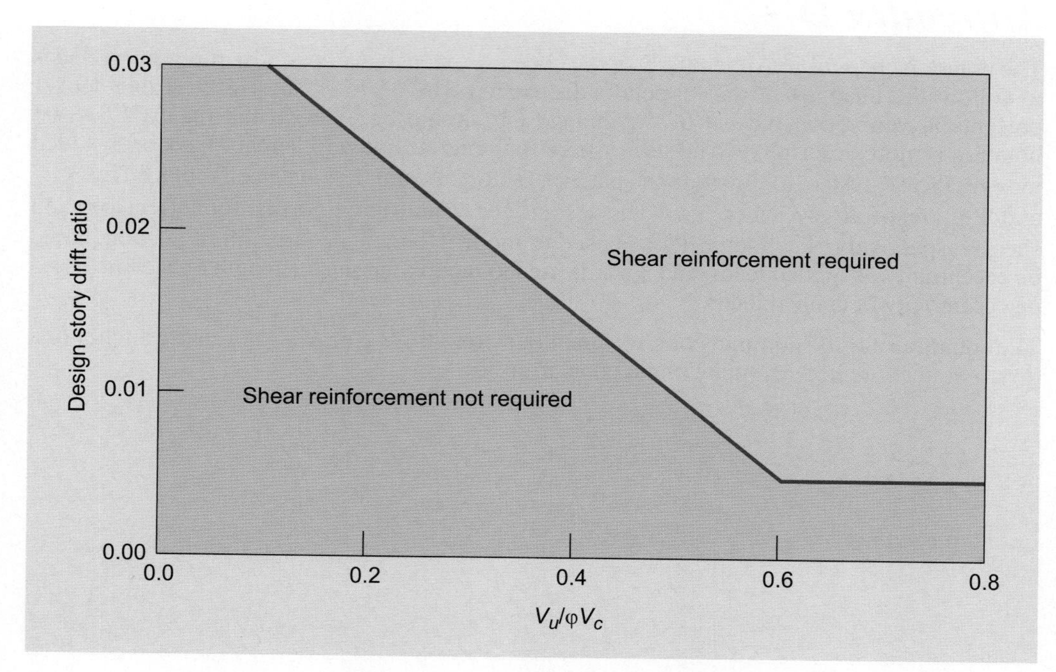

Figure 19A-2 Shear reinforcement requirement of ACI 318-05 Section 21.11.5

Photos provided courtesy of Decon U.S.A., Inc.

Figure 19A-3 Studrails as slab shear reinforcement

Appendix D

D.0 Notations The notation changes in Appendix D of the code are extensive and vitally important if one is to follow the changes in that appendix from ACI 318-02 to ACI 318-05. Table 19A-1 presents a comprehensive list of the changes. It should be evident that the ACI 318-05 notation is more descriptive. The subscripts *c* for concrete and *a* for anchor have been added in several cases. Also, to illustrate the pattern, factors Ψ_1 and Ψ_5 have been replaced by $\Psi_{ec,N}$ and $\Psi_{ec,V}$, respectively, where e_c, *N* and *V* stand for eccentricity, normal force (tension) and shear, respectively. The factor $\Psi_{ec,N}$ is used to modify the tensile strength of anchors based on eccentricity of applied loads, and $\psi_{ec,V}$, is used to modify the shear strength of anchors based on eccentricity of applied loads.

D.5.2.1 The equations for the nominal concrete breakout strength, N_{cb}, or N_{cbg}, of a single anchor or a group of anchors in tension are changed as follows:

$$N_{cb} = \frac{A_n}{A_{No}} \psi_2 \psi_3 N_b \qquad \text{(19A-3)}$$

is replaced by

$$N_{cb} = \frac{A_{nc}}{A_{Nco}} \psi_{ed,N} \psi_{c,N} \psi_{cp,N} N_b \qquad \text{(19A-4)}$$

and

$$N_{cbg} = \frac{A_n}{A_{No}} \psi_1 \psi_2 \psi_3 N_b \qquad \text{(19A-5)}$$

is replaced by

$$N_{cbg} = \frac{A_{nc}}{A_{Nco}} \Psi_{ec,N} \Psi_{ed,N} \Psi_{c,N} \Psi_{cp,N} N_b \qquad \text{(19A-6)}$$

Apart from a change of notation, a new modification factor $\psi_{cp,N}$ is added to each equation for reasons that need to be explained. ACI 318-02 Appendix D assumed that anchors with an edge distance equal to $1.5h_{ef}$ [h_{ef} = effective embedment depth of anchor] or greater developed the basic concrete breakout strength in tension. Test experience has since shown that many torque-controlled and displacement-controlled expansion anchors (see definition of expansion anchors in ACI 318-05 Section D.1) and some undercut anchors (see ACI 318-05 Section D.1) require an edge distance greater than $1.5h_{ef}$ to meet this requirement in uncracked concrete without supplementary reinforcement to control splitting. These types of anchors introduce splitting tensile stresses in the concrete during installation that are increased during load application and may cause a premature splitting failure. $\psi_{cp,N}$ is a new modification factor for these types of anchors to prevent splitting failure where supplementary reinforcement to prevent splitting is not present, and its value is specified in the new Section D.5.2.7.

D.5.2.2 The basic concrete breakout strength of a single anchor in tension in cracked concrete is given as

$$N_b = k_c \sqrt{f'_c} h_{ef}^{1.5} \qquad \text{(19A-7)}$$

WHERE

K_c = 24 for cast-in anchors, and

K_c = 17 for postinstalled anchors

ACI 318-05 has added:

> "The value of k_c for post-installed anchors shall be permitted to be increased above 17 based on ACI 355.2 product-specific tests, but shall in no case exceed 24."

The following is removed from RD.5.2.2:

> "When using k [now k_c] values from ACI 355.2 product approval reports, ψ_3 [now ψ_c] shall be taken as 1.0 because the published test results of the ACI 355.2 product approval tests provide specific k values for cracked or uncracked concrete."

The intent is to clarify the design of postinstalled anchors in cracked and uncracked concrete in the body of the code rather than in the commentary.

D.5.2.3

This section now states that where anchors are located less than $1.5h_{ef}$ from three or more edges, the value of h_{ef} used in Eqs. D-4 through D-11 must be the greater of $c_{a,\max}/1.5$ [$c_{a,\max}$ = the largest of the influencing edge distances that are less than or equal to 1.5 times the actual h_{ef}] and one-third of the maximum spacing between anchors within the group. It may be recalled that in ACI 318-02, it used to be just $c_{a,\max}/1.5$.

The limit on h_{ef} of at least one-third of the maximum spacing between anchors within the group prevents the designer from using a calculated strength based on individual breakout prisms for a group anchor configuration. ACI 318-05 Figure RD.5.2.2, reproduced here as Figure 19A-4, is useful in understanding the requirement of D.5.2.3. To visualize the requirement, move the concrete breakout surface, which originates at the actual h_{ef}, in a direction parallel to the applied tension toward the surface of the concrete. The value of h_{ef} used in Eqs. D-4 to D-11 is determined when either: (a) the outer boundaries of the failure surface first intersects a free edge, or (b) the intersection of the breakout surface between anchors within the group first intersects the surface of the concrete. Point A in Figure 19A-4 defines the intersection of the transported failure surface with the concrete surface and determines the value of h_{ef} to be used in the computation of anchor breakout strength.

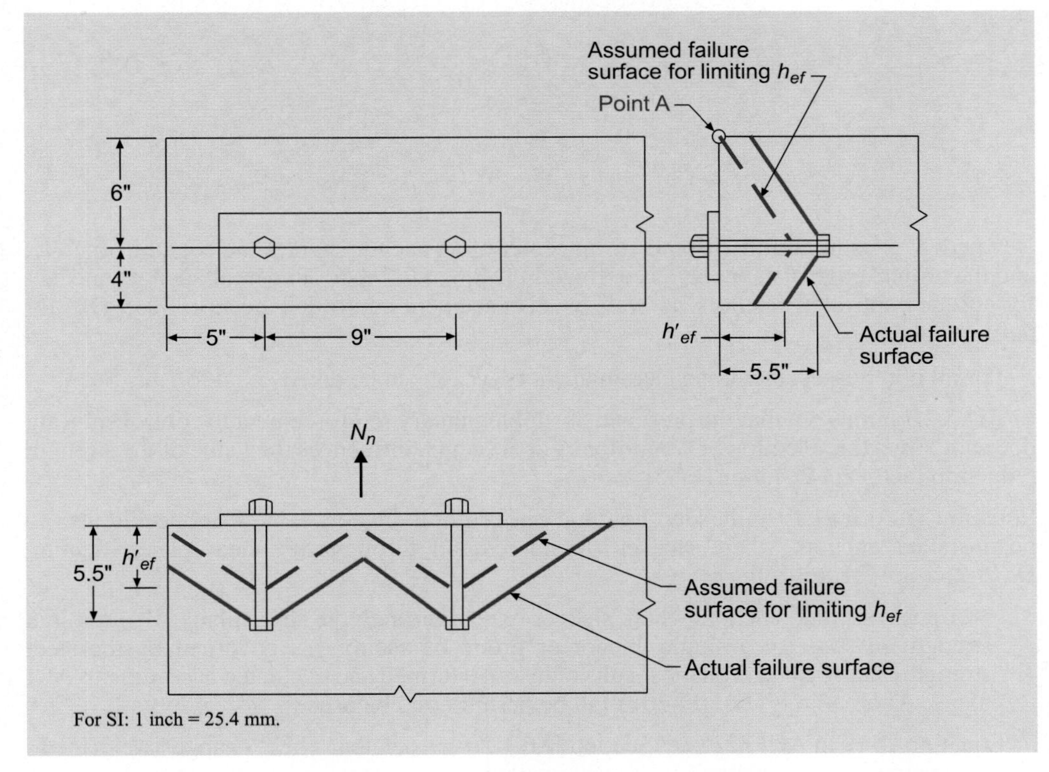

Figure 19A-4
Anchor in tension close to three or more edges

In Figure 19A-4, the actual h_{ef} is 5.5 in., but three edges are within $1.5h_{ef}$ or 8.25 in. from the end anchor. Therefore, the limiting value of h_{ef} (h'_{ef} in the figure) is the larger of $c_{a,\max}/1.5$ and one-third of the maximum spacing for an anchor group. This gives h'_{ef} = max (6/1.5, 9/3) = 4 in., which is to be used for the value of h_{ef} in Eqs. D-4 to D-11, including the calculation of A_{Nc} [projected concrete failure area of a single anchor or group of anchors, for calculation of strength in tension]; $A_{Nc} = (6 + 4)(5 + 9 + 1.5 \times 4) = 200$ in.^{2} Note that by ACI 318-02, h'_{ef} would also have been equal to 6/1.5 = 4 in. The new modification does not make any difference in this particular example until the spacing between the anchors exceeds $(6/1.5) \times 3 = 12$ in.

D.5.2.4 Figure RD.5.2.4, showing the definition of e'_N for a group of anchors loaded eccentrically in tension [distance between resultant tensile load and the centroid of the group], is significantly revised to add clarity and simplification.

D.5.2.6 ACI 318-05 has added the following requirements:

1. Where the value of k_c used in Eq. D-7 (Eq. 19A-7 shown above) is taken from the ACI 355.2 product evaluation report for postinstalled anchors qualified for use in cracked and uncracked concrete, the values of k_c and $\psi_{c,N}$ shall be based on the ACI 355.2 product-evaluation report.
2. Where the value of k_c used in Eq. D-7 is taken from the ACI 355.2 product evaluation report for postinstalled anchors qualified for use only in uncracked concrete, $\psi_{c,N}$ shall be taken as 1.0.

The intent once again is to clarify the design of postinstalled anchors used in cracked and uncracked concrete in the body of the code.

D.5.2.7 This is a new section that specifies the value of the new modification factor $\psi_{cp,N}$ for postinstalled anchors designed for uncracked concrete in accordance with D.5.2.6 without supplementary reinforcement to control splitting, as noted in the discussion of Section D.5.2.1. This factor is given by:

$$\Psi_{cp,N} = 1.0 \qquad \text{if } c_{a,min} \geq c_{ac} \qquad (19A\text{-}8)$$

$$\Psi_{cp,N} = \frac{c_{a,\min}}{c_{ac}} \geq \frac{1.5h_{ef}}{c_{ac}} \qquad \text{if } c_{a,min} < c_{ac} \qquad (19A\text{-}9)$$

where $c_{a,\min}$ is minimum distance from center of an anchor shaft to the edge of concrete, and the critical edge distance, c_{ac}, is defined in D.8.6 as $2.5h_{ef}$ for undercut anchors and $4h_{ef}$ for torque-controlled anchors as well as displacement-controlled anchors (see D.1 for definitions).

For all other cases, including cast-in anchors, $\psi_{cp,N}$ is to be taken equal to 1.0.

RD.5.2.7 points out that the presence of supplementary reinforcement to control splitting does not affect the selection of Condition A or B (which influences the value of the strength reduction factor) in D.4.4 or D.4.5.

D.6.1.2 Equation D-20 for cast-in headed bolt and hooked bolt anchors is now applicable also to postinstalled anchors where sleeves do not extend through the shear plane. Section D.6.1.2(c) now requires that

> "for post-installed anchors where sleeves extend through the shear plane, V_{sa} [nominal strength in shear of a single anchor or group of anchors as governed by the steel strength] shall be based on the results of tests performed and evaluated according to ACI 355.2. Alternatively, Eq. (D-20) shall be permitted to be used."

Equation D-19 in ACI 3218-02 Section D.6.1.2(c) is deleted. These changes are made to require testing if the contribution of postinstalled anchor sleeves to shear strength is to be considered.

D.6.2.1

Section D.6.2.1(c) now states:

> "for shear force parallel to an edge, V_{cb} or V_{cbg} [nominal concrete breakout strength in shear] shall be permitted to be twice the value of the shear force determined from Eq. (D-21) or (D-22), respectively, with the shear force assumed to act perpendicular to the edge and with $\psi_{ed,V}$ [shear strength correction factor for edge distance] taken equal to 1.0."

This is to clarify how to evaluate the shear breakout strength when anchors are loaded parallel to an edge.

RD.6.2.1 is revised as follows:

> "~~The assumption shown in the upper right example of Fig. RD.6.2.1(b), with the case for two anchors perpendicular to the edge, is a conservative interpretation of the distribution of the shear force on an elastic basis.~~ When using Eq. (D-22) for anchor groups loaded in shear, both assumptions for load distribution illustrated in examples on the right side of Fig. RD.6.2.1(b) should be considered because the anchors nearest the edge could fail first or the whole group could fail as a unit with the failure surface originating from the anchors farthest from the edge. If the anchors are welded to a common plate, when the anchor nearest the front edge begins to form a failure cone, shear load would be transferred to the stiffer and stronger rear anchor. For this reason, anchors welded to a common plate do not need to consider the failure mode shown in the upper right figure of Fig. RD.6.2.1(b). ~~For cases where nominal strength is not controlled by ductile steel elements, D.3.1 requires that load effects be determined by elastic analysis~~"

> "The case of shear force parallel to an edge is shown in Fig. RD.6.2.1(c). A special case can arise with shear force parallel to the edge near a corner. In the example of a single anchor near a corner (see Fig. RD.6.2.1(d)), the provisions for shear in the direction of the load should be checked in addition to the provisions for shear in the direction parallel to the edge. ~~where the edge distance to the side c_2 is 40 percent or more of the distance c_1 in the direction of the load, the shear strength parallel to that edge can be computed directly from Eq. (D-20) and (D-21) using c_1 in the direction of the load.~~"

These changes are intended to provide guidance for computing the nominal concrete breakout strength in shear for anchor groups and for anchors that are loaded parallel to an edge.

D.6.2.4

Section D.6.2.4 now states:

> "Where anchors are influenced by three or more edges, the value of c_{a1} [distance from center of an anchor shaft to the edge of concrete in the direction of applied shear] used in Eqs. D-23 through D-28 shall not exceed the greatest of: $c_{a2}/1.5$ [c_{a2} = distance from center of an anchor shaft to the edge of concrete in the direction perpendicular to c_{a1}] in either direction, $h_a/1.5$ [h_a = thickness of member in which an anchor is located, measured parallel to anchor axis] and one-third of the maximum spacing between anchors within the group."

It may be recalled that in 318-02, it used to be:

> ". . . edge distance c_1 [now c_{a1}] shall be limited to h [now h_a]/1.5."

The changes are made so that the overly conservative concrete breakout strengths in shear given by ACI 318-02 for anchors influenced by three or four edges would be more in accordance with test results. The limit on c_{a1} of at least one-third of the maximum spacing between anchors within the group keeps the calculated strength from being based on individual breakout prisms for a group anchor configuration. Figure RD.6.2.4, reproduced here as Figure 19A-5, is useful in understanding the requirement of D.6.2.4. To visualize the requirement, move the concrete breakout surface originating at the actual c_{a1} in the direction of the applied shear toward the surface of the concrete. The value of c_{a1} to be used in Eqs. D-21 to D-28 is determined when either: (a) the outer boundaries of the failure surface first intersect a free edge, or (b) the intersection of the breakout surfaces between anchors within

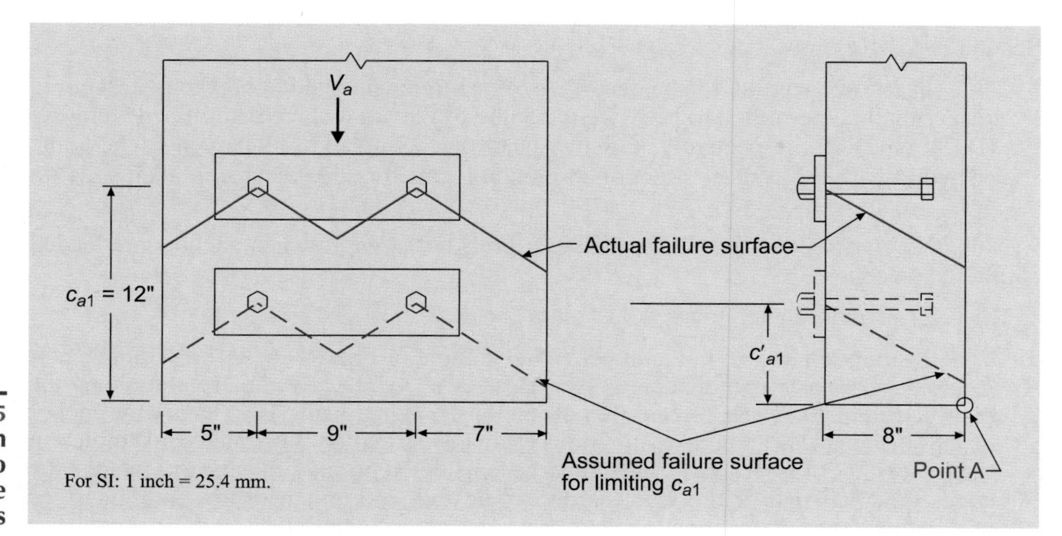

Figure 19A-5 Anchor in shear close to three or more edges

the group first intersects the surface of the concrete. Point A in Figure 19A-5 defines the intersection of the transported failure surface with the concrete surface, and determines the value of c_{a1} to be used in the computation of anchor breakout strength.

In Figure 19A-5, the actual c_{a1} is 12 in., but two orthogonal edges are within $1.5c_{a1}$ or 18 in. from the anchor group; c_{a2} = the larger of 5 in. and 7 in. = 7 in. and h_a = 8 in. Therefore, the limiting value of c_{a1} (c'_{a1} in the figure) is the largest of $c_{a2,\max}/1.5$, $h_a/1.5$ and one-third of the maximum spacing for an anchor group. This gives $c_{a1} = \max(7/1.5, 8/1.5, 9/3) = 5.33$ in., which is to be used for the value of c_{a1} in Eqs. D-21 to D-28, including the calculation of A_{Vc} [projected concrete failure area of a single anchor or group of anchors for calculation of strength in shear]; $A_{Vc} = (5 + 9 + 7)(1.5)(5.33) = 168$ in.^{2}, which is the cross-sectional area of the member. Note that by ACI 318-02, c'_{a1} would also have been equal to 8/1.5 = 5.33 in. The new modifications do not make any difference in this particular example until the spacing between the anchors exceeds (8/1.5) × 3 = 16 in. and/or the larger orthogonal edge distance c_{a2} exceeds 8 in.

D.6.3.1 In D.6.3.1, a new equation is added for the pryout strength of a group of anchors in shear:

$$V_{cpg} = k_{cp} N_{cbg} \tag{19A-10}$$

where k_{cp}, coefficient for pryout strength, is given in D.6.3.1 and N_{cbg}, nominal concrete breakout strength in tension of a group of anchors, is given by Eq. D-5.

D.8.6 A new Section D.8.6 is added requiring that the critical edge distance, c_{ac}, (see discussion of Section D.5.2.7 above), unless determined from tension tests in accordance with ACI 355.2,[78] shall not be taken less than:

$2.5h_{ef}$ for undercut anchors

$4h_{ef}$ for torque-controlled anchors

$4h_{ef}$ for displacement-controlled anchors

Table 19-2. ACI 318 Appendix D notation changes from 2002 to 2005 edition

Notation		
ACI 318-05	**ACI 318-02**	**Description**
A_{brg}	same	Bearing area of the head of stud or anchor bolt, in.^{2}, Appendix D.
A_{Nc}	A_N	Projected concrete failure area of ~~an~~ a single anchor or group of anchors, for calculation of strength in tension, in.^{2}, ~~as defined in~~ see D.5.2.1. ~~A_N shall not be taken greater than nA_{No}. See Fig. RD.5.2.1(b)~~, Appendix D.
A_{Nco}	A_{No}	Projected concrete failure area of ~~one~~ a single anchor, for calculation of strength in tension ~~when~~ if not limited by edge distance or spacing, in.^{2}, ~~as defined in~~ see D.5.2.1 ~~Fig. RD.5.2.1(a)~~, Appendix D.
A_{se}	same	Effective cross-sectional area of anchor, in.^{2}, Appendix D
A_{Vc}	A_V	Projected concrete failure area of ~~an~~ a single anchor or group of anchors, for calculation of strength in shear, in.^{2}, ~~as defined in~~ see D.6.2.1 and ~~A_V shall not be taken greater than nA_{Vo}. See~~ Fig. RD.6.2(b), Appendix D.
A_{Vco}	A_{Vo}	Projected concrete failure area of ~~one~~ a single anchor, for calculation of strength in shear, ~~when~~ if not limited by corner influences, spacing, or member thickness, in.^{2}, ~~as defined in~~ see D.6.2.1 ~~and see Fig. RD.6.2(a)~~, Appendix D.
	c	~~Distance from center of an anchor shaft to the edge of concrete, in., Appendix D.~~
c_{ac}		Critical edge distance required to develop the basic concrete breakout strength of a post-installed anchor in uncracked concrete without supplementary reinforcement to control splitting, in., see D.8.6, Appendix D.
$c_{a,\max}$	$c_{\max}$	~~The largest edge~~ maximum distance from center of an anchor shaft to the edge of concrete, in., Appendix D.
$c_{a,\min}$	$c_{\min}$	~~The smallest edge~~ minimum distance from center of an anchor shaft to the edge of concrete, in., Appendix D.
c_{a1}	c_1	Distance from the center of an anchor shaft to the edge of concrete in one direction, in.; ~~where~~ If shear ~~force~~ is applied to anchor, ~~c_1~~ c_{a1} is taken in the direction of the applied shear ~~force~~. If tension is applied to the anchor, c_{a1} is the minimum edge distance ~~See Fig. RD.6.2(a)~~, Appendix D.
c_{a2}	c_2	Distance from center of an anchor shaft to the edge of concrete in the direction ~~orthogonal~~ perpendicular to ~~c_1~~ c_{a1}, in., Appendix D.
d_o	same	Outside diameter of anchor or shaft diameter of headed stud, headed bolt, or hooked bolt, in., see ~~See also~~ D.8.4, Appendix D.
d'_o	same	Value substituted for d_o when an oversized anchor is used, in., see ~~See~~ D.8.4, Appendix D.
e_h	same	Distance from the inner surface of the shaft of a J- or L-bolt to the outer tip of the J- or L-bolt, in., Appendix D.
e'_N	same	~~Eccentricity of normal force on a group of anchors; the~~ distance between ~~the~~ resultant tension load on a group of anchors loaded in tension and the centroid of the group of anchors loaded in tension, in.; e_N' is always positive~~. See Fig. RD.5.2(b) and (c)~~, Appendix D.
e'_V	same	~~Eccentricity of shear force on a group of anchors; the~~ distance between ~~the~~ resultant shear load on a group of anchors loaded in shear in the same direction ~~point of shear force application~~ and the centroid of the group of anchors loaded in ~~resisting~~ shear in the same direction ~~of the applied shear~~, in.; e_V' is always positive, Appendix D.
f_c	same	Specified compressive strength of concrete, psi, Chapters 4, 5, 8-12, 14, ~~15,18-22~~ 18, 19, 21 22, Appendices A-D.
$\sqrt{f'_c}$	same	Square root of specified compressive strength of concrete, psi, Chapters 8, 9, 11, 12, 18, 19, 21, 22, Appendix D.
f_r	same	Modulus of rupture of concrete, psi, ~~See~~ see 9.5.2.3, Chapters 9, 14, 18, ~~Appendices~~ Appendix B~~, D~~.

(Continued)

Table 19-2. ACI 318 Appendix D notation changes from 2002 to 2005 edition (Cont"d)

Notation		
ACI 318-05	**ACI 318-02**	**Description**
	f_t	~~Calculated concrete tensile stress in a region of a member, psi, Appendix D.~~
f_{uta}	f_{ut}	Specified tensile strength of anchor steel, psi, Appendix D.
	f_{utsl}	~~Specified tensile strength of anchor sleeve, psi, Appendix D.~~
f_{ya}	f_y	Specified yield strength of anchor steel, psi, Appendix D.
h_a	h	Thickness of member in which an anchor is ~~anchored~~ located, measured parallel to anchor axis, in., Appendix D.
h_{ef}	same	Effective ~~anchor~~ embedment depth of anchor, in., see ~~See~~ D.8.5 ~~and Fig. RD.1~~, Appendix D
k_c	k	Coefficient for basic concrete breakout strength in tension, Appendix D.
k_{cp}	same	Coefficient for pryout strength, Appendix D.
ℓ_e	ℓ	Load bearing length of anchor for shear, ~~not to exceed $8d_o$,~~ in., see D.6.2.2, Appendix D. ~~= h_{ef} for anchors with a constant stiffness over the full length of the embedded section, such as headed studs or post-installed anchors with one tubular shell over the full length of the embedment depth, Appendix D.~~ ~~= $2d_o$ for torque controlled expansion anchors with a distance sleeve-separated from the expansion sleeve, Appendix D~~
n	same	~~Number of anchors in a group, Appendix D.~~ Number of items, such as strength tests, bars, ~~or~~ wires, monostrand anchorage devices, anchors, or shearhead arms ~~being spliced or developed along the plane of splitting~~, Chapters 5, 11, 12, 18, Appendix D.
N_b	same	Basic concrete breakout strength in tension of a single anchor in cracked concrete, ~~as defined in D.5.2.2,~~ lb, see D.5.2.2, Appendix D.
N_{cb}	same	Nominal concrete breakout strength in tension of a single anchor, ~~as defined in D.5.2.1,~~ lb, see D.5.2.1, Appendix D.
N_{cbg}	same	Nominal concrete breakout strength in tension of a group of anchors, ~~as defined in D.5.2.1,~~ lb, see D.5.2.1, Appendix D.
N_n	same	Nominal strength in tension, lb, Appendix D.
N_p	same	Pullout strength in tension of a single anchor in cracked concrete, ~~as defined in D.5.3.4 or D.5.3.5,~~ lb, see D.5.3.4 and D.5.3.5, Appendix D
N_{pn}	same	Nominal pullout strength in tension of a single anchor, as defined in D.5.3.1, lb, see D.5.3.1, Appendix D.
N_{sa}	N_s	Nominal strength of a single anchor or group of anchors in tension as governed by the steel strength, ~~as define in D.5.1.1 or D.5.1.2,~~ lb, see D.5.1.1 and D.5.1.2, Appendix D.
N_{sb}	same	Side-face blowout strength of a single anchor, lb, Appendix D.
N_{sbg}	same	Side-face blowout strength of a group of anchors, lb, Appendix D.
N_{ua}	N_u	Factored tensile force ~~load~~ applied to anchor or group of anchors, lb, Appendix D.
s	same	~~Anchor center to center spacing, in., Appendix D.~~ Center-to-center spacing of items, such as longitudinal reinforcement, transverse reinforcement, prestressing tendons, wires, or anchors, ~~spacing of shear or torsion transverse reinforcement in direction parallel ot longitudinal reinforcement,~~ in., Chapters ~~11,~~ 10-12, 17-~~18~~, 21, Appendix D.
	s_o	~~Spacing of the outer anchors along the edge in a group, in., Appendix D.~~
s_s	s	Sample standard deviation, psi, Chapter 5, Appendix D.
	t	~~Thickness of washer or plate, in., Appendix D.~~

(Continued)

Table 19-2. ACI 318 Appendix D notation changes from 2002 to 2005 edition (Cont"d)

Notation		Description
ACI 318-05	ACI 318-02	
V_b	same	Basic concrete breakout strength in shear of a single anchor in cracked concrete, ~~as defined in D.6.2.2 or D.6.2.3,~~ lb, see D.6.2.2 and D.6.2.3, Appendix D.
V_{cb}	same	Nominal concrete breakout strength in shear of a single anchor, ~~as defined in D.6.2.1,~~ lb, see D.6.2.1, Appendix D.
V_{cbg}	same	Nominal concrete breakout strength in shear of a group of anchors, ~~as defined in D.6.2.1,~~ lb, see D.6.2.1, Appendix D.
V_{cp}	same	Nominal concrete pryout strength of a single anchor, ~~as defined in D.6.3,~~ lb, see D.6.3, Appendix D.
V_{cpg}		Nominal concrete pryout strength of a group of anchors, lb, see D.6.3, Appendix D.
V_n	same	Nominal shear strength, lb, Chapters 8, 10, 11, 21, 22, Appendixees ~~C,~~ D.
V_{sa}	V_s	Nominal strength in shear of a single anchor or group of anchors as governed by the steel strength, ~~as defined in D.6.1.1 or D.6.1.2,~~ lb, see D.6.1.1 and D.6.1.2, Appendix D.
V_{ua}	V_u	Factored shear force ~~load~~ applied to a single anchor or group of anchors, lb, Appendix D.
ϕ	same	Strength reduction factor, see 9.3, Chapters 8-11, 13, 14, ~~17-19, 21, 22,~~ 17-22, Appendices A, B, C, D.
$\psi_{c,N}$	ψ_3	~~Modification factor, for strength in tension, to account for cracking, as defined in D.5.2.6 and D.5.2.7, Appendix D.~~ Factor used to modify tensile strength of anchors based on presence or absence of cracks in concrete, see D.5.2.6, Appendix D.
$\psi_{c,P}$	ψ_4	~~Modification factor, for pullout strength, to account for cracking, as defined in D.5.3.1 and D.5.3.6, Appendix D.~~ Factor used to modify pullout strength of anchors based on presence or absence of cracks in concrete, see D.5.3.6, Appendix D.
$\psi_{c,V}$	ψ_7	~~Modification factor, for strength in shear, to account for cracking, as defined in D.6.2.7, Appendix D.~~ Factor used to modify shear strength of anchors based on presence or absence of cracks in concrete and presence or absence of supplementary reinforcement, see D.6.2.7 for anchors in shear, Appendix D.
$\psi_{cp,N}$		Factor used to modify tensile strength of post-installed anchors intended for use in uncracked concrete without supplementary reinforcement, see D.5.2.7, Appendix D.
$\psi_{ec,N}$	ψ_1	~~Modification factor, for strength in tension, to account for anchor groups loaded eccentrically, as defined in D.5.2.4, Appendix D.~~ Factor used to modify tensile strength of anchors based on eccentricity of applied loads, see D.5.2.4, Appendix D.
$\psi_{ec,V}$	ψ_5	~~Modification factor, for strength in shear, to account for anchor groups loaded eccentrically, as defined in D.6.2.5, Appendix D.~~ Factor used to modify shear strength of anchors based on eccentricity of applied loads, see D.6.2.5, Appendix D.
$\psi_{ed,N}$	ψ_2	~~Modification factor, for strength in tension, to account for edge distances smaller than $1.5h_{ef}$, as defined in D.5.2.5, Appendix D.~~ Factor used to modify tensile strength of anchors based on proximity to edges of concrete member, see D.5.2.5, Appendix D.
$\psi_{ed,V}$	ψ_6	~~Modification factor, for strength in shear, to account for edge distances smaller than $1.5c_1$, as defined in D.6.2.6, Appendix D.~~ Factor used to modify shear strength of anchors based on proximity to edges of concrete member, see D.6.2.6, Appendix D.

REFERENCES

[73]*Preparation of Notation for Concrete*, ACI 104, American Concrete Institute, Farmington Hills, MI, 1971.

[74]*Test Method for Determining Density of Structural Lightweight Concrete*, ASTM C 567, American Society for Testing and Materials, West Conshohocken, PA, 2005.

[75]ACI Committee 440, *Guide for the Design and Construction of Concrete Reinforced with FRP Bars*, ACI 440.1R-03, American Concrete Institute, Farmington Hills, MI, 2003, 42 pp.

[76]ACI Committee 440, *Guide for the Design and Construction of Externally Bonded FRP Systems for Strengthening of Concrete Structures*, ACI 440.2R-02, American Concrete Institute, Farmington Hills, MI, 2002, 45 pp.

[77]AASHTO, *Standard Specifications for Highway Bridges*, 17th Edition, American Association for State Highways and Transportation Officials, Washington DC, 2001.

[78]*Qualification of Post-Installed Mechanical Anchors in Concrete and Commentary*, ACI 355.2/355.2R, American Concrete Institute, Farmington Hills, MI, 2004.

[79]*Structural Welding Code – Steel*, ANSI/AWS D1.1–02, American Welding Society, Miami, FL, 2002.

[80]*Minimum Design Loads for Buildings and Other Structures*, ASCE 7-02, American Society of Civil Engineers, New York, 2002.

[81]Saatcioglu, M., and Razvi, S.R., "Displacement-Based Design of Reinforced Concrete Column for Confinement," *ACI Structural Journal*, Vol. 99, No. 1, January-February 2002, pp. 3-11.

[82]Pessiki, S., Greybeal, B., and Mudlock, M., "Proposed Design of High-Strength Spiral Reinforcement in Compression Members," *ACI Structural Journal*, Vol. 98, No. 6, November-December 2001, pp. 799-810.

[83]Richart, F.E., Brandzaeg, A., and Brown, R.L., "The failure of Plain and Spirally Reinforced Concrete in Compression," *Bulletin No.190*, University of Illinois Engineering Experiment Station, Urbana, IL, 1929.

[84]Frosch, R.J., "Modeling and Control of Side Face Beam Cracking," *ACI Structural Journal*, Vol. 99, No. 3, May-June 2002, pp. 376-385.

[85]*PCI Design Handbook: Precast and Prestressed Concrete*, Precast/ Prestressed Concrete Institute, 4th Edition, Chicago, IL, 1992.

[86]Zia, P., and Hsu, T.T.C., "Design for Torsion and Shear in Prestressed Concrete Flexural Members," *PCI Journal*, Vol. 49, No. 3, May-June 2004, 34-42.

[87]Klein, G.J., "Design of Spandrel Beams," *PCI Specially Funded Research Project No. 5*, Precast/ Prestressed Concrete Institute, Chicago, IL, 1986

[88]Collins, M.P., and Mitchell D., "Shear and Torsion Design of Prestressed and Non-Prestressed Concrete Beams," *PCI Journal*, Vol. 25, No. 4, September-October, 1980 pp. 32-100.

[89]Bondy, K.B., "Moment Redistribution: Principles and Practice Using ACI 318-02," *PTI Journal*, Vol. 1, No.1, Post-Tensioning Institute, Phoenix, AZ, January, 2003, pp. 3-21.

[90]Lin, T.Y., and Thornton, K., "Secondary Moment and Moment-Redistribution in Continuous Prestressed Concrete Beams," *PCI Journal*, Vol. 17, No. 1, January-February, 1972, pp. 8-20.

[91]Collins, M.P., and Mitchell, D., *Prestressed Concrete Structures*, Response Publications, Canada, 1997, pp. 517-518.

[92]Megally, S., and Ghali, A., "Punching Shear Design of Earthquake Resistant Slab-Column Connections," *ACI Structural Journal*, Vol. 97, No. 5, September-October, 2000, pp. 720-730.

[93]Moehle, J.P., "Seismic Design Considerations for Flat Plate Construction," Mete A. Sozen Symposium: A Tribute from His Students; SP-162, American Concrete Institute, Farmington Hills, MI, 1996.

CHAPTER

19

Part 3 DESIGN EXAMPLES

This part of the chapter contains numerical examples, the purpose of which is to illustrate the application of certain code provisions that are relatively new and are therefore probably not well understood. The first two examples show the application of unified design, which was moved from Appendix B into the main body of the ACI 318 standard in its 2002 edition. The third example illustrates the emulative design of a special precast concrete moment frame, provisions for which were first introduced in the 2002 edition of ACI 318. Example 4 illustrates the allowable stress design of anchor bolts, provisions for which are not to be found in ACI 318, but are included in the IBC; the 1997 UBC was used as the source of the provisions. Example 5 illustrates the strength design of anchors, provisions for which are to be found in Appendix D of ACI 318, which was added to the standard for the first time in 2002. Example 6 illustrates the strong column-weak beam requirement that has been part of ACI 318 design provisions for special moment frames for a long time; the requirements, however, were significantly modified in the 1999 edition. Examples 7 and 8 illustrate the design of special reinforced concrete shear walls, which must be used in high seismic applications. The design provisions for these changed drastically between ACI 318-95 and ACI 318-99. The shear wall design procedure of the 1997 UBC for high seismic zones was used as a source by ACI 318-99. Finally, Example 9 illustrates the application of strut-and-tie models, provisions for which were first introduced in Appendix A to ACI 318-02.

Example 1 *Unified Design Example of a Doubly Reinforced Rectangular Beam Section*

Service Loads

Live load = 2.47 klf

Dead load = 1.05 klf, including self-weight

Material Properties

Concrete: $f_c' = 4000$ psi

Unit Weight = 150 pcf

Steel: $f_y = 60{,}000$ psi

Member Dimensions

30-ft simple span; Cross section restricted, for architectural reasons, to that shown in Figure 19E-1.1.

For tension-controlled behavior, $\varepsilon_t = 0.005$

From strain diagram, the depth of neutral axis for $\varepsilon_t = 0.005$ is:

$$c_{max} = \frac{0.003}{0.003 + 0.005} d_t = \frac{3 \times 21.5}{8} = 8.06 \text{ in.}$$

The depth of the equivalent stress block is

$$a_{max} = 0.85\, c_{max} = 0.85 \times 8.06 = 6.85 \text{ in.}$$

The compression force in the concrete equals the tension force in the steel:

$$0.85 f_c' a_{max} b = 0.85 \times 4 \times 6.85 \times 16 = 372.81 \text{ kips} = A_{smax} \times 60$$

$$A_{smax} = 372.81/60 = 6.21 \text{ in.}^2$$

The moment strength provided by the reinforcement A_{smax} is:

$$M_{nmax} = A_{s\,max} f_y \left(d_t - \frac{a_{max}}{2} \right) = 372.81 \left(21.5 - \frac{6.85}{2} \right)$$

$$= 6738.5 \text{ in.-kips} = 561.54 \text{ ft-kips}$$

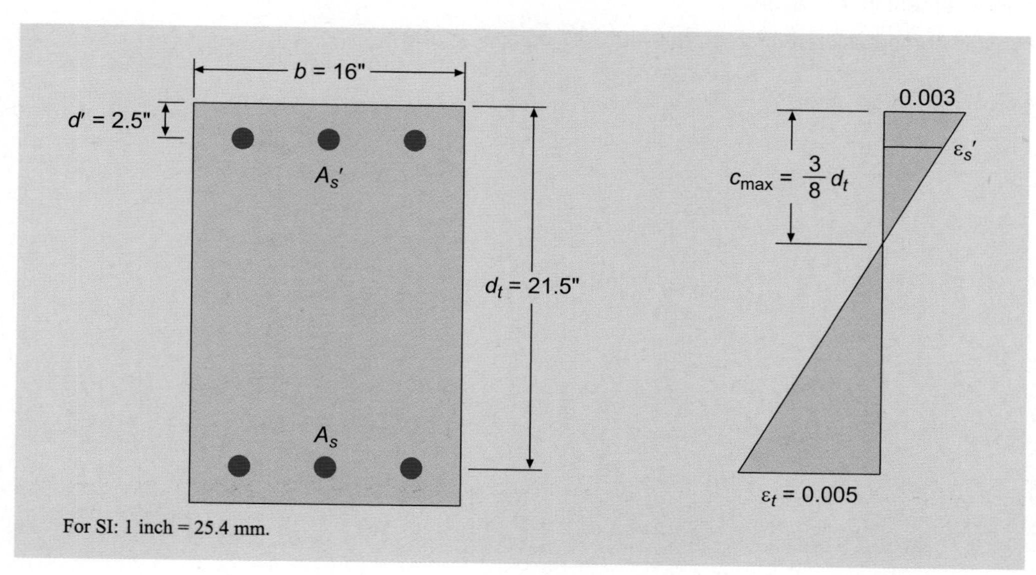

Figure 19E-1.1 Beam cross section and strain profile

Example 1—(continued)

The required moment strength is:

$$M_u = (1.2DL + 1.6\ LL)\frac{l^2}{8} = (1.2 \times 1.05 + 1.6 \times 2.47) \times \frac{30^2}{8}$$

$$= 5.212 \times \frac{30^2}{8} = 586.4 \text{ ft-kips}$$

The nominal moment strength

$M_n \geq M_u/\phi$ = 586.4/0.9 = 651.5 ft-kips,

Since ϕ = 0.9 for tension-controlled behavior.

Additional moment strength needed:

$$\Delta M_n = M_n - M_{nmax} = 651.5 - 561.54 = 90 \text{ ft-kips}$$

Additional tension steel required:

$$\Delta A_s = \frac{\Delta M_n}{(d_t - d')f_y} = \frac{90 \times 12}{(21.5 - 2.5) \times 60} = 0.95 \text{ in.}^2$$

Total tension steel required:

$$A'_{s,\,req} = A_{smax} + \Delta A_s = 6.21 + 0.95 = 7.16 \text{ in.}^2$$

Strain in the compression reinforcement:

$$\varepsilon'_s = \frac{c_{max} - d'}{c_{max}} \times 0.003 = \frac{8.06 - 2.5}{8.06} \times 0.003 = 0.00207 > \varepsilon_y$$

Therefore, $f'_s = f_y$

Compression steel required:

$$A'_{s,\,req} = \frac{\Delta M_n}{(d_1 - d')f_y} = \frac{90 \times 12}{(21.5 - 2.5) \times 60} = 0.95 \text{ in.}^2$$

Provide tension reinforcement with 6 #10 bars, A_s = 7.62 in.^{2}

They cannot be accomodated in one layer. Strength should be rechecked after arranging reinforcement in two layers.

Provide compression reinforcement with 2 #7 bars, A'_s = 1.20 in.^{2}

Example 2 *Design Axial Load-Moment Interaction Diagram of a Column Section Using Unified Design*

Material Properties

Concrete: $f_c' = 4{,}000$ psi

Unit weight = 150 pcf

Steel: $f_y = 60{,}000$ psi

Given Section

The column section under consideration is shown in Figure 19E-2.1. The layer of steel A_{s1} is taken as the extreme tensile steel.

1. The point (0, ϕM_n)

A_{s1} = 2.54 in.^{2} ;

A_{s2} = 2.54 in.^{2}

The location of the neutral axis is calculated through trial and error.

First, assume the location of the neutral axis coincides with the layer 2 steel. The depth of the neutral axis is then:

c = 2.635 in.

a $= 0.85 \times c = 2.24$ in.

The strain distribution across the section is shown in Figure 19E-2.2. The strain and stress in Layer 2 steel is:

$\varepsilon_{s2} = 0$

$f_{s2} = 0$

The strain in layer 1 steel is:

$$\varepsilon_{s1} = \frac{d_1 - c}{c} \times 0.003 = \frac{17.365 - 2.24}{2.24} \times 0.003 = 0.02 > \varepsilon_y$$

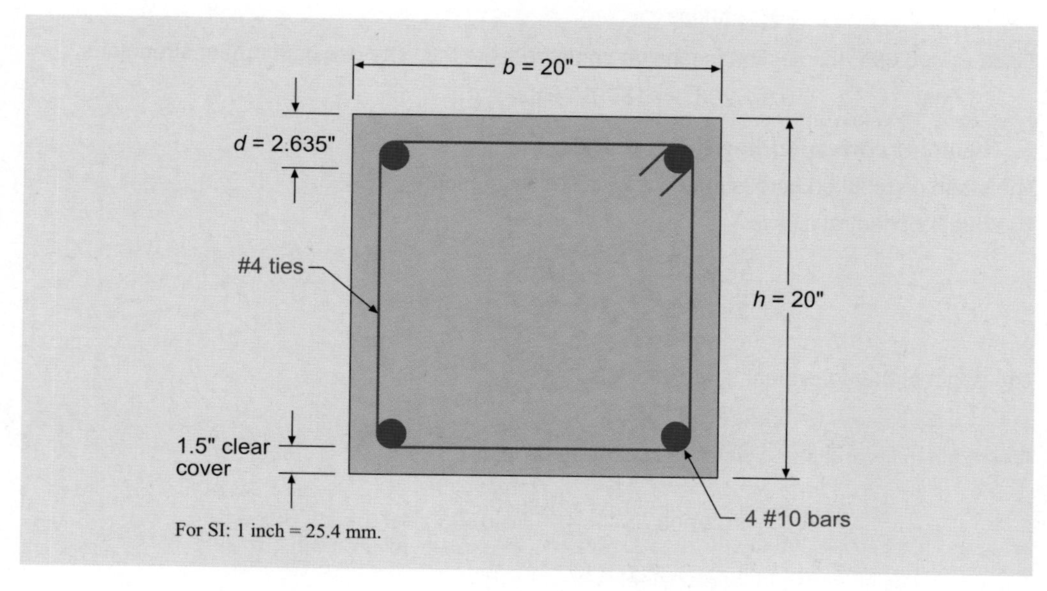

Figure 19E-2.1 **Details of the column section**

Example 2—(continued)

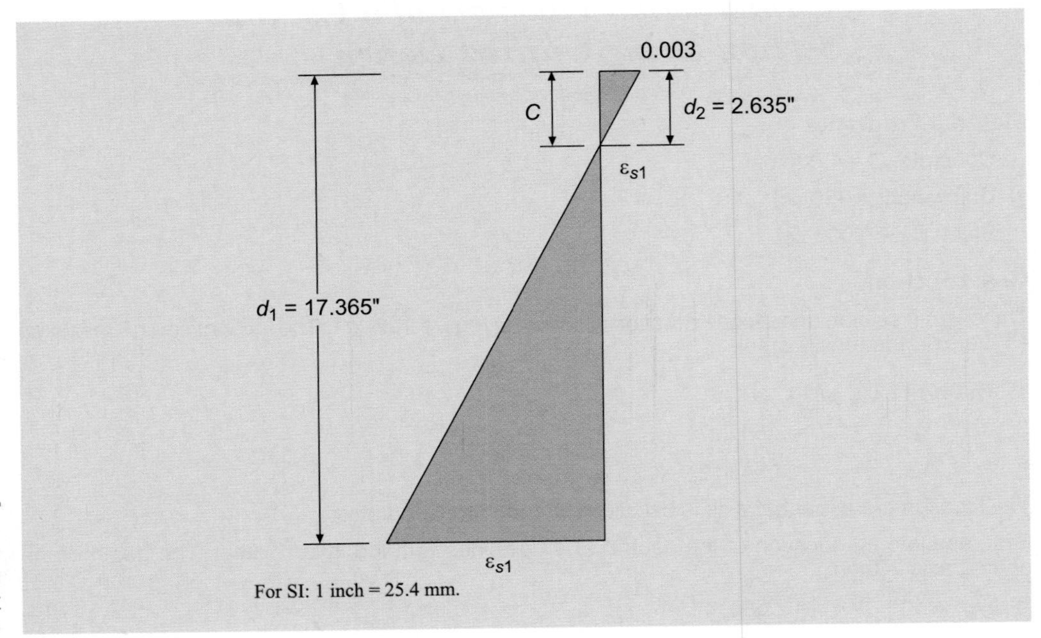

Figure 19E-2.2 Strain profile in the column section at (O,M_n)

The tension force in layer 1 steel is:

A_sf_y = 2.54 × 60 = 152.4 kips

The compression force in the concrete is:

$0.85f_c'ab$ = 0.85 × 4 × 2.24 × 20 = 152.3 kips

which is very close to the tension force. This indicates that the assumed neutral axis location is accurate enough.

The nominal moment strength is

M_n = A_sf_y(17.365 – 2.24/2) = 152.4 × 16.245 = 2476 in.-kips = 206.3 ft-kips

Since ε_{s1} > 0.005, the section is tension-controlled, ϕ = 0.9. The design moment strength is:

ϕM_n = 0.9 × 206.3 = 185.7 ft-kips

2. The point corresponding to $\varepsilon_s = 0.005$

The strain distribution across the section is shown in Figure 19E-2.3.

The depth of neutral axis is:

$$c = \frac{0.003}{0.003 + 0.005} d_t = \frac{3 \times 17.365}{8} = 6.51 \text{ in.}$$

The depth of the equivalent stress block is:

a = 0.85c = 5.54 in.

The strain in layer 2 steel, which is in compression, is:

$$\varepsilon_{s2} = \frac{c - d'}{c} \times 0.003 = \frac{6.51 - 2.635}{6.51} \times 0.003 = 1.79 \times 10^{-3} < \varepsilon_y$$

The compressive stress in layer 2 steel is:

f_{s2} = (1.79 × 10^{-3})(29,000) = 51.8 ksi

Example 2—(continued)

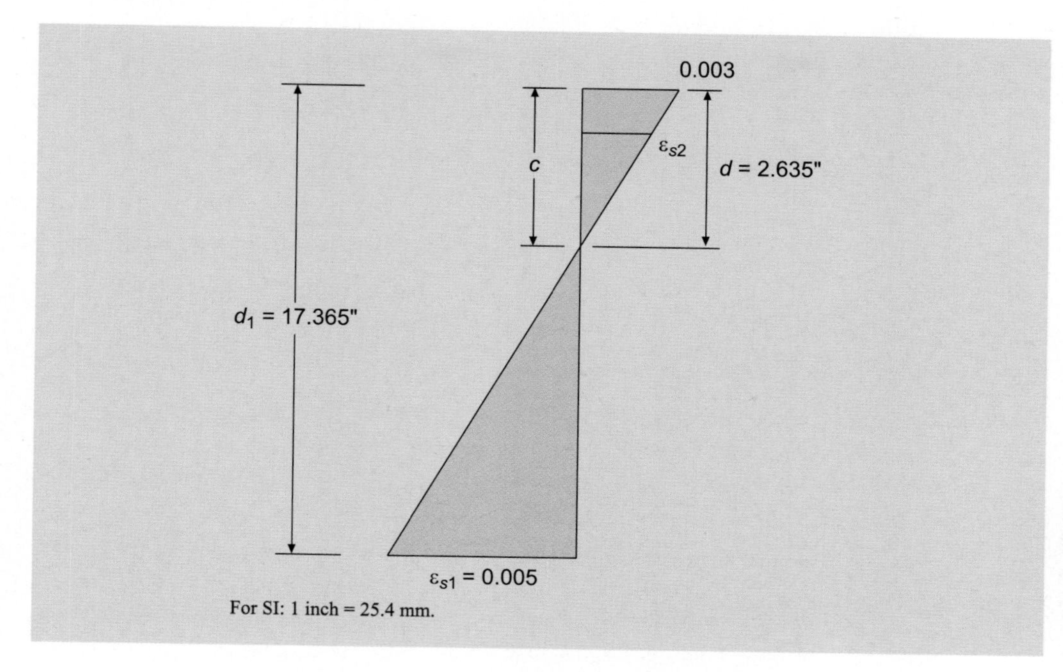

Figure 19E-2.3 Strain profile in the column section for $\varepsilon_{s1} = 0.005$

The total compressive force is:

$$\Sigma C = 0.85f'_c ab + A_{s2}(f_{s2} - 0.85f'_c)$$
$$= 0.85 \times 4 \times 5.54 \times 20 + (51.8 - 0.85 \times 4) \times 2.54$$
$$= 376.5 + 122.9 = 499.4 \text{ kips}$$

The tensile force from the layer 1 steel is:

$$T_{s1} = 2.54 \times 60 = 152.4 \text{ kips}$$

The nominal axial compressive strength in this case is:

$$P_n = 499.4 - 152.4 = 347 \text{ kips}$$

The nominal moment strength is

$$M_n = 376.5 \times (10 - 5.54/2) + 122.9 \times (10 - 2.635) + 152.4 \times (17.365 - 10)$$
$$= 2722 + 905 + 1122 = 4749 \text{ in.-kips} = 395.8 \text{ ft-kips}$$

Since $\varepsilon_{s1} = 0.005$, the section is tension-controlled, $\phi = 0.9$.

$$\phi P_n = 0.9 \times 347 = 312.3 \text{ kips}$$
$$\phi M_n = 0.9 \times 395.8 = 356.2 \text{ ft-kips}$$

3. The point corresponding to $\varepsilon_s = \varepsilon_y$

The strain distribution across the section is shown in Figure 19E-2.4.

The strain in the extreme tensile steel is:

$$\varepsilon_y = 2.07 \times 10^{-3}$$

The depth of neutral axis is:

$$c = \frac{0.003}{0.003 + 0.00207} d_t = 0.592 \times 17.365 = 10.28 \text{ in.}$$

$$a = 0.85c = 8.73 \text{ in.}$$

Example 2—(continued)

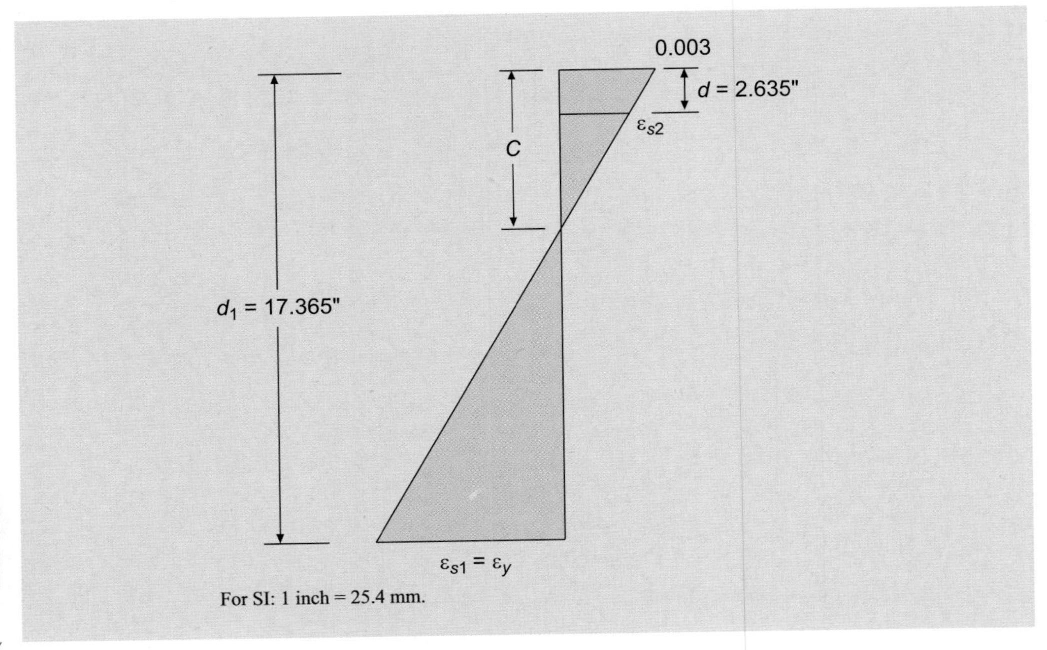

Figure 19E-2.4 Strain profile in the column section for $\varepsilon_s = \varepsilon_y$

The strain in layer 2 steel is:

$$\varepsilon_{s2} = \frac{c - d'}{c} \times 0.003 = \frac{10.28 - 2.635}{10.28} \times 0.003 = 2.23 \times 10^{-3} > \varepsilon_y$$

The stress in layer 2 steel is:

f_{s2} = 60 ksi

The total compressive force is:

$$\Sigma C = 0.85 f_c' ab + A_{s2}(f_{s2} - 0.85 f_c') = 0.85 \times 4 \times 8.73 \times 20 + (60 - 0.85 \times 4) \times 2.54 = 594 + 144 = 738 \text{ kips}$$

T_{s1} = 2.54 × 60 = 152.4 kips

The nominal axial compressive strength in this case is:

P_n = 738 – 152.4 = 585.6 kips

The nominal moment strength is

$$M_n = 594 \times (10 - 8.73/2) + 144 \times (10 - 2.635) + 152.4 \times (17.365 - 10)$$
$$= 3347 + 1061 + 1122 = 5530 \text{ in.-kips} = 460.8 \text{ ft-kips}$$

Since ε_{s1} = 0.002, the section is compression-controlled, ϕ = 0.65.

ϕM_n = 0.65 × 460.8 = 299.5 ft-kips

ϕP_n = 0.65 × 585.6 = 380.6 kips

4. The point corresponding to $\varepsilon_s = 0.5\varepsilon_y$

The strain distribution across the section is shown in Figure 19E-2.5.

The strain in the extreme tensile steel is

$0.5\varepsilon_y = 0.5(2.07 \times 10^{-3}) = 1.03 \times 10^{-3}$

Example 2—(continued)

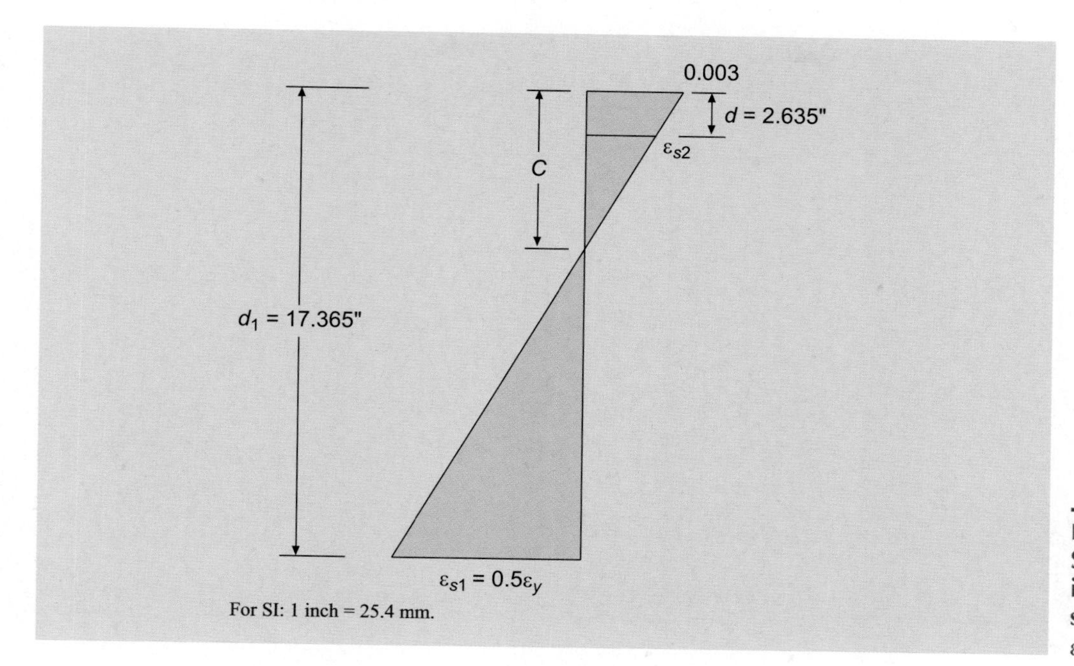

Figure 19E-2.5 Strain profile in the column section for $\varepsilon_s = 0.5\varepsilon_y$

The depth of neutral axis is:

$$c = \frac{0.003}{0.003 + 0.00103} d_t = 0.744 \times 17.365 = 12.91 \text{ in.}$$

$$a = 0.85c = 10.98 \text{ in.}$$

The strain in layer 2 steel is:

$$\varepsilon_{S2} = \frac{c - d'}{c} \times 0.003 = \frac{12.91 - 2.635}{12.91} \times 0.003 = 2.39 \times 10^{-3} > \varepsilon_y$$

The stress in layer 2 steel is:

$$f_{s2} = 60 \text{ ksi}$$

The total compressive force is:

$$\Sigma C = 0.85 f_c' ab + A_{s2}(f_{s2} - 0.85 f_c') = 0.85 \times 4 \times 10.98 \times 20$$

$$+ (60 - 0.85 \times 4) \times 2.54$$

$$= 746 + 144 = 890 \text{ kips}$$

$$T_{s1} = 2.54 \times (0.5 \times 60) = 76.2 \text{ kips}$$

The nominal axial compressive strength in this case is:

$$P_n = 890 - 76.2 = 813.8 \text{ kips}$$

The nominal moment strength is

$$M_n = 746 \times (10 - 10.98/2) + 144 \times (10 - 2.635) + 76.2 \times (17.365 - 10)$$

$$= 3364 + 1061 + 561 = 4986 \text{ in.-kips} = 415.5 \text{ ft-kips}$$

Since $\varepsilon_{s1} < 0.002$, the section is compression-controlled, $\phi = 0.65$.

$$\phi M_n = 0.65 \times 415.5 = 270 \text{ ft-kips}$$

$$\phi P_n = 0.65 \times 813.8 = 529 \text{ kips}$$

Example 2—(continued)

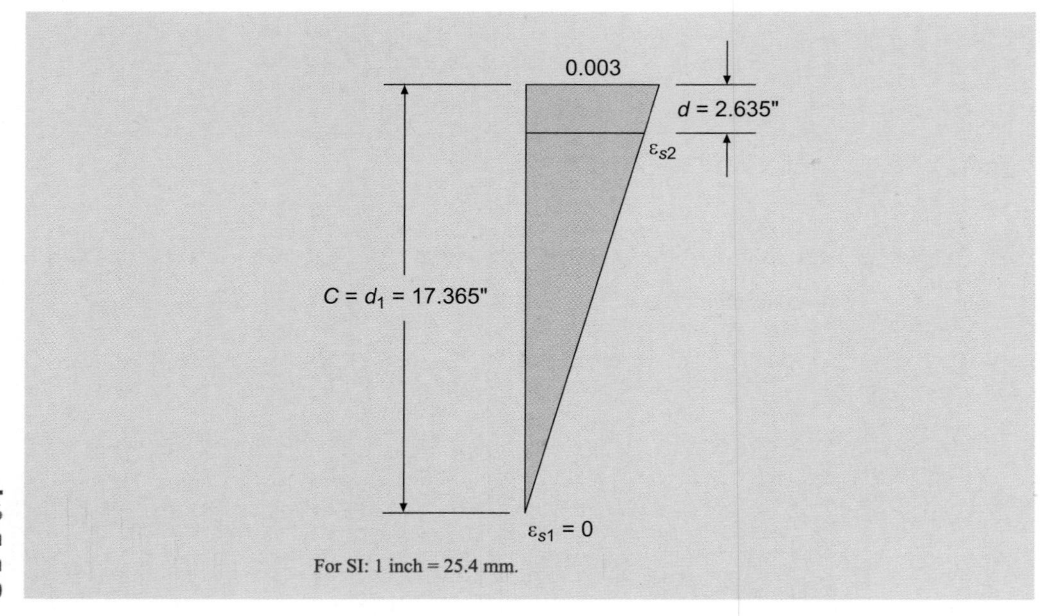

Figure 19E-2.6 Strain profile in the column section for $\varepsilon_s = 0$

5. The point corresponding to $\varepsilon_s = 0$

The strain distribution across the section is shown in Figure 19E-2.6.

The depth of neutral axis is:

$c = 17.365$ in.

$a = 0.85c = 14.76$ in.

The strain in layer 2 steel is

$$\varepsilon_{s2} = \frac{c-d'}{c} \times 0.003 = \frac{17.365-2.635}{17.365} \times 0.003 = 2.54 \times 10^{-3} > \varepsilon_y$$

The stress in layer 2 steel is:

$f_{s2} = 60$ ksi

The total compressive force is:

$$\Sigma C = 0.85 f_c' ab + A_{s2}(f_{s2} - 0.85 f_c')$$

$$= 0.85 \times 4 \times 14.76 \times 20 + (60 - 0.85 \times 4) \times 2.54$$

$$= 1004 + 144 = 1148 \text{ kips}$$

In this case, $T_{s1} = 0$.

The nominal axial compressive strength in this case is:

$P_n = 1148$ kips

The nominal moment strength is

$$M_n = 1004 \times (10 - 14.76/2) + 144 \times (10 - 2.635)$$

$$= 2630 + 1061 = 3691 \text{ in.-kips} = 307.6 \text{ ft-kips}$$

Since $\varepsilon_{s1} < 0.002$, the section is compression-controlled, $\phi = 0.65$.

$\phi M_n = 0.65 \times 307.6 = 200$ ft-kips

$\phi P_n = 0.65 \times 1148 = 746.2$ kips

Example 2—(continued)

6. Pure Compression (ϕP_n, *0*)

The nominal compressive strength with zero eccentricity is:

$$P_0 = 0.85f_c'(A_g - A_{st}) + f_yA_{st}$$
$$= 0.85 \times 4 \times (20 \times 20 - 2.54 \times 2) + 60 \times 2.54 \times 2$$
$$= 1647.5 \text{ kips}$$

The design compressive strength at zero eccentricity is:

$$\phi P_0 = 0.65 \times 1647.5 = 1071 \text{ kips}$$

The design compressive strength considering accidental eccentricity is:

$$\phi P_n = 0.80 \times 1071 = 856.7 \text{ kips}$$

The nominal and design strength interaction diagrams shown in Figure 19E-2.7 below can be drawn on the basis of the above calculations.

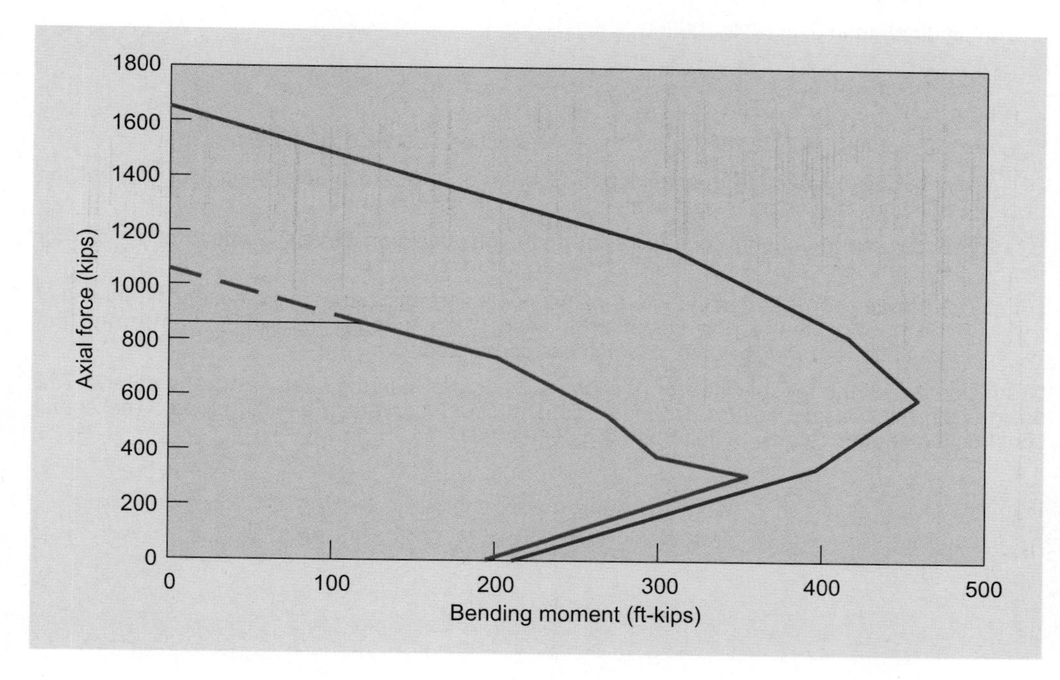

Figure 19E-2.7 Design and nominal strength interaction diagram for the column section

Example 3 *Design Example of a 12-Story Precast Frame Building Using Strong Connections*

A typical floor plan and elevation of the example building are shown in Figures 19E-3.1 and 19E-3.2, respectively. Figures 19E-3.1 and 19E-3.2 clearly show the precast elements out of which the building is to be constructed. The relevant design data are given below.

Service Loads

Live load = 50 psf

Superimposed dead load = 42.5 psf

Material Properties

Concrete: f_c' = 4000 psi [6000 psi for columns in the bottom six stories]

Unit weight = 150 pcf

Member Dimensions

Transverse beams:	24 × 26 in.
Longitudinal beams:	24 × 20 in.
Columns:	24 × 24 in.
Slabs:	7 in.

Figures 19E-3.1 and 19E-3.2 depict the three connections detailed in this example:

1. A strong connection near midspan of an interior beam on the third floor level of the building. The beam is part of an interior longitudinal frame.
2. A column-to-column connection at midheight between Levels 2 and 3 of an interior column stack that is part of an interior longitudinal frame.
3. A strong connection at the interface between a precast beam at the second floor level of the building that forms the exterior span of an exterior transverse frame, and the continuous corner column to which it is connected.

Grout sleeves for the mechanical connections are not shown in the sketches of the details. The designs did not consider reinforcement for construction loads (such as lifting load). The actual construction sequence may be left up to the contractor.

Example 3—(continued)

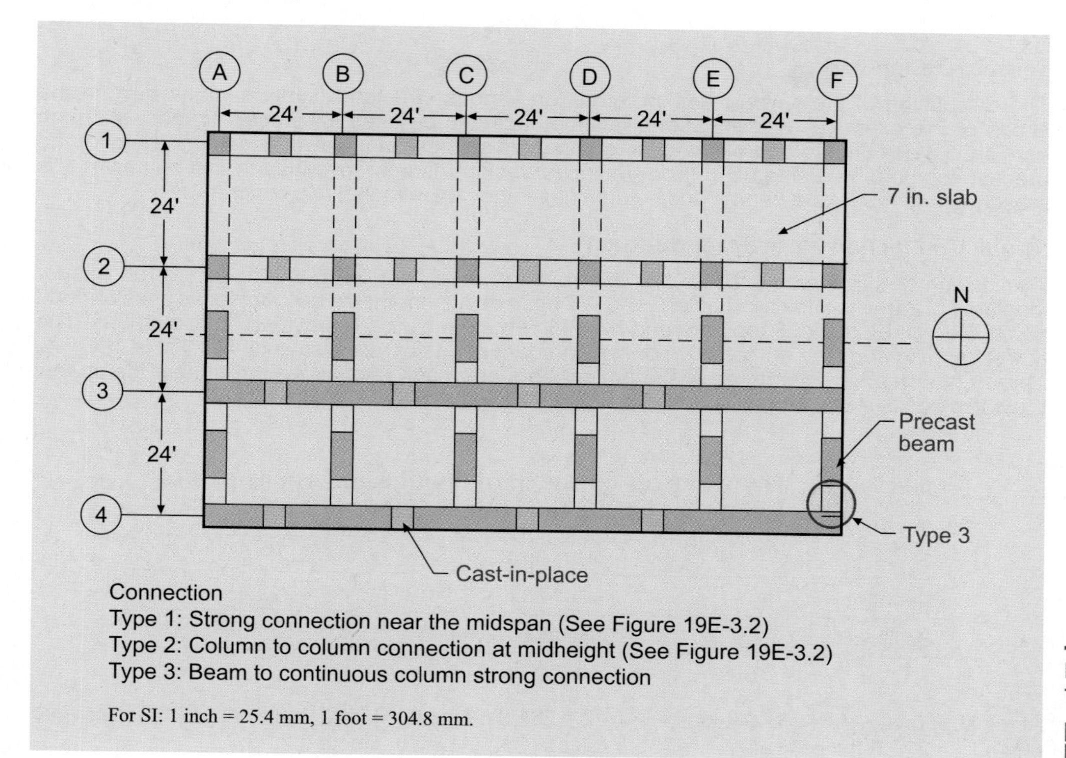

Figure 19E-3.1
Typical floor plan of example building

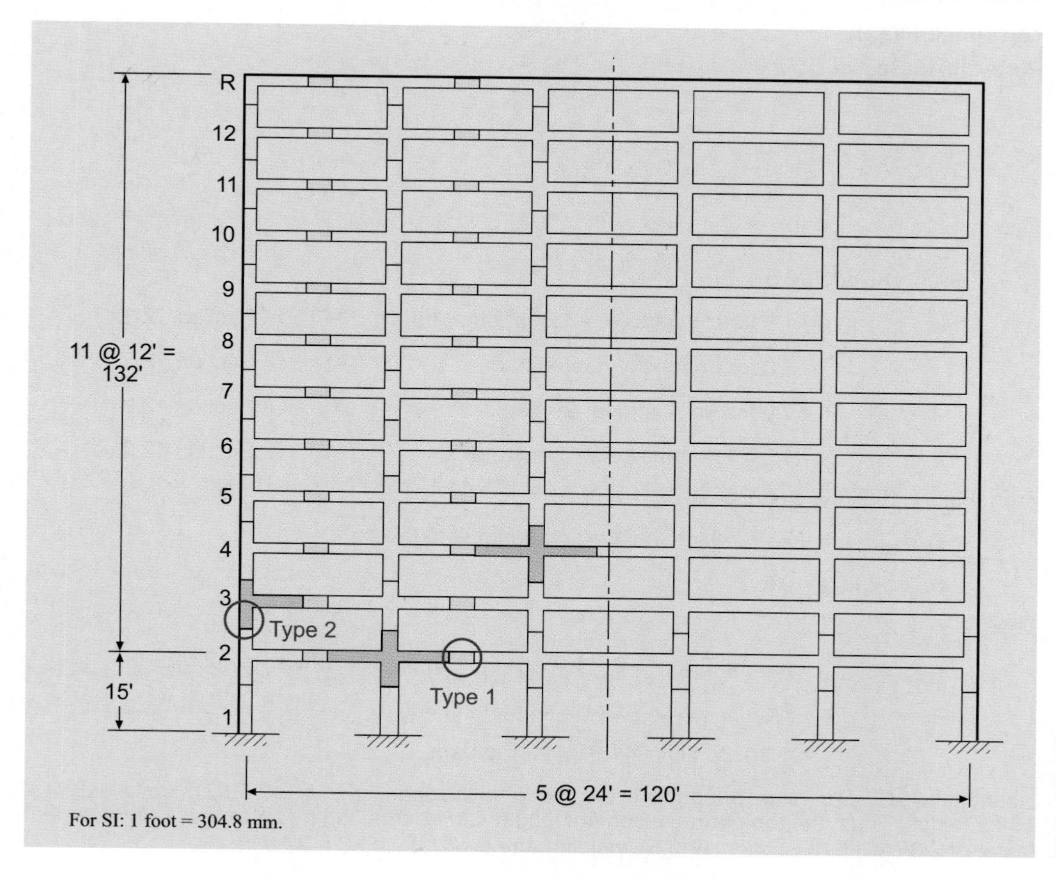

Figure 19E-3.2
Elevation of example building

Example 3—(continued)

Seismic Design Forces

The calculation of the seismic design forces on various structural components is beyond the scope of this example. Traditional analysis methods can be used for precast frames, although care should be taken to approximate the component stiffness in a way that is appropriate for the precast components being used. For emulation design (as it is described in this example) it is reasonable to model the beams and columns as if they were monolithic concrete.

Strong Connection Near Beam Midspan

The design bending moments for the beam at the third floor level that is part of an interior longitudinal frame is shown in Table 19E-3.1. These design moments account for all possible load combinations. Eight No. 9 top bars and five No. 9 bottom bars are provided at all supports. The corresponding negative and positive design moment strengths are also shown in Table 19E-3.1. Three of the top bars and three of the bottom bars are made continuous throughout the spans, providing positive and negative design moment strengths of 220.6 ft-kips.

Table 19E-3.1: **Design forces for the third floor beam forming part of an interior longitudinal frame of the building**

	M_u (ft-kips)	ϕM_n (ft-kips)
Negative	–510.8	–522.0
Positive	+311.7	+351.0

At the supports, $\phi M_n^+ = 351.0$ ft-kips > $\phi M_n^-/2 = 261.0$ ft-kips (OK) (ACI 318 Section 21.3.2.2). Also $(\phi M_n)_{min.} = 220.6$ ft-kips > $\phi M_n^-/4 = 130.5$ ft-kips (OK) (ACI 318 Section 21.3.2.2).

Lap splice length

Per ACI 318 Section 12.2.3:

$$\ell_d = \left[\frac{3}{40} \frac{f_y}{\sqrt{f'_c}} \frac{\psi_t \psi_e \psi_s \lambda}{\left(\frac{c_b + k_{tr}}{d_b} \right)} \right] d_b$$

$(c_b + K_{tr})/d_b \le 2.5$

ψ_t = 1.3 for top bars; ψ_t = 1.0 for other bars (ACI 318 Section 12.2.4)

ψ_e = 1.0 for nonepoxy-coated bars (ACI 318 Section 12.2.4)

ψ_s = 1.0 for No. 7 and larger bars (ACI 318 Section 12.2.4)

λ = 1.0 for normal-weight concrete (ACI 318 Section 12.2.4)

From Figure 19E-3.3, c_b = 2.56 in.

c_b/d_b = 2.27, which makes it reasonable to take:

$(c_b + K_{tr})/d_b = 2.5$

Thus:

$\ell_d = (3 \times 60{,}000 \times 1.0 \times 1.0 \times 1.0 \times 1.0)\, d_b/[40 \times (4000)^{0.5} \times 2.5]$

$= 28.5 d_b = 32.2$ in. for bottom bars, and

$= 1.3 \times 28.5 d_b = 41.8$ in. for top bars

Note that two No. 9 bars are adequate in the interior of the span, i.e., $\phi M_n = 150.3$ ft-kips > $\phi M_n^-/4 = 130.5$ ft-kips. Thus, the top bar development length can be reduced by an excess reinforcement factor of 2/3 (ACI 318 Section 12.2.5), yielding an $\ell_d = (2/3) \times 41.8 = 27.9$ in.

Example 3—(continued)

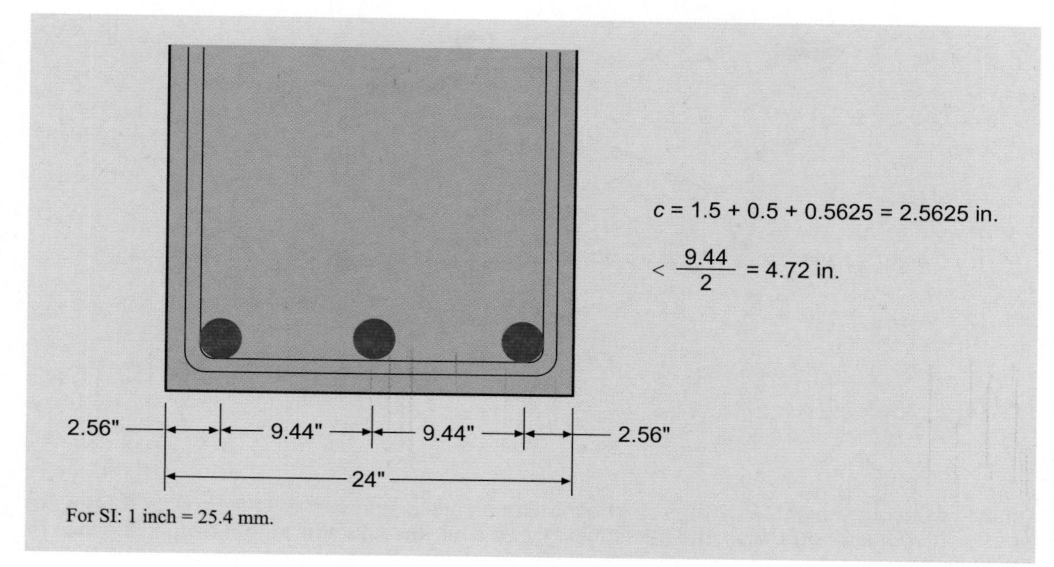

Figure 19E-3.3 Cross section of beam designed

Because all three top bars and all three bottom bars are spliced at the same location, Type B splices are to be used for both the top bars and the bottom bars. The Type B splice length = 1.3 × 32.2 = 41.9 in. for the bottom bars. Provide the same 42 in. (3 ft-6 in.) splice lengths for both the top and the bottom bars.

Reinforcing bar cutoff

M_{pr}^+ and M_{pr}^- are calculated with $\phi = 1.0$ and $f_y = 75$ ksi, ignoring compression steel.

For five No. 9 bottom bars:

$a = A_s f_y / 0.85 f'_c b = 5 \times 75/(0.85 \times 4 \times 24) = 4.6$ in.

$M_{pr}^+ = A_s f_y (d - a/2)$

$= 5 \times 75[17.44 - (4.6/2)]/12$

$= 473.1$ ft-kips

Where $d = 20 - 1.5$ (clear cover) – 0.5 (diameter of No. 4 stirrup) – 0.564 ($^1/_2$ diameter of No. 9 bar) = 17.44 in.

For eight No. 9 top bars:

$a = 8 \times 75/(0.85 \times 4 \times 24) = 7.4$ in.

$M_{pr}^- = 8 \times 75[17.44 - (7.4/2)]/12$

$= 688.2$ ft-kips

$w_D = 0.15 \times (7/12) \times 24$ (slab weight) + 0.0425×24 (superimposed DL) + $0.15 \times 24 \times 13/144$ (beam weight)

$= 3.45$ kips per ft at midspan

$0.9w_D = 0.9 \times 3.45 = 3.11$ kips per ft at midspan

Determine the cutoff point for five of the 8 No. 9 top bars:

The distance from the face of the interior support to where the moment under the loading considered equals ϕM_n^- (three No. 9) = 220.6 ft-kips is readily obtained by summing the moments about Section A-A (Figure 19E-3.4):

$$(x/2)(3.11x/11)(x/3) + 688.2 - 220.6 - 69.9x = 0$$

$$\text{or, } 0.0471x^3 - 69.9x + 467.6 = 0$$

Solving for x, $x = 6.91$ ft

Example 3—(continued)

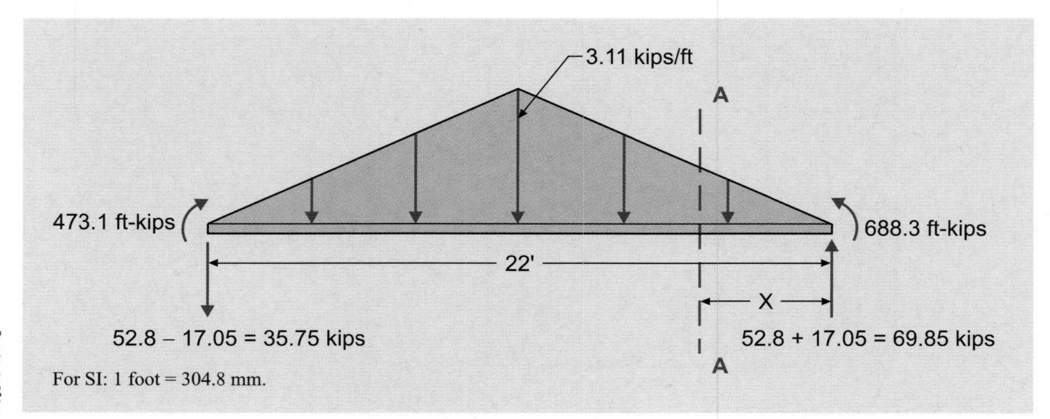

Figure 19E-3.4 Reinforcing bar cutoff

Thus, five out of the eight No. 9 bars can be terminated at a distance of 6.91 ft + d = 8.4 ft from the faces of interior supports. Also, the bars must extend a full development length beyond the face of the support:

$$\ell_d = \left[\frac{3}{40}\frac{f_y}{\sqrt{f'_c}}\frac{\psi_t\psi_e\psi_s\lambda}{\left(\frac{c_b + K_{tr}}{d_b}\right)}\right]d_b$$

$(c_b + K_{tr})/d_b \leq 2.5$

ψ_t	= 1.3 for top bars; ψ_t = 1.0 for other bars	(ACI 318 Section 12.2.4)
ψ_e	= 1.0 for nonepoxy-coated bars	(ACI 318 Section 12.2.4)
ψ_s	= 1.0 for No. 7 and larger bars	(ACI 318 Section 12.2.4)
	= 1.0 for normal-weight concrete	(ACI 318 Section 12.2.4)
K_{tr}	= 0 (conservative)	

$(c_b + K_{tr})/d_b = (1.35 + 0)/1.128 = 1.2 < 2.5$

Thus:

$\ell_d = (3 \times 60{,}000 \times 1.3 \times 1.0 \times 1.0 \times 1.0)d_b/[40 \times (4000)^{0.5} \times 1.2]$

$= 77d_b$

$= 77 \times 1.128 = 86.9$ in. $= 7.2$ ft < 8.4 ft

The total required length of the five No. 9 bars must be at least 8.4 ft beyond the face of the supports.

Flexural reinforcement shall not be terminated in a tension zone unless one or more of the conditions of ACI 318 Section 12.10.5 are satisfied. Because the point of inflection is approximately 10.7 ft from the face of the right support, which is greater than 8.4 ft, the five No. 9 bars cannot be terminated here unless one of the conditions of ACI 318 Section 12.10.5 is satisfied.

Check if the factored shear force V_u at the cutoff point does not exceed 2/3 of the design strength ϕV_n. Although No. 4 stirrups at an 8 in. spacing would satisfy shear requirements in this segment of the beam, No. 4 stirrups at a 6 in. spacing is provided to satisfy ACI 318 Section 12.10.5.1:

$\phi V_n = \phi(V_c + V_s)$

$= 0.75[2(4000)^{0.5} \times 24 \times 17.44 + (0.4 \times 60{,}000 \times 17.44/6)]/1000$

$= 92.0$ kips

Example 3—(continued)

$(2/3)\phi V_n$ = 61.3 kips > V_u = 60.0 kips at 8.4 ft from face of support

Because $(2/3)\phi V_n > V_u$, the cutoff point for the five No. 9 bars can be 8.4 ft beyond the face of the support.

The cutoff point for two of the five No. 9 bottom bars can be determined in a similar fashion. These bars can be cutoff at 8.4 ft from the face of the support as well, which is short of the splice closure.

Check connection strength

For strong connections, the requirements of ACI 318 Section 21.6.2(b) must be satisfied:

$\phi S_n \geq S_e$

Where

S_n = nominal flexural or shear strength of the connection

S_e = moment or shear at connection corresponding to development of probable strength at intended yield locations, based on the governing mechanism of inelastic lateral deformation, considering both gravity and earthquake load effects

At the connection, the spacing of the No. 4 hoops is 4 in., in accordance with ACI 318 Section 21.3.2.3. Thus,

ϕV_n = $\phi(V_c + V_s)$

= $0.75[2(4000)^{0.5} \times 24 \times 17.44 + (0.4 \times 60{,}000 \times 17.44/4)]/1000$

= 118.2 kips

Gravity load on beam:

$1.2w_D + 0.5w_L = (1.2 \times 3.45) + (0.5 \times 0.05 \times 24) = 4.74$ kips per ft

Maximum shear force V_e at the connection due to gravity and earthquake effects occurs at 9.125 ft from the face of the right support, and is equal to 61.0 kips < ϕV_n = 118.2 kips OK (see Figure 19E-3.5).

Maximum moment M_e at the connection due to gravity and earthquake effects occurs at 9.125 ft from the face of the left support and is equal to 174.0 ft-kips < ϕM_n = 220.6 ft-kips OK (see Figure 19E-3.5).

The strong connection near the beam midspan is illustrated in Figure 19E-3.6.

Column-to-Column Connection at Midheight

The design forces for the interior column between levels 2 and 3 (which is part of an interior longitudinal frame) are shown in Table 19E-3.2 for the load combination $1.2D + 0.5L + E$.

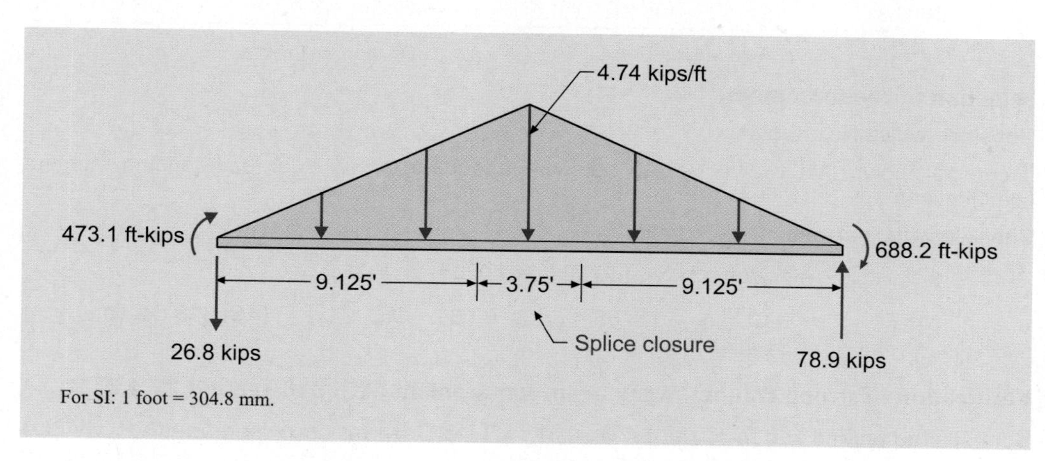

Figure 19E-3.5 Connection strength

Example 3—(continued)

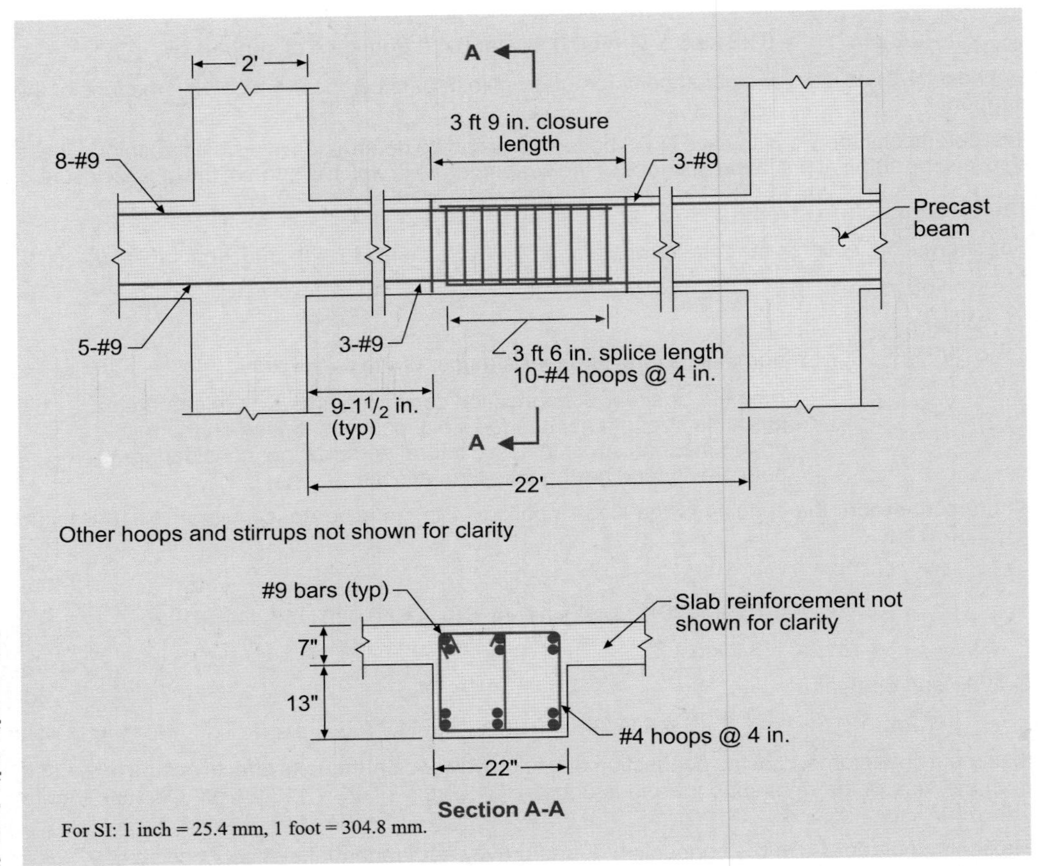

Figure 19E-3.6 **Beam-to-beam strong connection near midspan of third-floor beam forming part of interior longitudinal frame**

Table 19E-3.2. **Design forces for the interior column between the second and third floors, forming part of an interior longitudinal frame of the building**

Axial Load (kips)	Moment (ft-kips)		Shear (kips)
	Top	Bottom	
1609.8	–408.3	+467.5	70.7

Selection of reinforcement

Consider twelve No. 10 bars.

It can be shown that twelve No. 10 bars are adequate for all axial load-bending moment combinations.

Check longitudinal reinforcement ratio:

ρ_g $= A_{st}/bh = (12 \times 1.27)/24^2 = 0.0265$

ρ_{min} $= 0.01 < \rho_g = 0.0265 < \rho_{max} = 0.06$ OK (ACI 318 Section 21.4.3.1)

Verification of strong column, weak beam requirement (ACI 318 Section 21.4.2)

Between the second and third floor levels, $M_{nc} = 1182.3$ ft-kips corresponding to $P_u = 711.4$ kips. Similarly, between the third and fourth levels, $M_{nc} = 1168.4$ ft-kips corresponding to $P_u = 655.5$ kips. Thus,

Example 3—(continued)

ΣM_{nc} = 1182.3 + 1168.4 = 2350.7 ft-kips

The nominal flexural strength M_n of the beams framing into the column must include the slab reinforcement within an effective slab width equal to:

16 (slab thickness) + beam width = $(16 \times 7) + 24 = 136$ in.

Center-to-center spacing of the beams = $24 \times 12 = 288$ in.

Span/4 = $(24 \times 12)/4 = 72$ in. (governs)

The minimum required area of steel in the 72-in. effective width = $0.0018 \times 72 \times 7 = 0.91$ in.^{2}, which corresponds to five No. 4 bars @ 72/5 = 14.4 in. spacing. Because maximum bar spacing = $2h$ = 14 in., provide No. 4 @ 14 in. at both the top and bottom of the slab. Note that according to ACI 318 Figure 13.3.8, 100 percent of both the top and bottom reinforcement in the column strip must be continuous or anchored at the support.

From a strain compatibility analysis, negative M_{nb} = 736.0 ft-kips and positive M_{nb} = 459.0 ft-kips. Thus,

ΣM_{nb} = 736.0 + 459.0 = 1195.0 ft-kips

2350.7 ft-kips > $(6/5) \times 1195$ = 1434.0 ft-kips OK

The intent of this code provision is to prevent a story mechanism, rather than prevent local yielding in a column. The 6/5 factor is clearly insufficient to prevent column yielding if the adjacent beams both hinge. Therefore, confinement reinforcing is required in the potential hinge regions of a frame column.

Minimum connection strength

It can be shown from a strain compatibility analysis that M_{pr} for a column between the second and the third floor levels, corresponding to an axial load of 711.4 kips, is equal to 1244.1 ft-kips.

Also, as given above, M_{nc} = 1182.3 ft-kips corresponding to an axial load of 711.4 kips. From a strain compatibility analysis, $\varepsilon_t = 0.00223$ so that $\phi = 0.65 + (0.00223 - 0.002) \times 250/3 = 0.67$ (ACI 318 Section 9.3.2.2 and ACI 318 Fig. R9.3.2).

Column-to-column connection must have ϕM_n at least equal to $0.4M_{pr}$ (ACI 318 Section 21.6.2(d)). Thus,

ϕM_n = 0.67×1182.3

= 792.1 ft-kips > $0.4M_{pr} = 0.4 \times 1244.1$ = 497.6 ft-kips OK

Twelve No. 10 bars are adequate.

Splice all 12 bars at midheight, as shown in Figure 19E-3.7.

Column-Face Strong Connection in Beam

A strong connection is to be designed at the interface between a precast beam at the second floor level of the building that forms the exterior span of an exterior transverse frame, and the continuous corner column to which it is connected. The beam is reinforced with 5 No. 9 bars at the top and four No. 9 bars at the bottom at its ends. M_{pr}^+ and M_{pr}^- are calculated with $\phi = 1.0$ and f_y = 75 ksi ignoring compression steel.

For four No. 9 bars:

$a = A_s f_y / 0.85 f'_c b = (4 \times 1.0) \times 75/(0.85 \times 4 \times 24) = 3.68$ in.

$M_{pr}^+ = A_s f_y (d - a/2) = 4.0 \times 75(23.44 - 3.68/2)/12 = 540.0$ ft-kips

since d = 26 – 1.5 (clear cover) – 0.5 (diameter of No. 4 stirrup) – 0.5625 (diameter of No. 9 bar/2) = 23.44 in.

For five No. 9 bars:

$a = (5 \times 1.0) \times 75/(0.85 \times 4 \times 24) = 4.60$ in.

$M_{pr}^- = 5.0 \times 75(23.44 - 4.60/2)/12 = 660.6$ ft-kips

Assuming a 2 ft-6 in. cast-in-place closure, the beam span between critical sections is loaded as shown in Figure 19E-3.8.

Example 3—(continued)

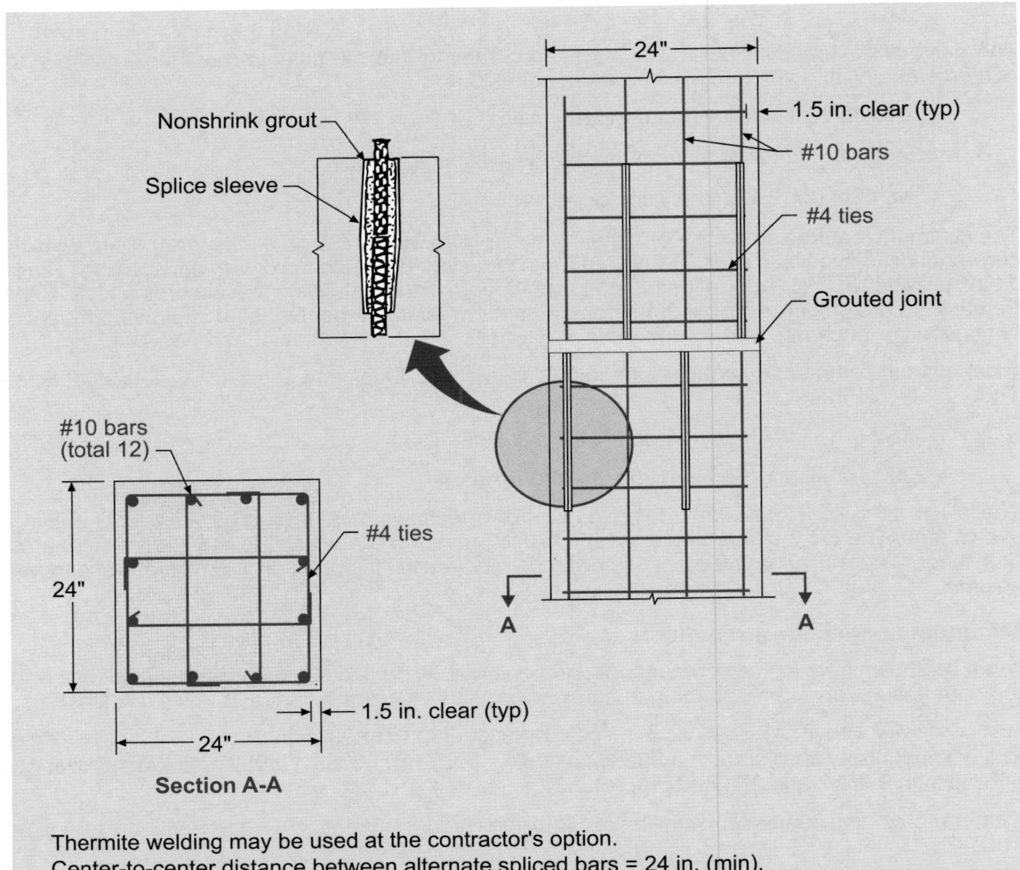

Figure 19E-3.7 Column-to-column strong connection at midheight of interior column between second and third floors, forming part of interior longitudinal frame

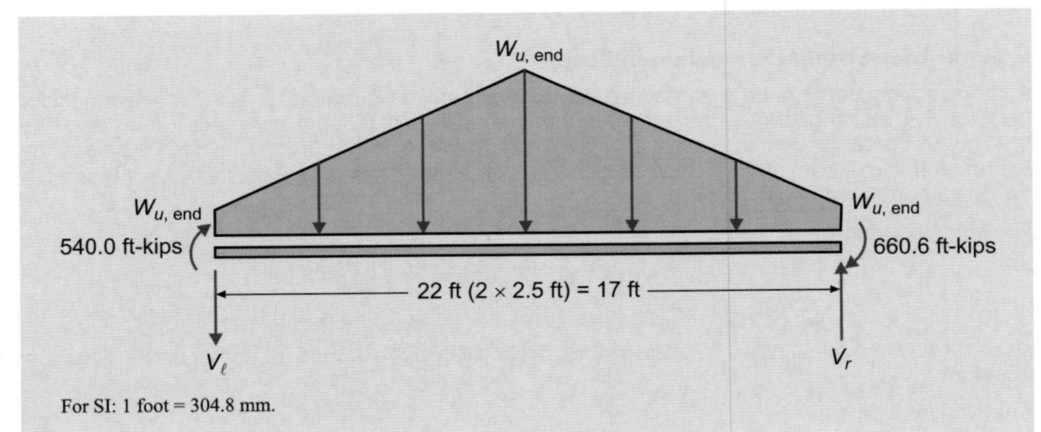

Figure 19E-3.8 Beam span between critical sections

Example 3—(continued)

Calculation of shear forces and bending moments at nonlinear action locations

The shear forces at the critical sections and the bending moments at the connections can be determined for the two governing load combinations as follows.

Load combination 1: $U = 1.2D + 0.5L + E$

w_D = 0.15 × (7/12) × 13 (slab weight) + 0.0425 × 13 (superimposed DL) + 0.15 × 24 × 19/144 (beam weight)

= 2.17 kips per ft at midspan

w_L = 0.05 × 13 = 0.65 kips per ft at midspan

$w_{u,\text{mid}}$ = (1.2 × 2.17) + (0.5 × 0.65) = 2.93 kips per ft at midspan

$w_{u,\text{end}}$ = 2.93 × 2.5/11 = 0.67 kips per ft at the ends of the beam segment

Summing moments about the left end of the beam:

$V_r(17) = (0.67 \times 17^2/2) + [(2.93 - 0.67) \times 17^2/4] + 540.0 + 660.6$

or, V_r = 85.9 kips

$V_\ell = 85.9 - (0.67 \times 17) - [(2.93 - 0.67) \times 17]/2 = 55.3$ kips

The bending moments at the faces of the left and right supports are:

$M^+_\ell = 540.0 + (55.3 \times 2.5) = 678.3$ ft-kips

$M^+_r = 660.6 + (85.9 \times 2.5) = 875.3$ ft-kips

Load combination 2: $U = 0.9D + E$

$W_{u,\text{mid}} = 0.9w_D = 0.9 \times 2.17 = 1.95$ kips per ft at midspan

$W_{u,\text{end}} = 1.95 \times 2.5/11 = 0.44$ kips per ft at the ends of the beam segment

Summing moments about the left end of the beam:

$V_r(17) = (0.44 \times 17^2/2) + [(1.95 - 0.44) \times 17^2/4] + 540.0 + 660.6$

or, V_r = 80.8 kips

$V_\ell = 80.8 - (0.44 \times 17) - [(1.95 - 0.44) \times 17]/2 = 60.5$ kips

The bending moments at the faces of the left and right supports are:

$M^+_\ell = 540.0 + (60.5 \times 2.5) = 691.1$ ft-kips

$M^+_r = 660.6 + (80.8 \times 2.5) = 862.6$ ft-kips

Thus, the governing bending moments at the faces of the supports are M^+_ℓ = 691.1 ft-kips and M^+_r = 875.3 ft-kips.

Strength design of connection

At the bottom of the connection, provide an additional four No. 9 bars to the four No. 9 bars and at the top of the section, provide an additional five No. 9 bars to the five No. 9 bars. From a strain compatibility analysis considering all of the reinforcement in the section:

$\phi M^+_n = 728.8$ ft-kips $> M^+_\ell = 691.1$ ft-kips OK

$\phi M^-_n = 888.2$ ft-kips $> M^+_\ell = 875.3$ ft-kips OK

For both the positive and negative design strengths, the section was determined to be tension-controlled.

Maximum reinforcement ratio = (10 × 1.0)/(24 × 22.33) = 0.019 < 0.025 OK

Anchorage and splices

Per ACI 318 Section 21.5.4.1:

$\ell_{dh} = f_y d_b/65(f'_c)^{0.5} = (60{,}000 \times 1.128)/[65 \times (4000)^{0.5}] = 16.5$ in.

The reinforcement details for the connection are shown in Figure 19E-3.9.

Example 3—(continued)

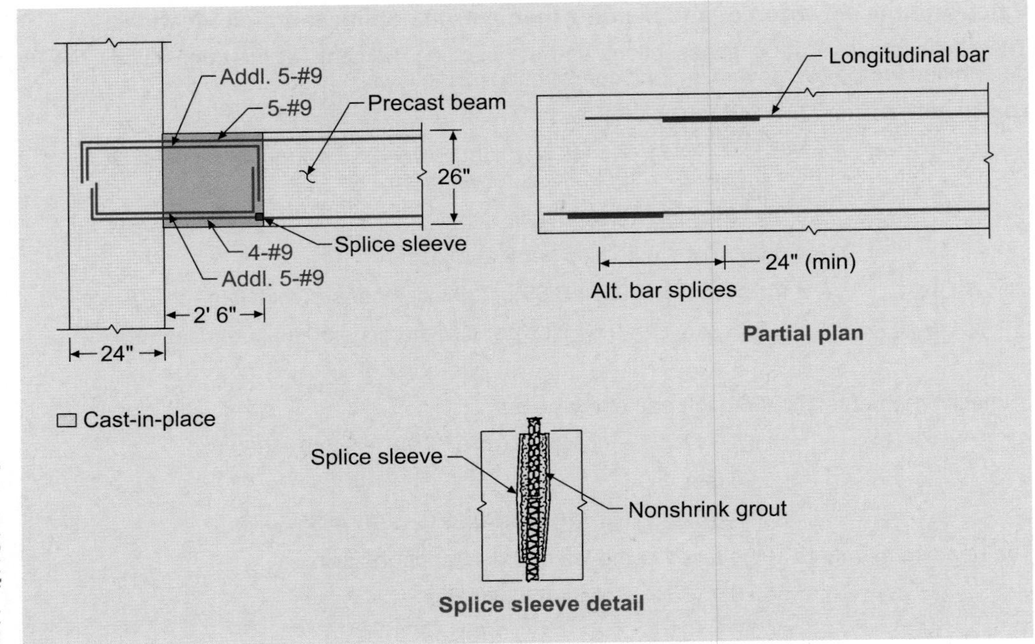

Figure 19E-3.9
Beam-to-continuous-column strong connection at exterior span of transverse frame at second floor level

Acknowledgement

This example has been adapted from:

Ghosh, S.K., Nakaki, S.D., and Krishnan, K., "Precast Structures in Regions of High Seismicity: 1997 UBC Design Provisions," *PCI Journal*, Vol. 42, No. 6, November – December 1997, pp. 76-93.

Example 4 *Allowable Stress Design Check on Anchor Bolt*

Given Data:

1 in. anchor bolt diameter

$f_c' = 3500$ psi

P_s = applied tension service load = 1500 lbs

V_s = applied shear service load = 2500 lbs

Edge distance = 5 in.

Spacing = 9 in.

Check Equation 19-1 of 2006 IBC Section 1911.2 for anchors subjected to combined tension and shear:

$(P_s/P_t)^{5/3} + (V_s/V_t)^{5/3} \leq 1$

P_t = allowable tension service load from 2006 IBC Table 1911.2

= 3250 lbs for $f_c' = 3000$ psi with edge distance = 6 in. and spacing = 12 in.

= 3650 for $f_c' = 4000$ psi with edge distance = 6 in. and spacing = 12 in.

From interpolation, $P_t = 3450$ psi for $f_c' = 3500$ psi with edge distance = 6 in. and spacing = 12 in.

V_t = allowable shear service load from 2006 IBC Table 1911.2

= 4500 lbs for $f_c' = 3000$ psi with edge distance = 6 in. and spacing = 12 in.

= 5300 lbs for $f_c' = 4000$ psi with edge distance = 6 in. and spacing = 12 in.

From interpolation, $V_t = 4900$ psi for $f_c' = 3500$ psi with edge distance = 6 in. and spacing = 12 in.

Reduction due to edge distance (ACI 318 Section 1911.3) = 1 – 5/6 = 0.17

Reduction due to spacing (ACI 318 Section 1911.3) = 1 – 9/12 = 0.25 governs

$P_t = 0.75 \times 3450 = 2588$ lbs

$V_t = 0.75 \times 4900 = 3675$ lbs

$(P_s/P_t)^{5/3} + (V_s/V_t)^{5/3} = (1500/2588)^{5/3} + (2500/3675)^{5/3}$

$= (0.58)^{5/3} + (0.68)^{5/3}$

$= 0.40 + 0.53 = 0.93 < 1$ OK

Example 5 *Anchoring to Concrete per Appendix D of ACI 318-05*

Determine the adequacy of the anchor bolt depicted in Figure 19E-5.1 according to the provisions of Appendix D of ACI 318-05. The anchor bolt is located away from the corner of the concrete footing to which it is attached.

Given data:

$^1/_2$ in. diameter ASTM F 1554 Grade 36 hex head anchor:

f_{uta} = 58 ksi, f_{ya} = 36 ksi, A_{se} = 0.142 in.^{2}, A_{brg} = 0.291 in.^{2}, elongation = 23% reduction of area = 40%

f'_c = 4000 psi

Service loads shown in Figure 19E-5.1 are due to wind

Factored Design Loads

From Section 9.2, for wind loads:

$N_{ua} = 1.6 \times 1100 = 1760$ lbs

$V_{ua} = 1.6 \times 425 = 680$ lbs

Design Tensile Strength, ϕN_n

- ***Steel strength, ϕN_{sa}***

From Equation (D-3): $\phi N_{sa} = \phi n A_{se} f_{uta}$

ASTM F 1554 Grade 36 steel is a ductile steel in accordance with the definition given in Section D.1, since elongation = 23% > 14%, and reduction of area = 40% > 30%.

ϕ = 0.75 for a ductile steel element subjected to tension loads (ACI 318 Section D.4.4(a) and Table 19E-5.1 below).

$\phi N_{sa} = 0.75 \times 1 \times 0.142 \times 58 \times 1000 = 6177$ lbs

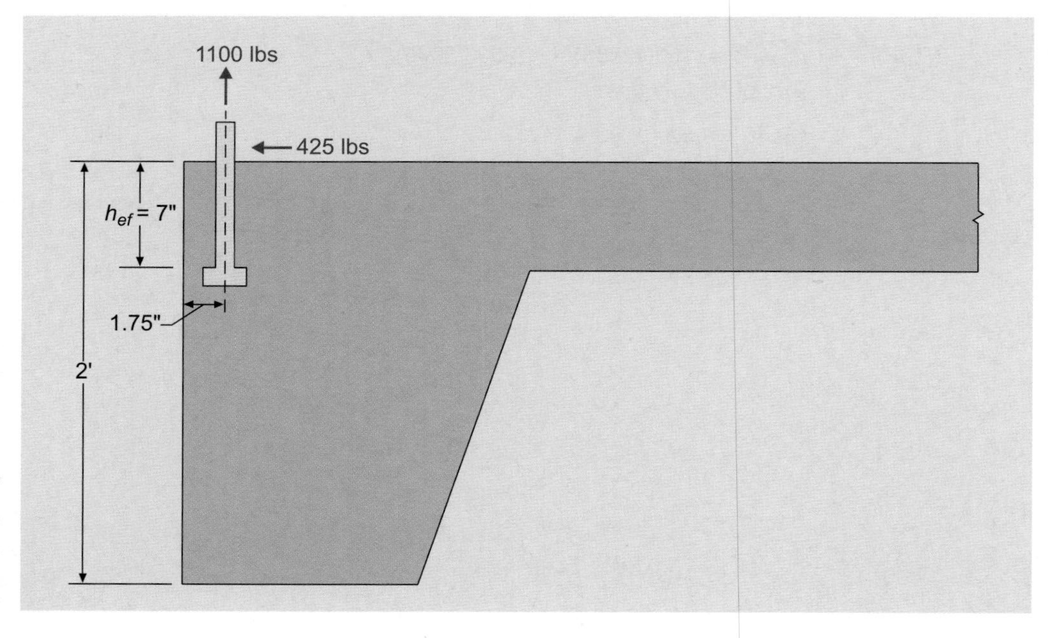

Figure 19E-5.1 Fastener subjected to tension and shear

Example 5—(continued)

Table 19E-5.1. Strength reduction factors of Appendix D

Strength Governed by					Strength Reduction Factor ϕ, for Use with Load Combinations in: Section 9.2	Appendix C
Ductile Steel Element						
Tension					0.75	0.80
Shear					0.65	0.75
Brittle Steel Element						
Tension					0.65	0.70
Shear					0.60	0.65
Concrete Breakout, Side-Face Blowout, Pullout, or Pryout*						
Shear	Breakout			Condition A	0.75	0.85
				Condition B	0.70	0.75
	Pryout			Condition B	0.70	0.75
Tension	Cast-in headed studs, headed bolts or hooked bolts		CB/SFB	Condition A	0.75	0.85
				Condition B	0.70	0.75
			P/P	Condition B	0.70	0.75
	Postinstalled anchors with category determined from ACI 355.2-01	Category 1	CB/SFB	Condition A	0.75	0.85
				Condition B	0.65	0.75
			P/P	Condition B	0.65	0.75
		Category 2	CB/SFB	Condition A	0.65	0.75
				Condition B	0.55	0.65
			P/P	Condition B	0.55	0.65
		Category 3	CB/SFB	Condition A	0.55	0.65
				Condition B	0.45	0.55
			P/P	Condition B	0.45	0.55

*CB/SFB = Concrete Breakout and Side Face Blowout
P/P = Pullout and Pryout

Example 5—(continued)

- ***Concrete breakout strength, ϕN_{cb}***

From Equation (D-4):

$$N_{cb} = \frac{A_{Nc}}{A_{Nco}} \psi_{ed,N} \psi_{c,N} \psi_{cp,N} N_b$$

Since no supplementary reinforcement has been provided (Condition B), ϕ = 0.70 (Section D.4.4(c)ii and Table 19E-5.1).

From Equation (D-6) and Figure 19E-5.2: $A_{Nco} = 9h_{ef}^2 = 9 \times 7^2 = 441$ in.^{2}

From Figure 19E-5.3: $A_{Nc} = (c_{a1} + 1.5h_{ef}) \times (2 \times 1.5h_{ef}) = (1.75 + 10.5) \times (2 \times 10.5) = 257.3$ in.^{2}

Since $c_{a,\min} = 1.75$ in. $< 1.5h_{ef} = 10.5$ in., use Equation (D-11) to determine $\psi_{ed,N}$:

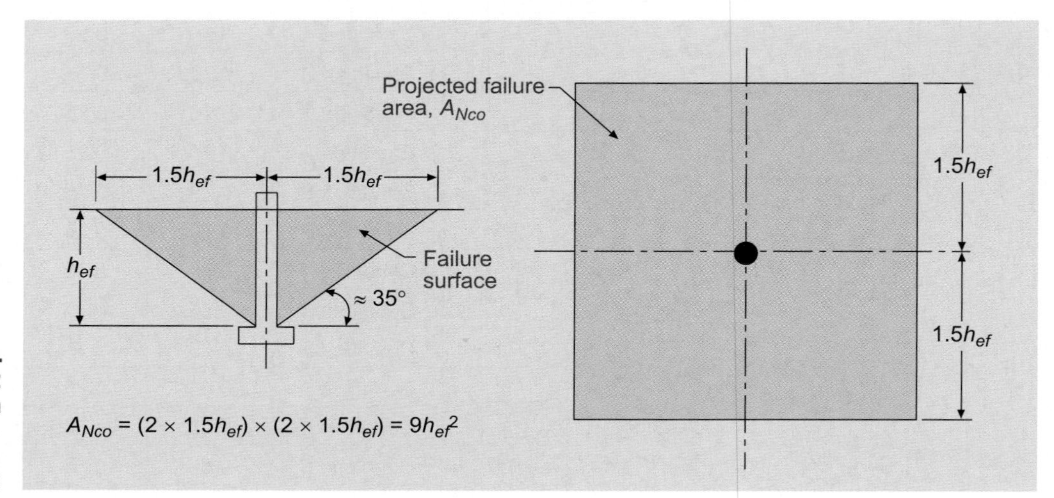

Figure 19E-5.2 Calculation of A_{Nco} per Section D.5.2.1 of ACI 318-05

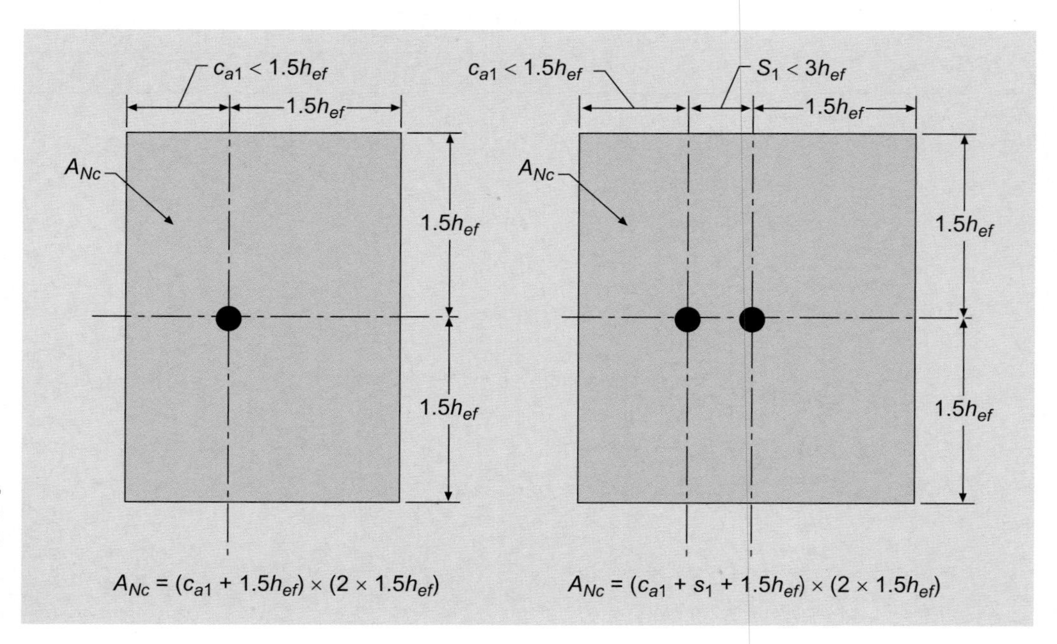

Figure 19E-5.3 Calculation of A_{Nc} per Section D.5.2.1 of ACI 318-05

Example 5—(continued)

$\psi_{ed,N}$ $= 0.7 + 0.3(c_{a,\min}/1.5h_{ef}) = 0.7 + 0.3(1.75/10.5) = 0.75$

Assuming that cracking will occur at the edge of the foundation at the location of the anchor, $\psi_{c,N}$ = 1.0 (Section D.5.2.6)

From Section D.5.2.7, for cast-in anchor: $\psi_{cp,N} = 1.0$

From Equation (D-7), for a cast-in anchor: $N_{cb} = k(f_c')^{0.5}(h_{ef})^{1.5} = 24 \times (4000)^{0.5}\,(7)^{1.5}$ = 28,112 lbs

ϕN_{cb} $= 0.70 \times (257.3/441) \times 0.75 \times 1 \times 1 \times 28{,}112 = 8611$ lbs

- ***Concrete pullout strength, ϕN_{pn}***

From Equation (D-14):

ϕN_{pn} $= \phi\psi_{c,P}N_p$

Because Condition B applies for pullout strength in all cases (Table 19E-5.1), $\phi = 0.70$

Assuming that cracking will occur at the edge of the foundation at the location of the anchor, $\psi_{c,P}$ = 1.0 (Section D.5.3.6)

From Equation (D-15):

N_p $= 8A_{brg}f_c' = 8 \times 0.291 \times 4000 = 9312$ lbs

ϕN_{pn} $= 0.70 \times 1.0 \times 9312 = 6518$ lbs

- ***Concrete side-face blowout strength, ϕN_{sb}***

Since c_{a1} = 1.75 in. < $0.4h_{ef}$ = 2.8 in., side-face blowout failure must be investigated (Section D.5.4.1).

From Equation (D-17): $\phi N_{sb} = \phi 160 c_{a1}(A_{brg})^{0.5}\,(f_c')^{0.5}$

Since no supplementary reinforcement has been provided (Condition B), $\phi = 0.70$ (Section D.4.4(c)ii and Table 19E-5.1).

ϕN_{sb} $= 0.7 \times 160 \times 1.75 \times (0.291)^{0.5} \times (4000)^{0.5} = 6687$ lbs

In accordance with Section D.4.1.2, ϕN_n is the lowest design strength from all of the above failure modes, i.e., $\phi N_n = \phi N_s = 6177$ lbs.

Design Shear Strength, ϕV_n

- ***Steel strength, ϕV_s***

From Equation (D-20):

ϕV_{sa} $= \phi n 0.6 A_{se}\,f_{uta}$

$\phi = 0.65$ for a ductile steel element subjected to shear loads (Section D.4.4(a) and Table 19E-5.1).

ϕV_s $= 0.65 \times 1 \times 0.6 \times 0.142 \times 58 \times 1000 = 3212$ lbs

- ***Concrete breakout strength, ϕV_{cb}***

From Equation (D-21) for the shear force applied perpendicular to the edge:

ϕV_{cb} $= \phi(A_{Vc}/A_{Vco})\psi_{ed,V}\psi_{c,V}V_b$

Because no supplementary reinforcement has been provided (Condition B), $\phi = 0.70$ (Section D.4.4(c)i and Table 19E-5.1).

From Figure 19E-5.4: $A_{Vco} = 4.5c_{a1}^2 = 4.5 \times 1.75^2 = 13.8$ in.^{2}

Because the member thickness > $1.5c_{a1}$ = 2.6 in. and the distance to an orthogonal edge c_{a2} > $1.5c_{a1}$, $A_{Vc} = A_{Vco} = 13.8$ in.^{2} (Section D.6.2.1) and $\psi_{ed,V} = 1.0$ (Equation D-27).

Assuming that cracking will occur at the edge of the foundation at the location of the anchor, $\psi_{c,V}$ = 1.0 for the case when no supplementary reinforcement has been provided (Section D.6.2.7).

From Equation (D-24): $V_b = 7(\ell_e/d_o)^{0.2}(d_o)^{0.5}(f_c')^{0.5}(c_{a1})$

Example 5—(continued)

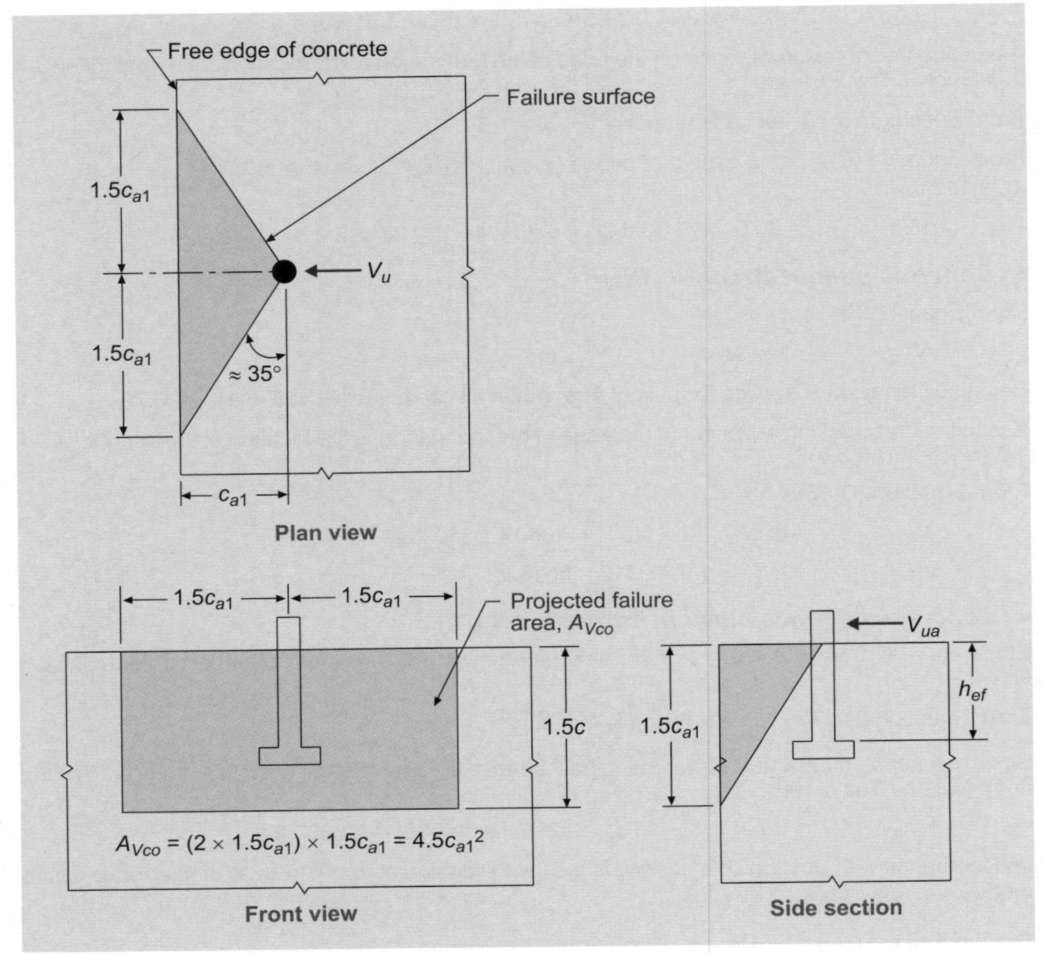

Figure 19E-5.4 Calculation of A_{Vco} per Section D.6.2.1 of ACI 318-05

ℓ_e $= h_{ef} = 7$ in. > $8d_o = 8 \times 0.5 = 4$ in., use $\ell_e = 4$ in.

V_b $= 7 \times (4/0.5)^{0.2} \times (0.5)^{0.5} \times (4000)^{0.5} \times (1.75)^{1.5} = 1099$ lbs

ϕV_{cb} $= 0.70 \times (1/1) \times 1.0 \times 1.0 \times 1099 = 769$ lbs

- ***Concrete pryout strength, ϕV_{cp}***

From Equation (D-29):

ϕV_{cp} $= \phi k_{cp} N_{cb}$

Since Condition B applies for pryout strength in all cases (Table 19E-5.1), $\phi = 0.70$

K_{cp} = 2.0, since $h_{ef} = 7.0$ in. > 2.5 in. (Section D.6.3.1)

N_{cb} = 8611/0.70 = 12,301 lbs

ϕV_{cp} $= 0.70 \times 2.0 \times 12{,}301 = 17{,}221$ lbs

In accordance with Section D.4.1.2, ϕV_n is the lowest design strength from all of the above failure modes, i.e., $\phi V_n = \phi V_{cb} = 769$ lbs.

Tension and Shear Interaction

Since $V_{ua} = 680$ lbs > $0.2\phi V_n = 0.2 \times 769 = 154$ lbs, full strength in tension is not permitted.

Example 5—(continued)

Since N_{ua} = 1760 lbs > 0.2ϕN_n = 0.2 × 6177 = 1235 lbs, full strength in shear is not permitted

Check combined shear and tension by equation (D-31):

$$(N_{ua}/\phi N_n) + (V_{ua}/\phi V_n) \leq 1.20$$

$$N_{ua}/\phi N_n = 1760/6177 = 0.29$$

$$V_{ua}/\phi V_n = 680/769 = 0.88$$

$$0.29 + 0.88 = 1.17 < 1.20$$ OK

Required Edge Distances Spacings and Thickness to Preclude Splitting Failure

Assuming that the anchor is untorqued, the minimum edge distances are the minimum cover requirements of Section 7.7 (Section D.8.2).

According to Section 7.7.1, the minimum cover = 1.5 in. for a 0.5 in. diameter bar (equivalent to a No. 4 bar) when the concrete is exposed to earth or weather. The clear cover to the anchor bolt shaft = 1.75 – (0.5/2) = 1.50 in., which is equal to the minimum required concrete cover. Note that the clear cover to the anchor bolt head (1.1875 in.) will be slightly less than 1.5 in., but the difference will still be within the minus $^3/_8$ in. tolerance allowed for cover per Section 7.5.2.1.

Example 6 *Relative Strengths of Columns And Beams at Joint*

Given Data:

Beams	= 34 × 24 in.
Columns	= 40 × 40 in. (interior)
C/C distance of columns	= 26 ft (in both directions)
Factored axial load	= 4480 kips on column C1 (Figure E19-4.1)
	= 4244 kips on column C2 (Figure E19-4.1)
Service dead load	= 3.36 kips/ft
Service live load	= 0.83 kips/ft
Main reinforcement in columns	= 36 No. 10 bars
Main reinforcement in beams	= 6 No. 8 bars (top)
	= 3 No. 8 bars (bottom)
Effective flange width	= 78 in. (Figure E19-4.2)
Specified compressive strength of concrete	= 4000 psi
Specified yield strength of reinforcement	= 60,000 psi

Other data are assumed as needed. See Figures E19-6.1 and E19-6.2 for details.

Figure E19-6.3 shows the P-M interaction diagram for the interior column.

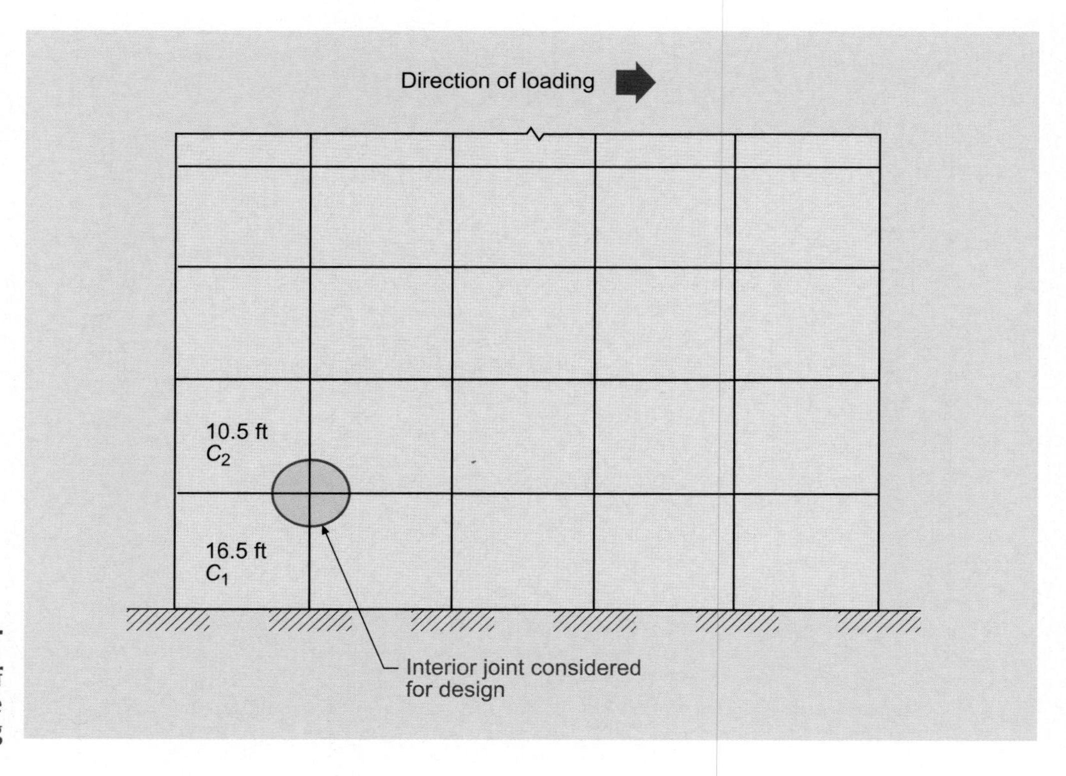

Figure 19E-6.1 Elevation of example building

Example 6—(continued)

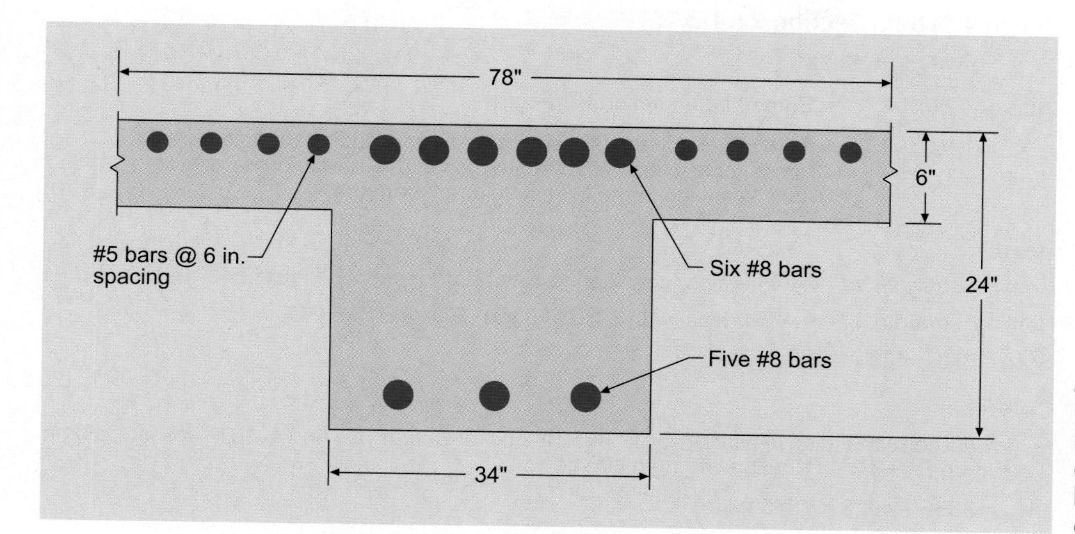

Figure 19E-6.2 Reinforcement details at the beam-column connection

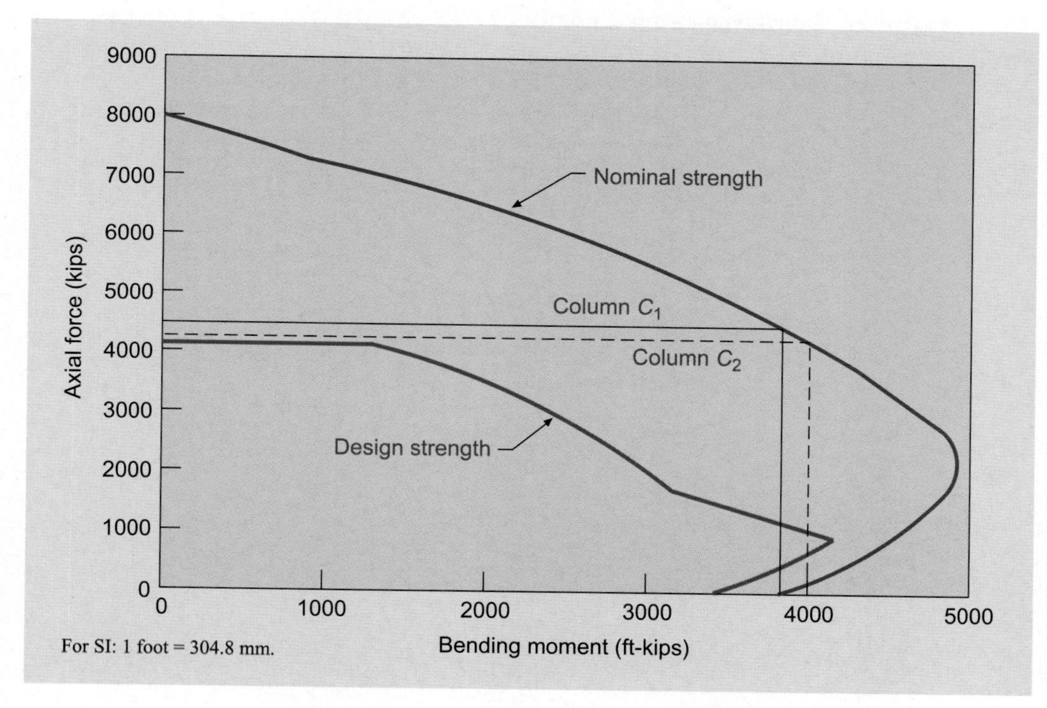

Figure 19E-6.3 Design and nominal strength interaction diagrams for interior columns C1

Example 6—(continued)

Per ACI 318-05 Section 21.4.2.2:

$\Sigma M_{nc} \geq (6/5)\ \Sigma M_{nb}$,

Where ΣM_{nb} = Sum of beam flexural strengths

ΣM_{nc} = Sum of column flexural strengths corresponding to factored axial forces, consistent with the direction of the lateral forces considered, resulting in the lowest flexural strengths.

Beams:

Nominal strength, M_n^- = 648 ft-kips (considering slab reinforcement) (Figure E19-4.2)

Nominal strength, M_n^+ = 438 ft-kips (with 3 No. 8 bars) (Figure E19-6.2)

SM_{nb} = 648 + 438 = 1086 ft-kips

Columns:

M_n (lowest corresponding to different P_u) = 3859 ft-kips for Column C1 and 4016 ft-kips for Column C2 (Figure E19-6.3 — Nominal strength curve)

ΣM_{nc} = 3859 + 4016 = 7875 ft-kips

Check Strength:

$\Sigma M_{nc}/\Sigma M_{nb}$ = 7875/1086 = 7.3 > 1.2 OK

Example 7 *Design of RC Shear Wall per IBC 2006*

Given Data:

Specified compressive strength of concrete f_c' = 4000 psi
Specified yield strength of reinforcement f_y = 60,000 psi
Structural system: Building Frame with Special Reinforced Concrete Shear Walls
Site Class: D
S_{DS} = 1.33g
S_{D1} = 0.93g
C_d = 5.0
Occupancy Category = I
R = 6
Importance Factor, I = 1.0
Redundancy factor, ρ = 1.0
Shear wall thickness, h = 12 in.
Seismic Design Category = E
Base shear, V = 1402 kips
Service-level axial dead load on shear wall = 316 kips
Service-level axial live load on shear wall = 34 kips

The internal forces that are due to lateral loads are shown in Table 19E-7.1. The displacement at the top of the building along the shear wall line was found to be 0.76 in. by elastic analysis under the code-prescribed forces.

Table E19-7.1. Summary of design axial force, shear force and bending moment for shear wall between grade and level 2

Loads	Symbol	Axial Force (kips)	Shear Force (kips)	Bending Moment (ft-kips)
Dead load	D	316	0	0
Live load	L	34	0	0
Lateral load	Q_E	38	382	16,855
Load Combinations				
1	$1.2D + 1.6L$	434	0	0
2	$1.2D + (\rho Q_E + 0.2S_{DS}D) + 0.5L$	518	382	16,855
3	$0.9D - (\rho Q_E + 0.2S_{DS}D)$	162	–382	–16,855
		Load Factor		
Load Combinations		***D***	***L***	Q_E
1		1.2	1.6	0
2		1.47	0.5	1
3		0.63	0	–1

Design displacement (δ_u) = 5.0 × 0.76 = 3.80 in.

Note: A shear wall is a structural wall. The terms are used interchangeably here.

Example 7—(continued)

Shear Wall Design

The design of one of the shear walls at the base of the structure shown in Figures E19-7.1 and E19-7.2 is illustrated in this example. Similar procedures may be followed to design the shear wall at the other floor levels. The design of shear walls by IBC 2006 follows the procedure in ACI 318-05.

Design loads

Table 19E-7.1 shows a summary of the axial force, shear force and bending moment at the base of the example shear wall based on different load combinations.

Required axial load strength, P_u = 518 kips

Required shear strength, V_u = 382 kips

Required flexural strength, M_u = 16,855 ft-kips

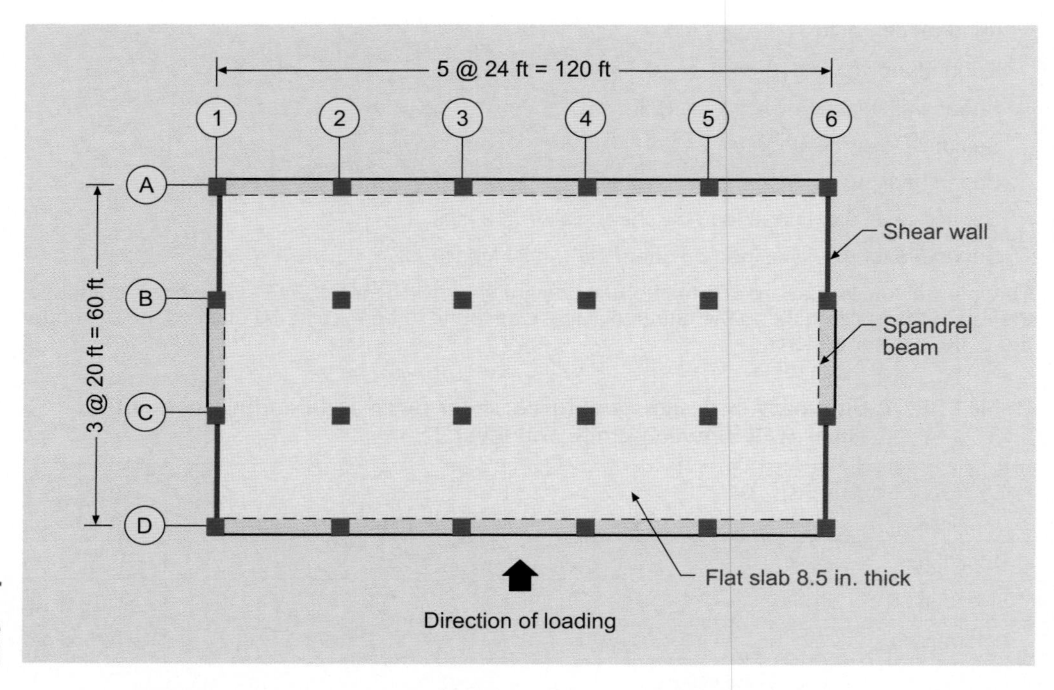

Figure 19E-7.1 Plan of example building considered

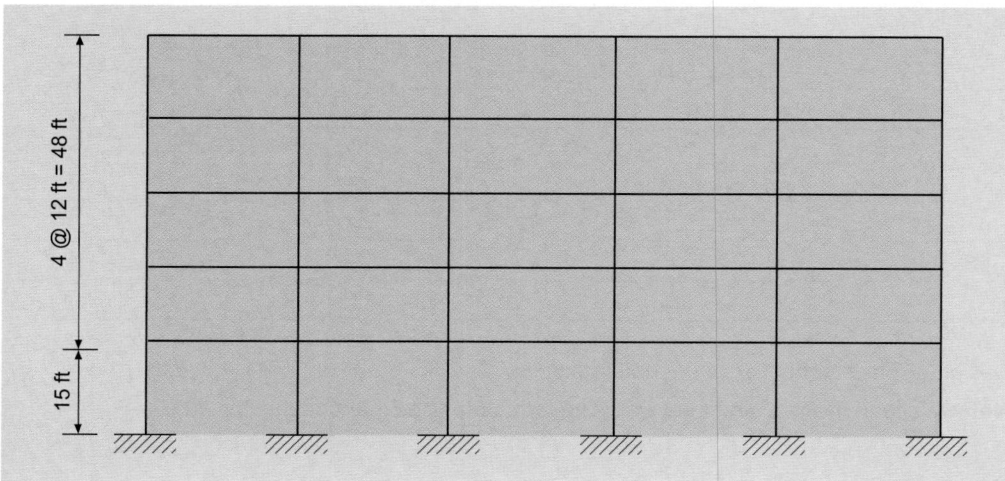

Figure 19E-7.2 Elevation of example building considered

Example 7—(continued)

Check strength under flexure and axial load

Draw P-M interaction diagram for the shear wall with assumed dimensions of wall and assumed longitudinal reinforcement in boundary elements and web. Check to see that all the points representing strength demand (from the three load combinations shown in Table 19E-7.1) are within the design strength interaction diagram.

In this example, the shear wall dimensions and reinforcement, as shown in Figure 19E-7.3, are considered.

Using 12 No. 8 bars in each boundary column, the reinforcement ratio for the boundary column is $(12 \times 0.79)/(18 \times 18) = 2.93\%$. This is high, but not excessive, and was judged to be acceptable.

Figure 19E-7.4 shows the P–M interaction diagrams for the example shear wall. As can be seen, all the points representing required strength are within the design strength curve.

One other quantity needs to be determined at this stage. That is the neutral axis depth, *c*, corresponding to the maximum axial force (in the presence of lateral force).

P_u = 518 kips

c = 26.3 in. (using an in-house computer program)

Design for shear

Height of the shear wall, h_w = 63 ft

Length of the shear wall, $\ell_w = 20 + 18/12 = 21.5$ ft

$h_w/\ell_w = 63/21.5 = 2.93$

ACI 318 Section 21.7.4.4

V_u must not exceed $\phi 8 A_{cv}(f'_c)^{0.5}$

$A_{cv} = 12 \times (20 \times 12 + 18) = 3096$ in.^{2}

Take $\phi = 0.75$, since a wall with $h_w/\ell_w = 2.93$ is not going to be governed by shear in its failure mode.

$\phi 8 A_{cv}(f'_c)^{0.5} = 0.75 \times 8 \times 3096\ (4000)^{0.5}/1000$

$= 1175$ kips > 382 kips (V_u) OK

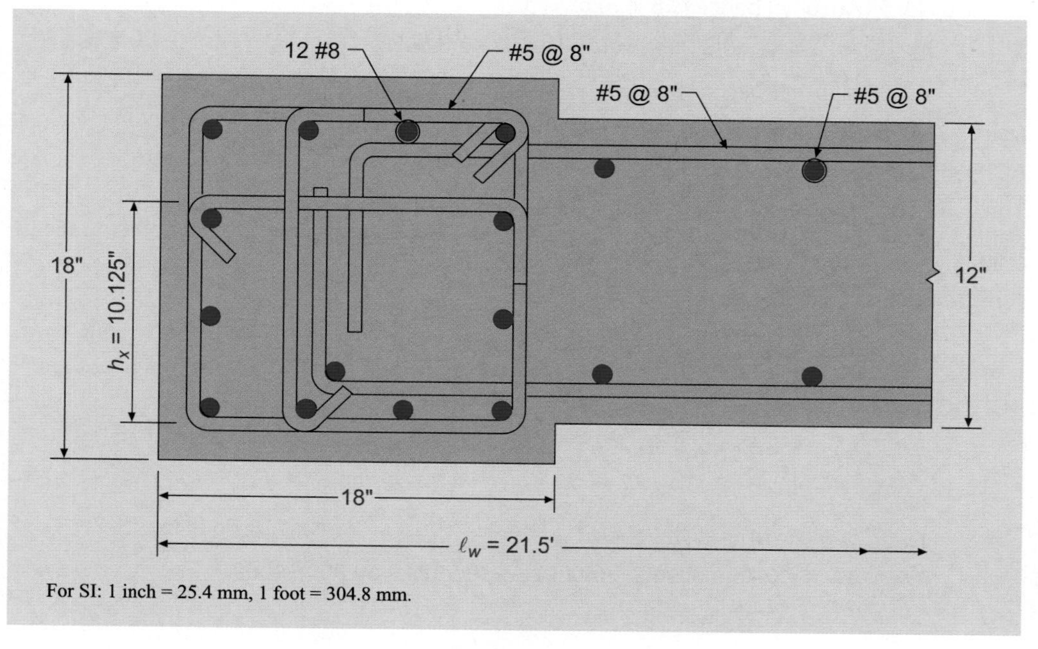

Figure 19E-7.3 Reinforcement details for a shear wall

Example 7—(continued)

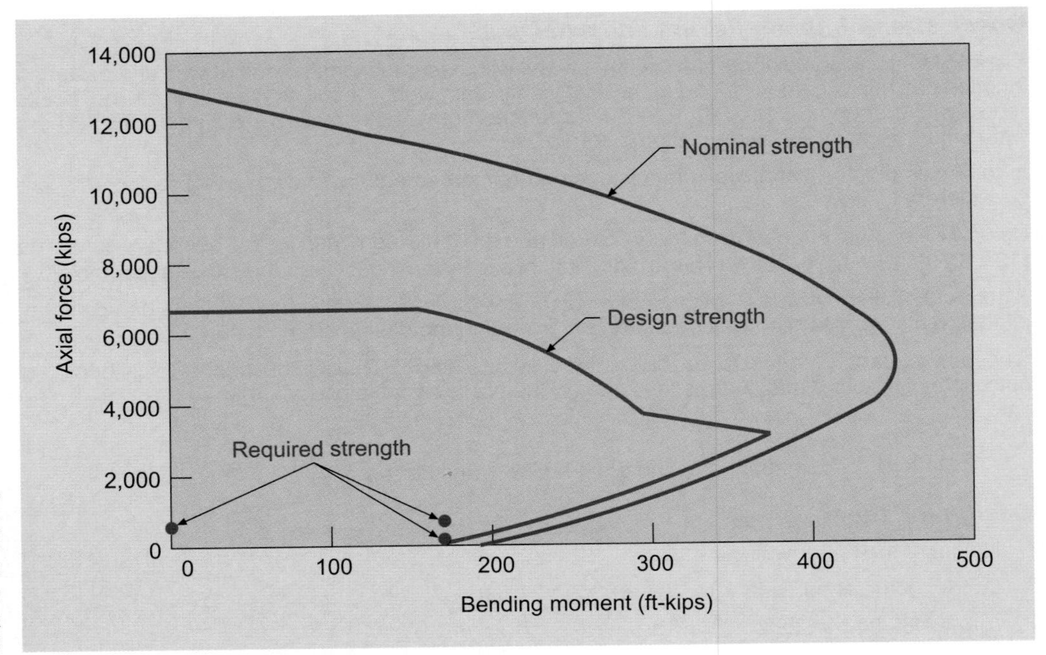

Figure 19E-7.4 Design and nominal strength interaction diagrams for a shear wall

ACI 318 Section 21.7.2.2

At least two curtains of reinforcement must be used if $V_u > 2A_{cv}(f_c')^{0.5}$

$2A_{cv}(f_c')^{0.5}$ $= 2 \times 3096(4000)^{0.5}/1000$

$= 392$ kips > 382 kips

Only one curtain needs to be provided. However, provide two curtains of reinforcement.

ACI 318 Section 21.7.2.1

For two No. 5 horizontal bars @ 18 in. o.c.

ρ_t $= (2 \times 0.31)/(12 \times 18) = 0.0029 > 0.0025$ OK

s $= 18$ in. ≤ 18 in. OK

Use two No. 5 horizontal bars @ 18 in. o.c.

ACI 318 Section 21.7.4.3

The vertical reinforcement ratio (ρ_ℓ) must not be less than the horizontal reinforcement ratio (ρ_t) if the ratio $h_w/\ell w < 2.0$. Since $h_w/\ell_w = 2.93 > 2.0$, use minimum reinforcement ratio of 0.0025.

Provide two No. 5 vertical bars @ 12 in. o.c.

ρ_ℓ $= 0.0043 > 0.0025$ OK
(21.7.2.1)

ACI 318 Section 21.7.4.1

For h_w/ℓ_w $= 2.93 > 2.0$

α_C $= 2$

V_n $= A_{cv}[\alpha_c (f_c')^{0.5} + \rho_t f_y]$ (ACI 318 Equation 21-7)

ϕV_n $= 0.75 \times 3096[2 \times (4000)^{0.5} + (0.0029 \times 60{,}000)]/1000$

$= 698$ kips > 382 kips OK

Example 7—(continued)

Design for flexure and axial loads (ACI 318 Section 21.7.5)

ACI 318 Section 21.7.5.1

Shear walls and portions of such walls subject to combined flexural and axial loads shall be designed in accordance with Sections 10.2 and 10.3, i.e., the provisions applied to columns. Boundary elements as well as the wall web are considered effective. Also, 10.3.6 and the nonlinear strain requirements of 10.2.2 do not apply.

Boundary elements of special RC structural walls (ACI 318 Section 21.7.6)

ACI 318 Section 21.7.6.1

The need for special boundary elements at the edges of shear walls needs to be evaluated in accordance with ACI 318 Section 21.7.6.2 (displacement-based approach) or ACI 318 Section 21.7.6.3 (stress-based approach). In this example, the displacement-based approach is followed.

ACI 318 Section 21.7.6.2(a): Displacement-based approach

Compression zones must be reinforced with special boundary elements where:

$c \geq c_r \quad = \ell_w/600(\delta_u/h_w)$ (ACI 318 Section Equation 21-8)

As computed earlier,

$c \quad = 26.3$ in.

$\ell_w \quad = 21.5$ ft

$H_w \quad = 63$ ft

$\delta_U \quad = 3.8$ in. along the wall line

$\delta_u/h_w \quad = 3.8/(63 \times 12) = 0.005 < 0.007$ Use $\delta_u/h_w = 0.007$

$c_r \quad = (21.5 \times 12)/(600 \times 0.007) = 61.4$ in. $> c = 26.3$ in.

Therefore, boundary zone details per ACI 318 Section 21.7.6.4 are not needed. However, for illustration purposes, the detailing will be shown in this example.

ACI 318 Section 21.7.6.2 (b): Height of special boundary element

The special boundary element reinforcement must extend vertically from the critical section a distance not less than the larger of ℓ_w or $M_u/4V_u$.

$\ell_w \quad = 21.5$ ft . . . governs

$M_u/4V_u \quad = 16{,}855/(4 \times 382) = 11$ ft

Shear wall special boundary element details (ACI 318 Section 21.7.6.4)

ACI 318 Section 21.7.6.4 (a): Length of special boundary element

Special boundary element must extend horizontally from the extreme compression fiber a distance not less than the larger of $c - 0.1\ell_w$ and $c/2$.

$c - 0.1\ell_w \quad = 26.3 - (0.1 \times 21.5 \times 12) = 0.5$ in.

$c/2 \quad = 26.3/2 = 13.2$ in. . . . governs

Since the length of needed special boundary element (= 13.2 in.) does not exceed the depth of boundary column (= 18 in.), the entire boundary column is confined.

ACI 318 Section 21.7.6.4 (c): Transverse reinforcement

Special boundary element transverse reinforcement must satisfy the requirements of ACI 318 Sections 21.4.4.1 through 21.4.4.3, except ACI 318 Equation (21-3) need not be satisfied.

Boundary column confinement

Minimum area of rectangular hoop reinforcement (ACI 318 Section 21.4.4.1(b))

$A_{sh} \quad = 0.09sb_c f'_c/f_{yt}$ (ACI 318 Equation 21-4)

As there are four layers of longitudinal reinforcement in the boundary column, minimum number of legs (hoops and ties) needed to support alternate bars is 3.

Example 7—(continued)

Maximum horizontal spacing of hoop or crosstie legs,

h_x = 2[18 – 2(1.5 + 0.625 + 0.5)]/3 + 1 + 0.625 = 10.125 in.

According to Section 21.4.4.2, the spacing of transverse reinforcement must not exceed (a) one-quarter of the minimum member dimension, (b) six times the diameter of the longitudinal reinforcement, and (c) s_x, as defined by Equation (21-5).

s_x = 4 in. $\leq 4 + (14 - h_x)/3 \leq 6$ in. (ACI 318 Equation 21-5)

s_x = 4 in. $\leq$ 5.3 in. $\leq$ 6 in.

s $\leq$ 5.3 in.

$\leq 6d_b = 6 \times 1.0$

= 6.0 in.

$\leq$ minimum plan dimension/4 = 4.5 in. governs

b_c = 18 – (2 × 1.5) – 5/8 = 14.4 in.

A_{sh} $\geq 0.09 \times 4.5 \times 14.4 \times 4/60 = 0.39$ in.^{2}

With one tie all around the longitudinal reinforcement and one crosstie in either direction (as shown in Figure 19E-7.3),

A_{sh} provided = 3 × 0.31 = 0.93 in.^{2} > 0.39 in.^{2} OK

Example 8 *Design of RC Shear Wall per IBC 2006*

Given Data:

Specified compressive strength of concrete f'_c = 4000 psi

Specified yield strength of reinforcement f_y = 60,000 psi

Structural system: Building Frame with Special Reinforced Concrete Shear Walls

Site Class: D

S_{DS}	= 1.00 g
S_{D1}	= 0.60 g
C_d	= 5
Occupancy Category	= I
R	= 6
Importance Factor, I	= 1.0
Redundancy Factor, ρ	= 1.0
Shear wall thickness, h	= 12 in.
Seismic Design Category	= D

Service-level axial dead load on shear wall = 918 kips

Service-level axial live load on shear wall = 111 kips

The internal forces that are due to lateral loads are shown in Table 19E-8.1. The displacement at the top of the building along the shear wall line was found to be 0.39 in. by elastic analysis under the code-prescribed lateral forces.

Design displacement $\delta_u = 5.0 \times 0.39 = 1.95$ in.

Table E19E-8.1. Summary of design axial force, shear force and bending moment for shear wall between grade and level 2

Loads	Symbol	Axial Force (kips)	Shear Force (kips)	Bending Moment (ft-kips)
Dead load	D	918	0	0
Live load	L	111	0	0
Lateral load	Q_E	0	530	23,325
Load Combinations				
1	$1.2D + 1.6L$	1279	0	0
2	$1.2D + (\rho Q_E + 0.2S_{DS}D) + 0.5L$	1341	530	23,325
3	$0.9D - (\rho Q_E + 0.2S_{DS}D)$	643	530	23,325

Load Combinations	Load Factor		
	D	L	Q_E
1	1.2	1.6	0
2	1.4	0.5	1
3	0.7	0	–1

Example 8—(continued)

Shear Wall Design

The design of the shear wall along line 4 at the base of the building shown in Figure 19E-8.1 is illustrated in this example. The design of shear walls by IBC 2006 follows the procedure in ACI 318-05.

Design loads

Table 19E-8.1 shows a summary of the axial force, shear force and bending moment at the base of the example shear wall based on different load combinations.

Required axial load strength, P_u = 1341 kips

Required shear strength, V_u = 530 kips

Required flexural strength, M_u = 23,325 ft-kips

Design for shear

Height of the shear wall, h_w = 54 ft

Length of the shear wall, ℓ_w = 23 ft

h_w/ℓ_w = 2.35 >2

ACI 318 Section 21.7.4.4

V_u must not exceed $\phi 8A_{cv}(f_c')^{0.5}$

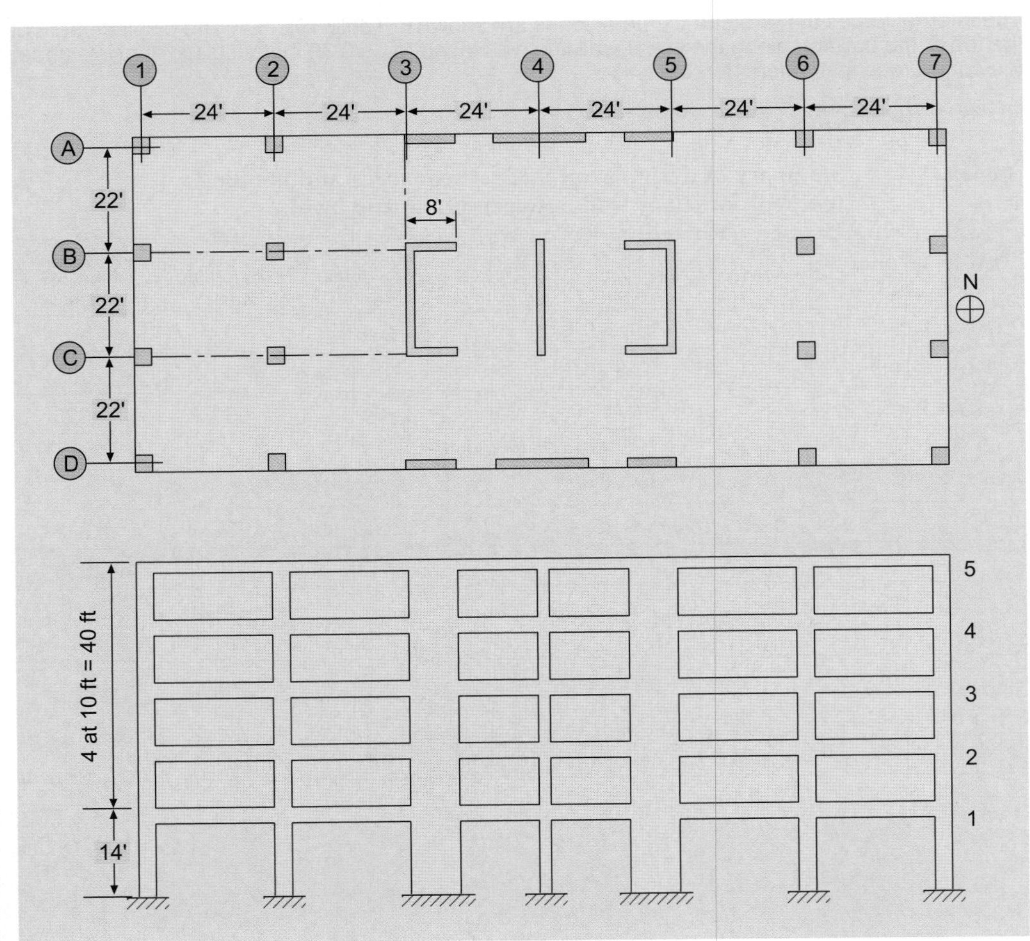

Figure 19E-8.1 Plan and elevation of example building

Example 8—(continued)

A_{cv} = 12 × 23 × 12 = 3312 in.^{2}

Take ϕ = 0.75, since a wall with h_w/ℓ_w = 2.35 is not going to be governed by shear in its failure mode.

$\phi 8A_{cv}(f_c')^{0.5}$ = 0.75 × 8 × 3312 × (4000)$^{0.5}$/1000

= 1257 kips > 530 kips (V_u) OK

ACI 318 Section 21.7.2.2

At least two curtains of reinforcement must be used if $V_u > 2A_{cv}(f_c')^{0.5}$

$2A_{cv}(f_c')^{0.5}$ = 2 × 3312 × (4000)$^{0.5}$/1000

= 419 kips < 530 kips

Therefore, two curtains of reinforcement must be used in the wall.

ACI 318 Section 21.7.2.1

For two curtains of No. 4 horizontal bars spaced at 12 in. o.c.

ρ_t = (2 × 0.20)/(12 × 12) = 0.0028 > 0.0025 OK

δ = 12 in. < 18 in. OK

Use two curtains of No. 4 horizontal bars @ 12 in. on center.

ACI 318 Section 21.7.4.3

The vertical reinforcement ratio (ρ_ℓ) must not be less than the horizontal reinforcement ratio (ρ_t) if the ratio h_w/ℓ_w < 2.0. Since h_w/ℓ_w = 2.35 > 2.0, use minimum reinforcement ratio of 0.0025.

Provide two curtains of No. 4 vertical bars @ 12 in. o.c.

ρ_ℓ = 0.0028 > 0.0025 OK

ACI 318 Section 21.7.4.1

For h_w/ℓ_w = 2.35 > 2.0

α_c = 2

V_n = $A_{cv}[\alpha_c (f_c')^{0.5} + \rho_\tau f_y]\alpha$ (ACI 318 Equation 21-7)

ϕV_n = 0.75 × (3312 [2 × (4000)$^{0.5}$ + (0.0028 × 60,000)]/1000

= 732 kips > 530 kips OK

Design for flexural and axial loads (ACI 318 Section 21.7.5)

ACI 318 Section 21.7.5.1

Shear walls and portions of such walls subject to combined flexural and axial loads shall be designed in accordance with Sections 10.2 and 10.3, i.e., the provisions applied to columns. Concrete and developed longitudinal reinforcement within the boundary elements and the wall web are considered effective. Also, 10.3.6 and the nonlinear strain requirements of 10.2.2 do not apply.

Draw the P-M interaction diagram for the shear wall with the assumed longitudinal reinforcement in the section. Check to make sure that all of the points representing strength demand (from the three load combinations shown in Table 19E-8.1) are within the design strength interaction diagram. In this example, two curtains of No. 4 bars @ 12 in. are provided in the web, and eight No. 11 bars are provided in four layers at each end of the shear wall. Figure 19E-8.2 shows the P-M interaction diagram for the example shear wall. As can be seen, the wall is adequate for the load combinations in Table 19E-8.1.

Boundary elements of special RC structural walls (ACI 318 Section 21.7.6)

ACI 318 Section 21.7.6.1

The need for special boundary elements at the edges of shear walls must be evaluated in accordance with ACI 318 Section 21.7.6.2 (displacement-based approach) or ACI 318 Section 21.7.6.3 (stress-based approach). In this example, the displacement-based approach is followed.

Example 8—(continued)

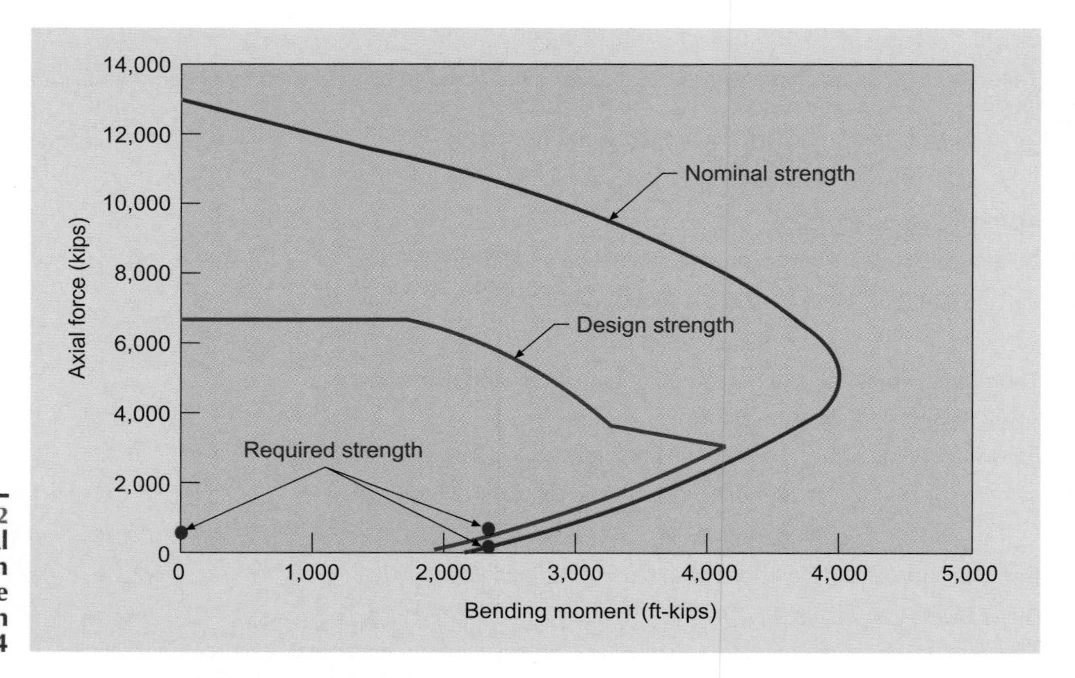

Figure 19E-8.2 Design and nominal strength interaction diagrams for the shear wall on column line 4

ACI 318 Section 21.7.6.2(a): Displacement-based approach

Compression zones must be reinforced with special boundary elements where:

$c \geq c_r \quad = \ell_{ww}/600(\delta_u/h_w)$ (ACI 318 Equation 21-8)

The neutral axis depth c corresponding to the maximum axial force (in the presence of lateral forces) needs to be determined:

For P_u = 1341 kips

ℓ = 54.1 in. (using an in-house computer program)

ℓ_w = 23 ft

h_w = 54 ft

δ_U = 1.95 in. along the wall line

δ_u/h_w = $1.95/(54 \times 12) = 0.003 < 0.007$ Use $\delta_u/h_w = 0.007$

c_r = $(23 \times 12)/(600 \times 0.007) = 65.7$ in. $> c = 54.1$ in.

Therefore, boundary zone details per ACI 318 Section 21.7.6.4 are not needed.

ACI 318 Section 21.7.6.5

Where special boundary elements are not required by ACI 318 Sections 21.7.6.2 or 21.7.6.3, the provisions of ACI 318 Section 21.7.6.5 must be satisfied. Boundary transverse reinforcement in accordance with ACI 318 Sections 21.4.4.1(c), 21.4.4.3 and 21.7.6.4 (a) must be provided if the longitudinal reinforcement ratio at the wall boundary is greater than $400/f_y$. The maximum longitudinal spacing of transverse reinforcement in the boundary must not exceed 8 in.

In this example, the longitudinal reinforcement ratio ρ at the wall boundary (as shown in Figure 19E-8.3) is:

ρ = $(8 \times 1.56)/[12 \times (11 + 2 \times 1.96)] = 0.0697 > 400/60{,}000 = 0.0067$

Therefore, provide transverse reinforcement in accordance with ACI 318 Sections 21.4.4.1(c), 21.4.4.3 and 21.7.6.4(a) at ends of wall.

Example 8—(continued)

ACI 318 Section 21.7.6.4(a) Length of boundary element

The boundary elements must extend horizontally from the extreme compression fiber a distance not less than the larger of the following:

$c - 0.1\ell_w = 54.1 - (0.1 \times 276) = 26.5$ in.

$c/2 = 54.1/2 = 27.1$ in. (governs)

Based on the 12 in. spacing of the No. 4 vertical bars in the web, provide confinement over a horizontal length of 30 in. at each end of the wall. The maximum horizontal spacing of the crossties is 14 in. according to ACI Section 21.4.4.

ACI 318 Section 21.7.6.5(b)

Since $V_u = 530$ kips $> A_{cv}(f'_c)^{0.5} = 210$ kips, horizontal reinforcement terminating at the edges of the wall must have a standard hook engaging the edge reinforcement, or the edge reinforcement must be enclosed in U-stirrups having the same size and spacing as, and spliced to, the horizontal reinforcement.

Reinforcement details for the shear wall are shown in Figure 19E-8.4.

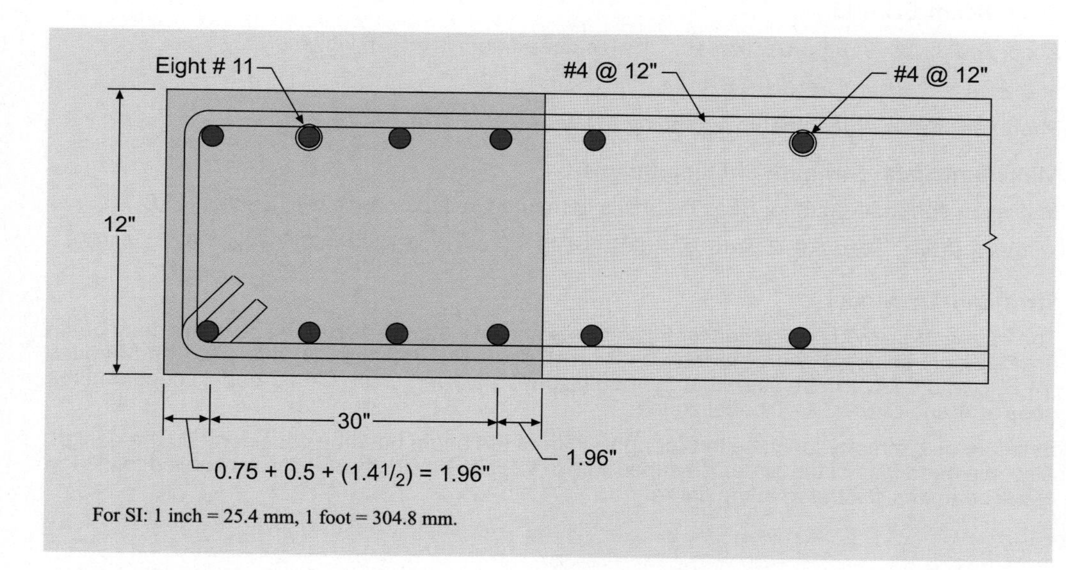

Figure 19E-8.3 Determination of reinforcement ratio at the end of the shear wall

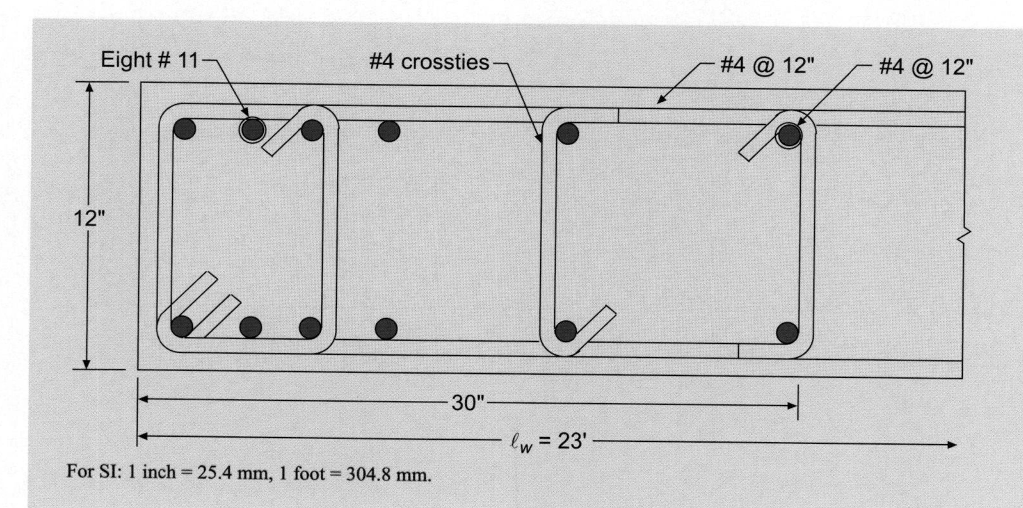

Figure 19E-8.4 Reinforcement details for the shear wall on column line 4

Example 9 *Strut-and-Tie Model per Appendix A of ACI 318-05*

Given Data:

Specified compressive strength of concrete f'_c = 4000 psi

Unit weight of concrete = 150 pcf

Specified yield strength of reinforcement f_y = 60,000 psi

P_D = 150 kips, P_L = 75 kips

Beam elevation and section are shown in Figure 19E-9.1.

Factored Loads and Reactions

Dead load of girder = (20 × 78/144) × 0.15 = 1.625 kips/ft

W_u = 1.2 × 1.625 = 1.95 kips/ft

P_u = (1.2 × 150) + (1.6 × 75) = 300 kips

Support reactions = (1.95 × 22/2) + 300 = 322 kips

Deep Beam Criteria

According to ACI 318-05 Section 10.7.1(a), a deep beam is one in which $\ell_n/h \leq 4$.

In this case, $\ell_n/h = 3 \times 6.5/6.5 = 3 < 4$

Therefore, this is a deep beam according to Section 10.7.1(a).

Maximum Shear Strength of Deep Beams

Maximum shear strength for deep beams is computed in accordance with Section 11.8.3:

$$\phi V_n = \phi 10(f'_c)^{0.5} b_w d = 0.75 \times 10 \times (4000)^{0.5} \times 20 \times (78 - 4.5)/1000 = 697 \text{ kips} > 322 \text{ kips OK}$$

Strut-and-Tie Model

The strut-and-tie model assumed for this beam is shown in Figure 19E-9.2. The entire deep beam is a D-region. The strut-and-tie model consists of 3 struts (AB, BC and CD), 1 tie (AD), and 4 nodes (A, B, C and D). It can be seen that the nodal zones at B and C are C-C-C nodal zones and the ones at A and D are C-C-T nodal zones.

Because of symmetry, only the left (or right) third of the beam need be considered in the design. Also, the dead load of the girder is lumped with the applied concentrated forces applied to the top of the beam, as illustrated in the figure.

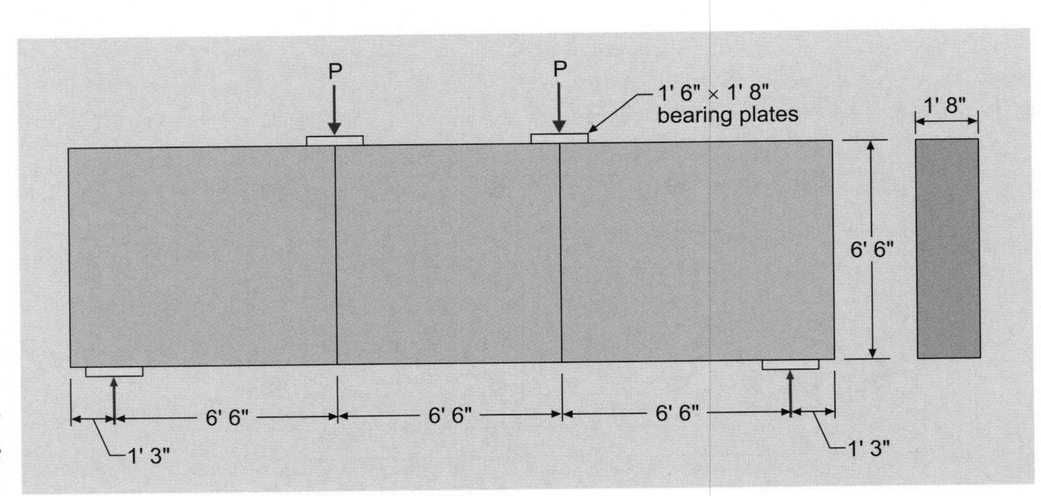

Figure 19E-9.1 Example of deep beam

Example 9—(continued)

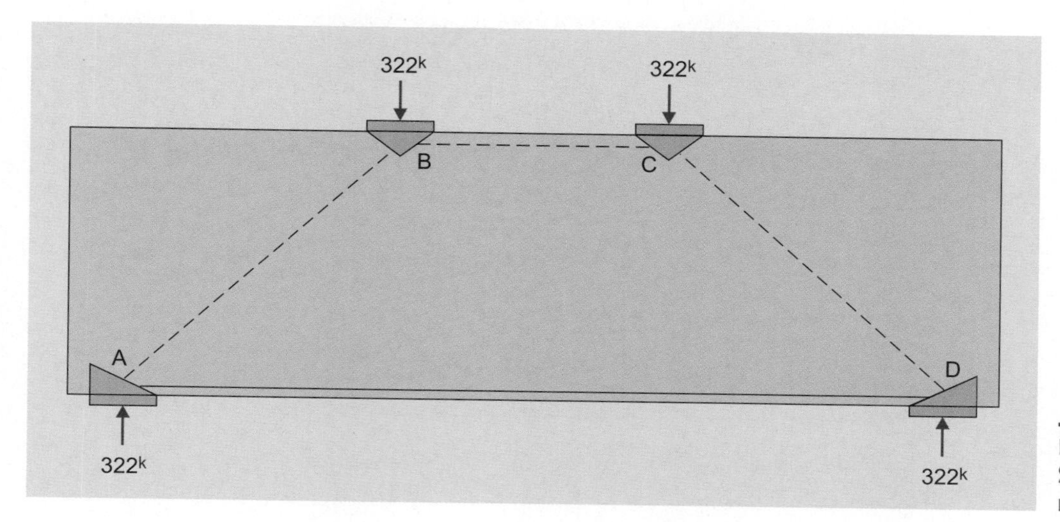

Figure 19E-9.2 Strut-and-tie model

Check bearing capacity at nodal zones

The bearing strength is given by Equation (A-7):

$$\phi F_{nn} = \phi f_{ce} A_{nz}$$

Assuming confining reinforcement is not provided in the nodal zone, the effective compressive strength of the concrete in the nodal zone f_{ce} can be determined from Equation (A-8):

$$f_{ce} = 0.85\beta_n f_c' = 0.85 \times 1.0 \times 4 = 3.4 \text{ ksi}$$

Where β_n = 1.0 in nodal zones bounded by struts or bearing areas, or both (Section A.5.2.1).

Therefore, for nodal zones B and C:

$$\phi F_{nn} = 0.75 \times 3.4 \times 18 \times 20 = 918 \text{ kips} > 322 \text{ kips} \qquad \text{OK}$$

Similarly, for nodal zones A and D:

$$\phi F_{nn} = 0.75 \times 0.85 \times 0.80 \times 4 \times 18 \times 20 = 734 \text{ kips} > 322 \text{ kips} \qquad \text{OK}$$

Where β_n = 0.80 in nodal zones anchoring one tie.

Position of nodes

The horizontal positions of nodes A and B coincide with the lines of action of the respective concentrated forces. The vertical position of these nodes must be either estimated or calculated based on the design strengths of the struts, tie and nodal zones. The latter of these two alternatives is used in this example.

The design strength of strut BC is

$$\phi F_{ns} = \phi f_{ce} A_{cs} \geq F_{BC}$$

$$= \phi 0.85\beta_s f_c' w_c b \text{ (strut)} \qquad \text{Equation (A-3)}$$

$$= \phi 0.85\beta_s f_c' w_c b \text{ (nodal zone)} \qquad \text{Equation (A-8)}$$

Because strut BC has a uniform cross section over its entire length, β_s = 1.0 per Section A.3.2.1. Also, β_n = 1.0, since the nodal zone at B is C-C-C (Section A.5.2.1). Therefore,

$$\phi F_{ns} = 0.75 \times 0.85 \times 1 \times 4 \times w_c \times 20 = 51w_c \geq F_{BC}$$

The design strength of tie AD is determined by Equation (A-6):

$$\phi F_{nt} = \phi A_{ts} f_y \geq F_{AD}$$

Also, the design strength of the nodal zone at A (C-C-T) is given by Equation (A-7):

$$\phi F_{nn} = \phi f_{ce} A_{nz} \geq F_{AD}$$

Example 9—(continued)

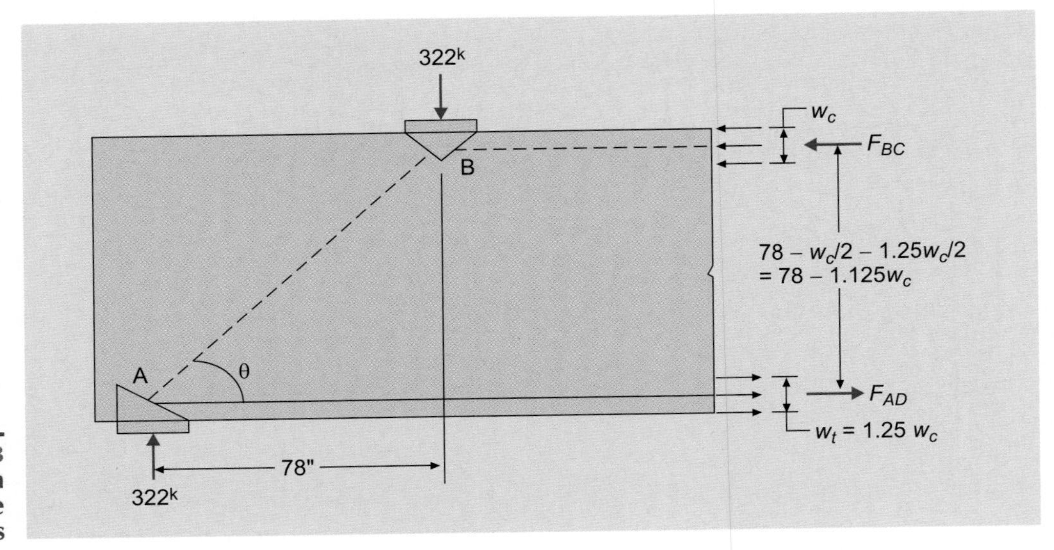

Figure 19E9.3 Determination of strut-and-tie widths

$= \phi 0.85 \beta_s f_c' w_c b$

$= 0.75 \times 0.85 \times 0.80 \times 4 \times w_t \times 20 = 40.8\ w_t \geq F_{AD}$

Where β_n = 0.80 in nodal zones anchoring one tie.

From the free-body diagram depicted in Figure 19E-9.3, it can be seen that $F_{BC} = F_{AD}$. Thus, equating the above equations,

$51w_c = 40.8w_t$, or $w_t = 1.25w_c$.

Summing moments about point A:

$322 \times 78 = F_{BC}(78 - 1.125w_c)$

$25{,}116 = 51w_c(78 - 1.125w_c)$

Solving the quadratic equation for w_c results in w_c = 7.0 in.

Then, $w_t = 1.25 \times 7 = 8.75$ in.

If these values are used for the widths of the strut and tie, the forces in these elements will be at their limits. Increase w_c to 7.5 in. and w_t to 9 in. to allow some margin.

The vertical position of node A is 9/2 = 4.5 in. from the bottom of the beam, and for node B, it is 7.5/2 = 3.75 in. from the top of the beam.

Design capacity of strut BC

The compressive force in strut BC = 322 × 78/ [78 – (7.5/2) – (9/2)] = 360 kips

Check capacity of strut BC:

$\phi F_{ns} = 51w_c = 51 \times 7.5 = 383$ kips > 360 kips OK

Reinforcement in tie AD

F_{AD} = 360 kips $\leq \phi F_{nt} = \phi A_{ts} f_y$

A_{ts} = 360/(0.75 × 60) = 8.0 in.^{2} > $200 b_w d/f_y$ = 4.9 in.^{2} (minimum reinforcement per Section 10.7.3)

Use two layers of four No. 9 bars (A_{ts} = 8.0 in.^{2})

Example 9—(continued)

Design capacity of strut AB

Angle of strut AB with respect to tie AD:

$\tan\theta$ = [78 – (7.5/2) – (9/2)]/78,

θ = 41.8° > 25° (Section A.2.5) OK

F_{AB} = 360/cos 41.8° = 483 kips

Width at top of strut:

$w_{ct} = \ell_b \sin\theta + w_c \cos\theta = 18\sin 41.8° + 7.5\cos 41.8° = 17.6$ in.

Width at bottom of strut:

$w_{cb} = \ell_b \sin\theta + w_t \cos\theta = 18\sin 41.8° + 9\cos 41.8° = 18.7$ in.

Design strength of the strut is based on the smaller of the two widths at the ends of the strut:

$\phi F_{ns} = \phi 0.85\beta_s f_c' w_{ct} b_s$

$= 0.75 \times 0.85 \times 0.75 \times 4 \times 17.6 \times 20 = 673$ kips > 483 kips OK

where $\beta_s = 0.75$ assuming that reinforcement satisfying Section A.3.3 is provided (Section A.3.1).

Anchorage of No. 9 tie bars

The two layers of four No. 9 tie bars must be properly anchored at both ends of the beam.

The critical section for development of the tie reinforcement is determined in accordance with Section A.4.3.2, which is illustrated in Figure 19E-9.4. The distance x from the edge of the bearing plate (nodal zone) to the critical section is determined from geometry:

tan 41.8° = 4.5/x, or x = 5 in.

Available length for straight bar development = 24 – 1.5 + 5 = 27.5 in., which is inadequate to develop a straight No. 9 bar.

The development length of a No. 9 bar with a standard 90-degree hook is determined from Section 12.5.2:

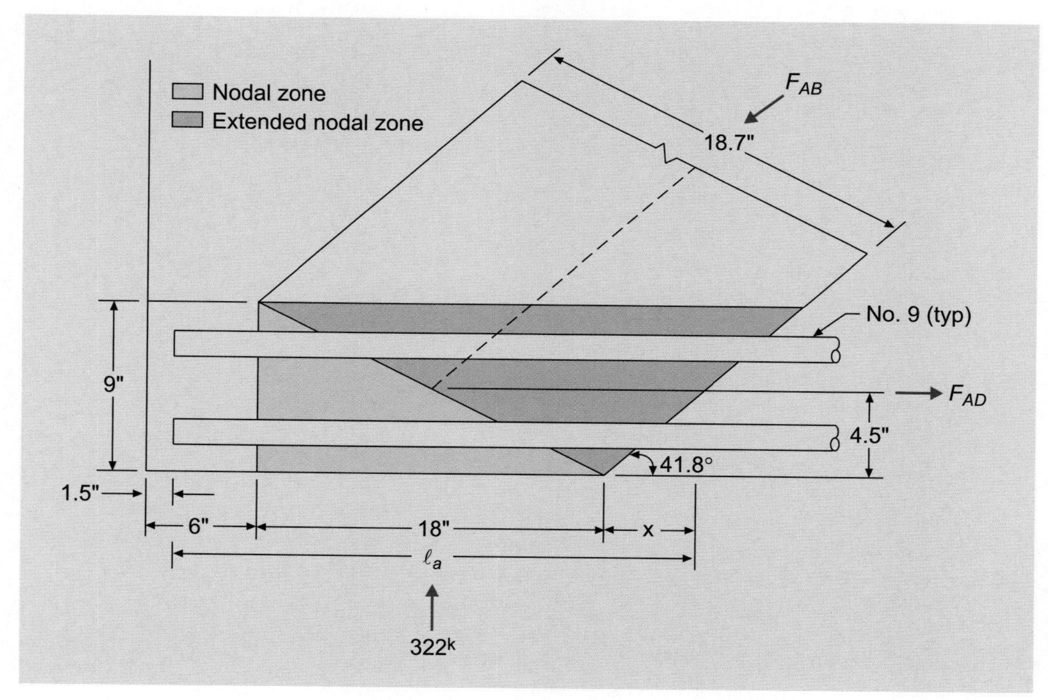

Figure 19E-9.4 Development of tie reinforcement in the extended nodal zone

Example 9—(continued)

ℓ_{dh} $= 0.02\psi_e \lambda f_y d_b/(f_c')^{0.5} = 0.02 \times 1.0 \times 1.0 \times 60{,}000 \times 1.128/(4000)^{0.5}$

$= 21.4$ in. < 27.5 in. OK

Horizontal and vertical reinforcement to resist splitting of diagonal struts

Provide horizontal and vertical reinforcement in accordance with Section A.3.3.

Horizontal reinforcement:

Try two No. 5 bars (one bar on each face) @ 12 in. o.c.

$\alpha_2 = \theta = 41.8°$

$A_{s2}\sin\alpha_2/b_s s_2 = (2 \times 0.31)\sin 41.8°/(20 \times 12) = 0.0017$

Vertical reinforcement:

Try two sets of two No. 4 overlapping ties @ 12 in. o.c.

$\alpha_1 = 90 - \alpha_2 = 48.2°$

$A_{s1}\sin\alpha_1/b_s s_1 = (4 \times 0.2)\sin 48.2°/(20 \times 12) = 0.0025$

Check Equation (A-4):

$0.0017 + 0.0025 = 0.0042 > 0.0030$ OK

CHAPTER

20

ALUMINUM

Chapter 20 covers the requirements for quality, design, fabrication and erection of aluminum structures. The IBC references two aluminum industry standards: *Aluminum Design Manual*: Part 1-A Aluminum Structures, Allowable Stress Design; Part 1-B Aluminum Structures, Load and Resistance Factor Design of Buildings and Similar Type Structures (ADM 1—00) and *Aluminum Sheet Metal Work in Building Construction*, Fourth Edition (ASM 35-0). As with other structural materials (concrete, masonry, steel and wood), the IBC references industry standards rather than transcribes the provisions directly into the code. The advantage to this approach is that the referenced standards are readily updated by revising the year in Chapter 35, Referenced Standards, thereby keeping the code current with the latest available industry standards.

CHAPTER

21

MASONRY

Before discussing the masonry provisions in the 2006 IBC, some historical background may be helpful. The masonry provisions in the 2000 IBC were an amalgamation of the provisions from the 1997 *Uniform Building Code* (UBC),[1] the *National Building Code* (NBC),[2] the *Standard Building Code* (SBC),[3] the NEHRP *Recommended Provisions for Seismic Regulations for New Buildings and Other Structures,*[4] ACI 530/ASCE 5/TMS 402 *Building Code Requirements for Masonry Structures*[5] and ACI 530.1/ASCE 6/TMS 602 *Specification for Masonry Structures.*[6] The 1997 UBC masonry provisions were the main resource for these provisions, although there are differences in content and format. The fact that the IBC adopts standards by reference has made the provisions more concise because many of the design provisions, test procedures and material standards are not included in the body of the code but are incorporated by reference. The national consensus standard for masonry referenced in the IBC is ACI 530/ASCE 5/TMS 402 *Building Code Requirements for Masonry Structures* and ACI 530.1/ASCE 6/TMS 602, *Specification for Masonry Structures*.

The UBC strength design provisions for masonry were developed from 1986 through 1994 by the Masonry Joint Ad Hoc Committee, which was a joint committee of the Structural Engineers Association of California (SEAOC) Code and Seismology Committees. The provisions were introduced into various editions of the UBC as follows:

1. Masonry shear walls and slender wall out-of-plane designs appeared in 1988.
2. Revisions of shear wall confinement trigger mechanism from stress based to strain based, and addition of wall-frame provisions appeared in 1991.
3. Reorganization of the chapter and addition of strength provisions for development, splices, anchorage and for beams, piers and columns appeared in 1994.

The masonry strength design provisions in the 1994 UBC were essentially complete. These provisions were essentially unchanged in the 1997 UBC. Many of the NEHRP provisions are based on the UBC strength provisions—some were incorporated in their entirety, whereas others were modified and updated with the results of the latest research by Technical Coordinating Committee for Masonry Research (TCCMAR) and others. The co-author of the 2000 *IBC Handbook—Structural Provisions*, Robert Chittenden, chaired SEAOC's Masonry Joint Ad Hoc Committee during the period of 1990 through 2000. Some of the SEAOC committee members are also members of NEHRP Technical Subcommittee 5 (TS-5).

In addition, consensus masonry source documents were produced by the Masonry Joint Standards Committee (MSJC), which is a joint committee of the American Concrete Institute (ACI), the American Society of Civil Engineers (ASCE), and The Masonry Society (TMS). These joint documents, formally known as ACI 530/ASCE 5/TMS 402 *Building Code Requirements for Masonry Structures*[5] and ACI 530.1/ASCE 6/TMS 602 *Specification for Masonry Structures*[6] are also referred to as the MSJC Code and the MSJC Specification. The terms "ACI 530/ASCE 5/TMS 402" and "MSJC Code" are used interchangeably in the commentary that follows.

Some of the SEAOC committee members were also members of MSJC. Because of the cross participation, which led to cross fertilization of ideas between the committees, the masonry provisions of the UBC, MSJC Code and NEHRP were converging.

Several of the source documents have extensive commentary. The ACI 530/ASCE 5/TMS 402 standard and the NEHRP provisions each have commentary incorporated directly into the document itself. The Masonry Society produced the *Commentary to*

Chapter 21 of the Uniform Building Code,[7] and Building Officials & Code Administrators International, Inc. (now ICC), produced the commentary to the NBC entitled *The National Building Code/1999—Commentary Volume 2.*[8]

Although the TMS Commentary is for the 1994 UBC, there were essentially no substantive changes in the masonry provisions between the 1994 and 1997 UBC.

This chapter will focus on the provisions within the IBC itself rather than repeat extensive commentary from these other documents. For extensive information on the origin of the various provisions in the 2000 IBC, refer to Table 21-1 of the *2000 IBC Handbook Structural Provisions*. The table shows the source of the various masonry provisions in the 2000 IBC. The reader is referred to the commentary for the cited source section for more extensive background on the specific provisions.

During the code development process that led up to the 2006 IBC, the trend was to adopt standards by reference rather than transcribe standards into the code itself. Accordingly, Chapter 21 decreased from 43 pages in the 2000 IBC to 29 pages in the 2003 IBC. This reduction in the number pages is due primarily to the increasing reliance on referencing the provisions in the ACI 530/ASCE 5/TMS 402 standard, rather than maintaining the provisions within the code itself.

The 2006 IBC references the 2005 edition of the ACI 530/ASCE 5/TMS 402 standard, which is divided into four parts:

Building Code Requirements for Masonry Structures (ACI 530-05/ASCE 5-05/TMS 402-05)

Commentary on Building Code Requirements for Masonry

Specification for Masonry Structures (ACI 530.1-05/ASCE 6-05/TMS 602-05)

Commentary on Specification for Masonry Structures

Because ACI 530/ASCE 5/TMS 402 contains a complete commentary on the provisions of the standard, the reader is referred to the commentary in the standard for a more detailed discussion of the specific provisions.

The majority of the code changes to both the 2003 and 2006 editions of the IBC were to make the masonry provisions in the code more consistent with the ACI 530/ASCE 5/TMS 402 standard. In some cases the code contains modifications to the requirements in the standard.

Section 2101 *General*

The masonry provisions govern the materials, design, construction and quality for masonry designed in accordance with working stress design, strength design or empirical design, and for glass masonry and masonry fireplaces and chimneys. Masonry veneer is covered in Chapter 14 of the IBC and Chapter 6 of the standard. This section contains the basic road map for the user. Section 2101.2 directs the user to specific sections depending on the specific design method or material being used. The requirements for construction documents in Section 2101.3 provide the minimum information to be presented in the design drawings and specifications. Table 1 shows the basic layout of Chapter 21 and how the section relates to the ACI 530/ASCE 5/TMS 402 standard.

Table 21-1.

Subject	IBC Section	ACI 530/ASCE 5/TMS 402 Section
General design requirements	2101	Chapter 1
Definitions	2102	Section 1.6
Construction materials	2103	Chapter 1
Construction requirements	2104	Chapter 1
Quality assurance	2105	Section 1.15
Seismic design	2106	Section 1.14
Allowable Stress Design	2107	Chapter 2
Strength Design	2108	Chapter 3
AAC masonry design	2101.2.2	Appendix A
Prestressed masonry	2101.2.3	Chapter 4
Empirical Design	2109	Chapter 5
Masonry veneer	1405	Chapter 6
Glass unit masonry	2110	Chapter 7
Masonry fireplaces	2111	—
Masonry heaters	2112	—
Masonry chimneys	2113	—

2101.2 **Design methods.** This section basically references the various sections within the code and in the standard depending on the specific design method and material used. Regardless of the design method used, all masonry construction must comply with the general requirements of Sections 2101 through 2104, as well as the applicable seismic design requirements in Section 2106. Section 2101.2 is broken down into six categories: allowable stress design, strength design, prestressed masonry, empirical design, glass unit masonry and masonry veneer. In most cases, the code references as well as modifies the provisions within the standard.

2101.2.1 **Allowable stress design.** Allowable stress design must comply with Sections 2106 and 2107. The term *working stress design* was changed to *allowable stress design* in the 2006 IBC to be consistent with the terminology used in the referenced standard and the load combinations in Section 1605.3. Section 2107 essentially references Chapter 1 of ACI 530/ASCE 5/TMS 402, which covers the general design requirements for masonry, and also modifies portions of Chapter 2 of the standard.

2101.2.2 **Strength design.** Strength design of masonry must comply with the provisions of Sections 2106 and 2108. Section 2108 references Chapter 1 of ACI 530/ASCE 5/TMS 402, which covers the general design requirements for masonry and modifies specific portions of Chapter 3 of the standard. The 2000 IBC contained provisions for strength design of masonry that were taken from the 1997 UBC because none existed in any national consensus standard. Provisions for strength design of masonry were incorporated into the 2002 edition of the ACI 530/ASCE 5/TMS 402 and were then deleted from the 2003 IBC.

AAC masonry is a relatively new building material in the United States, although its use is increasing. Autoclaved aerated concrete (AAC) masonry must comply with the provisions of Section 2106 and Chapter 1 and Appendix A of ACI 530/ASCE 5/TMS 402. Prior to its inclusion in the 2006 IBC, AAC masonry would need to be approved under the alternative materials, design and methods of construction provisions in Section 104.11. Now that it is included in the building code and the referenced standard, it will become an even more popular masonry construction method. It has been used in Europe for many years, where it is considered an effective and efficient building material. In approving the use of AAC in the 2006 IBC, a modification by the committee prohibits the use of AAC in the seismic force-resisting system of structures classified as Seismic Design Category B, C, D, E or F. The restriction on the use of AAC masonry in seismic force-resisting systems to seismic design category A structures is considered prudent until it has been cyclically tested and its seismic response characteristics can be evaluated by the Building Seismic Safety Council (BSSC).

Prestressed masonry must be designed in accordance with Chapters 1 and 4 of ACI 530/ASCE 5/TMS 402 and Section 2106. Special inspection is required during construction of prestressed masonry as described in Section 1704.5.

Empirically designed masonry must comply with the provisions of Sections 2106 and 2109 or Chapter 5 of ACI 530/ASCE 5/TMS 402.

Glass unit masonry must comply with the provisions of Section 2110 or Chapter 7 of ACI 530/ASCE 5/TMS 402.

Masonry veneer must comply with the provisions of IBC Chapter 14 or Chapter 6 of ACI 530/ASCE 5/TMS 402.

Note that the design method—whether allowable stress design, strength design, empirical design or prestressed masonry design—must also comply with the applicable seismic design requirements prescribed in Section 2106 based on the seismic design category of the building or structure. The seismic design category is determined in accordance with Section 1613.5.6.

Section 2102 *Definitions and Notations*

The masonry definitions in the 2000 IBC originated with UBC Section 2101.3 and NBC Section 2102.0. The sources for notation are UBC Section 2101.4 and Section 1.5 of ACI 530/ASCE 5/TMS 402. As the IBC evolved from the 2000 edition to the 2006 edition, several new definitions were added during the ICC code development process. Most of the new definitions related to new materials and construction methods or were clarifications of terminology. Most notable were the new definitions for prestressed masonry and the associated types of prestressed masonry shear wall definitions that were introduced into the 2003 IBC. The definition of Autoclaved Aerated Concrete (AAC) masonry was introduced into the 2006 IBC as well.

Some additional definitions were added to the 2006 IBC in order to be more consistent with definitions in the standard. For example, the terms *foundation pier*, *glass unit masonry* and *unreinforced (plain) masonry* were added to the code. Some minor clarifications were also made to the definitions. For example, the term *required strength* was relocated under the general term *strength* so that the three strength design terms *design strength*, *nominal strength* and *required strength* appear in one place.

Many of the notations were deleted from the 2003 IBC because they are either no longer used in the code or were only used in the strength design provisions of Section 2108. The majority of the strength design provisions in the 2003 IBC were deleted because a strength design procedure is included in Chapter 3 of the current ACI 530/ASCE 5/TMS 402 standard.

Section 2103 *Masonry Construction Materials*

This section contains minimum requirements for masonry construction materials. The 2006 IBC references over 250 ASTM Standards and Specifications, many of which apply to masonry construction materials.

2103.3 AAC masonry. AAC is autoclaved aerated concrete masonry that is a new material introduced into the 2006 IBC by reference to the 2005 edition of the ACI 530/ASCE 5/TMS 402 standard. It is defined in Section 2102 as a low-density cementitious product of calcium silicate hydrates, whose material specifications are defined in ASTM C 1386. See Sections 2102.1 and 2101.2.

2103.4 Stone masonry units. Stone masonry units must conform to the various applicable ASTM standards for exterior marble, limestone, granite, sandstone and slate building stone.

2103.5 Ceramic tile. References for ceramic tile mortar have been added based on requirements in the legacy model codes. During the development of the IBC, a frequent point of discussion concerned architectural materials such as ceramic tile and whether they belonged in structural chapters. For lack of a better place, the ceramic tile provisions were placed in Chapter 21, as in the other model codes. The code references the American National Standard Specifications for Ceramic Tile, ANSI A137.1.

2103.6 Glass unit masonry. The requirement for treating surfaces of the glass block in contact with mortar originated with 1997 UBC Section 2110.2. Reclaimed glass block units are not permitted.

2103.7 Second-hand units. Materials may be reused, provided that they meet all the code requirements for new materials. Note, however, that Section 104.9.1 restricts the use of used materials unless specifically approved by the building official. One should exercise caution in approving reuse of materials. The applicable material or design standards must be consulted to determine if reuse of used materials should be permitted. For example, glass block units cannot be reused as indicated in Section 2103.6.

This section allows the use of salvaged or *used* brick. Used bricks are often salvaged from the demolition of old buildings. Masonry units manufactured in the past may not have the same quality as masonry made to meet current standards. Caution should also be practiced when specifying used brick as a structural material. Even though the brick may appear clean, the pores in the bedding faces may be filled with cement paste, lime and other deleterious microscopic particles that may reduce the absorption properties of the brick, thereby adversely affecting the bond between the mortar and the masonry and reducing the mortar strength. Testing for the absorption rate of used masonry units in accordance with ASTM C 67 will give an indication of the bonding qualities of the used units with mortar. Used masonry units are best used as veneers, where the units are not relied upon for structural strength.

2103.8 Mortar. Mortar for use in masonry construction must conform to ASTM C 270 and the proportions in Table 2103.8(1) or the properties in Table 2103.8(2).

Table 2103.8(1). Mortars proportioned in accordance with this table should have a cube strength in excess of that required by ASTM C 270 for the various mortar types. Type M mortar is high-strength mortar having a minimum average 28-day compressive strength of 2,500 psi. Type M is suitable for general use. Type M mortar is recommended where maximum compressive strength is required such as in unreinforced masonry below grade. Type S mortar is recommended where a high lateral strength is required and is specifically recommended for reinforced masonry. The minimum average compressive strength for Type S mortar is 1,800 psi. Type N mortar is a medium strength mortar having a minimum average compressive strength of 750 psi. Type N mortar may be used where high compressive or lateral strength is not required. Type N is generally used in exposed masonry above grade where exposed to weather. Type O mortar is a low-strength mortar having a minimum average compressive strength of 350 psi. Type O mortar should only be used for nonbearing walls not exposed to severe weathering.

Table 2103.8(2). This table allows for the proportioning of mortar to meet the minimum requirements for 28-day compressive strength, minimum water retention and maximum air content.

Surface-bonding mortar. The specifications for premixed mortar, masonry units and other materials used in constructing dry-stack, surface bonded masonry, as well as the construction requirements, are referenced in this section. **2103.9**

Mortars for ceramic wall and floor tile. See comments on Section 2103.4 regarding placement of provisions for architectural materials in structural chapters. The requirements for electrically conductive dry-set mortars were deleted from the 2006 IBC because electrically conductive ceramic tile is no longer produced. **2103.10**

Mortar for AAC masonry. Mortar requirements for AAC masonry are covered in this section. Thin bed mortar requirements are covered in Section 2103.11.1, and mortar for leveling courses is covered in Section 2103.11.2. The term *thin-bed mortar* is a new definition in the 2006 IBC. Thin-bed mortar is used in AAC masonry construction with joints 0.06 inch or less. Mortar used for the leveling courses of AAC masonry must be Type M or S. **2103.11**

Grout. Grout proportion requirements are shown in Table 2103.12. The minimum grout strength requirement of 2,000 psi set forth in the model codes has been removed from the code provisions because ASTM C 476 permits grout to comply with the minimum strength requirement of 2,000 psi or to comply with the proportions shown in Table 2103.12. Grout batched in accordance with the proportions in Table 2103.12 will have a strength in excess of 2,000 psi. **2103.12**

Metal reinforcement and accessories. Requirements for metal reinforcement and accessories reflect updates made in the national consensus standards. Deformed reinforcement must meet the requirements in IBC Section 2103.13.1. A new standard has been approved for joint reinforcement and is referenced in IBC Section 2103.13.2. The various ASTM standards are listed in Table 21-2. **2103.13**

Table 21-2. ASTM standards for metal reinforcement and accessories

Reinforcement Type	ASTM Standard
Deformed reinforcing bars	A 615 (Billet steel) A 706 (Weldable) A 767 (Zinc coated) A 775 (Epoxy coated) A 996 (Rail and axle steel)
Masonry joint reinforcement	A 951
Deformed reinforcing wire	A 496
Wire fabric	A 185 (Plain steel welded wire fabric) A 497 (Welded deformed steel wire fabric)
Anchors, ties and accessories	A 36 (Structural steel) A 82 (Plain steel wire concrete reinforcement) A 240 (Stainless steel plate, sheet and strip) A 307 Grade A (anchor bolts) A 480 (Flat rolled stainless and heat-resisting steel plate, sheet and strip) A 1008 (Cold-rolled carbon steel sheet)
Prestressing tendons	A 421 (Wire) A 421 (Low-relaxation wire) A 416 (Strand and low-relaxation strand) A 722 (Bar)

Corrosion protection. The section provides requirements and standards for corrosion protection including mill galvanized, hot-dipped galvanized and epoxy coating methods. **2103.13.7**

2103.13.8 **Tests.** This code section originated with the NBC and SBC legacy model codes. The section requires tension and bending tests where unidentified reinforcement is to be used, but does not specify what standard test methods apply.

Section 2104 *Construction*

The provisions in IBC Section 2104 and the referenced standard, ACI 530.1/ASCE 6/TMS 602 provide minimum requirements to ensure the masonry is properly constructed consistent with the design methods used. More stringent construction requirements and referenced specifications than those contained in this code may be needed to satisfy aesthetic and architectural criteria. Such requirements should be written into the construction documents.

Many of the construction requirements that are contained in IBC Section 2104 are also contained in the ACI 530/ASCE 5/TMS 402 standard. The IBC structural committee decided to not repeat many of these requirements in the IBC insofar as they are already contained in the reference standard. The committee did, however, want certain minimum construction requirements related to placement of units and requirements for cold and hot weather construction. The committee's intent was that once users of the IBC become more familiar with the referenced standard, these construction provisions become redundant and may be deleted. In this way, conflicts between the provisions in the IBC and the referenced standard will be minimized.

2104.1 **Masonry construction.** This section cites the applicable code sections and references the construction provisions of ACI 530.1.

Section 2105 *Quality Assurance*

The quality assurance provisions in the 2000 IBC were modeled after those in the 1997 UBC and the ACI 530/ASCE 5/TMS 402 standard. Readers will find the requirements similar to those in the UBC, with a few exceptions as follows:

1. Compliance with the specified compressive strength of the masonry can no longer be verified by the Prism Test Record method that was permitted in the 1997 UBC. This procedure to ensure compliance was deleted because the masonry industry noted that it has rarely been used, and the requirements make it less likely to be used in the future. Moreover, such a procedure is not permitted in the ACI 530/ASCE 5/TMS 402 standard.
2. Design of masonry by the Allowable Stress Design method using one half of the allowable stresses to avoid the special inspection requirements is not permitted in the IBC. Although the UBC permitted half stress design of masonry for masonry construction without special inspection, this is not consistent with the ACI 530/ASCE 5/TMS 402 standard, which requires a minimum level of special inspection for all engineered masonry structures.

The IBC requires full allowable stress for masonry designed by the Allowable Stress Design Method along with mandatory special inspection. During the development of the IBC, it was noted that masonry damaged in earthquakes and high wind events often showed poor quality and workmanship, which contributed to the damage. Many of the masonry structures that failed showed evidence of missing reinforcement, missing grout or missing connectors, and various other improper construction techniques. Even when using half stresses for such designs, there would likely not be sufficient strength in poorly constructed masonry to support the expected wind loads or seismic demands. Special inspection requirements for masonry are covered in IBC Section 1704.5. IBC Section 2105 also requires testing of materials in accordance with applicable ASTM standards. Requirements for structural testing for seismic resistance and verification of masonry materials are covered in Section 1708 along with other structural testing requirements.

General. This section was adapted from Section 1.15 of ACI 530/ASCE 5/TMS 402, and Section 3.7 of ACI 530.1/ASCE 6/TMS 602. The quality assurance program is achieved through the special inspection and testing requirements covered in Chapter 17. **2105.1**

Section 2106 *Seismic Design*

IBC Section 2106 contains the minimum design requirements for masonry structures in seismic risk areas. The section was based on requirements in the 1997 *Uniform Building Code,* the MSJC Code and the NEHRP *Provisions*. The requirements are broken down for various seismic risk categories, with requirements being cumulative from lower seismic risk to higher seismic risk. The categories used to designate seismic risk are based on the concept of seismic design categories used in the 1997 NEHRP instead of seismic zones that were used in the UBC. Seismic design categories are based not only on the location of the structure in relation to known seismically active areas (spectral response accelerations), but also according to building importance (occupancy catagory) and the classification of the soil (site class) at the site of the structure. For more detailed information on the determination of seismic design category, refer to the discussion of Section 1613 and the NEHRP Commentary.

Seismic design requirements for masonry. General seismic design requirements are contained in this section. Prior to the 2006 IBC, it was not clear which masonry walls were considered part of the seismic force resisting system. The most common understanding included those walls that were specifically designed to resists seismic forces. A change in the 2006 IBC clarified which masonry walls are considered part of the seismic-force-resisting system by stating that all masonry walls, unless they are isolated on three edges from the in-plane motion of the basic structural system, must be considered part of the seismic-force-resisting system. This clarification is intended to address nonload-bearing walls that are connected to the lateral force resisting system and as a result would be subject to seismic forces. The new code language intends to prevent the incorporation of masonry elements that are incapable of resisting seismic demands that could be imposed on them during an earthquake. **2106.1**

Basic seismic-force-resisting systems. The basic seismic-force-resisting systems are described based on concepts from the 1997 NEHRP. The 2000 IBC listed five distinct types of shear wall systems that indicate the expected performance of the walls with various construction techniques. The five types of masonry shear wall systems are: ordinary plain masonry, ordinary reinforced masonry, detailed plain masonry, intermediate reinforced masonry and special reinforced masonry shear walls. These five shear wall types are assigned different seismic design coefficients, such as response modification coefficient R, based on the expected performance and ductility of the particular shear wall system involved. Certain shear wall types are required in each seismic region, and unreinforced shear wall types are not permitted in intermediate and high seismic risk regions, based on the seismic design category of the building. See Table 12.2-1 of ASCE 7-05. **2106.1.1**

In the 2003 IBC, the specific requirements for these shear wall types were removed from the code because Section 2106.1 references the 2002 edition of ACI 530/ASCE 5/TMS 402, which includes all of the design requirements for these shear wall systems.

In addition to the five types of conventionally reinforced shear wall systems, three new prestressed masonry shear wall systems were added to the 2003 IBC. These are: ordinary plain prestressed masonry, intermediate prestressed masonry and special prestressed masonry shear walls. All of the various types of masonry shear wall systems are defined under the term *shear wall,* in Section 2102.

In addition to referencing the applicable sections of ACI 530/ASCE 5/TMS 402, the code includes some specific requirements for the intermediate prestressed masonry and special prestressed masonry shear wall systems. When these systems were first introduced into the 2003 code, cyclic testing was still underway and results were not available. Consequently, the seismic design coefficients and height limitations for each system were proposed to be the same as the conventionally reinforced ordinary, intermediate and special reinforced masonry shear wall systems, respectively. The structural committee modified and the proposal by imposing various restrictions, which were approved as modified and are shown in Table 1617.6.2 of the 2003 IBC. In the 2006 IBC, which now references Table 12.2-1 of ASCE 7-05, the three distinct types were all combined into one category so that all prestressed masonry shear wall systems have the same seismic design coefficients and height limitations for the same type of building system. See item A12 of Table 12.2-1 for bearing wall buildings with prestressed masonry shear wall systems, and item B22 of Table 12.2-1 for building frame buildings with prestressed masonry shear wall systems. It should also be noted that in the 2006 IBC and ASCE 7-05, prestressed masonry shear wall systems are not permitted in Seismic Design Categories C, D, E and F.

Table 21-2 summarizes the requirements for each of the masonry shear wall systems described above.

2106.2 Anchorage of masonry walls. Because of the significant out-of-plane seismic forces that develop in concrete and masonry walls during seismic ground motion, the code requires them to be anchored to roofs and floors that provide lateral support for the wall. The anchorage must provide a positive and direct connection capable of resisting the design seismic forces but not less than a minimum strength level force of 280 per lineal foot of wall. Although *positive and direct* is not explicitly defined in the code, the idea is to transfer the lateral loads as directly as possible and not by indirect circuitous paths that are less reliable and more prone to failure.

Early editions of the legacy model codes required connections to floors and roofs parallel and perpendicular to wall to transfer design forces but not less than 200 pounds per lineal foot, which was an allowable stress design force. The strength level force is determined by 1.4 × 200 plf = 280 plf. Where anchor spacing exceeds 4 feet on center, the walls must be designed to resist bending horizontally between anchors. Obviously, where the anchors are used in masonry walls constructed of hollow units or cavity wall systems, they must be embedded in a reinforced grouted structural element within the wall. See the discussion under Section 1604.8.2.

2106.3 Seismic Design Category B. The requirements in Section 2106.2 of the 2000 IBC for Seismic Design Category A were deleted in the 2003 IBC because the section had no specific requirements related to seismic design. The code imposes some seismic related requirements on masonry structures beginning with Seismic Design Category B. Section 2106.3 references Section 1.14.4 of ACI 530/ASCE 5/TMS 402, which in turn imposes two additional requirements. Masonry partition walls, screen walls and other masonry elements that are not specifically designed to resist external loads other than the loads produced by their own weight are required be isolated from the rest of the structure so that the forces from the structure are not imparted to these elements. Any joints or connections between the structure and these isolated elements must be designed to accommodate the design seismic story drift.

2106.4 Additional requirements for Seismic Design Category C. Structures in Seismic Design Category C must conform to Section 1.14.5 of ACI 530/ASCE 5/TMS 402, the requirements for Seismic Design Category B described above, as well as the additional transverse reinforcing requirements described in this section. Columns that resist loads from discontinuous walls, and beams that resist loads from discontinuous walls or frames, require minimum transverse reinforcement. These requirements for strength and toughness are intended to prevent local failure or collapse in elements supporting discontinuous portions of the lateral-force-resisting system. Inherent in the code philosophy is the assumption that the inelastic demands on the structure will be well distributed throughout the lateral-force-resisting system. The value of the response modification factor, R, is based on this

Table 21-3. **Shear wall types and requirements**

Shear Wall System	Description and ACI530/ASCE 5/TMS 402 Section	Reinforcement Requirements	May be used in Seismic Design Category
Ordinary plain (unreinforcd) masonry shear wall	A masonry shear wall designed to resist lateral forces, neglecting stresses in reinforcement. Design in accordance with Section 1.14.2.2.1	None. Section must remain uncracked under design loads.	A and B
Detailed plain (unreinforced) masonry shear walls	A masonry shear wall designed to resist lateral forces, neglecting stresses in reinforcement. Design in accordance with Section 1.14.2.2.2	In accordance with Sections 1.14.2.2.2.1 and 1.14.2.2.2.2, $A_{s\ min}$ = 0.20 in^2 at corners, within 16 inches of each side openings, within 8 inches of wall ends, and maximum spacing of 120 inches on center. Horizontal reinforcement at bottom and top of wall openings and extend at least 24 inches or 40d_b past the opening, continuously at structurally connected floors and roofs and within 16 inches of the top of walls. Connections to floors and roofs parallel and perpendicular to wall to transfer design forces but not less than 200 pounds per lineal foot and spaced not more than 4 feet on center. Section must remain uncracked under design loads.	A and B
Ordinary reinforced masonry shear walls	A masonry shear wall designed to resist lateral forces, considering stresses in reinforcement. Section 1.14.2.2.3	In accordance with Sections 1.14.2.2.2.1 and 1.14.2.2.2.2 and as required by the design.	A, B and C
Intermediate reinforced masonry shear walls	A masonry shear wall designed to resist lateral forces, considering stresses in reinforcement. Section 1.14.2.2.4	In accordance with Sections 1.14.2.2.2.1 and 1.14.2.2.2.2 and as required by the design. Spacing of vertical reinforcement cannot exceed 48 inches on center.	A, B and C
Special reinforced masonry shear walls	A masonry shear wall designed to resist lateral forces, considering stresses in reinforcement Section 1.14.2.2.5	In accordance with Sections 1.14.2.2.2.1 and 1.14.2.2.2.2 and as required by the design. Maximum spacing of vertical and horizontal reinforcement must be the lesser of one-third the length of the shear wall, one-third the height of the shear wall or 48 inches. The minimum cross-sectional area of vertical reinforcement must be one-third of the required shear reinforcement. Shear reinforcement must be anchored around vertical reinforcing with standard hooks.	A, B, C, D, E and F
Ordinary plain prestressed masonry shear walls	A prestressed masonry shear wall designed to resist lateral forces, considering stresses in reinforcement. Chapter 4	In accordance with Chapter 4 of ACI 530/ASCE 5/TMS 402.	A and B
Intermediate prestressed masonry shear walls	A prestressed masonry shear wall designed to resist lateral forces, considering stresses in reinforcement. Section 1.14.2.2.4, Section 4.4.3 for flexural strength, Section 3.3.4.1.2 for shear strength and Sections 1.14.2.2.5, 3.3.3.5 and 3.3.4.3.2(c) for reinforcement.	In accordance with Section 2106.1.1.2. Flexural elements subjected to load reversals must be symmetrically reinforced, the nominal moment strength at any section along a member cannot be less than one-fourth the maximum moment strength, the cross-sectional area of bonded tendons is considered to contribute to the minimum reinforcement, tendons must be located in cells that are grouted the full height of the wall.	A and B

(Continued)

Table 21-3. **Shear wall types and requirements** (Cont'd)

Shear Wall System	Description and ACI530/ASCE 5/TMS 402 Section	Reinforcement Requirements	May be used in Seismic Design Category
Special prestressed masonry shear walls	A prestressed masonry shear wall designed to resist lateral forces, considering stresses in reinforcement. Section 1.14.2.2.5, Section 4.4.3 for flexural strength, Section 3.3.4.1.2 for shear strength and Sections 1.14.2.2.5(a), 3.3.3.5 and 3.3.4.3.2(c) for reinforcement.	In accordance with Section 2106.1.1.3. Flexural elements subjected to load reversals must be symmetrically reinforced, the nominal moment strength at any section along a member cannot be less than one-fourth the maximum moment strength, the cross-sectional area of bonded tendons is considered to contribute to the minimum reinforcement. All cells of the masonry wall must be fully grouted. Prestressing tendons must consist of bars conforming to ASTM A 722.	A and B

Note: The maximum spacing for reinforcement in an intermediate reinforced masonry shear wall is 4′–0″ c.c. This differs from the MSJC Code and NEHRP where the maximum spacing allowed is 10′–0″ c.c.

assumption, as well as the assumption of sufficient ductility and overstrength to meet the maximum anticipated seismic demands. Elements used to redistribute or transfer the effect of overturning forces and shears from stiff discontinuous elements are susceptible to increased localized or concentrated inelastic demands, which violates the above assumption, and may not achieve the required ductility. The requirement for a minimum amount of transverse reinforcement at a maximum spacing increases the maximum usable masonry strain and ductility in the element so that the element may meet the maximum seismic demand and distribute the overturning forces and shears without failure.

2106.5 Seismic Design Category D. Structures in Seismic Design D must conform to Section 1.14.6 of ACI 530/ASCE 5/TMS 402, the requirements for Seismic Design Category B and C, as well as the additional requirements of this section.

2106.5.1 Loads for shear walls designed by the allowable stress design method. For shear walls designed by the working stress design method, the in-plane shear from seismic forces determined by the working stress design method must resist 1.5 times the seismic forces required by Chapter 16. However, the 1.5 multiplier does not apply to the overturning moment.

2106.5.2 Shear wall shear strength. For shear walls with nominal shear strength in excess of the shear corresponding to the nominal flexural strength, two shear regions exist. The plastic hinge region is assumed to be between the base of the wall and a plane located at a distance equal to the length of the shear wall above the base. In the plastic hinge region, there is extensive cracking from flexure. Because of the extensive cracking, the capacity to develop shear resistance in the masonry is significantly reduced, and the shear strength of the wall is derived from the transverse reinforcement. Hence, the steel reinforcement should be designed to resist the entire shear force in plastic hinge regions, as required by this section. Note that the requirement applies to designs using either working stress design or strength design. For the region above the plastic hinge, the code requires nominal shear strength to be determined by the strength design procedure in accordance with Section 2108. The strength method must be used to determine the necessary amount of steel and spacing to meet the intent of this section. The restriction on the use of Type N mortar and masonry cement in the lateral-force-resisting system that was in the 2000 IBC is now part of the ACI 530/ASCE 5/TMS 402 standard. The sum of the vertical and horizontal reinforcement ratios must be at least 0.002 with a reinforcement ratio of not less than 0.0007 in each direction. The maximum spacing of reinforcement is 48 inches on center except for stack bond masonry walls, where the maximum spacing is 24 inches on center. Stack bond masonry must be fully grouted hollow open end units, hollow units with full head joints or solid units. Note that only special reinforced masonry walls are allowed in Seismic Design Category D. Refer to Table 12.2-1 of ASCE 7-05 for the seismic design coefficients and restrictions.

Seismic Design Category E or F. Structures in Seismic Design Category E or F must conform to Section 1.14.7 of ACI 530/ASCE 5/TMS 402, as well as the additional requirements specified for Seismic Design Categories B, C and D. The additional minimum reinforcing requirements for stack bond elements that are in the 2000 IBC are now part of the ACI 530/ASCE 5/TMS 402 standard. Stack bond used in elements that are not a part of the lateral force resisting system are required to have horizontal reinforcing ratio of at least 0.0015 with a maximum spacing of 24 inches on center and must be solid grouted and constructed of hollow open end units or two wythes of solid units. Stack bond masonry in elements of the lateral force resisting system are required to have horizontal reinforcing ratio of at least 0.0025 with a maximum spacing of 16 inches on center and must be solid grouted and constructed of hollow open end units or two wythes of solid units. Note that only special reinforced masonry shear walls are allowed in Seismic Design Category E and F. Refer to Table 12.2-1 of ASCE 7-05 for the seismic design coefficients and restrictions. **2106.6**

Section 2107 *Allowable Stress Design*

General. Section 2107 references the allowable stress design method in the MSJC Code with additional modifications that the ICC Structural Committee determined were appropriate. The modifications to the standard are in Sections 2107.2 through 2107.8. During the development of the 2000 IBC, the committee chose to reference the MSJC Code because the basic philosophy of the IBC is to adopt consensus standards for design procedures by reference wherever possible if a suitable consensus standard is available. The committee did this consistently throughout, with the exception of Section 2108 in the 2000 IBC, as no consensus standard existed for strength design of masonry at that time. The strength design procedure for masonry in Section 2108 of the 2000 IBC was based on the provisions in the 1997 UBC, and a version of these provisions was later incorporated in the 2002 edition of the MSJC code. **2107.1**

The allowable stress design provisions in the standard are based on the use of full design stresses only, assuming that all engineered structural masonry will have some level of special inspection. These minimum levels of special inspection have accordingly been incorporated into IBC Section 1704.5. The IBC and ACI 530/ASCE 5/TMS 402 standard do not allow the half stress design without special inspection that is permitted in the UBC.

There are minor differences between the MSJC Code and the UBC, e.g., slightly different modulus of elasticity values, but overall the procedures and allowable design values are quite consistent, allowing designers to easily make the transition from the UBC working stress design procedures to allowable stress design in the MSJC Code.

ACI 530/ASCE 5/TMS 402, Section 2.1.2, load combinations. Section 2.1.2.1 of ACI 530/ASCE 5/TMS 402 has load combinations that can be used when the legally adopted code does not specifiy load combinations. The section is deleted because the IBC includes complete allowable stress design load combinations in Section 1605.3. **2107.2**

ACI 530/ASCE 5/TMS 402, Section 2.1.3, design strength. This modification requires special inspection in accordance with IBC Chapter 17 rather than the requirements in the MSJC Code. The inspection requirements in Chapter 17 are much more comprehensive than those required in the MSJC Code. A code change to the 2000 IBC referenced the ACI 530/ASCE 5/TMS 402 standard but excluded Sections 2.1.2.1 (load combinations) and 2.1.3.4 (design strength). The intent was to exclude the load combinations and prohibit the use of the pseudo strength design method that is in the standard. A subsequent code change to the 2003 IBC further clarified the intent by adding Sections 2107.2 and 2107.3 to the 2006 IBC that specifically delete Sections 2.1.2.1 and 2.1.3.4 of ACI 530/ASCE 5/TMS 402. **2107.3**

2107.4 ACI 530/ASCE 5/TMS 402, Section 2.1.6, columns. This modification permits lightly loaded masonry columns in light-frame structures having a maximum area of 450 square feet and located in regions of relatively low seismicity to be reinforced with a single vertical reinforcing bar in each cell as long as the column can safely support the code required loads and deformations. In the MSJC Code, columns are defined by geometry, not by applied loads. Thus masonry members that have a certain geometry are classified as columns although they more closely resemble a flexural element with low axial load. Such columns are required by the MSJC Code to be reinforced with a minimum amount of vertical reinforcement and horizontal ties. Section 2107.4 exempts lightly loaded columns such as those used to support light-frame carport roofs, porches, sheds or similar structures that primarily experience axial tension and flexure in high wind events from the prescriptive requirements of the MSJC Code. This code section includes a list of prescriptive requirements for light-load columns that qualify: The nominal cross-sectional dimension of columns is 8 inches and the height cannot exceed 12 feet. Columns must be grouted solid and reinforced with one No. 4 bar centered in each cell. Roof systems must be adequately anchored to the columns to resist the design loads, and columns subjected to uplift loads must be adequately anchored to the footing. The code requires at least two No. 4 bars extending a minimum of 24 inches into the columns and bent horizontally a minimum of 15 inches in opposite directions into the footings. To achieve minimum stability, the total weight of the column and its footing must be at least 1.5 times the uplift load.

2107.5 ACI 530/ASCE 5/TMS 402, Section 2.1.10.7.1.1, lap splices. This modification requires the reinforcing bar splice lengths to be consistent for all masonry design methods (allowable stress design and strength design). The existing splice lengths in the MSJC code are based on developing the allowable stress in the bar by means of an assumed simplified bond mechanism. This method is unconservative for large bar sizes.

Both allowable stress and strength design methods assume that the reinforcing will have sufficient strength to resist the imposed forces. Splices of reinforcing bars must meet this same strength test; the same holds true for development length. In seismic regions, the force in any particular bar is indeterminate but may be at the yield strength of the bar. Required development and splice lengths are based on developing the yield strength of the reinforcing bar, including any apparent overstrength, without distress in the masonry, as is done in the strength design method.

The strength design provisions introduced in the 1994 UBC, Section 2108.2.2.6, contained more realistic requirements for development and splicing of reinforcing bars. These provisions were based on the research for development and splicing of reinforcing bars in concrete with adaptations as necessary to reflect that masonry does not have confinement reinforcement, as does concrete. Subsequent testing by the National Concrete Masonry Association validated the basic form of the equation used in the UBC but determined that the required splice length was conservative for small bars and unconservative for large bars. Hence, the equation in the 2000 IBC was modified to include a factor that increased splice and development lengths for bar sizes No. 8 and larger, and reduced the length for bar sizes No. 5 and smaller. The upper limit was removed from the equation, and the limit on cover depth for purposes of computation of the required splice and development length was increased from $3d_b$ to $5d_b$.

The splice length or development length required by the equation in the 2003 IBC, including the f factor, will develop 125 percent of the specified yield strength of the reinforcing bar. This allows for the likely overstrength of reinforcing bars and matches the requirements for welded or mechanical splices, which also must develop 125 percent of the specified yield strength of the bar.

During the development of the 2006 IBC, it was pointed out that the lap splice Equation 21-2 in the 2003 IBC produces lap splices that in some cases are unreasonably long and result in conditions that cannot be constructed in the field. During the development of the 2005 edition of the ACI 530/ASCE 5/TMS 402 standard, over half of the public comments were directed at this issue. A code change to the 2003 IBC modified the lap splice provisions to be essentially the same as the development length provision in the 1997 UBC (which originated with the 1985 UBC). The code change also increased lap splices by 50 percent in areas of high tensile demand (greater than 80 percent of the allowable steel tension stress *Fs*). The amendment to the standard in the 2006 IBC will in all probability be removed after further research resolves the issue and incorporates the new requirements into the standard.

The amendment in Section 2107.5 is only for splices. Although the same length is also needed to develop the strength of the reinforcing bar, the requirement for development length in the MSJC code Section 2.1.10.3 is not amended in the IBC.

The code language related to epoxy bars was added to the 2006 IBC because it is currently in the 2005 MSJC code and code users may incorrectly conclude that IBC Section 2107.5 replaces MSJC Section 2.1.10.7.1.1 entirely, which would result in the deletion of the language related to epoxy bars.

2107.6

ACI 530/ASCE 5/TMS 402, splice of reinforcement. The amendment to the standard is the last sentence of Section 2107.6, which requires bars larger than No. 9 to be spliced using mechanical connectors. The reason for the amendment is that the allowable stress design method of the standard allows reinforcing bar sizes up to No. 11. As noted in Section 2107.7, these large bar sizes are very difficult to lap splice without splitting the masonry. Using mechanical connectors can mitigate the splitting problem that occurs in ordinary lap splices. Hence, the code requires mechanical connectors for bar sizes larger than No. 9. The mechnical splice must develop 1.25 times the yield strength of the bar. This requirement is similar to the requirement for strength design in Section 3.3.3.4 of the standard.

2107.7

ACI 530/ASCE 5/TMS 402, maximum bar size. The amendment adds Section 2.3.6 to the standard that restricts the size of reinforcing bars used in walls. Placing large bars in masonry walls over-reinforces the section so that a brittle failure of the masonry is likely. It is difficult to develop lap splices for large reinforcing bars in thin masonry walls. Research shows that when the reinforcing bar size number is larger than the nominal thickness of the wall, the splice will fail by splitting of the wall and pullout of the bar before the strength of the bar is developed in the splice. The research shows that a rough rule of thumb for maximum bar size is (n-1), where n is the nominal thickness of the wall. Hence, the limit on bar size to one-eighth the nominal thickness. This requirement is similar to the requirement for strength design in Section 3.3.3.1 of the standard, except that Section 3.3.3.1 also imposes a maximum size limit of No. 9 on reinforcing bars.

2107.8

ACI 530/ASCE 5/TMS 402, maximum reinforcement percentage. The allowable stress design provisions of the the standard do not include maximum reinforcement ratio limits, although establishing criterion is under consideration by the MSJC. Until such time as maximum reinforcement ratio requirements are incorporated into the standard, it has been decided that special reinforced masonry shear walls that have relatively high inelastic demands should be subject to some reinforcement limitation. Thus, the amendment adds Section 2.3.7 to the standard, which imposes a maximum reinforcement ratio for special reinforced masonry shear walls. A minor modification in the 2006 IBC clarified that the requirement does not apply to the design of walls for out-of-plane loads.

Section 2108 *Strength Design of Masonry*

Section 2108 of the 2000 IBC has detailed requirements for the strength design of masonry. The 1999 edition of the ACI 530/ASCE 5/TMS 402 standard does not include a strength design procedure. A prime goal of the IBC structural committee was to reference national consensus design standards whenever possible rather than transcribe structural provisions into the IBC. However, in order to maintain the often preferred strength design method for masonry, the committee chose to incorporate the extensive strength design provisions within this section of the code. The strength design provisions in the 2000 IBC are based on those of the 1997 UBC. At that time, the structural committee encouraged the MSJC to develop and incorporate strength design provisions within the standard. This was achieved in the 2002 edition of the ACI 530/ASCE 5/TMS 402 standard, which contains comprehensive strength design provisions. As a result, the strength design provisions were removed from the 2003 IBC, and Section 2108 references Chapter 3 of the standard. The section also includes some modifications to the strength design provisions in the standard.

The provisions in Section 2108 of the 2000 IBC were originally based on the 1997 UBC strength design provisions for masonry, but the requirements for seismic design shear strength and anchor bolts were modified to more closely resemble the strength design provisions in the 1997 NEHRP *Provisions*. Many of the NEHRP *Provisions* are based on the UBC strength provisions but were modified and updated as a result of the latest seismic research. Additional changes were made to reconcile differences between the provisions in the two documents. The strength design provisions in Chapter 3 of the 2002 ACI 530/ASCE 6/TMS 402 were under development by masonry design experts for over 10 years. Many of the provisions in the IBC and in other documents such as NEHRP were based on the draft provisions that were being developed by the MSJC. Thus, many of the provisions in the 2005 MSJC are familiar to designers and are generally consistent with those in 2000 IBC Section 2108. An exception to the consistency between MSJC Chapter 3 and the 2000 IBC provisions is the intentional omission of masonry wall frame provisions. The MSJC decided to leave provisions for wall frames out of Chapter 3 of the standard because of considerable differences of opinion on the requirements for masonry wall frames and because very few buildings have been constructed using either the 2000 IBC or 1997 UBC provisions. Because the MSJC recognized that masonry moment wall frames are not commonly used in practice, the provisions were omitted from Chapter 3 until the need for such provisions arises and consensus provisions for wall frames can be developed.

There was a code change to the 2000 IBC that removed the strength design provisions from the code but intentionally left the masonry wall frame provisions in so that a designer who wanted to design a masonry wall frame system could do so. However, a subsequent code change deleted all of the strength design provisions from the code, including the masonry wall frame provisions. Consequently, there are no provisions for designing masonry wall frames in the 2006 IBC or the 2005 ACI 530/ASCE 6/TMS 402 standard. Therefore, such a design would probably best be done using Section 2108 of the 2000 IBC and approved under the alternative materials, design and methods of construction provisions in Section 104.11.

For a detailed discussion of the strength design provisions in the 2000 IBC, the reader is referred to the *2000 IBC Handbook—Structural Provisions*.

2108.1 **General.** The 2006 IBC references the strength design procedure in Chapter 3 of the 2005 edition of the ACI 530/ASCE 5/TMS 402 standard with some modifications. For the design of AAC masonry, the code references Appendix A of the standard.

2108.2 **ACI 530/ASCE 5/TMS 402, Section 3.3.3.3, development.** The required development length of reinforcement is determined by Equation 3-15 of the standard but cannot be less than 12 inches and need not be greater than 72 d_b. Equation 3-15 produces lap splices that are highly variable and in some cases are unreasonably long, resulting in conditions that are difficult to build in the field. During the development of the 2005 edition of the ACI 530/ASCE 5/TMS 402 standard, 148 public comments emphasized this problem. The 1997 UBC had a nearly identical equation, but the maximum lap splice length was capped at 52 bar-diameters. The 1997 UBC lap splices for Grade 60 reinforcement were determined by ASD in areas of high moment to be 1.5 times 48 bar diameters, or 72 bar-diameters. This modification to the standard essentially increases the 1997 UBC cap from 52 to 72 bar diameters to coordinate the ASD and strength design requirements. This modification will likely be removed from the IBC after further research resolves the situation and provisions are incorporated into the ACI 530/ASCE 5/TMS 402 standard.

2108.3 **ACI 530/ASCE 5/TMS 402, Section 3.3.3.4, splices.** Welded splices are not permitted in plastic hinge zones of intermediate or special reinforced masonry shear walls. Where used, welded splices must be of ASTM A706 steel reinforcement. This modification to the standard is based on a revision to the 2000 NEHRP *Provisions*. To achieve adequate performance, splices in reinforcing used in the lateral-force-resisting system that are subjected to high seismic strains must be capable of developing the full strength of the reinforcing steel. In order to be welded properly, the chemistry of the steel must have a limited carbon content as well as other elements such as sulfur and phosphorus. If the chemistry of the steel is not carefully controlled, the welding procedures in AWS D1.4, which are based on the steel chemistry, must be carefully adhered to in order to produce welds that develop the strength of the steel. If the carbon equivalent or the sulfur or phosphorus content is too high, the steel may not be weldable. Because the chemistry of reinforcing steel conforming to ASTM A615, A616 and A617 is not controlled and is often unknown, a quality weld that can develop the strength of the steel is not always possible. ASTM A706 steel has controlled chemistry and is always weldable. Welded splices are required to be able to develop only 125 percent of the specified yield strength of the spliced bars. However, because A615, A616, A617 as well as A706 bars can have actual yield strengths in excess of 125 percent of the specified yield strength, a code-conforming welded splice may fail before the spliced bars can yield. This would compromise the inelastic deformability of a structural member. Therefore, welded splices are prohibited within the potential plastic hinge region of members in structural systems in buildings assigned to the higher seismic design categories.

2108.4

ACI 530/ASCE 5/TMS 402, Section 3.3.3.5, maximum areas of flexural tensile reinforcement. Type 1 mechanical splices are not permitted to be used within a plastic hinge zone or within a beam-column joint of intermediate or special reinforced masonry shear walls. However, Type 2 mechanical splices are permitted in any location within a member. Type 1 splices may not be able to resist the stress levels that develop within the yielding region. Type 2 splices are required to develop the specified tensile strength of the spliced bars. ACI recommends that good detailing precludes the use of splices within regions subjected to potential yielding. However, if splices cannot be avoided, the designer should investigate and document the force-deformation characteristics of the spliced bar and the ability of the splice to meet the expected inelastic demand. See Sections 21.2.6.1 and 12.14.3.2 of ACI 318 for discussion of the types of mechanical splices.

This modification to the standard is based on a revision to the *2000 NEHRP Recommended Provisions*. Reinforcing steel is predominantly produced from remelted steel scrap. Because it is difficult to control the strength of the scrap steel, the resulting products tend to have a strength considerably higher than the specified yield strength. This is similar to the situation that occurred in structural steel where the actual yield strength can be much greater than the specifed yield strength. Because there is a lower limit but no upper limit on the yield strength other than for ASTM A706 bars, most reinforcing steel has a higher yield point than specified. Recent testing by the California Department of Transportation (CALTRANS) has shown that the over-strength can be as much as 60 percent over the specified strength. Splices in reinforcing steel used in the lateral force resisting system within plastic hinge zones and in beam-column joints are subjected to high seismic strains. In order to achieve adequate performance, these splices must be able to develop the full strength of the reinforcing. Therefore, Type 1 splices are prohibited within a plastic hinge zone or within a beam-column joint, but Type 2 splices are allowed because they are required to develop the specified tensile strength of the bar. Cyclic tests by CALTRANS of current splices meeting only the 125 percent requirement show that, in many cases, although the splices meet the 125 percent criterion, they cannot survive several excursions in the post yield range imposed by cyclic testing. Note that the 1999 edition of ACI 318 uses the terminology *mechanical splices*, which replaced *mechanical connections* as used in previous editions. This modification of the standard is intended to provide consistency between the IBC and ACI 318 with respect to mechanical splices.

Section 2109 *Empirical Design of Masonry*

The empirical design procedure for masonry is a prescriptive method of sizing and proportioning masonry structures using rules and formulas that were developed over many years. The procedure is based on experience and predates the engineering design methods. The empirical method was developed for use in smaller buildings with more interior walls and stiffer floor systems than are commonly built today. Gravity loads are assumed to be approximately centered on bearing walls and foundation piers, and the effects of reinforcement is neglected.

Section 2109 contains empirical design requirements for masonry that are nearly identical to those contained in Chapter 5 of ACI 530/ASCE 5/TMS 402. This section also contains requirements for adobe masonry construction that are similar to those contained in the *Standard Building Code* and the *Uniform Code for Building Conservation* (UCBC).

2109.1.1 **Limitations.** The restrictions on the use of empirical design were expanded in the 2006 IBC in order to make the IBC consistent with the 2005 edition of ACI 530/ASCE 5/TMS 402. The limitations on empirical design of masonry are generally based on the level of lateral load risk and are therefore driven by seismic design category, basic wind speed and building height. If a building structure exceeds the limitations prescribed in the section, then the building or structure must be designed in accordance with the engineering provisions covered in Section 2107 for allowable stress design or Section 2108 for strength design. Masonry foundation walls may be constructed in accordance with the prescriptive masonry foundation wall provisions covered in Chapter 18 in Section 1805.5.

Item 6 of this section is independent of the seismic design category or wind speed but rather imposes a restriction that reflects the underlying basis of empirical design. It restricts the use of the empirical design method to walls and piers that are loaded in such a way that the resultant of gravity loads lies within the middle third of wall thickness or pier cross section.

The section prohibits the use of empirical design for AAC masonry, which must be designed in accordance with the strength design procdure in Appendix A of the 2005 edition of the ACI 530/ASCE 5/TMS 402 standard.

Section 2110 *Glass Unit Masonry*

The section covers the empirical requirements for nonload-bearing glass unit masonry elements used in exterior or interior walls. Glass block cannot be used in fire walls, party walls, fire barriers or fire partitions, or for load-bearing construction.

Glass unit masonry provisions are similar to those contained in the UBC with additional updates and revisions based on the 1999 MSJC Code. See also the TMS commentary[7] and MSJC commentaries.[5, 6]

Section 2111 *Masonry Fireplaces*

The section covers masonry fireplaces and their foundations constructed of concrete or masonry. Table 2111.1 and Figure 2111.1, which summarizes the fireplace and chimney requirements, were deleted from the 2006 IBC because they are out of date and inconsistent with the provisions in the code. Note that Section 2111 covers requirements for masonry fireplaces, not masonry chimneys. Masonry chimney requirements are covered in Section 2113.

The fireplace and chimney provisions in Sections 2111 and 2113 originated with the three legacy model codes but were updated and revised to be consistent with the corresponding provisions in the IRC. Users of the 1997 UBC will find consistent seismic reinforcing and anchorage requirements (from UBC Section 3102.4.3 in IBC Sections 2111.3 and 2113.3), hearth requirements, firebox minimum dimensions, clearance requirements, flue sizing charts, etc. The format of the provisions, however, is dramatically different than the UBC, and numerous changes were made to achieve consistency between the IBC and the IRC.

Section 2112 *Masonry Heaters*

This section covers requirements for masonry heaters as defined in Section 2112.1. Section 2112 in the 2006 IBC was revised in its entirety. The purpose of the revisions was to coordinate the masonry heater provisions in the IBC and IRC and to reference the current ASTM and UL standards.

Section 2113 *Masonry Chimneys*

Masonry chimneys constructed of concrete or masonry. The section covers seismic anchorage and reinforcing, footing support and general construction requirements for masonry chimneys. The provisions are primarily based on the fireplace and chimney requirements of the three legacy model codes but were subsequently updated and revised in order to be consistent with the corresponding provisions in the IRC.

REFERENCES

[1]International Conference of Building Officials, *Uniform Building Code*, Whittier, CA, 1997, copyright held by International Code Council.

[2]Building Officials and Code Administrators International, *The BOCA National Building Code*, Country Club Hills, IL, 1993, 1996, 1999, copyright held by International Code Council.

[3]Southern Building Code Congress International, *Standard Building Code*, Birmingham, AL, 1994, 1997, 2000, copyright held by International Code Council.

[4]NEHRP, *NEHRP (National Earthquake Hazard Reduction Program) Recommended Provisions for Seismic Regulations for New Buildings and Other Structures*, Building Seismic Safety Council, Washington, DC, 1997 (2000).

[5]ACI 530/ASCE 5/TMS 402, *Building Code Requirements for Masonry Structures*, American Concrete Institute, Farmington Hills, MI, 2005.

[6]ACI 530.1/ASCE 6/TMS 602 , *Specification for Masonry Structures*, American Concrete Institute, Farmington Hills, MI, 2005.

[7]TMS, *Commentary to Chapter 21, Masonry, of the Uniform Building Code*, The Masonry Society.

[8]*The BOCA National Building Code—Commentary Volume 2*, International Code Council, Washington, DC, 1999.

BIBLIOGRAPHY

ASTM C 270-96a, *Standard Specification for Mortar for Unit Masonry*, American Society for Testing and Materials, West Conshohocken, PA, 1996.

ICC, *Handbook to the Uniform Building Code: An illustrative commentary*, International Code Council, Washington, DC, 1998.

ICC, *Proposed Changes to the Final Draft of the International Building Code*, International Code Council, Washington, DC, January 1999.

ICC, *1998 Proposed Changes to the First Draft of the International Building Code*, International Code Council, Washington, DC, April 1998.

ICC, *2003 International Building Code Update Resource Handbook*, International Code Council, Washington, DC, November, 2003.

ICC, *2006 IBC Code Change Resource Collection*, International Code Council, Washington, DC, June, 2006.

SEAOC, *Recommended Lateral Force Requirements and Commentary*, Structural Engineers Association of California, Sacramento, CA, 1996 (1999).

UBC-IBC Structural Comparison & Cross Reference, International Code Council, Washington, DC, 2000.

CHAPTER

22

STEEL

This chapter essentially gives the user a roadmap to the various design standards that apply to steel construction. Previous editions of the model codes often only transcribed parts of the design standards, causing confusion for code users. The 1997 *Uniform Building Code* (UBC) was the first edition of the UBC to begin the transition to adopting national consensus standards by reference similar to the *National Building Code* and the *Standard Building Code*. During the IBC code development process, this trend continued to the point where Chapter 22 of the 2006 IBC consists of less than three pages.

Section 2201 *General*

2201.1 Scope. Chapter 22 covers requirements for quality, design, fabrication and construction of steel structures including structural steel, joists, cable structures, storage racks and cold-formed steel.

Section 2202 *Definitions*

2202.1 Definitions. The definitions in Chapter 22 are the minimum necessary to apply the provisions in the code. An extensive code change during the 2002 code cycle added perforated shear wall provisions for cold-formed steel-framed walls, which resulted in several new definitions being added to Section 2202.1 of the 2003 IBC. These definitions were deleted in the 2006 IBC because the perforated shear wall provisions were replaced by the referenced standard, AISI *Standard for Cold-Formed Steel Framing-Lateral Design* (*AISI-Lateral*). The 2006 IBC only has three definitions: cold-formed steel construction, steel joist and structural steel member. All other definitions were deleted because they are covered in the various referenced steel standards.

It should be noted that the definition of light-frame construction in Section 202 pertains to repetitive wood framing but also includes light-gage steel framing members. This is important because in some cases the code has exceptions that apply to buildings of light-frame construction, which includes light-gage steel. For example, Section 1805.4.2 permits concrete footings supporting walls of light-frame construction to be designed in accordance with Table 1805.4.2, which is the familiar prescriptive foundation table that gives the width and thickness of continuous footings based on the number floors supported.

The 2000 and 2003 editions of the IBC included Section 2202.2, which defined engineering nomenclature (symbols) used in the code. The section was deleted from the 2006 IBC because engineering nomenclature and notation are defined in the various referenced steel standards.

Section 2203 *Identification and Protection of Steel for Structural Purposes*

Identification. All structural steel used for load carrying purposes must be properly identified in order to determine conformance with the appropriate specification and grade. Unidentified steel must be tested for conformity to the applicable standard. **2203.1**

Protection. Protection of structural steel and cold-formed steel structural members or panels from corrosion is required. Protection may be by painting or galvanizing. Painting of structural steel is required to comply with the requirements of AISC 360. **2203.2**

Section 2204 *Connections*

The 2003 IBC had separate sections covering requirements for bolting and welding. In the 2006 IBC, bolting and welding are consolidated into one section on connections.

Welding. This section does not directly reference the *AWS Structural Welding Code*. Rather, it requires that welding be performed in accordance with the applicable steel specification used for the design. The applicable standards are referenced in Section 2205 for structural steel, Section 2206 for steel joists, Section 2207 for steel cable structures, Section 2209 for cold-formed steel and Section 2210 for cold-formed steel light-framed construction. This is done because the various referenced standards either reference different editions of an AWS D1.1 specification, reference a different AWS specification, or the referenced standard includes its own welding requirements such as in the Steel Joist Institute (SJI) standards. For example, 1989 AISC ASD specification references AWS D1.1-88; the 1993 AISC LRFD specification references AWS D1.1-92; and the 1996 AISI specification for cold-formed steel references AWS D1.3-89. **2204.1**

This approach has both positive and negative impacts. Positive impacts result from a reduction in confusion as to what welding requirements apply. However, the negative impact is that necessary changes and the results of the latest research are not always incorporated respectively into the referenced welding specification. For example, the valuable lessons on structural welding learned from the SAC Joint Venture investigation[1,2] after the Northridge earthquake are not required by either the 1989 AISC ASD or the 1993 AISC LRFD.

The 2006 IBC references AISC 360-05, *Specification for Structural Steel Buildings*, which is a unified standard that replaced and combined the AISC ASD and AISC LRFD specifications into one document. The 2006 IBC also references AISC 341-05, *Seismic Provisions for Structural Steel Buildings*, including Supplement No. 1 dated 2006. The 2005 editions of AISC 360 and AISC 341 both reference the 2004 edition of the AWS D1.1 welding code.

Special inspection of welding is required as indicated in this section and in Section 1704.

Bolting. The philosophy of this section is similar to that used for welding requirements, as the design standards reference different editions of the bolting standards. For example, the 1989 AISC ASD specification references the *Specification for Structural Joints Using ASTM A325 or A490 Bolts* by the Research Council for Structural Connections (RCSC), whereas 1993 AISC LRFD specification references *Load and Resistance Factor Design Specification for Structural Joints Using A325 or A490 Bolts* by RCSC. **2204.2**

Bolt standards are not referenced directly in the code itself, but are referenced in the particular design standard. The applicable standards are referenced in Section 2205 for structural steel, Section 2206 for steel joists, Section 2209 for cold-formed steel and Section 2210 for cold-formed steel light-framed construction.

The 2006 IBC references the 2005 editions of AISC 360 (*Specification for Structural Steel Buildings*) and AISC 341 (*Seismic Provisions for Structural Steel Buildings*) which both reference the 2004 edition of the RSCS *Specification for Structural Joints Using ASTM A325 or A490 Bolts*.

Special inspection of the installation of high-strength bolts is required as indicated in this section and in Section 1704. Special inspection is required for installation and tightening of high-strength bolts regardless of the tightening method or whether the bolts are slip critical or simple shear-bearing connections. The only case where special inspection may not be required is where the connections are *designed* to use ordinary mild steel (ASTM A307) bolts but the engineer *specifies* high-strength bolts. This is similar to the exception for special inspection for concrete footings in Section 1704.4, which states, "the *structural design* of the footing is based on a specified compressive strength, f'_c, no greater than 2,500 pounds per square inch, regardless of the compressive strength *specified* in the construction documents or used in the footing construction."

Section 2205 *Structural Steel*

Structural steel must be designed in accordance with the referenced standard, AISC 360-05, which supersedes the previous LRFD, ASD and HSS specifications. AISC 360 **provides requirements for the design and construction of structural steel buildings and other structures, and incorporates *both* allowable stress design, and load and resistance factor design, methods. The design provisions for single angles and hollow structural sections are also included in the standard.** In addition to the new standard, AISC publishes AISC 325-05, the thirteenth major update of the AISC *Steel Construction Manual*. The manual combines the Allowable Stress Design, and Load and Resistance Factor Design methods and replaces both the 9th Edition ASD Manual and the 3rd Edition LRFD Manual, as well as much of the *HSS Connections Manual*. The manual contains the following specifications, codes and standards: 2005 AISC 360 *Specification for Structural Steel Buildings,* **2004 RCSC *Specification for Structural Joints Using ASTM A325 or A490 Bolts* and 2005 AISC *Code of Standard Practice for Steel Buildings and Bridges.***

2205.2 **Seismic requirements for steel structures.** The 2006 IBC references AIC 341—05, *Seismic Provisions for Structural Steel Buildings*, including Supplement No. 1 dated 2006. The AISC seismic provisions incorporate the latest requirements based on the NEHRP provisions[3] and the significant research results of the SAC Joint Venture investigation[1,2], which was initiated as a result of the extensive damage to welded steel moment frames in the 1994 Northridge earthquake. AISC 341 contains an extensive commentary and bibliography. The reader is referred to the commentary in AISC 341 for a detailed discussion on the seismic design provisions for steel buildings.

The AISC *Seismic Design Manual* includes printed versions of AISC 341-05 and AISC 358-05. AISC 358 covers prequalified connections for special and intermediate steel moment frames.

The requirements for design of structural steel structures resisting seismic forces are based on the seismic design category of the building. Requirements for Seismic Design Category A, B or C are in Section 2205.2.1, and requirements for Seismic Design Category D, E or F are in Section 2205.2.2. For a detailed discussion on determination of seismic design category, see commentary under Section 1613.

2205.2.1 **Seismic design category A, B or C.** Steel structures assigned to Seismic Design Category A, B or C may be of any construction permitted in Section 2205, and may be designed as "structural steel systems not specifically detailed for seismic resistance" using an *R* factor of 3. See Item H of Table 12.2-1 of ASCE 7. It should be noted that this category does not include cantilever column systems, which are covered under Item G in Table 12.2-1. As an alternate, structures in Seismic Design Category A, B or C may also be designed and detailed in accordance with AISC 341 using the appropriate *R* factor for the building system involved. In a sense, this is trading ductile detailing for design force. Using systems with higher *R* values produces lower lateral forces but requires more ductile detailing to ensure ductile response characteristics. Because the 2006 IBC references ASCE 7 for seismic design, the seismic design coefficients (such as the *R* factor) for the various types of building types and structural systems are given in Table 12.2-1 of ASCE 7.

Seismic design category D, E or F. Unlike structural steel systems in Seismic Design Category A, B or C, steel structures in Seismic Design Category D, E or F are required to be designed and detailed in accordance with AISC 341 using the appropriate *R* factor for the building system involved. **2205.2.2**

Seismic requirements for composite construction. The design and construction of composite steel and concrete structural systems resisting seismic forces are required to conform to the requirements of the AISC 360 and ACI 318 standards. The appropriate *R* factor from ASCE 7 may be used where the structure is designed and detailed in accordance with AISC 341. Composite systems in buildings in Seismic Design Category B or higher are required to be designed and detailed in accordance with AISC 341. AISC 341, Part II, covers design and construction of composite steel and concrete systems and uses the LRFD procedure. **2205.3**

Seismic design category D, E or F. Composite structures are permitted to be used in Seismic Design Categories D, E and F, subject to the limitations in Section 12.2.1 of ASCE 7, which covers requirements for the various types of structural systems and limitations. Section 2205.3.1 requires substantiating evidence that demonstrates that the proposed system performance meets the intent of AISC 341, Part II. The substantiating evidence is subject to review and approval of the building official. This section requires cyclic testing for composite elements and connections that are expected to sustain inelastic deformations under anticipated seismic loads. **2205.3.1**

Section 2206 *Steel Joints*

As in the three previous legacy model codes, steel joists are required to be designed, manufactured and tested in accordance with the specifications published by the Steel Joist Institute (SJI).

Joists are used primarily as gravity-load-carrying members. When joists are incorporated into the lateral-force-resisting system as collectors, chords and diaphragm ties, care must be taken to ensure that the joists are specifically designed and detailed to properly function as these elements. Normally, joists are designed only for gravity and wind uplift loads. A continuous load path is necessary for chords, collectors and diaphragm ties, and their connections. The typical seat on a steel joist is eccentric to the chord and is not ordinarily designed to carry the axial force and resulting moment through the seat when functioning as a collector or chord. Generally, special seats must be designed and manufactured, or the load path must be specifically directed through the chords by direct chord-to-chord connection. See Fisher[4] for methods of designing chord to chord connections.

A code change during the 2004/2005 code cycle created four new code sections within Section 2206 of the 2006 IBC that are intended to clarify the responsibilities of the registered design professional and the joist engineer/specialty structural engineer. The new sections pertain to construction drawings, calculations and certification. These new provisions require the steel joist industry to meet requirements similar to those for the pre-engineered wood truss industries. The new requirements for steel joists systems are similar to the requirements for Section 2303.4 for pre-engineered wood trusses and Section 1901.4 for precast concrete.

The new code sections in the 2006 IBC clarify the difference between joist placement and layout plans and open web steel joists and joist girder construction drawings. Joist placement and layout plans are generally recognized by the industry as not requiring an engineer's signature and seal, whereas open web steel joists and joist girder construction drawings do require an engineer's design, review and seal. The new sections clarify the responsibilities of the registered design professional and the joist engineer/specialty structural engineer. The basis of the specialty structural engineer designation is the American Society of Civil Engineer (ASCE)/Council of American Structural Engineers (CASE) Document 962 D 2003, "National Practice Guidelines for Specialty Structural Engineers." A specialty structural engineer is retained by the supplier or subcontractor who is responsible for design, fabrication and sometimes installation of engineered structural elements.

The code change proponent cited a position paper published in March of 2003 by the Steel Joist Institute (SJI) regarding signing and sealing of steel joist placement plans and pointed out that there is a potential for critical items that can only be designed and detailed by the joist engineer/specialty structural engineer to slip through the cracks because the SJI paper does not require a seal to be placed on the joist placement plans. Typically, these critical items are detailed and included on the joist placement plans but the joist engineer may not sign and seal the drawings citing the SJI position paper. In the opinion of the proponent, this creates the possibility that critical items could be designed, detailed and coordinated by designers and drafters and not by experienced and qualified professional engineers. Critical items that could be overlooked are 1) joist to joist or joist to girder connections; 2) compression chord bridging design and bridging connection details for cantilevered and uplift conditions; 3) compression chord design and detailing for conditions where the compression chord is not continuously braced; 4) special loading conditions; and 5) special configurations.

In approving the code change, the IBC structural committee felt that the code change clarifies the roles of the registered design professional and the steel joist manufacturer and will better serve the engineering community and industry.

2206.1 **General.** The design, manufacture and installation of open web steel joists and joist girders are required to conform to the one of the three Steel Joist Institute (SJI) standard specifications listed in this section: Open Web Steel Joists K-Series, Long span Steel Joists LH Series and Deep Long span Steel Joists DLH Series and Joist Girders. Seismic design must be in accordance with the provisions of Section 2205, which references ASCE 7 or Section 2210.5 for light-framed cold-formed steel construction. Section 2210.5 references the AISI *Standard for Cold-formed Steel Framing Lateral Design* (AISI-Lateral).

2206.2 **Design.** This section requires the registered design professional to indicate on the construction documents the steel joist or joist girder designations used as well as layout scheme, end support, anchorages, bridging, bridging termination connections and bearing connections that resist uplift and lateral loads. The construction documents must also include any special loads or conditions that are listed in the code section.

2206.3 **Calculations.** The steel joist/girder manufacturer is required to design the steel joists and/or steel joist girders to support the anticipated loads and load combinations prescribed by Chapter 16 and in accordance with the applicable SJI specifications. The registered design professional responsible for the design of the building may require joist/girder calculations to be prepared along with a cover letter bearing the seal and signature of the joist manufacturer's registered design professional. In addition to the standard steel joist or joist girder calculations, other items such as non-SJI standard bridging details and nonstandard connection details, field splices and joist headers are required to be submitted.

2206.4 **Steel joist drawings.** Steel joist placement plans that indicate the steel joist products specified on the construction documents are required to be submitted to the building department for review and approval. This code section is very specific as to what the joist placement plans are required to include. Note that the section specifically states that steel joist placement plans do not require the seal and signature of the joist manufacturer's registered design professional.

Certification. The steel joist manufacturer is required to submit a certificate of compliance stating that work was performed in accordance with approved construction documents and with the applicable SJI standard specifications. See also Section 1704.2.2 regarding exemption of special inspection where work is done by an approved fabricator. A certificate of compliance from the fabricator is required to be submitted to the building official stating that the work was performed in accordance with approved construction documents and with SJI standard specifications. **2206.5**

Section 2207 *Steel Cable Structures*

General. See the commentary in ASCE 19 for a detailed explanation of the structural requirements for cable structures. **2207.1**

Seismic requirements for steel cable. The seismic modifications to ASCE 19 that appear in this section originated with the NEHRP *Provisions*.[2] **2207.2**

Section 2208 *Steel Storage Racks*

The 2006 IBC references the 2002 edition of the Rack Manufacturers Institute (RMI) Specification for Design, Testing and Utilization of Industrial Steel Storage Racks. Use of the RMI specification is limited to the specific types of storage racks listed in this code section, which are the typical low to moderate height industrial racks mounted at grade as used in warehouse and warehouse retail applications. Seismic design of other types of racks not specifically included in the RMI specification must be designed in accordance with Section 15.5.3 of ASCE 7.

During the development of the 2003 NEHRP *Provisions*, the issue of seismic safety of steel storage racks was raised because of the recent proliferation of *big box retail* stores. As a result, FEMA funded the report, *Seismic Considerations for Steel Storage Racks Located in Areas Accessible to the Public* (FEMA 460). This resource is available from the Building Seismic Safety Council at www.bssconline.org.

Section 2209 *Cold-formed Steel*

General. This section references the 2001 edition of the *North American Specification for the Design of Cold-formed Steel Structural Members*, including 2004 Supplement (NAS—01) published by the American Iron and Steel Institute (AISI). The 1996 edition of the specification combined LRFD and ASD into one specification. The 1999 edition, published as Supplement No. 1 to the 1996 edition, further refined the specification and added some new provisions. Subsequently, a cooperative effort by the AISI Committee on Specifications for the Design of Cold-Formed Steel Structural Members, the Canadian Standards Association (CSA) Technical Committee on Cold Formed Steel Structural Members (S136) and the Camara Nacional de la Industria del Hierro y del Acero (CANACERO) in Mexico developed the 2001 Edition of the North American Specification for the Design of Cold-formed Steel Structural Members. The 2004 Supplement to the 2001 **2209.1**

North American Specification is the latest edition of the standard and is referenced in the 2006 IBC.

The NAS—01 standard and 2004 supplement were developed to assist engineers in the design of cold-formed steel structures in both commercial and residential construction. The 2004 supplement includes updates to the 2001 specification and incorporates the new provisions for the *Design of Cold-Formed Steel Structural Members Using the Direct Strength Method*. The direct strength method provides an alternative design approach for frequently used cold-formed sections such as channels and zees. The method permits engineers to design using gross properties and elastic buckling behavior of cross-sections to predict strength. With computer programs, this design method can simplify cold-formed steel design and better predict member behavior.

Cold-formed stainless steel structural systems are required to be designed in accordance with ASCE 8.

2209.2 Composite slabs on steel decks. Composite slabs of concrete and steel deck are required to be designed and constructed in accordance with ASCE 3. Composite slabs covered by this section are those constructed of concrete over metal deck forms where the slab acts compositely through the mechanism of mechanical interlock of the aggregate with deformations in the deck and bond to the deck. The steel deck acts like reinforcing steel in the positive moment region of the span and reduces the need for additional reinforcing bars or mesh. Where the loads are high, mechanical shear connectors, such as welded studs, are required.

Section 2210 *Cold-formed Steel Light-framed Construction*

During the code development process from the 2000 edition to the 2006 edition of the IBC, AISI developed several national consensus standards for cold-formed steel used in light-framed construction. The term *light-framed construction* is defined in IBC Section 202 as a type of construction whose vertical and horizontal structural elements are primarily formed by a system of repetitive wood or light-gage steel framing members. The AISI standards referenced in this section are:

- *North American Specification for the Design of Cold-formed Steel Structural Members, including 2004 Supplement* (NAS—01)
- *Standard for Cold-formed Steel Framing—General Provisions* (General—04)
- *Standard for Cold-formed Steel Framing—Header Design* (Header—04)
- *Standard for Cold-formed Steel Framing—Truss Design* (Truss—04)
- *Standard for Cold-formed Steel Framing—Wall Stud Design* (WSD—04)
- *Standard for Cold-formed Steel Framing—Lateral Design* (Lateral—04)
- *Standard for Cold-formed Steel Framing—Prescriptive Method for One- and Two-family Dwellings, including 2004 Supplemen*t (PM—01)

2210.1 General. This section references AISI-General and AISI-NAS standards for general requirements related to design, installation and construction of cold-formed steel framing. The standards give general requirements for residential and commercial construction, and prescriptive and engineered design. Examples include member identification and labeling through basic tolerances such as in-line framing. These are the base standards from which the other three framing standards discussed below were derived.

Headers. This section references AISI Header—04 for the design and installation of cold-formed steel headers. The two-part standard gives design professionals the required tools to design efficient built-up and L-shaped headers. **2210.2**

Trusses. This section references AISI Truss—04 for the design and installation of cold-formed steel trusses. The standard provides technical information and specification on cold-formed steel truss construction and applies to the design, quality assurance, installation and testing of cold-formed steel trusses. **2210.3**

Wall studs. This section references AISI WSD—04 for the design and installation of cold-formed steel wall studs. The standard provides requirements for the design and installation of structural and nonstructural walls in buildings. **2210.4**

Lateral design. This section references AISI Lateral—04 for the design of cold-formed steel light-framed shear walls and diaphragms used to resist wind and seismic loads. The standard contains design requirements for shear walls, diagonal strap bracing and diaphragms. **2210.5**

Prescriptive framing. This section references AISI PM—01 for the design and construction of cold-formed steel framing in detached one- and two-family dwellings and townhouses up to two stories in height. This standard is an updated version of previous CABO and IRC building code requirements as well as the previous 2000 edition of the standard. It incorporates latest developments such as the L-Header and an efficient design procedure for built-up headers. **2210.6**

REFERENCES

[1]SAC-95-02, *INTERIM GUIDELINES: Evaluation, Repair, Modification and Design of Steel Moment Frames*, SAC Joint Venture—A Partnership of the Structural Engineers Association of California, Applied Technology Council, and California Universities for Research in Earthquake Engineering, Sacramento, CA, 1995 (also known as FEMA 267).

[2]SAC-96-03, *INTERIM GUIDELINES: Advisory No. 1*, SAC Joint Venture—A Partnership of the Structural Engineers Association of California, Applied Technology Council, and California Universities for Research in Earthquake Engineering, Sacramento, CA, 1996 (also known as FEMA 267A).

[3]NEHRP, *NEHRP (National Earthquake Hazard Reduction Program) Recommended Provisions for Seismic Regulations for New Buildings and Other Structures*, Building Seismic Safety Council, Washington, DC, 1997 (2000).

[4]Fisher, James A., West, Michael, and Van de Pas, Julian, *Designing with Steel Joists, Joist Girders, and Steel Deck*, Vulcraft/Nucor, 1991.

BIBLIOGRAPHY

AISC LRFD-98, *Load and Resistance Factor Design Specification for Structural Steel Buildings*, American Institute for Steel Construction, Inc., Chicago, IL, 1998.

AISC HSS, *Specification for the Design of Hollow Structural Sections*, American Institute for Steel Construction, Inc., Chicago, IL, 1998.

ICC, *The BOCA National Building Code—Commentary Volume 2*, International Code Council, Washington, DC, 1999.

ICC, *Handbook to the Uniform Building Code: An illustrative commentary*, International Code Council, Washington, DC, 1998.

ICC, *Proposed Changes to the Final Draft of the International Building Code*, International Code Council, Washington, DC, January 1999.

ICC, *1998 Proposed Changes to the First Draft of the International Building Code*, International Code Council, Washington, DC, April 1998.

ICC, *2003 International Building Code Update Resource Handbook*, International Code Council, Washington, DC, November, 2003.

ICC, *2006 IBC Code Change Resource Collection,* International Code Council, Washington, DC, June, 2006.

SEAOC, *Recommended Lateral Force Requirements and Commentary*, Structural Engineers Association of California, Sacramento, CA, 1996 (1999).

SJI-1994, *Standard Specification, Load Tables, and Weight Tables for Steel Joists and Joist Girders*, Steel Joist Institute, Myrtle Beach, SC, 1994.

UBC-IBC Structural Comparison & Cross Reference, International Code Council, Washington, DC.

CHAPTER

23

WOOD

Chapter 23 in the 2000 IBC was an amalgamation of the 1997 *Uniform Building Code* (UBC), the 1996 *National Building Code* (NBC) and the 1997 *Standard Building Code* (SBC), with the addition of seismic provisions from the 1997 National Earthquake Hazard Reduction Program (NEHRP) *Recommended Provisions for Seismic Regulations for New Buildings and Other Structures*[1]. The provisions were selected from these source documents based on technical merit, necessity and clarity. New provisions were added where needed provisions did not exist.

The use of seismic design categories instead of seismic zones was a significant change from previous model codes. Seismic design categories are determined by a combination of use of the structure, as defined by the Seismic Use Group, and the potential seismic hazard at the site. The Seismic Use Group designation was deleted in the 2006 IBC, the Occupancy Category being used instead. See Table 1604.5 for Occupancy Category descriptions. The seismic hazard at the site is a combination of the seismic risk associated with potential ground motion and the site soil classification. See the discussion under Section 1613.5.2. The seismic zone designations in the legacy codes depend only on the location of the structure relative to areas of seismic ground motion hazard. See the commentary in Section 1613 for a more detailed discussion of earthquake loads, site class and determination of seismic design category.

Section 2301 *General*

2301.1 **Scope.** Chapter 23 covers materials, design construction and quality of wood buildings and structures.

The chapter is formatted into eight major sections that follow a logical format:

1. Section 2301—General
2. Section 2302—Definitions
3. Section 2303—Minimum Standards and Quality
4. Section 2304—General Construction Requirements
5. Section 2305—General Design Requirements for Lateral-Force-Resisting Systems
6. Section 2306—Allowable Stress Design
7. Section 2307—Load and Resistance Factor Design
8. Section 2308—Conventional Light-Frame Construction

In the 2003 IBC, the code user is referred to the 1997 *National Design Specification for Wood Construction* (NDS), published by the American Forest and Paper Association (AF&PA)[2] for allowable stress design (ASD), and to ASCE 16-95, *Standard for Load and Resistance Factor Design (LRFD) for Engineered Wood Construction*[3] for strength design. The AF&PA NDS is referenced in Section 2306 along with other national standards. In the 2003 IBC, ASCE 16-95 is referenced in Section 2307 for LRFD. One significant change to the 2006 IBC is that both Sections 2306 and 2307 now reference the 2005 edition of the AF&PA NDS because it is a dual format standard that includes both ASD and LRFD procedures.

Wood products addressed in Chapter 23 include boards, dimension lumber, posts, timbers, glued-laminated members, wood structural panels including plywood, composite panels, oriented strand board, particleboard, fiberboard, hardboard, prefabricated wood I-joist, structural composite lumber, laminated veneer lumber and parallel strand lumber. The chapter contains material specifications, quality requirements and design provisions, as well as empirical and prescriptive provisions for wood-frame construction.

Wood and wood-based products are also regulated in other chapters of the code. For example, Chapter 14 contains provisions for weather coverings for walls, as well as provisions for veneers and exterior trim. Wood and wood-based products used in fire-resistance-rated assemblies must also comply with Chapter 7. Wood and wood-based products used as interior finish on walls, ceilings and floors must comply with Chapter 8. Wood roof coverings are regulated in Chapter 15.

2301.2

General design requirements. There are three basic design methods permitted for wood structures. Allowable stress design in accordance with Section 2306, load and resistance factor design in accordance with Section 2307 and prescriptive conventional light-frame construction in accordance with Section 2308. Regardless of the design method used, the general design and construction requirements in Section 2304 apply as well. As an alternate to the prescriptive conventional construction provisions of Section 2308, an exception was added to the 2003 IBC that allows wood-frame buildings to be designed in accordance with the AF&PA *Wood Frame Construction Manual* (WFCM).

The first two design methods, ASD under Section 2306 and LRFD under Section 2307, are engineering methods, and the third is a prescriptive method. Engineering methods require the structure to be designed and detailed to resist all the applicable loads prescribed in Chapter 16. In contrast, the prescriptive provisions are essentially a collection of rules that anyone may follow without calculating any loads. If all of the rules are followed, the resulting structure is deemed to comply with the intent of the code. See Section 2308 for further discussion of the prescriptive conventional wood-frame construction provisions.

Prior to the 2006 IBC, the ASD method was based on NDS and the LRFD method was based on ASCE 16-95. The 2006 IBC references the 2005 NDS which is a dual format standard that contains both ASD and LRFD procedures and replaces the two other standards. The three design methods are discussed in more detail below.

1. **Allowable stress design.** Allowable stress design uses the load combinations of Section 1605.3 for the determination of strength. The designer has two options for ASD load combinations, the basic load combinations in Section 1605.3.1, or the alternative basic load combinations in Section 1605.3.2. It should be noted that all deformations and drifts from seismic load effects are determined using strength level forces in accordance with the requirements of Section 12.12 of ASCE 7. The seismic load effect, E, is not multiplied by 0.7 (or divided by 1.4) for determination of deformation and drift. The special seismic load combinations of Section 1605.4 are used for collector design for all structures assigned to Seismic Design Category B through F, except for wood-frame buildings braced entirely by light-frame shear walls. See discussion of Section 1605.4 for general application of the special seismic load combinations. Refer to Section 12.10.2.1 of ASCE 7.
2. **Load and resistance factor design.** The AF&PA/ASCE 16-95 *Standard for Load and Resistance Factor Design for Engineered Wood Construction* was first added to the 1997 UBC as a recognized standard. AF&PA/ASCE 16 was then incorporated into the 2000 IBC as a referenced standard in Section 2307. In the 2006 IBC, the LRFD procedure is included as part of the 2005 edition of the NDS; therefore, this section references AF&PA NDS instead. The applicable load combinations for strength design and LRFD are in Section 1605.2. Note that wood structures designed using the LRFD procedure are subject to the same general and lateral force design provisions as structures designed using ASD. See the 2005 NDS for LRFD requirements, strength properties and design values.
3. **Conventional light-frame wood construction.** The prescriptive conventional construction provisions are in Section 2308. Perhaps the most important aspect of the prescriptive conventional construction provisions are the restrictions and

limitations in Section 2308.2. A building or any portion of a building that does not conform to the limitations must be designed to resist all applicable loads of Chapter 16 in accordance with one of the engineering methods. See also Sections 2308.1.1 and 2308.4 for design of portions and elements. As with the engineering methods, buildings designed by the prescriptive conventional construction provisions of Section 2308 are also required to comply with the general construction requirements in Section 2304.

2301.3 Nominal sizes. Nominal lumber sizes, for example, 2 inches by 4 inches, are the sizes usually referred to or specified in this chapter. Actual or net dimensions, which are less than nominal dimensions, are used in structural calculations to determine member section properties, actual stresses and strength properties. The nominal and actual sizes of dimension lumber are established by Department of Commerce (DOC) Voluntary Product Standard PS 20, *American Softwood Lumber Standard*. The current edition referenced in the 2006 IBC is DOC PS 20-99, although DOC PS 20-05 is currently available from the National Institute of Standards and Technology (NIST). The nominal dressed sizes and section properties for sawn lumber and glued laminated timber are given in Tables 1A, 1B, 1C and 1D of the NDS. See Section 2302 for the definition of nominal size lumber.

Section 2302 *Definitions*

Definitions are used to clarify terms used in this chapter. Many of the definitions in this section are to clarify terms used in lateral-force-resisting systems.

New definitions in the 2006 IBC for prefabricated wood I-joist, structural composite lumber, laminated veneer lumber, parallel strand lumber and some modifications to the definition of composite panels, oriented strand board and plywood were made to coordinate the terms in the IBC with the current AF&PA NDS.

COLLECTOR. The collector collects shear from the diaphragm and delivers it to vertical lateral force resisting elements such as shear walls. The term *drag strut* is a colloquial expression that means *collector*. The IBC recognizes both horizontal and sloped (or nearly horizontal) diaphragms. The definition of collector applies to both horizontal and sloped diaphragms. The word *horizontal* is used to differentiate from vertical lateral force resisting elements such as a shear wall. In the legacy codes, the terms *horizontal diaphragm* and *vertical diaphragm* were used to describe diaphragms and shear walls. The new terminology in the IBC is to simply use the terms *diaphragm* and *shear wall*. For example, a code change in the 2006 IBC replaced the term *vertical diaphragm* with *shear wall* in Table 2308.9.3(4).

DIAPHRAGM, UNBLOCKED. A diaphragm is a horizontal or nearly horizontal (sloped) structural element that transmits lateral forces to the vertical resisting elements (walls or frames) of the lateral-force-resisting system. An unblocked diaphragm has edge nailing at the supported edges only. In an unblocked diaphragm the continuous panel joint is unblocked. In a blocked diaphragm all sheathing panel edges are supported by framing members or solid blocking members. Blocked diaphragms have continuity between the sheathing panel edges and therefore have significantly less deflection than unblocked diaphragms because of the stiffness developed by the continuity at the blocked panel edges.

The definition of a diaphragm existed in Chapter 23 of the 2000 IBC but not in Chapter 16. In the 2003 IBC the definition of a diaphragm was deleted from Chapter 23 and added to Chapter 16 along with the various types and elements of diaphragms. Chapter 23 of the 2006 IBC only has a definition of unblocked diaphragm. See Section 1602 for additional definitions pertaining to diaphragms. It should be noted that the definition of diaphragm in the 2006 IBC also includes horizontal bracing systems.

For the distribution of seismic or wind forces, the stiffness of the diaphragm relative to the stiffness of the vertical lateral-force-resisting elements (walls or frames) is the important parameter by which to classify the diaphragm. Refer to the commentary on Section 2305.2.5 for further discussion of diaphragm classifications.

GRADE (LUMBER). References DOC PS 20, *American Softwood Lumber Standard*, listed under DOC in Chapter 35. Voluntary Product Standard PS 20 is published by the U.S. Department of Commerce. The National Institute of Standards and Technology (NIST) administers the Department of Commerce Voluntary Product Standards program.

NATURALLY DURABLE WOOD. Includes decay resistant woods, which are redwood, cedar, black locust and black walnut, and termite resistant woods, which are redwood and eastern red cedar.

Naturally durable wood is the heartwood of a durable listed species. Note that only the heartwood of redwood and eastern red cedar are both decay and termite resistant.

PREFABRICATED WOOD I-JOIST. This definition was added to the 2006 IBC for structural members manufactured of sawn or structural composite lumber flanges and wood structural panel webs bonded together with exterior exposure adhesives in the form of an "I" cross-sectional shape.

SHEAR WALL. A shear wall is a vertical resisting element that is designed to resist lateral seismic and wind forces parallel to the plane of a wall. The definition of perforated shear wall and perforated shear wall segment were consolidated under the shear wall definition in the 2006 IBC. A perforated shear wall is a wood structural panel sheathed wall with openings that has not been specifically designed and detailed for force transfer around the openings. Perforated shear wall segments are sections of the shear wall with full-height sheathing that meets the height-to-width ratio limits specified in Section 2305.3.4. The terms *adjusted* and *unadjusted shear resistance* associated with perforated shear wall design were relocated to Section 2305.3.8.2 in the 2006 IBC.

STRUCTURAL COMPOSITE LUMBER. The definition of structural composite lumber was added to the 2006 IBC. The two types are laminated veneer lumber (LVL), which is composed of wood veneer sheet elements, and parallel strand lumber (PSL), which is composed of wood strand elements. Both LVL and PSL have wood fibers that are primarily oriented along the length of the member.

TREATED WOOD. Includes both fire-treated wood and wood treated to resist decay and termites.

WOOD STRUCTURAL PANEL. A panel manufactured from veneers, wood strands or wafers or a combination thereof bonded together with waterproof synthetic resins. In the 2003 IBC, the terms *composite panels*, *oriented strand board* (OSB) and *plywood* were added under the definition of *wood structural panel* for clarification. When used for structural purposes such as siding, roof and wall sheathing, subflooring, diaphragms and built-up members, wood structural panels must conform to the requirements for their type in DOC PS 1 or PS 2.

Section 2303 *Minimum Standards and Quality*

General. Structural lumber and wood structural panels are highly variable in strengths and other mechanical properties. The code requires that these materials (defined in the first paragraph of this section) conform to the applicable standards and grading rules specified in the code. Furthermore, the code requires that they be identified by a grade mark or be accompanied by a Certificate of Inspection issued by an approved agency. The grade mark is also required to be placed on the material by an approved agency (see labeling—Chapter 17). The proper use of a wood structural member cannot be determined unless it has been properly identified as to grade and species. Counterfeit grade stamps do occasionally appear on lumber in the field, and it is important that designers and enforcement personnel be **2303.1**

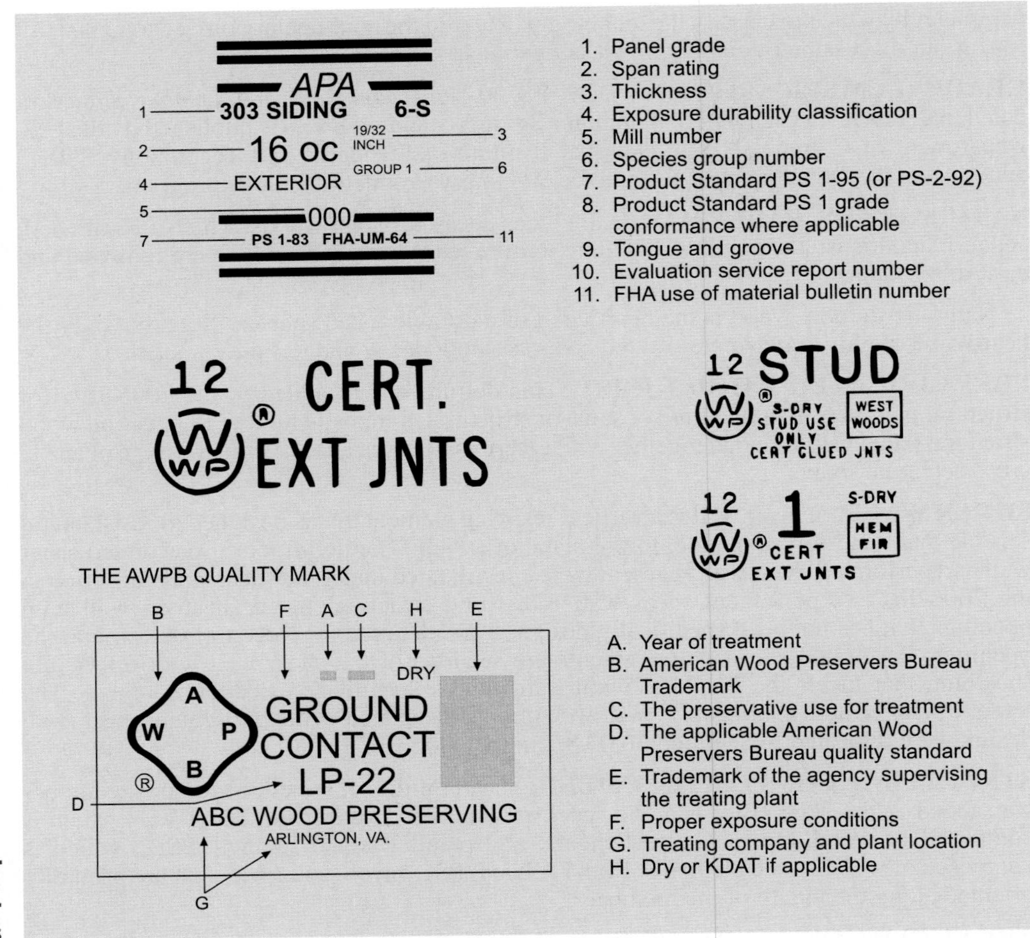

Figure 23-1
Typical lumber grade stamps

familiar with the grade-approved stamps. Examples of grade marks are shown in Figure 23-1.

Joist hangers are subject to the applicable requirements of this chapter as well as the requirements in Chapters 17 and 22.

2303.1.1 **Sawn lumber.** Lumber references the voluntary standard, *American Softwood Lumber Standard,* PS 20. Appendix R of the NDS references ASTM D 1990, ASTM D 245, ASTM D 2555, The Wood Handbook and PS20 for the classification, definition, methods of grading and development of design values for lumber. The NDS also references the various standard grading rule documents such as NLGA, NELMA, NSLB, SPIB, WCLB and RIS.

The changes in design values in the 1991 NDS for dimension lumber are based on the In-Grade Testing program conducted by the North American forest products industry. The program was carried out over an eight year period and involved the destructive testing of 70,000 pieces of lumber from 33 different species groups. A new test method standard, ASTM D4761, was also developed for the mechanical test methods used in the program. In addition, the standard practice, ASTM D1990, was developed for procedures used to establish design values for visually graded dimension lumber from test results obtained from in-grade test programs.

Approved end-jointed or edge glued lumber is presumed equivalent to solid sawed lumber of the same species and grade. The NDS permits the use of end-jointed lumber in Section 4.1.6 of that standard. When finger-jointed lumber is marked "STUD USE ONLY" or "VERTICAL USE ONLY" such lumber is limited to use where bending or tension

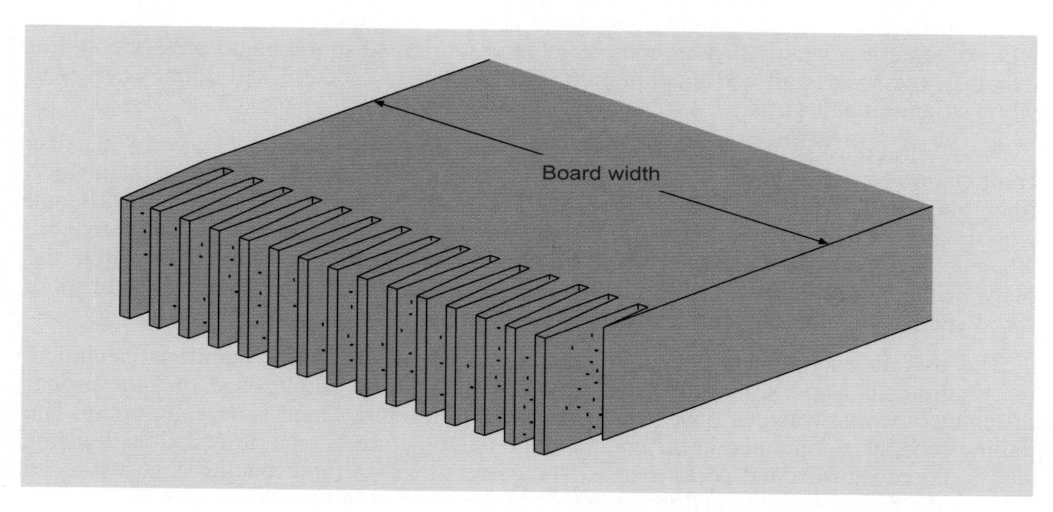

Figure 23-2 Finger-joint end joint

stresses are of short duration. The use of the term *approved* is intended to convey the need for quality control during the production of these glued products, and also to establish the qualification tests for the type of end joint used. Joints are tested for strength and for durability, and adhesive manufacturers test their products for durability. A technical discussion of finger-jointed lumber appeared in an article in the 1997 May/June issue of ICBO's *Building Standards Magazine* entitled "Wood in Construction: A Reliable System of Standards and Quality Control" by Sam W. Francis and David S. Collins. A similar technical article appeared in the SBCCI's *Southern Building Magazine* in March/April of 1995. An example of end-jointed lumber is shown in Figure 23-2, which illustrates a finger-jointed end joint.

2303.1.2

Prefabricated wood I-joists. The shear, moment and stiffness capacities of prefabricated wood I-joists must be established in accordance with ASTM D 5055. This standard also specifies that application details, such as bearing length and web openings, are to be considered in determining the structural capacity. Wood I-joists are manufactured out of sawn or structural composite lumber flanges and structural panel webs and are bonded together with exterior adhesives to form an "I" cross section. Wood I-joists are structural members typically used in floor and roof construction. The standard requires I-joist manufacturers to employ an independent inspection agency to monitor the procedures for quality assurance. The standard specifies that proper installation instructions accompany the product to the job site. The instructions are required to include weather protection, handling requirements and, where required, web reinforcement, connection details, lateral support, bearing details, web hole-cutting limitations and any special conditions.

A new definition for prefabricated wood I-joist was added to the 2006 IBC; Wood "I" joists are structural members manufactured of sawn or structural composite lumber flanges and wood structural panel webs bonded together with exterior exposure adhesives in the form of an "I" cross-sectional shape.

2303.1.3

Structural glued laminated timber. Glued laminated timbers are manufactured in accordance with ANSI/AITC A 190.1, which references several other AITC standards. ASTM D 3737 is the standard method for establishing allowable stresses for glued laminated timber. See AITC *Timber Construction Manual*.[4]

2303.1.4

Wood structural panels. Wood structural panels must conform to the voluntary product standards PS-1, for plywood, or PS-2, for oriented strand board (OSB) and other wood based structural use panels. PS-1 is the product standard for Construction and Industrial Plywood and PS-2 is the product standard for Performance Standard for Wood-based Structural-use Panels. Plywood, oriented strand board and composite panels are all wood structural panels covered under US DOC PS-1 and PS-2. Other than in the definition of wood structural panel in Section 2302, the code does not differentiate between the different types of structural panels such as composite panels, oriented strand board or plywood. For example, the shear wall and diaphragm Tables 2306.4.1, 2306.3.2 and 2306.3.1 refer to wood structural panels.

Wood structural panels include all-veneer plywood; composite panels consisting of a combination of veneer and wood-based material; and mat-formed panels that contain wood fiber only, such as OSB and waferboard. The primary distinction between OSB and waferboard is that the wood fibers in OSB are generally all oriented in the same direction, whereas the wood fibers in waferboard are oriented randomly in all directions within the plane of the board. Plywood is defined as panels made by cross laminating three or more wood veneers and joining the veneers together with glue.

The user is encouraged to obtain the referenced standards for additional technical information on wood structural panel products. It must be emphasized that the proper fastening of wood structural panels to the supporting structural frame is very important. The nailing schedules contained in the code and the manufacturer's recommendations must be strictly observed for good performance. The correct nail size and spacing is necessary to achieve the design strength and performance of the wood structural panel system.

DOC PS-1 has been developed as a guide and specification for the manufacturing of plywood intended for industrial and construction uses. DOC PS-2 is a consensus standard that has been developed as a specification of product performance for various grades of wood structural panels. Provisions in DOC PS-1 and DOC PS-2 define the requirements for structural-grade panels and give the requirements for sheathing and single floor-grade wood structural panels.

Wood structural panels manufactured in accordance with DOC PS-1 and DOC PS-2 are inspected and labeled to certify compliance by an approved agency. Examples are the American Plywood Association, Timber Engineering Company and Pittsburgh Testing Laboratories. The label identifies the grade and span ratings of the product. The inspection agencies maintain a continuous monitoring program designed to produce products that meet or exceed the applicable product standard. A number of tests, including deflection measurements, are required.

Besides the grades cited above, wood structural panels are also classified by exposure type:

1. Exterior—exterior type with a 100-percent waterproof glue line. Only the higher grades of veneers are allowed in exterior grades. Exterior rated panels are suitable for continuous exposure to weather.
2. Exposure 1—interior type made with waterproof exterior glue. Exposure 1 rated panels are suitable for extended construction exposure. The lower grades of veneers or strands used in the backs and interiors of Exposure 1 panels can affect the glue-line performance and cause delamination/deterioration during continuous exposure to weather.
3. Exposure 2—interior type made with interior glue. Exposure 2 rated panels are not suitable for exposure to weather.

Plywood is manufactured from more than 70 species of wood, which are divided into five groups in accordance with their stiffness and strength characteristics. Construction and industrial panel grades are generally identified under PS-1 in terms of face veneer grade or by a name indicating an intended use, such as APA-Rated Sturd-I-Floor. The plies may be of any species listed, except for panels designated Structural I and other special-use panels, which use only Group I species. The veneer grade defines the appearance in terms of natural, unrepaired growth characteristics (knots) and the number and size of repairs that may be made during manufacturing. The highest grades are N and A. The lowest grade is D. Grade D veneers may only be used for backs and inner plies of interior-use panels. Panels are also marked as sanded, unsanded and touchsanded.

OSB is manufactured from several species of wood. PS-2 sets forth performance requirements in terms of strength, stiffness and durability. OSB is manufactured to meet the strength, stiffness and durability requirements instead of being manufactured to a prescriptive recipe, as is plywood.

Wood structural sheathing and subflooring panels are classified as:

Rated Sheathing—Exterior

Rated Sheathing—Exposure 1

Rated Sheathing—Exposure 2

Structural I-Rated Sheathing—Exterior

Structural I-Rated Sheathing—Exposure 1

Wood structural panels intended for single-floor construction have limited voids in the inner plies in addition to the solid face veneer to prevent indentation caused by small concentrated loads. Single-floor panels intended for use under carpets and resilient flooring are classified as:

- Rated Sturd-I-Floor Exterior
- Rated Sturd-I-Floor Exposure 1
- Rated Sturd-I-Floor Exposure 2
- Rated Single Floor Exposure 2

Underlayment panels for use over subflooring are classified as:

- Underlayment Interior
- Underlayment Exposure 1
- C-C Plugged Exterior

Siding is manufactured as panels or as lapped siding and includes:

- 303-OL-MDO Exterior
- 303 Siding Exterior

Note that wood structural panels permanently exposed to weather, such as siding grades, must be exterior type. Panels that are interior type bonded with exterior glue (Exposure 1), are not allowed for siding applications or on the exposed underside of roof overhangs.

Fiberboard. Fiberboard is a smooth textured panel made up of natural fibers such as wood or cane. Fiberboard is used primarily as an insulating board and for decorative purposes but may also be used as wall or roof sheathing under the provisions of this section. Unlike particleboard, the cellulosic components of the fiberboard are broken down to individual fibers and molded to create the bond between the fibers. Other ingredients may be added during processing to provide or improve certain properties such as strength or water resistance, or to achieve specific surface finishes for decorative products. Fiberboard is used in most locations where panels are necessary, including wall sheathing, insulation of walls and roofs, roof decking, doors and interior finish. Fiberboard may not, however, be used for diaphragms. Certification of fiberboard products is performed by an approved agency. The material is generally labeled to indicate an intended use, strength values and flame resistance where applicable. **2303.1.5**

Fiberboard may be used as roof or wall insulation but is not intended for prolonged exposure to sunlight, wind, rain and snow. Where fiberboard is used as roof insulation, it must be protected with an approved roof covering to prevent water saturation and subsequent delamination and to avoid decay and destruction, caused by moisture, of the roofing bond. The section in the IBC pertaining to insulating roof decking was deleted from the 2006 IBC because fiberboard insulating roof decking is no longer manufactured.

Fiberboard is permitted without any fire-resistance treatment in the walls of all types of construction. When used in fire walls and fire separation walls, the fiberboard must be attached directly to a noncombustible base and protected by a tight-fitting, noncombustible veneer that is fastened through the fiberboard to the base. This will prevent the fiberboard from contributing to the spread of fire.

Fiberboard used in building construction must comply with ASTM C 208, *Specification for Cellulosic Fiber Insulating Board*. For several decades the fiberboard industry supported parallel ASTM and ANSI standards. During the last revision of ASTM C 208 the differences were resolved and the Board of Directors of the American Fiberboard Association voted to discontinue support of the ANSI standard in favor of ASTM C 208. Thus, the previous ANSI/AHA A194.1 standard was deleted in the 2006 IBC because the standard was withdrawn by ANSI and the fiberboard manufacturers no longer support it.

When used as structural sheathing, fiberboard must be identified by an approved agency.

2303.1.6 Hardboard. Hardboard is used as exterior siding and in interior locations for paneling and underlayment.

2303.1.7 Particleboard. Particleboard is a generic term for construction panels and products manufactured from cellulosic materials, usually wood, in the form of discreet pieces and particles as distinguished from fibers. The particles are combined with synthetic resins and other binders and bonded together under heat and pressure.

Particleboard is used as underlayment, siding and for shear walls. Particleboard used structurally for siding or shear walls must be stamped (labeled) M-S Exterior or M-2 Exterior. The “M” stands for medium density; the “2” designates the strength grade (grades range from 1 to 3); and “S” designates “special grade.” Particleboard designated M-S is medium density and has physical properties between an M-1 and M-2 designation. Both must be made with exterior glue.

Although similar in characteristics to medium-density Grade 1 particleboard, the particleboard intended for use as floor underlayment is designated “PBU” and has stricter limits on permitted levels of formaldehyde emission than those placed on Grade M particleboard. The particleboard intended for use as floor underlayment is not commonly manufactured with exterior glue, which could emit higher levels of formaldehyde than that permitted for “PSU” grade floor underlayment by ANSI A 208.1.

Particleboard underlayment is often applied over a structural subfloor to provide a smooth surface for resilient-finish floor coverings or textile floor coverings. The minimum $^{1}/_{4}$-inch thickness is suitable for use over panel-type subflooring. Particleboard underlayment installed over board or deck subflooring that has multiple joints should have a thickness of $^{3}/_{8}$ inch. Joints in the underlayment should not be located over the joints in the subflooring.

All particleboard underlayment with thicknesses of $^{1}/_{4}$ inch through $^{3}/_{4}$ inch should be attached with minimum 6d annular threaded nails spaced 6 inches on center on the edges and 12 inches on center for intermediate supports. See Table 2304.9.1.

2303.1.8 Preservative-treated wood. The applicable American Wood Preservers Association Standards are cited. Different preservative treatments are used depending on whether the wood is above ground or in contact with the ground. See commentary in Section 2304.11. ICC currently publishes a book containing all 24 AWPA standards that are referenced in both the 2003 and 2006 IBC and IRC.

2303.1.8.1 Identification. All wood required to be preservative treated by Section 2304.11 must be stamped (labeled) with the information listed in the section. There are no exceptions.

2303.1.8.2 Moisture content. Preservative treatments used in above-ground locations are water-borne salts. These salts may leach unless the wood is dried below a moisture content of 19 percent (i.e., dry) and covered with a protective material.

2303.1.9 Structural composite lumber. Structural composite lumber (SCL) is covered in the code because of its widespread use. Structural properties and strength capacities for SCL are set forth in manufacturers' literature and evaluation reports by ICC. A new definition for structural composite lumber was added to the 2006 IBC for structural members manufactured using wood elements bonded together with exterior adhesives. Two examples of structural composite lumber are laminated veneer lumber (LVL) and parallel strand lumber (PSL).

Reports prepared by approved agencies or evaluation reports published by the ICC Evaluation Service may be accepted as part of the evidence and data needed by the building official to form the basis of approval of a material or product. Such research reports supplement the resources of the building official and eliminate the need for the official to conduct a detailed analysis on every new product. The building official is not obligated to approve the evaluation reports issued by the model-code organizations, as the research reports are *advisory only* and are intended for technical reference. Note that evaluation reports are approved under the alternative materials, design and methods of construction provisions of Section 104.11.

Structural log members. A new Section 2303.1.10 was added to the 2006 IBC to provide structural capacity and grading requirements for logs used as structural members. In the past, the design of log structures could be challenging for both designer and building official because the building code contained no specific provisions that addressed structural capacity and grading requirements for logs used as structural members. All log structures require engineering, and the structural design values were approved under the alternative materials, design and methods of construction provisions in Section 104.11. This new section provides acceptable methods for establishing structural capacities of logs based on ASTM D3957 Standard Practices for Establishing Stress Grades for Structural Members Used in Log Buildings and specifies the requirement for a grading stamp or alternate certification of structural logs. At the time of this writing, the International Code Council (ICC) is in the process of developing a standard for log structures known as *Standard on the Design and Construction of Log Structures* (ICC400-2007). The goal of the new standard is to provide technical design and performance criteria that will facilitate and promote the design, construction and installation of safe reliable structures constructed of log timbers. It is intended to be used by design professionals, manufacturers, constructors, and building and other government officials, and be a referenced standard in future building codes. **2303.1.10**

Round timber poles and piles. A new Section 2303.1.11 referencing the ASTM standards for round timber poles and piles was added to the 2006 IBC to coordinate the requirement with Chapter 6 of the NDS. **2303.1.11**

Fire-retardant-treated wood. Fire-retardant-treated wood (FRTW) is defined as plywood and lumber that has been pressure impregnated with chemicals to improve its flame-spread characteristics beyond that of untreated wood. The principal objective of impregnating wood with fire-retardant chemicals is to produce a chemical reaction at certain temperature ranges. This chemical reaction reduces the release of certain intermediate products that contribute to the flaming of wood, and also results in the increased formation of charcoal and water. Some chemicals are also effective in reducing the oxidation rate for the charcoal residue. Fire-retardant chemicals also reduce the heat release rate of the FRTW when burning over a wide range of temperatures. This section gives provisions for the treatment and use of FRTW. However, the fire-retardant chemicals are generally quite corrosive. Corrosion-resistant fasteners must be used with FRTW. **2303.2**

The effectiveness of the pressure-impregnated fire-retardant treatment is determined by subjecting the material to tests conducted in accordance with ASTM E 84, with the modification that the test is extended to 20 minutes rather than 15 minutes. Under this procedure, a flame-spread index is established during the standard 10-minute test period. The test is continued for an additional 20 minutes. During this added time period, there must not be any significant flame spread. At no time must the flame spread more than $10^1/_2$ feet past the centerline of the burners. These criteria have been correlated with large-scale fire tests.

Labeling. Each piece of FRTW must be stamped (labeled). The labeling must show the performance of the material, including the 20-minute ASTM E84 test. The labeling must **2303.2.1**

state the strength adjustments, and conformance to the requirements for interior or exterior application.

The FRTW label must be distinct from the grading label to avoid confusion between the two. The grading label gives information about the properties of the wood before it is fire-retardant treated. The FRTW label gives properties of the wood after FRTW treatment. It is imperative that the FRTW label be presented in such a manner that it complements the grading label and does not create confusion over which label takes precedence.

The requirements for labeling fire-retardant-treated lumber and wood structural panels were expanded in the 2003 IBC, giving a list of specific requirements for the labeling.

2303.2.2 Strength adjustments. Several factors can significantly affect the physical properties of FRTW. These factors are the pressure treatment and redrying processes used, and the extremes of temperature and humidity that the FRTW will be subjected to once installed. The design values for all FRTW must be adjusted for the effects of the treatment and environmental conditions, such as high temperature in attic installations and humidity. The design adjustment values must be based on an investigation procedure, which includes subjecting the FRTW to similar temperatures and humidities. The procedure must be approved by the building official. The FRTW tested must be identical to that which is produced. The building official reviewing the test procedure should consider the species and grade of the untreated wood and conditioning of wood, such as drying before the fire-retardant-treatment process. A fire-retardant wood treater may choose to have its treatment process evaluated by the National Evaluation Services.

The FRTW is required to be labeled with the design adjustment values. These design adjustment values can take the form of factors that are multiplied by the original design values of the untreated wood to determine its allowable stresses, or new allowable stresses that have already been factored down in consideration of the FRTW treatment.

Two subsections were added to the 2003 IBC that prescribe specific strength adjustment requirements for treated wood structural panels and lumber. The effects of treatment and redrying after treatment and exposure to high temperatures and high humidities on the flexural properties of treated plywood and the design properties of treated lumber is required to be determined in accordance with ASTM standards D 5516 and D 5664. The section requires the manufacturer to publish allowable maximum loads and spans for service as floor and roof sheathing and the modification factors for roof framing for its particular treatment process. The section references the ASTM standard D 6841 to be used to evaluate the ASTM D 5664 test data.

2303.2.3 Exposure to weather, damp or wet locations. Some fire-retardant treatments are soluble when exposed to the weather or used under high-humidity conditions. When an FRTW product is to be exposed to weather conditions, it must be further tested in accordance with ASTM D 2898. The material is then subjected to the ASTM weathering test and retested after drying. There must not be any significant differences in the performance recorded before and after the weathering test.

2303.2.4 Interior applications. When an FRTW product is intended for use under high-humidity conditions, it must be further tested in accordance with ASTM D 3201. The material must demonstrate that when tested at 92-percent relative humidity, the moisture content of the FRTW does not increase to more than 28 percent. The label must show the test results.

2303.2.5 Moisture content. FRTW contains water-borne salts that are subject to leaching. The FRTW must be dried to the specified moisture contents after treating to minimize leaching. In addition, FRTW chemicals are quite corrosive to metal fasteners. Where the moisture content of the treated wood is too high, the corrosivity of the treated wood is even higher and contributes to greater corrosion of metal fasteners.

For wood that is kiln dried after treatment (KDAT), the kiln temperatures cannot exceed that used to dry the lumber and plywood that was submitted for the tests required by Section 2303.2.2.1 for plywood and 2303.2.2.2 for lumber.

Type I and II construction applications. Use of FRTW in Type I or II construction is limited to nonload-bearing partitions and exterior walls. **2303.2.6**

Hardwood and plywood. Hardwood plywood and decorative plywood is not used for structural purposes. The section references the American National Standard for Hardwood and Decorative Plywood. **2303.3**

Trusses. Metal-plate-connected trusses are typically constructed out of nominal dimension lumber with the metal-plate connectors placed on either the narrow or wide dimension of the lumber (4-inch by 2-inch lumber for floor trusses versus 2-inch by 4-inch lumber for roof trusses). NDS specifies the allowable design stresses for lumber, whereas while the Truss Plate Institute (TPI) *National Design Standard for Metal-Plate-Connected Wood Trusses* specifies how the allowable metal-plate design values are to be determined and how the maximum stresses in the truss elements are to be determined. **2303.4**

This section was revised in the 2006 IBC to more clearly define the requirements pertaining to metal-plate-connected wood trusses to achieve consistency with current design practice. Section 2303.4 of the 2006 IBC clarifies the provisions pertaining to metal plate connected wood trusses to be consistent with the current design practice and eliminates confusion regarding trusses submittals. The provisions include: general requirements for the design, manufacture and quality assurance of metal-plate-connected wood trusses; a clear definition of the truss designer; very specific and detailed requirements for truss design drawings; provisions for truss placement diagrams; requirements for the truss submittal package; requirements for truss member permanent bracing; and specific truss anchorage requirements. The revised section will improve the design, review and construction process related to metal-plate-connected wood truss systems.

Adequate bracing for trusses is critical. Lateral bracing requirements (e.g., brace points, bracing size, or strength and stiffness) should be specified by the truss designer. Temporary bracing should be left in place until permanent bracing is installed. All lateral bracing must be installed as assumed in the truss design so that the truss will have the same structural capacity for which it was designed. In any case, the individual truss member continuous lateral bracing locations are to be shown on the truss design drawings.

Permanent bracing must be installed in compliance with the truss industry's permanent bracing standard details that follow sound engineering practice. These details are usually provided by the component manufacturers to the building design professional as the projects are being designed. *The Building Component Safety Information (BCSI 1-03) Guide to Good Practice for the Handling, Installing & Bracing of Metal Plate Connected Wood Trusses* is a booklet produced by the Truss Plate Institute (TPI) and the Wood Truss Council of America (WTCA). It is the truss industry's new and improved guide for jobsite safety and truss performance and replaces the HIB-91 booklet from TPI.

The truss design drawings and specifications must be prepared by a registered design professional and must be provided to and approved by the building official prior to installation in the structure. It should be noted that the items listed in the code are the minimum requirements. The intent of adding the terms *truss design drawings* and *truss placement diagram* in the 2006 IBC was to minimize confusion that exists in the construction industry between a variety or terms that may mean the same thing, such as *construction documents*, *shop drawings*, etc. The term *truss placement diagram* is used by the truss industry and is very specific. These terms are intended to provide better clarity where truss submittals are concerned.

The truss design drawings are required to show permanent bracing, and the truss designer generally is the most knowledgeable concerning the required strength, stiffness and location of the bracing necessary to prevent buckling of the truss members. Bracing is necessary to resist buckling of the compression webs and chord under maximum gravity loads, as well as the uplift condition of dead load plus wind. Bracing may be necessary on the bottom chord at the first interior panel point to resist the wind uplift. The truss designer should specify either the strength and stiffness of braces, or the member size and grade of the braces (e.g., 2 by 4 DF #2) at the specified locations.

The truss designer should furnish complete calculations substantiating the size of all members and connector plate sizes. The truss calculations should indicate the combined stress index for members subjected to combined stresses from bending and axial compression and tension. The combined stress index should be less than one. The calculations are generally performed with a computer program; therefore, documentation may be required by the building official that substantiates the basis of the program used.

Section 2303.4.6 was added to clarify that the transfer of all design loads through the building structure and the connections of the trusses to the supporting structure to resist the loads are the responsibility of the registered design professional of the building.

Note that in the 2006 IBC, several sections pertaining to trusses were relocated from Section 2308 to 2304 for clarification, and consolidation, most notably Section 2303.4.1.7 pertaining to alterations to trusses, was relocated to Section 2303.4.1.7.

2303.5 **Test standard for joist hangers and connectors.** Section 1715.1 sets forth the test and acceptance criteria for joist hangers. Note that per Section 1715.1.4, only the design values based on the design limit states for the fasteners or the wood member may be increased for duration of load. If the hanger load is governed by the strength or deflection limit states of the steel, the design value may not be increased for duration of load. Evaluation reports and manufacturer's literature must provide the limit state data in order for the structure designer to comply with this requirement.

2303.6 **Nails and staples.** This section provides bending yield strength requirements for nails and staples. The bending yield strength requirements are those assumed in the NDS lateral strength tables, the UBC, NBC and SBC fastener schedules, and the model code evaluation reports. These strengths are standardized within the nail industry for engineered fasteners and are set forth in ASTM F 1667. Note that the NDS also requires that nails, staples and spikes must meet Federal Specification FF-N-105B, (1997), *Wire, Cut and Wrought Nails, Staples, and Spikes*. This specification also sets dimensional limits. A successful code change to the 2006 IBC added the shank length and diameter in parentheses in the tables for the various types of nails used in wood connections. See more complete discussion under Section 2304.9.1.

2303.7 **Shrinkage.** Two new sections were added to the 2003 IBC pertaining to shrinkage. Section 2303.7 requires the designer to consider the effects of cross-grain dimensional changes (shrinkage) in the vertical direction that can occur in lumber that was fabricated green. See Sections 2.7 and 2.11 of PS 20-99 for definitions of dry and green lumber: Dry lumber is lumber of less than nominal 5-inch thickness that has been seasoned or dried to a maximum moisture content of 19 percent. Green lumber is lumber of less than nominal 5-inch thickness that has a moisture content in excess of 19 percent. For lumber of nominal 5-inch or greater thickness (timbers), green is defined in accordance with the provisions of the applicable lumber grading rules. See also Section 2304.3.3.

Section 2304 *General Construction Requirements*

2304.1 **General.** The requirements of this section apply to ASD, LRFD *and* the conventional construction provisions.

2304.2 **Size of structural members.** Net dimensions are set forth in NDS. Refer to Tables 1A, 1B, 1C and 1D of the NDS for dressed lumber size and section properties of sawn lumber and glued laminated timber members.

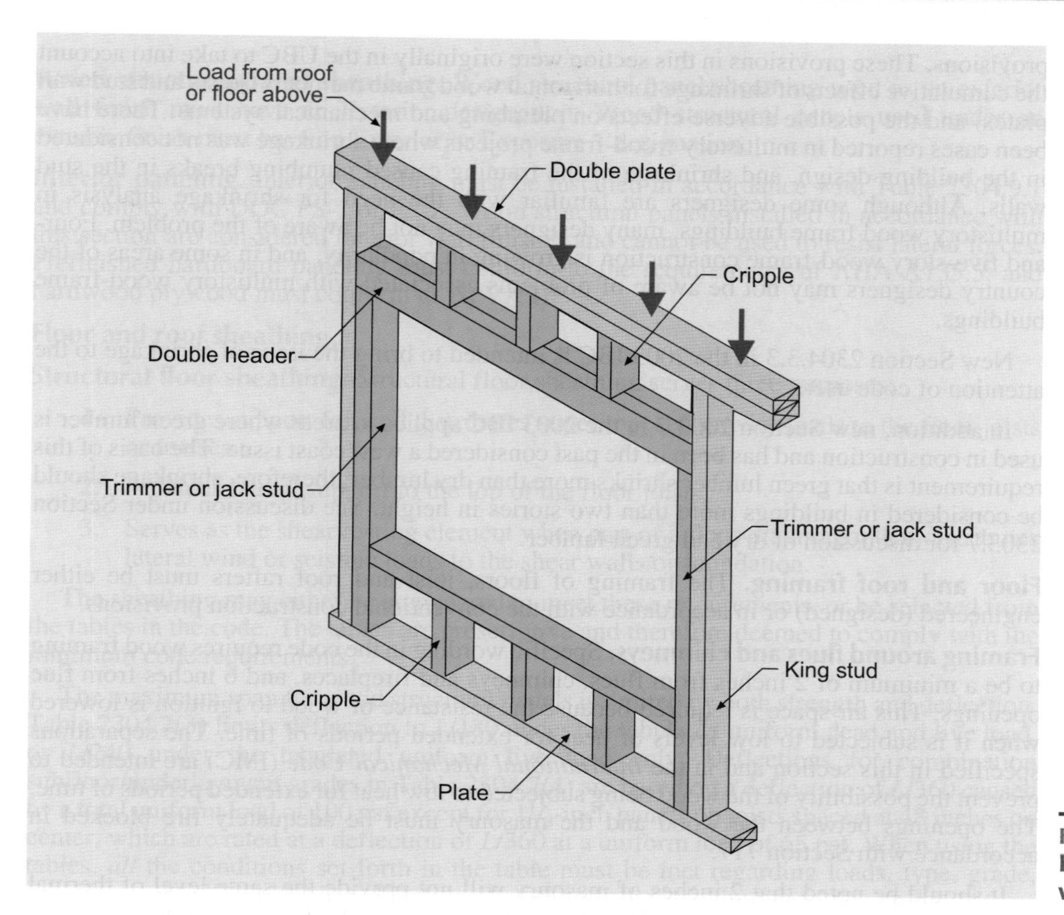

Figure 23-3
Headers over wall openings

Wall framing. Interior and exterior walls must be framed in accordance with conventional construction provisions unless a specific design is provided. **2304.3**

Bottom plates. Sill plates must be at least nominal 2x even if calculations by ASD or LRFD show that a smaller size is acceptable. This requirement applies both to engineered (designed) structures and those constructed in accordance with conventional construction provisions. **2304.3.1**

Framing over openings. Windows, doors, air-conditioning units and other service equipment require that openings be provided in wood-stud walls and partitions. Loads imposed above these openings must be transferred by a structural element above the opening to supports on both sides of the opening and then to a load-bearing wall or partition. In most wood-frame structures, these structural elements are composed of two pieces of 2-inch dimensional lumber plus a spacer or plate, or a solid 4x member. These elements must be fastened in accordance with Table 2304.9.1. Based on the span and the loading, some openings may require trusses, laminated veneer lumber, glued-laminated members or steel beams. Headers may be engineered or selected from Tables 2308.9.5 and 2308.9.6. Header tables are also published in a number of technical documents. All other headers must be engineered in accordance with Section 2301.2. In all cases, headers and their supports must be adequate to support the imposed loads (see Figure 23-3). **2304.3.2**

Shrinkage. New Section 2304.3.3 in the 2003 IBC imposes requirements for considering wood shrinkage. Wood-framed walls and bearing partitions cannot support more than two floors and a roof unless an analysis shows that there will be no adverse effects on the structure that are due to excessive shrinkage or differential movement caused by shrinkage. In drafting the 2000 IBC, the provisions pertaining to the effects of shrinkage on wood members were overlooked and should have been included in the general construction **2304.3.3**

IBC along with the addition of new Section 2306.1.4 incorporated the pertinent provisions of AITC 112-93 directly into the body of the code and deleted the reference to AITC 112-93. The title of Section 2304.8 was revised to indicate that the provisions cover all decking, including mechanically laminated and solid sawn decking. The capacity of lumber decking is arranged according to the various layup patterns described in Section 2304.8.2. The new Section 2306.1.4 gives the design capacity of lumber decking for flexure and deflection according to the formulas given in Table 2306.1.4.

2304.8.1 General. A laminated lumber floor or deck designed in accordance with this section provides the equivalent of a solid-wood deck and may be designed as if it were a solid-wood deck of the same thickness. However, the code requires a conservative design for bending strength in that continuous spans are designed using the moment coefficients for a simple span.

Because this type of floor or deck is labor intensive, it is not often used. It is intended primarily for heavy-timber construction. However, strength is generally not a controlling criterion with this type of deck; therefore, lower grades of lumber can usually be incorporated into the design.

One feature of the code related to construction of these decks that creates homogeneity is the requirement for nailing of adjacent laminations. Because the laminations are thoroughly side nailed and the nails are required to penetrate two and a half laminations, the floor or deck acts essentially as a homogeneous material.

2304.9 Connections and fasteners.

2304.9.1 Fasteners requirements. The code intends that Table 2304.9.1 provide minimum requirements for the number and size of nails connecting wood-framing members. This table accommodates the builder of nondesigned (conventional) construction, and also provides minimum fastening requirements for designed construction. Details such as end and edge distances and nail penetration are required to be in accordance with the applicable provisions of the NDS. Where required, corrosion-resistant fasteners must be either hot-dipped zinc-coated, aluminum alloy wire or stainless steel.

Table 2304.9.1 is comparable to the nailing tables in the other model codes. Footnote l, which requires 8d minimum nails for roof sheathing, corresponds to the 80 mph (fastest mile) entry in 1997 UBC Table 23-II-B-2 for wind uplift resistance.

Power driven fasteners, along with the typical sizes used, and staples are included in the table. The size designations in the table are common to all fastener manufacturers.

A successful code change to the 2006 IBC added the shank length and diameter in parentheses to the tables for the various types of nails used in wood connections.

It has often been reported that improper nail sizes have been used in wood frame building construction because the pennyweight system of specifying nail sizes is not universally understood. Code users sometimes focus on pennyweight (8d - 8 penny, 16d - 16 penny, etc.) and do not pay sufficient attention to the specific type of nail such as common, box, cooler, sinker, finish, etc. A typical example is substitution of box nails for common nails of the same pennyweight. The specific type of nail is critical because there can be significant differences in strength properties of connections nailed with nails of the same pennyweight but of different nail type. Specifying the nominal dimensions of nails in the fastening tables avoids confusion and reduces misapplications in nailed connections. The code change proponent expected some reluctance on the part of some in the building construction community to completely abandon the pennyweight system of designating nail sizes, so the code continues to maintain the pennyweight designations. Because nominal dimensions are not as subject to misinterpretation, the shank length and diameter in parentheses will prevent confusion and misapplication of the various types of nails used in wood connections.

2304.9.2 Sheathing fasteners. Fasteners should be driven flush with the surface of the sheathing, but not overdriven. Overdriving of fasteners can significantly reduce the shear capacity and ductility of the diaphragm or shear wall. For three-ply material, the strength reduction is significant if the fastener is overdriven through one ply.

If no more than 20 percent of the fasteners around the perimeter of the panel are overdriven by $^1/_8$ inch or less, then no reduction in shear capacity need be considered. If more than 20 percent of the fasteners around the perimeter are overdriven by any amount, or if any fasteners are overdriven by more than $^1/_8$ inch, then additional fasteners must be driven to maintain the desired shear capacity, provided that the additional fasteners will not split the substrate. For every two fasteners overdriven, one additional fastener should be driven. Panels with more than 20 percent of the fasteners overdriven greater than $^1/_8$ inch should be replaced.

Also, if the actual panel thickness is greater than the design panel thickness needed to resist the design shear, the panel shear capacity may be adequate without the driving of additional fasteners. For example, if the design required a $^{15}/_{32}$-inch-thick panel, but a $^{19}/_{32}$-inch-thick panel with all fasteners overdriven by $^1/_8$ inch is used for the sheathing, the panel is adequate because the net thickness is $^{15}/_{32}$ inch.

Joist hangers and framing anchors. Joist hangers and other framing hardware are usually evaluated by the model code agency's evaluation service. There are acceptance criteria for the performance of the hardware in terms of the strength limit states of the metal, wood and fasteners, as well as deflection limit states. **2304.9.3**

Test requirements for joist hangers are found in Section 1715.1. Only the design values based on the strength limit states for the fasteners or wood member (Items 4 or 5 in Section 1715.1.2) are allowed to be increased for duration of load per the NDS. No increase is allowed for loads limited by steel strength or deflection. ICC Evaluation Service reports and manufacturer's catalogs must provide the controlling limit strength of the connector in order for the designer to meet this requirement.

Other fasteners. Fasteners not specifically cited in the code are subject to building official approval. See Section 1703.4 for requirements related to research reports and acceptance of evaluation reports. **2304.9.4**

Fasteners in preservative-treated and fire-retardant-treated wood. The water-borne salts in preservative-treated and fire-retardant-treated wood are corrosive. Fasteners must be corrosion resistant when used with these materials. Corrosion-resistant fasteners are made of type 304 or 316 stainless steel, silicon bronze, copper or steel that has been hot-dipped or mechanically deposited zinc coated galvanized with a zinc coating of not less than 1.0 ounce per square foot. **2304.9.5**

The 2006 IBC introduced an alternate method for mechanically galvanizing, which is preferable to hot dipping for some types of fasteners as indicated in the exception this section. Class 55 was added to provide an equivalent amount of zinc as would be provided by the hot dip process in accordance with ASTM A 153. According to the American Galvanizers Association, mechanically plating to a thickness of 55 microns provides an equivalent coating to 1 ounce per square foot of hot-dipped galvanized zinc, which is what is provided for fasteners by ASTM A 153. Class 55 provides 55 microns of thickness.The section references ASTM A 53 or B 695 for coating weight requirements.

Electro-galvanized steel fasteners do not qualify as corrosion resistant; the zinc coating typically is about 0.1 ounce per square foot. Electro-galvanized nails are suitable only for occasionally wet locations such as for nailing composition shingles.

Load path. The code requires the load path to be continuous from the point of origin to the resisting element, which is generally from the roof to the foundation. A continuous load path for both gravity and lateral loads is necessary for adequate performance of the structure in response to superimposed vertical and lateral loads. This is especially critical in the case of high wind and earthquake lateral load demands. For example, visualize what happens to the wind suction load on a low-slope roof. The upward force that is not offset by the dead load of the roof elements must be resisted by dead load elsewhere. Similarly, where the structure is subjected to lateral earthquake loads, the inertial forces from the floor and roof diaphragms **2304.9.6**

must be effectively transferred to the vertical lateral-force-resisting elements, e.g., wood shear walls. A positive, properly detailed continuous load path is necessary to ensure the transfer of all gravity and lateral loads from the roof and floors down to the foundation and supporting soil.

2304.9.7 **Framing requirements.** Columns must be provided with full end bearing to transfer the loads or connections designed to resist the full compressive load. Connections must also be able to resist lateral and net uplift loads.

2304.10 **Heavy timber construction.** The provisions contain general provisions for column continuity, transfer of loads from beams to columns, other connection criteria, and requirements for structural anchorage and continuity. Minimum element size requirements for Type IV construction are found in Section 602.4.

2304.10.1 **Columns.** Columns must be continuous throughout all stories by concrete or metal caps, base plates, timber splice plates or other approved methods.

2304.10.1.1 **Column connections.** Girders and beams are require to be fitted around columns, and adjoining ends must be adequately tied to each other to transfer horizontal loads across the joints.

2304.10.2 **Floor framing.** Wall pockets or hangers are required where wood beams, girders or trusses are supported by masonry or concrete walls. Beams supporting floors are required to bear on girders or be supported by ledgers, blocks or hangers.

2304.10.3 **Roof framing.** Roof girders and alternate roof beams are anchored to supporting members with steel or iron bolts designed to resist vertical uplift of the roof.

2304.10.4 **Floor decks.** Floor decks and floor covering cannot extend closer than $^1/_2$ inch to walls with the space covered by a molding fastened to the wall.

2304.10.5 **Roof decks.** Where supported by a wall, roof decks must be anchored to walls by steel or iron bolts to resist uplift forces. This section in the 2000 IBC requires roof decks supported by a wall to be anchored to walls at intervals not exceeding 20 feet but did not specify the type or purpose of the anchorage. Code changes to the 2003 IBC added that the anchors are to resist uplift forces determined in accordance with Chapter 16 and the anchors must be steel or iron bolts of sufficient strength to resist vertical uplift of the roof.

2304.11 **Protection against decay and termites.** The provisions from the model codes were reorganized in the IBC for clarity. The provisions are grouped into the following areas:

1. Wood used above ground
2. Laminated timbers exposed to weather
3. Wood with ground contact
4. Supports for appurtenances
5. Termite protection
6. Retaining walls

The provisions of this section are intended to protect against decay and termite infestation. The provisions are based on the extensive material on biodeterioration of wood in the *Wood Handbook*.[5]

2304.11.2 **Wood used above ground.** Wood used above ground, if preservative treated, is usually treated with a water-borne preservative such as ammoniacal copper arsenate (ACA) or chromated copper arsenate (CCA) in accordance with AWPA U1 (Commodity Specifications A or F). The retention rates are lower than required for ground contact.

2304.11.2.1 **Joists, girders and subfloor.** There must be 18 inches clearance to joists and 12 inches clearance to wood girders if they are not of naturally durable or preservative-treated wood. See Figure 23-4.

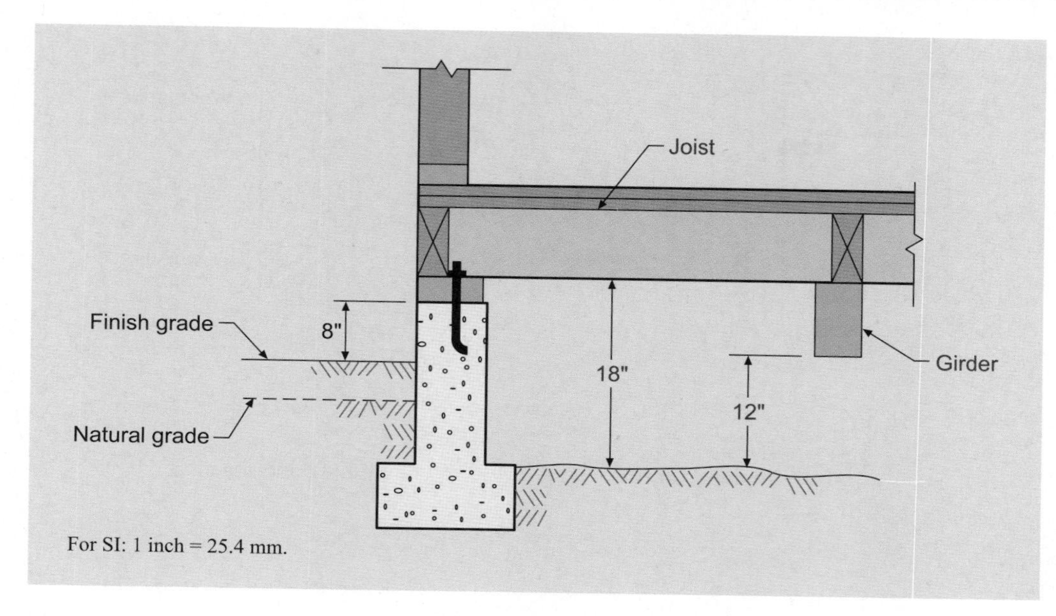

Figure 23-4 Under-floor clearance

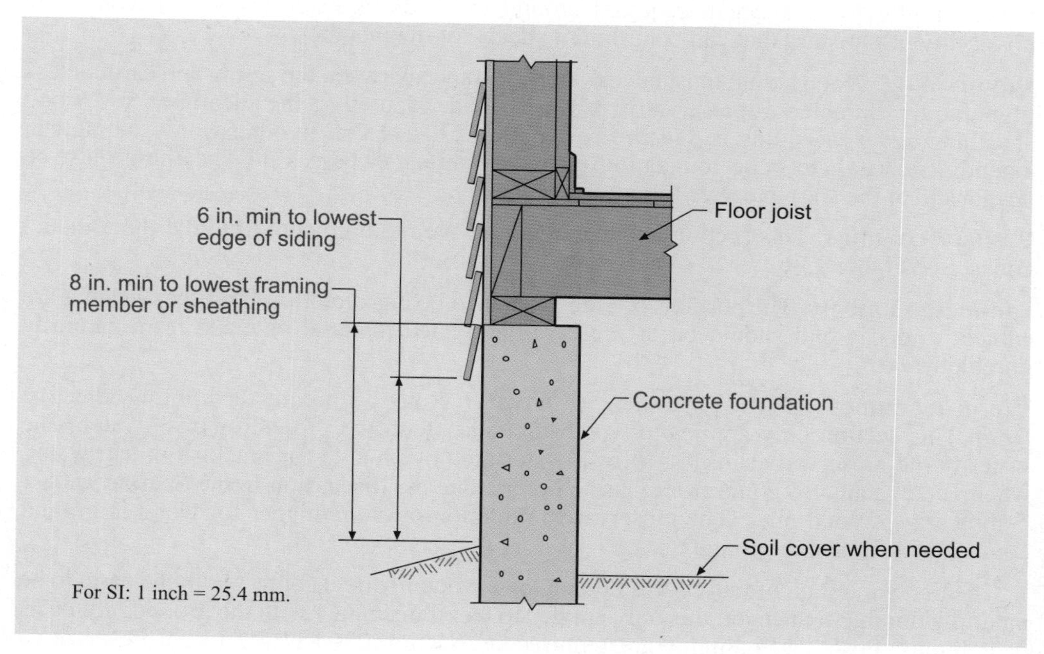

Figure 23-5 Clearance between wood framing, wood siding and earth

2304.11.2.2

Wood supported by exterior foundation walls. Framing, including sheathing (not siding), must have 8 inches of clearance from exposed earth if it is not naturally durable or preservative treated. See Figure 23-5. This section was retitled in the 2006 IBC to clarify that the section applies to wood framing resting on exterior foundation walls.

2304.11.2.3

Exterior walls below grade. These requirements were put in a separate subsection in the 2006 IBC to clarify that they apply to wood framing attached to the interior of exterior concrete or masonry foundation walls.

2304.11.2.4

Sleepers and sills. Concrete and masonry slabs that are in direct contact with the earth are very susceptible to moisture because of absorption of ground water. This can occur on interior slabs as well as at the perimeter. This section is intended to prevent use of untreated wood that may decay under such conditions. Concrete that is fully separated from the ground by a vapor barrier is not considered to be in direct contact with earth.

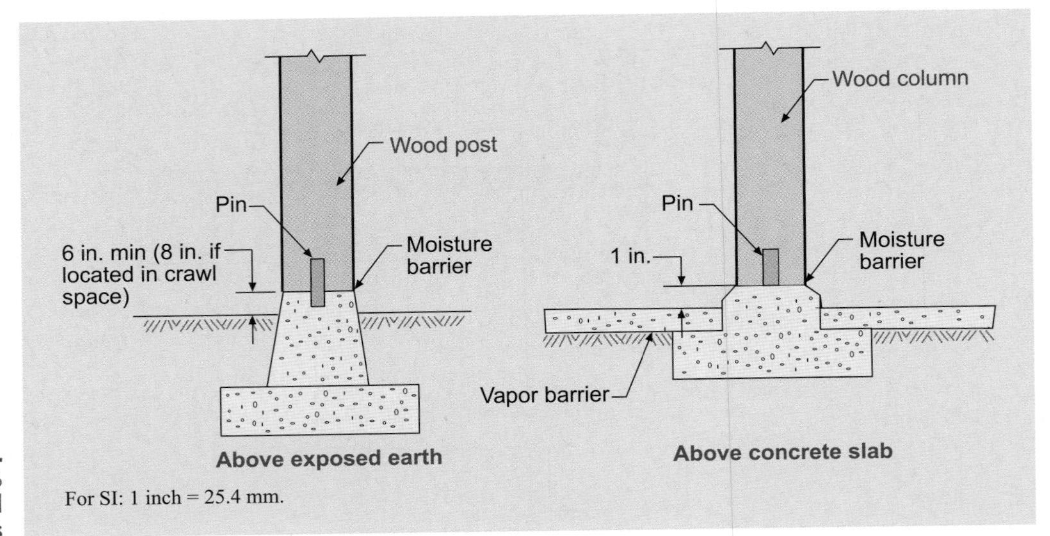

Figure 23-6 Posts and columns

2304.11.2.5 Girder ends. An airspace is required around the ends of wood girders to reduce the possibility of moisture that can contribute to decay of the member.

2304.11.2.6 Wood siding. Wood siding must have 6-inch clearance between the siding and earth unless of naturally durable or preservative-treated wood. Note that the clearance for wood sheathing under the siding is *8 inches* per Section 2304.11.2.2. In other words, siding can extend 2 inches below the foundation plate, or framing, whereas the sheathing must be terminated at the sill plate. See Figure 23-5.

2304.11.2.7 Posts or columns. This section is an amalgam of the requirements from the three model codes. See Figure 23-6.

2304.11.3 Laminated timbers. The portions of glue-laminated beams directly exposed to weather are subject to decay and should be of preservative-treated material or made from naturally durable wood.

2304.11.4 Wood in contact with the ground or fresh water. Note specifically the limiting adjective *fresh*. This section only applies to wood in contact with the ground or fresh water. The water-borne preservatives used for fresh water are not suitable for brackish or salt water, where attack can also come from marine borers. See the discussion in the Section 1808.1 commentary, Wood piles. The preservative retention rates are higher for wood in ground contact than for above-ground uses.

The first paragraph of this code section allows wood in direct contact with the earth to be naturally durable. However, this only applies to wood in contact with the ground, not posts or columns. Posts and columns are required by Section 2304.11.4.1 to be preservative treated.

2304.11.4.1 Posts or columns. Posts or columns embedded in concrete or embedded in earth, such as columns in a pole-supported structure, have no opportunity to dry and are subject to decay. Hence, they must be of preservative-treated wood.

2304.11.4.2 Wood structural members. Where wood framing is used to support floors and roofs that are moisture permeable, such as a concrete slab over a patio or a patio slab over a garage, the framing must be of pressure-treated wood or approved naturally durable species, unless the slab is separated from the framing by a waterproof membrane.

Although these wood framing members are not necessarily in direct contact with the ground, their exposure to moisture is similar to that of wood in direct contact with the ground. Therefore, the wood framing must be of naturally durable wood, or it must be preservative treated in accordance with AWPA C2, C9 or C22.

Supporting member for permanent appurtenances. Balconies, porches and other appurtenances are exposed to weather conditions and may not have protective overhangs. Water can collect in the joints and on the surfaces creating alternating cycles of wetting and drying conducive to decay. **2304.11.5**

Termite protection. Where termites are a significant hazard—for example, the southern states—floor framing must be preservative treated, naturally durable or have some other approved method of termite protection. Preservative treated wood must be in accordance with AWPA U1 for the species, product preservative and end use. **2304.11.6**

Wood used in retaining walls and cribs. Wood used in retaining walls, crib walls, bulkheads and other walls that retain or support the earth must, with only a few exceptions, be of preservative-treated wood specified as ground-contact-treated wood. **2304.11.7**

Attic ventilation. Refer to the discussion under Section 1203.2 in the *2006 IBC Handbook—Fire & Life Safety Provisions*. **2304.11.8**

Under-floor ventilation. Refer to the discussion under Section 1203.3 in the *2006 IBC Handbook—Fire & Life Safety Provisions*. **2304.11.9**

Long-term loading. Wood structural members are subject to long-term creep, which increases deflection, particularly where a high dead load is present. The 2000 and 2003 IBC do not permit wood members to permanently support the dead load of masonry or concrete, because additional deflection produced by long-term creep can cause severe cracking in the masonry or concrete. Section 2304.12 in the 2000 and 2003 IBC was a carryover from the 1997 UBC, which was intended to address concerns with long-term creep in wood members supporting concrete and masonry. The section consisted of a restriction against wood members being used to permanently support dead load of masonry or concrete followed by a list of exceptions to the restriction. **2304.12**

This provision pertaining to wood members supporting concrete and masonry was deleted in the 2006 IBC and replaced by a new section that references the design method for limiting long-term deflections in Section 3.5.2 and Appendix F of the NDS. Under sustained loading, wood members exhibit additional time dependent deformation (creep) which generally develops over long periods of time. The tabulated modulus of elasticity design values, *E*, in the *NDS* are intended to be used to estimate immediate deformation under load. Where dead loads or sustained live loads represent a relatively high percentage of total design load, creep is an appropriate design consideration, which is addressed within the NDS. The total deflection under long-term loading is estimated by increasing the initial deflection associated with the long-term load component by 1.5 for seasoned lumber or 2.0 for unseasoned or wet lumber or glued laminated timber. Section 3.5.2 and Appendix F of the NDS contain provisions to account for time dependent deformations known as creep.

Section 2305 *General Design Requirements for Lateral-Force-Resisting Systems*

These provisions are primarily based on a combination of 1997 UBC Section 2315 and 1997 NEHRP Sections 12.1 through 12.4.

There are separate sections for design of wood diaphragms (Section 2305.2) and wood shear walls (Section 2305.3). The term *diaphragm* refers exclusively to horizontal or nearly horizontal (sloping) elements such as floors and roofs; the term *shear walls* refers to the vertical resisting elements. In the 2006 IBC, the term *diaphragm* is defined in Section 1602, not in Chapter 23. Section 2302 contains the definition of unblocked diaphragm which is not in Section 1602, because the term pertains to wood frame systems only.

2305.1 General. Section 2305 contains requirements for wood shear walls and diaphragms resisting wind, seismic and other lateral loads. As an alternate to Section 2305, the 2006 IBC references the AF&PA *Special Design Provisions for Wind and Seismic* (SDPWS), which covers design and construction of wood members, fasteners and assemblies to resist wind and seismic forces. Like the 2005 edition of the AF&PA *National Design Specification (NDS) for Wood Construction*, the SDPWS is a dual format standard that contains both ASD and LRFD procedures. Although the SDPWS is an alternative to the provisions contained in Section 2305, the standard does not address all of the wood requirements that are included in the IBC, such as high-capacity diaphragms or staple-fastened shear walls.

Section 4.3.7 of the SDPWS includes a similar requirement that 3-inch nominal or wider framing and staggered nailing is required where the nominal unit shear capacity for seismic exceeds 700 plf in Seismic Design Categories D, E or F. This 700 plf nominal unit shear capacity corresponds to 350 plf (700/2) allowable unit shear for ASD and 560 plf (0.8 × 700) factored unit shear for LRFD. The ASD allowable unit shear capacity is determined by dividing the nominal unit shear by the reduction factor of 2.0, and the LRFD factored unit resistance is determined by multiplying the nominal unit shear capacity by ϕ_D of 0.80.

Structures use wood diaphragms as part of the lateral-force-resisting system to resist wind and seismic and other lateral loads. Wood diaphragms are often incorrectly assumed to be flexible. A wood diaphragm is usually flexible relative to a concrete diaphragm. A wood diaphragm may be stiff relative to a wood or gypsum board shear wall. In this case, the wood diaphragm is classified as rigid. However, a wood diaphragm is generally flexible relative to masonry or concrete shear walls.

Section 1602 of the 2000 IBC defines a diaphragm as flexible when the maximum lateral deformation of the diaphragm is more than two times the average story drift determined by comparing the deflection of the diaphragm itself with the deflection of the adjoining vertical resisting elements. Conversely, a diaphragm is considered rigid when the diaphragm deflection is equal to or less than two times the average story drift. These definitions apply to all building materials, not just wood. Defining whether the diaphragm is classified as flexible or rigid is necessary to determine how the lateral forces and torsional moments will be distributed to the vertical elements of the lateral-force-resisting system. Flexible diaphragms are assumed to have insufficient stiffness to distribute torsional moments.

Section 1602 of the 2006 IBC states that a diaphragm is flexible for the purpose of distribution of story shear and torsional moment where indicated by Section 12.3.1 of ASCE 7, as modified in Section 1613.6.1.

Section 1613.6.1 permits the assumption of flexible diaphragms under specific conditions. Diaphragms constructed of wood structural panels are permitted to be considered flexible, provided several conditions are met. See further discussion under Section 2305.2.5.

As an example, consider a 40-by-40 foot one-story Seismic Design Category D, Occupancy Category II building having a wood roof diaphragm with no openings, 10-foot plate height, Site Class D soil, and S_S = 0.75. The lateral-force-resisting elements are a 5-foot-long wood structural panel shear wall on each side (2:1 aspect ratio). The shear in the walls would be moderate at approximately 330 pounds per linear foot; assume $^{15}/_{32}$-inch sheathing with nailing at 4-inch spacing.

The shear wall deflection (primarily nail slip plus $^{1}/_{8}$-inch allowance for anchorage deflection) is about 0.30 inch. The diaphragm deflection is approximately 0.15 inch, assuming $^{3}/_{16}$-inch slip in each chord splice. The total deflection is 0.45 inches. The deflection of the diaphragm relative to the deflection of the shear walls is 0.15/0.30 = 0.5 < 2. Therefore, the diaphragm is stiff relative to the shear walls and classified rigid for lateral-force distribution. Note that the shear wall deflection above is twice that of the diaphragm.

If the wood shear walls were replaced by a 4-foot-long 8-inch CMU wall solidly grouted, the deflection of the wall would be approximately 0.05 inch. The deflection of the diaphragm relative to the shear walls is 0.15/0.05 = 3 > 2. Therefore, the diaphragm is considered flexible for purposes of distribution of lateral forces.

This simple, but realistic, example demonstrates that the common assumption that all wood diaphragms are flexible is incorrect. As another example, consider a typical two-story garden office structure with a width of 40 feet and a length of 120 feet using wood diaphragms. Masonry shear walls are used in the transverse direction at each end, and a wood shear demising wall is placed across the width of the building at mid-length. In the longitudinal direction, the walls are wood shear walls and glass curtain wall. Is the diaphragm flexible or rigid for purposes of distributing lateral forces? In the transverse direction, the diaphragms will likely be considered flexible relative to the shear walls, but in the longitudinal direction, the diaphragm will likely be considered rigid relative to the shear walls.

In previous model codes, wood-sheathed diaphragms have often been arbitrarily considered to be flexible by many registered design professionals and building officials. The 2000 IBC, the 1997 NEHRP, and the 1997 UBC codes stipulate that the determination of diaphragm rigidity or flexibility for determinating the method to distribute forces is dependent on the relative stiffness and deformations of the horizontal and vertical resisting elements. Wood-sheathed diaphragms in structures with wood frame shear walls with various types of sheathing may be relatively rigid compared with the vertical resisting system and, therefore, capable of transmitting the torsional moment caused by the lateral forces. A relative deformation of the diaphragm of two or more when compared with the vertical resisting system deformation under the same force is the benchmark used to define a diaphragm as being flexible.

See the 1999 SEAOC *Blue Book*[6] for a discussion of the many parameters necessary to classify the diaphragm as flexible or rigid. See also Example No. 1 in Volume II of the SEAOC *Seismic Design Manual*[7] for a demonstration of the necessary calculations.

Deformation compatibility is another area that must be considered by the engineer. Many of the failures in the Northridge earthquake were caused by the inability of the gravity load carrying system to sustain the lateral deformations and strains undergone by the lateral-force-resisting system. The code provisions intend for the structure to be adequately tied together for structural integrity. The gravity system relies on the lateral-load-resisting system for stability. Because the structure acts as a whole, the gravity system goes along for the ride as the lateral system resists the wind or earthquake demands. The maximum usable strain in the gravity elements must not be exceeded or there will be a high probability of collapse. The lateral resisting system must limit the deformations such that the maximum usable strains in the supported gravity system will not be exceeded. In other words, the deformation capability of the gravity system must be compatible with the deformations allowed by the lateral force resisting system. Thus, the term *deformation compatibility* is used.

Deformation compatibility provisions appeared in previous model codes but were often ignored by registered design professionals and building officials. Deformation compatibility requirements were enhanced in the 1997 UBC and the 1997 NEHRP in response to the damage caused by the 1994 Northridge earthquake. Those provisions were used as the basis for the provisions in the 2000 IBC Section 1617.6.4.3 for structures classified as Seismic Design Category D, E or F. The 2006 IBC references ASCE 7 for technical provisions related to seismic design. Refer to Section 12.12.4 for drift and deformation requirements in Seismic Design Categories D though F.

Deformation compatibility of connections between structural elements should be considered such that the deformation of each element and connection comprising the lateral-force-system is compatible with the deformations of the other lateral-force-resisting elements and connections and with the deformations of the gravity system and connections thereto. Particular care should be used to determine that the deformations are compatible with any brittle nonstructural elements such as masonry veneers.

The registered design professional should visualize the deformed shape of the structure to ensure that the connections provide the necessary ductility to meet the probable deformation demand placed on the structure. Unlike steel or other metal structures, wood is not a ductile material and virtually all the ductility achieved in the structure is from the metal connections. The intended failure mechanism of wood structures must be through the connections, including the nailing of structural panels; otherwise the failure will likely be brittle in nature. The philosophy of strong elastic columns and yielding beams cannot be projected from steel to wood structures. To enable a wood structure to deform and dissipate energy during a seismic event, specific elements must serve as the structural fuse in the structure and be ductile.

As an example of a compatibility issue, consider the deformation compatibility between a hold-down connector to the hold-down post and the edge nailing of shear wall sheathing to the hold-down post and adjacent bottom plate. Recent testing and observations from the Northridge earthquake have suggested that the hold-down post experiences notable displacement before significant load can be transferred through the hold-down connector. This is caused, among other things, by the oversizing of the bolt holes in the hold-down post and the deformation and rotation of the hold-down bracket. Anchor bolts connecting the bottom plate to the foundation below attempt to carry the shear wall uplift in tension as the hold-down post moves up. The sheathing, however, is nailed to both the bottom plate, which is held in place, and the tie-down post, which is being pulled up. The result is a large deformation demand being placed on the nails connecting the sheathing to the framing. This often results in nails pulling out of the sheathing at the hold-down postcorner, or the sill splitting from cross-grain tension and sometimes an unzipping effect in the nailing. This unzipping effect is caused by the sequential failure of the nails as the load is transferred to adjacent nails. Unzipping can ultimately affect a significant portion of the sheathing. The most effective solution currently available is to limit the slip and deformation at the hold-down post by using a very stiff nailed hold-down and ensuring that the hold-down bolt is pretensioned, in a sense. Use of bolted hold-downs should be avoided where high loads are placed on the hold-downs and deformation compatibility becomes a problem. At one point, SEAOC recommended setting the bottom of the hold down $^1/_2$ inch above the plate, drilling the stud holes and installing the stud bolts, then tightening the hold-down anchor bolt nut. This essentially tensions the chord stud. NEHRP recommends using only nailed hold-down connections.

NEHRP Section 12.3.4.4 requires the use of 3-inch by 3-inch by $^1/_4$-inch plate washers for all anchor bolts in shear walls regardless of seismic design category classification. The 2000 IBC requires 2-inch by 2-inch by $^3/_{16}$-inch plate washers only for structures classified Seismic Design Category D, E and F. A code change to the 2003 IBC changed the square plate washer size to 3 inch by 3 inch by $^1/_4$ inch thick to be consistent with the NEHRP provisions. However, a subsequent code change modified the minimum thickness from $^1/_4$ inch to 0.229 inches in order to allow the plate washers to be manufactured from cold rolled steel instead of $^1/_4$-inch hot rolled steel. The requirement for plate washers is discussed further in the commentary for Section 2305.3.11.

2305.1.1 Shear resistance based on principles of mechanics. When calculating the shear resistance of diaphragms or shear walls by the principle of engineering mechanics for seismic demands, the values of fastener strength and sheathing resistance should actually be based on the results of cyclic tests. The capacity determined from cyclic tests is likely to be lower than the capacity determined from monotonic testing. When determining values based on this method, the user should be aware that there are currently no accepted rational methods to calculate deflections for diaphragms and shear walls sheathed with materials other than wood structural panels fastened with nails. Therefore, if a rational method is to be used, the capacity of the fastener in the sheathing material should be validated by acceptable test procedures employing cyclic forces and displacements. Validation must include correlation between the overall stiffness and capacity predicted by principles of mechanics and that observed from test results. A diaphragm or shear wall sheathed with dissimilar materials on the two faces should be designed as a single-sided diaphragm or wall using the capacity of the stronger material and ignoring the shear capacity of the weaker material.

Framing. A diaphragm is a horizontal or nearly horizontal structural unit that acts as a deep beam or girder when flexible in comparison to its supports and as a plate when rigid in comparison to its supports. The analogy to a girder is appropriate because girders and diaphragms are made up as assemblies.[8,9] Sheathing acts as the *web* to resist the shear in diaphragms and is stiffened by the framing members, which also provide support for gravity loads. Flexure is resisted by the edge or boundary elements acting like *flanges* to resist induced tension or compression forces. The flanges may be the top plates, ledgers, bond beams or any other continuous element at the perimeter of the diaphragm. **2305.1.2**

The flange (chord) can serve several functions at the same time, providing resistance to loads and forces from different sources. When it functions as the tension or compression flange of the girder, it is important that the connection to the web be designed to accomplish the shear transfer. Because most diaphragm flanges consist of many pieces, it is important that the splices be designed to transmit the tension or compression occurring at the location of the splice and to recognize that the direction of application of seismic forces can reverse. The splices should be designed to minimize slip in the connection. A significant proportion of the diaphragm deflection is caused by deformation in the chord splice from the axial load. See the last term of Equation 23-1. It should also be recognized that the shear walls parallel to the flanges may be acting with the flanges to distribute the diaphragm shears. When seismic forces are acting at right angles to the direction considered previously, the flange then become a part of the reaction support system. It may function to transfer the diaphragm shear to the shear wall(s), either directly or as a collector (drag strut) between segments of shear walls that are not continuous along the length of the diaphragm. The term *diaphragm chord* is defined in Section 1602, and the term *collector* is defined in Section 2302.

Similar to the analogy for diaphragms, shear walls may be considered deep vertical cantilever beams. The flanges are subjected to tension and compression while the webs resist the shear. It is important that the flange members, splices at intermediate floors and the connection to the foundation be detailed and sized for the induced forces. The webs of diaphragms and shear walls often have openings. The transfer of forces around openings can be treated similarly to openings in the webs of steel girders. Members at the edges of openings have forces that are due to flexure and the higher web shear induced in them, and the resultant forces must be transferred into the body of the diaphragm beyond the opening, as required by this section. There is a similar requirement in Section 2305.1.3, Openings in shear panels, which states that "openings in shear panels which *materially* affect their strength shall be fully detailed on the plans and shall have their edges adequately reinforced to transfer all shearing stresses." The perforated shear wall method in Section 2305.3.8.2 allows unreinforced openings, provided the shear capacity is adjusted in accordance with Table 2305.3.8.2. For structures classified as Seismic Design Category D, E or F, all openings greater than approximately 14 inches by 14 inches should be reinforced. The additional cost of straps and blocking is insignificant in comparison to the improved performance under high seismic demands.

Collectors and their connections are critical to performance of the lateral-force-resisting system. The collector must collect the shears in the diaphragm in areas where there is no shear wall and deliver those forces to the shear walls. Premature failure of a collector or its connection may lead to complete structural failure. Section 1620.1.6 of the 2000 IBC (Seismic Design Category B and above) requires that collectors, collector splices and the connection of the collector to resisting elements have the strength to resist the special load combinations of IBC Section 1605.4; i.e., the maximum anticipated seismic force, E_m, unless the structure is braced entirely by light-frame shear walls. The 2006 IBC refers to Section 12.10.2 of ASCE 7-05. In other words, the collectors, splices and connections for a structure having a wood diaphragm and concrete tilt-up or masonry walls would have to be designed for the combinations using the maximum earthquake force, E_m, to ensure that they remain essentially elastic. The collectors, splices and connections for structures having a wood diaphragm and wood structural panel shear walls need only be designed to resist the load combinations of Section 1605.2 or 1605.3. See Sections 12.10.2 and 12.4.3.2 of ASCE 7-05. Note that in ASCE 7, the requirement applies to Seismic Design Category C through F.

Chords and collectors must be placed reasonably close to the plane of the diaphragm in order to function in accordance with the assumed behavior as a deep beam flange. If the chord or collector is too far from the plane of the diaphragm, shear lag will prevent the chord or collector from acting in accordance with the assumed behavior. For example, a top plate at the top of a parapet 3 feet above the plane of the roof diaphragm will not be effective as the diaphragm chord or collector.

2305.1.3 Openings in shear panels. Although the section requires that openings in shear walls that materially affect their strength have their edges reinforced, the meaning of the phrase *materially affect their strength* is not defined in the code. The perforated shear wall method in Section 2305.3.8.2 permits openings in shear walls without force transfer design, provided certain restrictions and adjustments are made. See commentary to Sections 2305.3.8.1 and 2305.3.8.2 for further discussion.

2305.1.4 Shear panel connections. Positive connections are required. The term *design forces* as used here may be confusing in regard to seismic loads. Design seismic forces determined from the provisions of Chapter 16 are strength level forces arising from the seismic load effect, *E*. When used for allowable stress design, the design force *E* is reduced in the ASD load combinations by 0.7 or *E*/1.4. See commentary to Section 2305.1.2 for forces required to attach collectors to shear walls.

Limiting the transfer of lateral forces by toenails or slant nails is necessary to minimize the failure mechanism of splitting caused by shrinkage of the blocking or closely spaced nails that provide a weakened plane. (See Figure 23-7). In addition, toenails may not be installed in the field with a high degree of precision.

The original shear value of 150 pounds/foot was allowed based on the spacing of 8d nails at 6 inches on center using the 1991 NDS nail values. A slant nail factor of 1.0 was used because the orientation where the nail is driven perpendicular to grain for both members is slant nailing. Shears exceeding 150 pounds/foot must be transferred by positive means such as face nailing to perpendicular blocking or framing anchors as shown in Figure 23-8. The limitation applies only to seismic forces in Seismic Design Category D, E or F.

The language of this section had a minor modification in the 2006 IBC. The 2003 IBC stated, "toenails shall not be used to transfer lateral forces in excess of 150 pounds per foot." This was changed in the 2006 IBC to read, "the capacity of toenail connections shall not be used when calculating lateral load resistance to transfer lateral earthquake forces in excess of 150 pounds per foot." The provision applies to the connections from diaphragms to shear walls, collectors or other elements, or from shear walls to other elements. The restriction on the use of toenails originated with the 1996 edition of the *Recommended Lateral Force Requirements and Commentary* of the Seismology Committee of the Structural Engineers Association of California (SEAOC), also known as the SEAOC *Blue Book*. It was later incorporated into the 1997 UBC. The intent was simple: If the design seismic load was less than or equal to 150 plf, the engineer could use toenails to provide the connection. If the design seismic load exceeded 150 plf, then toenails should not be used, and the engineer should specify some other method of connection such as a premanufactured metal connector. The requirement has been interpreted to mean that up to 150 plf of load could be resisted by the toenailed connection, and any additional load must be resisted by some other means. It should be noted that where ICC evaluation reports permit more than 150 plf through

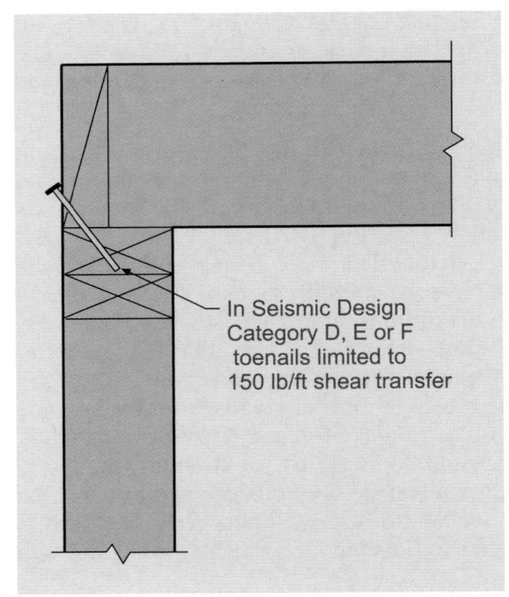

Figure 23-7 Slant nail shear transfer

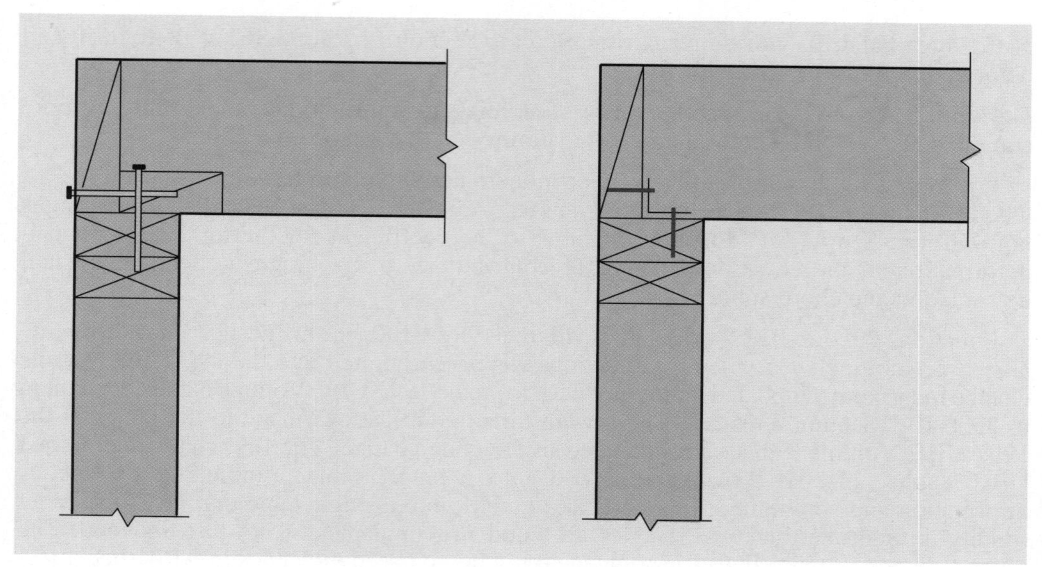

Figure 23-8 Positive shear transfer

toenailed band joists, these higher values were obtained by tests conducted through recognized acceptance criteria. ICC evaluation reports are approved under the alternative materials, design and methods of construction provisions in Section 104.11.

Wood members resisting horizontal seismic forces contributed by masonry and concrete walls. Because of the significant differences in in-plane stiffness between wood and masonry or concrete systems, the use of wood members to resist the seismic forces produced by masonry and concrete is not allowed in structures greater than one story in height. This is because of the probable torsional response a structure will exhibit. There are two exceptions where wood can be used as part of the lateral-load-resisting system in conjunction with concrete or masonry. The first exception is when the wood is in the form of a horizontal truss or diaphragm and the lateral loads do not produce significant rotation of the horizontal member. The torsion caused by the offset between the center of mass (CM) and center of rigidity (CR) should be less than the code-assumed accidental torsion; that is, the offset between the CM and the CR should be less than 5 percent of the building dimension perpendicular to the direction of loading. The second exception is for structures of two stories or less in height. In this case, the capacity of the wood shear walls should be sufficient to resist the magnitude of loads imposed. Five restrictions are imposed on structures using the second exception to ensure that the structural response will not include significant rotational effects and the drift will not cause failure of the masonry or concrete portions of the structure. **2305.1.5**

Wood members resisting seismic forces from nonstructural concrete or masonry. This section was added to the 2006 IBC to clarify that restrictions on wood frame construction resisting horizontal seismic forces apply to seismic forces contributed by masonry or concrete walls and not other types of masonry or concrete construction such as masonry veneer and concrete floors. The design of the lateral-force-resisting system is required to consider the effects of horizontal seismic forces contributed by masonry veneer or concrete floors. **2305.1.6**

Design of wood diaphragms. **2305.2**

General. The permissible deflection portion of the first sentence limits the deflection the same as does the language in Section 2305.3.1 for wood shear walls. The intent of the provision is to not exceed the more restrictive of the permissible deflection of attached or distributing elements or the drift limits of 2000 IBC Section 1617.3 less the deflections of the lateral-force-resisting system (shear walls). The drift limits noted in 2000 IBC Section **2305.2.1**

1617.3 are total drifts, not element drifts. Refer to Section 12.12 of ASCE 7-05 for drift and deformation limits of the 2006 IBC.

2305.2.2 Deflection. The first paragraph of this section refers to deformation compatibility. A more complete discussion on deformation compatibility is found in Section 2305.1.

Equation 23-1, based on the results of monotonic diaphragm tests, is only for blocked wood structural panels that are uniformly nailed. Adjustments must be made if the nailing is not uniform. See *ATC-7*[9] for modifications to the coefficient for the third term for nail deformation in the equation. The major contributors to diaphragm deflection are nail deformation and chord splice slip.

Equations 23-1 and 23-2 originated with the 1997 UBC, but the design values for e_n, G *and* t were not included when the equations were incorporated into the 2000 IBC. A code change incorporated updated tables of values from the 1997 UBC Volume 3 and added them to 2006 IBC Section 2305.2.2. There were three modifications made to the tables in the 1997 UBC Volume 3 that were included in the code change. The first change combined UBC Tables 23-2-H, 23-2-I and 23-3-J into a single table, eliminating extraneous information and combining separate G and t tables into a single table that gives the panel rigidity-through-the-thickness, Gt, for all wood structural panels, not just plywood. The second change is new values for oriented strand board based on values in APA's technical publication on APA *Performance-Rated Structural-Use Panels*. The values are recognized by the industry for all products meeting the criteria of either Voluntary Product Standards PS1 or PS2, both of which are referenced standards in the IBC. The third modification is the inclusion of the e_n values for 14-Ga x 2 inch staples, which are based on the original e_n equations in APA Lab Report RR-138. APA Lab Report RR-138 is the source document for the 4-term equation. The APA report describes the original derivation and gives the original equations for the development of the e_n values. Including the appropriate values in the 2006 IBC makes the equations useable and reduces the possibility for errors in the calculation process. It also provides a forum for future changes to these values in future editions of the code if necessary.

There is currently no rational method to calculate the deflection of an unblocked diaphragm. Monotonic racking tests indicate that the deflection of an unblocked diaphragm may be on the order of three times the deflection of blocked diaphragm.

2305.2.3 Diaphragm aspect ratios. The limits on diaphragm aspect ratios are empirically based on observations of earthquake damage. Diaphragms with aspect ratios less than the tabular limits appear to perform satisfactorily in most cases. The tabular values are similar to the limits in the NEHRP, except NEHRP limits the aspect ratio for double diagonal sheathing to 3:1.

2305.2.4 Construction. Although it was not explicitly stated in the 2003 IBC, the requirements of this section for exterior size and 4 by 8 panel size, except at boundaries and changes in framing, also apply to shear walls. A new Section 2305.3.3 was added to the 2006 IBC that prescribes the same requirements for shear walls as is required for diaphragms.

Although it is not clearly stated in the 2003 IBC, the minimum dimension of any sheet-type sheathing is 2 feet in any direction. The following additional language from the 1997 UBC was added to the 2006 IBC section, "where minimum sheet dimension shall be 24 inches unless all edges of the undersized sheets are supported by and fastened to framing members or blocking."

2305.2.4.1 Seismic Design Category F. In Seismic Design Category F, wood structural panels must be applied directly to the framing members. The use of gypsum wallboard between the wood structural panels and the framing members is prohibited because of the poor performance of nails in gypsum. NEHRP applies this restriction for fastening the wood structural panels directly to the diaphragm to structures assigned to Seismic Design Category E as well as Seismic Design Category F.

2305.2.5 Rigid diaphragms. Section 1602 of the 2000 IBC defines a diaphragm as being flexible when the maximum lateral deformation of the diaphragm is more than two times the average story drift determined by comparing the maximum deflection of the diaphragm to the corresponding maximum deflection of the adjoining vertical resisting elements of the lateral-force-resisting system. Conversely, a diaphragm will be considered rigid when the diaphragm deflection is equal to or less than two times the average story drift.

Section 1602 of the 2006 IBC states that a diaphragm is flexible for the purpose of distribution of story shear and torsional moment where indicated by Section 12.3.1 of ASCE 7, as modified in Section 1613.6.1. Section 1613.6.1 permits the assumption of flexible diaphragm under specific conditions. Diaphragms constructed of wood structural panels are permitted to be considered flexible, provided all of the following conditions are met: (1) Toppings of concrete or similar materials are not placed over wood structural panel diaphragms except for nonstructural toppings no greater than $1^1/_2$ inches thick; (2) Each line of vertical lateral-force-resisting elements complies with allowable story drift requirements of ASCE 7 Table 12.12-1; (3) Vertical lateral-force-resisting elements are light-framed walls sheathed with wood structural panels rated for shear resistance or steel sheets; and (4) Portions of wood structural panel diaphragms that cantilever beyond the vertical lateral-force-resisting elements are designed in accordance with Section 2305.2.5.

For flexible diaphragms, seismic forces should be distributed to the vertical resisting elements according to tributary area or simple beam analysis. Although rotation of the diaphragm may occur because lines of vertical elements have different stiffnesses, the diaphragm is not considered stiff enough to redistribute seismic forces through rotation. The diaphragm can be visualized as a single-span beam supported on rigid supports.

For diaphragms defined as rigid, rotational or torsional behavior is expected and results in redistribution of seismic forces to the vertical-force-resisting elements. It should be noted that the definition of rigid diaphragm is still in the 2006 IBC in Section 1602. Requirements for distribution of horizontal shear and torsional shear are in Section 12.8.4 of ASCE 7. Torsional response of a structure caused by irregular stiffness at any level within the structure can be a potential cause of failure. As a result, dimensional and diaphragm ratio limitations are provided for different categories of rotation. Additional requirements apply when the structure is deemed to have a torsional irregularity in accordance with Table 12.3-1 of ASCE 7.

To understand limits placed on diaphragms acting in rotation, it is helpful to consider two different categories of diaphragms. Category I includes rigid diaphragms that rely on force transfer through rotation to maintain stability and are thus torsionally irregular. An example would be an open-front structure with shear walls on the remaining three sides. For this category of structures, applicable limitations are set forth for wood diaphragms:

- Wood diaphragms shall not be used to resist forces contributed by masonry or concrete in structures over one story (see exceptions in Section 2305.1.5).
- The length of the diaphragm normal to the opening shall not exceed 25 feet (to perpendicular shear walls), and diaphragm *l/w* ratios are limited to 1.0 for one-story structures and 0.67 for more than one story.
- Additional limitations are necessary when rotation is significant enough to be considered a torsional irregularity.

Category II includes rigid diaphragms that have two or more supporting shear walls in each of the two perpendicular directions but, because the center of mass and center of rigidity do not coincide, there is redistribution forces to shear walls through rotation of the diaphragm. These can be further divided into Category IIA where the center of rigidity and mass are separated by a small portion of the structure's least dimension and the magnitude of the rotation is on the order of the accidental torsion specified in Section 12.8.4.2 of ASCE 7. For this level of rotation, Section 2305.1.5 Exception 1 might be considered applicable and, as a result, no particular limitations would be placed on diaphragm rotation for Category IIA. Category IIB, rigid diaphragms with eccentricities larger than those discussed in Section 12.8.4.2 of ASCE 7 should be subject to the following limitations for wood diaphragms:

- Diaphragm shall not be used to resist forces contributed by masonry or concrete in structures over one story.
- Additional limitations shall apply when rotation is significant enough to be considered a torsional irregularity.

Table 2305.2.3 provides limits for diaphragm ratios. Because flexible diaphragms have very little capacity for distributing torsional forces, further limitation of aspect ratios to less than the tabular maxima is used to limit diaphragm deformation such that rigid behavior will occur. The resulting deformation demand on the structure will then be limited. Where diaphragm ratios are further limited, exceptions permit higher ratios if calculations demonstrate that higher diaphragm deflections can be tolerated; i.e., deformation compatibility. In this case, it is important to determine the effect of diaphragm rigidity on the horizontal distribution and the ability of other structural and nonstructural elements to withstand the resulting element deformations.

Based on the above, open-front structures using wood diaphragms are severely limited in size to small buildings. The maximum dimensions are 25 feet deep by $37^1/_2$ feet wide. Similarly, structures with cantilevered diaphragms are allowed with the maximum dimension of the cantilever diaphragm limited to the lesser of two-thirds the diaphragm width or 25 feet. This limits the maximum plan dimension of the structure perpendicular to the cantilever to $37^1/_2$ feet.

The last class of structures allowed in this section are those with limited torsional irregularities. The diaphragm aspect ratios are limited such that the diaphragm can be assumed rigid if using wood shear walls. These structures could not exceed one story in height if a masonry or concrete structure (see Section 2305.1.5) and the capacity of the wood diaphragm is presumed adequate for those demands.

2305.3 Design of wood shear walls.

2305.3.1 General. A shear wall is a vertical structural unit that acts as a deep beam. The analogy to a beam is appropriate, as beams and shear walls are made up as assemblies. Sheathing and intermediate framing members act as the web to resist the shear in the shear wall. The sheathing is stiffened by the framing members, which also provide support for gravity loads. Flexure is resisted by the edge or boundary elements acting like flanges to resist induced tension or compression forces. The flanges are often called hold-down posts or chords.

The flange (chord) functions as the tension or compression flange of the beam. It is important that the connection to the web be designed to accomplish the shear transfer. Because multistory shear wall flanges consist of several pieces, it is important that the splices be designed to transmit the tension or compression occurring at the location of the splice and to recognize that the direction of application of seismic forces can reverse. The splices should be designed to minimize slip in the connection. A significant proportion of the shear wall deflection is caused by deformation in the chord splice from the axial load (see Equation 23-2, last term). It is important that the flange members, splices at intermediate floors, and the connection to the foundation be detailed and sized for the induced forces.

The webs of shear walls often have openings. The transfer of forces around openings can be treated similarly to openings in the webs of steel girders. Members at the edges of openings have forces that are due to flexure and the higher web shear induced in them, and the resultant forces must be transferred into the body of the diaphragm beyond the opening. Section 2305.3.8.2 permits unreinforced openings in shear walls (no force transfer design) with added restrictions outlined in the section. This method evolved into what is now referred to as the perforated shear wall method in the 2003 IBC and expanded further in the 2006 IBC.

The IBC has three types of shear wall design methods. The traditional method has full height shear wall segments that comply with the aspect ratio requirements and are usually restrained against overturning by hold-down devices. The second method considers the entire shear wall with openings in it, and the wall piers adjacent to the openings are wall segments. The method requires the forces around the openings to be designed and detailed. With this method, the hold down devices generally occur at the ends of the shear wall, not at each wall pier. See Figure 2305.3.5. The third and newest method is the perforated shear wall method, which is an empirical approach that does not require the forces adjacent to the openings to be addressed in the design, and no reinforcement at the openings is required. The perforated shear wall method specifically requires a hold-down device at the ends of perforated shear wall. See further discussion under Section 2305.3.8.2.

Note that only wood structural panel shear walls may be used for structures assigned to Seismic Design Category E or F. The uplift forces in the boundary elements must be resisted by either a portion of the structural dead load or by hold-down devices. See the commentary in Section 2305.3.6 for a discussion of hold-down requirements and considerations.

2305.3.2

Deflection. The "permissible deflection" in the first paragraph is limited by the drift limits of Section 12.12.1 and Table 12.12-1 of ASCE 7. See footnote c to Table 12.12-1 for deformation compatibility requirements for single-story buildings.

Equation 23-2 is limited to blocked wood structural panels with uniform nailing. There are currently no acceptable rational methods for determining the deflection of shear walls constructed of unblocked wood structural panels, single or double diagonally sheathed wood shear walls or shear walls constructed of particleboard, fiberboard, gypsum board or plaster. Thus, if using these materials, one must use methods other than rational calculations to satisfy the requirements for deformation compatibility and determination of whether the diaphragm is flexible or rigid relative to the shear wall. Except for diagonal sheathing, use of these alternative materials should be limited to the seismic design categories shown in Table 23-1 below.

Table 23-1. **Permitted shear wall material**

Material	Seismic Design Category
Wood structural panel (Sec. 2306.4.1)	A, B, C, D, E, F
Diagonally sheathed lumber (Sec. 2306.4.2)	A, B, C, D
Particleboard (Sec. 2306.4.3)	A, B, C
Fiberboard (Sec. 2306.4.4)	A, B, C
Gypsum board (Sec. 2306.4.5)	A, B, C, D[1]
Cement Plaster (Sec. 2306.4.5)	A, B, C, D[1]

[1]Subject to the limitations prescribed in Section 12.2.1 and Table 12.2-1 of ASCE 7. Building height limited to 35 feet.

Although their use is not restricted by the code, NEHRP Section 12.4.3.2 limits the use of particleboard, fiberboard, gypsum and plaster wall bracing to conventional construction in structures assigned to Seismic Design Category A, B or C. These alternative materials are not allowed in engineered construction in the NEHRP *Provisions*.

The capacity of shear walls must be based on experimental results as well as the observations of performance during earthquakes. The NEHRP does not allow shear walls sheathed with other than wood or wood-based structural-use panels to resist seismic lateral forces in engineered construction. Many of the NEHRP *Provisions* are a result of lessons learned from past earthquakes through observations of performance of structures sheathed with these materials during the Northridge earthquake and laboratory testing.

Equations 23-1 and 23-2 originated with the 1997 UBC, but the design values for e_n, G, and t were not included when the equations were incorporated into the 2000 IBC. A code change incorporated updated tables of values from the 1997 UBC Volume 3 and added them to 2006 IBC Section 2305.2.2. There were three modifications made to the tables in the 1997 UBC Volume 3 that were included in the code change. The first change combined UBC Tables 23-2-H, 23-2-I and 23-3-J into a single table, eliminating extraneous information and combining separate G and t tables into a single table that gives the panel rigidity-through-the-thickness, Gt, for all wood structural panels, not just plywood. The second change is new values for oriented strand board based on values in APA's technical publication on APA Performance-Rated Structural-Use Panels. The values are recognized by the industry for all products meeting the criteria of either Voluntary Product Standards PS1 or PS2, both of which are referenced standards in the IBC. The third modification is the inclusion of the e_n values for 14-Ga x 2 inch staples which are based on the original e_n equations in APA Lab Report RR-138. APA Lab Report RR-138 is the source document for the four-term equation. The APA report describes the original derivation and gives the original equations for the development of the e_n values. Including the appropriate values in the 2006 IBC makes the equations useable and reduces the possibility for errors in the calculation process. It also provides a forum for changes to these values in future editions of the code if necessary.

2305.3.3 **Construction.** In the 2003 IBC, Section 2305.2.4 contained information specific to diaphragms and shear walls as well as information common to both. The shear wall requirements in Section 2305.2.4 belong under Section 2305.3 instead of Section 2305.2. Section 2305.3.3 pertaining to construction of shear walls was added to the 2006 IBC for clarification. The reference to panels of $^1/_4$-inch thickness applies to shear walls only because the shear-wall tables have values for $^1/_4$-inch sheathing, but the diaphragm tables do not. Therefore, this language was relocated under Section 2305.3.3 in the 2006 IBC.

2305.3.4 **Shear wall aspect ratios.** Table 2305.3.4 limits the height to width ratio (h/w) for wood framed shear walls. The height-to-width ratio for single/sheathed shear walls is limited to 2:1 in accordance with the NEHRP *Provisions*.

Limits for fiberboard walls are also shown in the table. Wood structural panel shear walls are limited to 2:1 for seismic loads and $3^1/_2$:1 for other loads such as wind. Two footnotes were added to the 2003 IBC. For seismic loads, footnote "a" allows shear wall aspect ratios greater than 2:1 but not exceeding $3^1/_2$:1, provided the allowable shear value in Table 2306.4.1 is multiplied by 2w/h. Table 23-2 shows the factor 2w/h for various height-to-width ratios.

Table 23-2. **Reduction factors 2w/h**

Aspect Ratio h/w	Reduction Factor 2w/h
2.00	1.00
2.25	0.89
2.50	0.80
2.75	0.73
3.00	0.67
3.25	0.62
3.50	0.57

The code change that resulted in the 2w/h adjustment is based on a revision in the 2000 NEHRP *Provisions*. Many tests were conducted regarding the performance of wood-framed wall segments with narrow aspect ratios. The tests demonstrated that shear walls with narrow aspect ratios significantly affect stiffness and capacity when the aspect ratio exceeds 2:1. However, this does not preclude the use of narrow segments, provided the strength and stiffness effects are properly accounted for in the design. The 2w/h adjustment is based on a review of the shear wall test data and represents a conservative adjustment to the allowable shear design values for wood-framed shear walls. The maximum aspect ratio limit of 3.5:1 is maintained to ensure constructability and effective installation of hold-down devices.

The aspect ratio for wood-framed shear walls sheathed with gypsum board, gypsum lath or cement plaster were not incorporated into the 2000 IBC from the legacy model codes. This created some confusion among code users that these materials may not be permitted in wood-framed shear wall construction. Subsequently gypsum board, gypsum lath, cement plaster and footnote b were added to the 2003 IBC. The $1^1/_2$:1 aspect ratio shown in the table applies to unblocked construction. The aspect ratio is permitted to be increased to 2:1 when the wall is blocked construction as shown in Table 2306.4.5. The $1^1/_2$:1 ratio originated in all three legacy codes, and footnote b originated with the 1997 UBC.

Many clarification changes were made to the provisions for shear wall aspect ratios, shear wall heights and widths in the 2006 IBC. The changes include new definitions for perforated shear wall segment height, force transfer shear wall pier height and force transfer shear wall pier width. There are essentially three types of wood frame shear walls in the IBC: segmented (traditional) shear wall, shear walls with openings with force transfer detailed at the openings, and perforated shear walls without force transfer at the openings. The code change in the 2006 IBC clarified the requirements for the different shear wall types.

The terms *adjusted shear resistance* and *unadjusted shear resistance* were deleted in the 2006 IBC because they weren't actually definitions but technical requirements and were relocated to the technical provisions for design of perforated shear walls in Section 2305.3.8.2.2. Section 2305.3.3 was intended to limit the aspect ratio (height-to-width ratio) of segmented shear walls, force transfer piers and perforated shear walls but was not clear because the language in the 2003 IBC only specified aspect ratios for shear walls. The limitations on aspect ratio for shear walls shown in Table 2305.3.4 apply to segmental shear walls, full height segments in perforated shear walls and to wall piers in shear walls where the perimeter of openings are designed for force transfer. The changes in the 2006 IBC clarified the aspect ratios, height and width definitions and refer to Figure 2305.3.5 for clarification. Figure 2305.3.5(a) shows height and width for segmental shear walls and the segments of perforated shear walls. Figure 2305.3.5 shows height and width for shear walls with openings with force transfer around openings.

2305.3.5

Shear wall height definition. See Figure 2305.3.4(a) to understand the definitions. These definitions were excerpted from the NEHRP *Provisions* and are consistent with the dimensions illustrated in the code.

2305.3.5.1

Perforated shear wall segment height definition. This is a new section in the 2006 IBC, defining the height of a perforated shear wall segment, *h*, as the same as specified in Section 2305.3.5 for shear walls.

2305.3.5.2

Force transfer shear wall pier height definition. This is a new section in the 2006 IBC, defining the height, *h*, of a wall pier in a shear wall with openings designed for force transfer around openings as the clear height of the pier at the side of an opening as shown in Figure 2305.3.5(b).

2305.3.6

Shear wall width definition. The NEHRP *Provisions* recommend that the minimum width of a shear wall or shear wall segment be 2 feet. There are no cyclic test results for shear walls having a width less than 2 feet.

In the 2000 IBC the width of a shear wall was defined as the horizontal dimension of the shear wall sheathed between overturning restraints. That definition essentially made it impossible to meet a 2 to 1 aspect ratio with a typical wall consisting of only one 4 foot by 8 foot wood structural panel. The language in the 2003 IBC was changed to be consistent with the 2000 NEHRP *Provisions*. The revised definition for width identifies the *sheathed width* as the width dimension, and defines the width of a shear wall as the sheathed dimension of the shear wall in the direction of application of force.

2305.3.6.1

Perforated shear wall segment width definition. This is a new section in the 2006 IBC, defining the width of a perforated shear wall segment, *w*, as the width of full-height sheathing adjacent to openings in the perforated shear wall as shown in Figure 2305.3.5(a).

2305.3.6.2 Force transfer shear wall pier width definition. This is a new section in the 2006 IBC, defining the width, *w*, of a wall pier in a shear wall with openings designed for force transfer around openings as the sheathed width of the pier at the side of an opening as shown in Figure 2305.3.5(b).

2305.3.7 Overturning restraint. The net overturning moment at shear walls must be resisted by anchorage devices that are referred to as hold-downs, which are critical to satisfactory performance. The ultimate capacity of a shear wall is often controlled by the capacity of the hold-down rather than the shear capacity of the shear wall.

The allowable shear capacity of a shear wall is determined by dividing the minimum ultimate capacity by a safety factor of 2.5. Therefore, the ultimate capacity of the shear wall is at least 2.5 times the allowable shear. If the resistance from the combination of the hold-down capacity and the effective dead load is less than the boundary element uplift force associated with the ultimate shear capacity of the wall, then the wall capacity is limited by the hold-down capacity. The actual shear forces on the structure can easily be three times the allowable design force determined from the reduced base shear resulting from $E/1.4$. Thus, the wall will likely be loaded to its ultimate shear capacity when subjected to the design earthquake. Although not explicitly required by the code, hold-downs should be sized based on the lesser of the force determined using the special load combinations of Section 1605.4, $\Omega_o E$, or the ultimate shear strength of the wall so as to match the capacity of the hold-downs to the strength of the shear wall and prevent a weak link in the load path. The resistance of the hold-down may be based on the lesser of the ultimate capacity of the hold-down, the attachment of the hold-down to the boundary element, or the anchorage of the hold-down to the foundation or floor. The ultimate capacity for nails and bolts with $l/d \geq 4$ is approximately three times the allowable load. The ultimate capacity of an anchor bolt can be determined from the strength provisions for anchor bolts in ACI 318 using $\phi = 1.0$.

The code allows the deduction of effective dead load ($0.9D$) resistance from the uplift force determined from $E/1.4$. The hold-down is then designed for the net uplift force. This will result in under-capacity hold-downs as noted in the paragraph above. The uplift force should be determined using the capacity design approach for the shear force in the wall associated with $\Omega_o E$ prior to deduction of the effective dead load ($0.9D$). This will result in hold-downs with sufficient capacity to resist the ultimate shear capacity of the wall.

When using the above capacity approach, the foundation should, in turn, be designed for the ultimate capacity of the hold-down. The capacity of the foundation should be determined in accordance with ACI 318 using $\phi = 1.0$. Because most foundation widths and depths on low-rise buildings are determined by considerations other than foundation strength, this approach will seldom require any change in foundation dimensions, but may require a moderate increase in reinforcing steel.

Another issue for hold-down devices is the amount of deformation or slip allowed by the device. The purpose of the hold-down is to resist load and deflection in the form of uplift on the chord of the shear wall with zero slip. Hold-down devices should have zero slip relative to the chord before the uplift force is resisted. That is, the hold-down should resist uplift forces immediately, not only after a large amount of the force is resisted by the sheathing-to-plate nails because of slip.

Hold-down devices that permit significant vertical movement between the hold-down and the hold-down post can cause failure in the nails connecting the shear wall sheathing to the sill plate. High-tension forces and hold-down rotation caused by eccentricity may cause the bolts connecting the hold-down bracket to the hold-down post to pull through and split the hold-down post. Devices that permit such movement include heavily loaded one-sided bolted connections with small dimensions between elements resisting rotation that is due to eccentricity. Any device that uses over-drilled holes, such as most bolted connections, will also allow significant slip to occur between the device and the hold-down post before the load is effectively restrained. Both the NDS and the steel manual specify that bolt holes be over-drilled by not more than $^1/_{16}$ inch. This slip can cause much of the damage to the nails connecting the sheathing to the sill plate. Friction between the hold-down post and the device cannot be counted on to resist load because relaxation in the wood will cause a loss of clamping and, therefore, a loss in friction over time. This is why all hold-down qualification tests should be conducted with the bolts finger tightened as opposed to being tightened with a wrench.

Ideally, hold-down devices should exhibit zero slip as noted above. Examples of such devices are hold-downs attached to the hold-down post (chord) with nails and then bolted through the sill plate to the foundation or stud of a wall below with the anchor bolt tensioned to preload the hold-down.

Hold-downs should be qualified by cyclic testing. When nails, bolts or similar types of fasteners are used to connect the device to the chord, cyclic tests using sequential phased displacement test procedures, or the equivalent, must be required to show that no loss in wall capacity, ductility or stiffness is caused by the hold-down device. The bolts should be finger tightened, and friction between the tie-down device and the chord shall not be counted on to resist load. The tests should also simulate the deformation patterns expected in a wall assembly during a seismic event and use full wall assemblies.

One other item of consideration concerning shear walls and hold-downs is the eccentric force induced in the hold-down post (chord member). Hold-downs applied to one side of a chord stud can induce significant bending moments in the chord. The combined bending and tension in the chord can cause a brittle failure of the chord. Qualification testing of hold-downs for compliance with the City of Los Angeles interim standards[10] indicated that many manufactured hold-downs experienced a brittle failure of the chord at approximately two times the allowable loads; larger hold-downs (1-inch bolts) in some cases failed the chords prior to reaching the allowable load. The poor performance may be attributed to the manner in which the hold-downs were tested to establish allowable loads. Hold-down devices were typically tested in a steel jig rather than in an assembly resembling actual shear wall conditions. The allowable loads were based on the lower of the ultimate load divided by a factor of safety or the load calculated from the allowable bolt loads. To minimize the brittle chord failures, hold-downs should be qualified from assembly tests that resemble actual conditions of end use, and chords should be designed for combined tension plus induced moment from eccentricity. An alternative is to consider use of one hold-down on each side of the chord. See Reference 10 for a thorough discussion of the phenomena related to hold-down eccentricity.

In consideration of the above, the uplift force should be calculated using the distance between the centerline of the hold-down anchor and the centroid of the compression chord.

2305.3.8

Shear walls with openings. There is very little cyclic test data for shear walls with openings. Most of the provisions in this section are based on the principles of engineering mechanics or limited monotonic testing.

2305.3.8.1

Force transfer around openings. The minimum width or a wall pier (shear wall segment) should be 24 inches as recommended by NEHRP. This is also consistent with the minimum panel width of 24 inches required by Section 2305.2.4. Boundary members should extend past the opening far enough to develop the forces into the shear wall. See *ATC-7* for an example of the applicable mechanics and methods suitable for reinforcement of the opening. The analysis uses a Vierendel truss analogy.

2305.3.8.2

Perforated shear walls. There are essentially three methods for designing shear walls in wood-frame buildings: 1) the segmented approach simply ignores the openings and only includes full-height segments that are designed for the allowable shear values and may require hold-downs at each end of each segment, 2) the method where wall piers are designed and detailed with force transfer around the openings and 3) the perforated shear wall method that is described in Section 2305.3.8.2. The method used to design shear walls without force transfer around the openings is based on the perforated shear wall design method by AF&PA.[11]

The perforated shear wall design method of Section 2305.3.8.2 is a recently developed empirical method. This method recognizes the strength and stiffness contributed by sheathing above and below the wall openings. Historically, designers have used principles of mechanics to design wall areas above and below openings similar to coupling beams. This approach results in specific and often complicated detailing requirements that are difficult to inspect and construct. The perforated shear wall approach does not require special detailing for continuity around openings and thus is easier to inspect and construct in the field.

Traditional wood frame shear wall design uses full-height segments that comply with the aspect ratio requirements. The wall areas between segments that don't comply with aspect ratio are ignored and treated as openings. The sheathing above and below the openings, if present, is also ignored. In general, hold-down devices for overturning restraint were located at each end of the full height segments.

In 1981, Professor Hideo Sugiyama at the University of Tokyo, Japan, proposed an empirical equation to estimate the shear capacity and stiffness of shear walls with openings without intermediate overturning restraints. Sugiyama's equation serves as the basis for the perforated shear wall design method in the IBC. Sugiyama's equation and the resulting design procedure are discussed in several articles published by AF&PA. See reference 11. Full scale tests were done (Dolan[12] et al, 1996) that provided further verification of the empirical equation used as the basis of the procedure. The *Wood Frame Construction Manual* (WFCM) *for One- and Two-Family Dwellings – 1995 SBC High Wind edition* recognized the perforated shear wall without intermediate overturning restraint and unreinforced openings as an acceptable design method. Several refinements and clarification of the provisions were provided in the 2003 and 2006 IBC. The current procedure in the 2006 IBC has the limitations prescribed in Section 2305.3.8.2.1. Additional research is expected to modify the limitations and further expand the application of the method.

The key features of the perforated shear wall approach are 1) the full-height segments adjacent to the openings must comply with the aspect ratio limitations of Table 2305.3.4, 2) hold-downs need only be provided at the ends of the overall shear wall, not at the ends of each full height segment, 3) the capacity of the perforated shear wall is adjusted to be less than that of a segmented shear wall containing multiple segments and 4) full-height segments must be located at each end of the overall shear wall.

The method was originally incorporated into the 1995 edition of the *Wood Frame Construction Manual SBC High Wind Edition* and the 2000 IBC. The wording used in these documents created some confusion regarding proper application of the procedure. The language was improved in the 2000 NEHRP *Provisions* in an attempt to clarify the intent of the provisions. Code changes to the perforated shear wall provisions in the 2006 IBC provide clarification and more uniform application of the requirements.

There are several articles published by AF&PA pertaining to the origin and development of the perforated shear wall design approach. See reference 11.

2305.3.8.2.1 **Limitations.** Specific limitations were added to this section in the 2003 IBC. (1) Full-height segments are required at each end of a perforated shear wall. Openings are permitted to occur beyond the ends of the perforated shear wall, provided the width of the openings is not included in the width used to design the perforated shear wall. (2) The allowable shear used in the design from Table 2306.4.1 cannot exceed 490 plf. (3) Where there are out-of-plane offsets in a perforated shear wall, the portions of the wall on each side of the offset are considered separate perforated shear walls. (4) Collectors for shear transfer must be provided through the full length of the perforated shear wall. (5) A perforated shear wall must have uniform top of wall and bottom of wall elevations. Perforated shear walls not having uniform elevations at the top and bottom must be designed by other methods. (6) Perforated shear wall height, h, cannot exceed 20 feet.

2305.3.8.2.2 **Perforated shear wall resistance.** The resistance of perforated shear walls is calculated in accordance with this section. There were changes in the 2003 IBC that clarified the procedure for determining perforated shear wall resistance. The percentage of full-height sheathing is determined by calculating the sum of the widths of segments divided by the total width of the perforated shear wall, including openings. The maximum opening height is the maximum opening clear height that occurs in the perforated shear wall. If areas above and below an opening are unsheathed, then the height of the opening is the height of the wall. The unadjusted shear resistance is the allowable shear given in Table 2306.4.1 if the height-to-width ratio of the segments does not exceed 2:1 for seismic design and $3^1/_2$:1 for wind design. Where the height-to-width ratio of any perforated shear wall segments is greater than 2:1 but does not exceed $3^1/_2$:1, then the unadjusted shear resistance is multiplied by 2w/h for seismic design. The adjusted shear resistance is calculated by multiplying the unadjusted shear resistance given in Table 2306.4.1 by the shear resistance adjustment factors given in Table 2305.3.8.2. The section permits interpolation of adjustment factors in Table 2305.3.8.2 for intermediate percentages of full-height sheathing. The total perforated shear wall resistance is equal to the adjusted shear resistance times the sum of the widths of the perforated shear wall segments.

Anchorage and load path. The design requirements for perforated shear wall anchorage and load path detailing are given in Sections 2305.3.8.2.4 through 2305.3.8.2.8 or are based on recognized principles of mechanics. Wall framing, sheathing, sheathing attachment and fastener schedules must conform to the requirements of Section 2305.2.4 and Table 2306.4.1, except as specifically modified by these sections. **2305.3.8.2.3**

Uplift anchorage at perforated shear wall ends. Although hold-down devices to resist overturning forces are not required at wall segments, anchorage devices to resist uplift forces that are due to overturning must be provided at each end of the perforated shear wall. The uplift anchorage must comply with the general overturning restraint requirements of Section 2305.3.7, except that the section prescribes a minimum tension chord uplift force determined by Equation 23-3. **2305.3.8.2.4**

Anchorage for in-plane shear. The unit shear force, v, transferred into the top and out of the base of the perforated shear wall at full-height segments and into collectors connecting shear wall segments is determined by Equation 23-4. **2305.3.8.2.5**

Uplift anchorage between perforated shear wall ends. In addition to the uplift anchorage for the ends of the perforated shear wall required by Section 2305.3.8.2.4, the bottom plate of full-height segments are required to be anchored for a uniform uplift force, t, equal to the unit shear force, v, determined by Equation 23-4. **2305.3.8.2.6**

Compression chords. The section requires each end of each wall segment to be designed to resist a compression chord force, C, equal to the tension chord uplift force, T, determined by Equation 23-4. **2305.3.8.2.7**

Load path. A continuous load path to the foundation must be provided for uplift forces T and t, for shear forces, V and v, and for compression chord force, C. Although it may be obvious, this section requires elements resisting shear wall forces contributed by multiple stories to be designed to resist the sum of forces contributed by each story. **2305.3.8.2.8**

Deflection of shear walls with openings. The section gives the deflection of shear walls with openings to be taken as the maximum individual deflection of the shear wall segments determined in accordance with Equation 23-2 and divided by the applicable shear resistance adjustment factor from Table 2305.3.8.2. Note that Equation 23-2 applies to fully blocked shear walls that are uniformly fastened throughout. **2305.3.8.2.9**

Summing shear capacities. The capacities of dissimilar materials to either the same side, the opposite side or in the same line are not cumulative, because of incompatible stiffnesses and maximum usable strains. For example, the cracked stiffness of cyclically loaded gypsum is obviously much less than the stiffness of a wood structural panel subjected to the same seismic load. This holds true whether the dissimilar materials are in series or in parallel. Additionally, combinations of dissimilar materials have not been cyclically tested. **2305.3.9**

The section states that the shear values for the materials of the same thickness applied to both faces of the same wall are cumulative. Shear values for shear panels of different materials applied to the same face of the wall are not cumulative except as allowed in Table 2306.4.1. This means that the only case where two different materials can be combined on one face are wood structural panels applied over gypsum sheathing as specifically shown in Table 2306.4.1.

Since the 2000 IBC, the requirements for summing shear capacity underwent several clarifications. The 2000 IBC stated that where the material thicknesses are not equal, the allowable shear is equal to two times the shear capacity of the thinner material or the capacity of the thicker material, whichever is greater. Summing shear capacities of dissimilar materials applied to both faces or to the same wall line is not allowed. The exception for wind design permitted shear values for dissimilar materials applied to both faces of the same wall to be cumulative.

The language of the provision was found to be somewhat problematic. The code section intended to provide guidance on how to evaluate a shear wall with shear-resisting sheathing attached to opposite sides of the wall. The real intent of the section was to allow the use of two times the side of lesser *capacity* or the greater *capacity* side as the total *capacity* of the shear wall. The thickness of the sheathing materials is not the only variable that must be considered when evaluating the shear capacity of a given assembly. The nail size, the spacing, as well as the grade are equally important. It is the shear-resisting capacity of the sheathing on either side of the shear wall that must be evaluated, not just the thickness of the sheathing materials.

The problem can be illustrated in the following example: Assume a shear wall covered on one side with $^3/_8$-inch-thick Structural I sheathing fastened with 8d nails at 3 inches on center, which has a design capacity of 460 plf. The other side of the wall is covered with $^1/_2$-inch-thick sheathing with 8d nails at 6 inches on center, which has a design capacity of 260 plf. The way the section was written in the 2000 IBC, the greater of two times the thinner sheathing ($2 \times 460 = 920$ plf), or the capacity of the thicker side (260 plf), would be the allowable shear capacity of the wall. Thus, 920 plf would be allowed for this wall. This is incorrect and was not the intent of the code. In fact, 920 plf is greater than the cumulative capacities of both sides of the wall. The correct answer is obtained by using the proposed wording where the allowable capacity for the whole wall is the greater of two times the lesser capacity side or the greater capacity side. In this case, the greater of 520 plf (2×260) or 460 plf. The correct capacity of the two-sided wall in this example is two times the smaller shear capacity or the capacity of the stronger side, whichever is greater, which in this case is $2 \times 260 = 520$ plf. Notice that, in this example, the original wording yields a value that is 76 percent greater than the correct answer. An error of this magnitude is, needless to say, unacceptable.

Therefore, the language in the 2003 IBC was changed to clarify the wording while maintaining the original intent of the provision. The proposal also changed the word *both* to *opposite* in the exception. The use of *both faces* could be interpreted to mean that both materials had to be applied to each face. The word *opposite* faces is a better descriptor of what is intended—a different sheathing on each face. The result of these modifications are summarized as follows:

- Summing shear capacities of dissimilar materials applied to opposite faces is not allowed.
- Summing shear capacities of dissimilar materials in the same wall line is not allowed.
- The shear values for material of the same type and capacity applied to both faces of the same wall are cumulative.
- Where of the same material but the capacities are not equal, the allowable shear is two times the smaller shear capacity or the capacity of the stronger side, whichever is greater.
- For wind design, shear wall segments sheathed with wood structural panels, fiberboard structural sheathing or hardboard panel siding on one face and gypsum wallboard on the opposite face, the allowable shear capacity is equal to the sum of the sheathing capacities of each face separately.

2305.3.10

Adhesives. The provisions are based on assemblies having energy dissipation capacities that were recognized in setting the *R* factors. For diaphragms and shear walls utilizing wood framing, the energy dissipation is almost entirely due to nailbending. Fasteners other than nails and staples have not been extensively tested under cyclic load application. When adhesives have been tested in assemblies subjected to cyclic loading, they have had a brittle mode of failure. For this reason, adhesives are prohibited for wood-framed shear wall assemblies in high seismic areas, and only the tabulated values for nailed or stapled sheathing are allowed. Analysis and design of shear wall sheathing applied with adhesives are beyond the scope of the provisions. If one wished to use shear wall sheathing attached with adhesives as an alternate method of construction in accordance with Section 104.11, caution should be used.[12, 13] The increased stiffness will result in larger forces being attracted to the structure. The anchorage connections and adjoining assemblies must, therefore, be designed for these increased forces. Because of the brittle failure mode, these walls should be designed to remain elastic, similar to unreinforced masonry. The use of adhesives for attaching sheathing for diaphragms increases their stiffness, and could easily change the diaphragm response from flexible to rigid.

2305.3.11

Sill plate size and anchorage in Seismic Design Category D, E or F. Section 2305.3.11 contains special provisions for anchorage of foundation plates in buildings in Seismic Design Category D, E and F. Wood-frame shear wall tests demonstrate that the uplift movement at ends of shear walls that causes splitting of the sills can be reduced by installing square plate washers between the nut and sill. The 1997 and 2000 editions of the NEHRP recommends 3-inch by 3-inch by $^1/_4$-inch thick plate washers be used between the foundation sill plate and nut at shear walls and braced wall lines. The IBC *Final Draft*, which was used as the base document to develop the 2000 IBC, required 3-inch by 3-inch by $^1/_4$-inch thick plate washers because the requirement was based on the NEHRP *Provisions*. The size of the plate washers was changed from 3-inch by 3-inch by-$^1/_4$ inch to 2-inch by 2-inch by $^3/_{16}$-inch thick during the 2000 IBC code development process, which was consistent with the 1997 UBC. The reason given was lessons learned from the Northridge earthquake as well as adequate performance of sill plates tested with 2-inch by 2-inch by $^3/_{16}$-inch steel plate washers.

A code change to the 2003 IBC changed the square plate washer size to 3-inch by 3-inch by $^1/_4$-inch thick to be consistent with the NEHRP *Provisions*. However, a subsequent code change modified the minimum thickness from $^1/_4$ inch to 0.229 inches in order to allow the plate washers to be manufactured from cold rolled steel instead of $^1/_4$-inch hot-rolled steel. Cold-rolled steel of this thickness is more readily available, more economical and can be ordered with a hotdipped galvanized finish from the steel mill. Hot-rolled steel must be hot-dipped galvanized after fabrication. Some steel hardware manufacturers recommend that hotdipped galvanized plate washers be used where in contact with preservative-treated wood.

The hole in the plate washer is permitted to be diagonally slotted with a width up to $^3/_{16}$-inch wider than the bolt diameter with a length up to $1^3/_4$ inches, provided a standard cut washer is used between the square plate washer and the nut. The provision permitting slotted holes was taken directly from 2003 NEHRP *Provisions*.

Specific requirements for sill plate, anchorage and plate washers in Seismic Design Category D, E or F are based on the 1997 NEHRP *Provisions*. Shear values in this section are now given for both LRFD and ASD, insofar as LRFD is an accepted method of design.

The exception is often used in retrofit work where it is not practical to remove existing plates, and in new residential construction where short sections of wall with design shears greater than 350 pounds per linear foot would require all the sill plates to be increased to 3-inch members because of detailing difficulties encountered when wood floor framing bears on intermixed 3x and 2x plates.

Splitting of the bottom plate of the shear walls has been observed in tests as well as in structures subjected to earthquakes. Splitting of plates remote from the end of the shear wall can be caused by the rotation of individual sheathing panels inducing upward forces in the nails at one end of the panel and downward forces at the other. With the upward forces on the nails and a significant distance perpendicular to the wall to the downward force produced by the anchor bolt, high cross-grain bending stresses occur. Splitting can be reduced or eliminated by use of square plate washers sufficiently stiff to reduce the eccentricity and by using thicker sill plates. Thicker sill plates (3 inches nominal, 65 mm) are required for all shear walls for which 3-inch nominal (65 mm) framing is required. This helps prevent failure of the sill plate that is due to high lateral loading and cross-grain bending.

Section 2306 *Allowable Stress Design*

2306.1 Allowable stress design. The ANSI/AF&PA *National Design Specification (NDS) for Wood Construction* is adopted by reference without amendments. A major change for UBC users is the load duration factor for seismic, which was changed from 1.33 to 1.6 in the 1991 NDS. Although the 1997 UBC adopted the 1991 NDS, the duration factor for seismic and wind loads was amended in the UBC.

AITC 500, *Determination of Design Values for Structural Glued Laminated Timber* was deleted from the 2006 IBC because methods to determine design values for structural glued laminated timber are included in ASTM D3737, which is already a referenced standard in Chapter 35. AITC 500 is no longer being maintained and is no longer supported by the American Institute of Timber Construction.

The ANSI standard for Shallow Post Foundation Design, ANSI/ASAE EP486.1, was added to the 2003 IBC. The standard provides a design procedure for shallow post foundations that resist moments and lateral and vertical forces. The standard includes definitions, material requirements and design equations for designing post foundations. The Standard was developed by the ASAE Post and Pole Foundation Subcommittee, approved by the structures and Environment Division Standards committee and adopted by ASAE in March 1991. It was revised from 1992 to 1999. The 2006 IBC references the current 2000 edition.

2306.1.1 Joists and rafters. Although the section indicates that rafters may be designed using the span tables, the intent is that both joists and rafters are permitted to be designed by the *AF&PA Span Tables for Joists and Rafters*.

2306.1.2 Plank and beam flooring. Plank and beam flooring may be designed in accordance with the *AF&PA Wood Construction Data No. 4*.

2306.1.3 Treated wood stress adjustments. Several factors can significantly affect the physical properties of FRTW. These factors are the pressure treatment and redrying processes used, and the extremes of temperature and humidity that the FRTW will be subjected to once installed. The design values for all FRTW must be adjusted for the effects of the treatment and environmental conditions, such as high temperature and humidity in attic installations. The design adjustment values must be based on an investigation procedure that includes subjecting the FRTW to similar temperatures and humidities, and which has been approved by the building official. The FRTW tested must be identical to that which is produced. The building official reviewing the test procedure must consider the species and grade of the untreated wood, and conditioning of wood, such as drying before the fire-retardant-treatment process. A fire-retardant wood treater may choose to have its treatment process evaluated by ICC Evaluation Service.

The FRTW is required to be labeled with the design adjustment values. These design adjustment values can take the form of factors that are multiplied by the original design values of the untreated wood to determine its allowable stresses, or new allowable stresses that have already been factored down in consideration of the FRTW treatment.

2306.1.4 Lumber decking. The reference to AITC 112-93 in Section 2306.1 was removed from the 2006 IBC because AITC no longer maintains the standard. Although the American Institute of Timber Construction (AITC) no longer maintains the AITC 112-93 *Standard for Tongue-and-Groove Heavy Timber Decking*, design provisions for tongue-and-groove decking should still be available to designers, code users and building officials. Therefore, revisions to Section 2304.8 in the 2006 IBC along with the addition of new Section 2306.1.4 incorporated the pertinent provisions of AITC 112-93 directly into the body of the code and removed the reference to AITC 112-93. The title of Section 2304.8 was revised to indicate that the provisions cover all decking including mechanically laminated and solid sawn decking. The capacity of lumber decking is arranged according to the various layup patterns described in Section 2304.8.2. This new Section 2306.1.4 gives the design capacity of lumber decking for flexure and deflection according to the formulas given in Table 2306.1.4.

Wind provisions for walls. The allowable bending stress, F_b, in wood studs resisting wind loads may be increased in accordance with Table 2306.2.1 in lieu of using the repetitive member factor (C_r) increase of 1.15. **2306.2**

Wall stud bending stress increase. This increase recognizes that the stud and sheathing act compositely, which results in load sharing when properly designed and the appropriate blocking and fasteners are used. The wind pressures are transferred to the studs based on their relative stiffness to the sheathing. Because the minimum sheathing thickness is constant, the relative stiffness of the sheathing decreases as the stud depth increases; therefore, a higher proportion of load is carried by the stud as the stud depth increases. A minor change in the 2006 IBC clarified that the provision only applies to sawn lumber studs and the sheathing panel joints must occur over studs or blocking. **2306.2.1**

Wood diaphragms. The term *diaphragm* is only applied to horizontal or sloping panel elements that resist lateral forces. Vertical panel elements that resist lateral forces are termed *shear walls*. **2306.3**

Wood structural panel diaphragms. Allowable shears for these diaphragms may be taken from the table, or calculated by the principles of mechanics. The APA publication *Design/Construction Guide—Diaphragms*[8] provides a good example of the necessary calculations. For structures in Seismic Design Category D, E or F, the shear values should be based on cyclic tests or the tabular values and not calculated. **2306.3.1**

The values for allowable shear for nailed diaphragms are based on monotonic testing. The table also contains values for staples; these values are based on cyclic testing.

The minimum nail penetration requirements have been reduced from previous model code requirements where 8d and 10d nails are used. The minimum penetration for 10d nails in the 1997 UBC shear wall table is $1^5/_8$ inches, whereas in the IBC it is $1^1/_2$ inches. Studies undertaken using the European Yield Method (EYM) for calculating nail lateral values showed that the penetrations indicated in the tables in the legacy model codes were not necessary. The EYM methodology is included in the NDS. This change in minimum penetration permits the use of flat-wise blocking in diaphragms and shear walls where 10d nails are used.

When the framing lumber is of a species other than Douglas Fir-Larch or Southern Pine, the tabular values must be adjusted to account for the fastener behavior in framing members of other lumber species. The adjustments must be made in accordance with Footnote 1, but the resulting value may not exceed the tabular value.

High-load diaphragms shown in Table 2306.3.2 were added to the 2003 IBC. High-load diaphragms are required to have special inspection in accordance with Section 1704.6.1, as indicated in footnote "g" of Table 2306.3.2. The footnote was added to the table in the 2006 IBC so that code users are aware of the requirement for special inspection for high-load diaphragms.

The last footnote in the diaphragm and shear wall tables (Table 2306.3.1, Table 2306.3.2 and Table 2306.4.1) in the 2006 IBC was added to provide the factors to convert table values for shear loads of normal or permanent load duration as defined by the NDS. This was done in response to many inquiries over the years and a specific request by the Structural

Engineers Association of Washington. The proposed changes reflect the fact that the values in the tables have a built-in load duration factor of 1.6. As such, to convert to a normal load duration of 1.00, the tabular value must be multiplied by 0.63 ($1.6 \times 0.63 = 1.00$). To convert to a permanent load duration of 0.90, a factor of 0.56 must be used ($1.6 \times 0.56 = 0.90$). In addition, there were a number of editorial changes to Table 2306.3.2 in order for it to have the same format as other tables in Chapter 23.

2306.3.2 Shear capacities modifications. The allowable shear capacities for wood diaphragms may be increased 40 percent for wind design only. The wind loads determined by ASCE 7 have increased and recent research has provided a better understanding of how structures are affected by wind. Code-prescribed wind loads have been refined since the tables were developed. The tables use a 2.8 factor of safety. Wind loads are essentially monotonic and bounded; hence, a safety factor of 2 is sufficient for wind loading, resulting in a 40-percent increase in the tabulated values.

2306.3.3 Diagonally sheathed lumber diaphragms. Diagonally sheathed lumber diaphragms are seldom used in new construction, but are encountered in rehabilitation of older structures. The allowable shear values for diagonally sheathed lumber diaphragms are based on lateral design values for nails.

2306.4 Shear walls. Only wood structural panel shear walls are allowed in Seismic Design Category E or F. All edges of shear wall panels must be supported by studs or blocking.

2306.4.1 Wood structural panel shear walls. The allowable shear capacities for wood structural panel shear walls may be increased 40 percent for wind design only. Recent research has provided a better understanding of how structures are affected by wind. Code-prescribed wind loads have been refined because the tables use a factor of safety equal to 2.8. Wind loads are essentially monotonic and bounded; hence, a safety factor of 2.0 is sufficient for wind loading. This results in a 40 percent increase in the tabulated values.

IBC Section 2306.4.1 permits the tabular shear values in Table 2306.4.1 to be increased 40 percent for wind. It is simply a coincidence that this 1.4 increase is the same as 490/350 = 1.4, which is the factor used to convert strength level seismic loads to ASD level in the alternate allowable stress load combinations in Section 1605.3.2. The 1.4 increase factor for wind is to calibrate tabular values to a divisor of 2.0. The other 1.4 factor is used to scale code prescribed seismic loads from strength level down to service load (ASD) level. This is why the ASD load combinations in Sections 1605.3.1 and 1605.3.2 have 0.7E and E/1.4 for the seismic load effect. Note that the load factor for *E* in the strength design load combinations in Section 1605.2.1 is 1.0 because the code prescribed seismic load effect, *E*, is at strength level.

The 2006 IBC references the AF&PA *Special Design Provisions for Wind and Seismic* (SDPWS) which covers design and construction of wood members, fasteners and assemblies to resist wind and seismic forces. Table 4.3A of the *SDPWS* gives nominal unit shear capacities for wood-frame shear walls of wood-based panels. The nominal unit shear capacities for wind are given in a separate part of the table than the nominal unit shear capacities for seismic loads. The nominal unit shear capacities in the wind load part of the table are 1.4 times the values in the seismic load part of the table because they incorporate the 1.4 increase factor for wind. The 40 percent increase for wind is due to current thinking that a divisor for wind of 2.0 is acceptable rather than the 2.8 divisor used when the tables were originally developed. Note that further increases to the SDPWS nominal unit shear capacity for wind are not permitted.

Allowable shears for these shear walls may be taken from the table, or calculated by the principles of mechanics. For structures in Seismic Design Category D, E or F, the shear values should be based on cyclic tests or the tabular values, and not calculated.

The values for allowable shear are based on monotonic racking tests using aspect ratios of 1:1. The tabular values were reduced by as much as 25 percent by some jurisdictions as a result of the extensive damage to wood structures observed from the Northridge earthquake. Preliminary results from cyclic testing of wood structural panel diaphragms being conducted by APA indicate that some tabular values may need to be reduced by as much as 10 percent.

The table also contains values for staples, which are based on cyclic testing.

When the framing lumber is of a species other than Douglas Fir-Larch or Southern Pine, the tabular values must be adjusted to account for the fastener behavior in framing members of other lumber species. The adjustments must be made per Footnote 1, but the resulting value may not exceed the tabular value.

Footnotes e and f require 3-inch or wider framing members at adjoining panel edges where nail spacing at the panel edges is 2 inches or less. Recent earthquakes have shown that highly loaded shear walls with close nail spacing sustained major damage because of inadequate nail edge distance.

Footnote e applies to Seismic Design Categories D, E and F. All framing members receiving edge nailing, including sills and blocking, must be 3-inch nominal when the shear design value exceeds 350 pounds per lineal foot (ASD) or 490 pounds per lineal foot (LRFD). See the discussion of sill thickness in Section 2305.3.10.

Table 2306.4.1 of the 2003 IBC contains allowable shear values for allowable stress design (ASD) because Section 2306 applies to "Allowable Stress Design." Because Table 2306.4.1 is an ASD table, it is misleading to refer to 490 plf (LRFD) in footnote (i). Because this is an ASD table, the capacities should only be compared to wind and seismic loads from the allowable stress load combinations in Section 1605. Therefore, there is no need to refer to 490 plf (LRFD) in footnote i because Table 2306.4.1 only applies to the ASD procedure.

In the 2006 IBC, footnote i of Table 2306.4.1 was revised to no longer refer to 490 plf for load and resistance factor design (LRFD). The requirement was relocated to Section 2307, which covers requirements for LRFD. New Section 2307.1.1 requires wood structural panel shear walls in Seismic Design Category D, E or F where shear design values exceed 490 plf to have a single 3-inch nominal member at framing members receiving edge nailing from abutting panels. Note that two 2-inch nominal members fastened together to transfer shear between framing members in accordance with the NDS is an acceptable alternative.

Lumber sheathed shear walls. Diagonally sheathed shear walls use the same detailing and load requirements as diagonally sheathed diaphragms covered in Sections 2306.3.4 and 2306.3.5. **2306.4.2**

Particleboard shear walls. Particleboard shear walls should be used with caution in seismic force resisting systems because particleboard shear walls have not been cyclically tested. Note that type M-S or M-2 with exterior glue is required. Particle board shear walls have the same aspect ratio as wood structural panel shear walls in accordance with Table 2305.3.4. Particleboard shear walls are not allowed in buildings in Seismic Design Category D, E or F. The NEHRP does not allow particleboard shear walls in engineered construction, but does permit particleboard to be used as prescriptive braced wall panels in conventional construction. **2306.4.3**

Fiberboard shear walls. Fiberboard shear walls should only be constructed with a height-to-width ratio of $1^1/_2$:1 in accordance with Table 2305.3.4. For example, an 8-foot-high shear wall must be at least 5 feet 4 inches long. Note that the shear values for fiberboard shear walls are based on monotonic testing of panels with a height-to-width ratio of $1^1/_2$:1. **2306.4.4**

Fiberboard shear walls should be used with caution in seismic force resisting systems because they have not been cyclically tested. Fiberboard shear walls are not allowed in buildings in Seismic Design Category D, E and F. NEHRP Section 12.4.3.2 does not allow fiberboard shear walls in engineered construction, but does permit fiberboard to be used as prescriptive braced wall panels in conventional construction. Fiberboard shear walls are not permitted to resist lateral loads from concrete or masonry walls.

Table 2308.9.3(4) of the 2003 IBC was relocated to Section 2306.4.4 because the table contains design shear values for shear walls of fiberboard sheathing and is more appropriately located in Section 2306 under allowable stress design. The table was improperly placed by Section 2308, which contains prescriptive provisions for conventional light-frame construction that does not require design. The IBC structural committee modified the title of the table to use the term *shear walls* instead of *vertical diaphragms* to be consistent with current terminology.

2306.4.5 **Shear walls sheathed with other materials.** These tables for plaster and gypsum walls are relocated from Chapter 25. The detailed construction requirements for these walls are in Section 2306.4.5.1 and Chapter 25. Shear walls of these materials are rather brittle. The strength and stiffness of both plaster and gypsum shear walls decrease significantly under a limited number of cycles when subjected to seismic demands. Examination of the performance of structures subjected to strong shaking from the Northridge earthquake showed that these walls did not exhibit adequate strength or ductility. Therefore, these walls should be used to resist seismic demands only after careful consideration of the performance characteristics.

This section permits these gypsum and plaster shear walls to be used in Seismic Design Category A, B or C with no reduction in allowable shear; however, this section permits these shear walls to be used in Seismic Design Category D only with a 50-percent reduction in allowable shear, and prohibits their use in Seismic Design Category E or F. NEHRP Section 12.4.3.2 does not allow use of these wall types for engineered construction, nor does it allow their use for bracing of cripple walls in either engineered or conventional construction. NEHRP does, however, allow the use of these walls for conventional construction with increased shear panel lengths.

Footnote a in Table 2306.4.5 of the 2000 IBC required the maximum allowable shear values for stucco and gypsum board shear walls to be reduced by 50 percent for seismic loads in Seismic Design Category D, and they are not permitted to be used to resist seismic forces in Seismic Design Category E or F. This language originated with the 1997 UBC and was carried over during the initial development of Chapter 23 of the 2000 IBC. The 2003 IBC language in the section was changed to state that walls resisting seismic loads are subject to the limitations in Table 1617.6, which is consistent with the NEHRP and makes the code user aware of the restrictions on these materials when resisting seismic forces. Table 1617.6 of the 2003 IBC limits the use of lath, plaster and gypsum wallboard shear walls by assigning a response modification factor, R, of 2.0 rather than the R factor of 6.0 assigned to shear walls using wood structural panels. The table also restricts the height of the building to 35 feet in Seismic Design Category D, and prohibits the use of stucco and gypsum board shear walls to resist seismic forces in Seismic Design Category E or F. In the 2006 IBC, the majority of the technical provisions for seismic design were removed from the code and are in the ASCE 7 standard. The section in the 2006 IBC was changed so that walls resisting seismic loads are subject to the limitations in Section 12.2.1 and Table 12.2-1 of ASCE 7. Table 12.2-1 assigns a response modification factor of 2 and restricts the height of the building to 35 feet in Seismic Design Category D. Also, plaster and gypsum board shear walls are not permitted be used to resist seismic forces in Seismic Design Category E or F.

Section 2307 *Load and Resistance Factor Design*

This section in the 2000 and 2003 IBC references the *Load and Resistance Factor (LRFD) Standard for Engineered Wood Construction*, AF&PA/ASCE 16-95, for design of wood structures using the LRFD procedure. The 1997 NEHRP *Provisions* also reference this standard as the primary design procedure for engineered wood construction.The reference to the AF&PA/ASCE 16 LRFD standard was deleted from the 2006 IBC and replaced with a reference to the 2005 edition of the *National Design Specification (NDS) for Wood Construction*, which is a dual format specification that contains up-to-date provisions for both Allowable Stress Design (ASD) and LRFD. The 2005 edition of the NDS is the third revision of the NDS since the ASCE 16-95 LRFD standard was published. Many significant changes for wood member and connection design were introduced during the 1997, 2001 and 2005 editions of the NDS for the ASD procedure. These changes include revisions to member notching limits, design values, connection design provisions and member design for shear. The changes to the NDS are now part of the LRFD design procedure contained in the 2005 edition of the NDS.

Section 2308 *Conventional Light-Frame Construction*

Prescriptive conventional construction provisions originated with the repetitive light-frame wood construction provisions of the UBC. Early editions of the UBC had a section entitled "wood-joisted dwelling construction," which later became "light frame construction," which was later changed to "conventional construction provisions" in the 1970 UBC. The conventional construction provisions of the UBC have always been entirely prescriptive and were intended to apply to buildings constructed of repetitive light wood-framing members consisting of studs, joists and rafters.

The term *prescription* means, "the action of laying down authoritative rules or directions." The term *prescription* means "acquired by, founded on, or determined by prescription or by long-standing custom." Together these two definitions clearly describe the nature of the conventional wood frame construction provisions in the IBC: they are a set of *rules* based on long standing *custom*.

The underlying philosophy of prescriptive conventional construction was clearly defined in Section 2518 (a) of the 1970 UBC as follows, "The requirements contained in this section are intended for light-frame construction. *Other methods may be used provided a satisfactory design is submitted showing compliance with other provisions of this code*." Although the conventional construction provisions in the UBC were modified and expanded over the years, the introductory language of this section remained unchanged through all editions of the UBC up to and including the 1997 edition. The first sentence of IBC Section 2308 is essentially the same.

An essential feature of the conventional construction provisions is in this statement: "Other methods may be used provided a satisfactory design is submitted showing compliance with other provisions of this code." In other words, one need not conform to the restrictions, limitations and requirements of the provisions if a design is submitted to the jurisdiction that conforms to the engineering requirements of the code. In addition to Section 2518 (a), the 1970 UBC also included an exception in the engineering chapter, which stated, "Buildings or portions thereof that are constructed in accordance with the conventional light-framing requirements specified in Chapter 25 of this code shall be deemed to meet the requirements of this section." The 1997 UBC has the same exception, with the added phrase, "*Unless otherwise required by the building official* buildings or portions thereof that are constructed in accordance with the conventional light-framing requirements . . . " What these two sections mean is simple: a wood-frame building must either conform to all of the restrictions, limitations and requirements (rules) prescribed in the conventional construction provisions, or engineering must be provided that demonstrates compliance with the engineering requirements of the code. Where engineering is provided, the designer is required to determine all anticipated gravity and lateral loads that act on the structure and design and detail the various structural systems to resist these loads. This requirement is embodied in Section 1604.4 which states, "Load effects on structural members and their connections shall be determined by methods of structural analysis that take into account equilibrium, general stability, geometric compatibility and both short- and long-term material properties. Any system or method of construction to be used shall be based on a rational analysis in accordance with well-established principles of mechanics. Such analysis shall result in a system that provides a complete load path capable of transferring loads from their point of origin to the load-resisting elements."

Most of the seismic related provisions in this section originated with the 1997 UBC or are based on the 1997 NEHRP *Provisions*. New provisions have been added through the ICC code development process from the initial 2000 IBC to the current 2006 IBC.

The provisions of Section 2308 are based on experience gained over the last 60 years or more. In general, these types of provisions are reasonably easy to modify where experience shows they are inadequate. An example of such modification is the change in bracing requirements that resulted from experience gained in the 1971 San Fernando, 1987 Whittier and 1994 Northridge, California, earthquakes.

2308.1 General. The provisions of Section 2308 provide prescriptive construction details and methods for light-frame wood construction. Typical construction details as required by the provisions are shown in Figures 23-9 through 23-13. Light-frame wood construction consists of 2x construction with walls of 2-inch nominal thickness studs spaced at 16 or 24 inches on center, and roofs and floors framed of 2-inch nominal thickness rafters or joists spaced at 12, 16 or 24 inches on center. Studs are typically 2 by 4 or 2 by 6. Construction that meets the prescriptive provisions of Section 2308 is deemed to comply with the intent of the code without requiring engineering. Section 2308.2 contains the specific limitations for the conventional construction provisions.

In the 2006 IBC, detached one- and two-family dwellings and townhouses not more than three stories in height with a separate means of egress are specifically required to comply

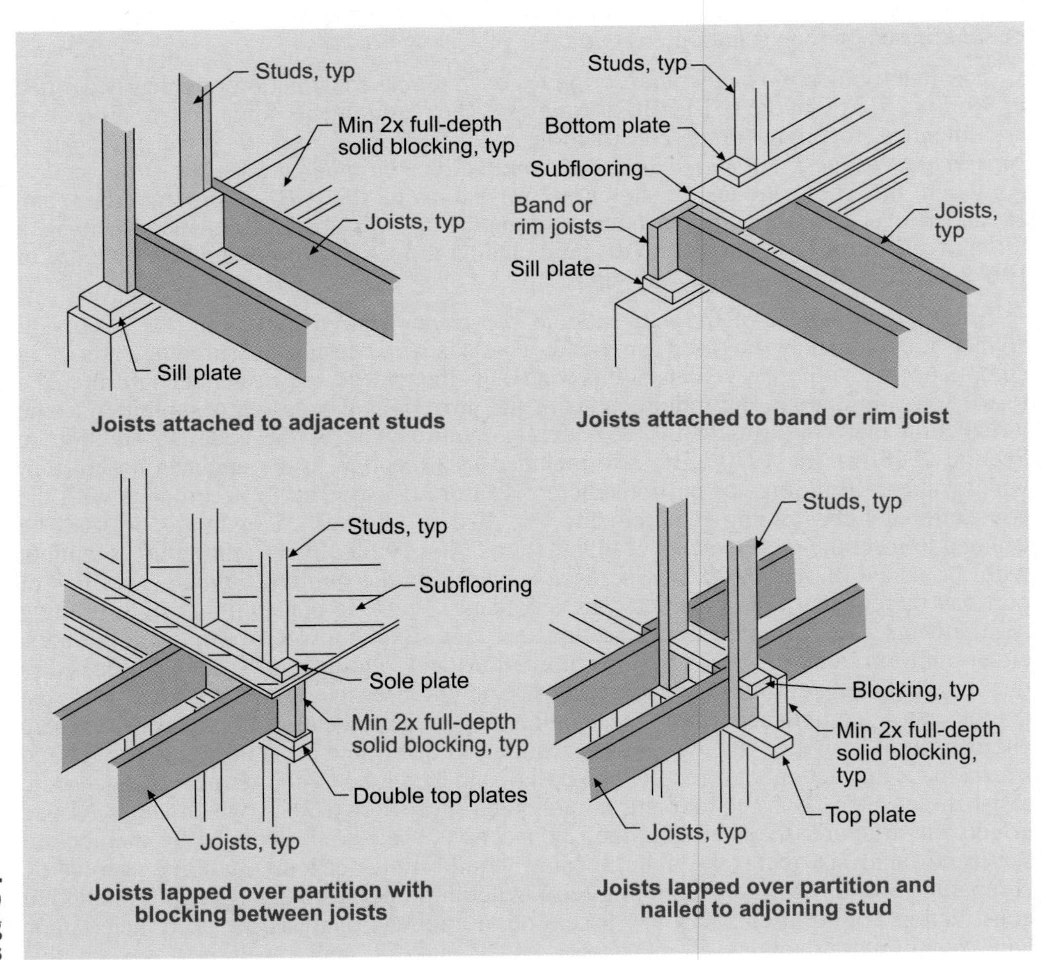

Figure 23-9
Typical framing details

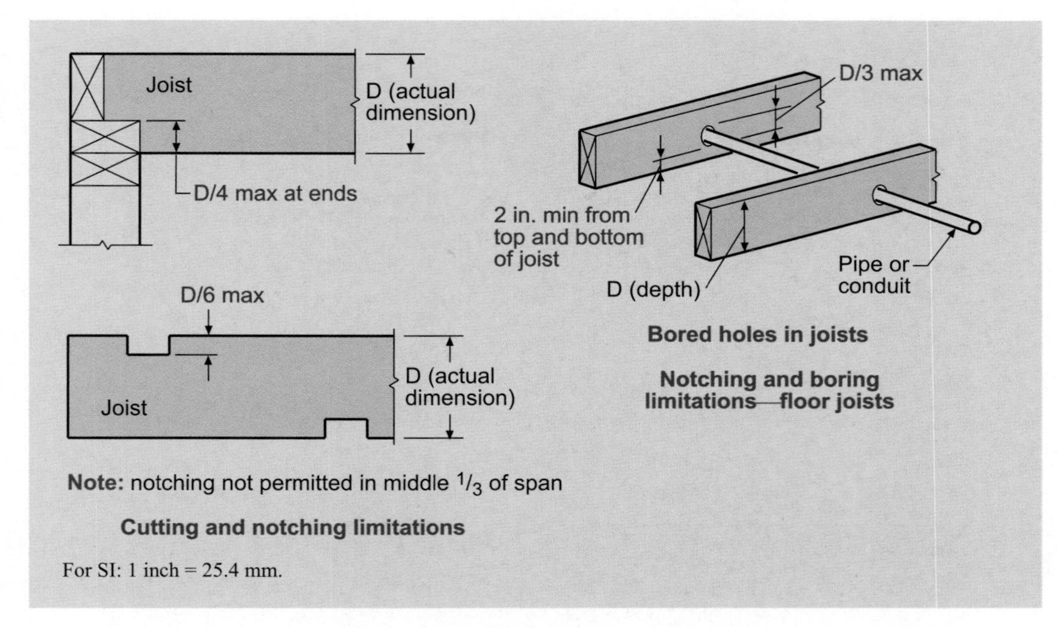

Figure 23-10
Typical details

with the *International Residential Code* (IRC). In addition, the definition of the term *townhouse* was added to Section 202 of the 2006 IBC because the term is used in this section. The language in the 2006 IBC is different from previous editions in that dwellings and townhouses are specifically required to conform to the IRC. However, Section R301.1.3 of the IRC requires structural elements that do not conform to the limits of the IRC to be designed in accordance with accepted engineering practice under the engineering provisions of the IBC. Because Section 2308.1 permits compliance with the AF&PA *Wood Frame Construction Manual* (WFCM), detached one- and two-family dwellings and townhouses must either conform to the IRC, the AF&PA WFCM or be designed in accordance with accepted engineering practice under the engineering provisions of the IBC.

The code change proponent indicated that the purpose of the code change was to make clear that Section 2308 is not an alternative for detached one- and two-family dwellings and townhouses, because Section 101.2 requires the use of the IRC for those buildings. Only buildings within the scope of the IBC and Section 2308.2 are permitted to utilize the conventional construction provisions of Section 2308. Another code change removed the story height exception for one- and two-family dwellings in Seismic Design Category C, D and E in Sections 2308.11 and 2308.12. The code change proponent stated that the story height exception for detached one- and two-family dwellings was removed because detached one- and two-family dwellings are within the scope of the IRC. Thus the 2006 IBC creates a peculiar situation: One- and two-family wood frame dwellings are not permitted to be designed by the conventional construction provisions in the IBC but must comply with either the IRC, the WFCM or be engineered. Yet hotels and multifamily dwellings (apartments) apparently are allowed to use the conventional construction provisions of Section 2308. This is a significant departure from the past because the conventional wood frame construction provisions of the IBC originated with the UBC, which were primarily intended for one- and two family dwellings, and hotels and multifamily dwellings (apartments) were usually engineered. At this point, the conventional wood-frame construction provisions in the 2006 IBC no longer have broad application.

Another change in the 2006 IBC pertains to portions of buildings that do not conform to the limitations for conventional construction. The section essentially expanded and clarified Section 2308.4.1, which requires portions of buildings of otherwise conventional construction that exceed the limits of Section 2308.2 to have those portions and the supporting load path be designed in accordance with accepted engineering practice and other provisions of the code. The code changes clarified the "design of portions"

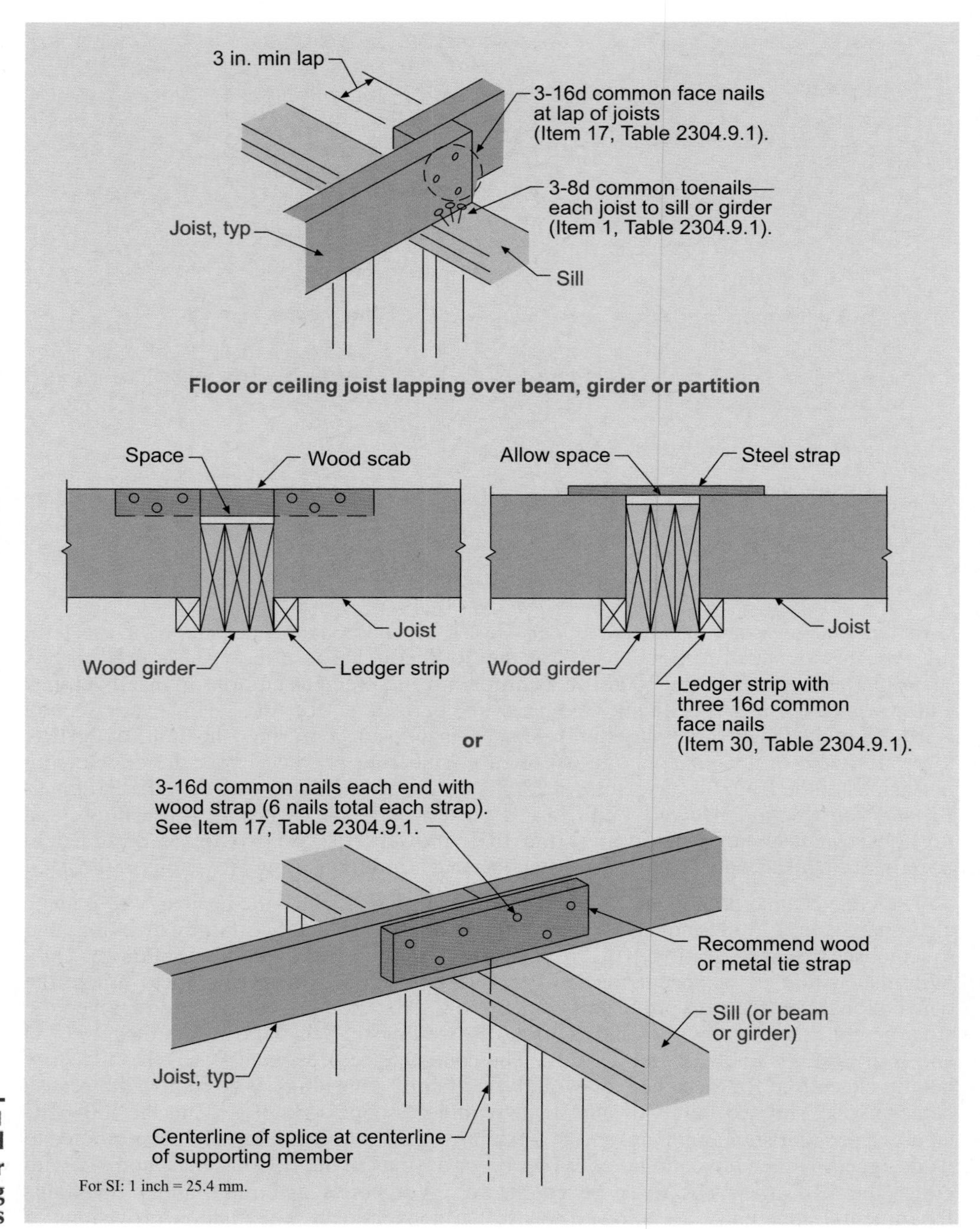

Figure 23-11 Typical details— floor or ceiling joists

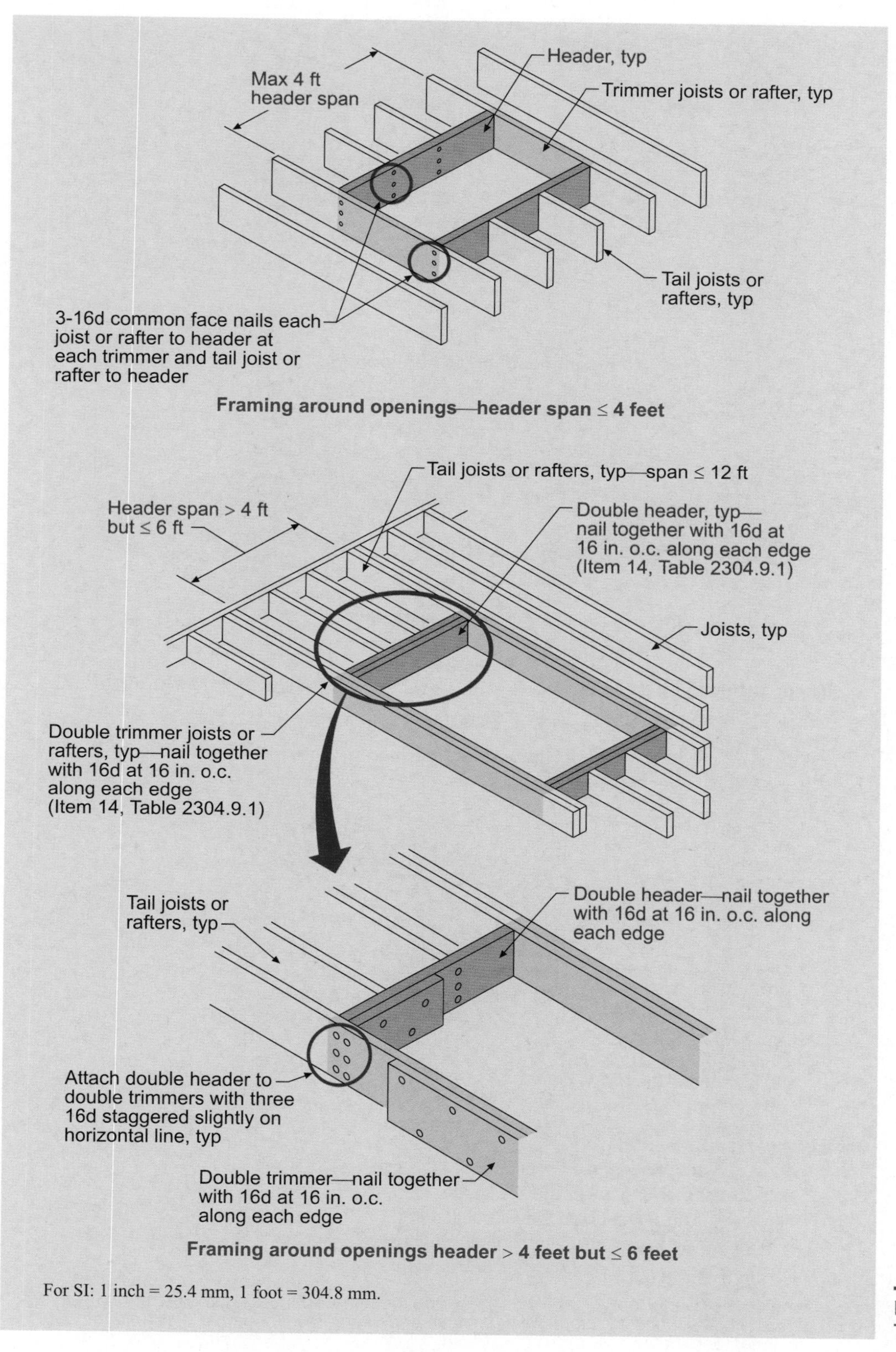

Figure 23-12
Typical details

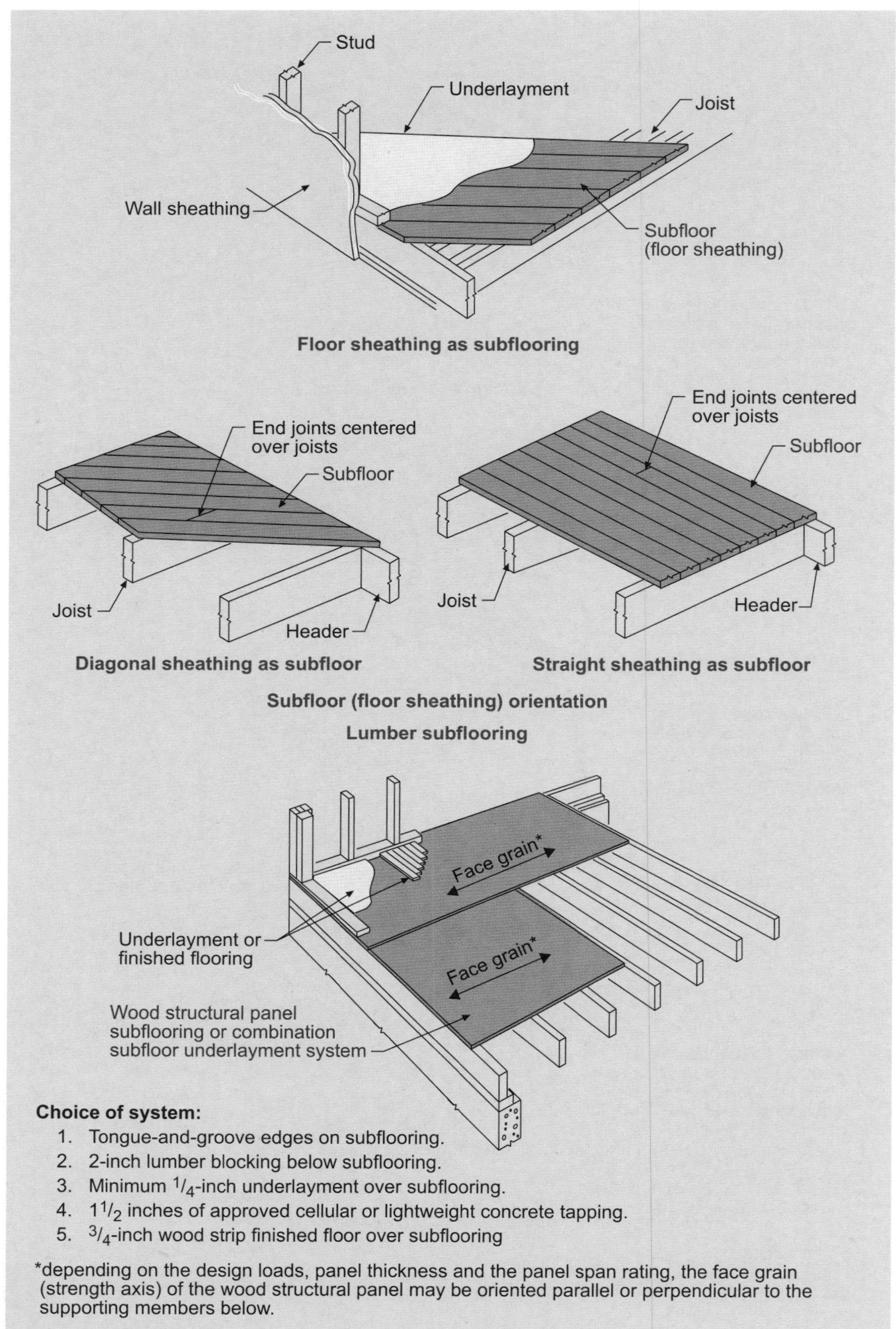

Figure 23-13 Plywood subfloor—typical details

requirements by distinguishing between *portions* and *elements*. The 2006 IBC clarifies that the code permits elements and members as well as rooms or a series of rooms to be engineered in an otherwise conventionally constructed building. In addition, the code change added the phrase "and supporting load path" to emphasize the importance of providing a continuous load path for the engineered portions or elements.

Portions exceeding limitations of conventional construction. This is a new section in the 2006 IBC pertaining to conventional buildings that have portions that exceed the limitations imposed by Section 2308.2. In this case, those portions as well as the supporting load path, must be designed in accordance with the engineering provisions of the code. In this section the term *portions* does not refer to structural elements but to parts of the building that contain volume and area such as a room or a series of rooms. See Section 2308.4 regarding design of elements or members that exceed the limitations of conventional construction. **2308.1.1**

Limitations. The structures to which conventional construction is applicable are described in this section. Additional requirements for structures classified as Seismic Design Category B or C are in Section 2308.11, and requirements for structures classified as Seismic Design Category D or E are in Section 2308.12. The conventional construction provisions are not permitted to be used for structures classified as Seismic Design Category F, which means engineering design is required for structures in Seismic Design Category F. **2308.1.2**

In Seismic Design Category D or E, cripple walls exceeding 14 inches are considered a story. Note that cripple walls exceeding 14 inches high in structures classified in any seismic design category are considered an additional story for purposes of bracing. See Sections 2308.9.4.1 and 2308.12.4. Typical conventional construction for structures with cripple walls use post and girder construction to support the floor joists in the interior of the structure.

Because interior braced wall lines may not necessarily extend down to the foundation, all lateral loads must be transferred through the cripple walls to the foundation. The cripple walls are heavily loaded, relative to the braced panel walls. The cripple walls thus are very important to satisfactory performance of the structure when subjected to lateral loads. By classifying the cripple walls as a story, higher bracing strength criteria are prescribed to resist the higher lateral loads. See, for example, Table 2308.12.4.

Section 2308.2 limits the average dead load to 15 psf for roofs and exterior walls, floors and partitions. In order to achieve uniformity, the 2006 IBC clarifies that the maximum dead load limit is for the combined roof and ceiling load. The 2006 IBC has two additional exceptions: (1) Stone or masonry veneer up to the lesser of 5 inches thick or 50 psf is permitted to a height of 30 feet above a noncombustible foundation with an additional 8 feet permitted for gable ends, and (2) concrete or masonry fireplaces, heaters and chimneys are permitted.

Basic wind speed greater than 100 mph (3-second gust). Where the basic 3-second-gust wind speed exceeds 100 mph (approximately 85 mph fastest mile wind speed), special details are necessary to resist the higher wind pressures and ensure load path continuity. The requirements and details are presented in the AF&PA WFCM or SBCCI SSTD 10 standards. **2308.2.1**

Buildings in Seismic Design Category B, C, D or E. Additional requirements for structures in Seismic Design Category B, C, D or E are covered in Sections 2308.11 and 2308.12. This section in the 2003 IBC included two exceptions to the requirements of Section 2308.11 that applied to detached one- and two-family dwellings classified in Seismic Design Category B or C. The exceptions were deleted in the 2006 IBC by the IBC structural committee in order to be consistent with other code changes that made it clear that Section 2308 is not intended to be an alternative for detached one- and two-family dwellings. **2308.2.2**

Braced wall lines. Braced wall lines are the lateral force resisting elements in conventional construction analogous to shear wall lines in engineered structures. The structural elements **2308.2.3**

must be well connected to ensure a continuous load path to effectively transfer lateral loads through the floor or roof systems to the braced wall panels within the braced wall lines. Braced wall panels may be any of the eight types described in Section 2308.9.3 subject to the requirements in Table 2308.9.3(1) as well as the specific seismic related requirements in Sections 2308.11 and 2308.12.

2308.3.1 **Spacing.** The 35-foot braced wall line spacing is from the NEHRP. Section 2308.12 reduces the spacing to 25 feet for buildings in Seismic Design Category D or E.

2308.3.2 **Braced wall panel connections.** The general intent of the provisions in this section is to provide prescriptive requirements for achieving a continuous load path for lateral loads from the roof and floor system to the braced wall panels. Recommended connection details for various framing conditions are shown in Figures 23-14 through 23-19. Figures 23-20 through 23-23 illustrate methods of transferring lateral loads from the roof system to braced wall panels.

These connection details are designed to transfer shears of approximately 100 – 200 pounds per foot, which is consistent with the allowable shear resistance of the braced panels.

2308.3.3 **Sill anchorage.** Where braced wall lines are required to be supported by continuous footings, sills must be anchored to the foundation with $^{1}/_{2}$-inch-diameter anchor bolts with a maximum spacing of 6 feet on center, or 4 feet on center for structures over two stories in height. Buildings in Seismic Design Category E require $^{5}/_{8}$-inch-diameter anchor bolts in accordance with Section 2308.12.9. See Section 2308.3.4 for braced wall line support requirements.

2308.3.4 **Braced wall line support.** Braced wall lines must be supported by continuous footings, except that interior braced wall lines need not be supported by continuous footings if the maximum plan dimension of the structure does not exceed 50 feet.

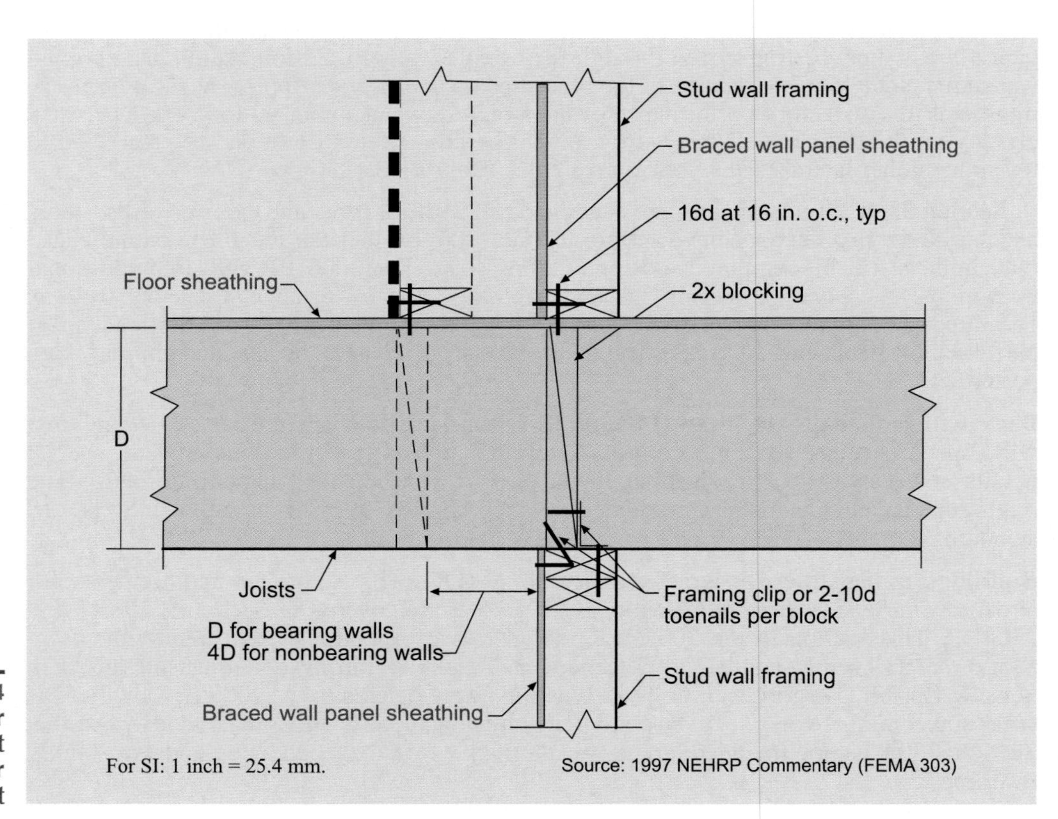

For SI: 1 inch = 25.4 mm.

Source: 1997 NEHRP Commentary (FEMA 303)

Figure 23-14 **Interior braced wall at perpendicular joist**

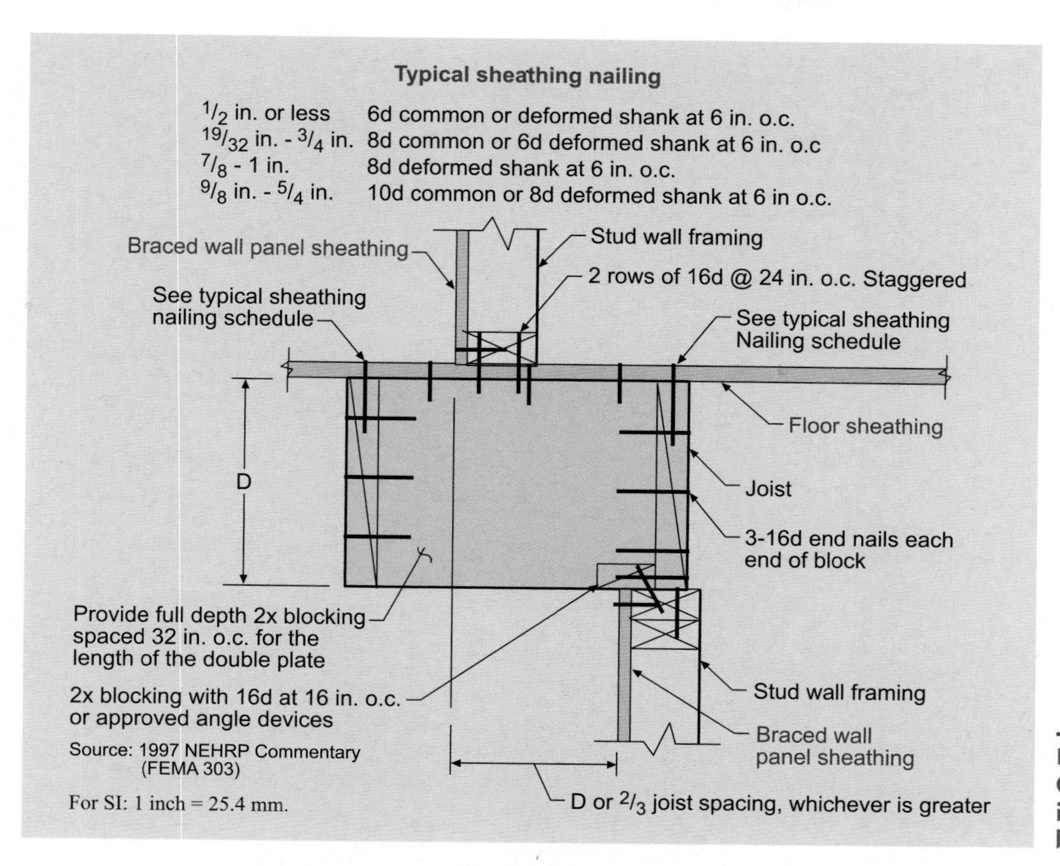

Figure 23-15
Offset at interior braced wall

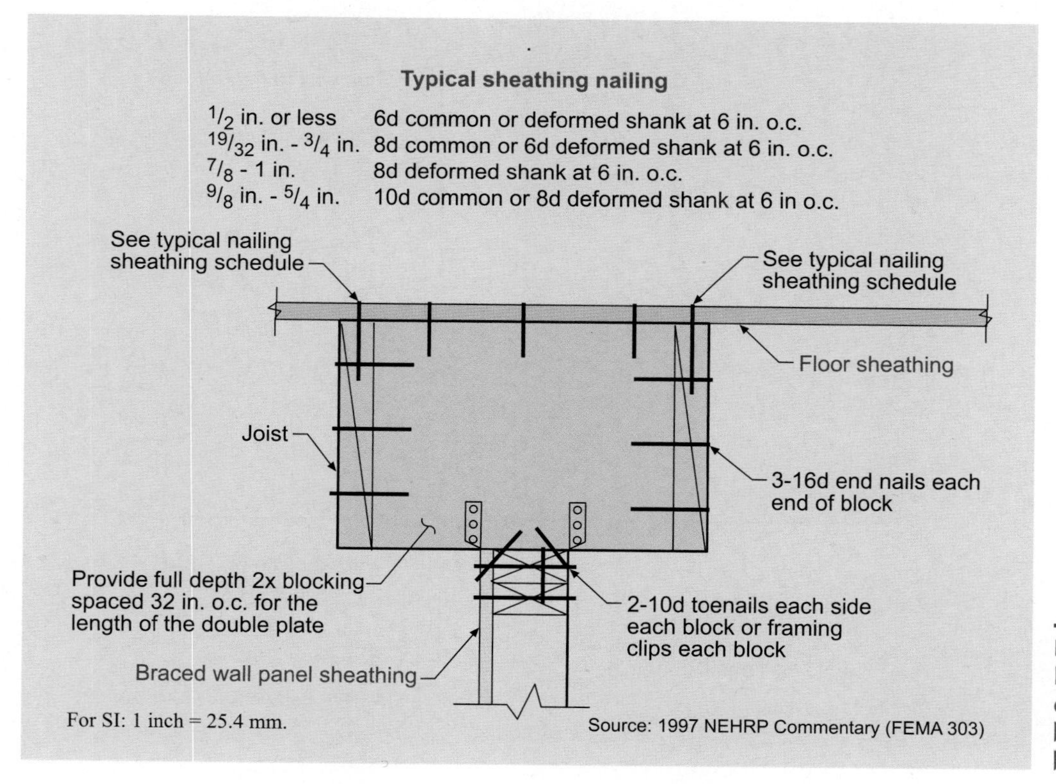

Figure 23-16
Diaphragm connection to braced wall below

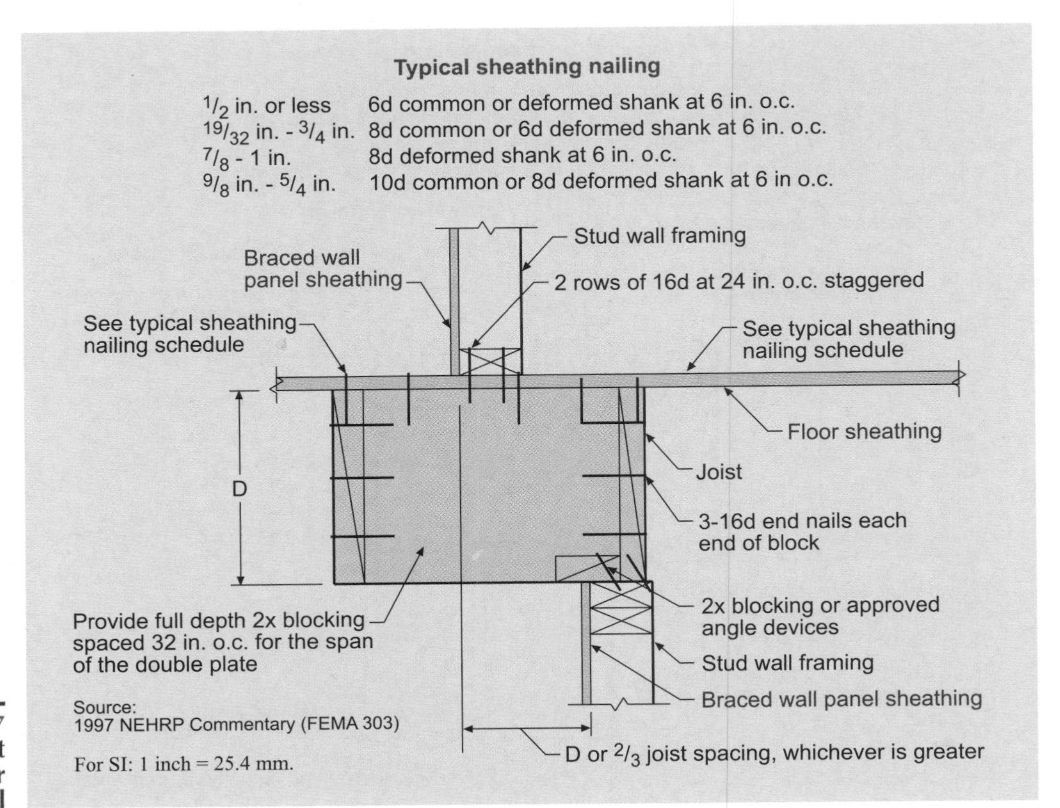

Figure 23-17 Offset at interior braced wall

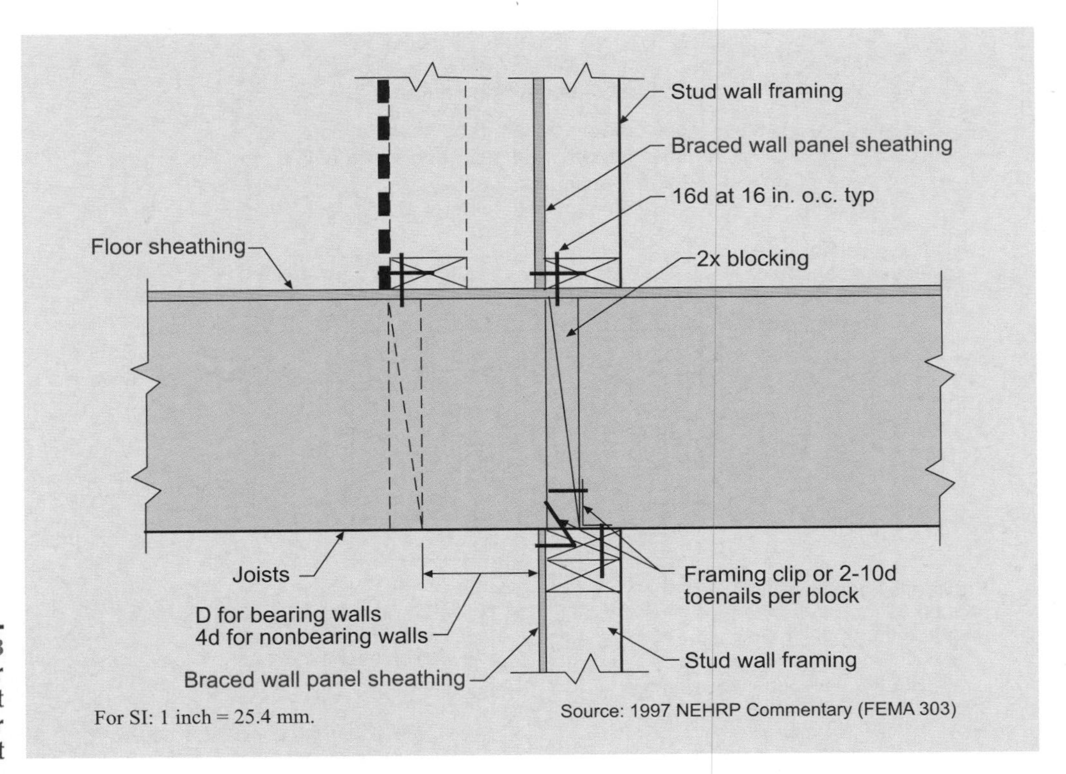

Figure 23-18 Interior braced wall at perpendicular joist

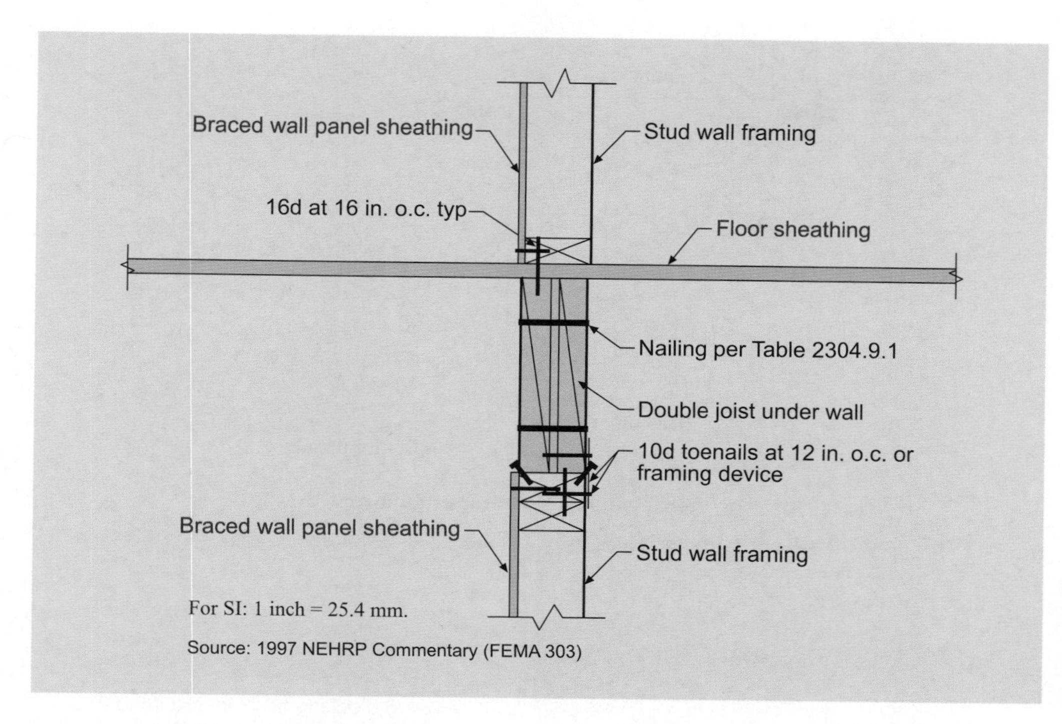

Figure 23-19 Interior braced wall at parallel joist

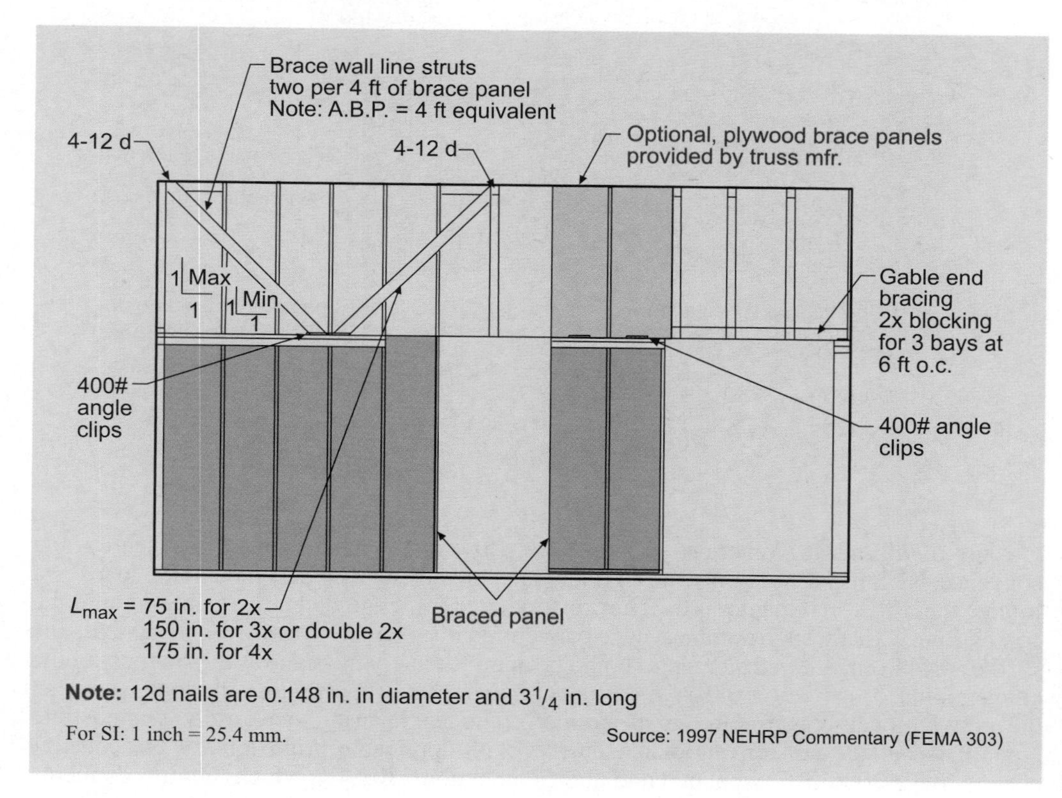

Figure 23-20 Suggested method for transferring roof diaphragm loads to braced wall panels

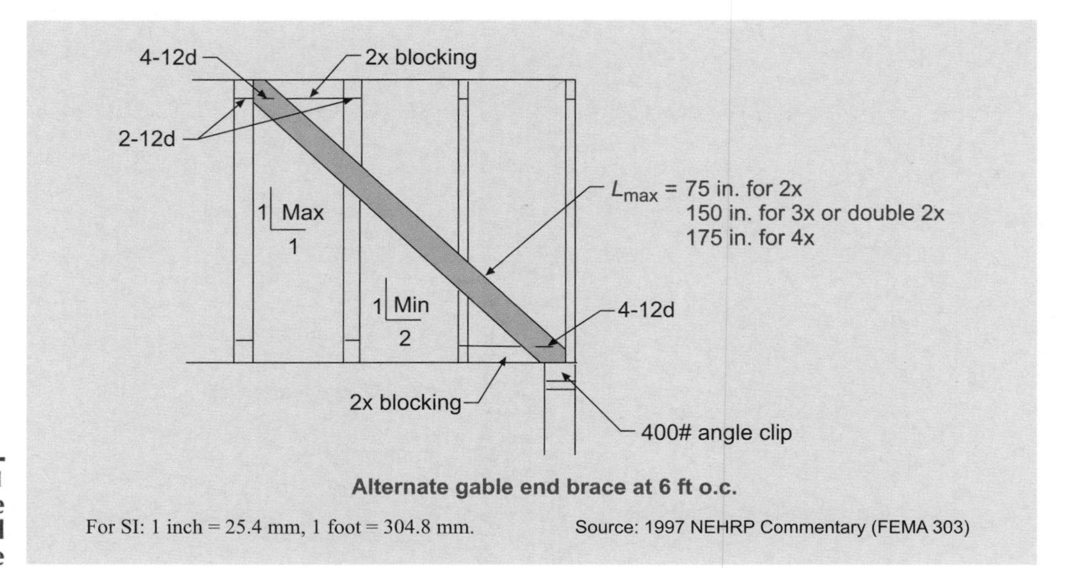

Figure 23-21 Alternate gable end brace

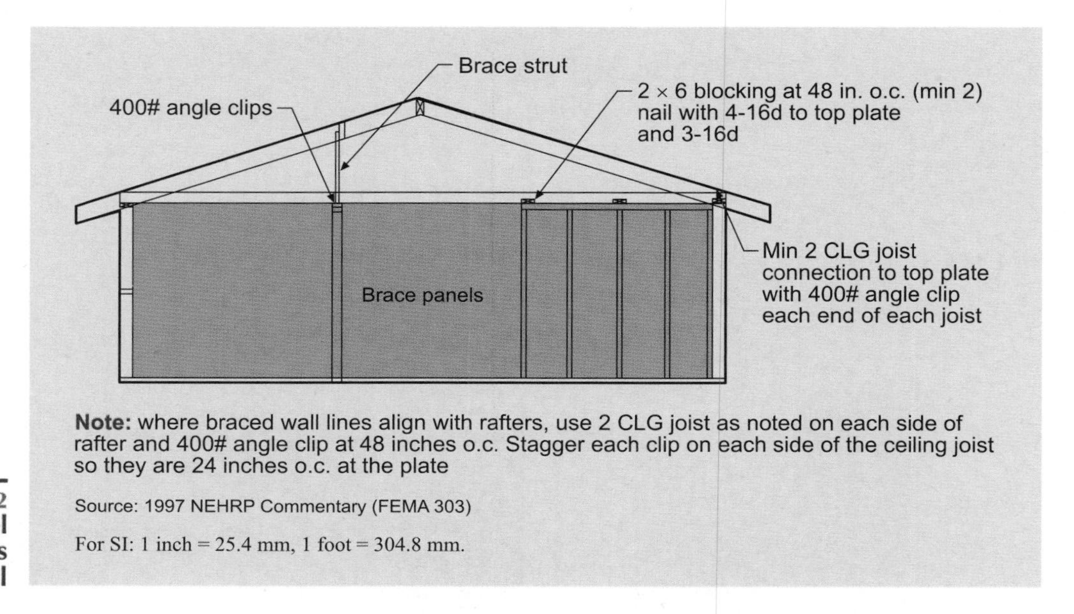

Figure 23-22 Wall parallel to truss bracing detail

2308.4 Design of elements. Where a structure consists of a combination of engineered and conventional construction, only the engineered elements must be designed to the loads and forces specified in Chapter 16. The performance of engineered elements should be compatible with the performance of a conventionally framed system. For example, the stiffness of the engineered elements should be approximately the same as the stiffness of the conventional construction unless a stiffness analysis for proportioning the loads between the conventional construction is performed. The engineering need only demonstrate compliance of the nonconventional elements with applicable provisions of the code. A change in the 2006 IBC expanded and clarified Section 2308.4.1, which requires elements of buildings of otherwise conventional construction that exceed the limits of Section 2308.2 to have those elements and the supporting load path be designed in accordance with accepted engineering practice and other provisions of the code. The code change clarified the "design of portions" (Section 2308.1.1) requirements by distinguishing between *portions* (Section 2308.1.1) and *elements* (Section 230.4). The 2006 IBC clarifies that the

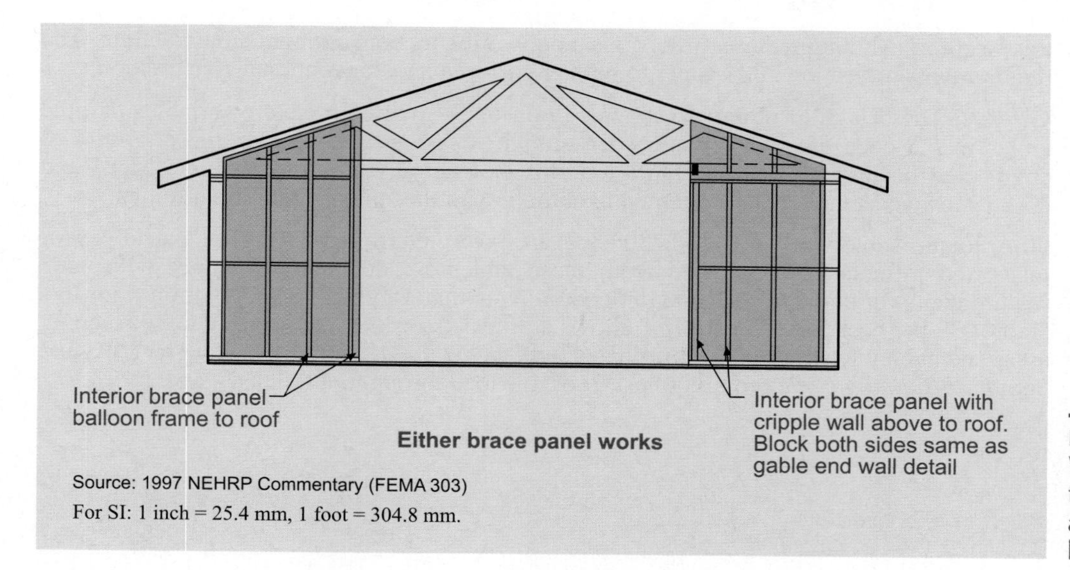

Figure 23-23 Wall parallel to truss alternate bracing detail

code permits elements and members as well as rooms or a series of rooms to be engineered in an otherwise conventionally constructed building.

2308.4.1

Elements exceeding limitations of conventional construction. When a conventionally constructed building contains structural elements that exceed the limits of Section 2308.2, the elements and their supporting load path must be designed in accordance with the engineering provisions of the code.

2308.4.2

Structural elements or systems not described herein. When a conventionally constructed building contains structural elements or systems that are not specifically covered by Section 2308, the elements and their supporting load path must be designed in accordance with the engineering provisions of the code. An example is floor and roof trusses.

2308.5

Connections and fasteners. The minimum connections and fastener requirements for conventional construction are the same as required for engineered construction. See Section 2304.9 and Table 2304.9.1 for minimum fastening requirements. See Table 2308.10.4.1 for rafter tie connections as a function of rafter slope and snow load. The number of 16d nails required as rafter tie connections per Table 2308.10.4.1 can be significantly greater than the minimum requirements of Table 2304.9.1.

Improper nail sizes have reportedly been used in wood frame building construction because the pennyweight system of specifying nail sizes is not universally understood. Code users often focus on pennyweight (8d - 8 penny, 16d - 16 penny, etc.) rather than pay proper attention to the specific style of nail such as common, box, cooler, sinker, finish, etc. A typical example is substitution of box nails for common nails of the same pennyweight. The specific style of nail is critical because there can be significant differences in strength properties of connections nailed with nails of same pennyweight but of different nail style. A code change to the 2006 IBC added the nominal dimensions of nails in the fastening tables in an effort to avoid confusion and help reduce misapplications in nailed connections. The code change proponent expected some reluctance on the part of some in the building construction community to completely abandon the pennyweight system of designating nail sizes, so the IBC continues to maintain the pennyweight system designations. Because nominal dimensions are not as subject to misinterpretation, the shank length and diameter was added in parentheses for the various styles of nails used in wood connections.

2308.6

Foundation plates or sills. Foundation requirements for light-frame construction are in Section 1805. Prescriptive requirements for foundation plates and anchorage are given in this section. Although these provisions apply only to conventional construction. The requirements in Section 2304 apply to both engineered and conventional construction.

2308.7 Girders. The design of girders to support floor loads involves so many variables that the tables in the code are only practical for simple cases. In many cases, girders must be engineered because of unusual configurations or loading conditions. Figures 23-24 and 23-25 show acceptable details deemed to comply with the splice and tie requirements.

2308.8 Floor joists. Tables 2308.8(1) and 2308.8(2) are excerpted from the AF&PA span tables for joists and rafters. Joists selected from these tables are deemed to comply with code requirements for both strength and deflection. Note that Table 2308.8(1) is for a floor live load of 30 psf because Table 1607.1 allows a 30 psf floor live load in residential sleeping areas and habitable attics, which originated with the *National Building Code*. The UBC and *Standard Building Code* require 40 psf floor live load throughout the residence.

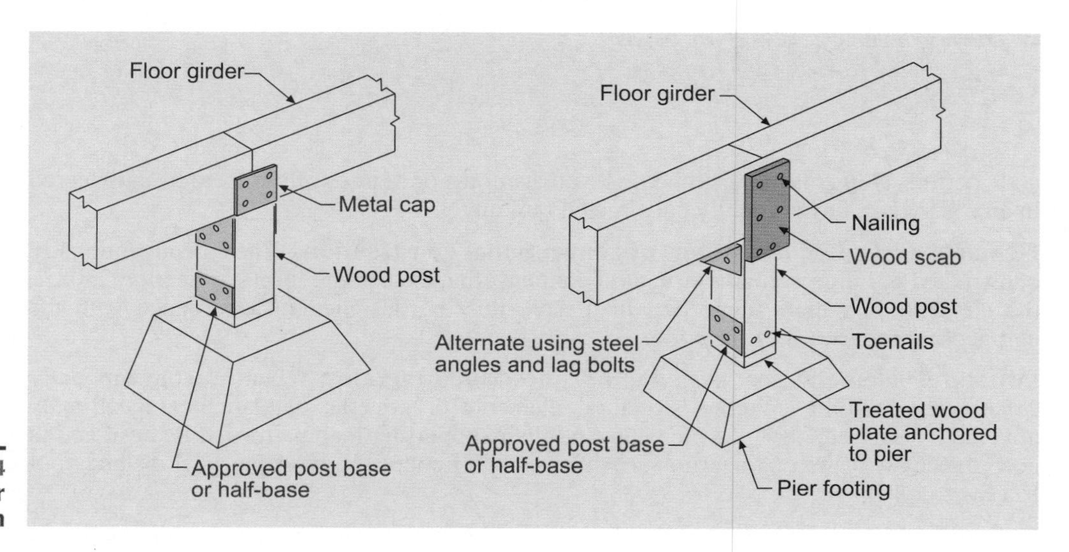

Figure 23-24
Post-to-girder connection

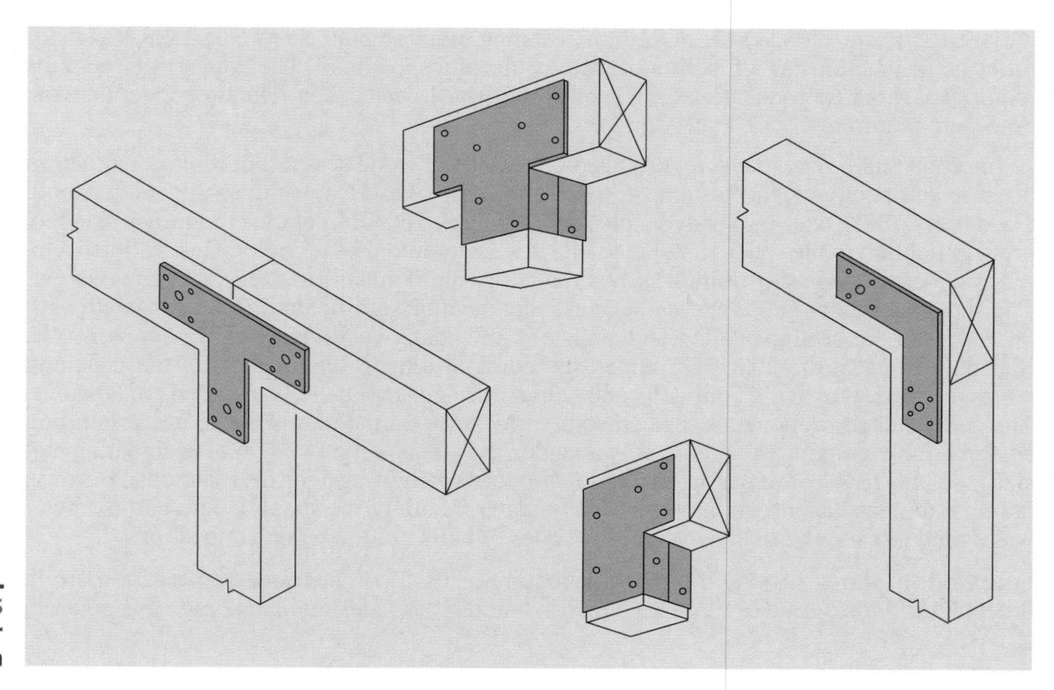

Figure 23-25
Post-to-girder connection

Bearing. Figures 23-26 and 23-27 illustrate minimum bearing requirements. **2308.8.1**

Framing details. Figure 23-28 depicts the provisions for support by solid blocking or other means, and joists framing into a girder. **2308.8.2**

Figure 23-29 illustrates the allowed notches or bored holes.

Figure 23-30 depicts the requirements for lapping and tying joists over a beam or partition.

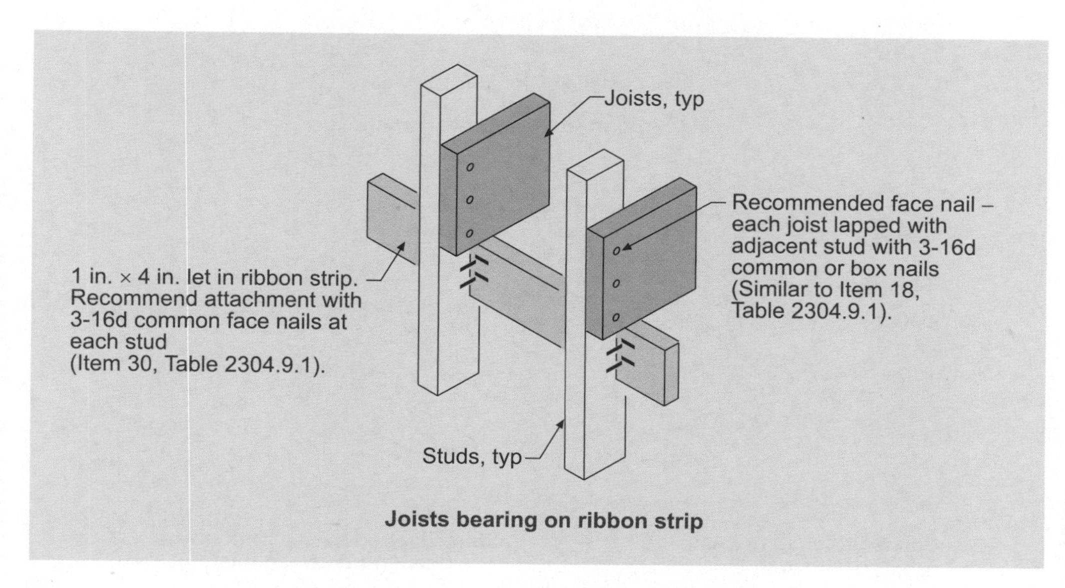

Figure 23-26 Bearing requirements

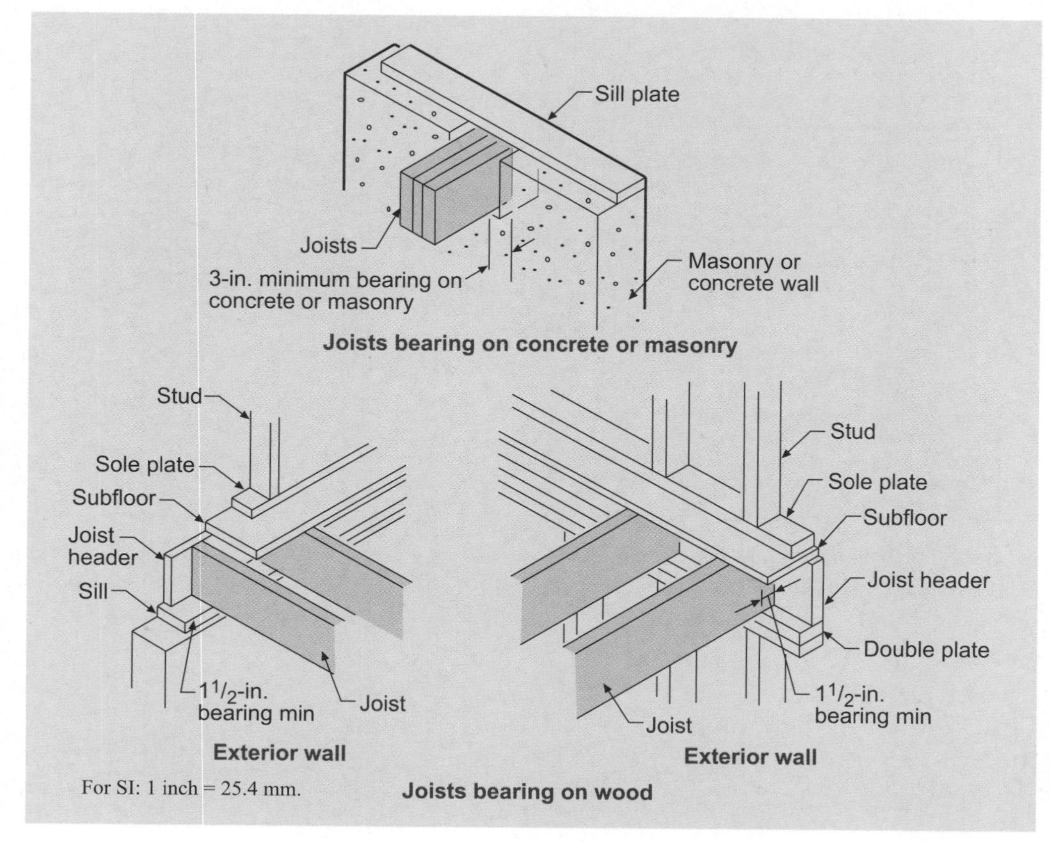

Figure 23-27 Bearing requirements

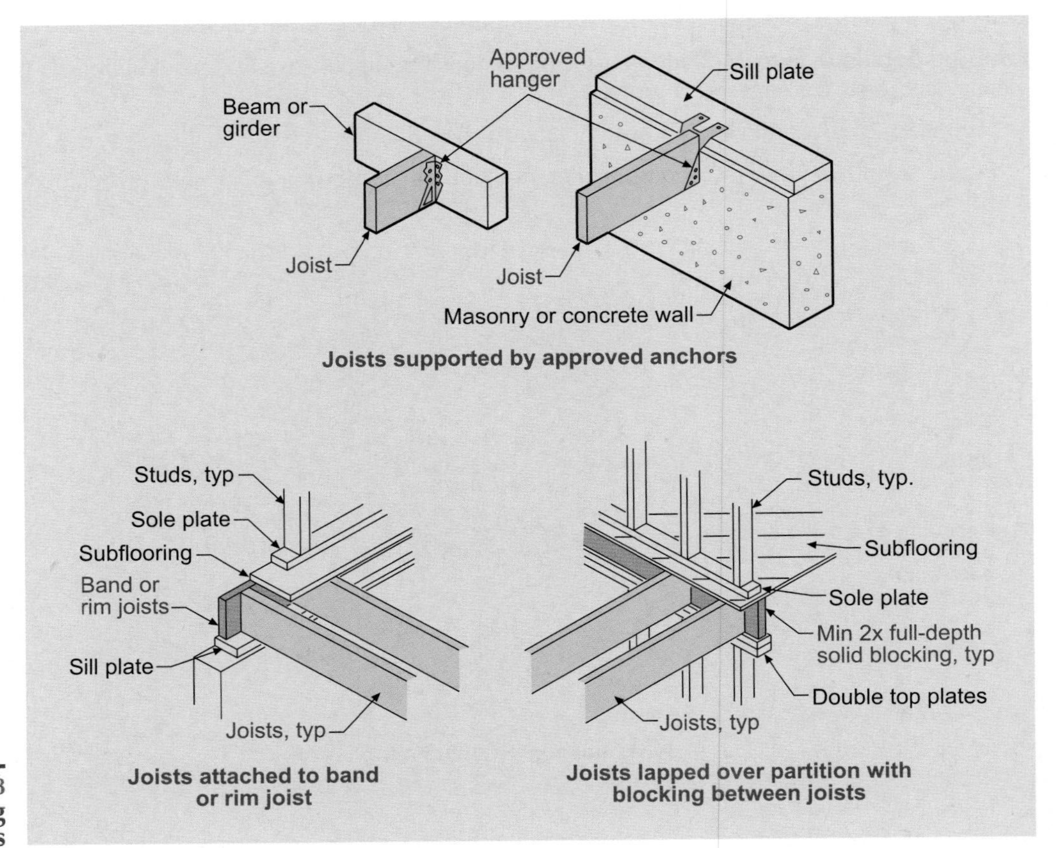

Figure 23-28 Framing details

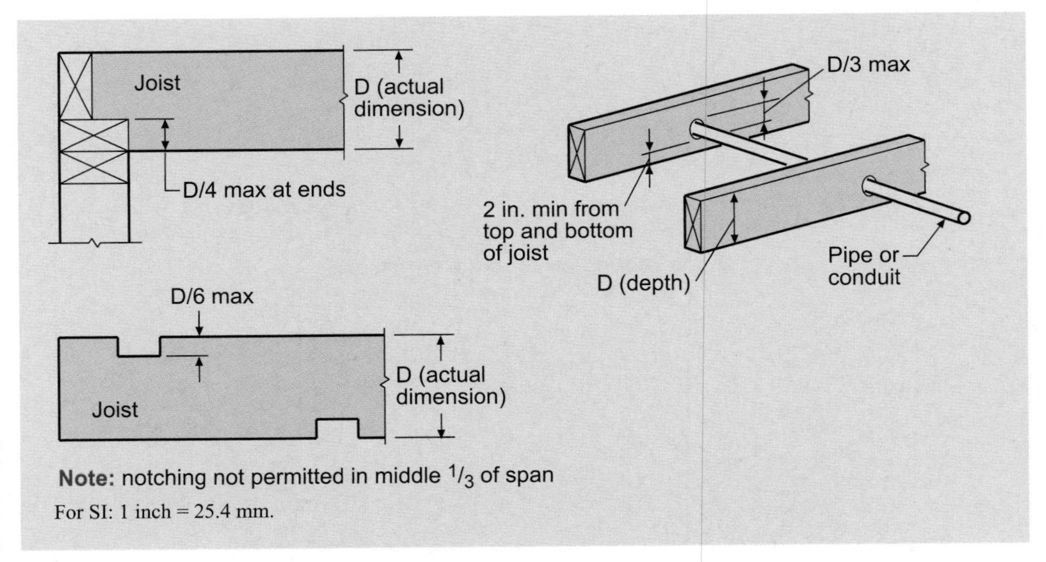

Note: notching not permitted in middle 1/3 of span

For SI: 1 inch = 25.4 mm.

Figure 23-29 Cutting, notching, or bored holes

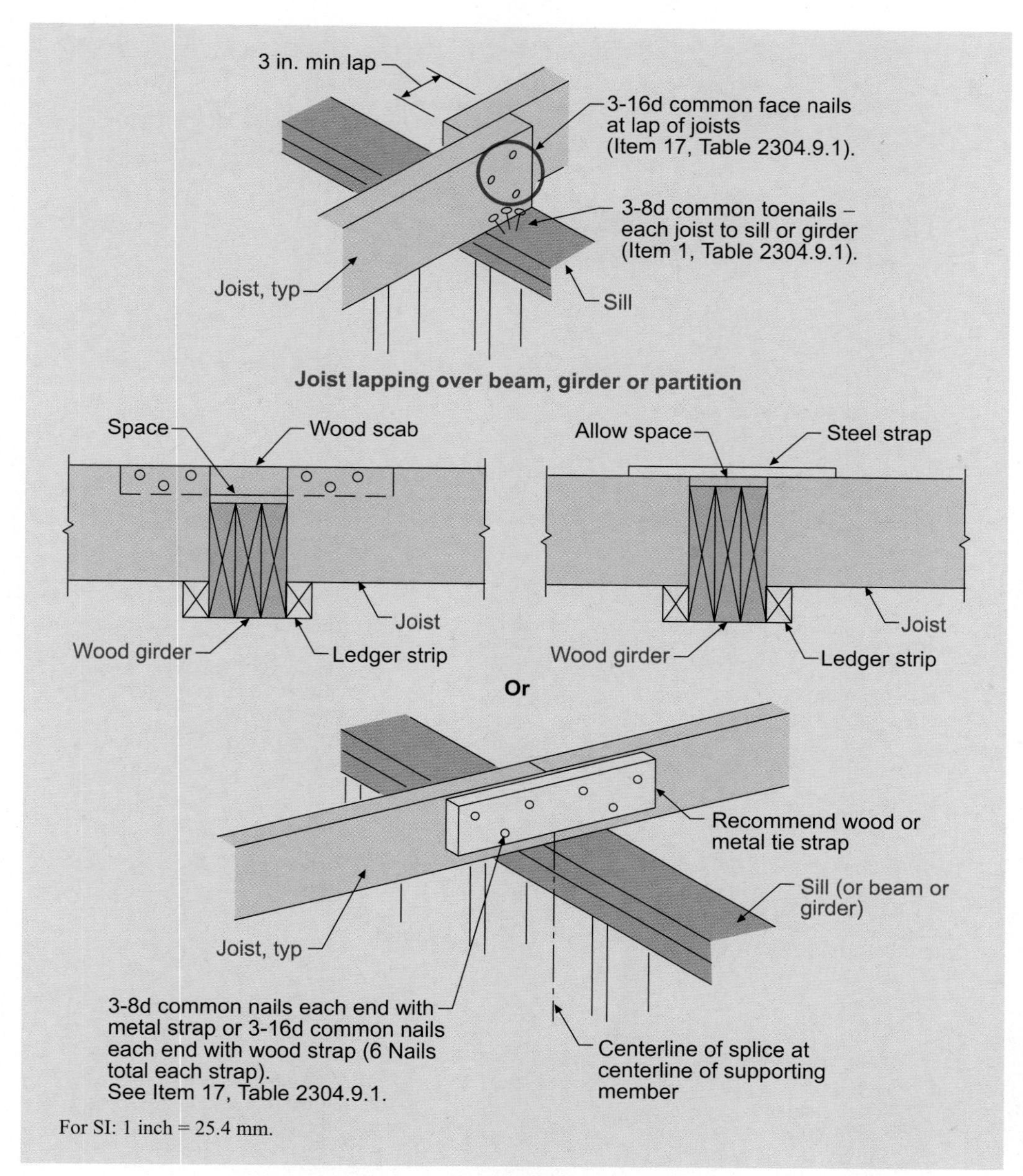

Figure 23-30
Floor joists tied over wood beam, girder or partition

Engineered wood products. This section specifically prohibits cuts, notches and holes bored in trusses, structural composite lumber, structural glue-laminated members or I-joists, unless specifically permitted by the manufacturer or where the registered design professional has considered their effects on the resistance capacity of the member. **2308.8.2.1**

Framing around openings. Figure 23-31 depicts acceptable framing details for openings not greater than 4 feet. Figure 23-32 depicts acceptable framing for openings greater than 4 feet but not greater than 6 feet. Figure 23-33 shows framing and hangers for openings greater than 6 feet. **2308.8.3**

Supporting bearing partitions. Figure 23-34 depicts the requirement for supporting bearing partitions. **2308.8.4**

Lateral support. Figure 23-35 depicts the lateral support requirements. **2308.8.5**

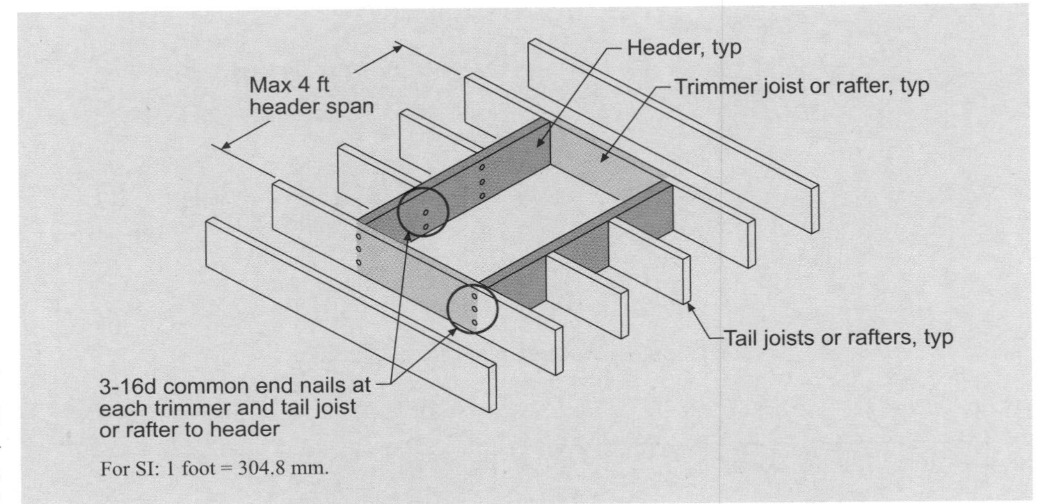

Figure 23-31 Framing around openings—Header span ≤ 4 feet

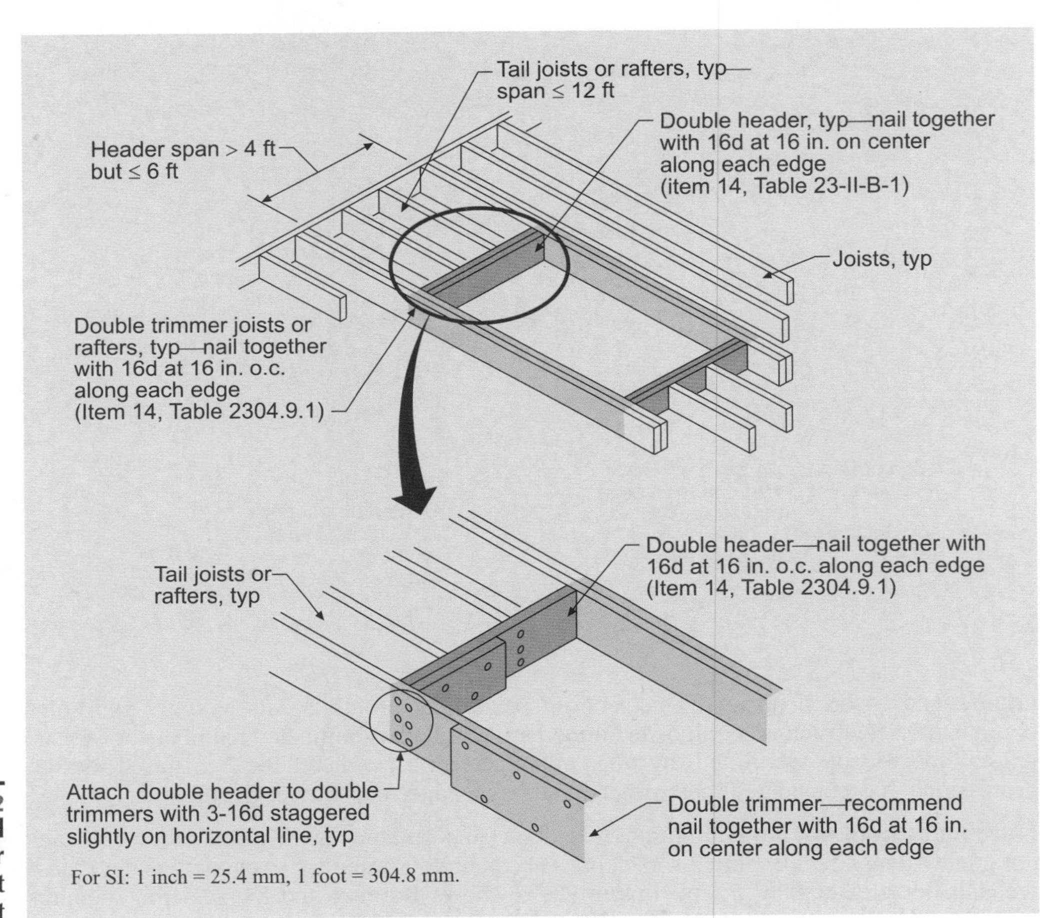

Figure 23-32 Framing around openings header > 4 feet but ≤ 6 feet

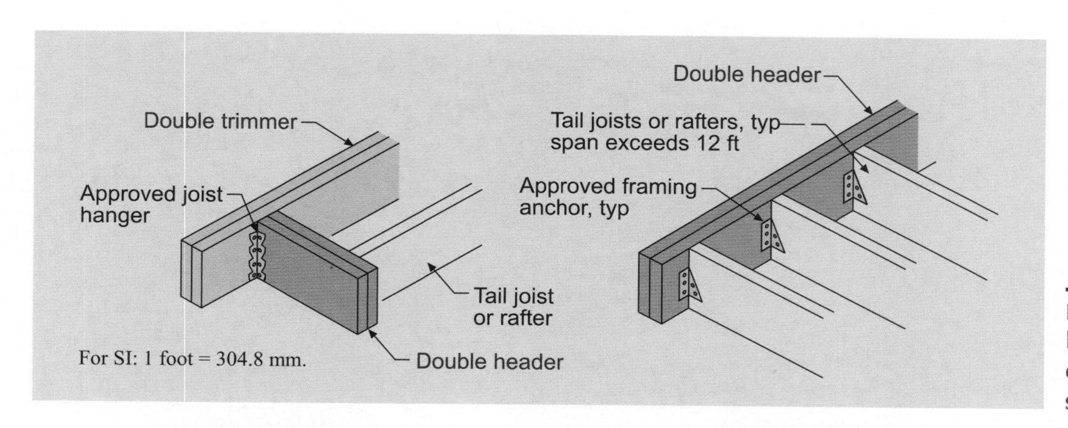

Figure 23-33
Framing around opening—Header span > 6 feet

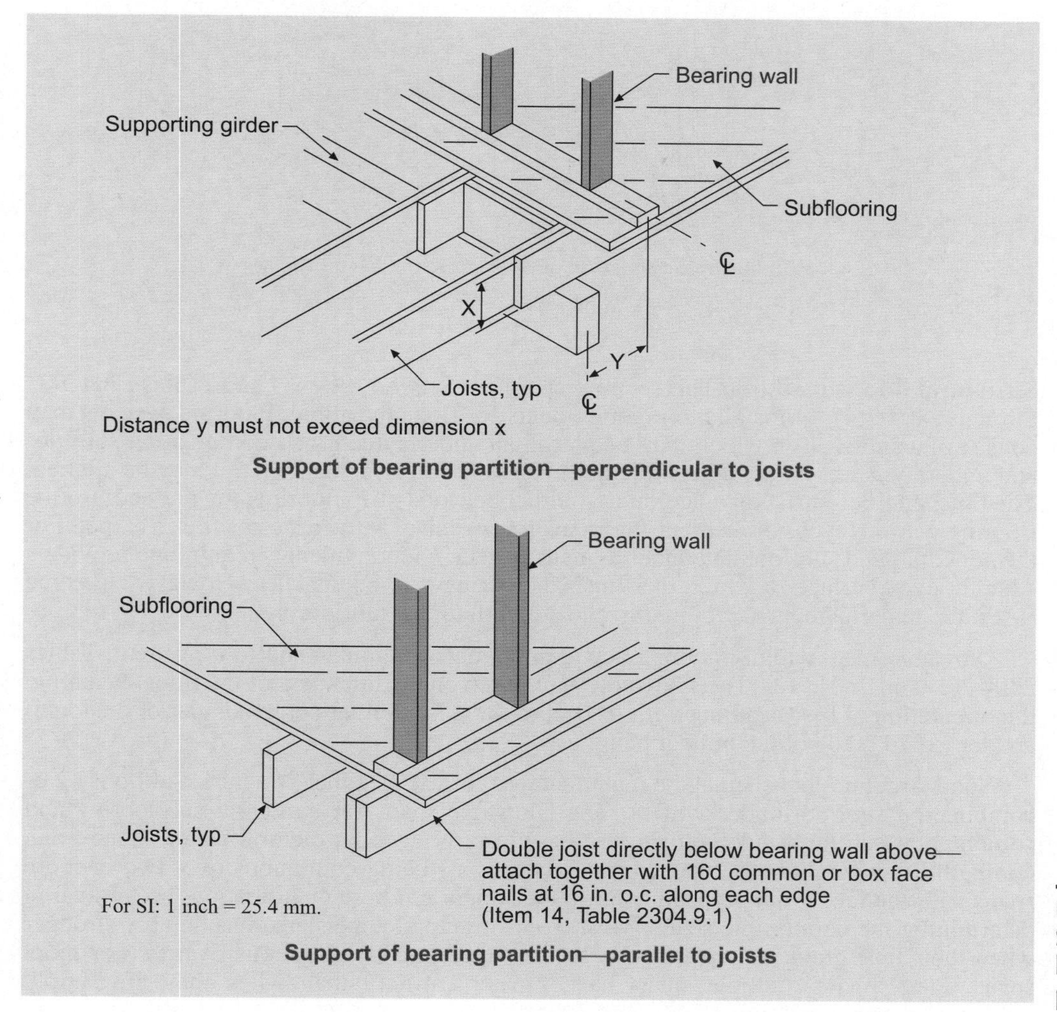

Figure 23-34
Supporting bearing partitions

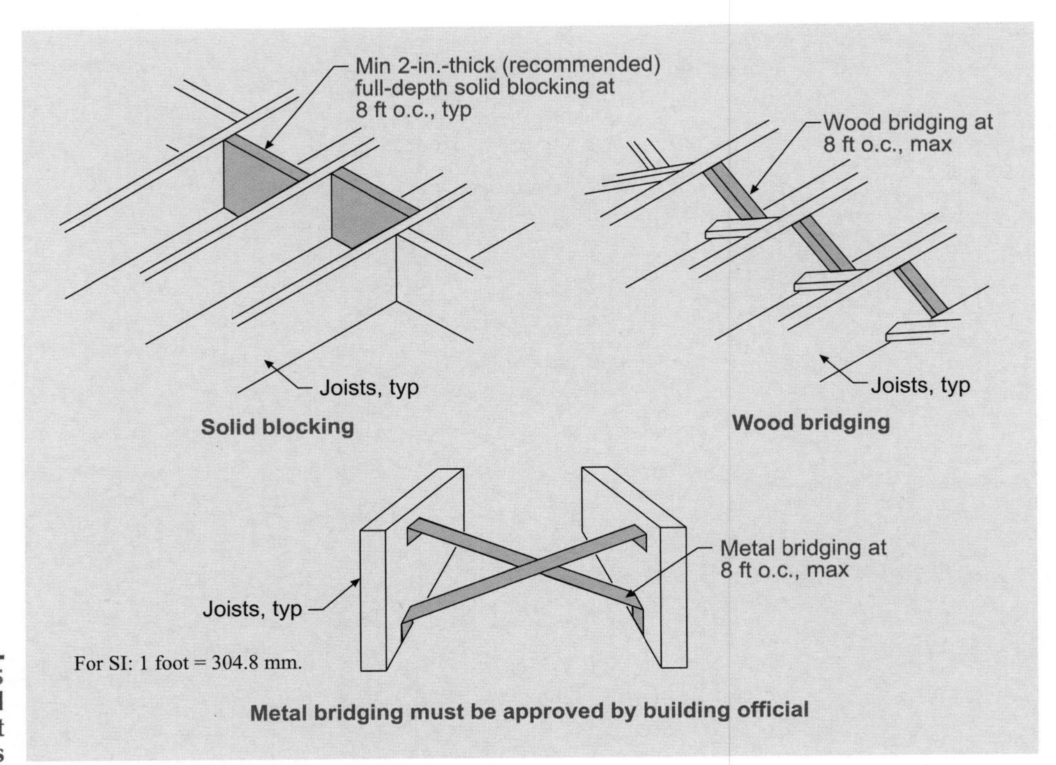

Figure 23-35 Lateral support requirements

2308.8.6 Structural floor sheathing. This section references Section 2304.7.1 and Tables 2304.7(1) through 2304.7(4) for installation requirements for floor sheathing. Flooring systems may consist of a subfloor on which may be placed an underlayment or a combination subfloor and underlayment system, upon which a finished floor-surfacing material may be applied. The finished floor-surfacing material may either be wood-strip flooring, tongue-and-groove flooring, various types of resilient floor coverings such as vinyl asbestos tile or carpet. For the noncombined subfloor and underlayment system, underlayment is required to provide a smooth, even surface to which the finished flooring will be attached. However, in some cases, the underlayment is required to add strength to the subflooring.

Allowable spans and minimum grade requirements for lumber subflooring are in Tables 2304.7(1) and 2304.7(2). The span tables are based on the thickness of the floor sheathing, the orientation of the sheathing with respect to the joists (either perpendicular or diagonal) and the board grade of the lumber being used.

Wood structural panels may be manufactured for use as either structural subflooring or combination subfloor-underlayment. The allowable spans for structural subflooring and combination subfloor-underlayment are based on the wood structural panel's face grain (strength axis) parallel to supporting members or its being continuous over two or more spans with the face grain perpendicular to the supports. These qualifications are critical in determining the permissible spans. Most wood structural panels are considerably stronger when their face grain is perpendicular to the supports and continuous over two or more spans. Panels with multiple spans have greater capacity than when they are simply supported between two joists.

To create a stiffer floor and prevent or minimize squeaking of the floor system after the building has been in use, the subfloor may be glued to the joists. This gluing prevents the relative movement between the panel and the joist that takes place when loads are placed on the floor, and provides additional stiffness.

Particleboard can be used as underlayment, structural subflooring or as combined subfloor-underlayment. Where used as underlayment, the code permits Type PBU

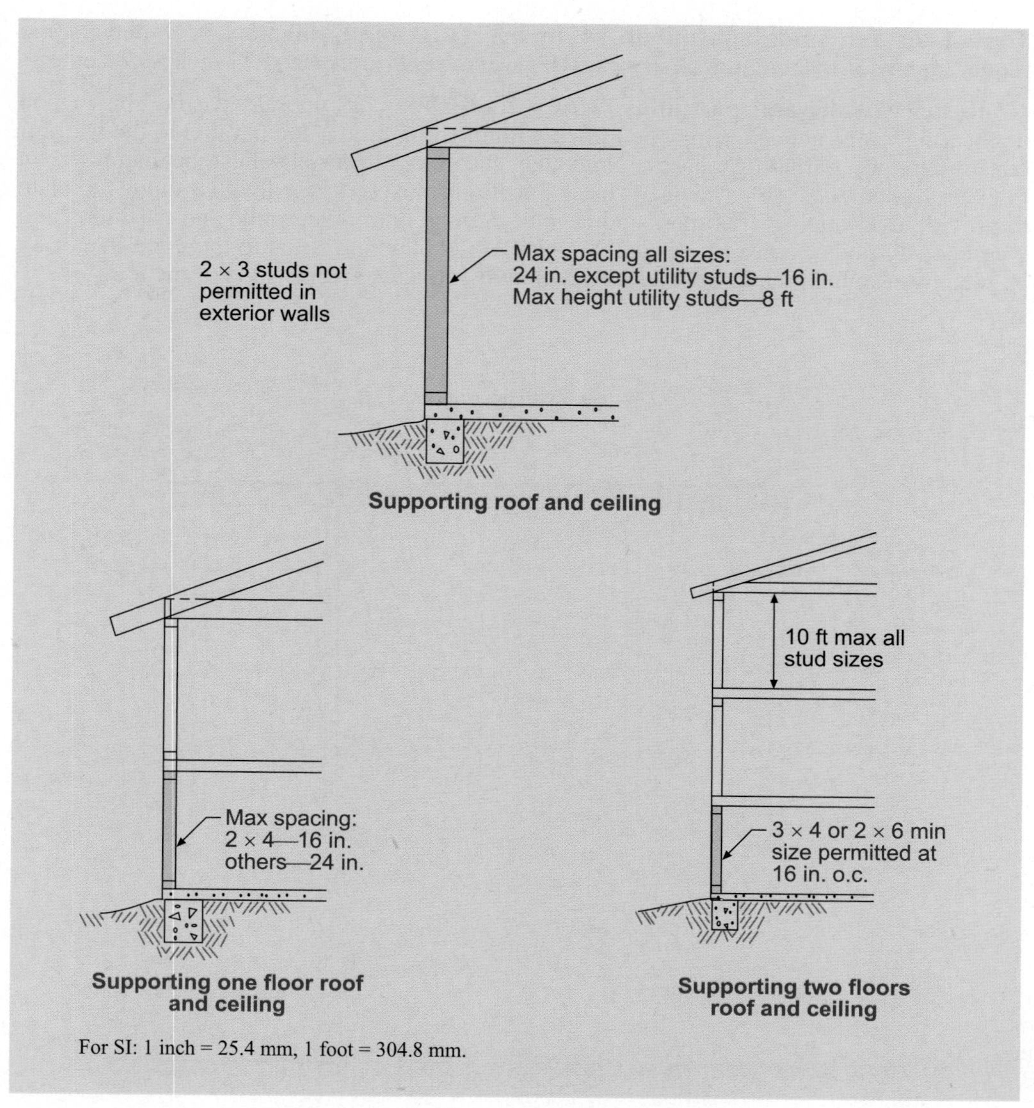

Figure 23-36 Stud requirements

particleboard in accordance with ANSI A208.1. Figure 23-13 depicts typical floor sheathing.

Wall framing. **2308.9**

Size, height and spacing. Figure 23-36 depicts stud requirements. **2308.9.1**

Framing details. Studs are required to be placed with their wide dimension perpendicular to the wall for maximum strength and stiffness. Not less than three studs are installed at each corner of exterior walls, although the exception allows two stud corners under specific conditions. **2308.9.2**

Top plates. Top plate splices require eight 16d face nails each side of the 4-foot lapped joint. 16d common nails are required by Item 10 of Table 2304.9.1. Single plate splices must have at least six 8d nails each side of the splice with a 3-inch wide by 6-inch long by 0.036 inch-thick steel strap. The NEHRP recommends this splice have 12 nails each side of the joint. **2308.9.2.1**

Figure 23-37 depicts the double top plate splice requirement; Figure 23-38 depicts the recommended single top plate splice.

2308.9.2.2 Top plates for studs spaced at 24 inches (610 mm). Figure 23-39 depicts the requirements for locating the joists or trusses where studs are spaced 24 inches on center.

2308.9.2.3 Nonbearing walls and partitions. This section allows for increased stud spacing on nonbearing walls, and studs may be turned with the long dimension parallel to the wall for construction of plumbing chases. Note that sheathing materials must be increased in thickness when wider stud spacing is used. Partitions must be capped with a single top plate to provide overlapping at corners and at intersections with other walls and partitions and continuously tied at joints by 16-inch long solid blocking and equal to the plate size or by $^1/_2$-inch by $1^1/_2$ -inch metal ties fastened with two 16d nails on each side of the joint.

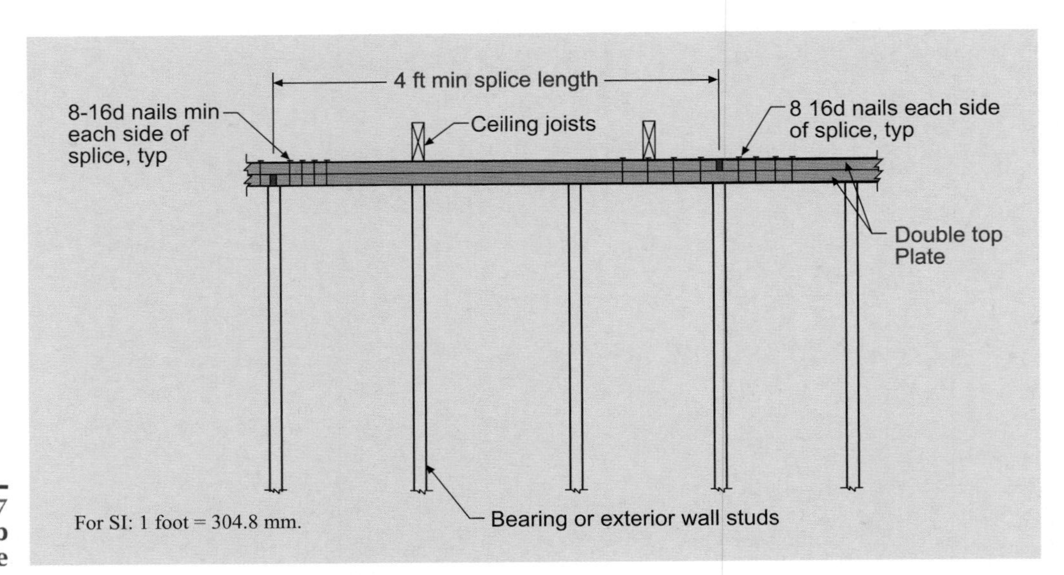

Figure 23-37 **Double top place splice**

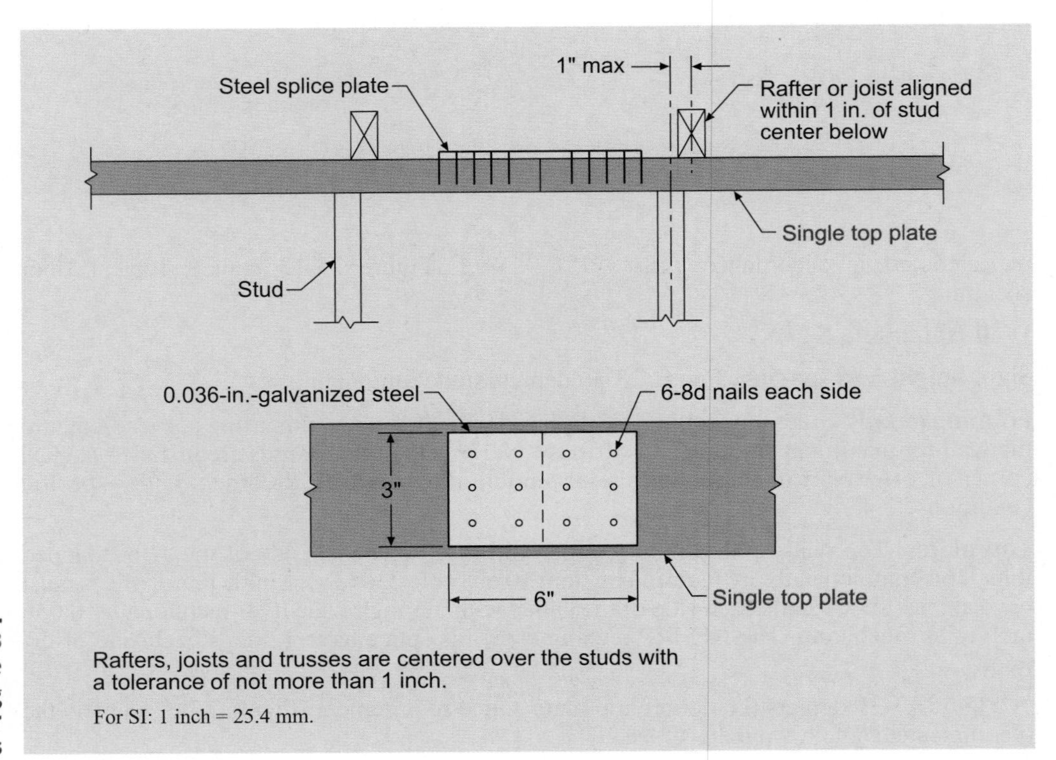

Figure 23-38 **Single top plate splice—bearing and exterior walls**

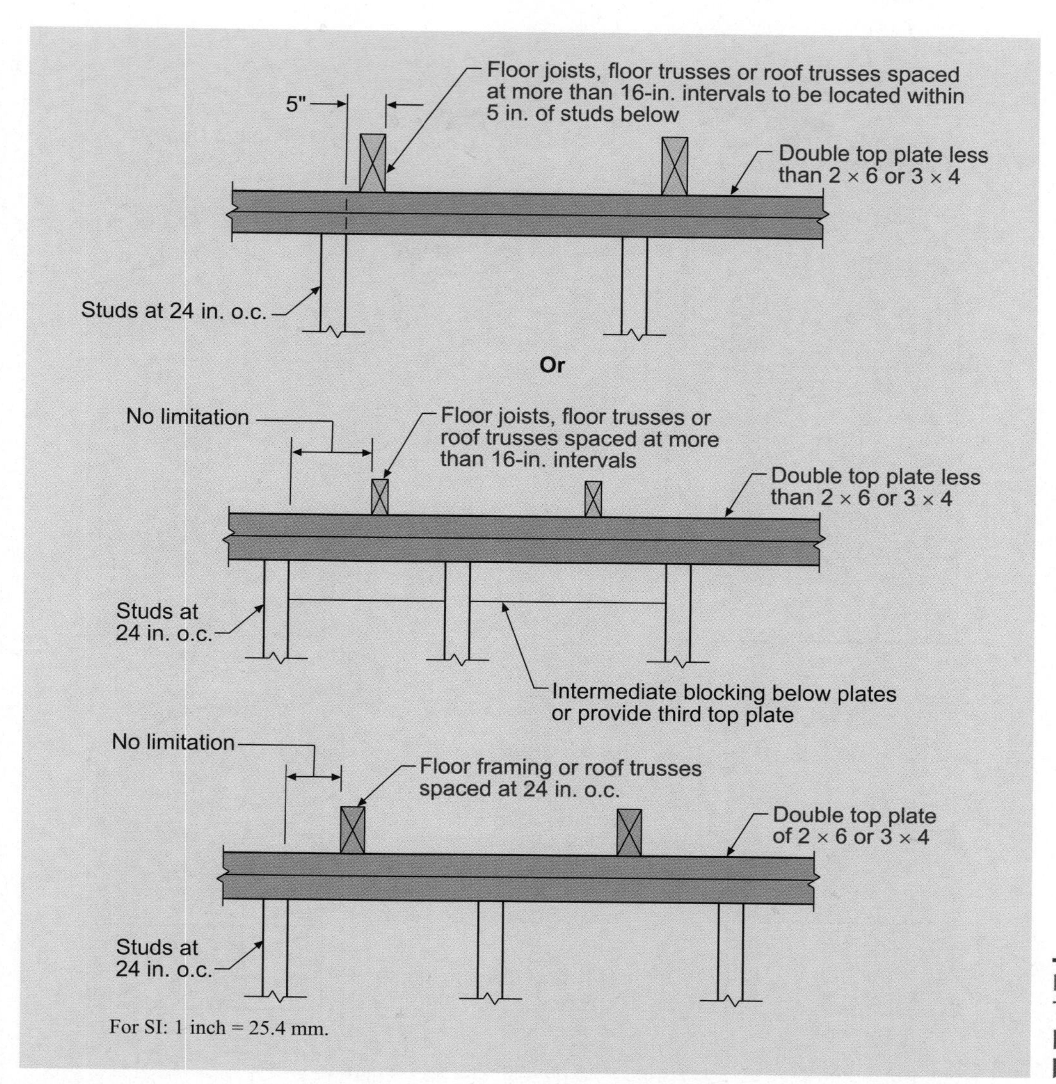

Figure 23-39 Top plate limitations—bearing

2308.9.3

Bracing. The bracing requirements for conventional construction are depicted in Figure 23-40. As required by Section 2308.3, braced wall lines are required at interior and exterior wall lines at the spacing of 25 or 35 feet, depending on the seismic design category assigned to the building. Prescriptive requirements for eight different braced wall panel types are described in this section.

Section 2308.9.3 of the 2000 and 2003 IBC requires that braced wall panels start not more than 8 feet from each end of a braced wall line. Section R602.10.1 of the 2003 IRC requires braced wall panels to begin no more than 12.5 feet from each end of a braced wall line. A code change to the 2003 IBC revised the maximum distance that braced wall panels are permitted to start from the end of a braced wall line to be 12.5 feet, which is consistent with the provisions in the IRC for SDC A, B and C. The requirement in the 2003 IBC that a designed collector be provided if the bracing begins more than 12.5 feet from the end of a braced wall line was deleted in the 2006 IBC because it made no sense within the context of the section. The code required that braced wall panels not start more than 8 feet from each end of a braced wall line yet required a designed collector to be provided if the bracing began more than 12.5 feet from the end of a braced wall line, which was not consistent. This was a peculiar requirement because the collector is an engineered component that must be designed to resist the governing lateral load, wind or seismic, and requires the determination

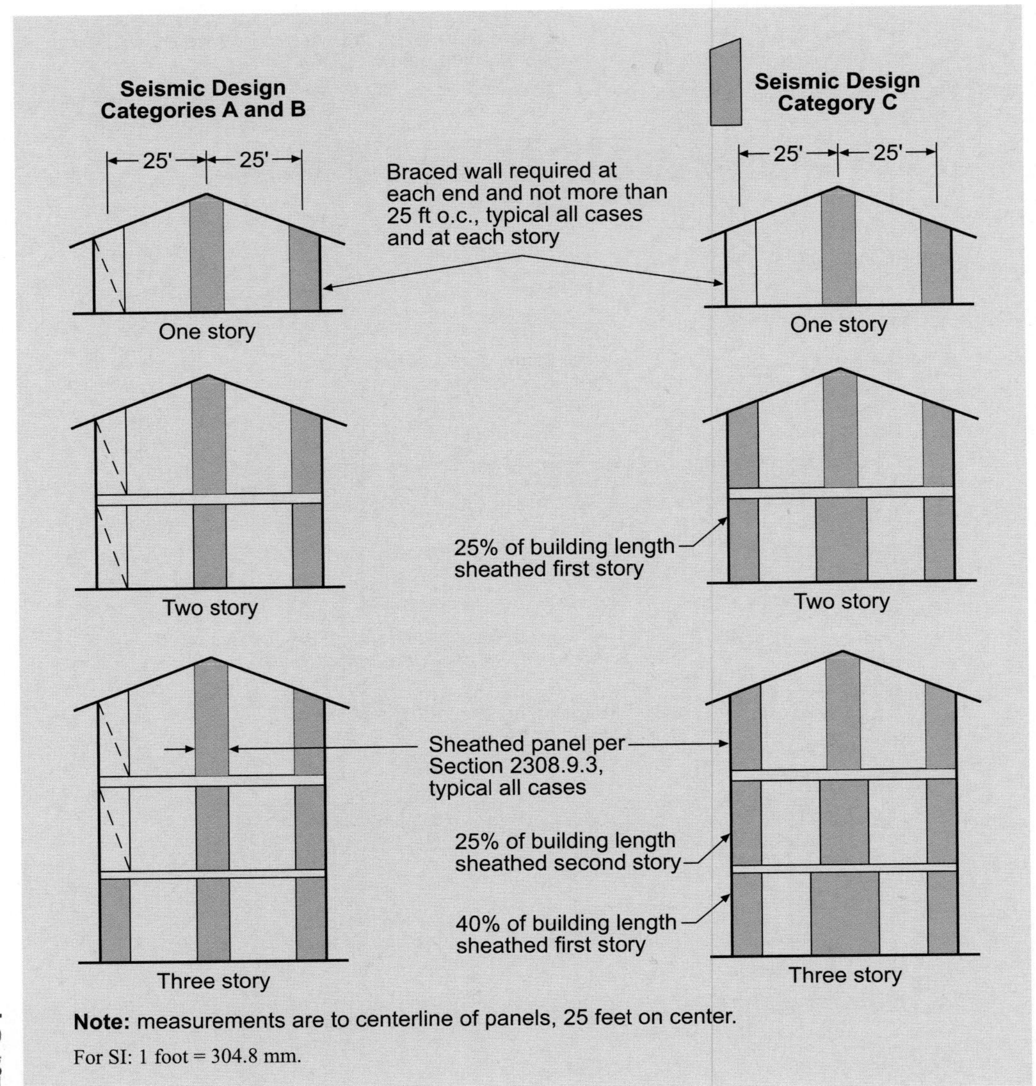

Figure 23-40
Wall bracing panel

and distribution of the governing load to properly design it. In addition, the result is an engineered component that is designed to transfer engineered lateral loads to prescriptive braced wall panels. The solution was to either provide prescriptive construction rules for the collector in various configurations or delete the provision from the code. A code change resolved the problem in the 2006 IBC by deleting the collector requirement. In order to conform to the prescriptive requirements, the code now requires braced wall panels to be located not more than 12.5 feet from the ends of braced wall lines with no reference to collector design. If braced wall panels do not meet this requirement, then the bracing system does not conform to prescriptive requirements and engineering is required. In this case, the design of the collector is an essential part of the engineering for the lateral-force-resisting system.

Figure 23-41 depicts this requirement.

Braced wall panels in a braced wall line must not be offset by more than 4 feet. Examples of acceptable locations for braced wall lines are illustrated in Figures 23-42 and 23-43.

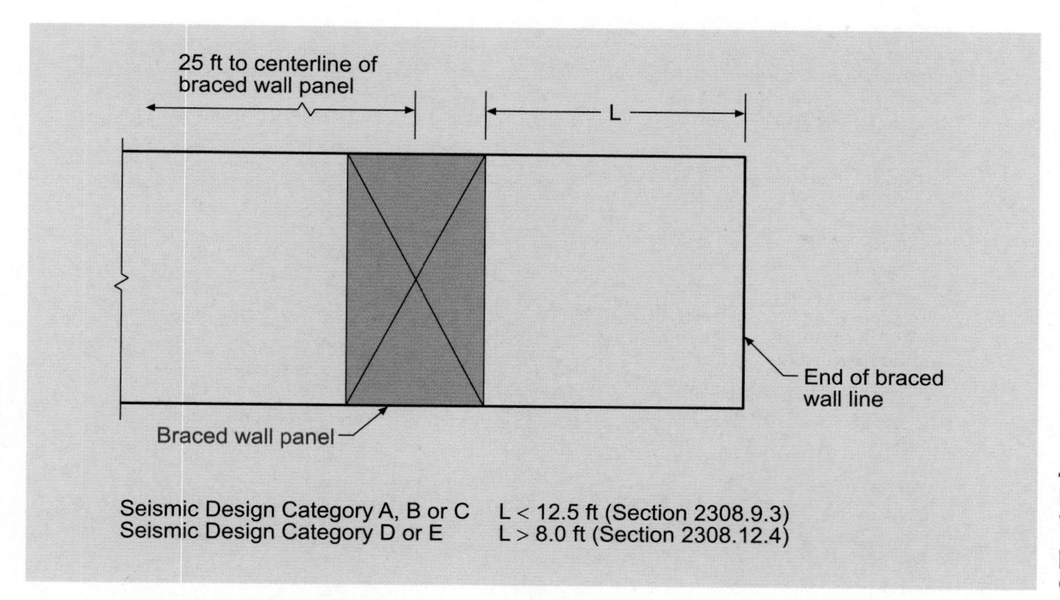

Figure 23-41 Wall bracing panel collector

Table 2308.9.3(1). This table is applicable only to structures classified as Seismic Design Category A, B or C. Structures classified as Seismic Design Category D or E have more stringent requirements, as shown in Table 2308.12.4. Table 2308.9.3(1) of the 2003 IBC states that braced wall panels must be "located at each end" yet Section 2308.9.3 permits braced wall panels to be located 12.5 feet from the end of a braced wall line. To resolve this, the language in Table 2308.9.3(1) of the 2006 IBC was revised to read "located in accordance with Section 2308.9.3."

2308.9.3.1

Alternate bracing. Braced panels in Section 2308.9.3 may be replaced by the alternate braced panels of this section, which were developed to solve the problem of having narrow walls adjacent to garage door openings. Note that the alternate braced panel is a complete assembly; therefore, all the requirements described in the section must be met. The alternate braced panel requirements are illustrated in Figure 23-44. At the present time, the alternate braced wall panel can be used on a one-story building or the first story of a two-story building. They cannot be used on the second story, because the code does not address the specific details and load path connections required for in-plane shear transfer and uplift anchorage to resist overturning forces at a second-story condition. See discussion under Section 2308.9.3.2 for requirements for alternate bracing wall panel adjacent to a door or window opening that are new in the 2006 IBC.

2308.9.3.2

Alternate bracing wall panel adjacent to a door or window opening. Section 2308.9.3.1 allows alternate braced wall panels to be used in lieu of braced wall panels described in Section 2308.9.3. The alternate braced wall panel in Section 2308.9.3.1 is required to be a minimum of 32 inches in length. Although this allows the alternate braced wall panel to fit in tight conditions in comparison to the 48-inch long braced wall panel, the 32 inches can be difficult to achieve when located adjacent to openings. This new alternative allows a reduction of the width of the full height segment of alternate braced wall panels to 16 inches wide for a one-story building, and to 24 inches wide for the first story of a two-story building. Because it is a prescriptive alternative bracing method, no engineering is required if constructed strictly in accordance with all of the requirements of the section as shown in Figure 23-45.

This so called "portal frame" alternative was first developed at the request of a building department in the State of Washington. Over the past decade, builders in the Pacific Northwest have been willing and technically capable of constructing these narrow *portal frames* in the field. The popularity of the detail grew until it was routinely permitted in a variety of jurisdictions in the Pacific Northwest. The original portal frame design was based

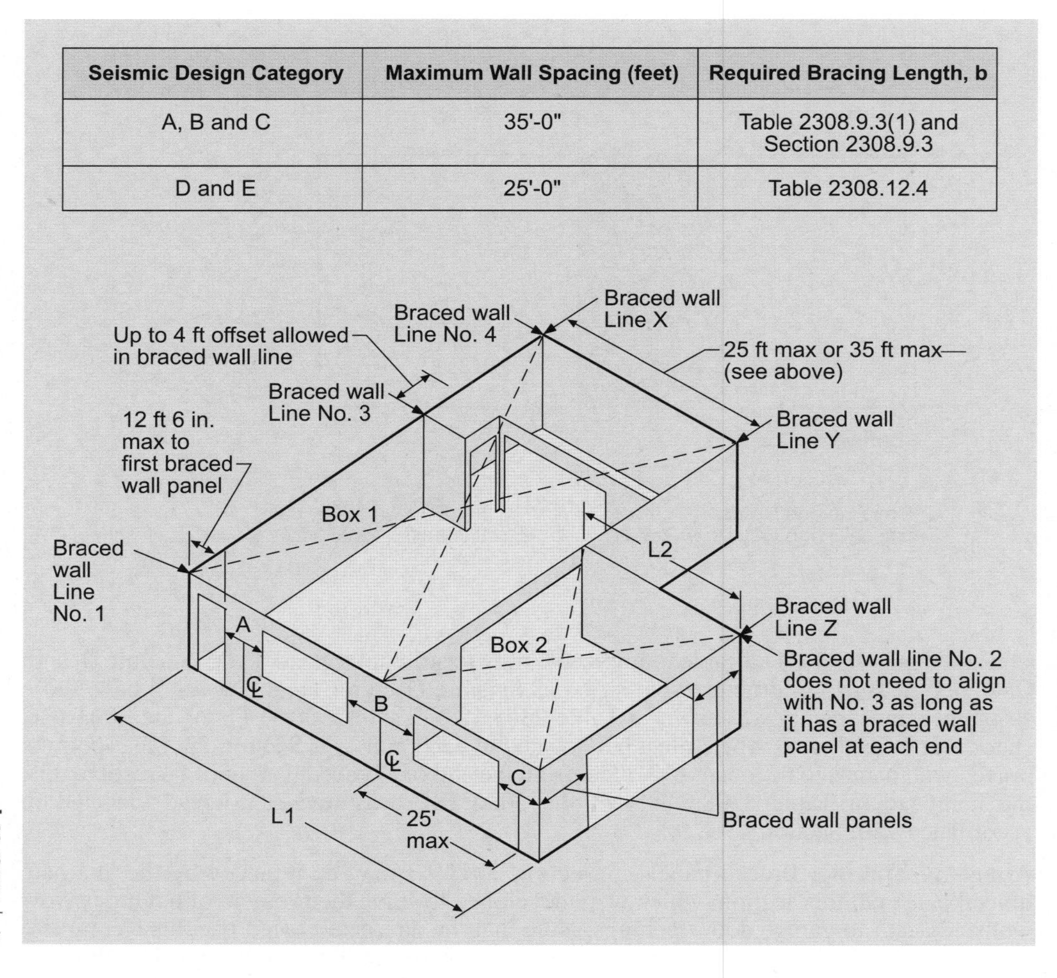

Seismic Design Category	Maximum Wall Spacing (feet)	Required Bracing Length, b
A, B and C	35'-0"	Table 2308.9.3(1) and Section 2308.9.3
D and E	25'-0"	Table 2308.12.4

Figure 23-42
Basic components of the lateral bracing system—one story

on monotonic testing. The APA Laboratory recently performed a series of cyclic tests using the Structural Engineers Association of Southern California (SEAOSC) cyclic testing protocol. During these tests, the single-sided/single-story 32-inch-wide alternate braced wall panel covered by Section 2308.9.3.1 was tested along with the 16-inch-wide proposed alternative portal frame system. The test results of the double-sided/two-story 32-inch-wide alternate braced wall panel was compared with the 24-inch-wide alternative portal frame. In both cases, the proposed alternative system performed significantly better in terms of both strength and stiffness than did the 32-inch alternative bracing panel permitted by the code.

The new alternative bracing method uses the header over the adjacent opening by running the header the length of the sheathed bracing panel to the first full-length stud. Where the sheathing and header overlap, the header is fastened to the full-height sheathing with a grid nailing pattern that provides a moment-resistant connection at the top. At the base of the sheathed section, embedded framing anchors nailed to the edge studs provide a moment resisting connection at the base. The framing anchors are sized to provide uplift and shear resistance capacity resulting in a reduction in anchoring requirements at the plate. Additional straps are required as shown in Figure 23-45.

2308.9.4 Cripple walls.

2308.9.4.1 Bracing. See the commentary for Section 2308.2. Cripple walls (sometimes referred to as foundation stud walls or knee walls) are stud walls usually less than 8 feet in height that rest on the foundation plate and support the first immediate floor above.

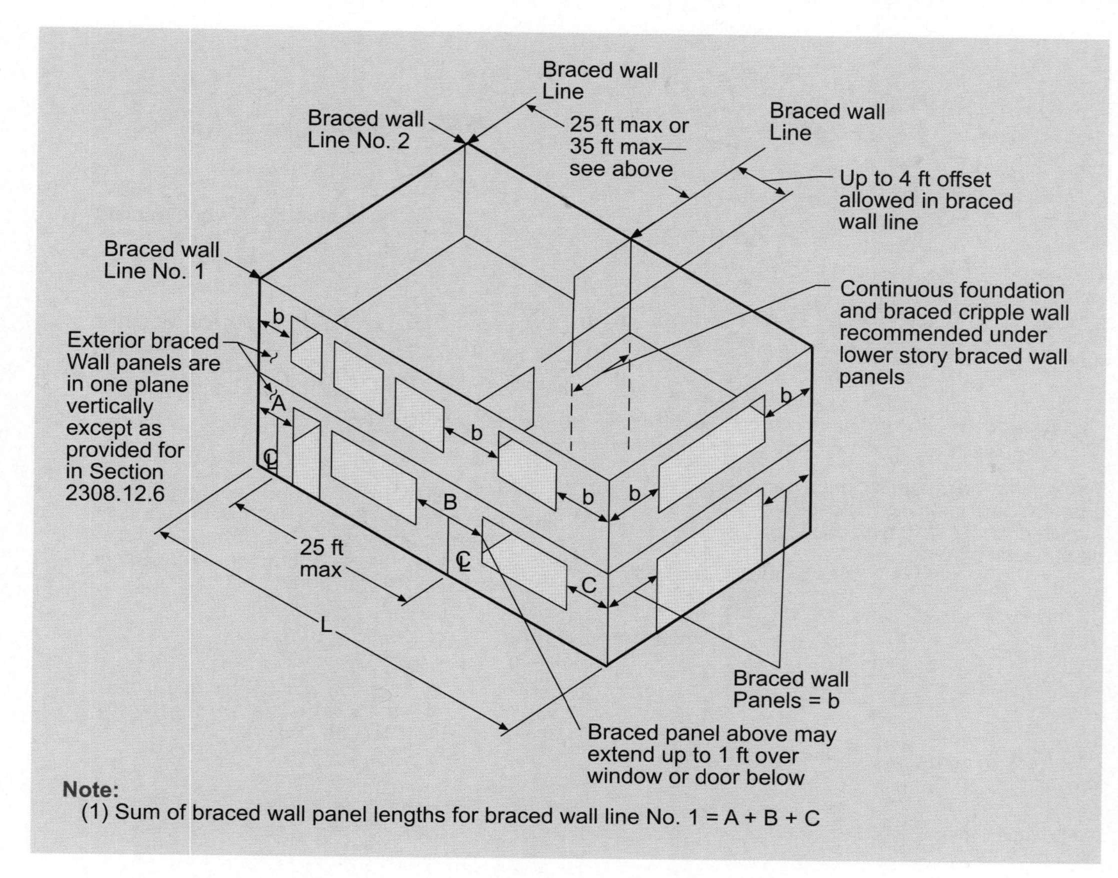

Figure 23-43 Basic components of the lateral bracing system—two stories

The code requires a minimum height of 14 inches for cripple wall studs, and this minimum is based on the length necessary to properly fasten the studs to the foundation wall plate and the double-wall plate above. Where the 14-inch minimum is not possible, the code then requires that the cripple wall be framed with solid blocking. Under this circumstance, the cripple wall studs, even though shorter than 14 inches in length, should be installed with wall plates as required by Section 2308.9.2 and with the solid blocking tightly fitted between each stud. This solid blocking performs three purposes: it provides a level uniform bearing surface for the support of the floor above, it transmits lateral forces from the floor to the foundation and it reduces the *racking* effects of the studs during a seismic or high-wind event. Wood structural panel sheathing may also be used to brace these walls, when adequate nailing is provided along the foundation sill and top plates.

Where cripple studs exceed 14 inches in height, the code has two requirements:

- Foundation stud walls having a stud height exceeding 14 inches shall be braced as a story in accordance with Table 2308.9.3(1) for Seismic Design Category A, B or C, and Table 2308.12.4 for Seismic Design Category D or E.
- Foundation stud walls exceeding 4 feet in height must be framed with studs having the size required for an additional story.

Thus, for a building that would be considered to be two stories in height with a crawl space beneath the first story but having foundation stud walls with the studs more than 4 feet in height, the code would require that the studs be framed with either 3 by 4 or 2 by 6 members, as would be required for the first story of a three-story building.

Openings in exterior walls. **2308.9.5**

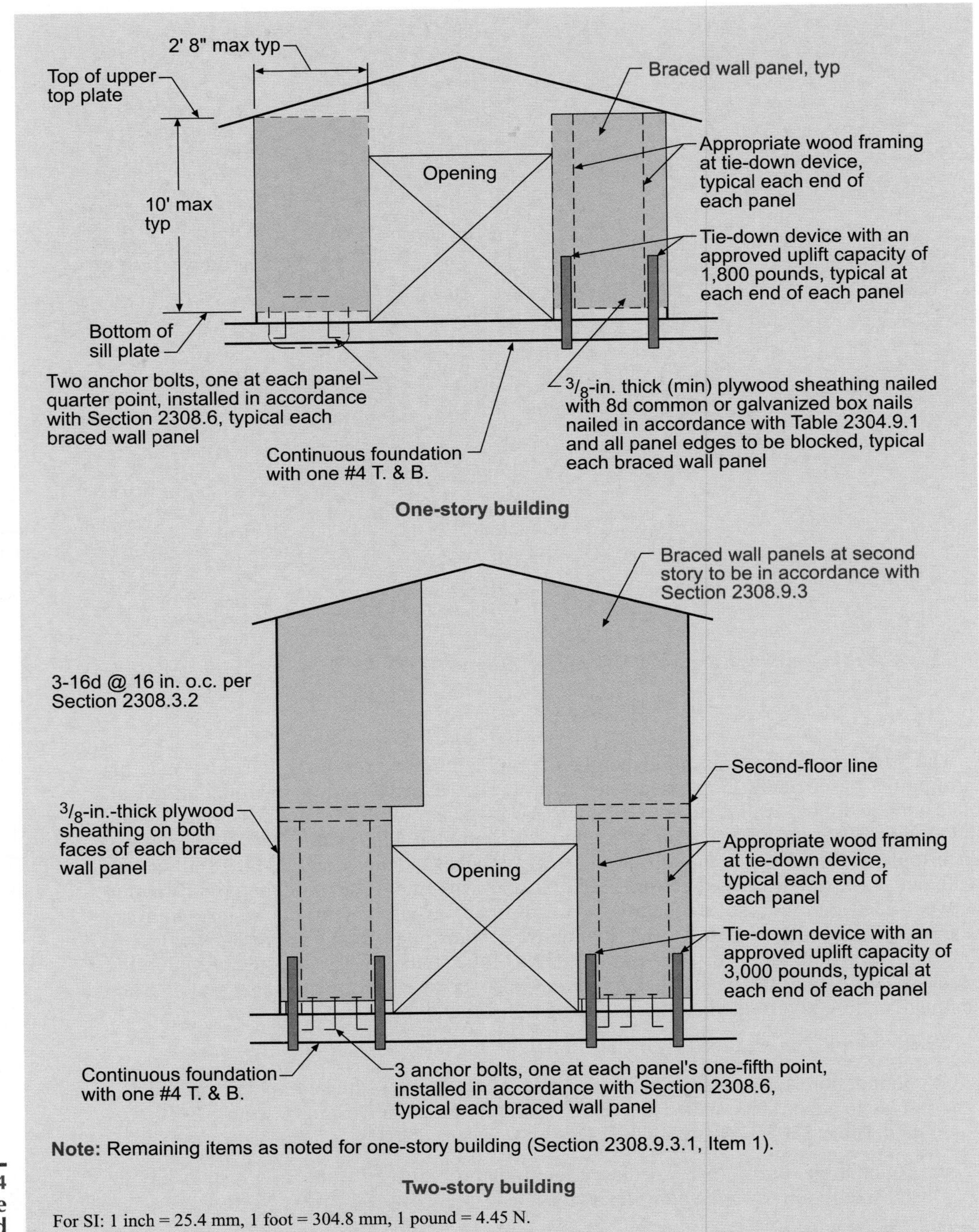

Figure 23-44 Alternate braced panels

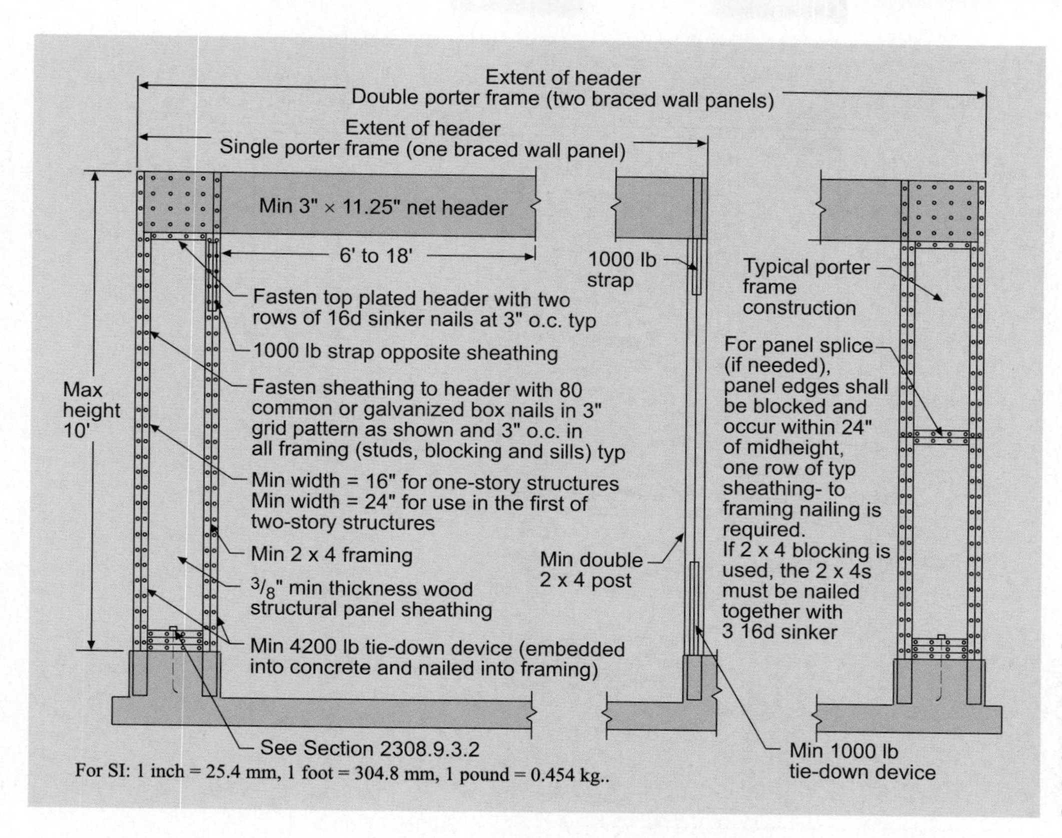

Figure 23-45
Alternate braced wall panel adjacent to a door or window opening

Headers. Headers selected in accordance with Table 2308.9.5 may be used only for structures that fit within the parameters of the table. Headers for other configurations or loading conditions must be designed in accordance with the engineering provisions of the code. Figure 23-46 illustrates header requirements. **2308.9.5.1**

Openings in interior-bearing partitions. Headers selected in accordance with Table 2308.9.6 may be used only for structures that fit within the parameters of the table. Headers for other configurations or loading conditions must be designed in accordance with the engineering provisions of the code. See Figure 23-45. **2308.9.6**

Pipes in walls. Figure 23-47 illustrates the requirements of this section for pipes in a wood-framed wall. **2308.9.8**

Bridging. Where stud partitions do not have adequate sheathing to brace the studs laterally in their weak (smaller) dimension, and the studs have a height-to-least-thickness ratio exceeding 50, the studs are required to have bridging or solid blocking with a minimum nominal thickness of 2 inches and a width the same as the studs. This blocking should be installed at heights that reduce the height-to-least-thickness ratio below 50. Use of 2 by 4 studs will require placement of the blocking when the maximum height of the wall is over 6 feet 3 inches ($1.5 \times 50 / 12 = 6.25$). **2308.9.9**

Cutting and notching. Figure 23-48 illustrates acceptable cutting or notching of studs. **2308.9.10**

Bored holes. Figure 23-49 illustrates acceptable dimensions and location of bored holes in studs. **2308.9.11**

Roof and ceiling framing. This section applies only to roofs with a minimum slope of 3:12 or greater. Where roofs have slopes less than 3:12, the horizontal thrust necessary to form a truss mechanism with the ridge board and ceiling joists or rafter ties becomes excessive. Thus, for roof slopes less than 3:12, the members supporting the rafters and ceiling joists **2308.10**

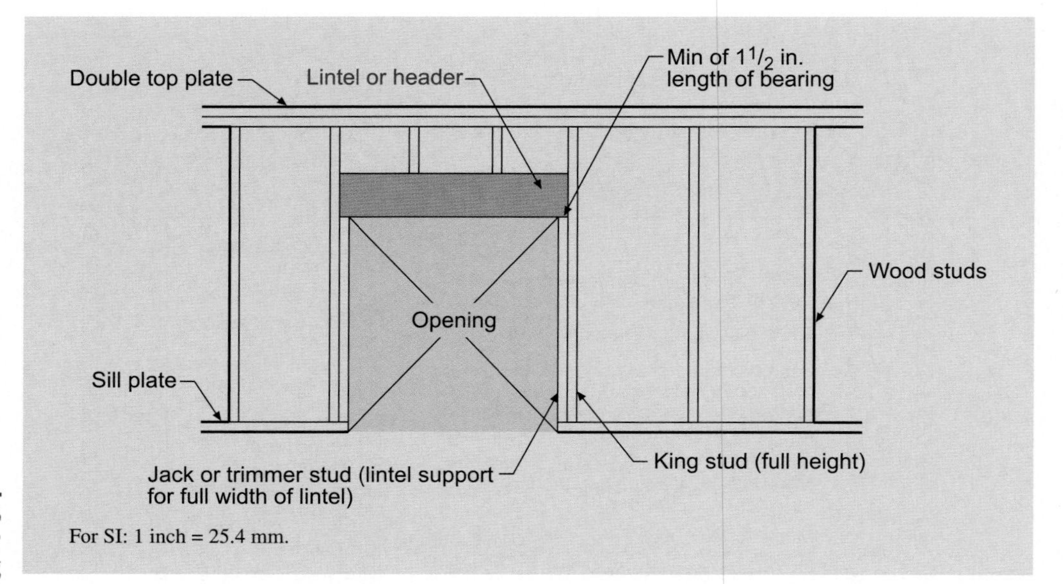

For SI: 1 inch = 25.4 mm.

Figure 23-46 Header over wall opening

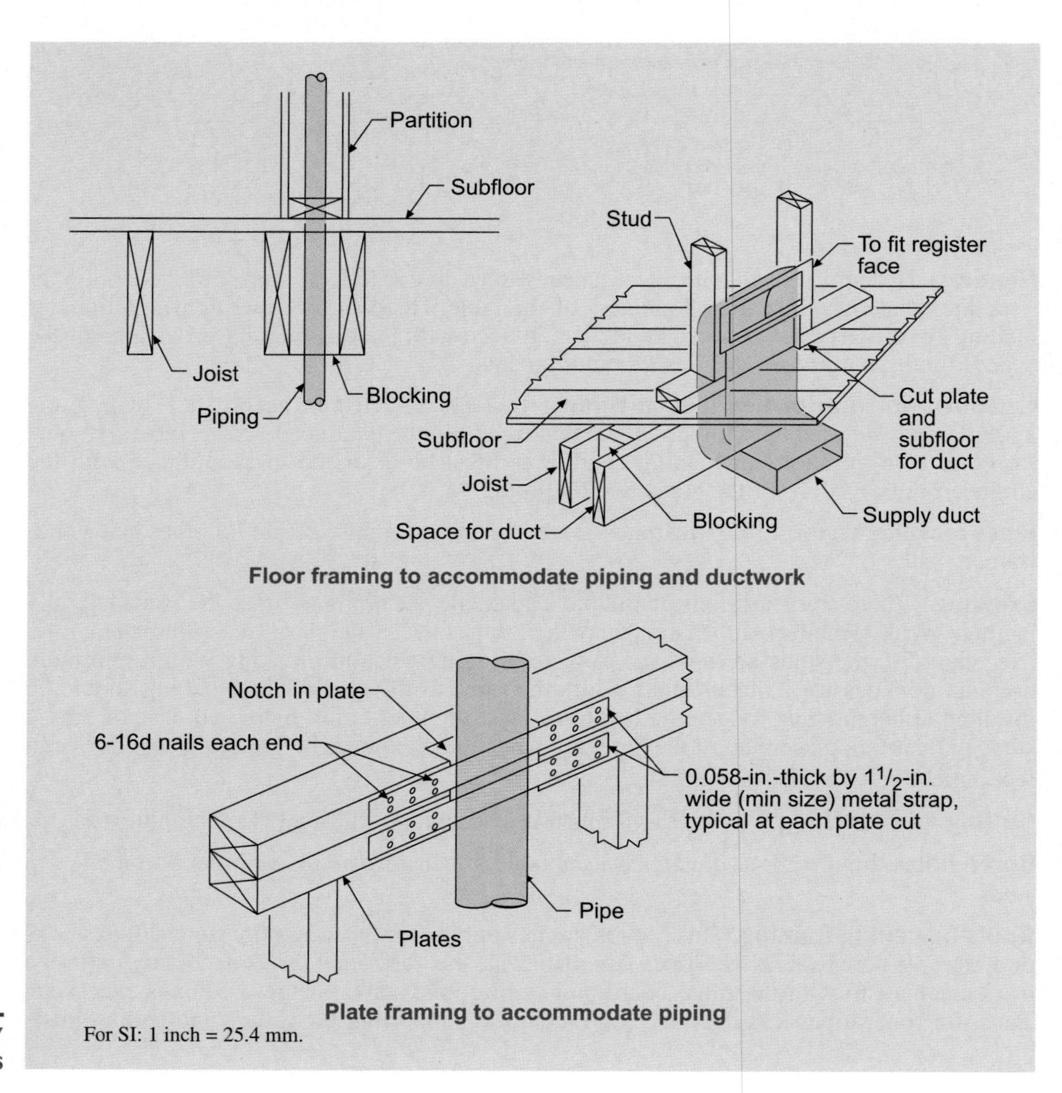

For SI: 1 inch = 25.4 mm.

Figure 23-47 Pipes in walls

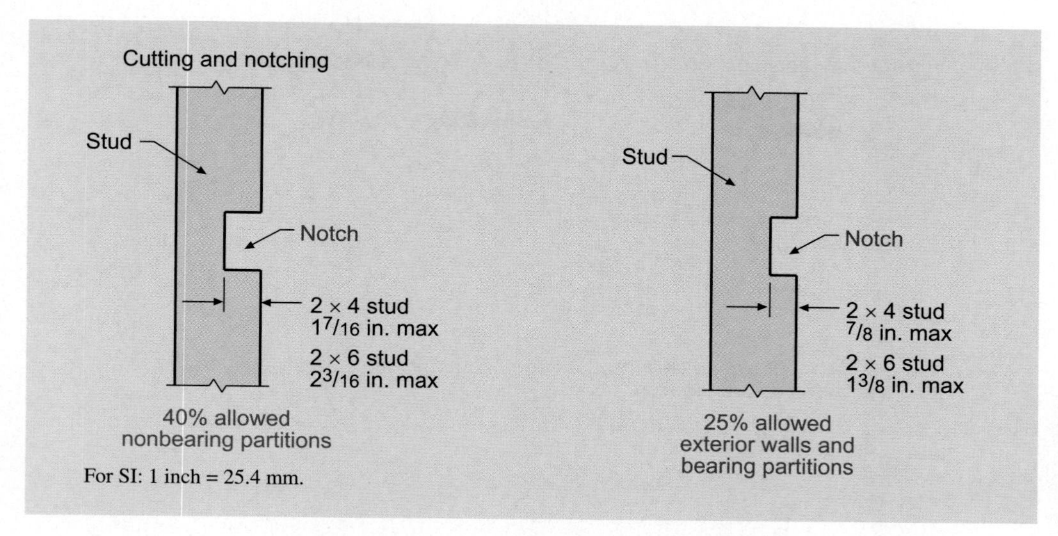

Figure 23-48 Cutting and notching of studs

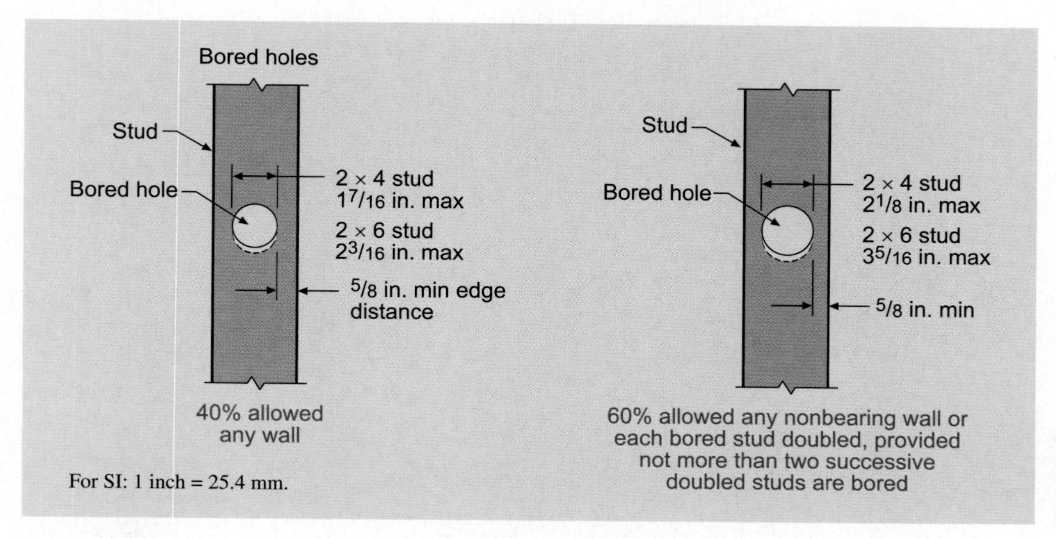

Figure 23-49 Bored holes in studs

such as the ridge beam, hip and valley rafters must be able to resist the gravity loads as beams instead of behaving like a truss. See Figures 23-50 and 23-51.

2308.10.1

Wind uplift. Wind suction pressures can cause considerable uplift forces on roof assemblies. The uplift loads must be positively transferred into the structure below to achieve sufficient gravity loads to resist the uplift. See also Section 2304.9.6, Load path, for additional commentary. Note that the uplift forces on roofs having slopes less than 3:12 will be larger than the forces for roofs with slopes equal to or greater than 3:12. Hence, although this section only applies to roofs with a slope greater than or equal to 3:12, ties of at least the same strength should be used on these lower slope roofs. The key features of the table are covered in the footnotes. The uplift loads are in pounds and are based on roof truss or rafter spacing of 24 inches on center. Other spacings must be adjusted. The table values have a built-in allowance for 10 psf roof dead load. The overhang loads are in pounds per foot of projection and based on 24-inch spacing. The overhang load must be added to the roof uplift load. The uplift loads are based on end zone component and cladding loads and allow reductions for connections away from the corner. The uplift loads are permitted to be reduced by 100 pounds for each full height wall above.

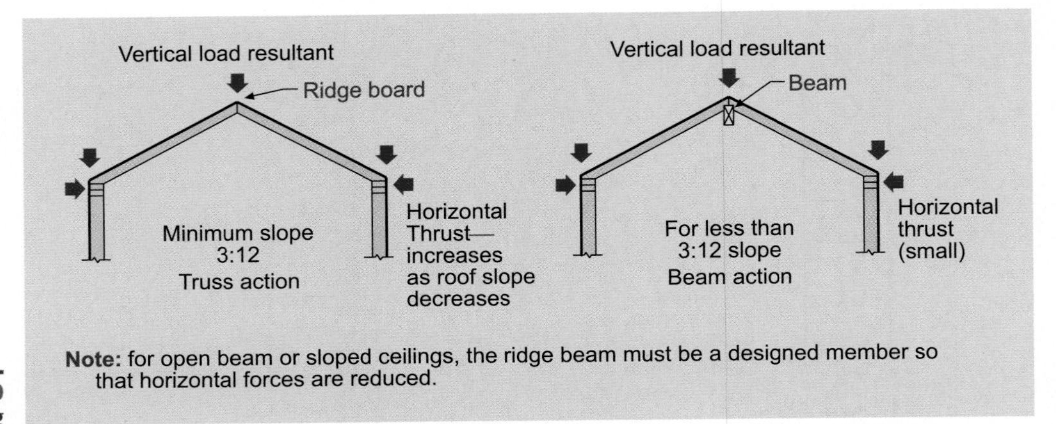

Figure 23-50 Roof framing

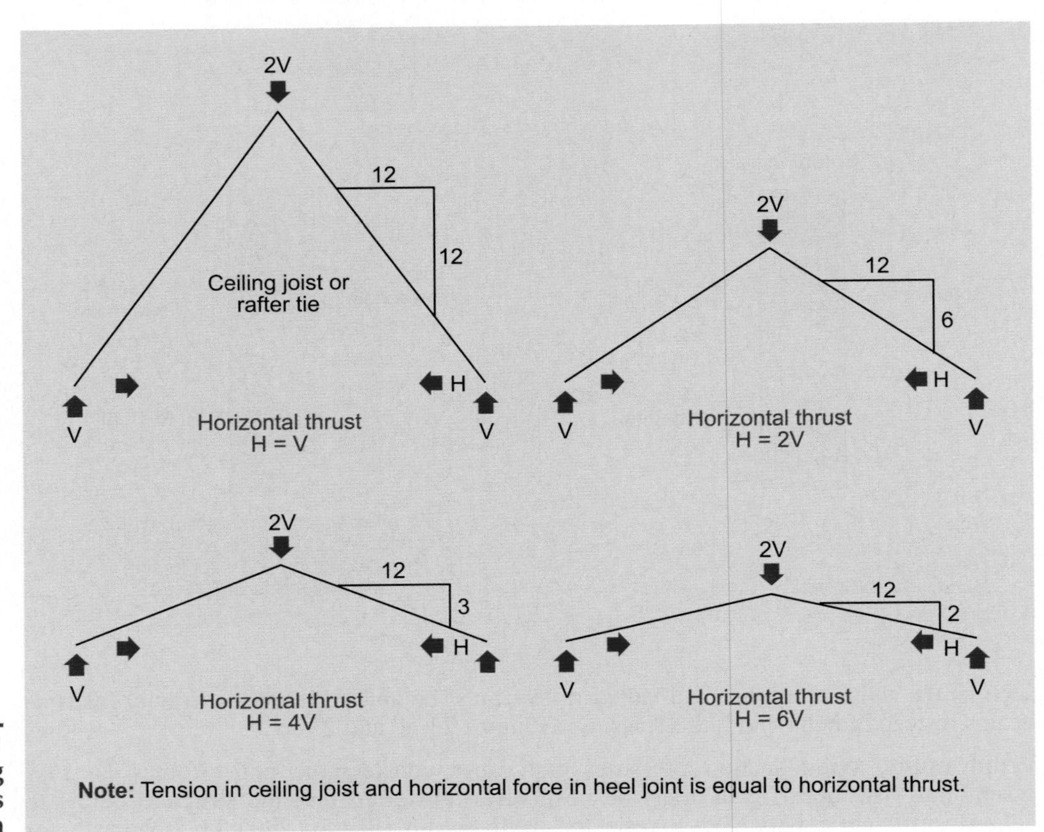

Figure 23-51 Roof framing thrusts—truss action

2308.10.2 Ceiling joist spans. Allowable spans for ceiling joists may be determined in accordance with the tables in the code. For other grades and species the section references *AF&PA Span Tables for Joists and Rafters*.

2308.10.3 Rafter spans. Allowable spans for rafters may be determined in accordance with the tables in the code. For other grades and species, the section references *AF&PA Span Tables for Joists and Rafters*.

2308.10.4 Ceiling joist and rafter framing. As noted in Section 2308.10 commentary, the roof rafters in conjunction with the ceiling joists or rafter ties form a simple truss. The ceiling joists or rafter ties resist the horizontal thrusts, as shown in Figure 23-51, in tension. Thus, the heel joint between the rafter and ceiling joist or rafter tie must have sufficient nailing to transfer the tension force to the ceiling joist and the ceiling joist splice must have sufficient nailing to resist the tension. Figure 23-52 illustrates the requirements. Nailing requirements for the heel joint are given Table 2308.10.4.1 based on roof slope and span, tie spacing, and either roof live load or ground snow load. Guidance and clarifications are given in the footnotes.

Ceiling joist and rafter connections. Ceiling joists and rafters must be nailed to each other and to the top wall plate in accordance with Tables 2304.9.1 and 2308.10.1. Ceiling joists must be tied over interior partitions and fastened to adjacent rafters to provide a continuous rafter tie across the building. Where ceiling joists are not parallel to rafters, a rafter tie must provide a continuous tie across the building at not more than 4 feet on center. In either case, the connections must be in accordance with Tables 2308.10.4.1 and 2304.9.1. Ceiling joists must have a minimum bearing of not less than 1 $^{1}/_{2}$ inches on the top plate at each end. **2308.10.4.1**

Notches and holes. This section contains prescriptive requirements for notching at the ends of rafters or ceiling joists, notches in the top or bottom of the rafter or ceiling joists, and bored holes in rafters or ceiling joists. **2308.10.4.2**

Framing around openings. This section contains prescriptive requirements for trimmer and header rafters used to frame around openings in roofs. **2308.10.4.3**

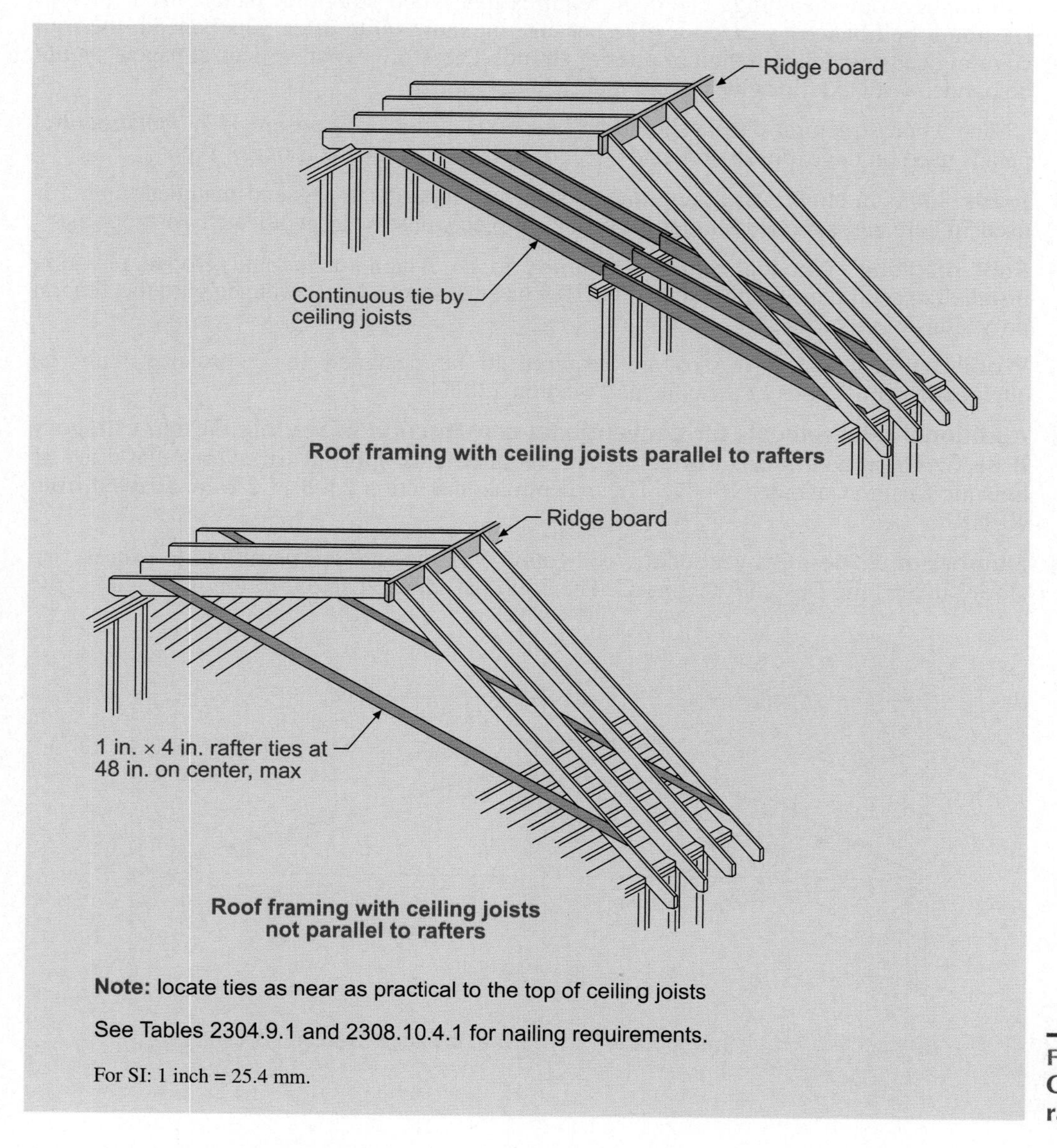

Figure 23-52 Ceiling and rafter framing

2308.10.5 Purlins. This section contains prescriptive provisions for purlin and strut bracing to reduce rafter spans. Purlins and struts are permitted to be installed to reduce the span of rafters within allowable span limits. Purlins are required to be supported by struts to bearing walls, which means a purlin that is braced to an interior wall creates a bearing wall. The maximum span of 2-inch by 4-inch purlins is 4 feet, and the maximum span of 2-inch by 6-inch purlins is 6 feet. However, in no case can the purlin be smaller than the supported rafter. Struts cannot be less than 2-inch by 4-inch members. The maximum unbraced length of struts is 8 feet and the minimum slope of the struts cannot be less than 45 degrees from the horizontal. Figure 53 illustrates purlin and strut requirements.

2308.10.6 Blocking. Roof rafters and ceiling joists must be supported laterally to prevent rotation and lateral displacement in accordance with Section 2308.8.5.

2308.10.7 Engineered wood products. Engineered wood products such as wood I-joists, glued-laminated timber and composite lumber cannot be notched or drilled except where specifically permitted by the product manufacturer or where the effects are specifically considered by the registered design professional.

2308.10.8 Roof sheathing. This section references the appropriate tables for wood structural panel and lumber roof sheathing. The code requires that wood structural panels used for roof sheathing be bonded by exterior glue because moisture quite often gets beneath the roof covering causing delamination of plies or strands Therefore, wood structural panels should be bonded with exterior glues to prevent delamination.

The wood structural panels should be labeled Exterior or Exposure 1. Wood structural panels used on the exposed underside of roof overhangs must be Exterior type.

2308.10.8.1 Joints. Joints in lumber sheathing must occur over supports unless end-matched lumber is used. Where end matched lumber is used, each piece must bear on at least two supports.

2308.10.9 Roof planking. Wood planking is required to be 2-inch tongue-and groove planking installed in accordance with Table 2308.10.9 or be designed in accordance with the general provisions of this code.

2308.10.10 Wood trusses. Wood trusses are required to be designed in accordance with the engineering provisions of the code and Section 2303.4.

2308.11 Additional requirements for conventional construction in Seismic Design Category B or C. These additional restrictions and requirements apply to structures classified as Seismic Design Category B or C. The exceptions in Section 2308.11.2 were adapted from NEHRP.

2308.11.1 Number of stories. Conventional wood-frame buildings are not permitted to exceed two stories in Seismic Design Category C. The 2006 IBC deleted an exception in the 2003 IBC

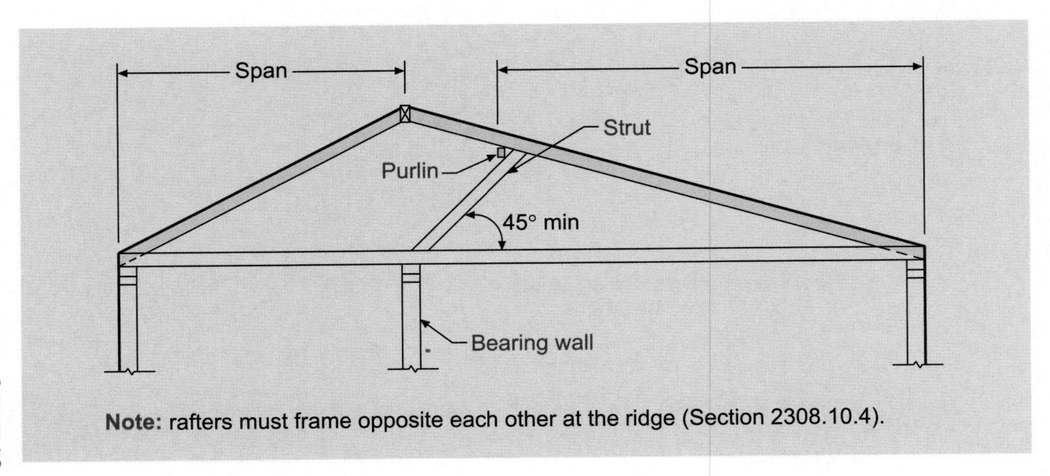

Figure 23-53 Rafter and purlin framing

that allowed an additional story for detached one- and two-family dwelling. The proponent reasoned that the story height exception for detached one- and two-family dwellings should be removed because detached one- and two-family dwellings are within the scope of the IRC.

Concrete or masonry. Section 2308.11.2 of the 2000 IBC imposes restrictions on the use of concrete or masonry walls and masonry veneer above the basement in Seismic Design Category B and C. A code change was proposed during the 2001 code cycle that removed these restrictions entirely but was not successful. This code change was subsequently modified by the committee, which resulted in an expansion of the provisions of Section 2308.11.2 as follows: **2308.11.2**

In Seismic Design Category B, masonry veneer is permitted to be used in three stories above grade if the lowest story is constructed of concrete or masonry walls. In addition, the wall bracing must be wood structural panels with a length that is increased to at least one- and one-half times the length required by Table 2308.9.3(1).

In Seismic Design Category B and C, masonry veneer is permitted to be used in two stories above grade where the lowest story has concrete or masonry walls.

In Seismic Design Categories B and C where the lowest story is not constructed of concrete or masonry walls, masonry veneer is permitted to be used in two stories above grade, provided the following criteria are met:

(1) The bracing required by Section 2308.9.3 must be wood structural panels with an allowable shear capacity of 350 plf minimum; (2) The bracing of the top story must be located at each end, at least every 25 feet o.c. and not less than 40 percent of the braced wall line. The bracing of the first story must be located at each end, at least every 25 feet o.c. and not less than 35 percent of the braced wall line; (3) 2000 pound capacity hold-down devices are required at the ends of braced wall panel segments from the second floor to first floor wall, and 3900 pound capacity hold-down devices are required from the first floor to the foundation; (4) Cripple walls are not permitted.

Framing and connection details. Provisions for stepped footings were adapted from NEHRP. NEHRP recommends that these provisions apply to all seismic design categories, but Seismic Design Category A is exempted in the IBC. **2308.11.3**

Anchorage. There are no special anchorage requirements in Seismic Design Category B or C. Braced wall lines are required to be anchored to the foundation in accordance with Section 2308.6. **2308.11.3.1**

Stepped footings. IBC Figure 2308.11.3.2 illustrates the requirements of this section. The phrase *lowest floor framing* means the framing of the lowest floor that is anchored directly to the foundation as shown on the left side of IBC Figure 2308.11.3.2. **2308.11.3.2**

Openings in horizontal diaphragms. Blocking and strapping are required to transfer shear around openings in diaphragms. IBC Figure 2308.11.3.3 illustrates the requirements. **2308.11.3.3**

Additional requirements for conventional construction in Seismic Design Category D or E. Additional restrictions and requirements are imposed on structures assigned to Seismic Design Category D or E, as prescribed in Sections 2308.12.1 through 2308.12.9. Conventional construction is not permitted in structures assigned to Seismic Design Category F. **2308.12**

Number of stories. Conventional wood-frame buildings are not permitted to exceed one story in Seismic Design Category D and E. The 2003 IBC had an exception that allowed an additional story for detached one- and two-family dwellings that was deleted in the 2006 IBC. The code change proponent indicated that the story height exception for detached one- and two-family dwellings was removed because detached one- and two-family dwellings are within the scope of the IRC. **2308.12.1**

Concrete or masonry. Masonry veneers cannot be used above the basement wall. This is also based on the NEHRP *Provisions*. **2308.12.2**

In the 2000 IBC, masonry veneer above the basement was prohibited in Seismic Design Category D and E. This restriction on the use of masonry veneer above the basement was removed in the 2003 IBC. An exception was added allowing masonry veneer to be used in the first story above grade plane in Seismic Design Category D, provided the following criteria are met: (1) Wall bracing must be wood structural panels with a minimum allowable shear capacity of 350 plf; (2) The bracing on the first story must be located at each end and at least every 25 feet on center and not less than 45 percent of the braced wall line; (3) 2100 pound capacity hold-down devices are required at the ends of braced walls from the first floor to the foundation; (4) Cripple walls are not permitted.

2308.12.3 Braced wall line spacing. Braced wall line spacing is limited to 25 feet maximum in Seismic Design Category D or E.

2308.12.4 Braced wall line sheathing. Braced wall panel lengths must be increased in accordance with Table 2308.12.4, depending on the value of the short period spectral response acceleration, S_{DS}. See Figure 23-42 for illustration of this requirement.

The second paragraph requires the length of the cripple wall braced panels to be $1^1/_2$ times the value in Table 2308.12.4 if the interior braced wall lines are not on a continuous foundation, such as interior braced wall lines supported by post-and-girder construction on precast piers. Note that interior braced wall lines must be on continuous foundations in accordance with Section 2308.3.4, unless the structure is not more than 50 feet in any plan dimension. Thus, a structure 40 feet by 60 feet would require continuous footings under the interior braced wall lines, whereas a structure that was 40 feet by 50 feet could use post-and-girder construction for support of the interior braced wall lines.

2308.12.5 Attachment of sheathing. Wall sheathing must be attached by the fasteners specified in the tables without substitution. Adhesives are not permitted to be used. Nails or staples yield when the braced panels are subjected to high seismic demand from ground motion and the panels exhibit ductile behavior. A panel attached by adhesives may behave in a brittle manner when subjected to high seismic demand.

2308.12.6 Irregular structures. Irregular structures in Seismic Design Category D or E cannot use conventional construction. The figures in the IBC graphically describe the various conditions to aid the code user.

The conditions describing irregular structures originated with the 1997 NEHRP *Provisions* and are very similar to the UBC conditions, except that the NEHRP *Provisions* apply the restrictions to Seismic Design Category C, D or E. IBC Figures 2308.12.6(1) through 2308.12.6(8) facilitate understanding the intent of the requirements.

Structures with geometric discontinuities in the lateral-force-resisting system sustain more earthquake and wind damage than structures without discontinuities. They have also been observed to suffer concentrated damage at the discontinuity location. For Seismic Design Category D or E, this section translates applicable irregularities from ASCE 7 Tables 12.3-1 and 12.3-2 into limitations on conventional light-frame construction. When a structure falls within the description of irregular, it is required that either the entire structure or the nonconventional portions be engineered in accordance with the engineered design provisions of the code. Although conceptually these are equally applicable to all seismic design categories, they are most critical in areas of high seismic risk, where damage caused by irregularities has repeatedly been observed in past earthquakes.

The engineered design of nonconventional portions in lieu of the entire structure is a common practice in some regions. The registered design professional must judge the extent of the portion required to be designed. This often involves design of the nonconforming element, force transfer into the element and a complete load path from the element to the foundation. A nonconforming portion may have enough of an impact on the behavior of a structure to warrant that the entire lateral-force-resisting system be engineered.

Item 1. This limitation is based on Item 4 of ASCE 7 Table 12.3-1 and applies when braced wall panels are offset out-of-plane from floor to floor. In-plane offsets are discussed in Item 3. Ideally braced wall panels should always stack above each other from floor to floor with the length stepping down at upper floors where less length of bracing panel is required.

Because cantilevers and setbacks are very often incorporated into residential construction, the exception offers rules by which limited cantilevers and setbacks can be considered conventional. Floor joists are limited to 2 by 10 (actual $1^1/_2$ by $9^1/_4$ inches) or larger and doubled at braced wall panel ends to accommodate the vertical overturning reactions at the end of braced wall panels. In addition, the ends of the cantilever are attached to a common rim joist to allow for redistribution of load. For rim joists that cannot run the entire length of the cantilever, the metal tie is intended to transfer vertical shear as well as provide a nominal tension tie. Limitations are placed on gravity loads to be carried by the cantilever or setback floor joists so that the joist strength will not be exceeded. The roof loads discussed are based on the use of solid sawn members where allowable spans limit the possible loads. Where engineered framing members such as trusses are used, gravity load capacity of the cantilevered or setback floor joists should be carefully evaluated.

Item 2. This limitation is based on Item 1 of ASCE 7 Table 12.3-1 and applies to open-front structures or portions of structures. The conventional construction bracing concept is based on using braced wall lines to divide a structure into a series of boxes of limited dimension, with the seismic force to each box being limited by the size. The intent is that each box be supported by braced wall lines on all four sides, limiting the amount of torsion that can occur. The exception, which permits portions of roofs or floors to extend past the braced wall line, is intended to permit construction such as porch roofs and bay windows, as illustrated in IBC Figure 2308.12.6(4). Walls that are not considered braced panels, i.e., with minimal lateral resistance, are allowed in areas where braced wall panels are prohibited. See the lower right-hand portion of IBC Figure 2308.12.6(4).

Item 3. This limitation is based on Item 4 of ASCE 7 Table 12.3-2 and applies when braced wall panels are offset in-plane. Ends of braced wall panels supported on window or door headers transfer large vertical reactions to headers that may not be of adequate size to resist these reactions. The exception permits a 4-foot extension of the braced wall panel over a nominal 4 by 12 or larger header on the basis that the vertical reaction will not result in critical shear or flexure. All other header conditions require an engineered design. Walls that are not considered braced panels, i.e., with minimal lateral resistance, are allowed over openings.

Item 4. This limitation results from observation of damage that is somewhat unique to split-level wood frame construction. If floors on either side of an offset move in opposite directions because of earthquake or wind loading, the short bearing wall in the middle becomes unstable and vertical support for the upper joists can be lost, resulting in a collapse. If the vertical offset is limited to a dimension equal to or less than the joist depth, then a simple strap tie directly connecting joists on different levels can be provided, and the instability eliminated.

Item 5. This limitation is based on Item 5 of ASCE 7 Table 12.3-1 and applies to nonperpendicular braced wall lines. When braced wall lines are not perpendicular to each other, further evaluation is needed to determine force distributions and the required lateral-support system.

Item 6. This limitation is based on Item 3 of ASCE 7 Table 12.3-1 and attempts to place a practical limit on openings in floors and roofs. Because stair openings are essential to residential construction and have long been used without any report of life-safety hazards resulting, these are felt to be acceptable conventional construction. See Section 2308.11.3.3 for detailing requirements for permitted openings.

NEHRP also considers a structure irregular if there are differences of more than 6 feet in height of the shear panels in a single story. Where the heights of a braced wall panel vary significantly, the stiffness and thus the distribution of lateral forces will also vary. The usual assumption of a uniform shear/foot will not be correct. The net result is a possible torsional irregularity. For example, if a structure on a hill is supported on 2-foot-high braced cripple wall panels on one side and 8-foot-high panels on the other, torsion and redistribution of forces will occur. Although the code does not currently require it, an engineered design is recommended to evaluate force distribution and provide adequate wall bracing and anchor bolting in this situation. This limitation applies specifically to walls from the foundation to the floor. Although gable-end walls have similar variations in wall heights, this has not been observed to be a significant concern in conventional construction.

2308.12.7 Anchorage of exterior means of egress components. Positive anchorage to the main structure is required for exterior egress balconies, exterior exit stairways and similar means of egress components. The anchorage must be spaced at no more than 8 feet on center. If anchorage to the main structure is not provided, then the egress component itself must be designed to resist lateral seismic forces. Toenails and nails in withdrawal do not perform well when subjected to cyclic seismic demand. See commentary for Section 2305.1.4.

2308.12.8 Steel plate washers. This requirement originated in response to the extensive splitting of wood sills during the Northridge earthquake. The splitting of sills was caused by a variety of factors including cross grain tension and oversized holes for anchor bolts, in addition to splitting from insufficient edge distance in heavily loaded walls. Subsequent cyclic testing of wood shear panel assemblies showed that use of 2-inch by 2-inch by $^3/_{16}$-inch, or larger, plate washers at the anchor bolts help prevent the splitting problem. See also the commentary pertaining to shear walls under Section 2305.3.10 for a more detailed discussion of the plate washer requirements. The *IBC Final Draft*, which was used as the base document to develop the 2000 IBC, required $3 \times 3 \times {}^1/_4$ inch thick plate washers, which was based on the 1997 NEHRP *Provisions*. The size of the plate washers was changed from $3 \times 3 \times {}^1/_4$ inch to $2 \times 2 \times {}^3/_{16}$ inch thick during the 2000 IBC code development process, which was consistent with the 1997 UBC. The reason given was adequate performance of sill plates tested with $2 \times 2 \times {}^3/_{16}$ inch steel plate washers. A code change to the 2003 IBC changed the square plate washer size to $3 \times 3 \times {}^1/_4$ inch thick to be consistent with the NEHRP *Provisions*. However, a subsequent code change modified the minimum thickness from $^1/_4$ inch to 0.229 inches in order to allow the plate washers to be manufactured from cold-rolled steel instead of $^1/_4$ inch hot-rolled steel. Cold-rolled steel of this thickness is more readily available, more economical and can be ordered with a hotdipped galvanized finish from the steel mill. Hot-rolled steel is not galvanized and must be galvanized after fabrication. Some steel hardware manufacturers recommend that galvanized plate washers be used where in contact with preservative-treated wood. The hole in the plate washer is permitted to be diagonally slotted with a width up to $^3/_{16}$ inch wider than the bolt diameter with a length up to $1^3/_4$ inches provided a standard cut washer is used between the square plate washer and the nut. The provision permitting slotted holes was taken directly from 2003 NEHRP *Provisions*.

2308.12.9 Anchorage in Seismic Design Category E. Anchor bolts for Seismic Design Category D are $^1/_2$-inch-diameter spaced at 48 inches on center instead of 6 feet on center. In Seismic Design Category E, $^5/_8$-inch-diameter anchor bolts spaced at 48 inches on center are required. The larger diameter bolts are required to transfer the increased in plane shear load from the anticipated increase in seismic demands.

REFERENCES

[1]NEHRP (1997, 2000, 2003), ***Recommended Provisions for Seismic Regulations for New Buildings and Other Structures***, Building Seismic Safety Council, Washington, DC.

[2]AF&PA, *ANSI/AF&PA: NDS-1997, National Design Specification for Wood Construction*, American Forest & Paper Association, Washington, DC.

[3]ASCE-16, *Standard For Load and Resistance Factor Design for Engineered Wood Construction:* AF&PA/ASCE 16-95, American Society of Civil Engineers, 1995, New York, NY.

[4]AITC, *Timber Construction Manual,* American Institute of Timber Construction, 2005, New York, NY, John Wiley and Sons, Inc.

[5]FPL, *Wood Handbook*, Reprinted from Forest Productions Laboratory General Technical Report FPL-GTR-113 with the consent of the USDA Forest Service, Forest Products Laboratory, Forest Products Society, 1998.

[6]SEAOC, *1999 Recommended Lateral Force Requirements and Commentary* (SEAOC Blue Book), Structural Engineers Association of California, 1999, ICC, Whittier, CA.

[7]SEAOC, *Seismic Design Manual, Volume II: Building Design Examples*, Structural Engineers Association of California, ICC, Whittier, CA.

[8]APA, *Design/Construction Guide—Diaphragms*, American Plywood Association, 1991, Tacoma, WA.

[9]ATC, *ATC-7: Guidelines for the Design of Horizontal Wood Diaphragms*, Applied Technology Council, 1981, Redwood City, CA.

[10]Nelson, Rawn F., and Hamburger, Ronald O., "Hold Down Eccentricity and the Capacity of the Vertical Wood Member," *Structure*, 1999.

[11]Stone, Jeffrey, Phillip Line and Brian Weeks (2000), Perforated Shear Wall Design, *Structural Engineer,* Civil Engineering News, INC., Alpharetta, GA.

[12]Dolan, J.D., and M.W. White, *Design Considerations for Using Adhesives in Shear Walls*, ASCE Journal of Structural Engineering 118(12): 3473-3480, 1992.

[13]Foschi, R.O., and A. Filiatrault, *Performance Evaluation of 3M Scotch Grip Wood Adhesive 5230 for the Static and Dynamic Design of Timber Shear Walls and Diaphragms,* Department of Civil Engineering Report, University of British Columbia, 1990, Vancouver, B.C.

BIBLIOGRAPHY

AF&PA, *Manual of Wood Construction: Load and Resistance Factor Design (LRFD),* American Forest & Paper Association. 1996, Washington, D.C.

APA, *Northridge California Earthquake, T94-5*, American Plywood Association, 1994. Tacoma, Washington.

APA, *Proposed Cyclic Testing Standard for Shear Walls*, American Plywood Association, 1996 Tacoma, Washington.

APA: Rose, J.D., *Preliminary Testing of Wood Structural Panel Shear Walls Under Cyclic (Reversed) Loading. Research Report 158*, American Plywood Association, 1996, Tacoma, Washington.

ASCE, *A Pre-Standard Report American Society of Civil Engineers*, ASCE Committee on Wood Load and Resistance Factor Design for Engineered Wood Construction, 1988. New York, New York.

ATC, *ATC-R-1: Cyclic testing of Narrow Plywood Shear Walls*, Applied Technology Council, 1995, Redwood City, California.

Breyer, Donald E., *Design of Wood Structures,* Third Edition, 1993, McGraw-Hill Book Company, New York, New York.

BOCA, *The BOCA National Building Code—Commentary Volume 2*, Building Officials & Code Administrators International, Inc., Country Club Hills, Illinois, 1999, copyright held by International Code Council.

BOCA, *National Building Code*, Building Officials & Code Administrators International, Inc., Country Club Hills, Illinois, 1996, copyright held by International Code Council.

CWC, *Wood Design Manual*, Canadian Wood Council, 1990, Ottawa, Canada.

CWC, *Wood Reference Handbook*, Canadian Wood Council, 1991, Ottawa, Canada.

CABO, *One- and Two-Family Dwelling Code*, Council of American Building Officials, 1995, Country Club Hills, Illinois, copyright held by International Code Council.

DofANAF, *Seismic Design for Buildings,* TM5-809-10 (Tri-Services Manual), Department of the Army, Navy and Air Force, 1992, U.S. Government Printing Office, Washington, D.C.

Dolan, J.D., 1994, *Proposed Test Method for Dynamic Properties of Connections Assemblies with Mechanical Fasteners*, ASTM Journal of Testing and Evaluation 22(6): 542-547.

Dolan, J.D., and Johnson, A.C., *Cyclic Tests of Long Shear Walls with Openings*, Virginia Polytechnic Institute and State University, Timber Engineering Report No. TE-1996-002, 1996 Blacksburg, Virginia.

Dolan, J.D., and Johnson, A.C., *Monotonic Tests of Long Shear Walls with Openings*, Virginia Polytechnic Institute and State University, Timber Engineering Report No. TE-1996-001, 1996 Blacksburg, Virginia.

EERI, *Northridge Earthquake Reconnaissance Report, Chapter 6, Supplement C to Volume 11, pp 125 et seq., Earthquake Spectra*, Earthquake Engineering Research Institute, 1996.

Faherty, Keith F., and T.G. Williamson, *Wood Engineering and Construction Handbook*, 1989 McGraw-Hill, New York, New York.

FPL, *Wood: Engineering Design Concepts*, Forest Products Laboratory, 1986, Materials Education Council, The Pennsylvania State University, University Park, PA.

Goetz, Karl-Heinz, Dieter Hoor, Karl Moehler, and Julius Natterer, *Timber Design and Construction Source Book: A Comprehensive Guide to Methods and Practice*, McGraw-Hill, 1989, New York, New York.

Hoyle and Woeste, *Wood Technology and Design of Structures.* Iowa State University Press, 1989.

HUD, *HUD Minimum Property Standards, Vol. I, II, and III*, U.S. Department of Housing and Urban Development, 1984, United States Government Printing Office Washington, D.C.

ICBO, *Uniform Building Code*, International Conference of Building Officials, 1994, Whittier, California, copyright held by International Code Council.

ICBO, *Handbook to the Uniform Building Code: An illustrative commentary*, International Conference of Building Officials, 1998, Whittier, California, copyright held by International Code Council.

Keenan, F.J., *Limit States Design of Wood Structures*, Morrison Hershfield Limited, 1986.

NEHRP, *Recommended Provisions for Seismic Regulations for New Buildings*, Building Seismic Safety Council, Washington, D.C., 1997, 2000, 2003.

NOAA, *San Fernando, California, Earthquake of February 9, 1971*, U.S. Department of Agriculture, National Oceanic and Atmospheric Administration, 1971, Washington, D.C.

Sherwood and Stroh, "Wood-Frame House Construction" in *Agricultural Handbook 73*. U.S. Government Printing Office, 1973,Washington, D.C.

Somayaji, Shan, *Structural Wood Design*. West Publishing Co., 1992, St. Paul, Minnesota.

Stalnaker, Judith J., and E.C. Harris, *Structural Design in Wood*, Second Edition, McGraw-Hill, 1996, New York, New York.

White, M.W. and J.D. Dolan, "Seismic Response of Timber Shear Walls. Part I: Aspect Ratios." Paper submitted for publication in *ASCE Journal of Structural Engineering*, 1996.

WWPA, *Western Wood Use Book*, Western Wood Products Association, Portland, Oregon, 1983.

Additional references are listed in John Peterson. 1983. "Bibliography on Lumber and Wood Panel Diaphragms," *Journal of Structural Engineering*, Vol. 109 No. 12. American Society of Civil Engineers, New York, New York.

Appendix

1

BACKGROUND TO SEISMIC STRENGTH DESIGN LOAD COMBINATIONS

ACI 318 through 1995, UBC through 1994

Strength design for reinforced concrete was first introduced into the 1956 edition of ACI 318 *Building Code Requirements for Reinforced Concrete*.[1.1] From the 1971 through the 1995 edition of the ACI 318 standard, the strength design load combinations involving earthquake forces remained unchanged from those shown in Item 1 of Table 1.1. The load combinations are meant to be used in conjunction with the strength reduction factors given in Chapter 9 of the ACI 318 standard. The *Uniform Building Code*, from its 1976 through its 1994 editions, modified the ACI strength design load combinations, for design in Seismic Zones 3 and 4, to those shown in Item 2 of Table 1.1. The strength reduction factors given in Chapter 9 of the ACI 318 standard were not modified in the UBC, thus increasing the margin of safety in comparison with ACI 318. When strength design for masonry was introduced into the UBC, the same seismic design load combinations as for concrete were adopted, along with strength reduction factors that were somewhat different from those of ACI 318 Chapter 9.

ANSI A58.1-82, ASCE 7-88

ANSI A58.1 (later ASCE 7) *Standard Minimum Design Loads for Buildings and Other Structures*,[1.2] in its 1982 edition, introduced a set of strength design load combinations that were different from those of ACI 318. The ANSI/ASCE load combinations were later adopted for load and resistance factor design (LRFD) of steel and subsequently also for LRFD of wood. Strength design for masonry, which was under development for quite some time and which was incorporated into the ACI 530-02/ASCE 5-02/TMS 402-02 masonry standard,[1.3] is also based on the ANSI/ASCE load combinations. The ANSI/ASCE strength design load combinations, as given in the 1988 edition of the ASCE standard,[1.2] are shown in Item 3 of Table 1.1. ASCE 7-88 did not anticipate the application of these load combinations to the strength design of concrete or masonry structures.

Appendix C to ACI 318-95, 318-99

ACI 318-95,[1.1] permitted, for the very first time, the proportioning of concrete structural elements by the ASCE 7 load combinations (seismic as well as nonseismic) "if the structural framing includes primary members of other materials . . ." Thus, a designer faced with the design of a building with a reinforced concrete shear wall core and steel framing outside of the core might choose to design the shear walls using the ASCE 7 rather than the ACI 318 load combinations. ASCE 7 gravity design load combinations provided lower design loads than the ACI 318 gravity design load combinations of that time. Although it was less clear how the design loads compared when it came to gravity plus lateral load (wind or earthquake) combinations, out of a desire to preserve the margin of safety in going from the ACI 318 to the ASCE 7 load combinations, a set of strength reduction factors having slightly lower values than the Chapter 9 ϕ-factors were developed and listed in Appendix C to ACI 318-95. The ASCE 7 load combinations, if chosen, had to be used with the strength reduction factors of Appendix C. ACI 318-99[1.1] did not change any of this.

ASCE 7-93, 7-95, 7-98

In all the load combinations mentioned so far, the earthquake effect, *E*, was the effect of service-level earthquake forces, as given, for instance, in the 1994 *Uniform Building Code* (UBC).[1.4] ASCE 7-93[1.2] introduced strength design load combinations involving strength-level earthquake forces, as shown in Item 4 of Table 1.1. Under Section A.9.11, Supplementary Provisions for Reinforced Concrete, ASCE 7-93 stated that the load combinations of ACI 318 were not applicable for the design of reinforced concrete to resist earthquake forces. Instead, ASCE 7 required that: "The load combinations for earthquake load in [ACI 318] shall be replaced with the load combinations of [ASCE 7-93] multiplied by the factor of 1.1, which accounts for an incompatibility between the ϕ-factors of [ACI 318] and the load factors of this document." Being a load standard, ASCE 7 did not want to extend itself into a modification of strength reduction factors. The multiplier of 1.1 was intended to provide the same level of safety as that produced by the reduced ϕ-factors of Appendix C to ACI 318-95. The seismic strength design load combinations and the stipulation concerning the multiplier 1.1 remained unchanged in ASCE 7-95. The seismic strength design load combinations did not change in ASCE 7-98; however, instead of applying a multiplier of 1.1 on the design loads, ASCE 7-98 required that the ϕ-factor in Appendix C to ACI 318-99 be used in computation with the design loads given by the seismic strength design load combinations.

Table 1-1. Seismic strength design load combinations in various codes and standards through the 2006 IBC

	(1) ACI 318-95, Ch. 9	(2) 1994 UBC	(3) ASCE 7-88
Service-level E	$U = 1.05D + 1.28L + 1.40E$	$U = 1.4(D + L + E)$	$U = 1.2D + 1.5E + (0.5L \text{ or } 0.2S)$
	$U = 0.9D + 1.43E$	$U = 0.9D \pm 1.4E$	$U = 0.9D - 1.5E$

	(4) ASCE 7-93, 7-95, & (5) 7-98	(6) 1997 UBC
Strength-level E	$U = 1.2D + 1.0E + 0.5L + 0.2S$	$U = 1.2D + 1.0E + (f_1L + f_2S)$
	$U = 0.9D + 1.0E$	$U = 0.9D \pm 1.0E$
	(7) ASCE 7-02, ASCE 7-05	**(8) 2000, 2003, 2006 IBC**
	$U = 1.2D + 1.0E + L + 0.2S$	$U = 1.2D + 1.0E + (f_1L + f_2S)$
	$U = 0.9D + 1.0E$	$U = 0.9D + 1.0E$

Notes (1), (4), (5), (7), (8) E can be positive or negative.

(1), (2) U was to be used in conjunction with the appropriate ϕ-factor from ACI 318-95 Ch. 9.

(3) ACI 318-95 and ACI 318-99 permitted the use of ASCE 7-88 U in conjunction with the appropriate ϕ-factor from Appendix C for the design of concrete members in structures of mixed construction only.

(4), (5) the load factor on L was required to be increased to 1.0 for garages, areas occupied as places of public assembly, and all areas where the live load is greater than 100 psf.

(5) U was to be used in conjunction with the appropriate ϕ-factor from ACI 318-95 Appendix C.

(4), (6) 1.1 U was to be used in conjunction with the appropriate ϕ-factor from ACI 318-88 (Revised 1992) or ACI 318-95 Ch. 9.

(7) The load factor on L is permitted to equal 0.5 for all occupancies in which unreduced live load is less than or equal to 100 psf, with the exception of garages or areas occupied as places of public assembly.

(7), (8) U is to be used in conjunction with the appropriate ϕ-factor from Ch. 9 of the referenced edition of ACI 318.

1997 UBC

The 1997 UBC adopted the ASCE 7-95 load combinations (same as those in ASCE 7-93) with two minor modifications. Instead of a live load factor of 0.5, which had to be increased to 1.0 in certain situations, the 1997 UBC prescribed a variable load factor, f_1, which was to be taken as 1.0 for floors in places of public assembly, for live loads in excess of 100 psf, and for garage live load. Similarly, instead of a fixed snow load factor of 0.2, the 1997 UBC prescribed a variable load factor, f_2, which was to be taken as 0.7 for roof configurations (such as saw tooth) that do not shed snow off the structure. The seismic strength design load combinations of the 1997 UBC are shown in Item 6 of Table 1.1. Exception 1 to Section 1612.2.1 of the 1997 UBC specified that load combinations that do not include seismic forces shall revert back to those of Section 1909.2 of ACI 318-95 for concrete design purposes. Exception 2 modified load combinations that included seismic forces, for concrete and masonry design, to:

$$U = 1.1\,[1.2D + 1.0E + (f_1L + f_2S)] \qquad \text{UBC (12-5a)}$$

$$U = 1.1\,[0.9D \pm 1.0E] \qquad \text{UBC (12-6a)}$$

The multiplier of 1.1 was based on the ASCE 7-95 approach.

The final draft of the *International Building Code* (IBC) did include the seismic strength design load combinations of the 1997 UBC. However, instead of adopting a multiplier of 1.1 on the design load, the IBC Final Draft, following ASCE 7-98, required: "For load combinations which include earthquake loads, the design strength shall be computed using the strength reduction factor, ϕ, listed in Appendix C [of ACI 318-95]."

It was not realized until actual designs were undertaken using the 1997 UBC that the multiplier of 1.1 should have never been applied to the UBC load combination (0.9D ± 1.0E), because this load combination was identical to the ACI 318-95 load combination (9-3), modified for inclusion of earthquake effects:

$$U = 0.9D + 1.43E \qquad \text{ACI (9-3a)}$$

or the 1994 UBC load combination:

$$U = 0.9D \pm 1.4E$$

if it was remembered that the E of the 1997 UBC was essentially 1.4 times the E of the 1994 UBC and ACI 318-95.

It was also questionable whether the 1.1 multiplier should apply to the load combination $[1.2D + 1.0\,E + (f_1L + f_2\,S)]$. Although a comparison of the ACI and the ASCE 7 load combinations readily revealed that ASCE 7 gravity load combinations definitely yielded lower design loads than the ACI 318 load combinations, it was far from clear if the same was true of the ASCE versus ACI gravity and lateral load combinations. Table 1.2 shows a comparison between design loads yielded by 1997 UBC Equation 12-5 and the corresponding gravity and earthquake load combination of ACI 318-95:

$$U = 0.75\,[1.4D + 1.7L + 1.87E] \qquad \text{ACI (9-2a)}$$

It can be seen that the comparison depends on three variables: the earthquake-to-dead (E/D), the snow-to-dead (S/D), and the live-to-dead (L/D) load (or load effect) ratios. The 1997 UBC Equation 12-5 yielded design loads that were over 10 percent lower than the corresponding design loads produced by ACI 318-95 Equation 9-2 (modified for earthquake effect) only when the live-to-dead load ratio exceeded 0.5, or even 0.6, which is not common in concrete structures.

The 1997 UBC Equation 12-5 would typically govern only the flexural design of moment frame beams. This was because the proportioning of reinforced concrete walls and columns is generally governed by the load combination yielding the minimum axial load in compression, or the maximum axial load in tension, which was typically Equation 12-6. A slight (up to 10 percent) underdesign of moment frame beams in flexure was of no concern. Underdesign in shear would have been a different matter, but that possibility was already precluded for beams of special as well as intermediate moment frames. A beam of a special moment frame must be designed for the maximum shear force that can develop in the beam, which is computed on the basis of moments equal to probable flexural strengths acting at the two ends. Also, the contribution of the concrete to shear strength is required to be neglected. A beam of an intermediate moment frame is also required to be designed for the maximum shear that can develop in it, which is computed on the basis of moments equal to nominal flexural strengths acting at the two ends.

In view of the above, it was recommended in References 1.5 and 1.6 that the 1.1 multiplier be not applied to the 1997 UBC Equation 12-5 or 12-6. In other words, Exception 2 to the 1997 UBC Section 1612.2.1 was recommended to be deleted.

ASCE 7-02, 7-05, 2000, 2003, 2006 IBC

For the same reasons as given above, the requirement of the working, the first and the final drafts of the IBC that the reduced strength reduction factors of Appendix C of ACI 318, rather than those of Chapter 9, be used in conjunction with the seismic strength design load combinations of the IBC, was proposed for deletion. The proposal was accepted, and the acceptance was reflected in the first (2000) edition of the IBC.

The seismic strength design load combinations have not changed from the 2000 to the 2003 to the 2006 IBC.

ASCE 7-02 dropped the ASCE 7-98 requirement: "where resistance to specified earthquake loads or forces E are included in design, the load combinations of Section 2.3 of ASCE 7-98 for strength design and the ϕ-factors of Appendix C of Ref. 9.9-1 [ACI 318-95] shall apply." ASCE 7-05 is the same in this regard as ASCE 7-02. In other words, the strength design load combinations in Section 2.3 of ASCE 7-02 and ASCE 7-05 are to be used in conjunction with the ϕ-factors, without modification, from Chapter 9 of ACI 318-02 and ACI 318-05, respectively.

Table 1-2. A comparison of seismic strength design load combinations of ASCE 7-95 (1997 UBC) and ACI 318-95

$$U(\mathrm{ACI}) = 1.05D + 1.28L + 1.43E$$

$$U(\mathrm{ASCE-7}) = 1.2D + 0.5L + 0.2S + 1.0E$$

$$\frac{U(\mathrm{ACI})}{U(\mathrm{ASCE}-7)} = \frac{1.05 + 1.28L/D + 1.0E/D}{1.2 + 0.5L/D + 0.2S/D + 1.0E/D}$$

L/D	*S/D* = 0.0	*S/D* = 0.1	*S/D* = 0.2	*S/D* = 0.5	*S/D* =1.0
***E/D* = 1**					
0.25	1.02	1.01	1.00	0.98	0.94
0.40	1.07	1.06	1.05	1.02	0.99
0.50	1.10	1.09	1.08	1.05	1.02
0.60	1.13	1.12	1.11	1.08	1.04
0.75	1.17	1.16	1.15	1.13	1.08
***E/D* = 2**					
0.25	1.01	1.01	1.00	0.08	0.96
0.40	1.05	1.04	1.04	1.02	0.99
0.50	1.07	1.06	1.06	1.04	1.01
0.60	1.09	1.08	1.08	1.06	1.03
0.75	1.12	1.12	1.11	1.09	1.06
***E/D* = 3**					
0.25	1.01	1.01	1.00	0.99	0.97
0.40	1.04	1.03	1.03	1.01	0.99
0.50	1.05	1.05	1.04	1.03	1.01
0.60	1.07	1.07	1.06	1.05	1.03
0.75	1.10	1.09	1.09	1.07	1.05

(Continued)

Table 1-2. A comparison of seismic strength design load combinations of ASCE 7-95 (1997 UBC) and ACI 318-95 (Cont'd)

E/D = 4					
L/D	*S/D* = 0.0	*S/D* = 0.1	*S/D* = 0.2	*S/D* = 0.5	*S/D* =1.0
0.25	1.01	1.00	1.00	0.99	0.97
0.40	1.03	1.03	1.02	1.01	0.99
0.50	1.04	1.04	1.04	1.03	1.01
0.60	1.05	1.05	1.05	1.04	1.02
0.75	1.08	1.07	1.07	1.06	1.01

REFERENCES

[1.1]ACI Committee 318, *Building Code Requirements for Structural* (previously Reinforced) *Concrete* (ACI 318) *and Commentary* (ACI 318R), American Concrete Institute, Detroit, MI, 1956, 1963, 1971, 1977, 1983, 1989, 1995, and Farmington Hills, MI, 1999, 2002, 2005.

[1.2]American Society of Civil Engineers, *ASCE Standard Minimum Design Loads for Buildings and Other Structures*, ASCE 7-88, ASCE 7-93, ASCE 7-95, ASCE 7-98 (also ANSI A58-55, ANSI A58.1-72, ANSI A58.1-82), New York, N. Y., 1990, 1993, 1995, and 2000, respectively.

[1.3]Masonry Standards Joint Committee, *Building Code Requirements for Masonry Structures* (ACI 530-95/ ASCE 5-95/ TMS 402-95), (ACI 530-99/ ASCE 5-99/ TMS 402-99), (ACI 530-02/ ASCE 5-02/ TMS 402-02), (ACI 530-05/ ASCE 5-05/ TMS 402-05), American Concrete Institute, Farmington Hills, MI (previously Detroit, MI), American Society of Civil Engineers, Reston, VA (Previously New York, N.Y.), The Masonry Society, Boulder, CO, 1995, 1999, 2002, 2005.

[1.4]Pacific Coast Building Officials Conference (later the International Conference of Building Officials), *Uniform Building Code*, Long Beach, CA (later Whittier, CA), 1935, 1952, 1967, 1970, 1973, 1976, 1979, 1982, 1985, 1988, 1991, 1994, 1997.

[1.5]Ghosh, S.K., "Design of Reinforced Concrete Buildings under the 1997 UBC", *Building Standards*, International Conference of Building Officials, May-June 1998, pp. 20-24.

[1.6]Ghosh, S.K., "Needed Adjustments in 1997 UBC," *Proceedings*, 1998 Convention, Structural Engineers Assocation of California, October 7-10, 1998, Reno-Sparks, NV, pp. T9.1-T9.15.

Appendix

2

BACKGROUND TO THE WIND LOAD PROVISIONS OF MODEL CODES AND STANDARDS

ASCE 7 through 1995 Edition

The following is adapted largely from References 2.4 and 2.5.

The first major improvements in the standardized wind load provisions of this country occurred with the publication of ANSI A58.1-1972.[2.6] Unfortunately, however, the new provisions never gained wide acceptance by the model code bodies and practicing engineers, primarily because they represented a quantum jump in sophistication in comparison with the codes of practice of that time. The ANSI A58.1-1972 provisions were also flawed with ambiguities, inconsistencies in terminology and a format that permitted certain provisions to be misinterpreted.

There was consensus among professional practitioners that the ANSI 1972 document needed revision, and a new subcommittee of ANSI A58.1 was appointed in 1976 to accomplish this task. The subcommittee reviewed all available research data and deliberated on various proposals advanced to improve and simplify the provisions. The draft produced by the subcommittee was subjected to the ANSI consensus process. Upon conclusion of that process, the revision was published as ANSI A58.1-1982.[2.6] The revised standard contained an innovative approach to wind loads for components and cladding of buildings. Wind load specification was based on understanding the aerodynamics of wind pressure in building corners, eaves and ridge areas, as well as the effects of area averaging on pressures.

In 1985, the American Society of Civil Engineers (ASCE) assumed responsibility for the committee that establishes design loads for buildings and other structures. The ASCE Committee reviewed the wind load provisions of ANSI A58.1-1982 and made minor changes and clarifications that were incorporated in ASCE 7-88, published in 1990. A revised version of ASCE 7-88 was published in ASCE 7-93. However, the wind load provisions were not changed and remained essentially the same as in ANSI A58.1-1982, until the publication of ASCE 7-95, which did make major changes in those provisions.

The most significant change in ASCE 7-95 was the use of 3-second gust speed, rather than the fastest-mile wind speed, as the basis of the basic wind speed used in design. The change necessitated revision of terrain and height factors, gust-effect factors and pressure coefficients for components and cladding. Additional significant changes and additions were as follows:

- Provisions were added for wind speed-up over isolated hills and escarpments (topographical effect).
- New provisions were added for full and partial loading on main wind-force resisting systems of buildings with a mean roof height greater than 60 feet (torsional loading effect).
- An alternative procedure was added for determining external loads on main wind-force resisting systems of low-rise buildings.
- Internal pressure coefficients were increased for partially enclosed buildings located in hurricane-prone regions.
- Pressure coefficients for components and cladding were added for hipped, stepped, multispan and sawtooth roofs.
- Velocity pressure exposure coeffecients were revised to be compatible with 3-second gust speed.

---

*A version of this appendix was published in References 2.1 and 2.2. A condensed version is also available in Reference 2.3.

- Gust effect factor procedures were unified for flexible and nonflexible buildings and structures. A new procedure was provided for calculating a gust effect factor.
- Pressures for components and cladding of buildings with a mean roof height less than 60 feet were reduced for buildings sited in Exposure B.

Other changes in tables and figures and in the wind tunnel test procedure incorporated the latest available technical information. The basic format of the wind provisions remained the same as in ASCE 7-88.

ASCE 7-98

The wind provisions of ASCE 7-98 contained several changes from those of ASCE 7-95 that are important to note:

1. The basic wind speed map in the 1998 standard was updated based on a new and more complete analysis of hurricane wind speeds. The new analysis included many more predictions for sites away from the coast, and it included the best available information regarding the rate at which hurricane winds degrade as they move inland. The most noticeable result was a significant reduction in wind speeds in inland Florida.
2. Wind directionality was now explicitly accounted for in the calculation of the wind loads; it was no longer just a component of the wind load factor. A new factor, K_d, was introduced, which varies depending on the type of structure.
3. A new Simplified Procedure for wind design was introduced for the very first time. The format of the entire wind section was changed to increase its ease of use. The wind provisions were organized by three distinct methods. Method 1 "Simplified Procedure," is restricted in use to simple diaphragm buildings that meet certain criteria. Method 2 "Analytical Procedure," may be used for essentially any building or structure. Method 2 contains provisions for buildings of all heights, including those commonly referred to as *low-rise buildings*. Method 3, "Wind Tunnel Procedure," includes provisions for conducting a wind tunnel test.
4. The definitions of Exposures C and D were changed slightly to allow the shorelines in hurricane-prone regions to be classified as Exposure C rather than Exposure D. The commentary pertaining to the exposure categories was expanded to provide more information and guidance to the user and now included photographs to assist in the understanding of the different exposure categories.

ASCE 7-02

In ASCE 7-02, the simplified design procedure, Method 1, of ASCE 7-98 was discarded. The simplified design procedure in Section 1609.6 of the 2000 IBC, with only a few relatively minor modifications, was adopted instead. This simplified procedure is based on the low-rise analytical procedure of ASCE 7 and bears strong resemblance to it. Its applicability is broader than that of the simplified design procedure in ASCE 7-98.

ASCE 7-02 required that a ground surface roughness within each 45-degree sector be determined for a distance upwind of the site. Three surface roughness categories were defined as shown in Table 2.1.

Three exposure categories were defined in terms of the three roughness categories, as shown in Table 2.2.

The former Exposure A (centers of large cities) was deleted.

Method 2, Analytical Procedure, for (MWFRS of) low-rise buildings was revised to provide clarification. The different load cases were clearly delineated.

New pressure coefficients were provided for determination of wind loads on domed-roof buildings.

Provisions for calculating wind loads on parapets were added.

The design load cases for the MWFRSs of buildings designed by the general analytical procedure (as distinct from the low-rise analytical procedure) were different in ASCE 7-98 than in ASCE 7-02. Consideration of wind-induced torsion was now required for all buildings, not just buildings having mean roof height exceeding 60 feet.

In the table of roof pressure coefficients for the design of the MWFRS by the general analytical procedure, a low suction coefficient of 0.18 was added for the windward roof in all cases where only a high suction coefficient was provided earlier. The intent of the new low suction coefficient is to require the roof to be designed for zero or a slightly positive (inward acting) pressure, depending upon whether the building is enclosed or partially enclosed, respectively.

Table 2-1. **Surface roughness categories of ASCE 7-02 and 7-05**

Surface Roughness Category	Description
B	Urban and suburban areas, wooded areas or other terrain with numerous closely spaced obstructions having the size of single-family dwellings or larger.
C	Open terrain with scattered obstructions having heights generally less than 30 ft. This category includes flat open country, grandstands and all water surfaces in hurricane-prone regions.
D	Flat, unobstructed areas and water surfaces outside hurricane-prone regions. This category includes smooth mud flats, salt flats and unbroken ice.

Table 2-2. **Exposure categories of ASCE 7-02 and 7-05**

Surface Roughness Category	Description
B	Surface Roughness B prevails in the upwind direction for at least 2630 ft or 10 times the building height (2600 ft or 20 times the building height in ASCE 7-05), whichever is greater.
C	All cases where Exposure B or D does not apply.
D	Surface Roughness D prevails in the upwind direction for at least 5000 ft or 10 times the building height (or 20 times the building height in ASCE 7-05), whichever is greater. Exposure D extends inland from the shoreline a distance of 660 ft or 10 times the building height (600 ft or 20 times the building height in ASCE 7-05), whichever is greater.

ASCE 7-05

Several changes are made in the set of conditions that must be met by a building for its MWFRS to be qualified to be designed by Method 1 - Simplified Procedure. The restriction that the building not be subjected to topographic effects is omitted. Topographic effects are now accounted for in the simplified design procedure by including a topographic effect factor in the calculation of the design wind pressure.

The conditions that must be met by a building for its components and claddings to be eligible to be designed by Method 1 are not changed, except that the restriction concerning topographic effects is lifted, as in the case of the MWFRS.

Simplified design wind pressures and net design wind pressures can now be calculated for basic wind speeds of 105, 125 and 145 miles per hour.

ASCE 7-05 now explicitly states that the basic wind speeds estimated from regional climatic data for special wind regions outside hurricane-prone areas can be lower than those given in ASCE 7-05 Figure 6-1. For estimation of basic wind speeds from regional wind data in special wind regions outside hurricane-prone areas, a minimum criterion is specified.

ASCE 7-02 required Exposure D to extend inland from the shoreline for a distance of 660 feet or 10 times the height of the building, whichever was greater. ASCE 7-05 requires Exposure D to extend into downwind areas of Surface Roughness B or C for a distance of 600 feet or 20 times the height of the building, whichever is greater. The multiplier of building height by which a certain terrain category has to extend in the upwind and the downwind direction of the building for qualification of an Exposure Category is changed from 10 to 20, as indicated above in the specific case of Exposure Category D. Other controlling distances are rounded off to the nearest 100 feet (Table 2.2).

A definition for eve height is added. Footnote 8 to Figure 6-10 (Low-Rise Analytical Procedure), which concerns delineation of boundary between windward zone pressures and leeward zone pressures, has been clarified.

Glazing in wind-borne debris regions that receives positive external pressure can no longer be treated as an opening for design purposes, instead of making it impact-resistant or protected.

Provisions for wind loads on parapets are updated. Values of the Combined Net Pressure Coefficient are updated from +1.8 and -1.1 to +1.5 and -1.0 for windward and leeward parapets, respectively. Application of the provisions to low-slope roofs has been clarified.

Design wind loads on free-standing walls and solid signs are revised.

Design wind loads on open buildings with monoslope roofs are revised. Design wind loads on open buildings with pitched or troughed roofs are provided for the first time.

New provisions are added for rooftop structures and equipment when the roof height of the building is less than 60 feet.

Fastest Miles versus 3-Second Gust Velocity of Wind

Wind-borne debris requirements are clarified as being applicable to Method 3 (Wind Tunnel Procedure). The requirements are the same as those for Method 2 (Analytical Procedure).

Wind is a turbulent flow, characterized by random fluctuations of velocity and pressure. If the instantaneous velocity of wind at a given point is recorded as a function of time on a chart, the result will look like that in Figure 2.1.

Because of its fluctuating nature, wind velocity needs to be studied statistically. A statistical property is the mean or average. Because the intensity of wind changes constantly, different averages are obtained by using different averaging times. For instance, while a 5-second average of the peak of a high wind may be 60 mph, the same wind averaged over an hour may be only 40 mph. This means that whenever a mean or average velocity of wind is desired, one must specify the averaging time.

The longest averaging time used for wind is a year, producing an annual or yearly average. For instance, the annual wind speed for Kansas City is about 11 mph. Although information on this speed is very important in wind energy utilization, it is useless for wind load on structures because in that case only high winds of short durations are of interest. The longest averaging time for high winds used in practice is one hour. In general, as the averaging time decreases, the maximum speed of the wind increases. Suppose V_T is the maximum wind speed based on an averaging time of T seconds, and V_H is the maximum hourly average. The relationship between V_T and V_H for open terrains is approximately

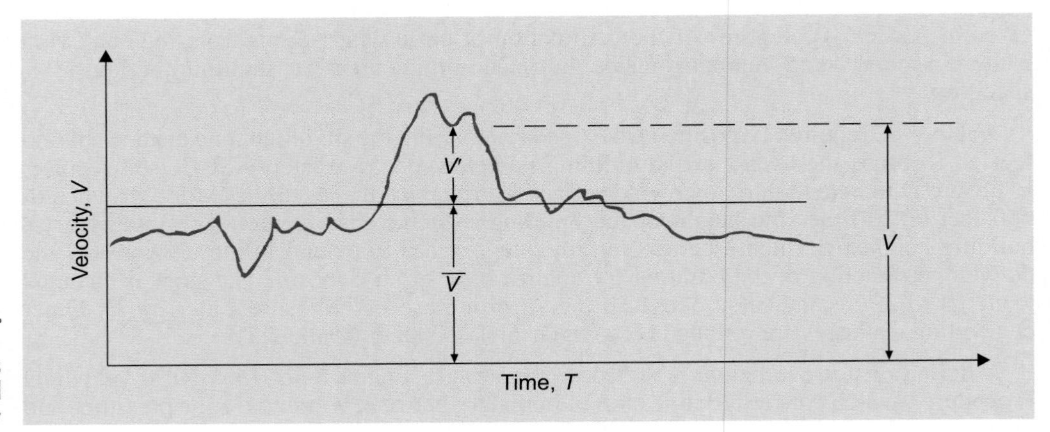

Figure 2-1 Fluctuations of wind velocity

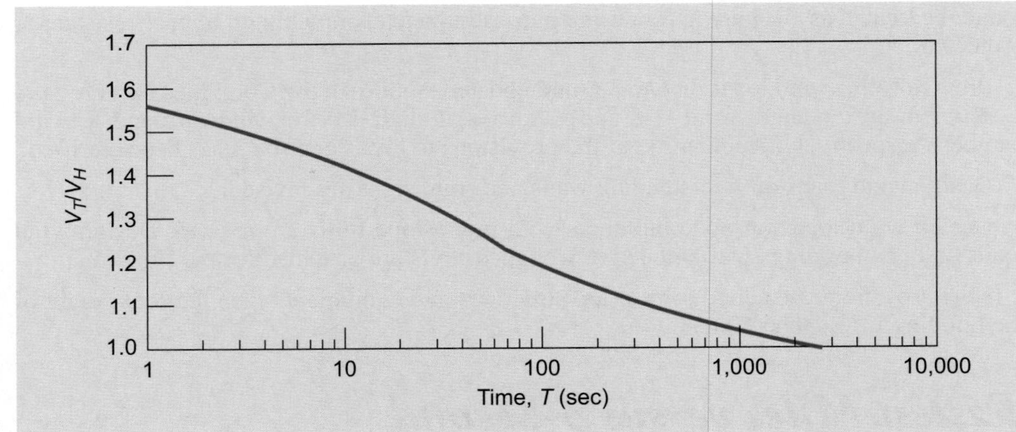

Figure 2-2 Relationship between hourly mean velocity of wind and velocity of wind averaged over a shorter period, T

given in Figure 2.2. From the figure, the 1-second gust speed above open terrains is almost 1.6 times the maximum hourly wind speed.

The mean wind speed used in ANSI A8.1-1972, ANSI A58.1-1982, ASCE 7-88 and ASCE 7-93 was the *fastest-mile wind*, which is the maximum wind speed, averaged over one mile of wind passing through a measuring instrument called an anemometer. The averaging time of the fastest-mile wind

$$T = \frac{3600}{V_{fm}}$$

where T is in seconds and the fastest-mile wind speed, V_{fm}, is in mph. For instance, if $V_{fm} = 60$ mph, the averaging time is T=3600/60 = 60 seconds. If V_{fm} is 120 mph, T decreases to 30 seconds. From the foregoing, the averaging time of the fastest-mile wind used in the design of structures, which normally uses a wind speed between 60 and 120 mph, is from 30 to 60 seconds. A fixed 3 seconds is the averaging time for the 3-second gust speed accepted as the basic wind speed in ASCE 7-95, ASCE 7-98, ASCE 7-02 and ASCE 7-05.

BOCA / National Building Code[2.7]

The BOCA/NBC adopted the wind design provisions of ANSI A58.1-1972 in its 1978 edition and retained them in the 1981 and 1984 editions. The revised wind design provisions of ANSI A58.1-1982 were adopted into the 1987 edition of the BOCA/NBC and were retained in the 1990 edition. The 1993 edition of the BOCA/NBC adopted the wind design provisions of ASCE 7-88, which were retained in the 1996 and 1999 editions. The wind design provisions of ASCE 7-95 were never adopted.

Standard Building Code[2.8]

The SBC adopted ANSI A58.1-1972 in the 1977 revisions to the 1976 SBC. The adopting language then appeared in the 1982 edition. Wind design using ANSI A58.1-1972 was permitted only for one- and two-story structures, provided the basic wind pressures from SBC Table 1205.1 were used. The 1982 SBC also adopted alternate wind load provisions (MBMA procedure) in Section 1206. This section was permitted to be used for the design of buildings with flat, single-slope and gable-shaped roofs with a mean roof height of 60 feet or less, provided the eave height did not exceed the least horizontal dimension of the building.

The 1985 edition had three procedures that could be used. Two of the procedures were contained in Section 1205, Wind Loads, and the third was in Section 1206, Alternate Wind Loads for Low-Rise Buildings. The first option allowed under Section 1205 was use of the provisions within the section. The second option permitted by Section 1205 was to use the wind design provisions of ANSI A58.1-1982, provided the basic wind pressures of Table 1205.1 were used. Table 1205.1 was based on the basic wind speed map of Figure 1205.1 (same as the 100-year mean recurrence interval basic wind speed map contained in ANSI A58.1-1972), which differed from the 50-year mean recurrence interval map in ANSI A58.1-1982.

The alternate wind load provisions of Section 1206 (MBMA procedures) were permitted to be used for the design of buildings with flat, single-slope and gable-shaped roofs with a mean roof height of 60 feet or less, provided the eave height did not exceed the least horizontal dimension of the building. Section 1206 contained its own basic wind speed map which was taken from ANSI A58.1-1982.

The 1988 SBC permitted any building or structure to be designed using the provisions of ANSI A58.1-1982. In addition, Section 1205.2 had provisions based on the MBMA procedures for buildings with flat, single-slope and gable-shaped roofs whose mean roof height was less than or equal to 60 feet. This edition did not require that the roof eave height be less than or equal to the least horizontal dimension of the building.

Section 1205.3 applied to buildings exceeding 60 feet in height but not more than 500 feet in height, provided the roof slope did not exceed 10 degrees or was not an arched roof. Buildings between 60 and 500 feet in height and not meeting these limitations, and all buildings over 500 feet in height, had to be designed according to ANSI A58.1-1982. The basic wind speed map within Section 1205 was the ANSI A58.1-1982 map.

The 1991 edition was essentially the same as the 1988 edition, except ANSI A58.1-1982 was updated to ASCE 7-88. The basic wind speed map within Section 1205 remained unchanged from the 1988 edition, because the basic wind speed map did not change within ASCE 7-88 from what was in ANSI A58.1-1982.

In the 1994 SBC, ASCE 7-88 was adopted by reference to apply to all buildings and structures. An exception continued to permit the MBMA procedures in Section 1606.2 to be used for buildings with flat, single-slope, hipped and gable-shaped roofs with mean roof heights not exceeding 60 feet or the least horizontal dimension of the building.

The 1997 edition was essentially the same as the 1994 edition, except that ASCE 7-88 was updated to ASCE 7-95. The basic wind speed map within Section 1606.2, Alternate wind loads for low-rise buildings, remained unchanged from the 1994 edition. It is necessary to point this out because the basic wind speed map of ASCE 7-95 was based on the 3-second gust wind speed.

The 1999 edition remained unchanged from the 1997 edition.

Uniform Building Code[2.9]

The following is adopted largely from SEAW *Wind Commentary to the Uniform Building Code*.[2.10]

Through its 1979 edition, the *Uniform Building Code* wind design provisions were based on the ANSI standard A58-1955.[2.6]

In 1976, the International Conference of Building Officials (ICBO) appointed a task force of its General Design Code Development Committee to update its wind load provisions. This was because of strong pressure to upgrade the simplified UBC wind design provisions at the same time the other model code writing bodies were considering adopting ANSI A58.1-1972.

Practicing engineers on the West Coast considered the ANSI provisions too complicated for practical engineering and lobbied for a simplified method of calculating wind forces. They were accustomed to determining wind pressures that were dependent only on the height of the building and the basic wind speed. These pressures were applied against the area within the building's projected vertical or horizontal outline. In contrast, the ANSI provisions required the use of six graphs to evaluate coefficients used in determining wind pressures. These pressures were applied normal to the face of the wall or roof and depended on the slope of the roof, internal pressure and other factors.

The wind provisions published in the 1982 edition of the UBC were based on those of ANSI A58.1-1972, although the calculation procedure was simplified. The changes proposed for ANSI A58.1-1982 were known at that time. The important changes proposed for ANSI A58.1-1982 were incorporated into the 1982 UBC. Few changes were made in the 1985 and 1988 UBC wind provisions.

The 1991 UBC made many changes in substance and format in its wind design provisions. There were further minor changes in the 1994 edition of the code, and no changes at all in the 1997 edition.

The 1991 UBC required wind pressures to be calculated in a manner similar to that of ASCE 7-88, except that only four coefficients needed to be determined. To simplify the calculation procedure, certain assumptions were made about the building. These assumptions limit the applicability of the UBC procedure.

The 1991 UBC Figure No. 23-1 was similar to the basic wind speed map published in ASCE 7-88 except that the contours and Special Wind Regions in the states of Washington and Oregon were different. The same figure appeared as Fig. 16-1 of the 1994 and the 1997 UBC.

In the 1991 edition, a new section was added that gathered many definitions, such as the three exposure categories, from the text and figures of the 1988 edition. To make the provisions consistent with the other code chapters, symbols and notations for wind were placed in a single location. Among the major new provisions was the addition of an exposure D terrain category for sites near large bodies of water.

Numerous changes were made to Tables 23-F, 23-G and 23-H (Tables 16-F, 16-G and 16-H, respectively, of later editions), bringing the 1991 UBC wind pressures more in line with ASCE 7-88. The original provisions of the 1982 UBC were conservative in their approaches and coefficients. For instance, they retained the traditional method of calculating wind forces, i.e., Method 2, and the pressure coefficients associated with basic wind speed were rounded up. Additionally, the forces associated with parts and portions, now called elements and components, were now keyed to tributary areas in the UBC.

The q_s and C_e factors were changed in the 1991 UBC to create a pressure profile equivalent to that of ASCE 7-88. The wind stagnation pressure, q_s, was extended to one decimal place instead of being rounded up to the next-highest integer.

The elements and components portions of Table 23-H (Table 16-H of later editions) were subdivided into coefficients at discontinuities and at areas away from these discontinuities.

Pressures at discrete elements were now based on tributary areas 10 square feet or smaller, between 10 and 100 square feet, and up to 1000 square feet. The term *corner* was now more clearly defined.

Instead of having separate tables of Importance Factors for wind and for earthquake, these were combined into Table 23-L. Tables 23-K and 23-L of the 1991 UBC were later combined into Table 16-K of the 1994 and 1997 UBCs. A seismic importance factor, I_p, for parts and positions of a structure, not previously part of Table 23-L, was included in Table 16-K of later editions.

The provisions for Glass and Glazing in UBC Chapter 54 (later Chapter 24) were modified, and the prescriptive provisions of Appendix Chapters 24 and 25 (later Appendix Chapters 21 and 23, respectively) were also extensively modified.

It should be obvious from the foregoing brief background segments that the wind provisions of the 2006 IBC, adopting Section 6 of ASCE 7-05 by reference with hardly any modification, and those of the 1997 UBC, which represented a simplified version of the wind load provisions of ASCE 7-88, represent different stages of evolution, and thus different levels of sophistication.

International Building Code

The first (2000) edition of the *International Building Code* (IBC)[2.11] adopted ASCE 7-98 for wind design. However, Method 1, Simplified Design of ASCE 7-98 was not adopted. Included in Section 1609.6 of the code was a different simplified design procedure, based on the low-rise analytical procedure (part of Method 2) of ASCE 7-98 and applicable only to simple diaphragm buildings, as defined in the code. For qualifying residential buildings, free of topographic effects, the SBCCI deemed-to-comply standard SSTD 10, *Standard for Hurricane Resistant Residential Construction*,[2.12] and the American Forest & Paper Association's (AF&PA) *Wood Frame Construction Manual* (WFCM)[2.13] were also allowed to be used. The 2000 IBC also added an alternative way of providing opening protection in one- and two-story buildings, included a conversion table between fastest-mile wind speed and three-second gust wind speed, and provided an optional design procedure for rigid tile roof coverings.

The second (2003) edition of the IBC adopted ASCE 7-02 for wind design. There was still a simplified design procedure, applicable to simple diaphragm buildings, in Section 1609.6 of the code. But it was now very close to Method 1, Simplified [Design] Procedure of ASCE 7-02, because (as mentioned earlier) ASCE 7-02 discarded Method 1 of ASCE 7-98, and adopted instead the simplified design procedure in Section 1609.6 of the 2000 IBC with some modifications. Qualifying residential buildings free of topographic effects could still be designed by SBCCI's SSTD10 or AF&PA's WFCM. The alternative way of providing opening protection in one- and two-story buildings, the conversion table between fastest-mile wind speed and three-second gust wind speed, and the optional design procedure for rigid tile roof coverings remained essentially unchanged.

ASCE 7-05 is adopted for wind design in the third (2006) edition of the IBC. Simplified wind design is no longer in the code; it is by reference to ASCE 7-05. Qualifying residential buildings free of topographic effects can still be designed by SBCCI's SSTD10 or AF&PA's WFCM. The alternative way of providing opening protection in one- and two-story buildings is retained in a modified form in the 2006 IBC. The conversion table between fastest-mile wind speed and three-second gust wind speed is revised. The optional design procedure for rigid tile roof coverings remains unchanged.

Discussion of Changes from ANSI A58.1-1972 to ASCE 7-05

Of all the changes from ANSI A58.1-1972 through ASCE 7-05, there are only a few that move in the direction of less conservatism in design. The first of these is the adoption of the low-rise analytical procedure in ASCE 7-95 as an alternative design approach for the MWFRS. This procedure can reduce design wind pressures significantly. Although generalizations are difficult because so many variables influence the determination of design wind pressures for a specific building, use of the alternate procedure can result in the total wind load being approximately 30 – 35 percent less than would be calculated using the primary procedure. Note that the low-rise analytical procedure was part of the *Standard Building Code* long before it was adopted by ASCE 7 and is based on comprehensive testing done at the University of Western Ontario, London.

In areas where the basic wind speed is low, the relative lack of conservatism of the low-rise procedure is mitigated somewhat by the requirement that all MWFRSs be designed for a minimum pressure of 10 psf applied to the area of the building projected onto a vertical plane. However, this provision is widely ignored and is not rigorously enforced by local jurisdictions. It needs to be taken more seriously by practitioners as well as local jurisdictions.

The second change was the introduction of the directionality factor K_d in ASCE 7-98. This led to a round-up of the wind load factor from 1.53 to 1.60 in strength design, which is conservative. This also decreased the design wind forces in Allowable Stress Design, which is widely used in the design of structures made of materials other than concrete.

The only other change possibly in the direction of less conservatism was the redrawing of the basic wind speed map in ASCE 7-98, which decreased the basic wind speeds in inland Florida. Obviously, when National Weather Service data indicate that a change is warranted, ASCE 7 has no reason to resist making that change.

By and large, the changes in ANSI A58.1/ASCE 7 have not been consistently in the direction of lower or higher design wind pressures. If there is a consistent trend to the changes, it is that the complexity of wind design has been steadily increasing. Much of the country has already experienced this complexity. It is being acutely felt in California where the 2006 IBC was adopted effective January 1, 2008. Wind design changed from that based on ASCE 7-88, meaningfully simplified (as incorporated in the 1997 UBC), to that based on ASCE 7-05, unsimplified.

1997 UBC versus IBC Comparison

Design wind forces at the various floor levels of an example concrete building, the plan and elevation of which are shown in Figures 2.3 and 2.4, respectively, were calculated using the general analytical procedure (Method 2) of ASCE 7-05 (which has been adopted into the 2006 IBC) and the wind design procedure of the 1997 UBC, which is a simplified version of that in ASCE 7-88. The building is assumed to be located in suburban Los Angeles (three-second gust wind speed of 85 mph), and the exposure category is assumed to be B. The simplification of the analytical procedure of the 1997 UBC was the result of a joint effort by the Structural Engineers Association of California (SEAOC) and the Structural Engineers Association of Washington (SEAW). It can be seen in Table 2.3 that the UBC procedure produces slightly, but not overly, conservative results, as it should. The efforts involved in the two cases were not comparable, with the ASCE 7-05 design taking considerably more time and being more complex (even though the different load cases in Figure 6-9 of ASCE 7-05, other than Load Case 1 were not even considered). The primary reason that accounts for the additional time is that the simplifications made by SEAOC/SEAW to the provisions of ASCE 7-88 are not available to the user of Method 2 of

ASCE 7-05. Also, as outlined in preceding sections, many complexities have been added to the wind design provisions of ASCE 7 between the 1988 and the 2005 editions. One example of the added complexity is the prescribed procedure for the computation of gust effect factors for flexible buildings. The example building being flexible, the gust effect factor had to be calculated. The calculation involves a large number of complex equations, and took an experienced engineer over an hour and a half to complete. Ironically, the factor turned out to be 0.87, which should be compared with the 0.85 prescribed for rigid buildings. Although no generalization is possible on the basis of one example, the UBC procedure, which has been in the *Uniform Building Code* since 1991, has been used in the design of a large population of structures located west of the Mississippi, in Indiana, and elsewhere. There is no record of distress that has been attributed to any deficiency in that design procedure.

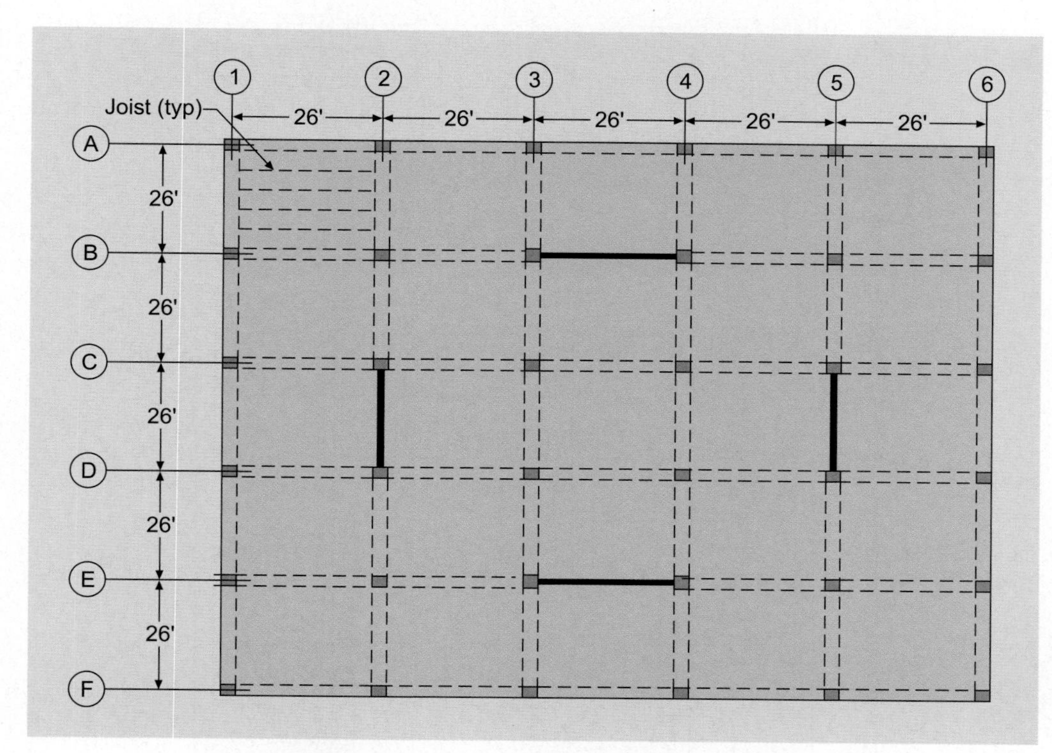

Figure 2-3
Plan of example concrete building

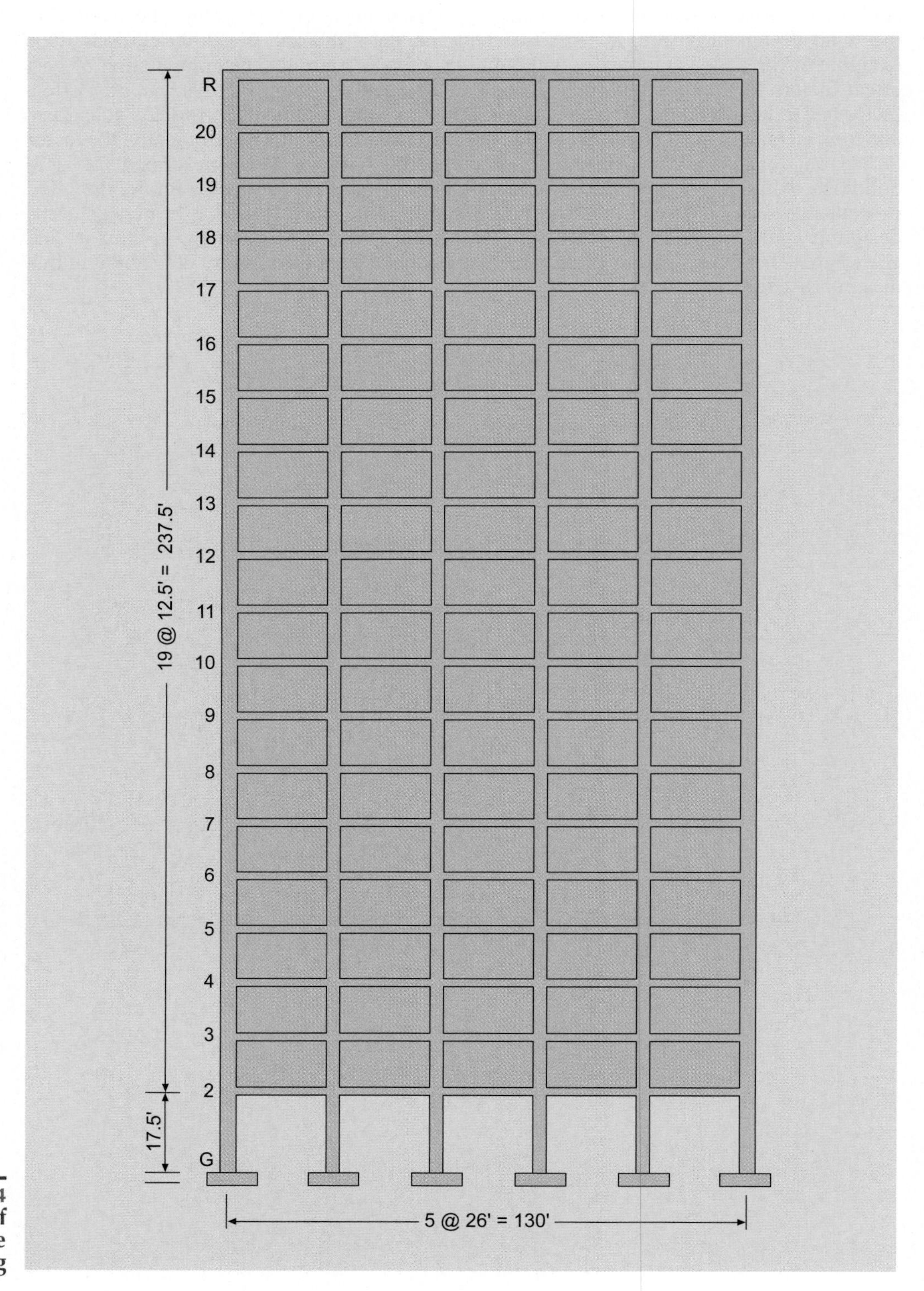

Figure 2-4
Elevation of example concrete building

Table 2-3. **Comparison of computed wind forces for example building.**

Floor Levels	Wind Forces (plf)		2006 IBC/1997 UBC
	1997 UBC	2006 IBC	
R	235	214	0.91
20	314	286	0.91
19	310	284	0.92
18	306	281	0.92
17	302	278	0.92
16	298	275	0.92
15	293	272	0.93
14	290	269	0.93
13	285	265	0.93
12	282	262	0.93
11	277	258	0.93
10	272	254	0.93
9	267	249	0.93
8	261	244	0.94
7	253	239	0.94
6	247	233	0.94
5	239	226	0.94
4	229	218	0.95
3	218	208	0.95
2	243	233	0.96

When the State of Oregon adopted the 2003 IBC as the basis of the 2004 *Oregon Structural Specialty Code*,[2.14] they made an amendment to the 2003 IBC allowing continued usage of the 1997 UBC wind design procedure (as adopted into the 1998 *Oregon Structural Specialty Code*). The State of Washington did not make a similar amendment when it adopted the 2003 IBC as the basis of the state code a few months ahead of Oregon. A simplification of the analytical procedures of ASCE 7-98 and -02 was under development by the Structural Engineers Association of Washington for quite some time. The simplified procedure has been published.[2.15] This procedure, however, does not appear ready for codification.

REFERENCES

[2.1]Ghosh, S.K., "The Evolution of Wind Provisions in Standards and Codes in the United States – Part 1," *Structural Engineer,* Zweig White Information Services, Skokie, IL, Web Link: www.GoStructural.com. December, 2006.

[2.2]Ghosh, S.K., "The Evolution of Wind Provisions in Standards and Codes in the United States – Part 2," *Structural Engineer,* Zweig White Information Services, Skokie, IL, January, 2007.

[2.3]Ghosh, S.K., "The Evolution of Wind Provisions in U.S. Standards and Codes," *Building Safety Journal*, International Code Council, Vol. 4, No. 6, December 2006

[2.4]Mehta, K.C., Marshall, R.D., and Perry, D.C., *Guide to the Use of the Wind Load Provisions of ASCE 7-88*, American Society of Civil Engineers, New York, NY, 1991.

[2.5]Mehta, K.C., and Marshall, R.D., *Guide to the Use of the Wind Load Provisions of ASCE 7-95*, American Society of Civil Engineers, New York, NY, 1998.

[2.6]American Society of Civil Engineers, *ASCE Standard Minimum Design Loads for Buildings and Other Structures*, ASCE 7-88, ASCE 7-93, ASCE 7-95, ASCE 7-98 (also ANSI A58-55, ANSI A58.1-72, ANSI A58.1-82), New York, NY, 1990, 1993, 1995 and 2000, respectively.

[2.7]Building Officials and Code Administrators International, *The BOCA National Building Code*, Country Club Hills, IL, 1978, 1981, 1984, 1987, 1990, 1993, 1996, 1999, copyright held by International Code Council.

[2.8]Southern Building Code Congress International, *Standard Building Code*, Birmingham, AL, 1979, 1982, 1985, 1988, 1991, 1994, 1997, 1999, copyright held by International Code Council.

[2.9]Pacific Coast Building Officials Conference (later the International Conference of Building Officials), *Uniform Building Code*, Long Beach, CA (later Whittier, CA), 1935, 1952, 1967, 1970, 1973, 1976, 1979, 1982, 1985, 1988, 1991, 1994, 1997, copyright held by International Code Council.

[2.10]Structural Engineers Association of Washington, *Wind Commentary to the Uniform Building Code*, 1991 Edition, Seattle, WA, 1993.

[2.11]International Code Council, *International Building Code*, Washington, DC, 2000, 2003, 2006.

[2.12]Southern Building Code Congress International, *Standard for Hurricane Resistant Residential Construction*, SSTD 10, Birmingham, AL, 1999, copyright held by International Code Council.

[2.13]American Forest and Paper Association, *Wood Frame Construction Manual for One- and Two-Family Dwellings*, WFCM, Washington, DC, 1996, 2001.

[2.14]Building Codes Division, *Oregon Structural Specialty Code*, Department of Consumer and Business Services, State of Oregon, West Salem, OR, 2004.

[2.15]Structural Engineers Association of Washington, *SEAW's Handbook of a Rapid Solutions Methodology for Wind Design*, SEAW RSM-03, Applied Technology Council, Redwood City, CA, 2004.

Appendix 3

BACKGROUND TO SEISMIC GROUND MOTION IN SEISMIC DESIGN

Background to the seismic zoning map of the 1988 UBC was presented by the author in Reference 3.1. It was compiled from a number of readily available sources: a 1969 paper by Algermissen,[3.2] the commentary portions of ASCE 7-88,[3.3] ATC 3-06,[3.4] and the 1988 SEAOC *Blue Book*.[3.5] That information is expanded and brought up to date, relying in part upon the commentary to the 1997 – 2003 NEHRP *Provisions*.[3.6]

Seismic Zones

The concept of seismic zones was first introduced into the 1935 edition of the *Uniform Building Code* (UBC).[3.7] The code included a map of the 11 western states, showing three different seismic zones of approximately equal seismic probability, with the most active being Zone 3, and the least active being Zone 1. The western coast of California was assigned to Zone 3, in part because of the historic seismicity of the region. Zone 2 included the central valley of California and extended into Nevada, Utah, Oregon and Montana. The remaining states were essentially in Zone 1. A seismic zonation factor equal to 1, 2 and 4, in Zones 1, 2 and 3, respectively, directly amplified the design base shear.

A seismic probability map of the United States was prepared in 1948 by Roberts and Ulrich with the advice of seismologists throughout the United States[3.8] and was issued by the Coast and Geodetic Survey (later the U.S. Geological Survey) in 1948. In 1949, the seismic probability map was revised such that the Charleston, South Carolina area was changed from Zone 3 to Zone 2, and a Zone 3 was set up for the Puget Sound region of Washington, which had formerly been included in Zone 2.[3.9] The revised map was adopted by the Pacific Coast Building Officials Conference (later ICBO) for inclusion in the 1952 edition of the *Uniform Building Code*. Subsequent editions of the UBC, up to and including the 1967 edition, incorporated this map with no changes.

The seismic zone map adopted into the 1970 and 1973 editions of the UBC evolved from the work of Algermissen during the 1960s[3.2] and is based on the maximum recorded intensity of shaking without regard to the frequency with which such shaking might occur. As originally published, this map had four zones (0, 1, 2 and 3), so that several areas in the Eastern United States were in the same zone as California. During the 1970s the map was modified and eventually adopted into the 1976 edition of the UBC with a Zone 4 in parts of California and Nevada, and with the Z factor for the remaining Zone 3 areas reduced. The boundary between Zones 3 and 4 was determined by proximity to the major fault systems. The boundary was set at 25 miles from a fault considered capable of generating an earthquake of magnitude 7.0 or greater and 15 miles from a fault that could generate an earthquake of magnitude between 6.0 and 7.0. The distances were based on the Schnabel-Seed attenuation curve[3.10] and a value of 0.3*g* for the peak acceleration at the boundary on a rock site. The seismic zone map of 1976 UBC remained unchanged through the 1985 edition of that code.

ATC 3-06

In 1976 Algermissen and Perkins published a new contour map for peak ground acceleration on rock, based on a uniform probability of occurrence throughout the 48 contiguous states.[3.11] The probability that the contoured peak acceleration would not be exceeded was given as 90 percent in 50 years. In developing this map, the first step was to delineate zones within which earthquakes may occur and establish for each zone the frequency of earthquakes with different magnitudes. Attenuation equations were selected for both the eastern and the western parts of the country. These several parameters served as input to a computer program that computed the frequency for different peak accelerations at all points of a gridwork covering the 48 states.

The Seismic Risk Committee of ATC 3 modified the Algermissen-Perkins map to make it more suitable for use in a document resembling a model code.[3.4] The concept of Effective Peak Acceleration (EPA), related to the damageability of ground shaking, was introduced, and certain small zones of very high peak acceleration were eliminated, partly on the basis that their retention would constitute microzoning, which the committee had been instructed to avoid. The contours of the Algermissen-Perkins map were smoothed, so as to avoid the appearance of great precision, and in some locales the contours were shifted on the basis of more recent knowledge. The result was a map that retained the basic principles and trends of the Algermissen-Perkins map but lacked the internal consistency of that map. It was estimated that the probability of not exceeding the contoured values of effective peak acceleration within 50 years was 80 percent to 95 percent. Maps of EPA for Alaska, Hawaii, Puerto Rico and several territories were drawn using the best available information from various sources as guidance.

It was decided during the initial stages of development of ATC 3 that ground-shaking regionalization maps should take into account the distance from anticipated earthquake sources. This decision reflects the observation that the higher frequencies in ground motion attenuate more rapidly with distance than the lower frequencies. Thus, at distances of 60 miles or more from a major earthquake, flexible buildings may be more seriously affected than stiff buildings. To accomplish the objectives of the decision, it proved necessary to use separate ground motion parameters and, therefore, to prepare a separate map for each.

The two parameters used to characterize the intensity of design ground shaking were the Effective Peak Acceleration (EPA) and the Effective Peak Velocity (EPV). To best understand the meaning of EPA and EPV, they should be considered normalizing factors for construction of smoothed elastic response spectra for ground motions of normal duration. The EPA is proportional to spectral ordinates for periods in the range of 0.1 second to 0.5 second, while the EPV is proportional to spectral ordinates at a period of about 1 second (Figure 3.1). The constant of proportionality (for a spectrum corresponding to 5 percent of critical damping) is set at a value of 2.5 in both cases.

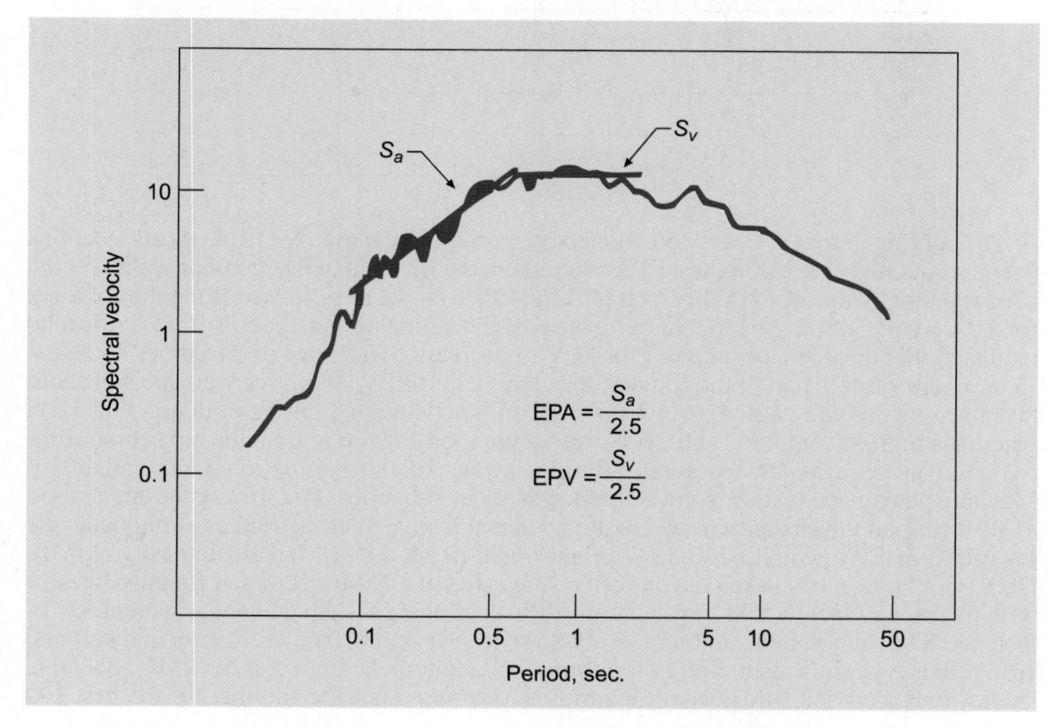

Figure 3-1
Schematic representation of effective peak acceleration and effective peak velocity

The EPA and EPV are related to peak ground acceleration and peak ground velocity, but are not necessarily the same as, or even proportional to, peak acceleration and velocity.

When very high frequencies are present in the ground motion, the EPA may be significantly less than the peak acceleration. This is consistent with the observation that chopping off the highest peak in an acceleration time history has very little effect upon the response spectrum computed from that motion, except at periods much shorter than those of interest in ordinary building practice. Furthermore, a rigid foundation tends to screen out very high frequencies in the free field motion. On the other hand, the EPV is generally greater than the peak velocity at large distances from a major earthquake. Ground motions increase in duration and become more periodic with distance. These factors tend to produce proportionately larger increases in the portion of the response spectrum represented by the EPV.

If an earthquake is of very short or very long duration, then it may be necessary to correct the EPA and EPV values to more closely represent the event. It is well documented that two motions having different durations but similar response spectra cause different degrees of damage—the damage being less for the shorter duration. In particular, there have been numerous instances where motions with very large accelerations and short duration have caused little or even no damage. Thus, when expressing the significance of a ground motion to design, it may be appropriate to decrease the EPA and EPV obtained from the elastic spectrum for a motion of short duration. On the other hand, for a motion of very long duration it may be appropriate to increase the EPA and EPV.

For ease in developing base shear formulas, it proved desirable to also express EPV by a dimensionless parameter (A_v), which is an acceleration coefficient. This parameter is referred to as the Effective Peak Velocity-Related Acceleration Coefficient. The relationship between EPV and A_v is given in Table 3.1 and may be expressed as follows:

$$A_v = \text{EPV (in./sec)} \times 0.4 / 12$$

Table 3-1. **Relationship between EPV and A_v**

Effective Peak Velocity (in./sec)	Velocity-Related Acceleration Coefficient, A_v
12	0.4
6	0.2
3	0.1
1.5	0.05

The ATC map for EPV was constructed by modifying the map for EPA. At all locations where a contour gives the highest EPA in a region, the EPV along that contour was set at the corresponding value of EPA. For example, the EPV was set at 12 in./sec along the contours for EPA = 0.4*g*. A study by McGuire,[3.12] based upon strong motion records from California, indicated that the distance required for EPV to decrease by a factor of 2 is about 80 miles. Thus, in the western part of the country, the contours for EPV = 6 in./sec were located about 80 miles outside the contours for EPV = 12 in./sec. Similarly, in Washington and Utah where the highest contour is at 0.2*g*, corresponding to EPV = 6 in./sec, the next contour for EPV = 3 in./sec was located about 80 miles away. The strong motion data available to McGuire were inadequate beyond a distance of about 100 miles. To estimate the attenuation of EPV beyond this distance, it was assumed that at large distances from an earthquake, the logarithm of EPV would be linearly proportional to Modified Mercalli Intensity (MMI). Data from large earthquakes in California suggested that MMI decreased roughly linearly with distance, which would translate into EPV continuing to halve at equal increments of distance. Thus subsequent contours were also spaced at about 80 miles. For the midwest and east, reliance was placed entirely on information about the attenuation of MMI. Available studies indicated that MMI decays logarithmically with distance and that for the first 100 miles from a large earthquake the attenuation is roughly the same as in the west. The implication is that the distance required for EPV to halve increases with distance. Thus, starting from the contour for EPV = 6 in./sec centered on southeastern Missouri, the contour for EPV = 3 in./sec would be 80 miles away, and the contour for EPV = 1.5 in./sec would be about 160 miles beyond that. It was also stipulated that a contour for EPV should never fall inside the corresponding contour for EPA. After these various rules were applied to produce a set of contours for EPV, considerable smoothing was done and contours were joined where they fell close together. The EPV map had neither the detailed theoretical basis nor the internal consistency of the Algermissen-Perkins map for peak acceleration.

ANSI A58.1 - 1982

The ATC 3-06 map for Effective Peak Velocity-Related Acceleration Coefficient was selected as the basis for the zoning map in ANSI A58.1-1982.[3.3] The contour map was converted into a zoning map as follows.

Table 3-2.

A_v	Zone
> 0.4	4
0.2 to 0.4	3
0.1 to 0.2	2
0.05 to 0.1	1
< 0.05	0

There were several reasons for the choice of the map for EPV over that for EPA. First, the contours for EPA are very closely spaced in some regions. This is especially true in California and Nevada; use of the EPA map or the Algermissen-Perkins map would have meant that both states would be subdivided into five zones. Second, since only one map was to be used for the sake of simplicity, it seemed appropriate to use the more conservative map. Finally, use of the map for EPV meant a more modest change in the zonation of the country from that given in the1979 UBC.[3.7]

The seismic zoning map in ANSI A58.1-1982, when compared with that in 1985 UBC, reduced the Z factor in many areas, especially in the eastern U.S.

ASCE 7-88

ASCE 7-88 used the same seismic zoning map as in ANSI A58.1-1982.

1988 through 1997 UBC

The 1987 SEAOC Seismology Committee collaborated with a number of other organizations and individuals to produce a zone map of the United States for use in the 1988 UBC. There is only one zone map, but the philosophy embodied in the two ATC maps was followed. In drawing the zone boundaries, both acceleration and velocity-related maps were consulted, and, if they disagreed, the one indicating the higher zone prevailed. The design base shear equation was modified so that the values of Z would correspond to the estimated values of effective peak acceleration. It was intended that the boundary between Zone 3 and Zone 4 correspond to a peak acceleration of 0.3*g* (or equivalent velocity), and the boundary between Zone 2 and Zone 3 correspond to 0.2*g*. The map is based principally on the ATC maps with modifications reflecting other sources of information, including the maps of acceleration and velocity published by Algermissen et al. in 1982.[3.13] The ATC maps and the maps by Algermissen et al. are based primarily on the historical record of earthquakes, which in California is short relative to the repeat time of major earthquakes even on the most active faults. To avoid the danger of missing areas that are potentially active but simply by chance have not shown activity during the short time of the historical record, geologic data were consulted, particularly data on fault slip rates. On the basis of geologic data, Zones 3 and 4 were extended in Southern California to encompass the Garlock fault. The modification was made with the intent to be consistent with the philosophy that the Z coefficient should correspond to ground motion values with a 10-percent probability of being exceeded in 50 years. The criterion was abandoned, however, in making many other modifications. It was abandoned in drawing the boundary between Zones 2 and 3 in California and Oregon. The boundary was drawn so as to exclude Zone 2 from California in in order to accommodate the desire to retain the structural detailing requirements of Zones 3 and 4 for the state. The criterion was also abandoned in the southwestern part of California and elsewhere to accommodate local opinion.

The 1988 seismic zoning map remained unchanged in the 1991 edition of the UBC. The 1994 UBC seismic zone map was somewhat different from the 1991 map. The notable changes from 1991 were as follows:

1. Parts of southern Arizona, including the city of Tucson, previously in Zone 1, were placed in Zone 2A.
2. Parts of the island of Oahu, including the city of Honolulu, previously in Zone 1, were placed in Zone 2A.
3. Western parts of the state of Oregon, previously in Zone 2B, were placed in Zone 3.
4. Western parts of the state of Washington, previously in Zone 2B, were placed in Zone 3.
5. The city of San Diego, previously in Zone 3, was placed in Zone 4.

Further changes in the seismic zones of the Hawaiian islands were made in the seismic zone map of 1997 UBC—the big island of Hawaii was assigned to Seismic Zone 4. Otherwise, the seismic zone map remained unchanged from the 1994 UBC.

1985 through 1994 NEHRP Provisions

The two maps developed during the ATC-3 project from a single map prepared by Algermissen and Perkins in 1976 were adopted into the 1985 edition of the NEHRP *Provisions* and were retained and used as the basis of seismic design through the 1994 edition.

In 1982, Algermissen and coworkers published a set of probabilistic maps for both acceleration and velocity using three different exposure times (thus three different levels of probability for excessive ground motions).[3.13] The basic procedure for generation of these new maps was not greatly different from that used for the 1976 map. The major difference was that the maps for ground velocity were prepared from basic data rather than being extrapolated from the acceleration map. The maps were thought to represent a step forward, and a modified version was incorporated into the 1988 edition of the NEHRP *Provisions* through an appendix for trial use and evaluation. Presented in the ground motion maps of the contiguous United States were the expected maximum horizontal acceleration and velocity in rock for exposure times of 50 and 250 years (average return periods for the expected ground motions of 474 and 2372 years, respectively). The mapped accelerations had a 90-percent probability of not being exceeded in the appropriate exposure times.

Subsequently, attenuation equations became available that permitted construction of response spectra directly by means of separate attenuation equations for spectral ordinates for a number of periods. Use of response spectra in design avoids many of the problems associated with the peak response parameters that were in widespread use. Because of these developments, interest turned toward formulation of design procedures that make use of these methods. Consequently, the use of the peak parameters was dropped from the "Appendix to Chapter 1" in the 1991 edition of the NEHRP *Provisions* in favor of furthering development of the spectral approach.

The hazard maps of the contiguous United States included for use with that appendix[3.14, 3.15] presented the spectral response accelerations at periods of 0.3 second and 1 second (as a percentage of the acceleration due to gravity), for an S_2 soil profile for exposure times of 50 and 250 years (average return periods of 474 and 2372 years, respectively). The mapped spectral response accelerations had a 90 percent probability of not being exceeded in the appropriate exposure times. More spectral ordinates could have been calculated, but Algermissen et al.[3.15] showed that the two ordinates mapped provided a reasonable approximation to a spectrum determined using all available ordinates.

Both the 50-year maps and the 250-year maps were updated by the USGS from the 1991 edition maps to reflect additional data and were included with the 1994 NEHRP *Provisions*. The Appendix to Chapter 1 of the 1994 NEHRP *Provisions* summarized unsuccessful attempts to develop so-called design value maps that were intended to replace the A_a and A_v maps in use since 1985. BSSC rejected the direct use of the 50-year spectral ordinate maps as design value maps for reasons stated in that appendix.

ASCE 7-93, 7-95

The seismic design provisions of ASCE 7-93, for the first time, became based on the 1991 NEHRP *Provisions*, including the maps of A_a and A_v. In fact, the 1993 edition of ASCE 7 was brought out solely for the purpose of updating the seismic design provisions. The seismic design provisions of ASCE 7-95 were updated to those based on the 1994 NEHRP *Provisions*. The maps did not change.

BOCA/NBC and SBC

The BOCA *National Building Code* (NBC) through its 1990 edition and the *Standard Building Code* (SBC) through its 1991 edition adopted the national loading standard, ASCE 7 or ANSI A58.1, for seismic design purposes. The 1993 edition of the BOCA/NBC and the 1994 edition of the SBC directly adopted seismic design provisions based on the 1991 NEHRP *Provisions*. The 1996 and 1999 editions of the BOCA/NBC and the 1997 and 1999 editions of the SBC retained 1991 NEHRP-based provisions for seismic design. However, all four of these documents allowed seismic design by ASCE 7-95 which had adopted seismic design provisions based on the 1994 NEHRP *Provisions*.

1997 NEHRP Provisions, ASCE 7-98, and 2000 IBC

As indicated above, the design ground motions in *legacy* model codes and in the NEHRP *Provisions* through its 1994 edition were based on an estimated 90-percent probability of not being exceeded in 50 years (about a 500 year mean recurrence interval or return period). This changed with the 1997 edition of the NEHRP *Provisions* and thus with ASCE 7-98 and the 2000 IBC.

Given the wide range in return periods for maximum-magnitude earthquakes in different parts of the Untied States (100 years in parts of California to 100,000 years or more in several other locations), the 1997 NEHRP *Provisions* focused on defining maximum considered earthquake ground motions for use in design. These ground motions may be determined in different manners depending on the seismicity of an individual region; however, they are uniformly defined as "the maximum level of earthquake ground shaking that is considered as reasonable to design buildings to resist." This definition facilitates the development of a design approach that provides approximately uniform protection against collapse throughout the United States.

It is widely recognized that the ground motion difference between 10-percent and 2-percent probabilities of being exceeded in 50 years in coastal California is typically smaller than the corresponding difference in inactive seismic areas such as the eastern and central United States. Figure 3.2, reproduced from the commentary to the 1997 NEHRP *Provisions*, plots the spectral acceleration at a period of 0.2 second, normalized at a

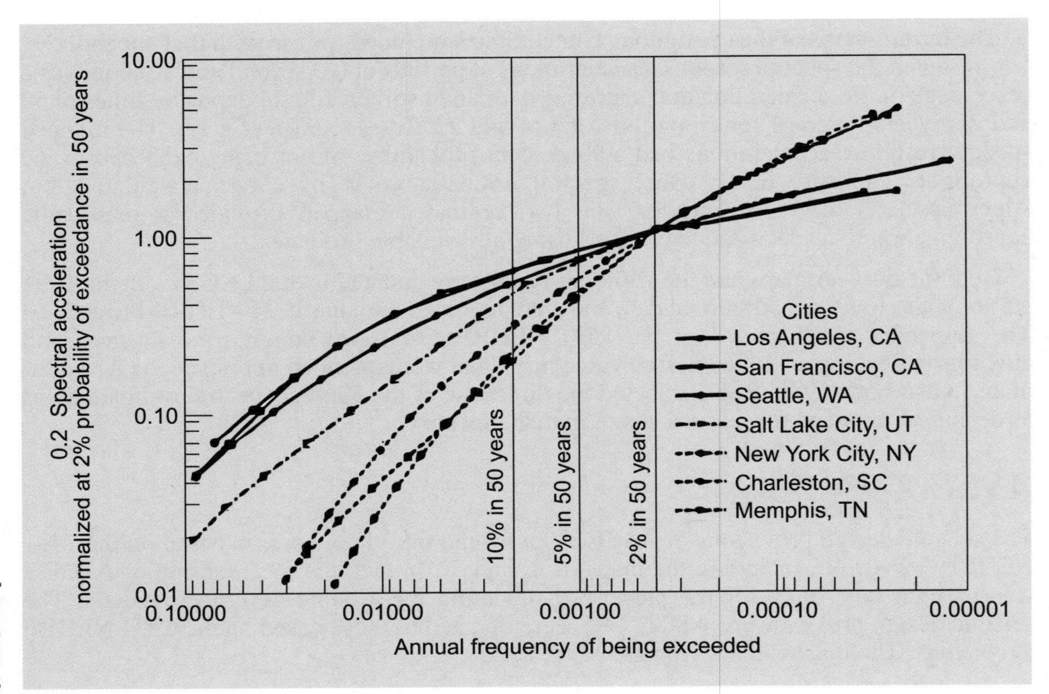

Figure 3-2
Hazard curves for selected cities

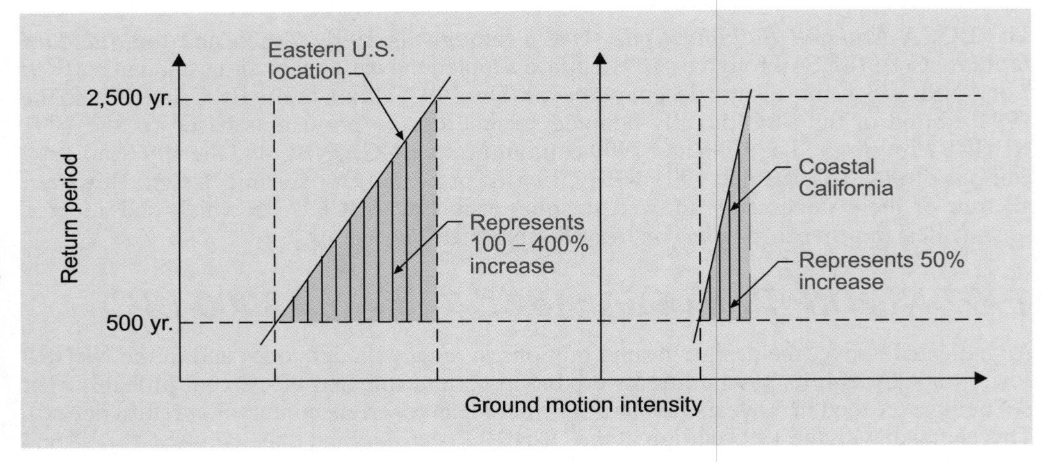

Figure 3-3
Comparison of earthquakes with long and short return periods in coastal California and eastern United States

2-percent probability of being exceeded in 50 years (10 percent in 250 years), versus the annual frequency of being exceeded.

The figure shows that in coastal California, the ratio between the 0.2-second spectral accelerations for the 2- and 10-percent probabilities of being exceeded in 50 years is about 1.5, whereas the ratio varies between 2.0 and 5.0 in other parts of the United States. This is schematically represented for added clarity in Figure 3.3.

The question therefore arose as to whether the definition of ground motion based on a constant probability for the entire United States would result in similar levels of seismic safety for all buildings.

In addressing the question it was recognized that seismic safety is the result not only of the design earthquake ground motion definition, but also of such critical factors as proper site selection, structural design criteria, analysis procedures, adequacy of detailing and quality of construction.

The seismic design provisions of the 1997 NEHRP *Provisions* are based on the assessment that if a building experiences a level of ground motion 1.5 times the design level of the 1994 and prior NEHRP *Provisions*, the building should have a low likelihood of collapse. Although quantification of this margin is dependent on the type of structure, detailing requirements, etc., the 1.5 factor was felt to be a conservative judgment.

As indicated above, in most U.S. locations, the 2-percent probability of ground motion values being exceeded in 50 years is more than 1.5 times those corresponding to a 10-percent probability within 50 years. This means that if the 10-percent probability of being exceeded in 50 years map were used as the design map and the ground motion corresponding to a 2-percent probability in 50 years were to occur, there would be a low confidence (particularly in the central and eastern U.S.) that buildings would not collapse, because of these larger ground motions. Such a conclusion for most of the U.S. was not acceptable. The only location where the above results seemed to be acceptable was coastal California (ground motion corresponding to a 2-percent probability of being exceeded in 50 years is about 1.5 times that corresponding to a 10-percent probability in 50 years) where buildings have experienced levels of ground shaking equal to and above the design value.

Probabilistic seismic hazard maps from U.S. Geological Survey for Coastal California indicate that the ground motion corresponding to a 10-percent probability of being exceeded in 50 years is significantly different (in most cases larger) than the design ground motion values contained in the 1994 NEHRP *Provisions* and in recent editions of the *Uniform Building Code*. One unique issue for coastal California is that the recurrence interval for the estimated maximum-magnitude earthquake is less than the recurrence interval represented by a 10-percent probability of being exceeded in 50 years. In other words, the recurrence interval for a maximum magnitude earthquake is 100 or 200 years versus 500 years.

Given that the maximum earthquake for many seismic faults in coastal California is fairly well known, a decision was made to develop a procedure that would use the best estimate of ground motion from maximum-magnitude earthquakes on seismic faults with higher probabilities of occurrence. For the purpose of the 1997 NEHRP *Provisions*, these earthquakes were defined as *deterministic earthquakes*. Following this approach and recognizing the inherent margin of 1.5 contained in the *Provisions*, it was determined that the level of seismic safety achieved in coastal California would be approximately equivalent to that associated with a 2- to 5-percent probability of being exceeded in 50 years for areas outside of coastal California. The use of the deterministic earthquakes to establish the maximum considered earthquake ground motions for use in design in coastal California resulted in a level of protection close to that implied in the 1994 NEHRP *Provisions*. Additionally, this approach resulted in less drastic changes to ground motion values for coastal California than the alternative approach of using probabilistic maps.

Based on the inherent margin contained in the NEHRP *Provisions*, the ground motion corresponding to a 2-percent probability of being exceeded in 50 years was selected as the maximum considered earthquake ground motion for use in design where the deterministic earthquake approach discussed above was not used.

To summarize the above discussions, the 1997 NEHRP *Provisions* maps reflected the following policy decisions that departed from past practice:

1. The maps defined the maximum considered earthquake ground motion for use in design procedures,
2. The use of the maps for design provided an approximately uniform margin against collapse for ground motions in excess of the design levels in all areas.
3. The maps were based on both probabilistic and deterministic seismic hazard evaluations.
4. The maps were response spectral ordinate maps and reflected the differences in short-period range of the response spectra for the areas of the United States and its territories with different ground motion attenuation characteristics and different recurrence times.

The various USGS probabilistic seismic hazard maps were combined with deterministic hazard maps by a set of rules (logic) to create the maximum considered earthquake ground motion maps that can be used to define response spectra for use in design. The United States and its territories were divided into:

1. Regions of negligible seismicity with very low probabilities of structural collapse,
2. Regions of low and moderate to high seismicity, and
3. Regions of high seismicity near known fault sources with short return periods.

The maximum considered earthquake ground motion maps for use in seismic design were developed by combining the regions as described below:

1. Where the mapped maximum considered earthquake ground motion values (based on a 2-percent probability of being exceeded in 50 years) for Site Class B, adjusted for the specific site conditions, were $\leq 0.25g$ for the short-period spectral response and $\leq 0.10g$ for the long-period spectral response, then the site was in a region of negligible seismicity, and a minimum lateral force design of 1 percent of the dead load of the structure was to be used in addition to the detailing requirements for Seismic Design Category A structures.
2. Where the maximum considered earthquake ground motion values (based on a 2-percent probability of being exceeded in 50 years) for Site Class B, adjusted for the specific site conditions, were greater than $0.25g$ for the short-period spectral response and $0.10g$ for the long-period spectral response, the maximum considered earthquake ground motion values (based on a 2-percent probability of being exceeded in 50 years, adjusted for the specific site conditions) were used until the values equaled the 1994 NEHRP *Provisions* ceiling design values increased by 50 percent (short-period = $1.50g$, long-period = $0.60g$). The ceiling design values were increased by 50 percent to represent the maximum considered earthquake ground motion values. This defined the sites in regions of low and moderate to high seismicity.
3. To shift from regions of low and moderate to high seismicity to regions of high seismicity with short return periods, the maximum considered earthquake ground motion values based on a 2-percent probability of being exceeded in 50 years were used until the values equaled the 1994 NEHRP *Provisions* ceiling design values increased by 50 percent (short-period = $1.50g$, long-period = $0.60g$). The ceiling design values were increased by 50 percent to represent maximum considered earthquake ground motion values. When 1.5 times the ceiling values were reached, they were used until the deterministic maximum considered earthquake map values of $1.5g$ (long-period) and $0.60g$ (short-period) were obtained. From there on, the deterministic maximum considered earthquake ground motion map values were used.

In some cases there were regions of high seismicity near known faults with return periods such that the probabilistic map values (2-percent probability of being exceeded in 50 years) exceeded the ceiling values of the 1994 NEHRP *Provisions* increased by 50 percent and were less than the deterministic map values. In these regions, the probabilistic map values were used for the maximum considered earthquake ground motions.

The basis for using past ceiling design values as the transition between the two regions was that experience had shown regularly configured, properly designed structures to have performed satisfactorily in past earthquakes.

The design level of ground motion was (and continues to be) 1/1.5 or 2/3 times the maximum considered earthquake ground motion. Figure 3.4 schematically illustrates the definition of the design earthquake.

In summary, the 1997 NEHRP *Provisions* used new procedures based on the use of spectral response acceleration rather than the traditional peak ground acceleration and/or peak ground velocity. The use of spectral ordinates and their relationship to building codes

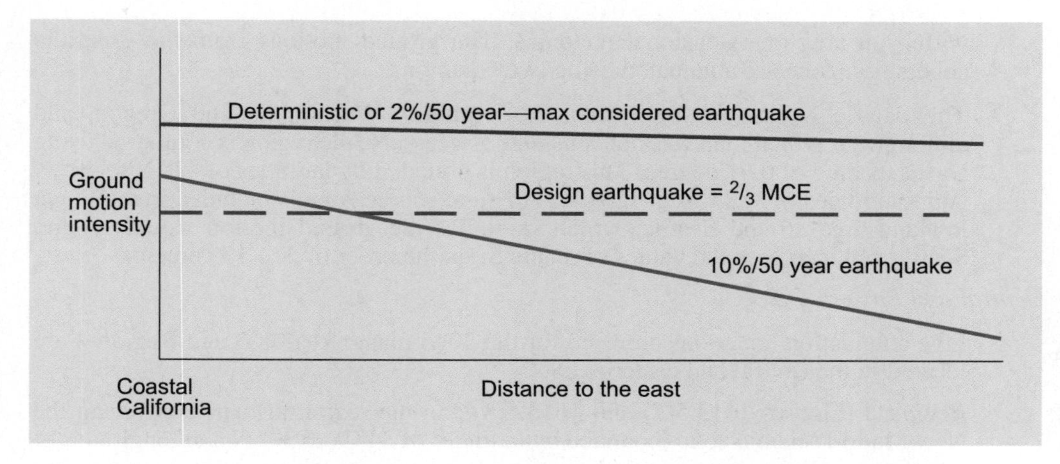

Figure 3-4 Definition of design earthquake—schematic

have been described by Leyendecker et al.[3.16] The spectral accelerations used in the design approach were obtained from combining probabilistic maps (Frankel et al.[3.17]) prepared by the U.S. Geological Survey (USGS) with deterministic maps using procedures developed by the Building Seismic Safety Council's Seismic Design Procedures Group (SDPG). The SDPG recommendations were based on the 1996 USGS probabilistic hazard maps with additional modifications based on review by the SDPG and the application of engineering judgment. The SDPG effort has sometimes been referred to as Project 97.

2000 NEHRP Provisions, ASCE 7-02, and 2003 IBC

The maps remained unchanged from the 1997 NEHRP *Provisions*, ASCE 7-98 and the 2000 IBC.

2003 NEHRP Provisions, ASCE 7-05, and 2006 IBC

Maps showing the contours of 5-percent damped 0.2-second and 1-second spectral response acceleration values for the maximum considered earthquake (MCE) ground motion have been updated for the 2003 NEHRP *Provisions*, ASCE 7-05 and the 2006 IBC. Figures 1615(1) through 1615(10) of the 2003 IBC have been replaced by Figures 1613.5(1) through 1613.5(14) of the 2006 IBC. The same maps are also included in Chapter 22 of ASCE 7-05. The changes for the 50 states and territories are discussed below.

48 CONTERMINOUS STATES: The maps of the 48 conterminous states are based on the 2002 USGS probabilistic maps that incorporate improved earthquake data in terms of updated fault parameters (such as slip rates, recurrence time and magnitude), additional attenuation relations (also referred to as ground motion prediction equations) and data calculated at smaller grid spacing as compared to the 1996 USGS probabilistic maps that formed the basis of the 2003 IBC and ASCE 7-02. The grid spacing used to contour the 2002 USGS probabilistic maps of the 48 states was 0.05 degree. For the MCE maps used in the 2006 IBC, five smaller regions were also contoured using a grid spacing of 0.01 degree. The small grid spacing makes interpolation in regions with numerous faults more accurate. The changes found in the 2006 IBC maps are discussed for the western U.S. along with its regions and the central and eastern U.S. along with its regions, as some of the changes apply to one but not the other.

Western U.S.

- The the ground motion values in the western U.S. are most affected by the attenuation equations, fault recurrence parameters, and fault geometry of the Cascadia subduction zone (located in Oregon and Washington). Five attenuation equations for shallow crustal faults were used, including one equation for faults within an area of extensional tectonics. The ground motions from the Cascadia subduction zone are attenuated using two equations.

- The Pacific Northwest region includes primarily Washington and Oregon, and although a regional map was not considered necessary, the region is contoured using a grid spacing of 0.01 degree. This region is bounded by latitudes of 41° N to 49° N and longitudes of 123° W to 125° W. An areal source zone is included in the Puget lowland area around Seattle, which smoothes the ground motion values around Seattle and increases the values of S_s and S_1 on the order of 5 to 10 percent.

Central and Eastern U.S.

- Five attenuation equations are used for the 2006 maps. Region 3 and Region 4 are located in the central and eastern U.S.

  Region 3 [Figures 1613.5(7) and 1613.5(8)]: In the region primarily covering the New Madrid seismic zone (bounded by latitudes of 34° N to 39° N and longitudes of 87° W to 92° W), a regional map was considered necessary because of steep changes in the ground motion. A deterministic area is included in the New Madrid seismic zone, marked by the shaded area in Figures 1613.5(7) and 1613.5(8), after a review of the New Madrid ground motions. Because MCE rules state that deterministic ground motions only govern over probabilistic ground motions if the deterministic values are lower, the inclusion of this deterministic area somewhat reduces the size of the high-ground-motion region near the New Madrid fault.

  Region 4 [Figure 1613.5(9)]: This regional map covers primarily South Carolina. The region is bounded by latitudes of 31° N to 35° N and longitudes of 77° W to 83° W. The ground motion values are calculated for a grid spacing of 0.05 degree instead of the smaller 0.01 degree. However a regional map was considered necessary because of the high gradient in the ground motion.

ALASKA: There are no differences in the maps between the 2003 and 2006 IBC for the state of Alaska.

HAWAII: There are no differences in the maps between the 2003 and 2006 IBC for the state of Hawaii.

PUERTO RICO AND THE U.S. VIRGIN ISLANDS: MCE ground motion maps [Figure 1613.5(13)] now provide contours of varying spectral values in Puerto Rico and the Virgin Islands, similar to those for the rest of the country, as opposed to the older practice of assigning a single value to the entire region. As a result of this update, most parts of Puerto Rico and the Virgin Islands have reductions in the mapped spectral values, except for the western one-third of Puerto Rico on the 0.2-sec map and the extreme southwestern part of Puerto Rico on the 1-sec map. The differences can be seen in Figures 16-10 and 16-11.

GUAM AND TUTUILLA: There are no differences in the maps between the 2003 and 2006 IBC for the territories of Guam and Tutuilla.

REFERENCES

[3.1]Ghosh, S.K., Domel, A.W., and Fanella, D.A., *Design of Concrete Buildings for Earthquake & Wind Forces,* Second Edition, Publication EB 113, Portland Cement Association, Skokie, IL, 1995

[3.2]Algermissen, S.T., "Seismic Risk Studies in the United States," Proceedings of the Fourth World conference on Earthquake Engineering, Santiago, Chile, 1969, Vol. I., pp. A-1.14 – A-1.27.

[3.3]American Society of Civil Engineers, *ASCE Standard Minimum Design Loads for Buildings and Other Structures,* ASCE 7-88 (formerly ANSI A58.1), New York, NY, 1990.

[3.4]Applied Technology Council, *Tentative Provisions for the Development of Seismic Regulations for Buildings,* ATC Publication ATC 3-06, NBS Special Publication 510, NSF Publication 78-8, U.S. Government Printing Office, Washington, DC, 1978.

[3.5]Seismology Committee, Structural Engineers Association of California, *Recommended Lateral Force Requirements and Commentary,* San Francisco, CA, 1988.

[3.6]Building Seismic Safety Council, *NEHRP (National Earthquake Hazards Reduction Program) Recommended Provisions for the Development of Seismic Regulations for New Buildings and Other Structures,* Washington, DC, 1997.

[3.7]Pacific Coast Building Officials Conference (later the International Conference of Building Officials), *Uniform Building Code*, Long Beach, CA (later Whittier, CA), 1935, 1952, 1967, 1970, 1973, 1976, 1979, 1982, 1985, 1988, 1991, 1994, 1997, copyright held by International Code Council.

[3.8]Roberts, E.B., and Ulrich, F.P., "Seismological Activities of the U.S. Coast and Geodetic Survey in 1948," Bulletin of the Seismological Society of America, Vol. 40, 1950, pp. 195 – 216.

[3.9]Roberts, E.B., and Ulrich, F.P., "Seismological Activities of the U.S. Coast and Geodetic Survey in 1949," Bulletin of the Seismological Society of America, Vol. 41, 1951, pp. 205 – 220.

[3.10]Schnabel, P.B., and Seed, H.B., "Acceleration in Rock for Earthquake in the Western United States," Bulletin of the Seismological Society of America, Vol. 63, 1973, pp. 501 – 516.

[3.11]Algermissen, S.T., and Perkins, D.M., *A Probabilistic Estimate of Maximum Acceleration in Rock in the Contiguous United States,* U.S. Geological Survey Open File Report 76-416, 1976, 45 pp.

[3.12]McGuire, R.K., *Seismic Structural Response Risk Analysis, Incorporating Peak Response Regression on Earthquake Magnitude and Distance,* Research Report R 74-51, Department of Civil Engineering, Massachusetts Institute of Technology, 1974, 371 pp.

[3.13]Algermissen, S.T., Perkins, D.M., Thenhaus, P.C., Hansen, S.L., and Bender, B.L., *A Probabilistic Estimate of Maximum Acceleration and Velocity in Rock in the Contiguous United States*, U.S. Geological Survey Open File Report 82-1033, 1982, 99 pp.

[3.14]Algermissen, S.T., Perkins, D.M., Thenhaus, P.C., Hansen, S.L., and Bender, B.L., *Probabilistic Earthquake Acceleration and Velocity Maps for the United States and Puerto Rico*, U.S. Geological Survey, Miscellaneous Field Studies Map MF-2120, 1990.

[3.15]Algermissen, S.T., Leyendecker, E.V., Bollinger, G.A., Donovan, N.C., Ebel, J.E., Joyner, W.B., Luft, R.W., and Singh, J.P., *Probabilistic Ground-Motion Hazard Maps of Response Spectral Ordinates for the United States*, Proceedings of the Fourth International Conference on Seismic Ordinates for the United States, Stanford University, Vol. II, 1991, pp. 687 – 694.

[3.16] Leyendecker, E.V., Perkins, D.M., Algermissen, S.T., Thenhaus, P.C., and Hanson, S.L., *USGS Spectral Response Maps and Their Relationships with Seismic Design Forces in Building Codes*, U.S. Geological Survey, Open-File Report 95-596, 1995.

[3.17] Frankel, A., Mueller, C., Bernard, T., Perkins, D., Leyendecker, E., Dickman, N., Hanson, S., and Hopper, M., *National Seismic Hazard Maps, June 1996: Documentation*, U.S. Geological Survey, Open-File Report 96-532, 1996.

Appendix

4

CONSIDERATION OF SITE SOIL CHARACTERISTICS

Seismic Design Provisions Prior to the 1994 NEHRP Provisions

The 1976 edition of the *Uniform Building Code,*[4.1] the seismic design provisions of which were based on the 1974 edition of the SEAOC *Blue Book*,[4.2] represents the beginning of relatively modern seismic design provisions in the United States.

An S coefficient was introduced into the design base shear formula of the 1976 UBC and the 1974 SEAOC *Blue Book* to account for the variability of soil conditions. The commentary to *Blue Book* editions prior to 1974 pointed out that "the absence of a soil factor, S, should not be interpreted as meaning that the effect of soil conditions on building response is not important." There had in fact been consideration of the soil-dependence of the design base shear in early editions of the UBC. According to Reference 4.3, "it is interesting to note that early seismic design codes contained provisions for increased design lateral forces for buildings sited on poor soils—provisions that were to disappear in 1949 [this reference is to the 1949 UBC], only to be reintroduced in 1974 [this reference is to the 1974 *Blue Book*]."

The approach used in the 1974 *Blue Book* was to consider the possibility of resonance between the structure and the soil by comparing the natural period of vibration of the structure, T, and the fundamental period of the site, T_s. The following expression, introduced into the 1976 UBC from the 1974 *Blue Book*, was retained in *Uniform Building Code* editions through 1985 (Figure 4.1):

Shortly after the 1974 *Blue Book* was completed, several studies appeared, which included some statistical treatment of strong motion response spectra recorded on different types of soil deposits.[4.4, 4.5, 4.6] By using these results it was possible to develop a more direct code formulation. Such a direct approach was adopted by ATC 3-06,[4.7] which considered three soil profile types to be different enough in seismic response to warrant separate

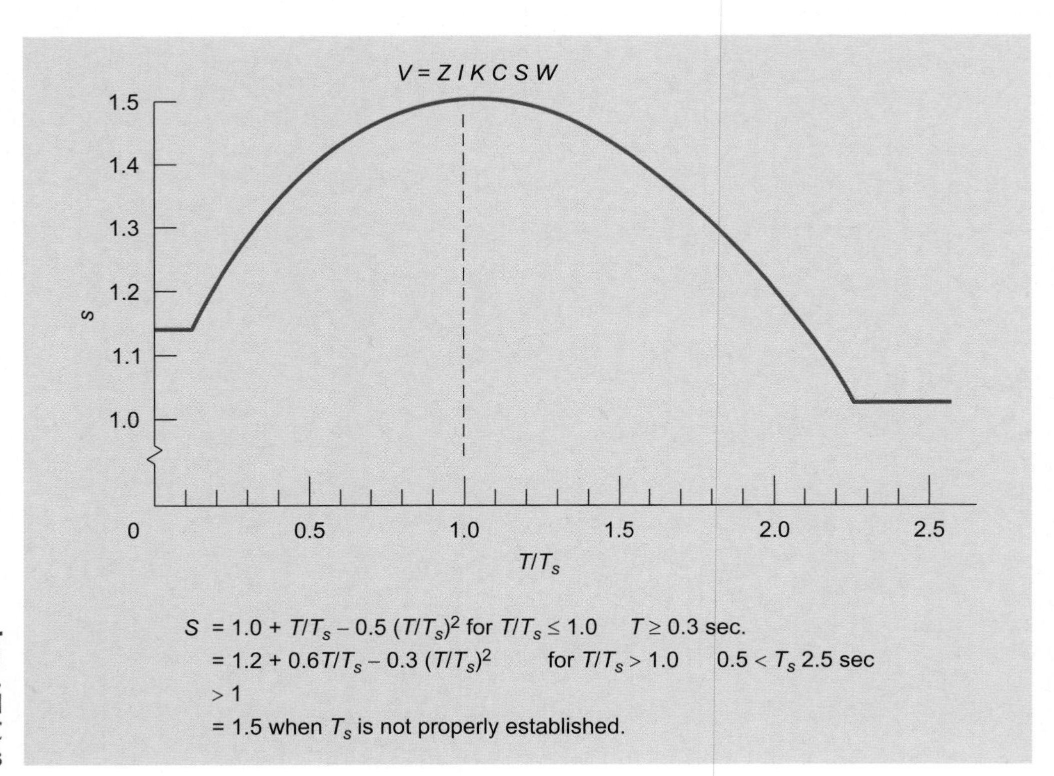

Figure 4-1 Site coefficient, S, of 1985 and earlier UBC editions

seismic coefficients (S factors). The direct approach was adopted as an alternative to the site-structure resonance approach in the 1988 UBC. Experience from the 1985 Mexico earthquake prompted the addition of a fourth soil profile type in the 1988 NEHRP *Provisions*[4.8] and the 1988 UBC. The four soil profile types are described in the 1988 NEHRP *Provisions* Commentary as follows:

1. Soil Profile Type S_1 – a soil profile with either (1) rock of any characteristic, either shale-like or crystalline in nature, that has a shear wave velocity greater than 2,500 ft/sec (762 m/sec), or (2) stiff soil conditions where the soil depth is less than 200 ft (61 m) and the soil types overlying the rock are stable deposits of sands, gravels or stiff clays.
2. Soil Profile Type S_2 – a soil profile with deep cohesionless or stiff clay conditions where the soil depth exceeds 200 feet and the soil types overlying rock are the stable deposits of sands, gravels or stiff clays.
3. Soil Profile Type S_3 – A soil profile containing 20- to 49-foot-thick soft- to medium-stiff clays with or without intervening layers of cohesionless soils.
4. Soil Profile Type S_4 – A soil profile characterized by shear wave velocity of less than 500 ft/sec (152 m/sec) containing more than 40 feet (12 m) of soft clays or silts.

The above definitions are close to those of the 1994 UBC.

1994 NEHRP Provisions

The need for improvement in codifying site effects was discussed at a 1991 National Center for Earthquake Engineering Research [NCEER] workshop[4.9] that made several general recommendations. A committee was formed during that workshop to pursue resolution of pending issues and develop specific code recommendations. The committee collected information, guided related research, discussed the issues and organized a November 1992 Site Response Workshop in Los Angeles.[4.10] This workshop discussed the results of a number of empirical and analytical studies and approved consensus recommendations that formed the basis of extensive modifications to the consideration of site effects in the 1994 NEHRP *Provisions*. The basis is summarized here from the 1994 NEHRP *Provisions* Commentary.

In the 1985 Mexico and 1989 Loma Prieta earthquakes, low maximum rock accelerations of 0.05g to 0.10g were amplified by factors of about 1.5 to 4 at sites containing soft clay layers ranging in thickness from a few feet to more than a hundred feet and having depths of rock up to several hundred feet. The average amplification factor for soft soil sites tends to decrease as the rock acceleration increases—from 2.5 to 3 at low accelerations to about 1.0 for a rock acceleration of 0.4g. This effect is directly related to the nonlinear strain behavior of the soil.

Low peak accelerations can be amplified several times at soil sites, especially those containing soft layers and where the rock is very deep. On the other hand, larger peak accelerations can be amplified to a lesser degree and can even be slightly deamplified at very high rock accelerations. In addition to peak rock acceleration, a number of factors including soil stiffness and layering play a role in the degree of amplification. One important factor is the impedance contrast between soil and underlying rock.

Earthquake records on soft to medium clay sites subjected to low acceleration levels indicate that the soil/rock amplification factors for long-period spectral accelerations can be significantly larger than those for peak accelerations. Furthermore, the largest amplification often occurs at the natural period of the soil deposit. In Mexico City in 1985, the maximum rock acceleration was amplified four times by a soft clay deposit that would have been classified as S_4, whereas the spectral amplitudes were about 15 to 20 times as large as on rock at a period near 2 seconds. In other parts of the valley where the clay is thicker, the spectral amplitudes at periods ranging between 3 and 4 seconds were amplified about 15 times, but the damage was less due to the low rock motion intensity at these very long periods. Records obtained at some soft clay sites during the 1989 Loma Prieta earthquake indicated a maximum amplification of long-period spectral amplitudes on the order of three to six times. At least at one site, the largest soil/rock amplification appeared to occur at the natural period of the soil deposit, similar to the occurrence in Mexico City.

Two Factor Approach. To summarize the above discussion, soil sites generally cause a higher amplification of rock spectral accelerations at long periods than at shorter periods and, for a service level of shaking ($A_a \approx A_v \approx 0.4$; see Appendix 3 for definitions of A_a and A_v), the shorter-period amplification or deamplification is small. However, short-period accelerations, including the peak acceleration, can be amplified several times, especially at soft sites subject to low levels of shaking. The latter evidence suggested a two-factor approach sketched in Figure 4.2. In this approach, adopted in the 1994 NEHRP *Provisions*, the short-period plateau, of height proportional to A_a, is multiplied by a short-period site coefficient, F_a, and the curve proportional to A_v/T is multiplied by a long-period site coefficient, F_v. Both F_a and F_v depend on the site conditions and on the level of shaking, defined in the 1994 NEHRP *Provisions* by the A_a and A_v coefficients, respectively.

Strong-motion records obtained from the Loma Prieta earthquake provided important quantitative measures of the in-site response of a variety of geological deposits to damaging levels of shaking. The data provided empirical estimates of the site coefficients F_a and F_v as a function of mean shear wave velocity for input ground motion near $0.1g$. Extrapolation of amplification estimates at the $0.1g$ level as derived from the Loma Prieta data had, of necessity, to be based on laboratory and theoretical modeling considerations because few or no strong-motion records have been obtained at higher levels of motion, especially on soft soil deposits. Fortunately, it was found that the functional relationship between the logarithms of amplification and mean shear wave velocity is a straight line.

Site Class Definition. In view of the above, the site categories in the 1994 and subsequent NEHRP *Provisions* are defined in terms of the average shear wave velocity in the top 100 feet of the profile, v_s. If the shear wave velocities are available for a site, they should be used.

However, in recognition of the fact that in many cases the shear wave velocities are not available, alternative definitions of the site categories are also included in the 1994 and subsequent NEHRP *Provisions*. They use the standard penetration resistance for cohesionless soil layers and the undrained shear strength for cohesive soil layers. These

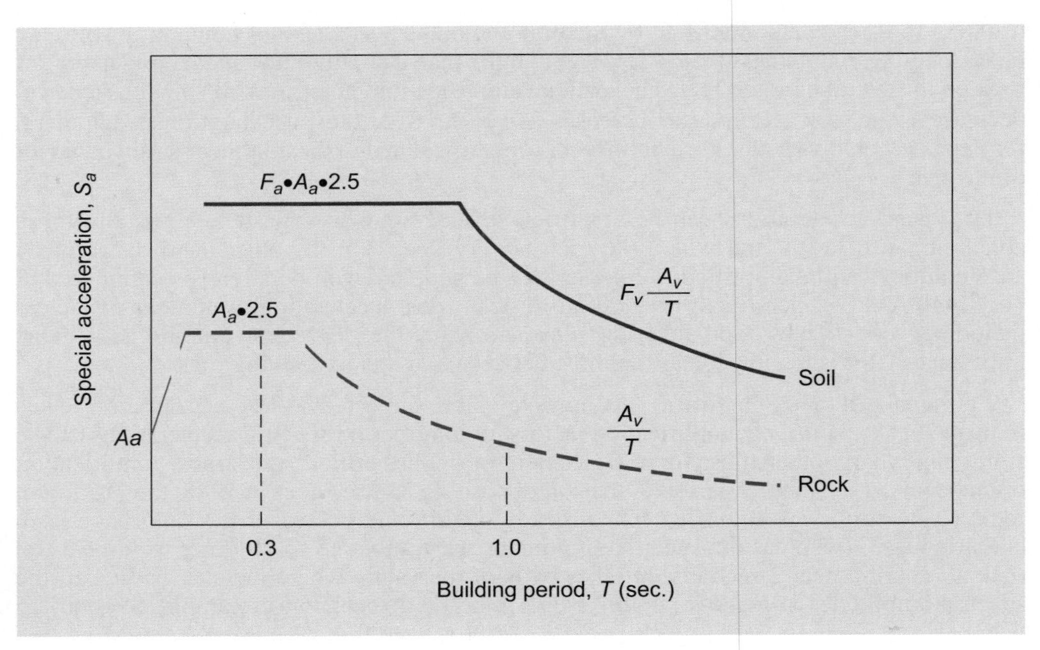

Figure 4-2
Two-factor approach to local site response

alternative definitions are rather conservative insofar as the correlation between site amplification and these geotechnical parameters is less certain than that with v_s. There will be cases where the values of F_a and F_v will be smaller if the site category is based on v_s rather than on geotechnical parameters. Also, the 1994 and subsequent NEHRP *Provisions* Commentary cautions the reader not to interpret the site category definitions as implying any specific numerical correlation between shear wave velocity on the one hand and standard penetration resistance or shear strength on the other.

Studies have indicated that earthquake motions on outcrops of hard rock tend to be smaller than on outcrops of regular rock by 10 to 40 percent at both short and long periods (except at very small periods under about 0.2 second where the reverse may be true). On the basis of these studies and observations, the 1994 and subsequent NEHRP *Provisions* incorporate the difference between regular rock, B, and hard rock of $v_s > 5000$ ft/sec by defining a new hard rock site category, A, and assigning to it site factors $F_a = F_v = 0.8$.

Dynamic Site Response Analysis

In the 1994 and subsequent NEHRP *Provisions*, coefficients F_a and F_v are not provided for Site Class F soils and site-specific geotechnical investigations, and dynamic site response analyses are required for these soils. The exception is that for structures having a fundamental period of vibration equal to or less than 0.5 second, values of F_a and F_v for liquefiable soils may be determined assuming liquefaction does not occur. The exception is provided because ground motion data obtained in liquefied soil areas during earthquakes indicate that short-period ground motions are attenuated because of liquefaction, whereas long-period ground motions may be amplified. The following text, added for the first time to the 2000 NEHRP *Provisions* Commentary, is being reproduced for the benefit of the reader from the 2003 NEHRP *Provisions* Commentary (with the permission of the Building Seismic Safety Council) because no guidance on the topic is readily available. Components of a dynamic site response analysis include: (1) modeling the soil profile; (2) selecting rock motions to input into the soil profile; and (3) conducting a site response analysis and interpreting the results.

1. Modeling the soil profile: Typically, a one-dimensional soil column extending from the ground surface to bedrock is adequate to capture first-order site-response characteristics. However, two- to three-dimensional models may be considered for critical projects when two or three-dimensional wave propagation effects may be significant (e.g., in basins). The soil layers in a one-dimensional model are characterized by their total unit weights, shear wave velocities from which low-strain (maximum) shear moduli may be obtained and by relationships defining the nonlinear shear stress-strain relationships of the soils. The required relationships for analysis are often in the form of curves that describe the variation of shear modulus with shear strain (modulus reduction curves) and by curves that describe the variation of damping with shear strain (damping curves). In a two- or three-dimensional model, compression wave velocities or moduli or Poissons ratios are also required. In an analysis to estimate the effects of liquefaction on soil site response, the nonlinear soil model must also incorporate the buildup of soil pore water pressures and the consequent effects on reducing soil stiffness and strength. Typically, modulus reduction curves and damping curves are selected on the basis of published relationships for similar soils.[4.11–4.16] Site-specific laboratory dynamic tests on soil samples to establish nonlinear soil characteristics can be considered where published relationships are judged to be inadequate for the types of soils present at the site. The uncertainty in soil properties should be estimated, especially the uncertainty in the selected maximum shear moduli and modulus reduction and damping curves.
2. Selecting input rock motions: Acceleration time histories that are representative of horizontal rock motions at the site are required as input to the soil model. Unless a site-specific analysis is carried out to develop the rock response spectrum at the site, the Maximum Considered Earthquake (MCE) response spectrum for Site

Class B rock can be defined using the general procedure described in 2003 NEHRP *Provisions* Section 3.3 (IBC Section 1613.5). For hard rock (Site Class A), the spectrum may be adjusted using the site factors in Tables 3.3-1 and 3.3-2 [IBC Tables 1613.5.3(1) and 1613.5.3 (2)]. For profiles having great depths of soil above site class A or B rock, consideration can be given to defining the base of the soil profile and the input rock motions at a depth at which soft rock or very stiff soil of Site Class C is encountered. In such cases, the MCE rock response spectrum may be taken as the spectrum for Site Class C defined using the site factors in Tables 3.3-1 and 3.3-2. Several acceleration time histories, typically at least four, recorded during earthquakes having magnitudes and distances that significantly contribute to the site seismic hazard should be selected for analysis. The U.S. Geological Survey results for de-aggregation of seismic hazard (website address: http://earthquake.usgs.gov/research/hazmaps/interactive/) can be used to evaluate the dominant magnitudes and distances contributing to the hazard. Prior to analysis, each time history should be scaled so that its spectrum is at the approximate level of the MCE rock response spectrum in the period range of interest. It is desirable that the average of the response spectra of the suite of scaled input time histories be approximately at the level of the MCE rock response spectrum in the period range of interest. Because rock response spectra are defined at the ground surface rather than at a depth below a soil deposit, the rock time histories should be input to the analysis as outcropping rock motions rather than at the soil-rock interface.

3. Site response analysis and results interpretation: Analytical methods may be equivalent linear or nonlinear. Frequently used computer programs for one-dimensional analysis include the equivalent linear program SHAKE[4.17] and nonlinear programs DESRA-2,[4.18] MARDES,[4.19] SUMDES,[4.20] D-MOD,[4.21] and TESS,[4.22] and DESRAMUSC.[4.23] For analysis of liquefaction effects on site response, computer programs incorporating pore water pressure development (effective stress analyses) must be used (e.g., DESRA-2, SUMDES, D-MOD, TESS and DESRAMUSC). Response spectra of output motions at the ground surface should be calculated and the ratios of response spectra of ground surface motions to input outcropping rock motions should also be calculated. Typically, an average of the response spectral ratio curves is obtained and multiplied by the MCE rock response spectrum to obtain the MCE soil design response spectrum. Sensitivity analyses to evaluate effects of soil property uncertainties should be conducted and considered in developing the design response spectrum.

ASCE 7-05

In ASCE 7-05, in the tables giving site coefficients F_a (Table 11.4-1) and F_v (Table 11.4-2), the reader is referred to Section 11.4.7 for site coefficients for Site Class F. The 1994 through the 2003 editions of the NEHRP *Provisions* simply require site-specific geotechnical information and dynamic site response analyses to be performed, which was also the case with ASCE 7-98 and ASCE 7-02.

ASCE 7-05 Section 11.4.7 now clearly specifies exactly when a site-specific ground motion procedure is required, although it is always permitted to be used. A *site response analysis* per Section 21.1 is required for Site Class F soils, unless the exception to Section 20.3.1 is applicable. The *site response analysis* is a new site-specific ground motion procedure introduced in ASCE 7-05. This new method determines the maximum considered earthquake (MCE) response spectrum of the building site by first starting with a bedrock response spectrum and then working upward to convert it to a response spectrum on the ground surface using the properties of the soil layer above the bedrock. A *ground motion hazard analysis* is required for seismically isolated structures and structures with damping systems on sites that are *near-source* (i.e., where S_1 is greater than or equal to $0.6g$). The *ground motion hazard analysis* title is used for the site-specific procedure included in ASCE

7-02 and in the 1994 through the 2003 editions of the NEHRP *Provisions* and is discussed in the previous section.

REFERENCES

[4.1]Pacific Coast Building Officials Conference (later the International Conference of Building Officials), *Uniform Building Code*, Long Beach, CA (later Whittier, CA), 1935, 1952, 1967, 1970, 1973, 1976, 1979, 1982, 1985, 1988, 1991, 1994, 1997, copyright held by International Code Council.

[4.2]Seismology Committee, Structural Engineers Association of California, *Recommended Lateral Force Requirements and Commentary,* San Francisco (later Sacramento), CA, 1974, 1988, 1996, 1999.

[4.3]Applied Technology Council, *A Critical Review of Current Approaches to Earthquake-Resistant Design,* Publication ATC 34, Redwood City, CA, 1995.

[4.4]Mohraj, B.J., "A Study of Earthquake Response Spectra for Different Geologic Conditions," Bulletin of the Seismological Society of America, Vol. 66, No.3, June 1976, pp. 915 – 935.

[4.5]Seed, H.B., Ugas, C., and Lysmer, J., "Site Dependent Spectra for Earthquake-Resistant Design," Bulletin of the Seismological Society of America, Vol. 66, No.1, February 1976, pp. 221 – 224.

[4.6]Kiremidjian, A.S., and Shah, H.C., "Probabilistic Site Dependent Response Spectra," Journal of the Structural Division, Proceedings ASCE, Vol. 106, No. ST1, January 1980, pp. 69 – 86.

[4.7]Applied Technology Council, *Tentative Provisions for the Development of Seismic Regulations for Buildings,* ATC Publication ATC 3-06, NBS Special Publication 510, NSF Publication 78-8, U.S. Government Printing Office, Washington, DC, 1978.

[4.8]Building Seismic Safety Council, *NEHRP (National Earthquake Hazards Reduction Program) Recommended Provisions for the Development of Seismic Regulations for New Buildings (and Other Structures),* Washington, DC, 1985, 1988, 1991, 1994, (1997, 2000, 2003).

[4.9]Whitman, R., Editor, *Proceedings of the Site Effects Workshop: October 24-25, 1991,* Report NCEER - 92-0006, National Center for Earthquake Engineering Research, Buffalo, NY, 1992.

[4.10]Martin, G.M., Editor, *Proceedings of the NCEER/SEAOC/BSSC Workshop on Site Response During Earthquakes and Seismic Code Provisions, November 18-20, 1992,* University of Southern California, Los Angeles, 1994.

[4.11]Seed, H.B., and Idriss, I.M., *Soil Moduli and Damping Factors for Dynamic Response Analyses,* Report No. EERC 70-10, University of California, Berkeley, Earthquake Engineering Research Center, 1970.

[4.12]Seed, H.B., Wong, R.T., Idriss, I.M., and Tokimatsu, K., "Moduli and Damping Factors for Dynamic Analyses of Cohesionless Soils," Journal of Geotechnical Engineering, ASCE, Vol. 112, No. 11, 1986, pp. 1016 – 1032.

[4.13]Sun, J.I., Golesorkhi, R., and Seed, H.B., *Dynamic Moduli and Damping Ratios for Cohesive Soils,* Report No. UCB/EERC-88/15, University of California, Berkeley, Earthquake Engineering Research Center, 1988.

[4.14]Vucetic, M., and Dobry, R., "Effect of Soil Plasticity on Cyclic Response," Journal of Geotechnical Engineering, ASCE, Vol. 117, No. 1, 1991, pp. 89 – 107.

[4.15]Electric Power Research Institute, *Guidelines for Determining Design Basis Ground Motions,* Report No. EPRI TR-102293, Electric Power Research Center, Palo Alto, CA, 1993.

[4.16]Kramer, S.L., *Geotechnical Earthquake Engineering,* Prentice Hall, NJ, 1996.

[4.17]Idriss, I.M., and Sun, J.I., User's Manual for SHAKE91, Center for Geotechnical Modeling, Department of Civil and Environmental Engineering, University of California, Davis, 1992, 13 pp. (plus Appendices).

[4.18]Lee, M.K.W., and Finn, W.D.L., DESRA-2, *Dynamic Effective Stress Response Analysis of Soil Deposits with Energy Transmitting Boundary including Assessment of Liquefaction Potential,* Soil Mechanics Series No. 36, Department of Civil Engineering, University of British Columbia, Vancouver, Canada, 1978, 60 pp.

[4.19]Chang, C.-Y., Mok, C.M., Power, M.S., and Tang, Y.K., *Analysis of Ground Response at Lotung Large-Scale Soil-Structure Interaction Experiment Site,* Report No. NP-7306-SL, Electric Power Research Institute, Palo Alto, CA, 1991.

[4.20]Li, X.S., Wang, Z.L., and Shen, C.K., SUMDES, *A Nonlinear Procedure for Response Analysis of Horizontally-Layered Sites Subjected to Multi-Directional Earthquake Loading,* Department of Civil Engineering, University of California, Davis, 1992.

[4.21]Matasovic, N., *Seismic Response of Composite Horizontally-Layered Soil Deposits,* Ph.D. Dissertation, Civil and Environmental Engineering Department, University of California, Los Angeles, 1993, 452 pp.

[4.22]Pyke, R.M., *TESS: A Computer Program for Nonlinear GroundResponse Analyses.* TAGA Engineering Systems & Software, Lafayette, CA, 1992.

[4.23] Qiu, P., *Earthquake-Induced Nonlinear Ground Deformation Analyses*, Ph.D. Dissertation, University of Southern California, Los Angeles, 1998.

INDEX

Note: Chapter and section numbers shown in italics are chapters and sections in ASCE 7-05 (not the IBC) discussed in Chapter 16 of this publication.

A

B

C

D